Hypoparathyroidism

Maria Luisa Brandi
Edward Meigs Brown
Editors

Hypoparathyroidism

Springer

Editors
Maria Luisa Brandi, MD, PhD
Unit of Metabolic and Bone Diseases
University of Florence
Florence
Italy

Edward Meigs Brown, MD
Division of Endocrinology,
Diabetes and Hypertension
Harvard Medical School and
Brigham and Women's Hospital
Boston, MA
USA

ISBN 978-88-470-5865-1 ISBN 978-88-470-5376-2 (eBook)
DOI 10.1007/978-88-470-5376-2
Springer Milan Heidelberg New York Dordrecht London

Softcover reprint of the hardcover 1st edition 2015

Printed on acid-free paper

Springer is part of Springer Science+Business Media (www.springer.com)

Foreword

Hypoparathyroidism, manifesting as tetany, was first encountered in the late nineteenth century as a postoperative complication of total thyroidectomy, but, as described by Mannstadt and Potts in the first chapter of this volume, the connection between parathyroid glands, post-thyroidectomy tetany, and hypocalcemia took years to be elucidated. Isolation and characterization, first from bovine and then from human parathyroid glands, of the biologically relevant agent parathyroid hormone (PTH) took decades more. The pioneering work of the late Gerald Aurbach in this endeavor (as cited by Mannstadt and Potts) and in many other seminal studies that illuminated our understanding of the causes of hypoparathyroidism is especially fitting in this volume, to which so many of Aurbach's collaborators and fellows and even fellows of those fellows have contributed.

The comprehensive scope of this volume reflects the fact that studies of hypoparathyroidism go well beyond a simple delineation of the consequences of hormone deficiency. Part I, "Anatomy and physiology of the parathyroid glands," depicts (a) the complicated embryology and development of the parathyroids, (b) the intricate homeostatic mechanisms involving not only PTH but also vitamin D and FGF23 that maintain normal calcium and phosphate metabolism, (c) the relationship between PTH and PTH-related peptides (PTHrP), and (d) the signal transduction mechanisms that govern Ca++ regulation of PTH secretion as well as PTH action on its target organs. Two particularly notable features of signal transduction relevant to PTH are the unique calcium-sensing receptor (CaSR), which in parathyroid cells transduces the signal from extracellular ionized calcium to changes in PTH secretion, and the role of G proteins in coupling both the CaSR and the PTH receptor to downstream intracellular effectors. The Gs protein, which couples not only PTH receptors but many other hormone receptors to stimulation of the second messenger, cyclic AMP, is the subject of its own chapter, given its relevance to a unique form of hypoparathyroidism discussed below.

Part II, "Conditions of hypoparathyroidism," beginning with considerations of epidemiology and clinical presentation, proceeds to detailed descriptions of the genetic and acquired diseases that result in PTH deficiency as either an isolated manifestation or as part of a syndromic constellation of other abnormalities. One of the remarkable features of hypoparathyrodism is that until recently, it represented the sole hormone deficiency disorder for which cognate hormone replacement therapy is not the standard of care. Treatment with vitamin D (in one of its several forms) and calcium supplements, while effective in restoring normocalcemia, is not, strictly speaking, physiological. Chapters discussing "conventional treatment" (i.e., vitamin D and calcium supplements) and the more physiological replacement approaches with the 1–34 fragment of PTH or the intact 1–84 amino acid hormone provide an important perspective on the potential changes in practice that may ensue.

Mannstadt and Potts, in the first chapter, pay tribute appropriately to one of endocrinology's greatest figures, Fuller Albright. Without benefit of modern tools such as radioimmunoassay, he astutely recognized that a subset of subjects with hypoparathyroidism suffered from resistance to PTH action rather than true deficiency of the hormone. Hence, he termed the disorder in these subjects *pseudohypoparathyroidism* (PHP), the first description of a hormone resistance syndrome. Analogous disorders were later recognized for most other peptide and also steroid hormones showing how studies of hypoparathyroidism provided a paradigm that influenced our understanding of endocrine disease in general. Perhaps fittingly from a historical perspective, following Albright's initial description, studies by Aurbach and colleagues localized the site of hormone resistance in PHP as proximal to cyclic AMP generation. Studies by Aurbach's fellows and others eventually defined the defect in "classic" PHP (characterized by the phenotypic appearance termed *Albright's hereditary osteodystrophy*) as loss of function mutations in the gene encoding the alpha subunit of the Gs protein. Thus, the first described hormone resistance disorder also became the first human disease recognized to be caused by a G protein mutation. The complex transcriptional regulation of the Gs alpha gene involving multiple transcripts, each with differential parental imprinting, also turned out to have a correlate in human phenotypic expression called pseudoPHP. All of these conditions are well described in Part III of this volume, "Functional Hypoparathyroidism."

This brief overview should make clear that the editors of this volume, Maria Luisa Brandi and Edward M. Brown, have succeeded in bringing together a broad array of experts from a number of fields, each of whom has contributed to a volume that in aggregate is surely greater than the sum of its parts. From the most basic aspects of biochemistry, molecular biology, and physiology to depiction at the molecular, genetic, and clinical level of a wide variety of disorders, the chapters in this volume define our current understanding of hypoparathyroidism.

Allen M. Spiegel
Department of Medicine,
and Montefiore-Einstein Center for Cancer Care,
Montefiore Medical Center,
Albert Einstein College of Medicine,
Bronx, NY, USA

Preface

This volume is the first devoted solely to hypoparathyroidism and is intended for those who must deal with this disorder. The introduction of a book on a field of clinical medicine today requires considerable justification. In the field of hypoparathyroidism, the recent growth of medical knowledge in general has created the need for a new volume fully dedicated to this topic.

That there has been a recent explosion of knowledge on the subject of hypoparathyroidism cannot be doubted, and much of the acceleration in its pace can be attributed to new insights into etiopathogenesis, epidemiology, and substitutive therapy. PubMed offers about 7,100 references to hypoparathyroidism since 1926, with over 1,200 published in the past 5 years. It is now a pleasure to record that the intensified research in hypoparathyroidism has matured to the point that a handbook can be fully dedicated to an area that is enjoying a spectacular increase in interest among basic researchers and clinicians alike.

The information made available in the 38 chapters of this volume has been brought together from widely diverse sources, and, in some instances, is reported here for the first time. Many subjects have been presented both in broad outline and in more comprehensive detail in different chapters to meet the differing requirements of the target audiences. The book is designed for use in the clinic and in the basic science laboratory connected to the clinic. Every effort has been made to make available sufficient basic and clinical knowledge to satisfy the reader's curiosity about each of these aspects.

This book is dedicated to our mentor, Gerald D. Aurbach, MD, whose pioneering studies on parathyroid hormone led to dramatic advances in our understanding of some hereditary diseases of calcium metabolism including pseudohypoparathyroidism. Gerry was not only a fine physician and an elegant scientist but also a lover of classical music and an avid fan of the Washington Redskins football team. Through his wisdom and daring insights, he showed us the way in science and in life. This book honors him as a great scientist but also as a gentle, wise, and supportive person, loved and respected by everyone.

Credits for this book are many. The first of these goes to the authors of the chapters, all leaders in their respective fields, for it is the success of their endeavors that forms the basis of and the reason for this publication. Credit is also due to the Springer team for their formidable effort in attending to the many needs that were part of the making of the handbook on hypoparathyroidism. The foreword to the book was kindly provided by Dr. Allen M. Spiegel, another of Gerry's students, who has contributed greatly to our understanding of calcium metabolism.

Florence, Italy — Maria Luisa Brandi
Boston, MA, USA — Edward Meigs Brown

Contents

Part II Conditions of Hypoparathyroidism

Contributors

Murat Bastepe Endocrine Unit, Department of Medicine, Massachusetts General Hospital and Harvard Medical School, Boston, MA, USA

Katherine Benson Department of Endocrinology, Concord Repatriation General Hospital, Sydney, NSW, Australia

John P. Bilezikian Metabolic Bone Disease Unit, Columbia University Medical Center, College of Physicians and Surgeons, Columbia University, New York, NY, USA

Maria Luisa Brandi Bone Metabolic Diseases Unit, Department of Surgery and Translational Medicine, University of Florence, Florence, FI, Italy

Edward Meigs Brown Division of Endocrinology, Diabetes and Hypertension, Department of Medicine, Brigham and Women's Hospital, Boston, MA, USA

Luisella Cianferotti Unit of Bone and Mineral Metabolism, Department of Surgery and Translational Medicine, University of Florence, Florence, Italy

Bart L. Clarke Division of Endocrinology, Diabetes, Metabolism, and Nutrition, Department of Internal Medicine, Mayo Clinic College of Medicine, Rochester, MN, USA

David E.C. Cole Departments of Laboratory Medicine and Pathobiology, Medicine, and Genetics (D.E.C.C.), Sunnybrook Health Sciences Centre, University of Toronto, Toronto, ON, Canada

Mark Stuart Cooper Department of Endocrinology, Concord Repatriation General Hospital, Sydney, NSW, Australia

Natalie E. Cusano Metabolic Bone Disease Unit, Columbia University Medical Center, College of Physicians and Surgeons, Columbia University, New York, NY, USA

Gordon B. Cutler Jr. Gordon Cutler Consultancy, LLC, Deltaville, VA, USA

Pierre D'Amour Department of Medicine, CHUM Saint-Luc Hospital, Montréal, Québec, Canada

David W. Dempster Department of Pathology and Cell Biology, College of Physicians and Surgeons of Columbia University, New York, NY, USA

Department of Regional Bone Center, Helen Hayes Hospital, West Haverstraw, NY, USA

Francesca M. Elli Endocrinology and Diabetology Unit, Department of Clinical Sciences and Community Health, University of Milan, Fondazione IRCCS Ca' Granda Ospedale Maggiore Policlinico, Milan, Italy

Sergio Fabbri Department of Surgery and Translational Medicine, University of Florence, Florence, FI, Italy

Thomas J. Gardella Endocrine Unit, Massachusetts General Hospital, Harvard Medical School, Boston, MA, USA

Francesco Giudici Department of Surgery and Translational Medicine, University of Florence, Florence, Italy

Francesca Giusti Bone Metabolic Diseases Unit, Department of Surgery and Translational Medicine, University of Florence, Florence, Italy

David Goltzman Calcium Research Laboratory and Department of Medicine, McGill University Health Centre, Royal Victoria Hospital, Montreal, QC, Canada

Departments of Medicine and Physiology, McGill University, Montreal, QC, Canada

Ana Rita Gomez Department of Surgery and Translational Medicine, University of Florence, Florence, FI, Italy

Geoffrey N. Hendy Departments of Medicine, Physiology, and Human Genetics (G.N.H.), McGill University, Montreal, QC, Canada

Calcium Research Laboratory and Hormones and Cancer Research Unit, Royal Victoria Hospital, Montreal, QC, Canada

Eli Hershkovitz Pediatric Endocrinology Unit, Soroka Medical Center and Faculty of Health Sciences, Ben Gurion University of the Negev, Beer Sheva, Israel

Giancarlo Isaia Gerontology and Bone Metabolic Disease Section, Department of Medical Sciences, University of Torino, Torino, Italy

Harald Jüppner Endocrine Unit and Pediatric Nephrology Unit, Massachusetts General Hospital and Harvard Medical School, Boston, MA, USA

Andrew C. Karaplis Division of Endocrinology, Department of Medicine and Lady Davis Institute for Medical Research, Jewish General Hospital, McGill University, Montreal, QC, Canada

E. Helen Kemp Department of Human Metabolism, Faculty of Medicine, Dentistry and Health, University of Sheffield, The Medical School, Sheffield, South Yorkshire, UK

Aliya Khan Department of Medicine, McMaster University, Hamilton, ON, Canada

Christopher S. Kovacs Faculty of Medicine – Endocrinology, Obstetrics and Gynecology, and BioMedical Sciences, Health Sciences Centre, Memorial University of Newfoundland, St. John's, NL, Canada

Henry M. Kronenberg Endocrine Unit, Massachusetts General Hospital, and Harvard Medical School, Boston, MA, USA

Agnès Linglart Department of Pediatric Endocrinology, Center of Reference for Rare Disorders of the Calcium and Phosphorus Metabolism, Paris-Sud Hospital and Paris-Sud University, Le Kremlin Bicêtre, France

Michelangelo Mancuso Neurological Clinic, University of Pisa, Pisa, Italy

Nancy R. Manley Department of Genetics, University of Georgia, Athens, GA, USA

Michael Mannstadt Endocrine Unit, Massachusetts General Hospital, Harvard Medical School, Boston, MA, USA

Giovanna Mantovani Endocrinology and Diabetology Unit, Department of Clinical Sciences and Community Health, University of Milan, Fondazione IRCCS Ca' Granda Ospedale Maggiore Policlinico, Milan, Italy

Margherita Marchetti Gerontology and Bone Metabolic Disease Section, Department of Medical Sciences, University of Torino, Torino, Italy

Laura Masi Metabolic Bone Diseases Unit, University Hospital AOU-Careggi University of Florence, Florence, Italy

Department of Surgery and Translational Medicine, University of Florence, Florence, Italy

Piergiorgio Messa Department of Nephrology, Urology and Renal Transplant, Fondazione Ca' Granda IRCCS Ospedale Maggiore Policlinico, Milan, Italy

Deborah M. Mitchell Endocrine Unit, Massachusetts General Hospital, Harvard Medical School, Boston, MA, USA

M. Andrew Nesbit Centre for Biomedical Sciences, University of Ulster, Coleraine, County Londonderry, Northern Ireland, UK

Daniele Orsucci Neurological Clinic, University of Pisa, Pisa, Italy

Paola Divieti Pajevic Endocrine Unit, Massachusetts General Hospital, and Harvard Medical School, Boston, MA, USA

Ruti Parvari Shraga Segal Department of Microbiology, Immunology and Genetics, Faculty of Health Sciences, Ben Gurion University of the Negev, and National Institute of Biotechnology Negev, Beer Sheva, Israel

John T. Potts Jr. Clinical Medicine, Massachusetts General Hospital, Boston, MA, USA

Lars Rejnmark Department of Endocrinology and Internal Medicine, Aarhus University Hospital, Aarhus C, Denmark

René Rizzoli Division of Bone Diseases, Faculty of Medicine, Geneva University Hospitals, Geneva 14, Switzerland

Mishaela R. Rubin Metabolic Bone Disease Unit, Columbia University Medical Center, College of Physicians and Surgeons, Columbia University, New York, NY, USA

James E. Sanders Hypoparathyroidism Association, Idaho Falls, ID, USA

Dolores M. Shoback Department of Medicine, University of California, San Francisco, San Francisco, CA, USA

Metabolism and Endocrinology Section, San Francisco Department of Veterans Affairs Medical Center, San Francisco, CA, USA

Endocrine Research Unit, San Francisco Department of Veteran Affairs Medical Center, San Francisco, CA, USA

Gabriele Siciliano Neurological Clinic, University of Pisa, Pisa, Italy

Jim Sliney Jr. Division of Endocrinology, Columbia University Medical Center, New York, USA

Oren Steen Department of Medicine, McMaster University, Hamilton, ON, Canada

Francesco Tarsitano Obstetrics and Gynecology Department, A.O.U. Federico II, Napoli, Italy

Marina Tarsitano Department of Genetics, Biochemical Laboratory, Salerno, Italy

Susanne Thiele Division of Pediatric Endocrinology and Diabetes, Department of Pediatrics and Adolescent Medicine, University of Luebeck, Luebeck, Schleswig-Holstein, Germany

Francesco Tonelli Department of Surgery and Translational Medicine, University of Florence, Florence, Italy

Andrea Vitale Department for Exercise Science and Research, Parthenope University, Naples, Italy

Anthony P. Weetman Department of Human Metabolism, Faculty of Medicine, Dentistry and Health, University of Sheffield, The Medical School, Sheffield, South Yorkshire, UK

Marc N. Wein Endocrine Unit, Massachusetts General Hospital, and Harvard Medical School, Boston, MA, USA

Lee S. Weinstein Metabolic Diseases Branch, National Institute of Diabetes and Digestive and Kidney Diseases, National Institutes of Health, Bethesda, MD, USA

Jean-Louis Wémeau Department of Endocrinology and Metabolic Diseases, Marc Linquette Endocrinological Clinic, Claude Huriez Hospital, CHRU, Lille Cedex, France

Amber Wheeler Department of Medicine, University of California, San Francisco, San Francisco, CA, USA

Metabolism and Endocrinology Section, San Francisco Department of Veterans Affairs Medical Center, San Francisco, CA, USA

Karen K. Winer National Institutes of Health, Eunice Kennedy Shriver National Institute of Child Health and Human Development/PGNB, Bethesda, MD, USA

Part I

Anatomy and Physiology of the Parathyroid Glands

The History: From Ivar Sandström to the Sequence of Parathyroid Hormone

1

Michael Mannstadt and John T. Potts Jr.

1.1 Introduction

The history of the discovery of the parathyroid glands (Table 1.1) and the evolution of knowledge about parathyroid hormone (PTH) are rewarding for the insights it provides into the basic physiology and pathophysiology of bone and mineral metabolism and the diseases hypoparathyroidism and hyperparathyroidism. The early history illustrates the surprising contradictions and confusion that can occur in scientific investigations along the pathway to a clearer understanding of biology and disease. This chapter will focus on hypoparathyroidism consistent with the theme of this volume.

1.2 Discovery of the Glands

Credit for the discovery of the parathyroid glands clearly belongs to Ivar Sandström (1852–1889), who found this new organ in the dog in 1877 when he was a medical student in Uppsala, Sweden. Later, as a temporary research assistant in the Department of Anatomy, he systematically and thoroughly investigated these glands further and identified this novel organ also in cats, oxen, horses, rabbits, and finally humans. He examined 50 corpses and found all four glands in most of them. Although the relationship to the thyroid gland was unsolved, it was Sandström who gave them the name "glandulae parathyreoideae," parathyroid glands.

> Although both of the aforementioned kinds of glands [the parathyroids versus accessory thyroid glands] could with equal reasons claim the name of accessory thyroid glands, a special name seems to be required for those which are the subject of this paper [the parathyroids], both with regard to the essentially different structure and on account of the fact that this kind of gland [the parathyroids] is constant in its occurrence [referring to his careful work in dogs, cats, horses, and rabbits as well as 50 human postmortem subjects] while the other one [accessory thyroids] is extremely variable. I therefore suggest the use of the name *Glandulae parathyreoideae*; a name in which the characteristic of being bye-glands to the thyroid is expressed. [1, 2]

He described in great detail the variable size, form, and color of the four glands found in humans, their vascular supply, and their microscopic appearance. He had, of course, no knowledge as to the function of this new organ. In a style that is somewhat different from today's rigid scientific language, he writes

> Concerning the physiological importance of these glands for the organism, we are not able, from reasons that are quite apparent, to allow ourselves even to make a guess. [1, 2]

M. Mannstadt, MD (✉)
Endocrine Unit, Massachusetts General Hospital, Harvard Medical School, Thier 10, 50 Blossom St, Boston, MA 02114, USA
e-mail: mannstadt@mgh.harvard.edu

J.T. Potts Jr., MD
Clinical Medicine, Massachusetts General Hospital, 55 Fruit Street, BAR-516, Boston, MA 02114, USA
e-mail: potts.john@mgh.harvard.edu

M.L. Brandi, E.M. Brown (eds.), *Hypoparathyroidism*,
DOI 10.1007/978-88-470-5376-2_1, © Springer-Verlag Italia 2015

Table 1.1 Short history of parathyroid hormone

1880	*Sandström* identifies parathyroids in humans
1900	Parathyroids first recognized as functionally distinct from thyroid (*Vassale and Generali*) Acceptance of evidence mixed
1910–1925	Vital role established: removal causes tetany Function debated over control of calcium vs. detoxifying function (guanidine toxicity)
1925	Endocrine function established by *Collip* Parathyroid extracts reverse tetany
1929–1952	*Albright* describes idiopathic hypoparathyroidism, pseudohypoparathyroidism (PHP), and pseudo-PHP
1959	*Era of chemical biology begins* *Aurbach* isolates and purifies PTH intact by use of organic solvents
1971–1975	Structure and synthesis of PTH Definitive research and clinical uses
1987	Clinically useful immunoassays of PTH Laboratory diagnosis of hypoparathyroidism possible
1990	*Full impact of molecular biology unfolds* Receptor cloned Genetic defects in hypoparathyroidism defined Genetic manipulations define PTH function in vivo in rodents

The inability to publish his work in a major medical journal was a great disappointment to him and prevented his international recognition. In 1880, he published his discovery "On a New Gland in Man and Several Mammals" in a local Swedish journal [1, 2]. It turned out to be his only scientific publication. Although Sandström received two local Swedish prizes for his discovery, international recognition was not achieved before his death and not until decades later. His personal life story was a tragic one. He had apparently inherited a propensity to depression and although he continued his studies and even finished his medical degree in 1887, he was hospitalized several times during that period. He frequently expressed disappointment at his lack of recognition and failure to be permitted further opportunity to work as an investigator. He shot and killed himself in 1889 at the age of 37.

The glands were also found by others before him, but had not been carefully and systematically examined. Sir Richard Owen probably first recognized them around 1852 in the Indian rhinoceros [3]. His publication, which was in all likelihood not accessible to Ivar Sandström, describes the glands in one single sentence:

> A small compact yellow glandular body was attached to the thyroid at the point where the veins emerge. [3]

The systematic description of the parathyroids in humans and in several other species and therefore their potential significance for human physiology and disease clearly begin with Sandström. The entertaining and highly readable monograph by Jörgen Nordenström [4] recalls the early history and explains the unusual circumstances that led to a postmortem examination of the rhinoceros in the nineteenth century. More tellingly, it chronicles the tragic story of Sandström who never received full credit for the significance of his painstaking work.

1.3 Clarification of the Separate Anatomy of the Parathyroids

Two lines of observation, one based on clinical experience and the other on animal experiments, led, but surprisingly quite slowly, to the identification of the key role played by the parathyroids. Clinical reports discussed the poorly understood complex set of symptoms that we now recognize as hypocalcemic tetany in patients operated on for thyroid disease after extensive (probably often complete) thyroidectomy. Mortality from thyroidectomy was so high (as great as 40 %) that the famous Austrian surgical pioneer Theodor Billroth (1829–1894) stopped thyroid surgery for several years. In some cases, tetany was the cause of death. Nordenström in his monograph provides some of these dramatic examples of patients who developed tetany and sometimes death after thyroidectomy [4]. The physicians were at a loss to understand, let alone treat, patients in whom tetany developed. He notes that, because of the typical spasm of the hands that can be seen in shoemakers, the condition came to be called Schusterkrampf (shoemaker's cramp).

There were many false steps, and constant controversy, on the way to a full appreciation of the role of the parathyroids, but not until the early part of the twentieth century was it recognized that it was inadvertent removal of the parathyroids and not the thyroid that caused tetany in patients undergoing thyroidectomy (see below). Even after others eventually duly noted the work of Sandström, it was not clear that the parathyroids were a separate organ system rather than embryonic thyroid glands and/or an accessory part of the thyroid. This confusion in animal experiments in part arises from the presence of multiple glands as well as their variable location particularly what we now term the inferior parathyroids, which may be intrathyroidal.

In a comprehensive and scholarly review of the experimental studies, Boothby [5] observes (but in retrospect not entirely correctly) that Sandström believed that the glands were likely accessory thyroid tissue. Baber [6] in 1881 clearly described these glands under the name "undeveloped portions of the thyroid." (Baber was apparently not aware of Sandström's earlier paper). Horsley [7] in 1886 correctly deduced as a result of careful experimentation that these tissues recognized by Sandström and Baber were not undeveloped tissue of the thyroid, but separate organs. Horsley demonstrated that after partial thyroidectomy, while the "undeveloped tissue" (the parathyroids) did not show any enlargement or conversion to thyroid tissue, the remaining thyroid tissue did immediately hypertrophy.

In 1891, Gley reported his findings that experimental thyroidectomy in animals often resulted in tetany [8, 9]. However, he incorrectly deduced that the external parathyroids were indeed embryonic thyroid tissue. He arrived at this deduction because he correctly noted that tetany did not develop if these external parathyroids were spared, but he noted that these glands doubled in size after removal of the thyroid. Therefore, he promulgated the view that they were then taking on the function of the thyroid without realizing, of course, that the internal parathyroids had been removed with the bulk of the thyroid tissue, and this led to hyperplasia of the remaining parathyroids. He persisted in this view despite the observations of Horsley.

It was the work of Vassale and Generali published in 1896 that disagreed with Gley's contention that the glands were embryonic thyroid rests and established that they were special organs distinct from the thyroid. In a series of papers resulting from careful work [10, 11], they demonstrated that removal of all four parathyroid glands caused tetany even if significant amounts of thyroid tissue were preserved, whereas total thyroidectomy did not cause tetany if at least one parathyroid gland was spared.

Still, others as cited by Boothby contributed to the work that finally established the glands as essential to prevent tetany. Especially notable was the work of the great Viennese pathologist Jakob Erdheim (1874–1937) in 1904–1906. Through postmortem observations in patients who died of tetany after thyroidectomy, Erdheim established with painstaking care that the parathyroids were totally absent. Erdheim also undertook experimental studies in rats, which normally have only two parathyroid glands, which are readily visible. Using cautery, Erdheim was able to destroy various portions of these glands without damaging the thyroid. Complete removal of all parathyroid tissue with preservation of the thyroid resulted in tetany similar to the earlier work of Vassale and Generali.

Erdheim even provided what could have been an early clue to the role of the glands in calcium metabolism through his observations in his parathyroidectomized animals. He demonstrated that the tooth discoloration that developed in some of the surviving rats (teeth constantly grow in all rats) was due to the sudden cessation of calcium deposition coincident with their loss of the parathyroid glands (hypoparathyroid state) and subsequent low blood calcium levels [12].

1.4 Physiological Role of the Glands

Intense debate, surprising in retrospect, centered on the cause of the tetany and the role of these vital glands. Although the true explanation, severe hypocalcemia, was carefully documented by a number of investigators, others concluded that the principal function of the glands was detoxification. One of the main reasons the detoxification theory could survive so long was because attempts to treat animals with extracts of the parathyroid glands was not effective in reversing the tetany. We can now appreciate that obtaining parathyroid hormone from parathyroid extracts was unsuccessful at the time.

William MacCallum and his coworkers beginning in 1908 were strong proponents of the view that the parathyroid glands were somehow involved in control of blood calcium. In 1909, MacCallum and his colleague Carl Voegtlin were able to demonstrate that infusions of calcium completely reversed the symptoms of cramps that dogs suffered after removal of their parathyroids [13]. They also measured blood calcium levels and reported that they were lower than normal in parathyroidectomized dogs. They concluded

> Tetany occurs spontaneously in many forms and may also be produced by the destruction of the parathyroid glands… The injection of a solution of a salt of calcium into the circulation of an animal in tetany promptly checks all the symptoms and restores he animal to an apparently normal condition. [13]

However, they were unable to reverse the tetany by administration of extract of the glands. Others also demonstrated that calcium would reverse the tetany in experimental animals after parathyroidectomy [14]. Confusion developed, however, when Koch stated in 1912 that there were high levels of methyl guanidine found in the urine of animals with tetany after parathyroidectomy [15, 16]. A few years later, Paton demonstrated that administration of guanidine or methyl guanidine could apparently cause symptoms characteristic of tetany in rats [17].

In closing his very scholarly review summarizing the field as of 1921, Boothby concluded:

- Removal of all parathyroid tissue in animals causes tetany and death; the younger the animals, the worse the problem. (Some noted that herbivores were more resistant.)
- Preservation of small amounts of parathyroid tissue prevents or greatly minimizes the tetany.
- The parathyroids have a function separate from that of the thyroid – their only relationship is an anatomic proximity.
- Their function remains unclear. It seems to be concerned with calcium metabolism or guanidine metabolism or both. Nonetheless, administration of large amounts of calcium is usually of benefit in lessening the symptoms in patients suffering tetany after thyroid surgery.
- Reported cases of idiopathic tetany are not necessarily related to the parathyroids, and the association of tetany with the function of the parathyroids is only firm in humans after extensive thyroid surgery.

It is evident that these early workers were imaginative investigators who learned much with what today might be considered rudimentary tools. Sometimes progress was stalled for years. Investigators, as well as clinicians, were not often aware of some innovative findings lacking the rapid access to medical information available today. Sometimes, innovative findings that did not fit into the paradigms of the day were rejected or ignored.

1.5 The Parathyroids Are Endocrine Glands

A definitive series of experiments by James Collip (1892–1965) in 1925 resolved the controversy about the function of the glands [18]. Collip prepared hot hydrochloric acid extracts of the parathyroid glands; an approach that he correctly hypothesized was needed to free the active substance from other stromal components of the gland and to render it soluble. He showed that these acid extracts of the parathyroid gland would completely relieve the tetany that followed parathyroidectomy in experimental animals and in humans [19]. Thereby, he established that the parathyroids are endocrine glands that secreted a hormone, PTH. Another author, Adolph Hansen, reported a similar acid extraction procedure in 1924 [20, 21] and claimed priority for the discovery although his efforts to demonstrate biological actions with his extract were at best inconsistent, so the bulk of the credit belongs to Collip in the opinion of the present authors.

The availability of biologically active extracts of parathyroid hormone made available by pharmaceutical firms such as Lilly immediately attracted the interest of clinical investigators, who administered the preparations in clinical investigation in patients to better understand the etiology and pathophysiology of such conditions as idiopathic tetany. Prior to the availability of active preparations of PTH, the state of knowledge in the field was as summarized above by Boothby [5], namely, that it was unproven whether idiopathic tetany could be due to a failure of the parathyroid glands. Leading clinical investigators in several institutions, most notably Fuller Albright (1900–1969) and his colleagues in the endocrine group at Massachusetts General Hospital (MGH) used these clinically available preparations (termed parathormone) to reverse hypocalcemia in hypoparathyroidism. (Beyond the scope of this chapter is the use of these preparations that led clinical investigators to discover the first patient with overactivity of the parathyroids in the United States). Administration of these PTH preparations to patients with idiopathic hypoparathyroidism confirmed the diagnosis by the demonstration of prompt phosphaturia achieved with what was then termed the Ellsworth-Howard test [22]. The brilliant observation of Albright and colleagues led further to the identification of a form of hormone resistance to parathyroid hormone as the cause in some patients with apparent hypoparathyroidism. The investigators demonstrated a failure of the extracts to promote phosphaturia in certain patients with additional striking phenotypic features, later termed Albright's osteodystrophy, leading them to clarify the entity of pseudohypoparathyroidism (PHP) [23] and later (foreshadowing the delineation of the role of gene imprinting in hereditary disorders many years later) the entity of pseudo-pseudohypoparathyroidism [24]. Their remarkable foresight obtained on clinical grounds alone linked the two diseases with similar phenotypic features, the former PHP with hormone resistance and the latter pseudo-PHP devoid of hormone resistance per se.

The successful extraction of PTH from the glands created problems that blocked further progress toward fully characterizing the structure of parathyroid hormone. When techniques for protein structural analysis became available (following the seminal work of Sanger who determined the structure of insulin [25]), there was interest in applying the techniques to parathyroid hormone. The Collip hot acid extraction method had an undesired side effect (as we understand the issue in retrospect). The hormonal peptide was not only liberated and solubilized, but also cleaved at multiple sites (most likely at asparagine or aspartate acid sites within the sequence) giving a multiplicity of peptides of varying length with a low yield of any one. In a 1954 report, for example, Handler et al. [26] summarized their frustration at the inability to use the techniques then available for purification. They stated

> 1) the active material in the gland… may be a large protein which in the course of isolation is degraded into fractions of varying size each of which still has activity, or 2) the active material may not be a large molecule at all, but instead a small molecule which adheres to each one of the fractions. [26]

The problem was compounded because the method of monitoring purification, the bioassay, was itself difficult. The assay in use at that time involved injections of purified preparations into parathyroidectomized rats to raise the blood calcium concentration. The precision of the technique was much less than that of an enzymatic assay. The field remained stalled until a breakthrough development in 1959.

1.6 Era of Chemical Biology

In 1959, Aurbach reported a new technique that solved the problem and resulted in purification of the intact, native polypeptide [27]. By using organic solvents like hot acid, he liberated the peptide in an active form but without producing multiple cleavage products. Later Rasmussen and Craig confirmed his results using an analogous technique [28].

With continued advances in protein sequencing techniques, which became available in the late 1950s and early 1960s, two independent groups determined the structure of PTH, first of bovine hormone, in 1970 [29, 30]. Accumulation of sufficient amounts of parathyroid tissue was possible using cows and other large animal species used for meat consumption by scientists working with slaughterhouses and the meat production industry. Only several years later, after laborious accumulation of sufficient material from human parathyroid glands that were available as the byproduct of surgically removed parathyroid tumors, could the structure of human hormone itself be approached and ultimately completely solved by 1978 [31].

It was hypothesized that a molecule comprising the first 34 residues might be sufficiently long to be biologically active. This somewhat arbitrarily chosen peptide length was based on the deduced amino acid sequence of PTH, on the reports that hot acid produced active fragments, and on considerations of peptide synthesis techniques then available. Successful reports of full biological activity of PTH(1–34), first for the bovine hormone in 1971 [32] and then later the human in 1974 [33], confirmed that the structure of the compound had been accurately deduced and even more importantly provided a material for definitive animal and clinical use.

Availability of highly purified parathyroid hormone and active synthetic fragments made it also possible to develop improved immunoassays based on the principles clarified by Ekins in 1980 [34]. He championed the use of double antibody methods or so-called sandwich assays. Much of the circulating parathyroid hormones are fragments, most of them biologically inactive [35]. These fragments were often detected in the earlier radioimmunoassay techniques. Overall, as noted in earlier reviews [36], this caused a lack of precision in the results with these earlier assay techniques (see also Chap. 4). The introduction of an effective double antibody assay in 1987 [37] greatly improved the detection capacity of the assays such that the low levels of PTH seen in patients with hypoparathyroidism could be readily distinguished from normal levels making it possible to accurately confirm by laboratory techniques the presence of hypoparathyroidism. The even greater advance in this instance (but beyond the scope of this chapter) was to greatly improve the capacity of the assays to discriminate between the diagnosis of primary hyperparathyroidism (elevated levels of PTH) and hypercalcemia of malignancy (low levels of PTH) (see also chapter 4).

1.7 Era of Molecular Biology

The wide availability of the powerful techniques of molecular biology accelerated progress leading to the successful cloning of the receptor for the hormone in 1991 [38] (see Chap. 9). Parallel advances in cell biology from many fields provided improved techniques that permitted a much clearer delineation of critical steps in hormone action in target cells (especially in bone and kidney) using the cloned receptor and synthesized fragments of PTH and introduced the current era of the molecular biology of parathyroid hormone [39].

As will be reviewed in Chaps. 16, 17, 18, 19, 20, and 21, the powerful techniques of molecular biology have aided in characterizing the many genetic defects responsible for hypoparathyroidism [40]. They include, but are not limited to, the rare loss-of-function mutations in the PTH gene itself or in transcriptions factors key to the development of the parathyroid glands (such as GCM2 and GATA3) and mutations in the AIRE gene leading to inherited forms of autoimmune hypoparathyroidism (APECED). The importance of the molecular diagnosis for patient care is illustrated by the autosomal-dominant hypocalcemia

(ADH), common among the inherited forms of hypoparathyroidism, which is caused by activating mutations in the calcium-sensing receptor. Patients with ADH are particularly prone to hypercalciuria and nephrocalcinosis, therefore rendering the molecular diagnosis important for the treating physician. Mutations in the gene encoding the guanine-binding protein G11 have recently been identified as a cause of hypoparathyroidism [41, 42], demonstrating the power of genetics in shedding light on important signaling pathways in the parathyroid glands. Molecular biology also clarified the mechanisms of resistance to PTH in pseudohypoparathyroidism. Mutations in *GNAS*, the gene encoding the alpha subunit of Gs, or methylation changes at the GNAS locus are responsible for this imprinted disorder (see Chaps. 10, 32, 33, 34, and 35). In addition, the greater understanding of the molecular actions of parathyroid hormone has led to such advances as a long-acting form of parathyroid hormone termed LA-PTH [43] which has potential as a hormone replacement therapy for hypoparathyroidism, one of the few endocrine deficiency states heretofore not treated by replacement with the missing hormone. Clinical investigators have successfully demonstrated that treatment with PTH(1–34) and PTH(1–84) is a possible therapy for patients with hypoparathyroidism (see Chaps. 30 and 31). Recently, the first randomized, placebo-controlled phase 3 clinical trial using human recombinant PTH(1–84) was successfully completed [44]. PTH replacement therapy for hypoparathyroidism, which addresses the underlying defect, could therefore become a practical reality in the not too distant future.

References

1. Sandström I (1880) Om en ny körtel hos menniskan och åtskilliga däggdjur. Upsala Lakareforen Forh 15:441–471
2. Seipel CM (1938) An english translation of Sandström's Glandulae Parathyreoideae. Bull Inst Hist Med 6:179–222
3. Owen R (1862) On the anatomy of the Indian rhinoceros. Trans Zool Soc Lond 4:31–58
4. Nordenström J (2013) The hunt for the parathyroids. Karolinska Institute University Press, Stockholm
5. Boothby WM (1921) The parathyroid glands: a review of the literature. Endocrinology 5:403–440
6. Baber EC (1881) Researches on the minute structure of the thyroid gland. Philos Trans 172:577–608
7. Horsley V (1892) Remarks on the function of the thyroid gland: a critical historical review. Br Med J 1(215–219):265–268
8. Gley E (1891) Sur les fonctions du corps thyroide. Comptes Rendus Soc Biol Paris 43:841–842
9. Gley E (1891) Sur la toxicité des urine des chiens thyroîdectomisés. Contribution a l'étude des fonctions du corps thyroïde. Comptes Rendus Soc Biol Paris 3:366–368
10. Vassale G, Generali F (1896) Sur les effets de l'extirpation des glandes parathyréoïdes. Arch Ital Biol 25:459–464
11. Vassale G, Generali F (1896) Fonction parathyroidienne et fonction thyroidienne. Arch Ital Biol 33: 154–155
12. Erdheim J (1911) Über die Dentinverkalkung im Nagezahn bei der Epithelkörperchentransplantation. Frankfurt Z Pathol 7:295–347
13. MacCallum WG, Voegtlin C (1909) On the relation of tetany to the parathyroid glands and to calcium metabolism. J Exp Med 11:118–161
14. Parhon C, Urechie CS (1907) Untersuchungen über den Einfluss den die Calcium and Sodiumsalze auf den Verlauf der experimentellen Tetanie ausüben. Neurol Centralbl 26:1099
15. Koch WF (1913) Toxic bases in the urine of parathyroidectomized dogs. J Biol Chem 15:43–63
16. Koch WF (1915) The physiology of the parathyroid glands. J Labor Clin Med 1:299–315
17. Paton DN, Findlay L, Burns D (1914–1915) On guanidine or methylguanidine as a toxic agent in the tetany following parathyroidectomy. J Physiol 49:17–18
18. Collip JB (1925) The extraction of a parathyroid hormone which will prevent or control parathyroid tetany and which regulates the level of blood calcium. J Biol Chem 63:395–438
19. Collip JB, Leitch DB (1925) A case of tetany treated with parathyrin. Can Med Assoc J 15:59–60
20. Hanson AM (1923) An elementary chemical study of the parathyroid glands of cattle. Mil Surg 53:280–284
21. Hanson AM (1924) Experiments with active preparations of parathyroid other than that of desiccated gland. Mil Surg 53:701–718
22. Albright F, Ellsworth R (1929) Studies on the physiology of the parathyroid glands: I. Calcium and phosphorus studies on a case of idiopathic hypoparathyroidism. J Clin Invest 7:183–201
23. Albright F, Burnett CH, Smith PH (1942) Pseudohypoparathyroidism – an example of 'Seabright-Bantam syndrome': report of three cases. Endocrinology 30:922–932
24. Albright F, Forbes AP, Henneman PH (1952) Pseudo-pseudohypoparathyroidism. Trans Assoc Am Phys 65:337–350

25. Sanger F (1949) The terminal peptides of insulin. Biochem J 45:563–574
26. Handler P, Cohn DU, Dratz AF (1954) Metabolic interrelations with special reference to calcium. In: 5th Josiah Macy conference. Progress Associated, Inc, New York
27. Aurbach GD (1959) Isolation of parathyroid hormone after extraction with phenol. J Biol Chem 234:3179–3181
28. Rasmussen H, Craig L (1959) Purification of parathyroid hormone by use of countercurrent distribution. J Am Chem Soc 81:5003
29. Niall HD, Keutmann H, Sauer R, Hogan M, Dawson B, Aurbach G, Potts J Jr (1970) The amino acid sequence of bovine parathyroid hormone I. Hoppe Seylers Z Physiol Chem 351:1586–1588
30. Brewer HB Jr, Ronan R (1970) Bovine parathyroid hormone: amino acid sequence. Proc Natl Acad Sci U S A 67:1862–1869
31. Keutmann HT, Sauer MM, Hendy GN, O'Riordan LH, Potts JT Jr (1978) Complete amino acid sequence of human parathyroid hormone. Biochemistry 17:5723–5729
32. Potts JT Jr, Tregear GW, Keutmann HT, Niall HD, Sauer R, Deftos LJ, Dawson BF, Hogan ML, Aurbach GD (1971) Synthesis of a biologically active N-terminal tetratriacontapeptide of parathyroid hormone. Proc Natl Acad Sci U S A 68:63–67
33. Tregear GW, van Rietschoten J, Greene E, Niall HD, Keutmann HT, Parsons JA, O'Riordan JL, Potts JT Jr (1974) Solid-phase synthesis of the biologically active N-terminal 1–34 peptide of human parathyroid hormone. Hoppe Seylers Z Physiol Chem 355:415–421
34. Ekins R (1980) More sensitive immunoassays. Nature 284:14–15
35. Potts JT (2005) Parathyroid hormone: past and present. J Endocrinol 187:311–325
36. Jüppner H, Potts JT Jr (2002) Immunoassays for the detection of parathyroid hormone. J Bone Miner Res 17(Suppl 2):N81–N86
37. Nussbaum SR, Zahradnik RJ, Lavigne JR, Brennan GL, Nozawa-Ung K, Kim LY, Keutmann HT, Wang CA, Potts JT Jr, Segre GV (1987) Highly sensitive two-site immunoradiometric assay of parathyrin, and its clinical utility in evaluating patients with hypercalcemia. Clin Chem 33:1364–1367
38. Jüppner H, Abou-Samra AB, Freeman M et al (1991) A G protein-linked receptor for parathyroid hormone and parathyroid hormone-related peptide. Science 254:1024–1026
39. Gardella T, Jüppner H, Brown E, Kronenberg H, Potts J Jr (2010) Parathyroid hormone and parathyroid hormone -related peptide in the regulation of calcium homeostasis and bone regulation. In: DeGroot L, Jameson J (eds) Endocrinology, 6th edn. W. B. Saunders Co., Philadelphia
40. Thakker RV (2001) Genetic developments in hypoparathyroidism. Lancet 357:974–976
41. Nesbit MA, Hannan FM, Howles SA, Babinsky VN, Head RA, Cranston T, Rust N, Hobbs MR, Heath H 3rd, Thakker RV (2013) Mutations affecting G-protein subunit alpha11 in hypercalcemia and hypocalcemia. N Engl J Med 368:2476–2486
42. Mannstadt M, Harris M, Bravenboer B, Chitturi S, Dreijerink KM, Lambright DG, Lim ET, Daly MJ, Gabriel S, Jüppner H (2013) Germline mutations affecting Galpha11 in hypoparathyroidism. N Engl J Med 368:2532–2534
43. Maeda A, Okazaki M, Baron DM et al (2013) Critical role of parathyroid hormone (PTH) receptor-1 phosphorylation in regulating acute responses to PTH. Proc Natl Acad Sci U S A 110:5864–5869
44. Mannstadt M, Clarke BL, Vokes T et al (2013) Efficacy and safety of recombinant human parathyroid hormone (1–84) in hypoparathyroidism (REPLACE): a double-blind, placebo-controlled, randomised, phase 3 study. Lancet Diabetes Endocrinol 1:275–283

2 Embryology of the Parathyroid Glands

Nancy R. Manley

2.1 Introduction

The parathyroid glands develop from the pharyngeal pouches, transient endodermal outpocketings that also form the thymus and ultimobranchial bodies in vertebrates. The parathyroids vary in number and final location in different vertebrates, including in humans and mice. Despite its importance in calcium physiology, the molecular regulators and cellular events underlying parathyroid organogenesis have only recently begun to be elucidated, in part due to their small size, nondescript shape, and variable locations. Recent work has identified some of the key molecular regulators of parathyroid organogenesis, including the transcription factors GCM2, GATA3, and TBX1, and the sonic hedgehog (SHH) signaling pathway, and the morphogenetic events leading to their development have begun to be defined. The parathyroid glands develop from a shared initial organ primordium with the thymus glands, leading to interesting connections between these two organs with diverse functions. Finally, a recent study has shown that parathyroid cell fate may be unstable during late fetal development. Further understanding of the mechanisms controlling parathyroid specification and embryonic development could contribute to better understanding of parathyroid biology and improved treatment for hypoparathyroidism in humans.

2.2 Anatomy of Parathyroid Organogenesis

Parathyroids originate from the posterior pharyngeal pouches, transient bilateral endodermal outpocketings that form from the pharynx during embryogenesis. The number of parathyroids and which pouches they originate from is species specific; humans and birds (chickens) have four parathyroids arising from the 3rd and 4th pharyngeal pouches (pp) [1–4], while mice have two parathyroids that come from the 3rd pp [5]. Nearly all of the information we have regarding parathyroid organogenesis has come from studies in mice, facilitated by the identification of the early regulator of parathyroid differentiation, *glial cells missing 2*, or *Gcm2* [6]. The expression of *Gcm2* throughout parathyroid organogenesis allowed the tracking of parathyroid-fated cells throughout embryonic development and has been a key to the recent developments in understanding parathyroid organogenesis.

Initial parathyroid organogenesis is closely linked to thymus organogenesis – these organs arise from different regions of the same pouches, and during development they undergo a series of morphogenetic events to form separate organs (reviewed in [7]). The initial parathyroid domain forms in the dorsal-anterior region of the pp and

N.R. Manley
Department of Genetics, University of Georgia,
500 DW. Brooks Drive, Athens, GA 30602, USA
e-mail: nmanley@uga.edu

M.L. Brandi, E.M. Brown (eds.), *Hypoparathyroidism*,
DOI 10.1007/978-88-470-5376-2_2, © Springer-Verlag Italia 2015

the forming pp-derived organ primordia, the ventral domain of which constitutes the developing thymus. These primordia must (in mice and humans) detach from the pharynx via localized apoptosis [7, 8]. The thymus and parathyroid domains separate from each other by less well-understood mechanisms, likely involving both differential cell adhesion, involvement of the surrounding neural crest cells (NCCs), and physical forces derived in part from thymus migration [5, 9]. Current evidence suggests that while the thymus lobes actively migrate, via the activity of the NCC-derived capsule [10]; the parathyroids do not themselves migrate but are "dragged" along by the migrating thymus lobes until the separation process is complete. This process introduces variability in their final locations, most often near the lateral aspects of the thyroid gland, but can be nearly anywhere in the neck region.

As a result of this connection during early organogenesis, the thymus and parathyroids have been often studied together and have been suggested to have functional overlap as well. These issues are discussed at the end of this chapter; the majority of this chapter will focus on the current knowledge of the molecular regulation of parathyroid cell fate specification and differentiation during organogenesis.

2.3 Molecular Regulators of Initial Parathyroid Specification

2.3.1 Transcription Factors

Because of their small size, variable location, and rather indistinct shape, little was known about parathyroid gland organogenesis until the identification of the early parathyroid marker *Gcm2*. *Gcm2* encodes a transcription factor related to the *glial cells missing* gene, originally identified in *Drosophila* as a molecular switch between neural and glial cell fate (reviewed in [11]. Although *Gcm2* does not have this same function in mammals, it plays a critical role in parathyroid development [12]. However, *Gcm2* expression does not appear to specify parathyroid cell fate or define the parathyroid domain during initial organogenesis. In the absence of *Gcm2*, the parathyroid domain (or at least a domain that expresses some parathyroid-related genes) appears to be specified at E10.5. This domain then undergoes rapid and coordinated apoptosis at about E11.5–12 [13]. Thus, other transcription factors and signaling pathways must specify parathyroid fate. While several candidates have been identified, the transcriptional network that specifies cell fate, and directly or indirectly upregulates *Gcm2* expression, has still not been clearly articulated.

A suite of genes including *Hoxa3*, *Pax1,9*, *Eya1*, and *Six1,4* have been proposed to constitute a Hox-Pax-Eya-Six network that controls early pouch patterning and organogenesis. While single and double mutants for these genes generally result in parathyroid agenesis or severe hypoplasia, the exact structure of such a network and whether these genes act individually or in concert to affect parathyroid fate specification is less clear. The first to be identified was *Hoxa3* [14]. Null mutants are aparathyroid and athymic, and due in part to the classical role of HOX proteins in specifying regional identity, the prevailing model has been that *Hoxa3* specifies 3rd pp identity and patterning [15]. However, recent evidence has demonstrated that in *Hoxa3* mutants, *Gcm2* is expressed its normal domain, but at very low levels indicating that *Hoxa3* upregulates *Gcm2* but is not required to specify parathyroid fate [16, 17]. Whether this regulation is direct or indirect is unknown; however, evidence from $Hoxa3^{+/-}Pax1^{-/-}$ mutants suggests that *Hoxa3* may work with the paired box transcription factor PAX1. *Pax1* single mutants have normal initial *Gcm2* expression, but do not maintain it, resulting in significant parathyroid hypoplasia [18]; this phenotype is exacerbated in $Hoxa3^{+/-}Pax1^{-/-}$ compound mutants. *Eya1* and *Six1,4* have also been shown to be required for *Gcm2* expression, and mutants result in loss through apoptosis. As loss of *Gcm2* itself is sufficient to cause apoptosis, it is possible that the effects of all of these genes, either individually or as a pathway or network, are mediated by their effect (direct or indirect) on *Gcm2* expression.

The two best candidates for transcriptional regulators that specify parathyroid fate are TBX1 and GATA3, both of which are expressed in the parathyroid domain in the 3rd pp and have been implicated in regulating *Gcm2*. *Tbx1* expression is correlated spatially and temporally with *Gcm2*, and its expression in the 3rd pp is unaffected in *Gcm2* null mutant mice [13], indicating that it acts upstream of, or in parallel to, Gcm2. However, recent work from the author and collaborators has shown that ectopic expression of *Tbx1* in the 3rd pp outside the parathyroid domain is not sufficient to induce *Gcm2* expression [19], and *Tbx1* null mutants do not form the caudal pouches at all [20]. Thus, it is unclear whether TBX1 plays any specific role in parathyroid specification or organogenesis and, if so, whether it regulates *Gcm2* expression directly or indirectly. In contrast, GATA3 has been shown to directly bind to the *Gcm2* promoter region and upregulate its expression, and *Gcm2* levels are reduced even in heterozygotes [21]. Whether GATA3 plays a role in organ fate specification is less clear. *Gata3*$^{+/-}$ heterozygotes have fewer *Gcm2-expressing* cells, suggesting that GATA3 could affect cell fate [21]. However, this possibility has not been directly investigated.

The final candidate gene identified so far is *Sox3*. Human mutations in Sox3 are associated with hypoparathyroidism, and *Sox3* is expressed in the 3rd pp and developing parathyroids in mice [22]. However, no direct connection has so far been made between *Sox3* and *Gcm2* expression or other aspects of parathyroid organogenesis, so its specific role is still unknown. Thus, while all of these transcription factors have been shown to affect organogenesis and patterning, the identity of the direct targets for these transcription factors and clear evidence for a role in specifying parathyroid cell fate, as opposed to promoting *Gcm2* expression, is lacking.

2.3.2 Signaling Pathways

While transcriptional regulators generally act cell autonomously, signaling pathways can act either within or between tissues to influence cell fate and/or differentiation. Thus, signals that specify parathyroid fate could be expressed either within the endoderm or in the adjacent NCC mesenchyme, and there is evidence for both. Three signaling pathways, SHH, BMP4, and FGF8/10, have been implicated as positive or negative regulators of parathyroid fate in the 3rd pp in mice and are discussed below. All of them are expressed within the endoderm. However, data from *Splotch* mutant mice, which have a deficiency in NCCs, have shown that the size of the parathyroid domain within the pouch is in part determined by signals from the surrounding NCCs [23]. Thus, signals coming from either or both cell types during patterning could influence the location and size of the parathyroid domain within the endoderm.

The earliest identified signaling pathway to influence parathyroid fate within the pouch endoderm is *sonic hedgehog* (SHH). *Shh* null mutant mice fail to establish a prospective parathyroid domain or express *Gcm2*, and thymus fate spreads to encompass the entire pouch [24]. However, there are conflicting data on whether SHH is acting directly within the endoderm or indirectly (either from adjacent endoderm or through a NCC-mediated mechanism) to establish parathyroid fate [24, 25]. Intriguingly, *Tbx1* is known to act downstream of SHH signaling in heart development [26], raising the possibility that SHH acts in part through inducing *Tbx1* in this case as well. However, gain of function studies in the author's lab, in which ectopic SHH signaling in other domains of the 3rd pp in mouse embryos induced *Tbx1*, but not *Gcm2*, indicate that this pathway is not sufficient to turn on *Gcm2* outside the normal parathyroid domain [19]. These data indicate that either other SHH targets, or additional signals or pathways, may be required to fully induce the parathyroid pathway.

The fibroblast growth factor (FGF) signaling pathway has also been implicated in suppressing parathyroid fate and/or differentiation. The main *Fgf* gene implicated in 3rd pp patterning and development in mice is Fgf8, but as *Fgf8* null mutants fail to form the caudal pouches, loss of function approaches is limited. However, members of the sprouty (*Spry*) class of FGF inhibitors are expressed in the 3rd pp in mice, and mutations in these genes cause enhanced and ectopic FGF signaling throughout the pouch at E10.5 and later [7]. In *Spry1,2* double mutants, parathyroid size is reduced, and Gcm2 expression is delayed, indicating that excessive FGF signaling can suppress parathyroid specification and differentiation. This effect was suppressed by reducing the dosage of *Fgf8*, which is normally expressed in the ventral endoderm and off by E11.5. However, FGF10 is also expressed in the NCC mesenchyme adjacent to the dorsal domain, so some of the effect of FGF signaling on the parathyroid domain may come from FGF10. These results suggest that the effects of FGF signaling on parathyroid organogenesis may occur quite early and from both the within the endoderm and from the NCC mesenchyme, to restrict parathyroid fate to the most dorsal domain of the pouch.

The last signaling pathway that has been implicated in parathyroid fate specification is the BMP pathway, specifically BMP4. The role of BMP4 is less clear, as there is evidence for both a positive and a negative role. Like *Fgf8*, *Bmp4* expression is not expressed in the parathyroid domain but is restricted to the ventral thymus domain. In the SHH null, *Bmp4* expansion throughout the 3rd pp is coincident with loss of the parathyroid domain and expansion of thymus fate. Furthermore, the expression of the BMP inhibitor Noggin in the NCC mesenchyme surrounding the dorsal parathyroid domain suggests that suppressing BMP signaling is important for parathyroid fate or differentiation. Taken together, these data have been interpreted to indicate a SHH-BMP mutual antagonism in establishing parathyroid and thymus cell fate in the 3rd pp [5]. However, evidence from chick showed that inhibition of BMP signaling (via ectopic Noggin) suppressed *Gcm2* expression, at least at early stages of pouch development, suggesting that BMP signaling is at least transiently a positive regulator of *Gcm2* expression and parathyroid differentiation in this system. Thus, the role of BMP signaling in parathyroid fate specification and/or differentiation, and whether there are species-specific differences in this process, will require further investigation.

2.4 Differentiation and Survival of Parathyroids: *Gcm2*

Once the parathyroid domain is established, upregulation of *Gcm2* expression is necessary and sufficient for parathyroid differentiation and survival. *GCM2* is also known to be important in human parathyroid development, as both dominant negative [27] and loss-of-function [28] *GCM2* alleles are associated with hypoparathyroidism in humans (see also Chap. 14). In the *Gcm2* null mutant mouse, the parathyroid domain is specified, as evidenced by normal expression of the parathyroid-associated genes *Tbx1*, *Ccl21*, and *Casr* (*calcium-sensing receptor*) in the dorsal domain at E10.5 and failure of the thymus domain to expand into this region [13]. However, these cells fail to upregulate *parathyroid hormone* (*Pth*) at E11.5 and undergo coordinated apoptosis soon after, by E12.5. GCM2 also works with the transcription factor MAFB to upregulate *Pth* gene expression [29]. *MafB* mutation also affects parathyroid separation from the thymus and may itself be regulated by GCM2.

Thus, upregulation of *Gcm2* is a critical step in early parathyroid differentiation and survival. *Gcm2* continues to be expressed in parathyroids after the early stages of differentiation, and the loss of parathyroids after downregulation of *Gcm2* expression in *Hoxa3* and *Pax1* mutants suggests that it may still be required for parathyroid survival at least during fetal development. However, in the absence of conditional deletion of *Gcm2* at later stages, it is not clear if it is required for parathyroid maintenance once they are established.

2.5 The Thymus-Parathyroid Connection

2.5.1 Do the Thymus and Parathyroids Have Overlapping Functions?

The primary functions of the thymus and parathyroid glands are quite distinct, with the thymus playing a critical role in producing T cells and parathyroids controlling calcium physiology through the production of PTH. However, the physical connection between the thymus and parathyroid organs during early organogenesis has led to reports that these organs may indeed have overlapping functions.

The original report of the *Gcm2* null mutant phenotype received attention not only because it was the first gene to specifically be required only for parathyroid organogenesis but also because of the conclusion that the thymus could act as a secondary source of PTH [12]. This conclusion was based on survival of a significant proportion of *Gcm2* null mutants, even in the absence of parathyroid glands, their report of low levels of serum PTH in the absence of parathyroids, and on the observation that removing the thyroid and parathyroids together from wild-type mice did not cause lethality, which removing the thymus as well caused rapid death (presumably due to lack of PTH). As the parathyroids had been thought to be the sole source of physiological PTH, this was considered a significant finding with potential implications for human health [30]. A more recent study, based on this conclusion, reported the ability to generate and isolate parathyroid-like cells from thymic epithelial cells, as an initial effort to produce parathyroid cells for transplant [31].

While this report was consistent with the common origin of the thymus and parathyroids in the 3rd pp, work from the author's lab showed that this conclusion was not entirely accurate [2]. Instead, the PTH thought to be produced by the thymus was produced by authentic parathyroid cells that remain attached to the thymus during normal organogenesis. This study showed that the process of thymus-parathyroid separation is inefficient and "messy," leading to small clusters of parathyroid cells remaining associated with the thymus and numerous small clusters of parathyroid cells throughout the neck region in addition to the primary parathyroid glands. These "ectopic" thymus-associated parathyroid cells are the likely source of PTH in the original *Gcm2* null paper and also call into question the identity of the parathyroid cells that were thought to have been generated from thymus cells in the 2011 study [31], as these could have been parathyroid cells already present in the thymus.

While the thymus does not have true parathyroid-like function, the parathyroid domain during initial organogenesis does have a transient thymus-related function. At E11.5, prior to the separation of the two organs, the parathyroid domain expresses *Ccl21*, a chemokine that contributes to initial immigration of lymphoid progenitors to the thymus, which is important in early thymus organogenesis [32, 33]. Therefore, while the thymus doesn't appear to have any parathyroid function, the parathyroid domain does help recruit lymphoid cells to the thymus, at least during initial organogenesis.

2.5.2 Stability of Parathyroid Cell Fate

The presence of small clusters of parathyroid cells throughout the neck in both mice and humans, as a consequence of normal development, also has another unusual consequence. In about half of mice and in a substantial percentage of humans, these remnants of the organ separation process can downregulate the parathyroid program and transdifferentiate in a thymus fate, forming small cervical thymi [34, 35]. In addition, the author's lab has recently shown that about 25 % of these cervical thymi have previously differentiated as parathyroid, including prior expression of *Pth* [36]. These parathyroid-derived cervical thymi (pCT) generate T cells with a specific functional phenotype that could have implications for the function of the immune system in individuals with pCT [36]. While the mechanisms by which this cell fate switch occurs are unknown, parathyroid fate appears to stabilize at about the newborn stage, after which the frequency of cervical thymi remains constant. This "window of opportunity" for parathyroid cells to downregulate the parathyroid program and transdifferentiate to a thymus fate suggests that there is an underlying instability in parathyroid fate during a specific temporal window during the late fetal stage.

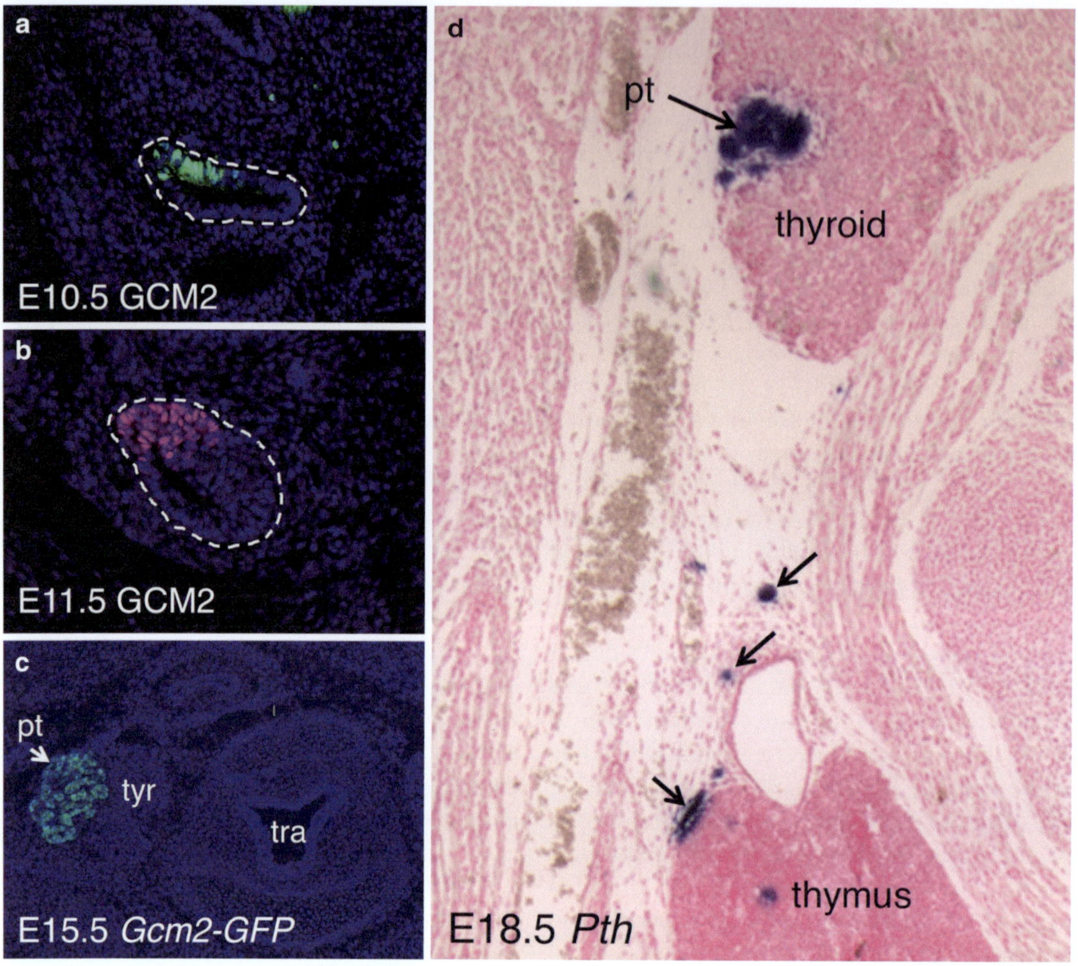

Fig. 2.1 Parathyroid organogenesis in the mouse embryo. (**a**, **b**) Saggital sections of mouse E10.5 (**a**) and E11.5 (**b**) embryos stained with an antibody recognizing GCM2. At these stages, GCM2 (*pink* or *green*) marks the dorsal-anterior domain of the 3rd pharyngeal pouch-derived organ primordium (*outlined* in *white dashed line*); the remainder of the pouch becomes thymus. (**c**) By E15.5, the parathyroid (*pt*) has separated from the thymus and is usually located near the lateral aspects of the thyroid lobes (*tyr*). In this panel, tra, trachea. *Gcm2* expression is shown using a GCM2-EGFP transgene. (**d**) At E18.5 and after birth, the main parathyroid gland (*pt*) is usually located at or within the thyroid gland, here identified by in situ hybridization with a probe for *Pth*. However, small clusters of parathyroid cells are present throughout the neck between the main parathyroid and the thymus gland (*arrows*), remnants of the process of organ separation and migration

This phenomenon is not just an oddity of development that may affect the immune system. Understanding how cell fate is stabilized is important to the issue of therapeutic stem cell-based interventions in general and to the generation of parathyroid cells for transplant in particular. Parathyroid cells are excellent targets for generation of differentiated cells for transplant from ES or iPS cells. Further investigation of this apparently inherent but transient instability, and how it is resolved during development, could provide important keys to future efforts to generate parathyroid cells for transplant.

Acknowledgments The author thanks Jena Chojnowski and Zhijie Liu for generating the images in Fig. 2.1. The author also thanks John O'Neil for providing helpful comments on the manuscript.

References

1. Gilmour J (1937) The embryology of the parathyroid glands, the thymus, and certain associated rudiments. J Pathol 45:507–522
2. Liu Z, Farley A, Chen L, Kirby BJ, Kovacs CS, Blackburn CC et al (2010) Thymus-associated parathyroid hormone has two cellular origins with distinct endocrine and immunological functions. PLoS Genet 6(12):e1001251
3. Okabe M, Graham A (2004) The origin of the parathyroid gland. Proc Natl Acad Sci U S A 101(51): 17716–17719
4. Neves H, Dupin E, Parreira L, Le Douarin NM (2012) Modulation of Bmp4 signalling in the epithelial-mesenchymal interactions that take place in early thymus and parathyroid development in avian embryos. Dev Biol [Research Support, Non-US Gov't] 361(2):208–219
5. Gordon J, Manley NR (2011) Mechanisms of thymus organogenesis and morphogenesis. Development 138(18):3865–3878
6. Kim J, Jones BW, Zock C, Chen Z, Wang H, Goodman CS et al (1998) Isolation and characterization of mammalian homologs of the Drosophila gene glial cells missing. Proc Natl Acad Sci U S A 95(21): 12364–12369
7. Gardiner JR, Jackson AL, Gordon J, Lickert H, Manley NR, Basson MA (2012) Localised inhibition of FGF signalling in the third pharyngeal pouch is required for normal thymus and parathyroid organogenesis. Development 139(18):3456–3466
8. Gordon J, Wilson VA, Blair NF, Sheridan J, Farley A, Wilson L et al (2004) Functional evidence for a single endodermal origin for the thymic epithelium. Nat Immunol 5(5):546–553
9. Gordon J, Patel SR, Mishina Y, Manley NR (2010) Evidence for an early role for BMP4 signaling in thymus and parathyroid morphogenesis. Dev Biol 339(1):141–154
10. Foster KE, Gordon J, Cardenas K, Veiga-Fernandes H, Makinen T, Grigorieva E et al (2010) EphB-ephrin-B2 interactions are required for thymus migration during organogenesis. Proc Natl Acad Sci U S A 107(30):13414–13419
11. Hashemolhosseini S, Wegner M (2004) Impacts of a new transcription factor family: mammalian GCM proteins in health and disease. J Cell Biol [Research Support, Non-US Gov't Review] 166(6): 765–768
12. Gunther T, Chen ZF, Kim J, Priemel M, Rueger JM, Amling M et al (2000) Genetic ablation of parathyroid glands reveals another source of parathyroid hormone. Nature 406(6792):199–203
13. Liu Z, Yu S, Manley NR (2007) Gcm2 is required for the differentiation and survival of parathyroid precursor cells in the parathyroid/thymus primordia. Dev Biol 305(1):333–346
14. Chisaka O, Capecchi MR (1991) Regionally restricted developmental defects resulting from targeted disruption of the mouse homeobox gene *hox-1.5*. Nature 350:473–479
15. Manley NR, Condie BG (2010) Transcriptional regulation of thymus organogenesis and thymic epithelial cell differentiation. Prog Mol Biol Transl Sci 92:103–120
16. Chen L, Zhao P, Wells L, Amemiya CT, Condie BG, Manley NR (2010) Mouse and zebrafish Hoxa3 orthologues have nonequivalent in vivo protein function. Proc Natl Acad Sci U S A 107(23):10555–10560
17. Chojnowski J, Trau H, Masuda K, Manley N (2013) Complex tissue specific roles for HOXA3 during thymus and parathyroid development. Paper presented at the international society for developmental biology conference, Cancun, June 2013
18. Su D, Ellis S, Napier A, Lee K, Manley NR (2001) Hoxa3 and pax1 regulate epithelial cell death and proliferation during thymus and parathyroid organogenesis. Dev Biol 236(2):316–329
19. Manley N, Bain V, Gordon J, Gutierrez K, Cardenas K, Richie E (2012) Regulation of thymus and parathyroid organ fate specification by Shh and Tbx1. Paper presented at the mouse molecular genetics conference, Asilomar, Sept 2012
20. Jerome LA, Papaioannou VE (2001) DiGeorge syndrome phenotype in mice mutant for the T-box gene, Tbx1. Nat Genet 27(3):286–291
21. Grigorieva IV, Mirczuk S, Gaynor KU, Nesbit MA, Grigorieva EF, Wei Q et al (2010) Gata3-deficient mice develop parathyroid abnormalities due to dysregulation of the parathyroid-specific transcription factor Gcm2. J Clin Invest 120(6):2144–2155
22. Bowl MR, Nesbit MA, Harding B, Levy E, Jefferson A, Volpi E et al (2005) An interstitial deletion-insertion involving chromosomes 2p25.3 and Xq27.1, near SOX3, causes X-linked recessive hypoparathyroidism. J Clin Invest 115(10): 2822–2831
23. Griffith AV, Cardenas K, Carter C, Gordon J, Iberg A, Engleka K et al (2009) Increased thymus- and decreased parathyroid-fated organ domains in Splotch mutant embryos. Dev Biol 327(1):216–227
24. Moore-Scott BA, Manley NR (2005) Differential expression of Sonic hedgehog along the anterior-posterior axis regulates patterning of pharyngeal pouch endoderm and pharyngeal endoderm-derived organs. Dev Biol 278(2):323–335
25. Grevellec A, Graham A, Tucker AS (2011) Shh signalling restricts the expression of Gcm2 and controls the position of the developing parathyroids. Dev Biol [Research Support, Non-US Gov't] 353(2):194–205
26. Garg V, Yamagishi C, Hu T, Kathiriya IS, Yamagishi H, Srivastava D (2001) Tbx1, a DiGeorge syndrome candidate gene, is regulated by sonic hedgehog during pharyngeal arch development. Dev Biol 235(1):62–73

27. Mannstadt M, Bertrand G, Muresan M, Weryha G, Leheup B, Pulusani SR et al (2008) Dominant-negative GCMB mutations cause an autosomal dominant form of hypoparathyroidism. J Clin Endocrinol Metabol [Case Reports Research Support, NIH, Extramural Research Support, Non-US Gov't] 93(9):3568–3576
28. Ding C, Buckingham B, Levine MA (2001) Familial isolated hypoparathyroidism caused by a mutation in the gene for the transcription factor GCMB. J Clin Invest 108(8):1215–1220
29. Kamitani-Kawamoto A, Hamada M, Moriguchi T, Miyai M, Saji F, Hatamura I et al (2011) MafB interacts with Gcm2 and regulates parathyroid hormone expression and parathyroid development. J Bone Miner Res [Research Support, Non-US Gov't] 26(10):2463–2472
30. Balling R, Erben RG (2000) From parathyroid to thymus, via glial cells. Nat Med [News] 6(8):860–861
31. Woods Ignatoski KM, Bingham EL, Frome LK, Doherty GM (2011) Directed trans-differentiation of thymus cells into parathyroid-like cells without genetic manipulation. Tissue Eng Part C Methods 17(11):1051–1059
32. Liu C, Saito F, Liu Z, Lei Y, Uehara S, Love P et al (2006) Coordination between CCR7- and CCR9-mediated chemokine signals in prevascular fetal thymus colonization. Blood 108(8):2531–2539
33. Liu C, Ueno T, Kuse S, Saito F, Nitta T, Piali L et al (2005) The role of CCL21 in recruitment of T-precursor cells to fetal thymi. Blood 105(1):31–39
34. Dooley J, Erickson M, Gillard GO, Farr AG (2006) Cervical thymus in the mouse. J Immunol 176(11):6484–6490
35. Terszowski G, Muller SM, Bleul CC, Blum C, Schirmbeck R, Reimann J et al (2006) Evidence for a functional second thymus in mice. Science 312(5771):284–287
36. Li J, Liu Z, Xiao S, Manley NR (2013) Transdifferentiation of parathyroid cells into cervical thymi promotes atypical T-cell development. Nat Commun [Research Support, NIH, Extramural] 4:2959

3 PTH and PTH-Related Peptides

Giancarlo Isaia and Margherita Marchetti

3.1 Introduction

Parathyroid hormone (PTH) is an 84-amino acid polypeptide endocrine hormone that is produced by the parathyroid glands and secreted into the circulation in response to low calcium levels [1]. PTHrP is a polypeptide that was originally isolated as the factor responsible for humoral hypercalcemia of malignancy. Subsequently, it became apparent that PTHrP is a critical developmental paracrine factor, and it is nearly ubiquitously produced and secreted by normal and malignant cells. Both peptides hold clinical interest for their capacities to control calcium/phosphate homeostasis and bone metabolism [2].

3.2 PTH and PTHrP: Genes, Structures, and Biosynthesis

PTH is a peptide comprised of a single 84-amino acid chain, which is synthesized and secreted by the parathyroid glands. The amino terminus (residues 1–34) is highly conserved and is important for the biological activity of the molecule. The PTH human gene is localized on chromosome 11p15 and consists of three exons. In the rough endoplasmic reticulum of the parathyroid cells, PTH is synthesized as a 115-amino acid polypeptide (prepro-PTH), which is first cotranslationally cleaved in the ER to pro-PTH (90 amino acids) and then to the mature, biologically active PTH molecule (84 amino acids) in the Golgi apparatus. The N-terminal cleaved pre-sequence is rich in hydrophobic amino acids that are necessary for transport of the nascent polypeptide chain into the endoplasmic reticulum, while the basic pro-peptide directs accurate cleavage of pro-PTH into the mature 1–84 molecule [3]. The homology among species is high in the region that encodes prepro-PTH, and substantial homology also is retained in the gene flanking regions, introns, and mRNA UTRs. The transcription of the prepro-PTH gene is in a relatively suppressed state under normocalcemic conditions. This posttranscriptional regulation is dependent upon binding of protective trans-acting factors to a specific element in the PTH mRNA 3′-UTR. The molecular weight of PTH(1–84) is 9.425 Da. The biosynthetic process is estimated to take less than 1 h. After its synthesis, PTH(1–84) is stored in secretory vesicles. In the secretory vesicles some C-terminal fragments are produced, especially at high extracellular calcium concentrations, which reduce the fraction of secreted PTH that is PTH(1–84). Under hypocalcemic conditions the stored PTH is released by exocytosis within seconds and, in a few hours, gene transcription increases. If hypocalcemic conditions persist for days, parathyroid cells start to replicate and became hyperplastic.

G. Isaia (✉) • M. Marchetti
Gerontology and Bone Metabolic Disease Section, Department of Medical Sciences, University of Torino, Corso Bramante 88/90, 10126 Torino, Italy
e-mail: giancarlo.isaia@unito.it; margherita.marchetti@hotmail.it

M.L. Brandi, E.M. Brown (eds.), *Hypoparathyroidism*,
DOI 10.1007/978-88-470-5376-2_3, © Springer-Verlag Italia 2015

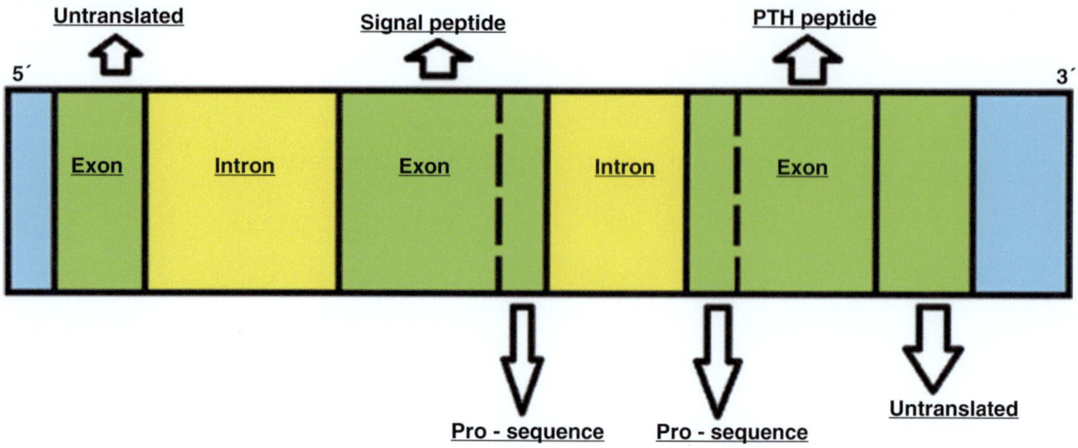

Fig. 3.1 Structure of human PTH gene

The mammalian and avian PTH genes include two introns, which are removed by RNA splicing while the final mature RNA is being generated, that divide the gene into the three exons encoding, respectively:

1. The 5′ untranslated region (UTR)
2. The signal peptide
3. The mature peptide and the 3′ UTR (Fig. 3.1)

Study of parathyroid hormone-related protein (PTHrP) began in 1987 as the culmination of a 40-year search for the humoral factor responsible for the most common paraneoplastic syndrome, humoral hypercalcemia of malignancy. Soon afterward came the discovery that PTHrP is a ubiquitously expressed protein that is essential for life. It is a classical paracrine peptide hormone that undergoes extensive posttranslational processing before its secretion. The several secreted forms of PTHrP have a broad range of effects in many organs involving development, survival, and function [4]. Eight of the first 13 N-terminal amino acids of PTHrP are identical to those in PTH, and the three-dimensional structures of residues 13–34 of PTH and PTHrP are strikingly similar; this part of the two molecules is responsible for the binding and activation of their common receptor, the PTH/PTHrP receptor, type 1 (PTHR1) [5].

Despite their N-terminal homology and their calciotropic properties, PTH and PTHrP are the products of separate genes located on distinct chromosomes. It is believed that both chromosomes originated from a single ancestral gene and were generated as distinct entities through tetraploidization events [6]. Human PTHrP is encoded by a single gene on chromosome 12p12.1-11.2, which has regions homologous to chromosome 11p15. With the exception of the short N-terminal region, the structure of PTHrP is not closely related to that of PTH. The regions of the chromosomes containing the two hormones have similar banding patterns and contain related genes, such as the A and B isoforms of lactate dehydrogenase, Sox 5 and 6, and the Hand K-ras genes [7, 8]. The PTHrP gene is more complex than the human PTH gene: it consists of 9 exons, and alternative splicing generates up to 12 transcripts, which encode three separate isoforms of 139, 141, or 173 amino acids. Expression of the PTHrP gene is regulated by many hormones and growth factors. The multiple products of its posttranslational processing, including glycosylation, combined with the short half-life of PTHrP mRNA and the multiple biological activities contained within PTHrP, equip it ideally to function as a paracrine effector with a developmental focus. Combined with the susceptibility of PTHrP to posttranslational modification through proteolysis and the generation of several constituent peptides, this increased complexity highlights the potential versatility of PTHrP as a paracrine regulator [9].

PTHrP is usually undetectable in the circulating blood of normal subjects, but is produced in a paracrine/autocrine fashion during fetal and adult life by a number of normal cells and tissues in which it is believed to play an expanding number of physiological roles through these autocrine/paracrine pathways. PTHrP is synthesized and expressed by various tissues, such as blood vessels, smooth muscles, growth plate chondrocytes, bone, keratinocytes, mammary gland, placenta, kidney, pancreas, and neuronal and glial tissues [10].

3.3 Regulation, Metabolism, and Catabolism of PTH and PTHrP

Once secreted, PTH is rapidly cleared from plasma through uptake principally by the liver and kidney, where PTH(1–84) is cleaved into amino- and carboxyl-terminal fragments that are then cleared by the kidney (see also Chap. 4). Carboxyl-terminal fragments of PTH can be found in the blood together with PTH(1–84), and they can also be secreted from the parathyroid glands. Peripheral metabolism of PTH does not appear to be regulated by physiologic states (high versus low calcium, etc.); hence peripheral metabolism of hormone, although responsible for rapid clearance of secreted hormone, appears to be a high-capacity, metabolically invariant catabolic process [11]. The plasma concentration of PTH(1–84) is 10–55 pg/ml. Circulating immunoreactive PTH in normocalcemic subjects includes:

- PTH(1–84)—5–30 %
- C-terminal fragments—70–95 %
- N-terminal fragments—a small percentage

In normocalcemic conditions, PTH(1–84) is about 20 % of total circulating PTH molecules; during hypocalcemia PTH increases to 33 % and decreases to 4 % under hypercalcemic conditions. PTH(1–84) has a plasma half-life of 2–4 min. In comparison, the C-terminal fragments, which are cleared principally by the kidney, have half-lives that are five to ten times longer [12, 13].

PTH secretion is mostly regulated by the level of serum ionized calcium and the concentration of 1,25-dihydroxyvitamin D3. Extracellular and resultant intracellular magnesium deficiency can modify parathyroid function too, inhibiting PTH secretion, while high serum magnesium concentrations, well above the physiological range, also inhibit PTH secretion. Calcium ions interact with a calcium sensor, the extracellular calcium-sensing receptor (CaSR), a G protein-coupled receptor (GPCR). This receptor is a member of a distinctive subfamily of the GPCR superfamily (family C or 3) that is characterized by large extracellular domains suitable for "clamping" small-molecule ligands. Stimulation of the receptor by high calcium levels suppresses PTH secretion. The receptor is expressed by the parathyroid glands and the calcitonin-secreting cells (C cells) of the thyroid, as well as in other sites such as the brain and kidney. Genetic evidence has revealed a key biologic role for the CaSR in regulating parathyroid gland responsiveness to calcium, inhibiting PTH secretion, PTH gene expression, and parathyroid cellular proliferation as well as enhancing intracellular degradation of PTH(1–84). The CaSR also promotes renal calcium clearance.

Additional factors that participate in PTH regulation are the 1,25-dihydroxyvitamin D3, the serum phosphate concentration, and fibroblast growth factor 23 (FGF23). 1,25-dihydroxyvitamin D3 reduces expression of the PTH gene by inhibiting its transcription and decreases parathyroid cellular proliferation. 1,25-dihydroxyvitamin D3, in contrast, inhibits transcription of the CaSR gene. In contrast to 1,25-dihydroxyvitamin D, phosphate loading increases both expression of the PTH gene (by promoting stability of prepro-PTH mRNA) and parathyroid cellular proliferation. Recent data indicate that fibroblast growth factor 23 (FGF23) is also an important regulator of parathyroid function, suppressing both PTH gene expression and PTH secretion.

In contrast to PTH, PTHrP is a paracrine factor expressed throughout the body. It is a classical paracrine regulator that undergoes extensive posttranslational processing before secretion, and it is normally a secretory protein that enters the endoplasmic reticulum (ER) under the direction

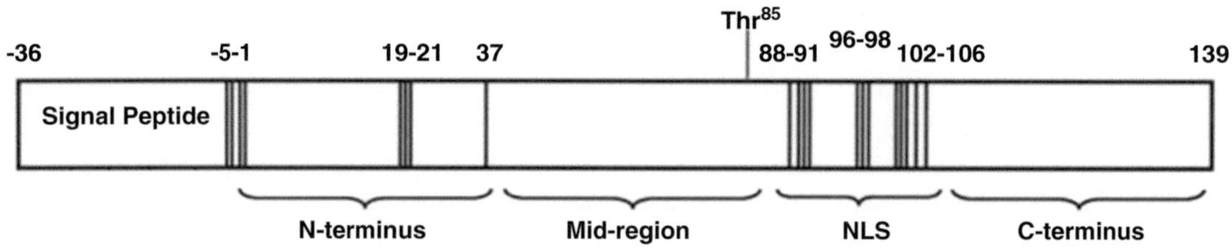

Fig. 3.2 Structure of the PTHrP protein, showing the signal peptide, amino (N)-terminus, mid region, nuclear localization signal (*NLS*), and carboxy (C)-terminus

of its signal peptide (Fig. 3.2) during its translation on ribosomes [14]. PTHrP is regulated by a variety of agents affecting its expression and secretion by many different cell types. An increase in PTHrP mRNA is observed rapidly and transiently following exposure of cells to serum, growth factors, and phorbol esters through mechanisms including increased gene transcription and enhanced mRNA stability. PTHrP can be ligated efficiently to multiple ubiquitin moieties. The ubiquitin-dependent proteolytic pathway is involved in regulating the metabolic stability of intracellular PTHrP, and this regulation may be an important mechanism for modulating its effects on cell growth and differentiation. Indeed, posttranslational control of PTHrP abundance may be defective in cancer cells [15].

3.4 Receptor and Biological Effects of PTH and PTHrP

Because there is significant homology within their N-termini, with 9 amino acid residues out of their first 13 amino acids being identical, PTH and PTHrP can bind to and activate their common receptor, the PTHR1, with equal affinity. The PTH/PTHrP receptor is a G protein-coupled receptor with 7 transmembrane-spanning domains and is encoded by a multi-exonic gene. The PTH/PTHrP receptor or PTH1R is expressed on target cells for PTH and PTHrP, such as osteoblasts in the bone and renal tubular cells in the kidney. The PTH1R is also expressed at lower levels in a number of other tissues, in which it mediates a large array of nontraditional paracrine and autocrine functions in response to locally produced PTHrP. Thus, some evidence has been provided for the existence of a different receptor for N-terminal PTHrP in keratinocytes, insulinoma cells, lymphocytes, and squamous carcinoma cell lines.

PTH has multiple functions. Its main activity is in the fine regulation of the concentration of calcium in the blood circulation, modulating movement of calcium into and out of the bone and renal tubular reabsorption of calcium so as to maintain serum calcium concentration within a narrow range. PTH has multiple actions on the bone, some direct and some indirect. It acts directly on osteoblasts, which then activate osteoblastic bone resorption and osteoclastogenesis. In the kidney PTH stimulates the conversion of 25-hydroxyvitamin D (25[OH]D) to 1,25-dihydroxyvitamin D3 (1,25[OH]2D3), its active metabolite, thereby enhancing calcium absorption in the gut. Thus PTH acts indirectly at the gastrointestinal tract through its effects on the 1-hydroxylation of 25-hydroxyvitamin D [16]. PTH also enhances renal calcium reabsorption and promotes renal phosphate excretion.

PTHrP can perturb calcium/phosphate homeostasis and bone metabolism under pathological conditions, i.e., when large tumors, especially of the squamous cell type, lead to massive overproduction of the hormone. However, normal circulating levels of PTHrP are negligible, and PTHrP is probably unimportant in normal calcium homeostasis in human adults. However, mice with a targeted disruption in the PTHrP gene show a lethal defect in bone development, thus demonstrating its importance in normal skeletal physiology. PTHrP is known to be a critical regulator of cellular and organ growth, development, migration, differentiation, and survival and of epithelial calcium ion transport in a variety of tissues. PTHrP is normally produced in many tissues and acts in those sites in a paracrine manner. In addition to these well-recognized and classical autocrine/paracrine roles, PTHrP has been observed to have intracrine actions as well, entering the nucleus under the direction of a nuclear localization signal (NLS) [17–19]. Because it interacts with the type 1 PTHR, injection of PTHrP produces hypercalcemia in experimental animals. There are only three identified circumstances in which PTHrP species are present in the circulation and act in an endocrine manner:

1. Fetal life, where PTHrP regulates maternal-to-fetal placental calcium transport.
2. Lactation, in which PTHrP is produced in the breast and reaches the circulation under the control of the CaSR. In this situation, the breast epithelial cells have been proposed to act as an "accessory parathyroid," increasing PTHrP release when blood calcium is low and vice versa. The increase in PTHrP when maternal calcium is reduced is thought to increase blood calcium concentration in the mother by mobilizing calcium from the bone and reducing its loss in the urine. In addition to regulating PTHrP release, the CaSR promotes transport of calcium into the milk, thereby ensuring adequate calcium in the milk for the newborn child when maternal calcium is sufficient. Thus, PTHrP is important not only during fetal but also newborn life for normal calcium and bone metabolisms.
3. The humoral hypercalcemic syndrome, in which PTHrP is produced by tumors and stimulates bone resorption.

There remains at present time no convincing evidence of biologically relevant circulating PTHrP levels otherwise in normal humans. The majority of the actions of PTHrP occur in a paracrine/autocrine manner, particularly in fetal development and physiology. Several examples are as follows [20, 21]:

- Stimulates bone resorption and also anabolic functions in the bone, when administered intermittently.
- Coordinates chondrocyte maturation, differentiation, and apoptosis to maintain the orderly growth of the long bones during development; regulates endochondral bone development.
- Ensures tooth eruption, by resorption of the alveolar bone to allow passage of the newly developed tooth. Tooth eruption requires the spatial coordination of bone cell activity. Osteoclasts must resorb the bone overlying the crown of the tooth to allow it to emerge, and osteoblasts must form the bone at the base of the tooth to propel it upward out of the crypt. PTHrP is normally produced by stellate reticulum cells, and it signals to dental follicular cells to promote the formation of osteoclasts above the crypt. In the absence of PTHrP, these osteoclasts do not appear, eruption fails to occur, and the teeth become impacted.
- Regulation of fetal mineral homeostasis; it works together with PTH to support the normal fetal blood calcium concentration.
- During lactation embryonic mammary development leading to nipple formation and branching morphogenesis. PTHrP might also participate in adolescent ductal morphogenesis.
- May modulate implantation of the fertilized ovum and retention of the embryo as well as relaxing the uterus and vascular smooth muscle; it inhibits oxytocin-stimulated activity during pregnancy and prevents preterm labor.
- Modulates trophoblastic growth and differentiation.
- During lactation induced by prolactin, it is released into the mother's bloodstream from the breast epithelial cells, as noted above, where it promotes calcium transport from blood to milk, increases mammary blood flow, and regulates maternal and neonate Ca-Pi metabolism.
- Acts as a vasodilator in resistance vessels, decreases vascular tone and blood pressure, and is believed to regulate regional and systemic hemodynamics.
- Regulates smooth muscle cell proliferation.
- PTHrP secreted in response to vasoconstrictor and mechanical stretching on smooth muscle cell mediates myorelaxant effects in a number of smooth muscle-containing organs. It also decreases vascular tone and blood pressure.
- Regulates keratinocyte differentiation, delaying terminal differentiation of hair follicles, epidermal keratinocytes, keratinization, and apoptosis. Decreases the number of hair follicles through epithelial-mesenchymal interactions.
- Delays beta cell apoptotic death, increases proliferation of human beta cells in the pancreas, and improves glucose-stimulated insulin secretion.

References

1. Clemens TL, Cormier S (2001) Parathyroid hormone-related protein and its receptors: nuclear functions and roles in the renal and cardiovascular systems, the placental trophoblasts and the pancreatic islets. Br J Pharmacol 134:1113–1136
2. Pioszak AA, Parker NR et al (2009) Structural basis for parathyroid hormone-related protein binding to the parathyroid hormone receptor and design of conformation-selective peptides. J Biol Chem 284: 28382–28391
3. Murray TM, Rao LG (2005) Parathyroid hormone secretion and action: evidence for discrete receptors for the carboxyl-terminal region and related biological actions of carboxyl- terminal ligands. Endocr Rev 26:78
4. Burtis WJ, Wu T et al (1987) Identification of a novel 17,000-dalton parathyroid hormone-like adenylate cyclase-stimulating protein from a tumor associated with humoral hypercalcemia of malignancy. J Biol Chem 262:7151–7156
5. Pinheiro P, Cardoso J et al (2010) Gene structure, transcripts and calciotropic effects of the PTH family of peptides in Xenopus and chicken. BMC Evol Biol 10:37
6. Gensure RC, Ponugoti B (2004) Identification and characterization of two parathyroid hormone-like molecules in zebrafish. Endocrinology 145:1634–1639
7. Gensure RC, Gardella TJ (2005) Parathyroid hormone and parathyroid hormone-related peptide, and their receptors. Biochem Biophys Res Commun 328: 666–678
8. Canario AVM, Rotllant J (2006) Novel bioactive parathyroid hormone and related peptides in teleost fish. FEBS Lett 580:291–299
9. McCauley LK, Martin TJ (2012) Twenty-five years of PTHrP progress: from cancer hormone to multifunctional cytokine. J Bone Miner Res 27:1231–1239
10. Datta NS, Abou-Samra AB (2009) PTH and PTHrP signaling in osteoblasts. Cell Signal 21:1245–1254
11. Potts JT Jr (2005) Chapter 347. Diseases of the parathyroid gland and other hyper- and hypocalcemic disorders. In: Harrison's internal medicine. Mc Graw Hill, New York
12. D'Amour P, Räkel A et al (2006) Acute regulation of circulating parathyroid hormone (PTH) molecular forms by calcium: utility of PTH fragments/PTH(1–84) ratios derived from three generations of PTH assays. J Clin Endocrinol Metab 91:283
13. Jüppner H, Abou-Samra AB et al (1991) A G protein-linked receptor for parathyroid hormone and parathyroid hormone-related peptide. Science 254:1024
14. Fiaschi-Taeschi NM, Stewart AF (2003) Minireview: parathyroid hormone-related protein as an intracrine factor—trafficking mechanisms and functional consequences. Endocrinology 144(2):407–411
15. Meerovitch K, Wing S (1997) Preproparathyroid hormone-related protein, a secreted peptide, is a substrate for the ubiquitin proteolytic system. J Biol Chem 272:6706–6713
16. Al-Azem H, Khan AA (2012) Hypoparathyroidism best practice & research clinical endocrinology & metabolism. Best Pract Res Clin Endocrinol Metab. 26:517–522
17. Bhatia V, Saini MK et al (2009) Nuclear PTHrP targeting regulates PTHrP secretion and enhances LoVo cell growth and survival. Regul Pept 158(1–3):149–155
18. Richard V, Luchin A (2003) Quantitative evaluation of alternative promoter usage and 30 splice variants for parathyroid hormone-related protein by real-time reverse transcription-PCR assay. Clin Chem 49:1398–1402
19. Sellers RS, Luchin AI (2004) Alternative splicing of parathyroid hormone-related protein mRNA: expression and stability. J Mol Endocrinol 33:227–241
20. Wysolmerski JJ (2012) Parathyroid hormone-related protein: an update. J Clin Endocrinol Metab 97(9):2947–2956
21. Dean T, Vilardaga JP et al (2008) Altered selectivity of Parathyroid Hormone (PTH) and PTH-Related Protein (PTHrP) for distinct conformations of the PTH/PTHrP receptor. Mol Endocrinol 22(1):156–166

PTH Assays and Their Clinical Significance

4

Pierre D'Amour

4.1 Introduction

The development of three generations of PTH assays has been necessary to understand the diversity of circulating molecular forms of PTH. The first PTH radioimmunoassay was described in 1963 [1]; the second-generation PTH immunoradiometric assay (IRMA) was described in 1987 [2], 24 years later; and the third-generation PTH IRMA was described in 1999 [3], 12 years later. Each generation has contributed to the description of circulating PTH immunoheterogeneity.

The first generation of PTH RIA used multivalent antibodies raised against parathyroid extracts of various species, more or less purified PTH(1–84) preparations, and eventually synthetic fragments representative of various regions of the PTH molecule. Epitopes recognized by these assays were mid-carboxyl-terminal, carboxyl-terminal, and rarely amino-terminal. Tracers evolved from ^{125}I-bPTH(1–84) to region-specific ^{125}I-synthetic fragments in the most sophisticated assays [4, 5]. These assays were initially responsible for the description of circulating PTH immunoheterogeneity [6, 7]. PTH composition in the basal state was demonstrated to be 20 % PTH(1–84) and 80 % carboxyl-terminal fragments missing an amino-terminal structure [8]. In rats injected with ^{125}I-bPTH(1–84), these fragments started their structure at positions 34, 37, 40, and 43 [8] but at positions 34, 37, 38, and 45 in man [9].

The second generation of PTH IRMA became available in 1987 [2]. It was commercialized by Nichols Institute. It uses a carboxyl-terminal antibody linked to a solid phase and a labeled amino-terminal antibody to reveal hPTH(1–84). Initially believed to react only with hPTH(1–84), this assay was demonstrated to also react with large carboxyl-terminal fragments possessing an amino-terminal structure, called non-(1–84) PTH fragments [10]. Their composition started at amino acids 4, 7, 10, and 15 with the major fragment starting at position 7 [11]. They represented 20 % of the immunoreactivity detected by a second-generation PTH assay but only 5 % of the immunoreactivity detected by a first-generation assay [12]. HPLC had to be used to separate PTH(1–84) from these C-fragments [12]. Because these fragments represent 5 % of circulating PTH, one has to reduce the amount of hPTH(1–84) estimated to be present in the sample to 15–20 % [12].

The first third-generation PTH assay was described in 1999 [3]. This assay was supposed to react only with hPTH(1–84) until we demonstrated that it also reacts with N-PTH [13]. N-PTH is believed to be phosphorylated on serine 17 [14]. It represents 7–8 % of circulating PTH detected by a second-generation assay with a 13–20 epitope [12] but only 2 % of circulating

P. D'Amour, MD
Department of Medicine,
CHUM Saint-Luc Hospital,
Montréal, Québec H3C 3J7, Canada
e-mail: pierre.damour@umontreal.ca

M.L. Brandi, E.M. Brown (eds.), *Hypoparathyroidism*,
DOI 10.1007/978-88-470-5376-2_4, © Springer-Verlag Italia 2015

PTH detected by a first-generation assay [12]. Overproduction of N-PTH has been described in primary and secondary hyperparathyroidism as well as in parathyroid carcinoma [15–21].

4.2 Origin of Circulating PTH Molecular Forms

PTH molecular forms are generated in part by the peripheral metabolism of PTH(1–84) in the liver [22, 23] or directly secreted by the parathyroid glands [7, 10, 23]. The amount of PrePro PTH mRNA controls PTH(1–84) synthesis [24]. Both 1,25$(OH)_2$D and calcium concentration exert a negative control on the amount of PrePro PTH mRNA, while phosphate concentration exerts a positive control [25, 26]. 1,25$(OH)_2$D acts though the vitamin D receptor presents in the parathyroid cells [27], while calcium and phosphate influence proteins that either reduce or enhance the stability of PrePro PTH mRNA, increasing or decreasing, respectively, the level of PrePro PTH mRNA [26].

The fate of newly synthesized PTH(1–84) is influenced by calcium concentration. In the presence of hypocalcemia, most of PTH(1–84) remains intact, but hypercalcemia influences its degradation into fragments [28]. In one degradative pathway, secretory granules fuse with lysosomes and cause complete PTH degradation [28]. In the second, they are degraded by intravesicular cathepsins B and D to generate C-PTH fragments [29].

Small and large C-PTH fragments are generated by the liver from PTH(1–84) [8, 23] in a calcium-independent manner [30]. Kidneys degrade filtered PTH(1–84) and C-PTH fragments [22]. This results in the accumulation of small and large C-PTH fragments in terminal renal failure [10, 31].

In the parathyroid venous blood in calves, plasma calcium influences the relative proportion of PTH(1–84) and of C-fragments. Hypocalcemia leads to mainly PTH(1–84) with few C-PTH fragments. Hypercalcemia on the other hand decreases PTH secretion and favors a high C-PTH fragment/PTH(1–84) ratio [32]. Again in calves, a sigmoidal relationship exists between calcium and parathyroid hormone secretion rate [33], and there is a non-suppressible fraction of PTH secretion mainly composed of C-PTH fragments [34].

Dogs maintained on a low calcium-vitamin D-deficient diet increased PTH(1–84) secretion relative to C-PTH fragments [35]. Over 2 years, they developed a fivefold increase in their parathyroid function with a decrease in their C-PTH fragments/PTH(1–84) ratio [36] with a slight decrease in the C-PTH fragments/PTH(1–84) ratio. When maintained on a calcium-sufficient diet for 1 month, which was supplemented with 1,25$(OH)_2$D IV 0.25 μg twice a day, serum calcium became normal, 1,25$(OH)_2$D remained elevated, and I-PTH decreased to a normal value, while the C-PTH/I-PTH ratio remained elevated at 14.8 [36]. Over the next 22 months, the situation remained the same with a normal diet supplemented with vitamin D. A decrease in the set point of PTH regulation by calcium was also observed with a very high C-PTH fragment/PTH(1–84) ratio of 11.4 [36]. Dogs injected with small doses of 1,25$(OH)_2$D, which had no influence on their serum calcium over 1 month, decreased I-PTH by 40 % without any change in the C-PTH levels resulting in high C-PTH fragments/PTH(1–84) ratio [37]. One month after removing half the parathyroid glands of the dogs, we observed that I-PTH was decreased less than C-PTH fragments, reducing the C-PTH fragments/PTH(1–84) ratio [37].

4.3 Molecular Forms of PTH in Specific Clinical Conditions

4.3.1 Primary Hyperparathyroidism

A set-point error in PTH secretion that increases the serum calcium level at which PTH secretion is suppressed characterizes primary hyperparathyroidism [38]. The exact cause of this "rightward" shift remains uncertain. The evolution of the C-PTH/I-PTH ratio under hypocalcemia or hypercalcemia remains normal but at a higher set-point level [38]. In some cases, the set-point error is related to a decreased expression of the extracellular calcium-sensing receptor [39], while rare cases have been reported in which the set point is normal and PTH secretion is increased but is limited to the non-suppressible fraction, with a high C-PTH/I-PTH ratio similar to cases of non-parathyroid hypercalcemia [40]. Low circulating levels of 25(OH)D can be found in patients with primary hyperparathyroidism [41–49] and have been associated with larger parathyroid tumors [50]. This is explained both by an inadequate supply of vitamin D [51] and increased degradation of 25(OH)D related to the stimulatory effect of hypercalcemia and high $1{,}25(OH)_2D$ levels on 24-hydroxylase activity. Histomorphometric features of the bone are influenced by vitamin D deficiency in primary hyperparathyroidism [52], and vitamin D repletion increases bone mineral density [53].

Table 4.1 Principal causes of hypoparathyroidism

Low-PTH-level hypoparathyroidism
Parathyroid destruction
Surgery
Autoimmune (isolated or polyglandular)
Cervical irradiation
Infiltration by metastasis or systemic diseases
Sarcoidosis, amyloidosis, hemochromatosis, Wilson's disease, thalassemia
Reduced parathyroid function
Hypomagnesemia
PTH gene defects
Activating calcium-sensing receptor mutations
Parathyroid agenesis
DiGeorge syndrome
Kenny-Caffey syndrome
Isolated X-linked hypoparathyroidism
Mitochondrial neuropathies
High-PTH-level hypoparathyroidism
Pseudohypoparathyroidism
Hypomagnesemia

4.3.2 Hypoparathyroidism

The principal causes of hypoparathyroidism are illustrated in Table 4.1 (see also Chap. 14). Hypoparathyroidism is associated with increased bone mineral density at the lumbar spine, hip, and radius sites [54–57]. Bone biopsy histomorphometric analysis shows greater cancellous bone volume, trabecular width, and cortical width compared with age- and sex-matched controls [55]. Greater bone surface density, trabecular thickness, trabecular number, and connectivity density are observed in microcomputed tomography in comparison with matched controls [58]. Markers of bone turnover are in the lower half of the normal range or frankly low [58]. Dynamic skeletal indices, including mineralizing surface and bone function, are profoundly suppressed in hypoparathyroid subjects on double-tetracycline labeling of biopsy specimens [59, 60]. Postsurgical hypoparathyroidism is associated with an increased risk of renal complications and hospitalizations related to seizures but not with an increased risk of cardiac arrhythmias or cardiovascular disease or death [61] (see also Chap. 22). PTH assays performed during surgery can predict hypoparathyroidism after surgery even in patients with MEN-type 1 [62–65].

Results of cryopreserved parathyroid autografts have been disappointing in one study [66] and slightly better in another study [67]. Various forms of therapy have been used to treat hypoparathyroidism mostly initially based on administration of $1{,}25(OH)_2D$ and calcium [65]. But more recently, both PTH(1–34) [68–71] and PTH(1–84) [59, 60, 72–75] have been used to treat primary hypoparathyroidism and have induced marked changes in bone turnover and structure [59, 73].

4.3.3 Non-parathyroid Hypercalcemia

Normocalcemic individuals made acutely hypercalcemic have an I-PTH level similar to patients with chronic non-parathyroid hypercalcemia but maintain an elevated C-PTH level in the absence of renal failure [76]. They have an elevated C-PTH/I-PTH ratio which is further increased when renal failure is present. This is related to an adaptation to chronic hypercalcemia with the production of more C-PTH fragments. It also explains the overlap of C-PTH values between mild primary hyperparathyroidism and nonparathyroid hypercalcemia with the majority of C-PTH assays [76, 77].

4.3.4 Secondary Hyperparathyroidism Related to Vitamin D Deficiency

Kawahara et al. demonstrated that 25(OH)D and not 1,25(OH)$_2$D were involved in the negative regulation of PTH mRNA via the existence in the PT-r parathyroid cell line of 1α-hydroxylase activity [78]. Björkman et al. demonstrated that the response of PTH to vitamin D supplementation was not only determined by the baseline PTH levels and change in the vitamin D status but also by age and mobility of the patient [79]. PTH decreases quite linearly during vitamin D supplementation [80]. This is in accord with several papers in the literature [81, 82].

4.3.5 Renal Failure

An accumulation of small and large carboxyl-terminal fragments is observed as renal failure progresses [10, 23, 31]. The kidney has a major role in C-PTH fragments disposal and this is absent in end-stage renal disease [8]. Small and large carboxyl-terminal fragments represent 95 % of circulating PTH in end-stage renal disease [10]. The composition of PTH changes as renal failure progresses: in hypocalcemic patients with severe secondary hyperparathyroidism, the C-PTH fragments/PTH(1–84) ratio is lower than it is in patients with mild secondary hyperparathyroidism, reflecting an adaptation to hypocalcemia [83] even in advanced renal failure [10]. Bone turnover was assessed by the PTH(1–84)/large C-PTH fragments ratio in patients with end-stage renal disease with success in one study [84] but without success in another [85]. In this last study, the lack of this relationship could be ascribed to 4 patients out of 34 [85]. A variation of the PTH(1–84)/large C-PTH fragments ratio has been observed as a function of the assays selected [86]. A decreased biological activity of PTH(1–84) has been linked to the accumulation of C-PTH fragments in renal failure [87]. An identical performance of second- and third-generation PTH assays has been observed in the diagnosis of renal failure [88].

The Allegro intact PTH assay of Nichols Institute was used to establish the KIDOQI clinical practice guidelines for PTH values in renal failure patients [89]. This assay is no longer available and has to be replaced by other assays. Souberbielle has demonstrated that the Elecsys PTH assay of second generation gave values very similar to the Nichols Allegro intact PTH assay over a large range of values [90].

Conclusion

Three generations of PTH assays have been used to characterize circulating molecular forms of PTH with success. Molecular forms described included small C-PTH fragments missing an N-structure with first-generation PTH assays [8], while non-(1–84) PTH fragments or large carboxyl-terminal fragments with a partial N-structure were described with the second-generation PTH assays [10, 11], and finally N-PTH, a phosphorylated form of PTH(1–84) with third-generation PTH assays [13].

References

1. Berson SA, Yalow RS, Aurbach GD, Potts JT (1963) Immunoassay of bovine and human parathyroid hormone. Proc Natl Acad Sci U S A 49:613–617
2. Nussbaum SR, Zahradnik RJ, Lavigne JR, Brennan GL, Nozawa-Ung K, Kim LY, Keutmann HT, Wang CA, Potts JT Jr, Segre GV (1987) Highly sensitive two-site immunoradiometric assay of parathyrin, and its clinical utility in evaluating patients with hypercalcemia. Clin Chem 33:1364–1367
3. John MR, Goodman WG, Gao P, Cantor TL, Salusky IB, Jüppner H (1999) A novel immunoradiometric assay detects full-length human PTH but not amino-terminally truncated fragments: implications for PTH measurements in renal failure. J Clin Endocrinol Metab 84:4287–4290
4. Segre GV, Tregear G, Potts JT Jr (1975) Development and application of sequence-specific radioimmunoassay for analysis of the metabolism of parathyroid hormone. In: O'Malley BW, Hardman JG (eds) Methods on enzymology, vol 37. Academic, New York, pp 38–66

5. D'Amour P, Labelle F, Lazure C (1984) Comparison of four different carboxylterminal tracers in a radioimmunoassay specific to the 68–84 region of human parathyroid hormone. J Immunoassay 5:183–204
6. Berson SA, Yalow RS (1968) Immunochemical heterogeneity of parathyroid hormone in plasma. J Clin Endocrinol Metab 28:1037–1047
7. D'Amour P, Labelle F, Lecavalier L, Plourde V, Harvey D (1986) Influence of serum Ca concentration on circulating molecular forms of PTH in three species. Am J Physiol 251:E680–E687
8. Segre BV, D'Amour P, Potts JT (1976) Metabolism of radioiodinated bovine parathyroid hormone in the rat. Endocrinology 99:1645–1652
9. Zhang CX, Weber BV, Thammavong J, Grover TA, Wells DS (2006) Identification of carboxyl-terminal peptide fragments of parathyroid hormone in human plasma at low-picomolar levels by mass spectrometry. Anal Chem 78:1636–1643
10. Brossard JH, Cloutier M, Roy L, Lepage R, Gascon-Barré M, D'Amour P (1996) Accumulation of a non-(1–84) molecular form of parathyroid hormone (PTH) detected by intact PTH assay in renal failure: importance in the interpretation of PTH values. J Clin Endocrinol Metab 81:3923–3929
11. D'Amour P, Brossard JH, Rousseau L, Nguyen-Yamamoto L, Nassif E, Lazure C, Gauthier D, Lavigne JR, Zahradnik RJ (2005) Structure of non-(1–84) PTH fragments secreted by parathyroid glands in primary and secondary hyperparathyroidism. Kidney Int 68:998–1007
12. D'Amour P, Räkel A, Brossard JH, Rousseau L, Albert C, Cantor T (2006) Acute regulation of circulating parathyroid hormone (PTH) molecular forms by calcium: utility of PTH fragments/PTH(1–84) ratios derived from three generations of PTH assays. J Clin Endocrinol Metab 91:283–289
13. D'Amour P, Brossard JH, Rousseau L, Roy L, Gao P, Cantor T (2003) Amino-terminal form of parathyroid hormone (PTH) with immunologic similarities to hPTH(1–84) is overproduced in primary and secondary hyperparathyroidism. Clin Chem 49:2037–2044
14. Rabbani SA, Kremer R, Bennett HP, Goltzman D (1984) Phosphorylation of parathyroid hormone by human and bovine parathyroid glands. J Biol Chem 259:2949–2955
15. Räkel A, Brossard JH, Patenaude JV, Albert C, Nassif E, Cantor T, Rousseau L, D'Amour P (2005) Overproduction of an amino-terminal form of PTH distinct from human PTH(1–84) in a case of severe primary hyperparathyroidism: influence of medical treatment and surgery. Clin Endocrinol (Oxf) 62:721–727
16. Tanaka M, Itoh K, Matsushita K, Matsushita K, Fujii H, Fukagawa M (2005) Normalization of reversed bio-intact-PTH(1–84)/intact-PTH ratio after parathyroidectomy in a patient with severe secondary hyperparathyroidism. Clin Nephrol 64:69–72
17. Arakawa T, D'Amour P, Rousseau L, Brossard JH, Sakai M, Kasumoto H, Igaki N, Goto T, Cantor T, Fukagawa M (2006) Overproduction and secretion of a novel amino-terminal form of parathyroid hormone from a severe type of parathyroid hyperplasia in uremia. Clin J Am Soc Nephrol 1:525–531
18. Boudou P, Ibrahim F, Cormier C, Sarfati E, Souberbielle JC (2006) Unexpected serum parathyroid hormone profiles in some patients with primary hyperparathyroidism. Clin Chem 52:757–760
19. Rubin MR, Silverberg SJ, D'Amour P, Brossard JH, Rousseau L, Sliney J Jr, Cantor T, Bilezikian JP (2007) An N-terminal molecular form of parathyroid hormone (PTH) distinct from hPTH(1 84) is overproduced in parathyroid carcinoma. Clin Chem 53:1470–1476
20. Caron P, Maiza JC, Renaud C, Cormier C, Barres BH, Souberbielle JC (2009) High third generation/second generation PTH ratio in a patient with parathyroid carcinoma: clinical utility of third generation/second generation PTH ratio in patients with primary hyperparathyroidism. Clin Endocrinol (Oxf) 70:533–538
21. Cavalier E, Daly AF, Betea D, Pruteanu-Apetrii PN, Delanaye P, Stubbs P, Bradwell AR, Chapelle JP, Beckers A (2010) The ratio of parathyroid hormone as measured by third- and second-generation assays as a marker for parathyroid carcinoma. J Clin Endocrinol Metab 95:3745–3749
22. D'Amour P, Segre GV, Roth SI, Potts JT Jr (1979) Analysis of parathyroid hormone and its fragments in rat tissues: chemical identification and microscopical localization. J Clin Invest 63:89–98
23. Nguyen-Yamamoto L, Rousseau L, Brossard JH, Lepage R, Gao P, Cantor T, D'Amour P (2002) Origin of parathyroid hormone (PTH) fragments detected by intact-PTH assays. Eur J Endocrinol 147:123–131
24. Habener JF, Rosenblatt M, Potts JT Jr (1984) Parathyroid hormone: biochemical aspects of biosynthesis, secretion, action, and metabolism. Physiol Rev 64:985–1053
25. Silver J, Russell J, Sherwood LM (1985) Regulation by vitamin D metabolites of messenger ribonucleic acid for preproparathyroid hormone in isolated bovine parathyroid cells. Proc Natl Acad Sci 82:4270–4273
26. Kilav R, Silver J, Naveh-Many T (2001) A conserved cis-acting element in the parathyroid hormone 3′-untranslated region is sufficient for regulation of RNA stability by calcium and phosphate. J Biol Chem 276:8727–8733
27. Wecksler WR, Ross FP, Mason RS, Posen S, Norman AW (1980) Biochemical properties of the 1 alpha, 25-dihydroxyvitamin D_3 cytoplasmic receptors from human and chick parathyroid glands. Arch Biochem Biophys 201:95–103
28. Setoguti T, Inoue Y, Shin M (1988) Electron-microscopic studies on the threshold value of calcium concentration for the release of storage granules and the acceleration of their degradation in the rat parathyroid gland. Cell Tissue Res 251:531–539
29. Hashizume Y, Waguri S, Watanabe T, Kominami E, Uchiyama Y (1993) Cysteine proteinases in rat parathyroid cells with special reference to their correlation with parathyroid hormone (PTH) in storage granules. J Histochem Cytochem 41:273–282

30. Bringhurst FR, Stern AM, Yotts M, Mizrahi N, Segre GV, Potts JT Jr (1989) Peripheral metabolism of [35S] parathyroid hormone in vivo: influence of alterations in calcium availability and parathyroid status. J Endocrinol 122:237–245
31. Segre GV, D'Amour P, Hultman A, Potts JT Jr (1981) Effects of hepatectomy, nephrectomy, and nephrectomy/uremia on the metabolism of parathyroid hormone in the rat. J Clin Invest 67:439–448
32. Mayer GP, Keaton JA, Hurst JG, Habener JF (1979) Effects of plasma calcium concentration on the relative proportion of hormone and carboxyl fragments in parathyroid venous blood. Endocrinology 104:1778–1784
33. Mayer GP, Hurst JG (1978) Sigmoidal relationship between parathyroid hormone secretion rate and plasma calcium concentration in calves. Endocrinology 102:1036–1042
34. Mayer GP, Habener JF, Potts JT Jr (1976) Parathyroid hormone secretion in vivo. Demonstration of a calcium-independent nonsuppressible component of secretion. J Clin Invest 57:678–683
35. Cloutier M, D'Amour P, Gascon-Barré M, Hamel L (1990) Low calcium diet in dogs causes a greater increase in parathyroid function measured with an intact hormone than with a carboxylterminal assay. Bone Miner 9:179–188
36. Cloutier M, Gascon-Barré M, D'Amour P (1992) Chronic adaptation of dog parathyroid function to a low-calcium-high-sodium-vitamin D-deficient diet. J Bone Miner Res 7:1021–1028
37. Cloutier M, Gagnon Y, Gascon-Barré M, Brossard JH, D'Amour P (1997) Adaptation of parathyroid function to intravenous 1,25-dihydroxyvitamin D3 or partial parathyroidectomy in normal dogs. J Endocrinol 155:133–141
38. Brossard JH, Whittom S, Lepage R, D'Amour P (1993) Carboxyl-terminal fragments of parathyroid hormone are not secreted preferentially in primary hyperparathyroidism as they are in other hypercalcemic conditions. J Clin Endocrinol Metab 77:413–419
39. Hannan FM, Thakker RV (2013) Calcium-sensing receptor (CaSR) mutations and disorders of calcium, electrolyte and water metabolism. Best Pract Res Clin Endocrinol Metab 27:359–371
40. D'Amour P, Weisnagel J, Brossard JH, Ste-Marie LG, Rousseau L, Lepage R (1998) Functional evidence for two types of parathyroid adenoma. Clin Endocrinol (Oxf) 48:593–601
41. Grey A, Lucas J, Horne A, Gamble G, Davidson JS, Reid IR (2005) Vitamin D repletion in patients with primary hyperparathyroidism and coexistent vitamin D insufficiency. J Clin Endocrinol Metab 90:2122–2126
42. Beyer TD, Chen EL, Nilubol N, Prinz RA, Solorzano CC (2007) Short-term outcomes of parathyroidectomy in patients with or without 25-hydroxy vitamin D insufficiency. J Surg Res 143:145–150
43. Untch BR, Barfield ME, Dar M, Dixit D, Leight GS Jr, Olson JA Jr (2007) Impact of 25-hydroxyvitamin D deficiency on perioperative parathyroid hormone kinetics and results in patients with primary hyperparathyroidism. Surgery 142:1022–1026
44. Silverberg SJ (2007) Vitamin D deficiency and primary hyperparathyroidism. J Bone Miner Res 22:V100–V104
45. Grubbs EG, Rafeeq S, Jimenez C, Feng L, Lee JE, Evans DB, Perrier ND (2008) Preoperative vitamin D replacement therapy in primary hyperparathyroidism: safe and beneficial? Surgery 144:852–858
46. Isidro ML, Ruano B (2009) Biochemical effects of calcifediol supplementation in mild, asymptomatic, hyperparathyroidism with concomitant vitamin D deficiency. Endocrine 36:305–310
47. Tucci JR (2009) Vitamin D therapy in patients with primary hyperparathyroidism and hypovitaminosis D. Eur J Endocrinol 161:189–193
48. Wagner D, Xia Y, Hou R (2013) Safety of vitamin D replacement in patients with primary hyperparathyroidism and concomitant vitamin D deficiency. Endocr Pract 19:420–425
49. Souberbielle JC, Bienaimé F, Cavalier E, Cormier C (2012) Vitamin D and primary hyperparathyroidism (PHPT). Ann Endocrinol 73:165–169
50. Rao DS, Honasoge M, Divine GW, Phillips ER, Lee MW, Ansari MR, Talpos GB, Parfitt AM (2000) Effect of vitamin D nutrition on parathyroid adenoma weight: pathogenetic and clinical implications. J Clin Endocrinol Metab 85:1054–1058
51. Bollerslev J, Marcocci C, Sosa M, Nordenström J, Bouillon R, Mosekilde L (2011) Current evidence for recommendation of surgery, medical treatment and vitamin D repletion in mild primary hyperparathyroidism. Eur J Endocrinol 165:851–864
52. Stein EM, Dempster DW, Udesky J, Zhou H, Bilezikian JP, Shane E, Silverberg SJ (2011) Vitamin D deficiency influences histomorphometric features of bone in primary hyperparathyroidism. Bone 48:557–561
53. Kantorovich V, Gacad MA, Seeger LL, Adams JS (2000) Bone mineral density increases with vitamin D repletion in patients with coexistent vitamin D insufficiency and primary hyperparathyroidism. J Clin Endocrinol Metab 85:3541–3543
54. Laway BA, Goswami R, Singh N, Gupta N, Seith A (2006) Pattern of bone mineral density in patients with sporadic idiopathic hypoparathyroidism. Clin Endocrinol (Oxf) 64:405–409
55. Rubin MR, Dempster DW, Zhou H, Shane E, Nickolas T, Sliney J Jr, Silverberg SJ, Bilezikian JP (2008) Dynamic and structural properties of the skeleton in hypoparathyroidism. J Bone Miner Res 23:2018–2024
56. Abugassa S, Nordenström J, Eriksson S, Sjödén G (1993) Bone mineral density in patients with chronic hypoparathyroidism. J Clin Endocrinol Metab 76:1617–1621
57. Touliatos JS, Sebes JI, Hinton A, McCommon D, Karas JG, Palmieri GM (1995) Hypoparathyroidism counteracts risk factors for osteoporosis. Am J Med Sci 310:56–60

58. Rubin MR, Dempster DW, Kohler T, Stauber M, Zhou H, Shane E, Nickolas T, Stein E, Sliney J Jr, Silverberg SJ, Bilezikian JP, Müller R (2010) Three dimensional cancellous bone structure in hypoparathyroidism. Bone 46:190–195
59. Rubin MR, Dempster DW, Sliney J Jr, Zhou H, Nickolas TL, Stein EM, Dworakowski E, Dellabadia M, Ives R, McMahon DJ, Zhang C, Silverberg SJ, Shane E, Cremers S, Bilezikian JP (2011) PTH(1–84) administration reverses abnormal bone-remodeling dynamics and structure in hypoparathyroidism. J Bone Miner Res 26:2727–2736
60. Sikjaer T, Rejnmark L, Rolighed L, Heickendorff L, Mosekilde L, Hypoparathyroid Study Group (2011) The effect of adding PTH(1–84) to conventional treatment of hypoparathyroidism: a randomized, placebo-controlled study. J Bone Miner Res 26:2358–2370
61. Underbjerg L, Sikjaer T, Mosekilde L, Rejnmark L (2013) Cardiovascular and renal complications to postsurgical hypoparathyroidism: a Danish nationwide controlled historic follow-up study. J Bone Miner Res 28:2277–2285
62. Cavicchi O, Piccin O, Caliceti U, Fernandez IJ, Bordonaro C, Saggese D, Ceroni AR (2008) Accuracy of PTH assay and corrected calcium in early prediction of hypoparathyroidism after thyroid surgery. Otolaryngol Head Neck Surg 138:594–600
63. Ezzat WF, Fathey H, Fawaz S, El-Ashri A, Youssef T, Othman HB (2011) Intraoperative parathyroid hormone as an indicator for parathyroid gland preservation in thyroid surgery. Swiss Med Wkly 141:w13299
64. Nilubol N, Weisbrod AB, Weinstein LS, Simonds WF, Jensen RT, Phan GQ, Hughes MS, Libutti SK, Marx S, Kebebew E (2013) Utility of intraoperative parathyroid hormone monitoring in patients with multiple endocrine neoplasia type 1-associated primary hyperparathyroidism undergoing initial parathyroidectomy. World J Surg 37:1966–1972
65. Cayo AK, Yen TW, Misustin SM, Wall K, Wilson SD, Evans DB, Wang TS (2012) Predicting the need for calcium and calcitriol supplementation after total thyroidectomy: results of a prospective, randomized study. Surgery 152:1059–1067
66. Borot S, Lapierre V, Carnaille B, Goudet P, Penfornis A (2010) Results of cryopreserved parathyroid autografts: a retrospective multicenter study. Surgery 147:529–535
67. Cohen MS, Dilley WG, Wells SA Jr, Moley JF, Doherty GM, Sicard GA, Skinner MA, Norton JA, DeBenedetti MK, Lairmore TC (2005) Long-term functionality of cryopreserved parathyroid autografts: a 13-year prospective analysis. Surgery 138(6):1033–1040; discussion 1040–1041
68. Winer KK, Sinaii N, Peterson D, Sainz B Jr, Cutler GB Jr (2008) Effects of once versus twice-daily parathyroid hormone 1–34 therapy in children with hypoparathyroidism. J Clin Endocrinol Metab 93:3389–3395
69. Winer KK, Sinaii N, Reynolds J, Peterson D, Dowdy K, Cutler GB Jr (2010) Long-term treatment of 12 children with chronic hypoparathyroidism: a randomized trial comparing synthetic human parathyroid hormone 1–34 versus calcitriol and calcium. J Clin Endocrinol Metab 95:2680–2688
70. Gafni RI, Brahim JS, Andreopoulou P, Bhattacharyya N, Kelly MH, Brillante BA, Reynolds JC, Zhou H, Dempster DW, Collins MT (2012) Daily parathyroid hormone 1–34 replacement therapy for hypoparathyroidism induces marked changes in bone turnover and structure. J Bone Miner Res 27:1811–1820
71. Winer KK, Zhang B, Shrader JA, Peterson D, Smith M, Albert PS, Cutler GB Jr (2012) Synthetic human parathyroid hormone 1–34 replacement therapy: a randomized crossover trial comparing pump versus injections in the treatment of chronic hypoparathyroidism. J Clin Endocrinol Metab 97:391–399
72. Rubin MR, Sliney J Jr, McMahon DJ, Silverberg SJ, Bilezikian JP (2010) Therapy of hypoparathyroidism with intact parathyroid hormone. Osteoporos Int 21:1927–1934
73. Cusano NE, Rubin MR, McMahon DJ, Zhang C, Ives R, Tulley A, Sliney J Jr, Cremers SC, Bilezikian JP (2013) Therapy of hypoparathyroidism with PTH(1–84): a prospective four-year investigation of efficacy and safety. J Clin Endocrinol Metab 98:137–144
74. Rejnmark L, Sikjaer T, Underbjerg L, Mosekilde L (2013) PTH replacement therapy of hypoparathyroidism. Osteoporos Int 24:1529–1536
75. Sikjaer T, Amstrup AK, Rolighed L, Kjaer SG, Mosekilde L, Rejnmark L (2013) PTH(1–84) replacement therapy in hypoparathyroidism: a randomized controlled trial on pharmacokinetic and dynamic effects after 6 months of treatment. J Bone Miner Res 28:2232–2243
76. Benson RC Jr, Riggs BL, Pickard BM, Arnaud CD (1974) Radioimmunoassay of parathyroid hormone in hypercalcemic patients with malignant disease. Am J Med 56:821–826
77. Lufkin EG, Kao PC, Heath H 3rd (1987) Parathyroid hormone radioimmunoassays in the differential diagnosis of hypercalcemia due to primary hyperparathyroidism or malignancy. Ann Intern Med 106:559–560
78. Kawahara M, Iwasaki Y, Sakaguchi K, Taguchi T, Nishiyama M, Nigawara T, Tsugita M, Kambayashi M, Suda T, Hashimoto K (2008) Predominant role of 25OHD in the negative regulation of PTH expression: clinical relevance for hypovitaminosis D. Life Sci 82:677–683
79. Björkman M, Sorva A, Tilvis R (2009) Responses of parathyroid hormone to vitamin D supplementation: a systematic review of clinical trials. Arch Gerontol Geriatr 48:160–166
80. Lips P (2001) Vitamin D deficiency and secondary hyperparathyroidism in the elderly: consequences for bone loss and fractures and therapeutic implications. Endocr Rev 22:477–501

81. Björkman M, Sorva A, Risteli J, Tilvis R (2007) Vitamin D supplementation has minor effects on parathyroid hormone and bone turnover markers in vitamin D-deficient bedridden older patients. Age Ageing 37:25–31
82. Vieth R, Ladak Y, Walfish PG (2003) Age-related changes in the 25-hydroxyvitamin D versus parathyroid hormone relationship suggest a different reason why older adults require more vitamin D. J Clin Endocrinol Metab 88:185–191
83. Brossard JH, Lepage R, Cardinal H, Roy L, Rousseau L, Dorais C, D'Amour P (2000) Influence of glomerular filtration rate on non-(1–84) parathyroid hormone (PTH) detected by intact PTH assays. Clin Chem 46:697–703
84. Monier-Faugere MC, Geng Z, Mawad H, Friedler RM, Gao P, Cantor TL, Malluche HH (2001) Improved assessment of bone turnover by the PTH-(1–84)/large C-PTH fragments ratio in ESRD patients. Kidney Int 60:1460–1468
85. Salusky IB, Goodman WG, Kuizon BD, Lavigne JR, Zahranik RJ, Gales B, Wang HJ, Elashoff RM, Jüppner H (2003) Similar predictive value of bone turnover using first- and second-generation immunometric PTH assays in pediatric patients treated with peritoneal dialysis. Kidney Int 63:1801–1808
86. Koller H, Zitt E, Staudacher G, Neyer U, Mayer G, Rosenkranz AR (2004) Variable parathyroid hormone(1–84)/carboxylterminal PTH ratios detected by 4 novel parathyroid hormone assays. Clin Nephrol 61:337–343
87. Slatopolsky E, Finch J, Clay P, Martin D, Sicard G, Singer G, Gao P, Cantor T, Dusso A (2000) A novel mechanism for skeletal resistance in uremia. Kidney Int 58:753–761
88. D'Amour P (2008) Lessons from a second- and third-generation parathyroid hormone assays in renal failure patients. J Endocrinol Invest 31:459–462
89. National Kidney Foundation (2003) KIDOQI clinical practice guidelines for bone metabolism and disease in chronic kidney disease. Am J Kidney Dis 42:S1–S202
90. Souberbielle JC, Boutten A, Carlier MC, Chevenne D, Coumaros G, Lawson-Body E, Massart C, Monge M, Myara J, Parent X, Plouvier E, Houillier P (2006) Inter-method variability in PTH measurement: implication for the care of CKD patients. Kidney Int 70:345–350

Control of Parathyroid Hormone Secretion by Extracellular Ca^{2+}

5

Edward M. Brown

5.1 Introduction

The chapter addresses the control of PTH secretion by extracellular Ca^{2+} (Ca^{2+}_o), which is a critical component of the homeostatic system maintaining nearly constant levels of Ca^{2+}_o and phosphate. It will cover not only the control of the secretion of preformed PTH by Ca^{2+}_o but also the regulation of other parameters of parathyroid function contributing to the overall secretory response to changes in Ca^{2+}_o. These are listed in Table 5.1 and include above and beyond rapid Ca^{2+}_o-induced changes in the release of stored PTH, its intracellular degradation, the biosynthesis of full-length PTH [PTH(1–84)], parathyroid cellular proliferation, and, perhaps, apoptosis. In aggregate, these determine the overall, minute-to-minute rate of PTH secretion from the total mass of parathyroid cells, which is the rate of secretion from each parathyroid cell summed over the total parathyroid cellular mass in any given individual.

In addition to Ca^{2+}_o, phosphate, and 1,25 dihydroxyvitamin D_3 [$1,25(OH)_2D_3$], fibroblast growth factor 23 (FGF23) also regulate parathyroid function (for reviews, see [1–4]), with phosphate stimulating it, while $1,25(OH)_2D_3$ and FGF23 inhibit it. The regulatory roles of these three factors are covered in Chaps. 6 and 8 and will only be alluded to briefly here.

E.M. Brown, MD
Division of Endocrinology, Diabetes
and Hypertension, Department of Medicine,
Brigham and Women's Hospital,
Room 223A, EBRC223A, 221 Longwood Ave.,
Boston, MA 02115, USA
e-mail: embrown@partners.org

5.2 Role of PTH in Maintaining Mineral Ion Homeostasis In Vivo

Figures 5.1 and 5.2 show how PTH contributes to regulating Ca^{2+} and phosphate homeostasis. Ca^{2+}_o is kept within a narrow range of ± ~1–2 % of its normal level in any given person, even though the normal level of Ca^{2+}_o in the population as a whole is ~8.5–10.5 mg/dl, that is, ± ~10 % [6]. The normal level of serum phosphorus varies over a broader range of ~2.5–4.5 mg/dl (e.g., ± ~30 %). Deviation of the levels of these two ions substantially above or below their normal limits can have severe clinical sequelae, as detailed elsewhere in this volume. Therefore, a homeostatic system maintaining their circulating levels within their respective normal ranges is critical.

Table 5.1 Parameters modulating the overall secretory rate of normal parathyroid glands

1. Alterations in the minute-to-minute secretion of preformed PTH
2. Alterations in the rate of PTH degradation
3. Alterations in preproPTH gene expression
4. Alterations in parathyroid cellular proliferation
5. Alterations in apoptosis of parathyroid cells

M.L. Brandi, E.M. Brown (eds.), *Hypoparathyroidism*,
DOI 10.1007/978-88-470-5376-2_5, © Springer-Verlag Italia 2015

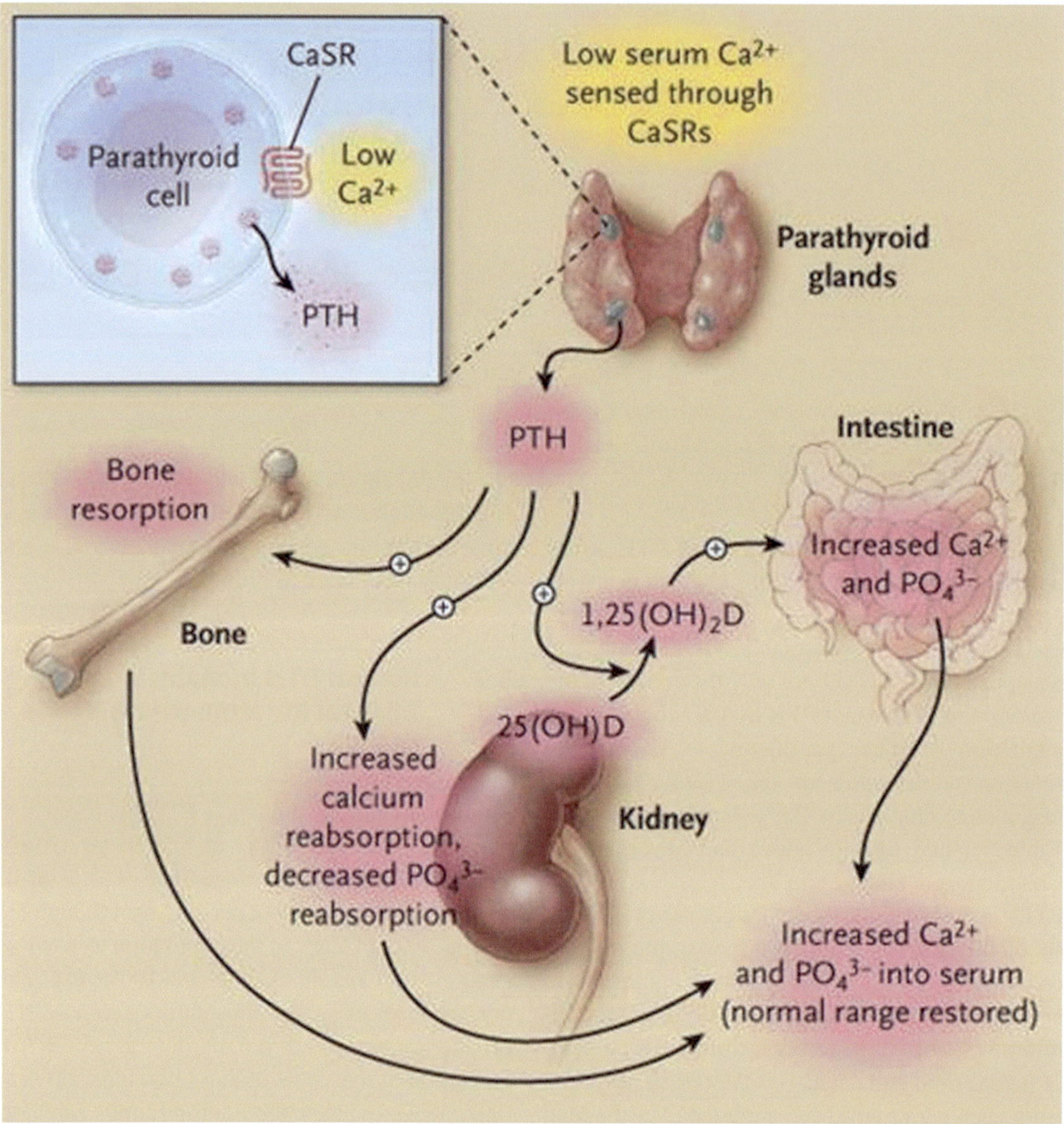

Fig. 5.1 Schematic representation of the major hormones and tissues participating in extracellular Ca^{2+} (Ca^{2+}_o) homeostasis. When defending against hypocalcemia—as illustrated here—parathyroid chief cells secrete more PTH, which modulates renal function so as to promote phosphaturia, enhance Ca^{2+} reabsorption, and increase $1,25(OH)_2D_3$ synthesis. The latter enhances intestinal Ca^{2+} and phosphate absorption. $1,25(OH)_2D_3$ and PTH act jointly to promote net release of skeletal Ca^{2+}. Increased movement of Ca^{2+} into the extracellular fluid (ECF) from bone and intestine, combined with diminished Ca^{2+} excretion, serve to normalize Ca^{2+}_o. See text for details (From Shoback [5], with permission)

Figure 5.1 demonstrates the homeostatic response to hypocalcemia. A key component is the parathyroid cell's capacity to sense small (~1–2 %) alterations in Ca^{2+}_o from its normal concentration and to respond with changes in PTH secretion that restore normocalcemia. The extraordinary sensitivity of the parathyroid cell to alterations in Ca^{2+}_o derives from the steep inverse sigmoidal relationship between PTH release and Ca^{2+}_o, with slight changes in Ca^{2+}_o producing large alterations in PTH secretion (Fig. 5.2).

PTH is a key Ca^{2+}_o-elevating hormone, as is $1,25(OH)_2D_3$. Four actions of PTH restore normocalcemia in the defense against hypocalcemia [5, 6, 8] (for review, see Chaps. 8 and 11): (1) PTH increases renal Ca^{2+} reabsorption in both the cortical thick ascending limb of Henle's loop and the distal convoluted tubule of the kidney. (2)

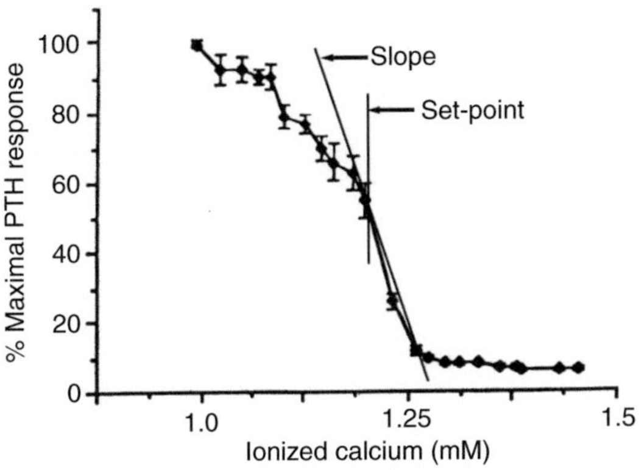

Fig. 5.2 Steep inverse sigmoidal curve relating Ca^{2+}_o and PTH in vivo. Serum PTH was measured by immunoradiometric assay while hyper- or hypocalcemia was induced in normal subjects by infusing calcium or ethylenediaminetetraacetic acid (EDTA), respectively. The steep slope of this relationship, which is a key component for maintaining Ca^{2+}_o within a very narrow range in normal persons in vivo, is illustrated, along with the set-point (the serum Ca^{2+} that produces half of the maximal suppression of PTH), which participates in "setting" the serum Ca^{2+} concentration. Two further parameters describing the curve (not shown) are the minimal and maximal levels of PTH secretion at high and low serum Ca^{2+}, respectively. See text for details (Reproduced in modified form with permission from Brown [7])

It enhances synthesis of $1,25(OH)_2D_3$ from $25(OH)D_3$ in the proximal tubule, which then augments intestinal Ca^{2+} absorption and phosphate through separate transport systems. (3) It increases net Ca^{2+} and phosphate release from bone. PTH's actions on bone are thought to involve alterations in bone turnover (i.e., reductions in bone formation and/or increases in bone resorption) and/or mobilization of a pool of soluble mineral ions at the bone surface [9, 10]. (4) Finally, PTH promotes phosphaturia, an action facilitating excretion of any excess phosphate originating from $1,25(OH)_2D_3$-induced GI absorption and/or PTH- or $1,25(OH)_2D_3$-stimulated release from the bone. The hyper- and hypophosphatemia in patients with hypo- or hyperparathyroidism [5, 11], respectively, document PTH's importance in maintaining phosphate homeostasis, notwithstanding the key roles that FGF23 plays in phosphate and to a lesser extent Ca^{2+}_o homeostasis (see below).

If normocalcemia is not restored through the series of events just noted, the parathyroid glands have several adaptive responses to hypocalcemia that enhance their secretory capacity beyond just the rapid secretion of preformed PTH, which lasts for 1–2 h [12, 13]. Table 5.1 and Fig. 5.3 list these responses and depict the approximate time courses of their contributions to the overall secretory response. The initial release of stored, preformed PTH occurs within seconds to minutes. It is followed by reduced intracellular degradation of PTH starting within 20 min [15–18], which increases the cellular content of PTH(1–84) [15]. A rise in PTH mRNA, occurring as early as an hour or less and lasting for weeks or longer, then ensues [19]. Finally, an increase in parathyroid cellular proliferation takes place within ~2 days and can persist indefinitely [6, 19–21]. The time courses of these various adaptive changes, lasting from seconds to years or more, ensure both immediate and sustained responses, without the presence of any "windows" of time that lack an enhanced secretory rate. For example, reduced degradation of PTH occurs prior to the depletion of preformed hormone, ensuring that more PTH(1–84) is available for secretion until expression of the preproPTH gene increases and, later, stimulation of parathyroid cellular proliferation commences.

Over the past 10 years, it has become clear that an additional homeostatic system is essential for preserving phosphate homeostasis and, to some extent, Ca^{2+}_o homeostasis. A central element is fibroblast growth factor 23 (FGF23), a phosphaturic hormone secreted predominantly by osteocytes [22] (See Chap. 8). Figure 5.4 depicts the principal ways in which FGF23 contributes to phosphate homeostasis. Note the homeostatic events occurring upon phosphate loading, which enhances FGF23 release. FGF23, in turn, acts to normalize the serum phosphate concentration, primarily via its powerful phosphaturic effect [1]. FGF23 also suppresses renal production of $1,25(OH)_2D_3$, thereby diminishing release and absorption of phosphate from the bone and intestine, respectively. Because $1,25(OH)_2D_3$ increases FGF23, there is a negative feedback loop, whereby FGF23 reduces $1,25(OH)_2D_3$ synthesis, which decreases FGF23 production further. Since both increased serum phosphate and reduced $1,25(OH)_2D_3$ stimulate parathyroid function [23], the resulting increase in circulating PTH, combined with a rise in FGF23, further stimulates renal phosphate loss. Note in Fig. 5.4 that every effect of FGF23, phosphate, PTH, and $1,25(OH)_2D_3$ on one another constitutes negative feedback loops. For example,

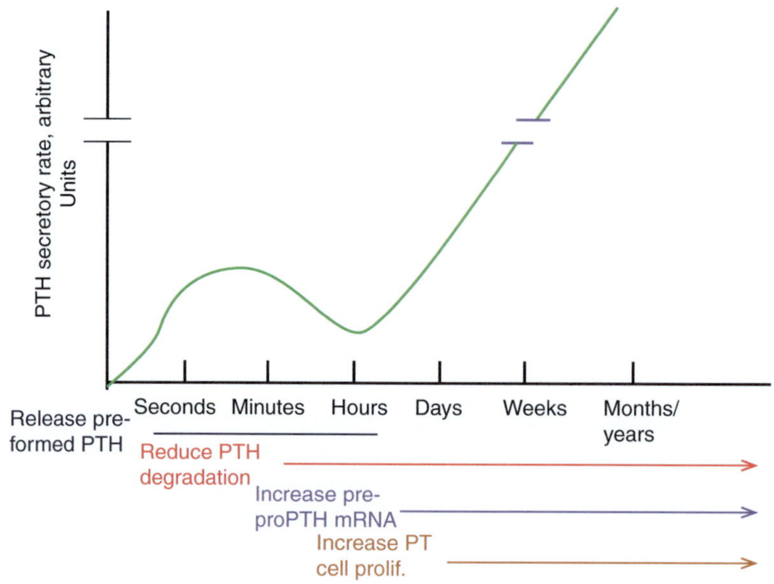

Fig. 5.3 Schematic timeline for the sequence of events enabling progressive elevations in the secretory rate of PTH in response to hypocalcemia over a time frame ranging from seconds to months or years. There is an initial increase in the secretion of stored PTH, which wanes over ~90 minutes. Beginning at around 20-30 mins, there is a reduction in intracellular degradation of PTH, which sustains PTH secretion at a level above baseline from 90 mins until there are further increases in hormone secretion resulting from increased synthesis owing to an elevation in preproPTH mRNA and stimulation of parathyroid cellular proliferation. Of the other key regulators of parathyroid function, increases in phosphate and decreases in 1,25(OH)2D3 both raise the level of preproPTH and enhance parathyroid proliferation, while a decrease in FGF23 increases PTH secretion within minutes and elevates the level of preproPTH within ~40 minutes to an hour. See text for details

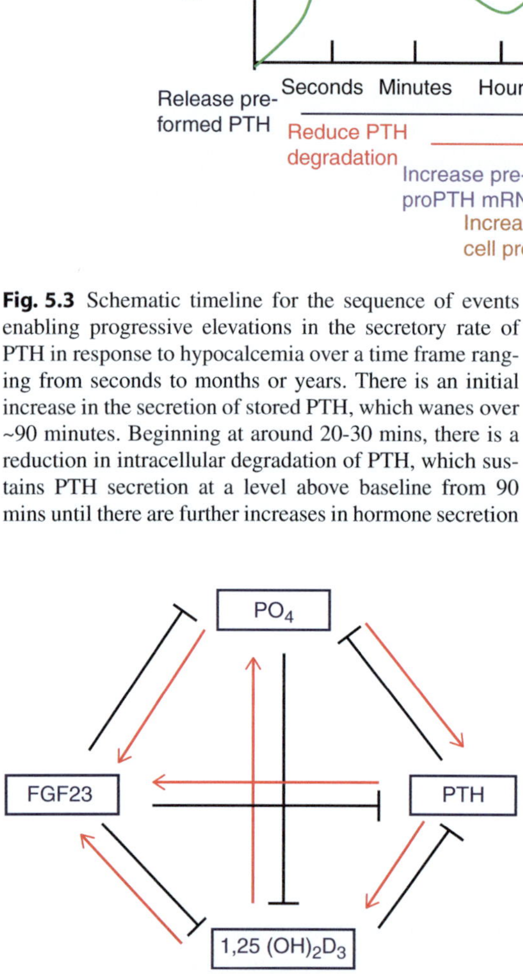

Fig. 5.4 Diagram showing the interactions that take place between extracellular phosphate and the hormones regulating it as part of the maintenance of phosphate homeostasis. *Red arrows* show actions that elevate the levels of the hormones shown or of extracellular phosphate. *Black lines* ending in perpendicular bars illustrate actions that reduce the levels of the indicated hormones or phosphate. This phosphate-regulating homoeostatic system interacts with that controlling Ca^{2+}_o homeostasis through direct effects of PTH on renal excretion of phosphate, synthesis of $1{,}25(OH)_2D_3$ and of FGF23, the direct actions of FGF23 on PTH and $1{,}25(OH)_2D_3$, and the direct effect of $1{,}25(OH)_2D_3$ on PTH. See text for details

hyperphosphatemia enhances both FGF23 and PTH release [4, 20, 24], and both promote hypophosphatemia via their phosphaturic effects. The resultant lowering in phosphate then decreases production of both PTH and FGF23. In considering the regulation of parathyroid function by Ca^{2+}_o, the complex interactions between serum Ca^{2+}, phosphate, PTH, FGF23, and $1{,}25(OH)_2D_3$ should be borne in mind to appreciate more fully the overall control of mineral ion homeostasis, a topic covered in recent reviews [1, 4, 25] and in Chap. 8.

5.3 Mechanisms That Determine the Overall Rate of PTH Secretion

5.3.1 Production of PTH-Containing Secretory Vesicles

PTH is released by exocytosis of PTH-containing secretory vesicles. The overall regulation of PTH(1–84) secretion from each parathyroid cell involves alterations in the synthesis and packaging of PTH in these vesicles, the degradation of PTH within them, the rate of release of PTH via exocytosis, and, to some extent, degradation of these vesicles prior to their secretion. Therefore, a description of how PTH is produced, packaged, and then secreted or, alternatively, broken down is the focus of the rest of this chapter.

After synthesis of preproPTH messenger RNA (mRNA) and its transport into the cytoplasm, the nascent peptide chain of preproPTH is directed into the lumen of the rough endoplasmic reticulum (ER) by the pre-segment of the preproPTH peptide with concomitant cleavage of the pre-segment over 10–15 min, leaving proPTH within the ER [18, 26]. ProPTH transits the Golgi within 5–10 min and is converted to PTH by the enzyme furin. The newly formed PTH likely resides within the so-called immature, prosecretory granules, which lack the dense core of the mature secretory granules [26]. Newly produced PTH is available for secretion within 20–30 min after its initial synthesis as preproPTH. Newly produced PTH has three possible fates—secretion, degradation, or conversion to a storage pool of PTH [18]. Stored PTH presumably resides in dense core secretory granules. Conversion to the storage pool occurs with a $t_{1/2}$ of several hours. Stored PTH can remain in secretory granules or, like newly formed PTH, be secreted or degraded. Seventy to eighty percent of initially synthesized PTH is degraded—a seemingly wasteful scenario [18]. However, as noted later in the context of the regulation of overall PTH production, modulation of PTH(1–84) degradation and associated changes in intact hormone secretion provide an effective regulatory mechanism.

What controls PTH secretion from the newly synthesized and stored pools of PTH? The use of dispersed parathyroid cells [27, 28] made it possible to investigate cellular mechanisms underlying the control of PTH secretion by Ca^{2+}_o and other factors. These studies showed that agents activating adenylate cyclase, such as beta-adrenergic agonists, promote PTH release solely from the storage pool [17, 29]. Therefore, the release of PTH by such agents is self-limited in vivo and in vitro, lasting for 20–30 min [13, 16]. Inducing hypocalcemia in vivo promotes a substantial, four- to sevenfold increase in the rate of PTH secretion lasting ~1–2 h [30]. The secretory rate then decreases to a lower level ~2–3-fold above the basal rate [30] (Fig. 5.3). Inducing hypocalcemia, in contrast to agents activating adenylate cyclase, mobilizes PTH from both the newly synthesized and stored pools [17].

The remainder of this chapter addresses in more detail the mechanisms controlling the overall rate of PTH secretion in response to changes in Ca^{2+}_o, its principal physiological regulator, and how this is matched to the body's physiological needs. The cellular and molecular mechanisms underlying the adaptive responses to changes in Ca^{2+}_o are depicted in Table 5.1 and are discussed in the temporal order in which they are activated in response to hypocalcemia.

5.3.2 The Physiology of Ca^{2+}_o-Regulated PTH Secretion

There is a steep inverse sigmoidal relationship between Ca^{2+}_o and PTH release that can be described by four parameters, the maximal and minimal rates of PTH secretion, the midpoint or "set point" of the curve, and its slope [31] (Fig. 5.2). The set point is related to the level at which Ca^{2+}_o is set in vivo, although the latter tends to be at a level of Ca^{2+}_o where PTH secretion is ~25 % of maximal rather than at the set point per se [31]. This gives the parathyroid cell a substantial secretory reserve with which to respond to a hypocalcemic stress. A rightward shift in set point is a key contributor to the hypercalcemia in PTH-dependent hypercalcemia [31]. Conversely, a leftward shift causes the parathyroid cell to be overly sensitive to Ca^{2+}, as, for example, in hypoparathyroidism caused by activating mutations in the calcium-sensing receptor (CaSR) or in its downstream G protein ($G\alpha_{11}$) [6, 32]. The curve's steep slope assures large changes in PTH in response to small alterations in Ca^{2+}_o, thereby contributing importantly to the narrow range within which Ca^{2+}_o is maintained in vivo.

5.4 Mechanisms by Which Ca^{2+}_o Regulates the Diverse Aspects of Parathyroid Function That Determine the Overall Rate of PTH Release

5.4.1 Molecular Basis for Ca^{2+}_o-Sensing

The CaSR is the molecular mechanism underlying most, if not all, of the effects of Ca^{2+}_o on the various aspects of parathyroid function detailed below [33–35]. Some appreciation of the structure and function of the CaSR and how it functions is essential for this discussion. The CaSR was first isolated from bovine parathyroid in 1993 [36]. Before it was cloned, substantial indirect evidence supported the concept that a G protein-coupled receptor mediated some, if not all, of the actions of Ca^{2+}_o on the parathyroid [7, 37, 38]. Exposing bovine parathyroid cells to elevated Ca^{2+}_o, for example, activates phospholipase C (PLC) increasing the cytosolic calcium concentration (Ca^{2+}_i) both via calcium release from intracellular stores (via IP_3 produced by PLC) and influx of extracellular Ca^{2+} [35, 39]. These responses closely resemble those produced by known G protein-coupled receptors (GPCRs) stimulating PLC via the pertussis toxin-insensitive G proteins, G_q and G_{11}. Furthermore, high-Ca^{2+}_o concentrations evoke a pertussis toxin-sensitive inhibition of adenylate cyclase, resembling the actions of other GPCRs inhibiting adenylate cyclase through the inhibitory G protein, G_i [7, 40]. This evidence prompted the use of expression cloning in *Xenopus laevis* oocytes to isolate the CaSR [36], which is a member of the family C GPCRs [41]. This family includes, in addition to the CaSR, 8 metabotropic glutamate receptors (mGluRs), 2 $GABA_B$ receptors, and receptors for sweet substances, odorants, and pheromones [41]. The human CaSR's predicted structure features a large, 612 amino acid extracellular domain (ECD) [42], a 250 amino acid transmembrane domain (TMD) containing the 7 transmembrane helices that are characteristic of the GPCRs, and, lastly, a 216 amino acid carboxy (C)-terminal tail (C-tail).

The cell surface CaSR functions as a dimer. The two monomers are linked by non-covalent hydrophobic interactions as well as by intermolecular disulfide bonds involving cysteines 129 and 131 of each monomer [43, 44]. The cell surface CaSR desensitizes relatively little after prolonged or repeated exposure to high Ca^{2+}_o, at least in parathyroid cells. A recent study demonstrated a lack of a Ca^{2+}_o-induced increase in a comparatively rapid baseline rate of internalization of the CaSR ($T_{1/2} \sim 15$ min) [45]. In contrast, a second documented a significant increase in internalization at high Ca^{2+}_o, a discrepancy not yet resolved [46]. Notably, both studies used CaSR-transfected HEK293 cells, which may or may not mimic faithfully the parathyroid cell per se. Regarding control of the CaSR's forward trafficking to the cell membrane, the first study noted above described a novel mechanism, called ADIS (*A*gonist-*D*riven *I*nsertional *S*ignaling) that produces agonist-dependent trafficking of intracellular CaSR to, rather than away from, the cell surface at high Ca^{2+}_o, thereby increasing the level of cell surface CaSR [45]. This mechanism has been postulated to ensure the CaSR's persistent expression on the cell surface, thus enabling it to continuously monitor and respond to changes in Ca^{2+}_o.

Molecular modeling strongly suggests that the CaSR's monomeric ECD forms a bilobed, Venus flytrap (VFT)-like structure with a crevice between the lobes [47, 48] (Fig. 5.5). The CaSR has marked positive cooperativity in binding Ca^{2+}_o presumably owing to its several Ca^{2+}_o binding sites on each monomer [48]. The best-characterized Ca^{2+}_o binding site lies between the two lobes of each VFT [48, 49]. Based on the known structures of glutamate-free and glutamate-bound mGluR ECDs, the cleft in the CaSR ECD is thought to open without agonist and to close upon binding Ca^{2+}_o. Thus, Ca^{2+}_o binding likely assists the transition of the inactive CaSR ECD to its active conformation, which then causes conformational changes of the CaSR's TMD and intracellular domains that likely initiate signal transduction, but remain to be elucidated. In contrast to Ca^{2+}_o, positive allosteric activators of the CaSR, for example, cinacalcet, which are elaborated upon later, bind within the CaSR's TMD.

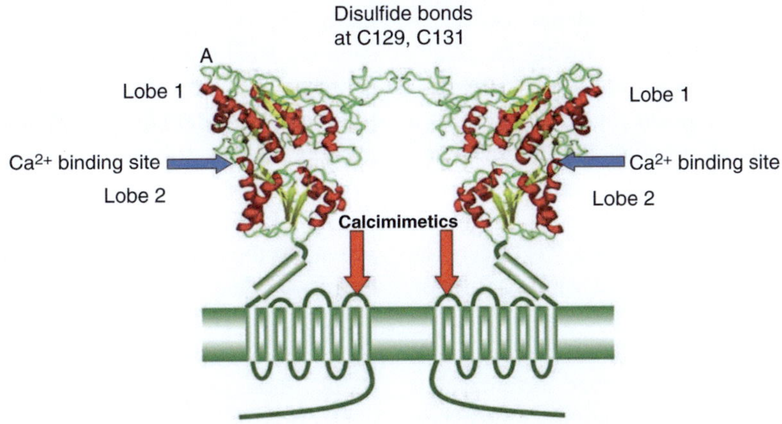

Fig. 5.5 Proposed structure of the human CaSR. Shown are the two disulfide-linked monomers of the extracellular domain (ECD) of the cell surface CaSR, each of which has two lobes that are predicted to fold into a structure resembling a Venus flytrap (VFT). The latter is postulated to close following binding of Ca^{2+} in the crevice between the lobes, thereby activating the CaSR. The *lower part* of the figure depicts the seven transmembrane helices that are characteristic of the superfamily of G protein-coupled receptors (GPCRs) and transduce the $Ca^{2+}{}_o$ signal from the ECD to the G proteins and diverse intracellular effector systems to which the CaSR is linked. Red segments of the receptor's ECD are alpha helices. The key calcium-binding site in the crevice between the two lobes of each monomer is shown, while the separate binding site for calcimimetics resides within the receptor's transmembrane domain. See text for details (Reproduced in modified form from Huang et al. [14], with permission)

The CaSR, after activation by $Ca^{2+}{}_o$, stimulates the G proteins, $G_{q/11}$, G_i, and $G_{12/13}$, which activate PLC, inhibit adenylate cyclase, and stimulate Rho kinase, respectively [35, 39]. In addition to directly inhibiting adenylate cyclase via G_i, the CaSR can also decrease cAMP indirectly by virtue of the PLC-mediated increases in $Ca^{2+}{}_i$, which, as a result, reduce the activity of Ca^{2+}-inhibitable forms of adenylate cyclase and/or stimulate phosphodiesterase [50]. Additional CaSR-regulated signaling pathways include mitogen-activated protein kinases (MAPKs) [e.g., extracellular signal-regulated kinase 1/2 (ERK1/2), p38 MAPK, and c-jun N-terminal kinase (JNK)]; phospholipases A_2 and D; and the epidermal growth factor (EGF) receptor (for recent reviews, see [35, 39]).

5.4.2 The CaSR Is the Mediator of the Regulation of PTH Secretion by $Ca^{2+}{}_o$

After its cloning, it was possible to identify those aspects of parathyroid function that are CaSR regulated. The CaSR's involvement in the control of PTH secretion by $Ca^{2+}{}_o$ was proven by showing that calcimimetic CaSR activators, for example, the second-generation compounds, NPS R-467 and R-568, or the third-generation drug, cinacalcet, rapidly inhibits PTH release in vivo and in vitro [51]. Calcimimetics are low molecular weight positive allosteric modulators of the CaSR that sensitize it to $Ca^{2+}{}_o$ (i.e., reduce the EC_{50} for activation of the receptor by $Ca^{2+}{}_o$). They do not act in the absence of $Ca^{2+}{}_o$ and require the presence of ~0.5–1.0 mM $Ca^{2+}{}_o$ to do so [52]. Calcimimetics also increase the cell surface expression of both wild-type CaSR [53] and some naturally occurring inactivating mutants (see below) [54], by serving as "pharmacochaperones" and/or by promoting ADIS. Cinacalcet has been approved by the FDA for three indications: (1) as a treatment for severe secondary hyperparathyroidism in patients being dialyzed for chronic kidney disease, (2) as medical therapy for hypercalcemia and hyperparathyroidism in patients with parathyroid cancer, and (3) as a treatment for severe primary hyperparathyroidism in whom surgery is not an option [52, 55, 56].

Further evidence for the CaSR's essential role in $Ca^{2+}{}_o$-regulated PTH secretion has been provided by the impaired CaSR-mediated inhibition of PTH secretion in individuals heterozygous or homozygous for inactivating mutations of the

CaSR, who have mild to moderate or severe hypercalcemia, respectively, owing to Ca^{2+}_o "resistance" [6]. Similarly, mice with "knockout" of the CaSR exhibit severe hypercalcemia as a result of severely dysregulated CaSR-mediated inhibition of PTH secretion [57, 58].

The signaling pathways and molecular mechanisms mediating CaSR-regulated PTH secretion remain elusive. A mouse model with inactivation of both G_q and G_{11} [59] has a phenotype similar to the mice with global knockout of the CaSR described above [57]. Moreover, heterozygous inactivating mutations of $G\alpha_{11}$ produce mild hypercalcemia and "resistance" to inhibition of PTH secretion by Ca^{2+}_o [32]. G_q and/or G_{11} must, therefore, be important mediators of the CaSR-regulated PTH secretion, presumably by stimulating PLC and modulating key intracellular signaling pathways participating in the high-Ca^{2+}_o-induced, CaSR-mediated inverse control of PTH secretion. The identity of these mediators remains incompletely understood. Ca^{2+}_o- and CaSR-evoked arachidonic acid production by PLA_2 is a candidate mediator [60], perhaps by being converted to metabolites of the 12- and 15-lipoxygenease pathways [61]. ERK1/2 has also been implicated in pathological parathyroid tissue [62]. In most secretory cells, a rise in Ca^{2+}_i stimulates exocytosis via stimulus-secretion coupling. The absence of SNAP-25 in parathyroid cells [63], a key component of the machinery mediating Ca^{2+}-induced exocytosis, has been suggested as a possible contributor to the paradoxical inhibition rather than the stimulation of PTH secretion at high Ca^{2+}_o. High levels of Ca^{2+}_o markedly polymerize the actin-based cytoskeleton under the plasma membrane of dispersed bovine parathyroid cells, perhaps physically blocking exocytosis, as PTH-containing vesicles may be immobilized in this cytoskeleton [64].

5.4.3 The CaSR Can Modulate PTH Secretion Independent of Changes in Ca^{2+}_o

There are several ways of activating the CaSR without increasing Ca^{2+}_o. For example, calcimimetics, by sensitizing the CaSR to Ca^{2+}_o and increasing its cell surface expression [56, 65], suppress PTH release despite the accompanying decrease in Ca^{2+}_o. In addition, interleukin-1β [66] and interleukin-6 [67] increase CaSR expression, which may be a contributor to the hypocalcemia encountered in inflammatory states and severe illness [68]. Moreover, because activating the CaSR can upregulate both its own expression and that of the vitamin D receptor (VDR) [65, 69] while activating the (VDR) upregulates its own expression and that of the CaSR, there is the possibility of synergistic interactions between the two receptors in regulating their target tissues.

Several agents activate the CaSR in addition to Ca^{2+}_o and calcimimetics. Polycationic agonists, including other di- (e.g., Mg^{2+}) and trivalent cations (Gd^{3+}) as well as organic polycations, for example, neomycin, are called type 1 agonists [33]; by definition they do not require Ca^{2+}_o to activate the receptor. They bind to the CaSR's ECD, unlike the calcimimetics described earlier, which are termed type 2 allosteric activators and bind to the CaSR's TMD. High Mg^{2+}_o, as encountered during magnesium infusion in the treatment of preeclampsia or eclampsia, inhibits PTH release and enhances renal Ca^{2+}_o excretion by activating the CaSR in the parathyroid and kidney, respectively, thereby causing the accompanying hypocalcemia and hypercalciuria [70] (for review, see Chap. 7). In addition to calcimimetics, several L-amino acids, particularly aromatic amino acids [71], can also function as type 2 allosteric activators. In contrast to the calcimimetics that bind to the CaSR's TMD, these amino acids interact with a site adjacent to the binding site for Ca^{2+}_o in the crevice between the ECD's two lobes and sensitize the receptor to Ca^{2+}_o. Aromatic amino acids, formed when a protein meal is digested [72], and high Ca^{2+}_o are potent stimulants of gastrin and gastric acid secretion as well as release of cholecystokinin from the small intestine [72, 73]. Therefore, in the GI tract, amino acids may serve as important activators of the CaSR, permitting the CaSR to monitor the levels of two key classes of nutrients, mineral ions (i.e., Ca^{2+} and Mg^{2+}), on the one hand, and amino acids/proteins, on the other.

5.4.4 Control of PTH Degradation by Ca^{2+}_o and the CaSR

PTH circulates as a complex mixture of intact PTH(1–84) and various carboxy (C)-terminal fragments [74], as described in greater detail in Chap. 4. Most of the C-terminal fragments have been cleaved within their amino-termini at amino acid residues 34 [PTH(34–84)] and 37 [PTH(37–84)] [15, 75–77]. These represent about 70–75 % of total immunoreactive PTH in the circulation during normocalcemia [78]. Because they lack much of PTH's N-terminus, these C-terminal fragments are inactive and are not recognized by the two-antibody "intact" sandwich assays for PTH. These assays are widely used in clinical practice and are described as "second-generation" immunoradiometric assays [79]. Immunoreactive species of PTH recognized by them must have both C-terminal as well as N-terminal epitopes proximal to amino acid 34. Large circulating fragments of PTH cleaved closer to their N-termini have been recognized more recently, especially PTH(7–84) [74]. These large PTH C-fragments are detected by second-generation immunoradiometric assays (which are not, therefore, truly "intact" assays), but not by more recently developed third-generation assays utilizing N-terminal antibodies recognizing epitopes at the extreme N-terminus (e.g., within residues 1–6) (third generation, so called "whole" PTH assays [80]). Consequently, third-generation assays are much more specific for PTH(1–84), but are not clearly better than second-generation assays in the clinical setting [80].

Large N-terminally truncated PTH fragments comprise about 5 % of total circulating immunoreactive PTH in normal subjects and are substantially less abundant than shorter fragments, such as PTH(37–84), which comprise 70–75 % of circulating PTH. PTH(1–84) represents 20–25 % of total PTH during normocalcemia and increases during hypocalcemia and declines during hypercalcemia [78]. None of the N-terminally truncated species of PTH stimulate the type 1 PTH receptor (PTHR1) in the bone and kidney [81] (see Chaps. 8 and 9).

The various C-fragments of PTH in blood are generated, at least in part, in parathyroid cells via enzymatic cleavage of PTH(1–84) within secretory vesicles, most likely by cysteine proteases, for example, cathepsins D and H [18, 77, 81]. Ca^{2+}_o modulates the intraglandular degradation of PTH, with more degradation at high Ca^{2+}_o and less at low [15, 75, 76]. The fragments generated at high Ca^{2+}_o are released from parathyroid cells in vitro as rapidly as 20 min following incubation in high-Ca^{2+}_o medium [16, 18]. Reduced breakdown of PTH at low Ca^{2+}_o likely contributes to the fact that PTH levels in vivo are higher than baseline after 1–2 h of hypocalcemia, despite depletion of stored PTH over this time frame [13, 82]. This increased rate of secretion of PTH(1–84) at low Ca^{2+}_o owing to its reduced degradation precedes the rise in PTH synthesis and secretion resulting in the increased expression of the preproPTH gene. The impact of Ca^{2+}_o-induced modulation of PTH degradation changes the percentage of total immunoreactive PTH that is intact PTH(1–84) during hypo- and hypercalcemia to 33 and ~10 %, respectively [78]. The CaSR mediates the Ca^{2+}_o-elicited changes in PTH degradation, as activation of the CaSR by cinacalcet, as with high Ca^{2+}_o, elevates the ratio of PTH fragments to PTH(1–84) in vivo [83].

5.4.5 Role of the CaSR in Regulating PTH Gene Expression

In addition to inhibiting PTH secretion, calcimimetics reduce preproPTH mRNA [84], documenting unequivocally the CaSR's role in the regulation in expression of this gene. Ca^{2+}_o-induced changes in preproPTH mRNA result from alterations in mRNA stability rather than in its transcription [84]. The 3′ untranslated region (UTR) of preproPTH mRNA has a conserved adenosine uridine (AU)-rich (AUR) sequence element, which enables regulation of mRNA stability by interacting with appropriate protein binding partners (Fig. 5.6) (for review, see [24]). Cytosolic proteins from the parathyroid glands of hypocalcemic rats interact with the AU-rich element (ARE), thereby stabilizing preproPTH mRNA and increasing its expression. Two proteins that bind to the 3′-ARE and increase its stability are the so-called AU-rich factor

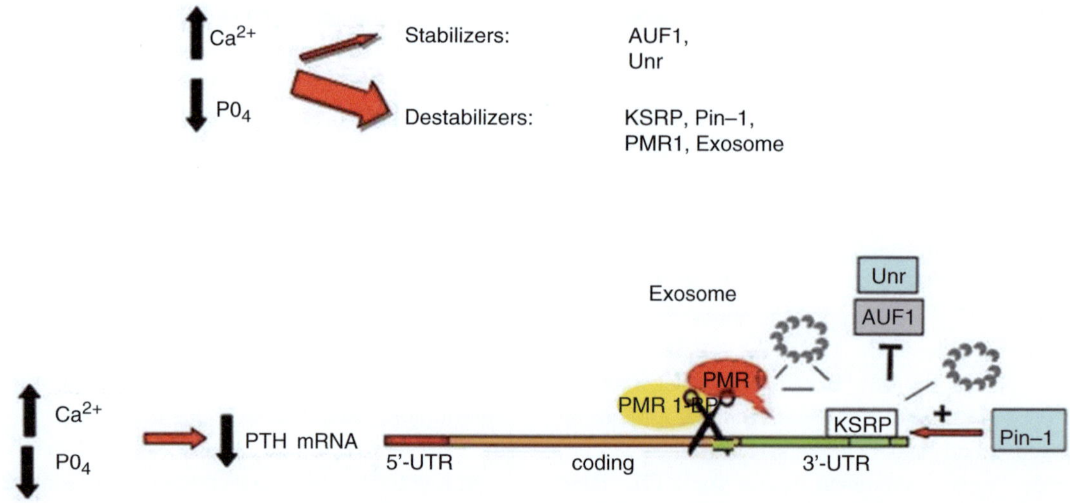

Fig. 5.6 Mechanism by which elevated levels of $Ca^{2+}{}_o$, acting via the CaSR, decrease the expression of the mRNA for preproPTH. Reductions in phosphate have similar actions. The stability of the mRNA is enhanced in the presence of low $Ca^{2+}{}_o$ (or high phosphate) by the binding of the AU-rich binding factor (AUF-1) and the protein called "Upstream of N-Ras" (Unr) to the ARE in the 3' UTR of the preproPTH gene. A key destabilizer of the mRNA is KSRP, which in the presence of high $Ca^{2+}{}_o$ or low phosphate, binds to the same site in the 3'-untranslated region of the PTH gene as AUF1 and displaces the latter. The peptidylprolyl isomerase, Pin-1, then activates KSRP, and the latter recruits the endonuclease, PMR1, and the exosome, which cleaves the mRNA internally. See text for details. Reproduced in modified form from Nechama M. et al. [85], with permission.

(AUF-1) and a second protein, Upstream of N-ras (Unr) [24]. Later studies showed that the decreased stability of preproPTH associated with dissociation of AUF-1 from the ARE is a consequence of the binding of an mRNA decay-promoting protein, called KH splicing regulatory protein (KSRP), to the same ARE. Pin-1, a petidylprolyl isomerase, then activates KSRP, thereby recruiting an endonuclease, PMR1, and the so-called exosome, which cleaves mRNAs internally [24, 85]. Following activation of the CaSR by a calcimimetic in rats with experimentally induced renal insufficiency, activation of calmodulin (CaM) and protein phosphatase 2B posttranslationally modifies AUF-1, which decreases its binding to the ARE, thereby decreasing the stability and pari passu preproPTH mRNA expression [86]. Prolonged exposure of parathyroid cells to low or high $Ca^{2+}{}_o$ increases or decreases, respectively, the machinery required for protein synthesis, that is, ER, Golgi apparatus, etc. [87, 88]. How the parathyroid cell recognizes the need for this increased biosynthetic capacity (i.e., producing more ER and Golgi rather than simply increasing preproPTH mRNA) is unclear.

5.4.6 $Ca^{2+}{}_o$ Regulates Parathyroid Cellular Proliferation via the CaSR

Proliferation of parathyroid cells during chronic hypocalcemia or vitamin D deficiency enhances the total secretory capacity of the parathyroid glands beyond that achieved solely by releasing preformed hormone, decreasing PTH degradation, and increasing preproPTH mRNA expression and is an important element in the defense against hypocalcemia. The CaSR's mediatory role in regulating parathyroid cellular proliferation is proven by the striking parathyroid enlargement in humans and mice with homozygous inactivation of the CaSR gene [57, 89] as well as the suppression by calcimimetics of the parathyroid growth in rat models of secondary hyperparathyroidism (SHPT) [90].

Experimental models of SHPT, particularly that produced by a high-phosphate diet in rats with experimentally induced renal impairment, have provided most of the available data addressing how $Ca^{2+}{}_o$ regulates parathyroid

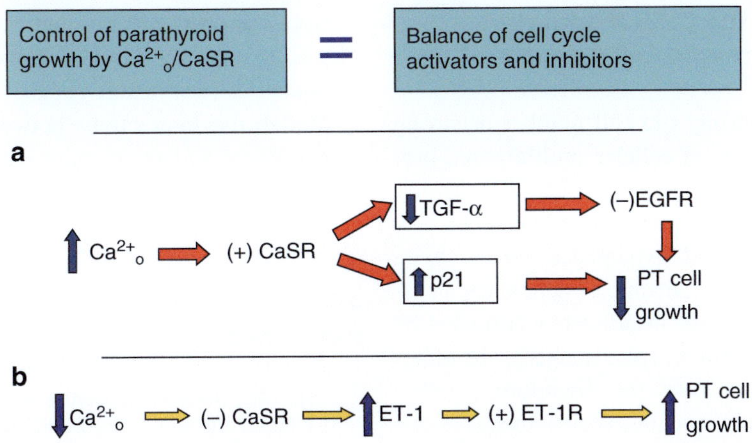

Fig. 5.7 Mechanisms underlying the regulation of parathyroid cellular proliferation by Ca^{2+}_o. (A) High Ca^{2+}_o (acting via the CaSR) decreases the expression of TGF-α, which is an activator of the pro-proliferative epidermal growth factor receptor (EGFR), and also decreases expression of the cyclin dependent kinase inhibitor, p21. (B) In addition to the actions of Ca^{2+}_o in (A), low Ca^{2+}_o upregulates the expression of endothelin-1 (ET-1), which activates the endothelin receptor, ET-1R, thereby stimulating parathyroid cellular proliferation. (+) depicts increased activity, and (-) indicates reduced activity. See text for details.

cellular hyperplasia [91]. Studies in normal animals are more difficult as there is minimal parathyroid cell division in normal parathyroid glands. The increased proliferation of parathyroid cells during SHPT is caused, at least in part, by an autocrine/paracrine loop that involves transforming growth factor-alpha (TGF-α) and its receptor, the epidermal growth factor receptor (EGFR) [92] (Fig. 5.7). TGF-α-EGFR signaling regulates progression from G1 to S in the cell cycle via the reciprocal effects on the stimulatory cyclin D1 and inhibitors of cyclin-dependent kinases, particularly p21 [93]. Elevations in cyclin D1 promote entry into mitosis and the inhibitors reduce it. Whether EGFR signaling occurs in normal parathyroid glands is unclear, however, since normal parathyroid cells express EGFR but not TGF-α or EGF [94].

Activation of EGFR activates ERK1/2, which elevates cyclin D1 expression in the rat model of SHPT described above [95]. The EGFR also transactivates the cyclin D1 gene [96]. The EGFR mediates the concomitant stimulation of parathyroid cellular proliferation, since an EGFR tyrosine kinase inhibitor (erlotinib) inhibited the parathyroid growth and associated increases in proliferating cell nuclear antigen (PCNA) and TGF-α in the SHPT model fed a high-phosphate diet [95].

In these rats, a high-Ca^{2+} diet elevated p21 expression and reduced the high-phosphate-evoked rise in TGF-α content (Fig. 5.7) [93, 97]. The high-Ca^{2+}_o-induced decrease in TGF-α is reminiscent of high-Ca^{2+}_o-evoked suppression of PTH release, suggesting that the CaSR may mediate this action of high Ca^{2+}_o. Subsequent studies showed coordinate increases in TGF-α and EGFR, while the SHPT was developing in this model and demonstrated that the increase in this signaling pathway further elevated TGF-α levels, producing a positive feedback loop [95].

Endothelin-1 (ET-1) is another paracrine regulator that may participate in the regulation of parathyroid cellular proliferation by Ca^{2+}_o (Fig. 5.7). In rats fed with a low-Ca^{2+} diet for 8 weeks, the number of proliferating parathyroid cells increased and was accompanied by an increase in immunostaining for ET-1 [98]. Administration of the endothelin receptor antagonist, bosentan, at the time when the low-calcium diet was instituted prevented the increase in cellular proliferation, supporting an important role of ET-1 in this low Ca^{2+}_o-induced parathyroid hyperplasia.

5.4.7 Does the CaSR Modulate Parathyroid Cell Apoptosis?

In addition to changes in cell number occurring through alterations in cellular proliferation, programmed cell death (apoptosis) can also modify cell number. It is a long-standing clinical observation that parathyroid hyperplasia caused by a reversible cause (i.e., vitamin D deficiency, phosphate loading, or renal insufficiency corrected by renal transplantation) spontaneously involutes slowly and incompletely [8]. Therefore, parathyroid cell apoptosis is not an efficient means of disposing unneeded parathyroid cells and, therefore, reducing overall secretory capacity [99]. Data obtained more recently using calcimimetics in diverse models of parathyroid dysfunction and a variety of methods for quantifying apoptosis are conflicting. For instance, one study showed no effect of cinacalcet on parathyroid apoptosis in experimental SHPT in rats [100]. In another study using a similar experimental model, NPS R-568 enhanced apoptosis, albeit at very high concentrations (10^{-4} M) of uncertain physiological relevance [101]. Given the potential therapeutic importance of promoting apoptosis with a calcimimetic or other agent in various forms of hyperparathyroidism, further studies are needed on this issue.

Conclusions

This chapter has focused on the effects of $Ca^{2+}{}_o$ on various aspects of parathyroid function determining the overall secretory rate at any given moment in time, for example, release of preformed stored of PTH, intracellular degradation of PTH, de novo synthesis of PTH, parathyroid cellular proliferation, and, perhaps, apoptosis. These $Ca^{2+}{}_o$-evoked, CaSR-mediated actions should be viewed in the context of the partially overlapping actions of $1,25(OH)_2D_3$, phosphate, and/or FGF23 on various aspects of parathyroid function (also see Chap. 8). How these four regulatory factors interact in their actions on the various aspects of parathyroid function discussed here in the wide variety of physiological and pathophysiological circumstances encountered in vivo are, in most cases, not well understood. The reader, in perusing the literature in this field from the past, present, and future, would do well to not only focus on the individual components reviewed here but also attempt to synthesize the interactions between the various regulators and their actions on the different aspects of parathyroid function contributing to the overall rate of PTH secretion over time.

References

1. Hu MC, Shiizaki K, Kuro-o M, Moe OW (2013) Fibroblast growth factor 23 and Klotho: physiology and pathophysiology of an endocrine network of mineral metabolism. Annu Rev Physiol 75:503–533
2. Brown AJ, Slatopolsky E (2007) Drug insight: vitamin D analogs in the treatment of secondary hyperparathyroidism in patients with chronic kidney disease. Nat Clin Pract Endocrinol Metab 3:134–144
3. Naveh-Many T, Nechama M (2007) Regulation of parathyroid hormone mRNA stability by calcium, phosphate and uremia. Curr Opin Nephrol Hypertens 16:305–310
4. Silver J, Rodriguez M, Slatopolsky E (2012) FGF23 and PTH–double agents at the heart of CKD. Nephrol Dial Transplant 27:1715–1720
5. Shoback D (2008) Clinical practice. Hypoparathyroidism. N Engl J Med 359:391–403
6. Brown EM (2007) Clinical lessons from the calcium-sensing receptor. Nat Clin Pract Endocrinol Metab 3:122–133
7. Brown EM (1991) Extracellular Ca2+ sensing, regulation of parathyroid cell function, and role of Ca2+ and other ions as extracellular (first) messengers. Physiol Rev 71:371–411
8. Bringhurst FR, Demay MB, Kronenberg HM (1998) Hormones and disorders of mineral metabolism. In: Wilson JD, Foster DW, Kronenberg HM, Larsen PR (eds) Williams textbook of endocrinology. W.B. Saunders, Philadelphia, pp 1155–1209
9. Huan J, Martuseviciene G, Olgaard K, Lewin E (2007) Calcium-sensing receptor and recovery from hypocalcaemia in thyroparathyroidectomized rats. Eur J Clin Invest 37:214–221
10. Neuman WF (1982) Blood: bone equilibrium. Calcif Tissue Int 34:117–120
11. Ward BK, Magno AL, Walsh JP, Ratajczak T (2012) The role of the calcium-sensing receptor in human disease. Clin Biochem 45:943–953
12. Schwarz P, Sorensen HA, McNair P, Transbol I (1993) Cica-clamp technique: a method for quantifying parathyroid hormone secretion: a sequential citrate and calcium clamp study. Eur J Clin Invest 23:546–553
13. Jung A, Mayer GP, Hurst JG, Neer R, Potts JT Jr (1982) Model for parathyroid hormone secretion and metabolism in calves. Am J Physiol 242:R141–R150
14. Huang Y, Zhou Y, Yang W, Butters R, Lee HW, Li S, Castiblanco A, Brown EM, Yang JJ (2007)

Identification and dissection of Ca(2+)-binding sites in the extracellular domain of Ca(2+)-sensing receptor. J Biol Chem 282:19000–19010
15. Morrissey JJ, Hamilton JW, MacGregor RR, Cohn DV (1980) The secretion of parathormone fragments 34–84 and 37–84 by dispersed porcine parathyroid cells. Endocrinology 107:164–171
16. Brown EM, Leombruno R, Thatcher J, Burrowes M (1985) The acute secretory response to alterations in the extracellular calcium concentration and dopamine in perifused bovine parathyroid cells. Endocrinology 116:1123–1132
17. Morrissey JJ, Cohn DV (1979) Regulation of secretion of parathormone and secretory protein protein-I from separate intracellular pools by calcium, dibutyryl cyclic AMP, and (l)-isoproterenol. J Cell Biol 82:93–102
18. Morrissey JJ, Cohn DV (1979) Secretion and degradation of parathormone as a function of intracellular maturation of hormone pools. Modulation by calcium and dibutyryl cyclic AMP. J Cell Biol 83:521–528
19. Naveh-Many T, Friedlaender MM, Mayer H, Silver J (1989) Calcium regulates parathyroid hormone messenger ribonucleic acid (mRNA), but not calcitonin mRNA in vivo in the rat. Dominant role of 1,25-dihydroxyvitamin D. Endocrinology 125: 275–280
20. Denda M, Finch J, Slatopolsky E (1996) Phosphorus accelerates the development of parathyroid hyperplasia and secondary hyperparathyroidism in rats with renal failure. Am J Kidney Dis 28:596–602
21. Goodman WG, Quarles LD (2008) Development and progression of secondary hyperparathyroidism in chronic kidney disease: lessons from molecular genetics. Kidney Int 74:276–288
22. Schaffler MB, Cheung WY, Majeska R, Kennedy O (2014) Osteocytes: master orchestrators of bone. Calcif Tissue Int 94:5–24
23. Demay MB (2013) Physiological insights from the vitamin D receptor knockout mouse. Calcif Tissue Int 92:99–105
24. Silver J, Naveh-Many T (2009) Phosphate and the parathyroid. Kidney Int 75:898–905
25. Lavi-Moshayoff V, Wasserman G, Meir T, Silver J, Naveh-Many T (2010) PTH increases FGF23 gene expression and mediates the high-FGF23 levels of experimental kidney failure: a bone parathyroid feedback loop. Am J Physiol Renal Physiol 299: F882–F889
26. Habener JF, Amherdt M, Ravazzola M, Orci L (1979) Parathyroid hormone biosynthesis. Correlation of conversion of biosynthetic precursors with intracellular protein migration as determined by electron microscope autoradiography. J Cell Biol 80:715–731
27. Morrissey JJ, Cohn DV (1978) The effects of calcium and magnesium on the secretion of parathormone and parathyroid secretory protein by isolated porcine parathyroid cells. Endocrinology 103:2081–2090
28. Brown EM, Hurwitz S, Aurbach GD (1976) Preparation of viable isolated bovine parathyroid cells. Endocrinology 99:1582–1588
29. Brown EM, Gardner DG, Windeck RA, Hurwitz S, Brennan MF, Aurbach GD (1979) ß-adrenergically stimulated adenosine 3′,5′-monophosphate accumulation in and parathyroid hormone release from dispersed human parathyroid cells. J Clin Endocrinol Metab 48:618–626
30. Schwarz P (1993) Dose response dependency in regulation of acute S-PTH(1–84) release in normal humans: a citrate and calcium infusion study. Scand J Clin Lab Invest 53:601–605
31. Brown EM (1983) Four parameter model of the sigmoidal relationship between parathyroid hormone release and extracellular calcium concentration in normal and abnormal parathyroid tissue. J Clin Endocrinol Metab 56:572–581
32. Nesbit MA, Hannan FM, Howles SA, Babinsky VN, Head RA, Cranston T, Rust N, Hobbs MR, Heath H 3rd, Thakker RV (2013) Mutations affecting G-protein subunit alpha11 in hypercalcemia and hypocalcemia. N Engl J Med 368:2476–2486
33. Brown EM, MacLeod RJ (2001) Extracellular calcium sensing and extracellular calcium signaling. Physiol Rev 81:239–297
34. Magno AL, Ward BK, Ratajczak T (2011) The calcium-sensing receptor: a molecular perspective. Endocr Rev 32:3–30
35. Conigrave AD, Ward DT (2013) Calcium-sensing receptor (CaSR): pharmacological properties and signaling pathways. Best Pract Res Clin Endocrinol Metab 27:315–331
36. Brown EM, Gamba G, Riccardi D, Lombardi M, Butters R, Kifor O, Sun A, Hediger MA, Lytton J, Hebert SC (1993) Cloning and characterization of an extracellular Ca(2+)-sensing receptor from bovine parathyroid. Nature 366:575–580
37. Nemeth E, Scarpa A (1987) Rapid mobilization of cellular Ca^{2+} in bovine parathyroid cells by external divalent cations. J Biol Chem 202:5188–5196
38. Shoback DM, Chen TH, Lattyak B, King K, Johnson RM (1993) Effects of high extracellular calcium and strontium on inositol polyphosphates in bovine parathyroid cells. J Bone Miner Res 8:891–898
39. Chakravarti B, Chattopadhyay N, Brown EM (2012) Signaling through the extracellular calcium-sensing receptor (CaSR). Adv Exp Med Biol 740: 103–142
40. Chen C, Barnett J, Congo D, Brown E (1989) Divalent cations suppress 3′,5′-adenosine monophosphate accumulation by stimulating a pertussis toxin-sensitive guanine nucleotide-binding protein in cultured bovine parathyroid cells. Endocrinology 124:233–239
41. Brauner-Osborne H, Wellendorph P, Jensen AA (2007) Structure, pharmacology and therapeutic prospects of family C G-protein coupled receptors. Curr Drug Targets 8:169–184
42. Garrett JE, Capuano IV, Hammerland LG, Hung BC, Brown EM, Hebert SC, Nemeth EF, Fuller F (1995) Molecular cloning and functional expression of human parathyroid calcium receptor cDNAs. J Biol Chem 270:12919–12925

43. Fan GF, Ray K, Zhao XM, Goldsmith PK, Spiegel AM (1998) Mutational analysis of the cysteines in the extracellular domain of the human Ca2+ receptor: effects on cell surface expression, dimerization and signal transduction. FEBS Lett 436:353–356
44. Pidasheva S, Grant M, Canaff L, Ercan O, Kumar U, Hendy GN (2006) Calcium-sensing receptor dimerizes in the endoplasmic reticulum: biochemical and biophysical characterization of CASR mutants retained intracellularly. Hum Mol Genet 15: 2200–2209
45. Grant MP, Stepanchick A, Cavanaugh A, Breitwieser GE (2011) Agonist-driven maturation and plasma membrane insertion of calcium-sensing receptors dynamically control signal amplitude. Sci Signal 4:ra78
46. Nesbit MA, Hannan FM, Howles SA, Reed AA, Cranston T, Thakker CE, Gregory L, Rimmer AJ, Rust N, Graham U, Morrison PJ, Hunter SJ, Whyte MP, McVean G, Buck D, Thakker RV (2013) Mutations in AP2S1 cause familial hypocalciuric hypercalcemia type 3. Nat Genet 45:93–97
47. Hu J, Spiegel AM (2003) Naturally occurring mutations in the extracellular Ca2+-sensing receptor: implications for its structure and function. Trends Endocrinol Metab 14:282–288
48. Huang Y, Zhou Y, Castiblanco A, Yang W, Brown EM, Yang JJ (2009) Multiple Ca(2+)-binding sites in the extracellular domain of the Ca(2+)-sensing receptor corresponding to cooperative Ca(2+) response. Biochemistry 48:388–398
49. Silve C, Petrel C, Leroy C, Bruel H, Mallet E, Rognan D, Ruat M (2005) Delineating a Ca2+ binding pocket within the venus flytrap module of the human calcium-sensing receptor. J Biol Chem 280:37917–37923
50. Geibel J, Sritharan K, Geibel R, Geibel P, Persing JS, Seeger A, Roepke TK, Deichstetter M, Prinz C, Cheng SX, Martin D, Hebert SC (2006) Calcium-sensing receptor abrogates secretagogue- induced increases in intestinal net fluid secretion by enhancing cyclic nucleotide destruction. Proc Natl Acad Sci U S A 103:9390–9397
51. Fox J, Lowe SH, Petty BA, Nemeth EF (1999) NPS R-568: a type II calcimimetic compound that acts on parathyroid cell calcium receptor of rats to reduce plasma levels of parathyroid hormone and calcium. J Pharmacol Exp Ther 290:473–479
52. Nemeth EF, Steffey ME, Hammerland LG, Hung BC, Van Wagenen BC, DelMar EG, Balandrin MF (1998) Calcimimetics with potent and selective activity on the parathyroid calcium receptor. Proc Natl Acad Sci U S A 95:4040–4045
53. Huang Y, Breitwieser GE (2007) Rescue of calcium-sensing receptor mutants by allosteric modulators reveals a conformational checkpoint in receptor biogenesis. J Biol Chem 282:9517–9525
54. Leach K, Wen A, Cook AE, Sexton PM, Conigrave AD, Christopoulos A (2013) Impact of clinically relevant mutations on the pharmacoregulation and signaling bias of the calcium-sensing receptor by positive and negative allosteric modulators. Endocrinology 154:1105–1116
55. Brown EM (2010) Clinical utility of calcimimetics targeting the extracellular calcium-sensing receptor (CaSR). Biochem Pharmacol 80:297–307
56. Nemeth EF, Shoback D (2013) Calcimimetic and calcilytic drugs for treating bone and mineral-related disorders. Best Pract Res Clin Endocrinol Metab 27:373–384
57. Ho C, Conner DA, Pollak MR, Ladd DJ, Kifor O, Warren HB, Brown EM, Seidman JG, Seidman CE (1995) A mouse model of human familial hypocalciuric hypercalcemia and neonatal severe hyperparathyroidism [see comments]. Nat Genet 11:389–394
58. Chang W, Tu C, Chen T, Liu B, Elalieh H, Dvorak M, Clemens T, Kream B, Halloran B, Bikle D, Shoback D (2007) Conditional knockouts in early and mature osteoblasts reveals a critical role for Ca2+ receptors in bone development. J Bone Miner Res 22:S79, Abst 1284
59. Wettschureck N, Lee E, Libutti SK, Offermanns S, Robey PG, Spiegel AM (2007) Parathyroid-specific double knockout of Gq and G11 alpha-subunits leads to a phenotype resembling germline knockout of the extracellular Ca2±sensing receptor. Mol Endocrinol 21:274–280
60. Bourdeau A, Souberbielle J-C, Bonnet P, Herviaux P, Sachs C, Lieberherr M (1992) Phospholipase-A2 action and arachidonic acid metabolism in calcium-mediated parathyroid hormone secretion. Endocrinology 130:1339–1344
61. Bourdeau A, Moutahir M, Souberbielle JC, Bonnet P, Herviaux P, Sachs C, Lieberherr M (1994) Effects of lipoxygenase products of arachidonate metabolism on parathyroid hormone secretion. Endocrinology 135:1109–1112
62. Corbetta S, Lania A, Filopanti M, Vincentini L, Ballare E, Spada A (2002) Mitogen-activated protein kinase cascade in human normal and tumoral parathyroid cells. J Clin Endocrinol Metab 87: 2201–2205
63. Lu M, Forsberg L, Hoog A, Juhlin CC, Vukojevic V, Larsson C, Conigrave AD, Delbridge LW, Gill A, Bark C, Farnebo LO, Branstrom R (2008) Heterogeneous expression of SNARE proteins SNAP-23, SNAP-25, Syntaxin1 and VAMP in human parathyroid tissue. Mol Cell Endocrinol 287:72–80
64. Quinn SJ, Kifor O, Kifor I, Butters RR Jr, Brown EM (2007) Role of the cytoskeleton in extracellular calcium-regulated PTH release. Biochem Biophys Res Commun 354:8–13
65. Mendoza FJ, Lopez I, Canalejo R, Almaden Y, Martin D, Aguilera-Tejero E, Rodriguez M (2009) Direct upregulation of parathyroid calcium-sensing receptor and vitamin D receptor by calcimimetics in uremic rats. Am J Physiol Renal Physiol 296: F605–F613
66. Nielsen PK, Rasmussen AK, Butters R, Feldt-Rasmussen U, Bendtzen K, Diaz R, Brown EM,

Olgaard K (1997) Inhibition of PTH secretion by interleukin-1 beta in bovine parathyroid glands in vitro is associated with an up-regulation of the calcium- sensing receptor mRNA. Biochem Biophys Res Commun 238:880–885
67. Canaff L, Zhou X, Hendy GN (2008) The proinflammatory cytokine, interleukin-6, up-regulates calcium-sensing receptor gene transcription via Stat1/3 and Sp1/3. J Biol Chem 283:13586–13600
68. Murphey ED, Chattopadhyay N, Bai M, Kifor O, Harper D, Traber DL, Hawkins HK, Brown EM, Klein GL (2000) Up-regulation of the parathyroid calcium-sensing receptor after burn injury in sheep: a potential contributory factor to postburn hypocalcemia. Crit Care Med 28:3885–3890
69. Maiti A, Beckman MJ (2007) Extracellular calcium is a direct effecter of VDR levels in proximal tubule epithelial cells that counter-balances effects of PTH on renal Vitamin D metabolism. J Steroid Biochem Mol Biol 103:504–508
70. Cholst I, Steinberg S, Tropper P, Fox H, Segre G (1984) The influence of hypermagnesemia on serum calcium and parathyroid hormone levels in human subjects. N Engl J Med 310:1221–1225
71. Conigrave AD, Mun HC, Lok HC (2007) Aromatic L-amino acids activate the calcium-sensing receptor. J Nutr 137:1524S–1527S; discussion 1548S
72. Conigrave AD, Mun HC, Brennan SC (2007) Physiological significance of L-amino acid sensing by extracellular Ca(2+)-sensing receptors. Biochem Soc Trans 35:1195–1198
73. Hira T, Nakajima S, Eto Y, Hara H (2008) Calcium-sensing receptor mediates phenylalanine-induced cholecystokinin secretion in enteroendocrine STC-1 cells. FEBS J 275:4620–4626
74. D'Amour P (2006) Circulating PTH molecular forms: what we know and what we don't. Kidney Int Suppl 102:S29–33
75. Mayer GP, Keaton JA, Hurst JG, Habener JF (1979) Effects of plasma calcium concentration on the relative proportion of hormone and carboxyl fragments in parathyroid venous blood. Endocrinology 104:1778–1784
76. Hanley D, Takatsuki K, Sultan J, Schneider A, Sherwood L (1978) Direct release of parathyroid hormone fragments from functioning bovine parathyroid glands in vitro. J Clin Invest 62:1247–1254
77. Segre GV, Niall HD, Sauer RT, Potts JT Jr (1977) Edman degradation of radioiodinated parathyroid hormone: application to sequence analysis and hormone metabolism in vivo. Biochemistry 16: 2417–2427
78. D'Amour P, Rakel A, Brossard JH, Rousseau L, Albert C, Cantor T (2006) Acute regulation of circulating parathyroid hormone (PTH) molecular forms by calcium: utility of PTH fragments/PTH(1–84) ratios derived from three generations of PTH assays. J Clin Endocrinol Metab 91:283–289
79. Nussbaum SR, Potts J Jr, Wang CA, Zahradnik R, Lavigne JR, Kim L, Segre GV (1987) A highly sensitive two-site immunoradiometric assay for parathyroid hormone (PTH) and its clinical utility in the evaluation of patients with hypercalcemia. Clin Chem 33:1364–1367
80. Gao P, Scheibel S, D'Amour P, John MR, Rao SD, Schmidt-Gayk H, Cantor TL (2001) Development of a novel immunoradiometric assay exclusively for biologically active whole parathyroid hormone 1–84: implications for improvement of accurate assessment of parathyroid function. J Bone Miner Res 16:605–614
81. Friedman PA, Goodman WG (2006) PTH(1–84)/PTH(7–84): a balance of power. Am J Physiol Renal Physiol 290:F975–F984
82. Schwarz P, Sorensen HA, Transbol I, McNair P (1992) Regulation of acute parathyroid hormone release in normal humans: combined calcium and citrate clamp study. Am J Physiol 263:E195–E198
83. Valle C, Rodriguez M, Santamaria R, Almaden Y, Rodriguez ME, Canadillas S, Martin-Malo A, Aljama P (2008) Cinacalcet reduces the set point of the PTH-calcium curve. J Am Soc Nephrol 19:2430–2436
84. Nechama M, Ben-Dov IZ, Silver J, Naveh-Many T (2009) Regulation of PTH mRNA stability by the calcimimetic R568 and the phosphorus binder lanthanum carbonate in CKD. Am J Physiol Renal Physiol 296:F795–F800
85. Nechama M, Peng Y, Bell O, Briata P, Gherzi R, Schoenberg DR, Naveh-Many T (2009) KSRP-PMR1-exosome association determines parathyroid hormone mRNA levels and stability in transfected cells. BMC Cell Biol 10:70
86. Levi R, Ben-Dov IZ, Lavi-Moshayoff V, Dinur M, Martin D, Naveh-Many T, Silver J (2006) Increased parathyroid hormone gene expression in secondary hyperparathyroidism of experimental uremia is reversed by calcimimetics: correlation with post-translational modification of the trans acting factor AUF1. J Am Soc Nephrol 17:107–112
87. Raisz L (1965) Regulation by calcium of parathyroid growth and secretion in vitro. Nature 44:103–110
88. Roth SI, Raisz LG (1966) The course and reversibility of the calcium effect on the ultrastructure of the rat parathyroid gland in organ culture. Lab Invest 15:1187–1211
89. Egbuna OI, Brown EM (2008) Hypercalcaemic and hypocalcaemic conditions due to calcium-sensing receptor mutations. Best Pract Res Clin Rheumatol 22:129–148
90. Wada M, Furuya Y, Sakiyama J-i, Kobayashi N, Miyata S, Ishii H, Hagano N (1997) The calcimimetic compound NPS R-568 suppresses parathyroid cell proliferation in rats with renal insufficiency. J Clin Invest 100:2977–2983
91. Finch JL, Lee DH, Liapis H, Ritter C, Zhang S, Suarez E, Ferder L, Slatopolsky E (2013) Phosphate restriction significantly reduces mortality in uremic rats with established vascular calcification. Kidney Int 84:1145–1153
92. Arcidiacono MV, Sato T, Alvarez-Hernandez D, Yang J, Tokumoto M, Gonzalez-Suarez I, Lu Y,

Tominaga Y, Cannata-Andia J, Slatopolsky E, Dusso AS (2008) EGFR activation increases parathyroid hyperplasia and calcitriol resistance in kidney disease. J Am Soc Nephrol 19:310–320
93. Dusso AS, Pavlopoulos T, Naumovich L, Lu Y, Finch J, Brown AJ, Morrissey J, Slatopolsky E (2001) p21(WAF1) and transforming growth factor-alpha mediate dietary phosphate regulation of parathyroid cell growth. Kidney Int 59:855–865
94. Gogusev J, Duchambon P, Stoermann-Chopard C, Giovannini M, Sarfati E, Drueke TB (1996) De novo expression of transforming growth factor-alpha in parathyroid gland tissue of patients with primary or secondary uraemic hyperparathyroidism. Nephrol Dial Transplant 11:2155–2162
95. Cozzolino M, Lu Y, Sato T, Yang J, Suarez IG, Brancaccio D, Slatopolsky E, Dusso AS (2005) A critical role for enhanced TGF-alpha and EGFR expression in the initiation of parathyroid hyperplasia in experimental kidney disease. Am J Physiol Renal Physiol 289:F1096–F1102
96. Lin SY, Makino K, Xia W, Matin A, Wen Y, Kwong KY, Bourguignon L, Hung MC (2001) Nuclear localization of EGF receptor and its potential new role as a transcription factor. Nat Cell Biol 3:802–808
97. Cozzolino M, Lu Y, Finch J, Slatopolsky E, Dusso AS (2001) p21WAF1 and TGF-alpha mediate parathyroid growth arrest by vitamin D and high calcium. Kidney Int 60:2109–2117
98. Kanesaka Y, Tokunaga H, Iwashita K, Fujimura S, Naomi S, Tomita K (2001) Endothelin receptor antagonist prevents parathyroid cell proliferation of low calcium diet-induced hyperparathyroidism in rats. Endocrinology 142:407–413
99. Wang Q, Palnitkar S, Parfitt AM (1996) Parathyroid cell proliferation in the rat: effect of age and of phosphate administration and recovery. Endocrinology 137:4558–4562
100. Colloton M, Shatzen E, Miller G, Stehman-Breen C, Wada M, Lacey D, Martin D (2005) Cinacalcet HCl attenuates parathyroid hyperplasia in a rat model of secondary hyperparathyroidism. Kidney Int 67:467–476
101. Mizobuchi M, Ogata H, Hatamura I, Saji F, Koiwa F, Kinugasa E, Koshikawa S, Akizawa T (2007) Activation of calcium-sensing receptor accelerates apoptosis in hyperplastic parathyroid cells. Biochem Biophys Res Commun 362:11–16

Phosphate Control of PTH Secretion

6

Piergiorgio Messa

6.1 Introduction

Phosphate, chiefly in association with calcium, is the main component of the mineralized phase of the bone, where it is mainly present in a crystalline form (hydroxyapatite). In addition to this essential function, phosphate is involved in a large series of basic biological processes, such as energy metabolic pathways, nucleic acid metabolism, cell signalling, activity of a large number of proteins involved in metabolic pathways, control of the transport of organic or inorganic compounds through the cellular membrane, and so forth [1–3].

6.2 General Notes on Phosphorus Metabolism

6.2.1 Main Functions and Body Distribution of Phosphorus

The total phosphate content of a medium-size human body approximates 650 g (12.0 g/kg of body weight), about 85 % of it being within the bone, 14 % in the intracellular space, and the remaining 1 % in the plasma. The intracellular phosphorus is mainly found in the form of organic compounds (nucleoside phosphates, phosphoproteins, phospholipids, 2,3-diphospho-glycerate, etc.), with a proportion of 10–100 to 1 as compared with the intracellular inorganic phosphate content, with the two components being in strict reciprocal equilibrium. In turn, the intracellular inorganic phosphorus equilibrates with extracellular phosphate, with a ratio of around 0.6. Phosphorus in blood circulates in part as inorganic phosphate (mainly associated with inorganic cations such as calcium, magnesium, and sodium) accounting for the usual normal range of 2.5–4.2 mg/dl, while a consistent amount of phosphate (8–13 mg/dl), which is not usually measured, is present as phosphoric esters or as a lipid compound [4, 5]. The variable content of phosphate in these different bodily pools and its global balance are under a complex and integrated control system, which involves a number of transport systems expressed in many tissues and organs and a large number of hormonal and paracrine/autocrine factors. On the other hand, phosphate per se can also modulate all these transport pathways and the related regulatory factors.

6.2.2 Factors Controlling Phosphate Metabolism

When looking at the issue of phosphate homeostasis, it is worth distinguishing whether we are dealing with the phosphate content in the body

P. Messa
Department of Nephrology, Urology and Renal Transplant, Fondazione Ca' Granda IRCCS Ospedale Maggiore Policlinico, Via Commenda 15, Milan 20122, Italy
e-mail: Piergiorgio.messa@policlinico.mi.it

M.L. Brandi, E.M. Brown (eds.), *Hypoparathyroidism*,
DOI 10.1007/978-88-470-5376-2_6, © Springer-Verlag Italia 2015

pool(s) or with the serum phosphorus levels. The bodily content of phosphate is determined by its global external balance, which is mainly regulated by the net flux through the intestine, on the one hand, and the renal excretion of phosphate, on the other. The intestinal absorption of phosphate is mainly dependent on dietary phosphorus content; active transcellular transport across the intestinal wall, which occurs through the sodium-phosphate cotransporter type 2b (Na-P 2b) and to a lesser extent the PiT-1 transporter; and on passive paracellular transport, which is mainly dependent on the phosphate concentration gradient between the intestinal lumen and the interstitial space. These absorptive processes are principally localized within the duodenum and jejunum. It has also been acknowledged that a minor amount of phosphate (approximately 200 mg/day) is secreted (in the pancreatic juice and intestinal secretion), so that the net intestinal absorption of this mineral is the result of the actively and passively absorbed phosphate minus the secreted amount. It is accepted that the net intestinal phosphate absorption approximates 60–70 % of its dietary intake, which is more or less 700–1,100 mg/day for a usual western dietary intake [6, 7]. The other component which regulates body content of phosphate is its excretion by the renal route. The renal phosphate excretion is dependent on the amount of phosphate filtered through the glomerulus, which is proportional to the glomerular filtration rate, and the amount of the filtered load reabsorbed by the proximal tubule, mainly regulated by the activity of the Na-Pi 2a and Na-Pi 2b cotransporters [8]. In a normal individual, in metabolic and phosphate balance, the renal excretion of phosphorus exactly matches its net intestinal absorption.

When we deal with the control of the serum phosphorus concentration, in addition to the above described mechanisms determining the bodily content of phosphate, other factors are involved that control the fluxes into and out of the bone and the changes in the distribution of inorganic phosphate between the intracellular and extracellular body compartments. Some of these processes, which are mainly responsible for the transport into and out of cells, have been demonstrated to depend on the activity of specific phosphate transporters (PiT-1 and PiT-2) almost ubiquitously expressed at the cell membrane [9–13] (see also Chap. 11). Furthermore, in the physiological control of serum phosphorus level, a critical role has been focused on the modulation of the distribution of phosphate within and outside of the miscible pool(s) in the bone. In the last few years, a huge amount of data has highlighted a key role of osteocytes in orchestrating these control mechanisms. In fact, these cells, far from being inert and quiescent as previously believed, have been shown to play a central role in the modulation of bone remodelling and mineral metabolism, by regulating the activities of the other bone cells (osteoblasts, osteoclasts), by functioning as mechanosensory cells, and by playing a true endocrine role through their secretion of circulating hormones (e.g., FGF23, sclerostin, osteocalcin) involved in mineral metabolism and in other putative functions [14, 15] (see also Chap. 11).

Overall, most of the mechanisms listed above are involved in the setting of both the balance and serum level of phosphorus and are, in turn, tightly controlled by hormonal, autocrine, and paracrine factors. Table 6.1 lists the most relevant of these factors and their main mechanisms of action.

6.2.3 Phosphorus Control of Its Controlling Factors

Many, if not all, of the metabolic pathways and hormonal/paracrine/autocrine factors involved in the regulation of either the balance and/or the serum levels of phosphate can be directly or indirectly modulated by the dietary phosphate intake and/or the serum phosphorus levels. Therefore, phosphate can self-regulate its own balance and serum concentration. In fact, increased dietary intake of phosphorus stimulates FGF23 and PTH synthesis and secretion, reduces vitamin D levels, and induces a decrease of both renal tubular and intestinal phosphate transport, while opposing effects having been shown during dietary phosphate deprivation [21–23]. In the following paragraphs, the putative mechanisms underlying the reciprocal control between phosphate and its regulatory factors will be described in more

Table 6.1 Factors controlling inorganic phosphate metabolism: short description of the main sites (bone, intestine, renal) and mechanisms of action, with the relevant references

Controlling factor	Site of action: Intestine	Site of action: Kidney	Site of action: Bone	Global effect on phosphate metabolism	Relevant references
Vitamin D	↑ Pi absorption	↑ Pi tubular reabsorption	↑ Bone turnover	↑ Serum Pi levels Positive Pi balance	[16]
PTH	Indirectly (↑vit D) ↑ Pi abs	↓ Pi tubular transport	↑ Bone turnover	NRF: ↓ serum Pi levels Neutral/neg. Pi balance IRF: =/↑ serum Pi levels Neutral/pos. Pi balance	[17]
FGF23/Klotho	Direct and indirect (↓vit D) ↓ Pi abs	↓ Pi tubular transport (via FGFR-1 and −4)	Direct and indirect (↓PTH) Reduced bone turnover	↓Serum Pi levels Negative Pi balance	[18]
Phosphatonins (DMP-1, MEPE, sFRP-4)	Not completely defined	↓ Pi tubular transport	Not completely defined	↓Serum Pi levels Negative Pi balance	[19]
Putative undefined intestinal factor(s)	? ↓ Pi absorption	?	?	↓ Serum Pi levels; maintenance of neutral Pi balance (?)	[20]

Notes: *Pi* inorganic phosphate, *PTH* parathyroid hormone, *NRF* normal renal function, *IRF* impaired renal function, *FGF23* fibroblast growth factor 23, *FGFR* FGF23 receptor, *DMP1* dentin matrix protein 1, *MEPE* matrix extracellular phosphoglycoprotein, *sFRP-4* secreted frizzled related protein 4

detail; however, it can be said that all of these mechanisms suggest the presence of some cellular mechanism(s) that senses phosphate. Having said that, it is also worth pointing out that no clear demonstration of the existence of such a mechanism has been so far produced (see below) nor is it clear whether this hypothesized mechanism senses phosphate concentration or phosphate intake or both.

In what follows, we will focus on the mechanism(s) by which phosphate and PTH reciprocally control one another, addressing in more detail the topic of this chapter, namely, the control of parathyroid function by phosphate.

6.2.3.1 PTH-Mediated Phosphate Control

As already mentioned, the pathways by which PTH affects phosphate metabolism are manifold (Fig. 6.1). PTH mainly acts through its receptor (PTH/PTH-related peptide receptor), which is mainly expressed in the renal tubular cells and in bone cells (osteoblasts and osteocytes). At the renal tubular level, PTH promotes the synthesis of the 1-α-hydroxylase enzyme (CYP27B1), which converts 25-OH-vitamin D to its most active compound (1,25-$(OH)_2$-vitamin D), the main hormonal factor stimulating the intestinal absorption of calcium and phosphorus (see also Chap. 11). At the same time, PTH, by increasing the intracellular levels of cAMP, inhibits the expression at the membrane level of the renal proximal tubular cells of Na-Pi 2a and Na-Pi 2c cotransporters, reducing the tubular absorption of phosphate with a consequent increase in its urinary excretion (see also Chap. 11). Furthermore, PTH, in addition to its multiple anabolic and catabolic effects on bone cells, can also stimulate the flux of phosphate outside the bone by inducing the resorption of the bone mineral phase by activating osteoclasts

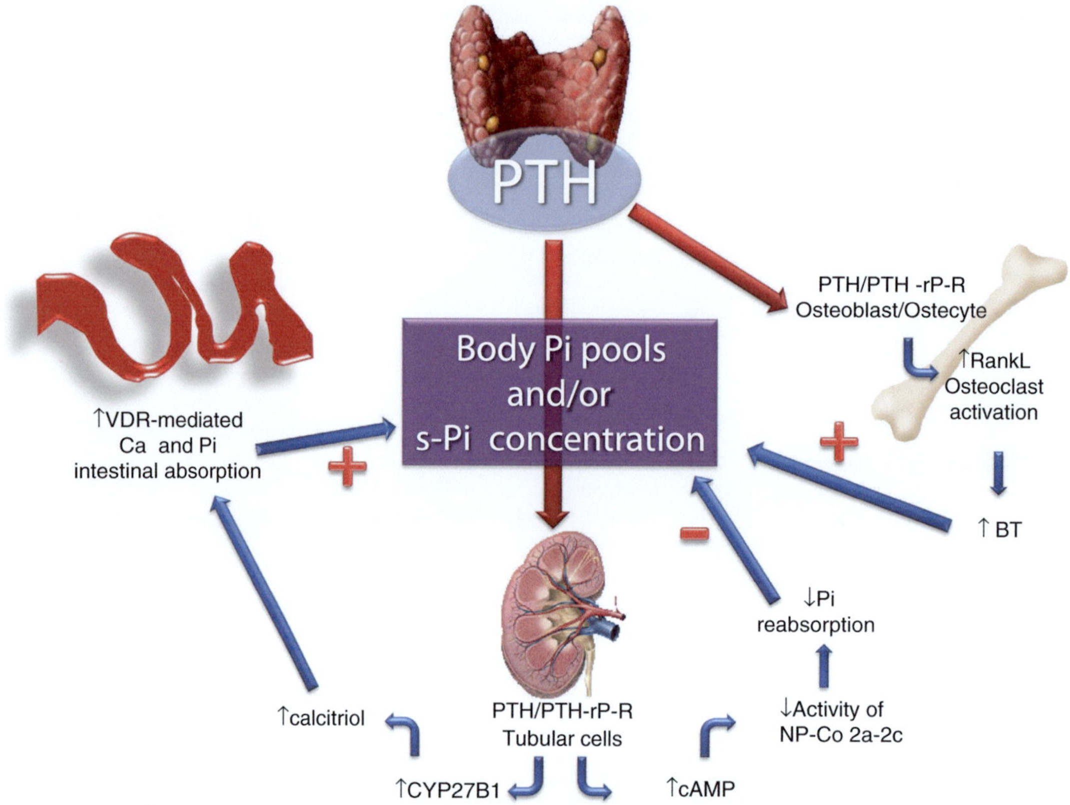

Fig. 6.1 Main mechanisms by which PTH controls phosphorus metabolism. *Pi* inorganic phosphate, *PTH/PTHrP-R* PTH-PTH-related peptide receptor, *RANK-L* ligand of receptor activator of nuclear factor kappa-B, *BT* bone turnover, *NP-Co 2a-2c* sodium-phosphate cotransporter 2a and 2c, *cAMP* cyclic adenosine monophosphate, *CYP27B1* cytochrome 27B1 (25-hydroxyvitamin D_3 1-alpha-hydroxylase), *VDR* vitamin D receptor

through increased RANKL expression by both osteoblasts and osteocytes [24].

Among all the abovementioned effects of PTH, the most potent one with respect to phosphate control is its inhibition of the renal tubular transport of phosphate, at least while renal function is preserved. In fact, hypophosphatemia is the most common feature of the conditions characterized by either primary or secondary hyperparathyroidism in patients with normal or near normal renal function. In contrast, when renal function is either acutely or chronically impaired, increased PTH levels, a usual finding in these conditions, are mainly associated with hyperphosphatemia, due to the reduced capacity of the kidney to eliminate phosphate, in the face of an almost unchanged intestinal absorption rate and increased resorption of phosphate from bone associated with the increased PTH levels.

6.2.3.2 Phosphate-Mediated Control of Parathyroid Function

In turn, it has long been accepted that phosphate itself can regulate PTH secretion by some complex and as yet not fully elucidated feedback mechanisms. There is also widespread evidence supporting the notion that phosphate-dependent PTH control can take place via both indirect and direct mechanisms. In the following paragraphs, these two different types of pathways will be dealt with separately. However, it would be worth underlining at the outset that most evidence of a putative phosphorus control of PTH secretion derives from experimental and clinical studies carried out in conditions mainly characterized by reduced renal function. So, most of these results cannot be easily generalized to the physiological or different pathological conditions.

6.2.4 Indirect Effects of Pi on PTH

The first known mechanism by which phosphate indirectly controls PTH levels is related to the reciprocal inverse relationship linking the changes in the serum concentrations of phosphate and calcium, by a physicochemical mechanism, with an increase and a decrease of phosphate producing a decrease and an increase in calcium, respectively [25, 26]. The consequent changes in calcium concentration, acting through the specific calcium sensor (CaSR), a G-protein-coupled receptor expressed at the plasma membrane levels of parathyroid cells, are the most potent stimuli for controlling the release of PTH from the secretory granules and its synthesis within the parathyroid glands and the rate of parathyroid cellular proliferation (see Chap. 5 dedicated to this topic in this book). So, in conditions of acute and, to a lesser extent, chronic increase in phosphorus levels, calcium levels tend to be reduced, inducing a stimulation of PTH release and secretion and in the long run, also stimulating parathyroid cell growth.

Another way by which phosphate can indirectly induce changes of PTH levels is related to the control of the enzyme CYP27B1 (25-OH-vitD-1-α-hydroxylase), which is involved in the synthesis of the most active vitamin D metabolite, 1,25-dihydroxyvitamin D (calcitriol), principally in the renal proximal tubule. Increased phosphate levels and possibly increased phosphorus intake have been demonstrated to be associated with reduced calcitriol synthesis and, consequently, with increased levels of PTH. In fact, there is much experimental and clinical evidence demonstrating that calcitriol controls PTH levels by both indirect and direct mechanisms. The best recognized indirect mechanism consists of the activation of the active transcellular transport of calcium at the intestinal level, which by increasing serum calcium levels downregulates PTH secretion. Another well-recognized indirect mechanism of calcitriol is related to its activation of bone cells (direct osteoblast/osteocyte activation and indirect osteoblast-mediated osteoclast activation); this is followed by increased bone turnover, which would be expected to be followed by increased release of calcium from its skeletal stores and inhibition of PTH release [27]. A third mechanism by which calcitriol can indirectly induce PTH release is related to the recently recognized stimulatory effect of calcitriol on FGF23 synthesis, which probably involves activation of FGF23 gene transcription through both direct and indirect VDRE-mediated mechanisms [28, 29]. In turn, both in vivo and in vitro studies have demonstrated that FGF23 reduces PTH mRNA levels and its secretion, through its interaction with its specific receptor (FGFR-1c) and the coreceptor Klotho, expressed in parathyroid cells [30, 31]. This inhibitory effect of FGF23 on PTH is mainly evident in conditions of relatively normal renal function, while it no longer functions normally in patients with hyperparathyroidism secondary to impaired renal function, probably due to a progressive loss of both FGFR-1c and Klotho expression in parathyroid cells [32]. Finally, as has long been recognized, calcitriol can directly inhibit PTH synthesis by a genomic effect mediated by its specific receptor (VDR), which is abundantly expressed in the parathyroid [33–36]. Also, the inhibitory effect of calcitriol tends to be dampened in chronic kidney disease patients, particularly in more advanced stages of the disease, due to a progressive reduction of VDR expression in parathyroid cells [37].

There are conflicting data on the possibility that phosphate per se plays a role in stimulating FGF23 secretion or synthesis and consequently affecting PTH levels in this way. On the one hand, it has been reported that both increased dietary intake of phosphate and an increase in its serum levels are associated with increased levels of FGF23, though the exact mechanism through which this effect might be put into play is not clear [21, 38, 39]. On the other hand, other authors failed to find any major change in FGF23 levels after modifying the dietary content of phosphate or after IV phosphate infusion

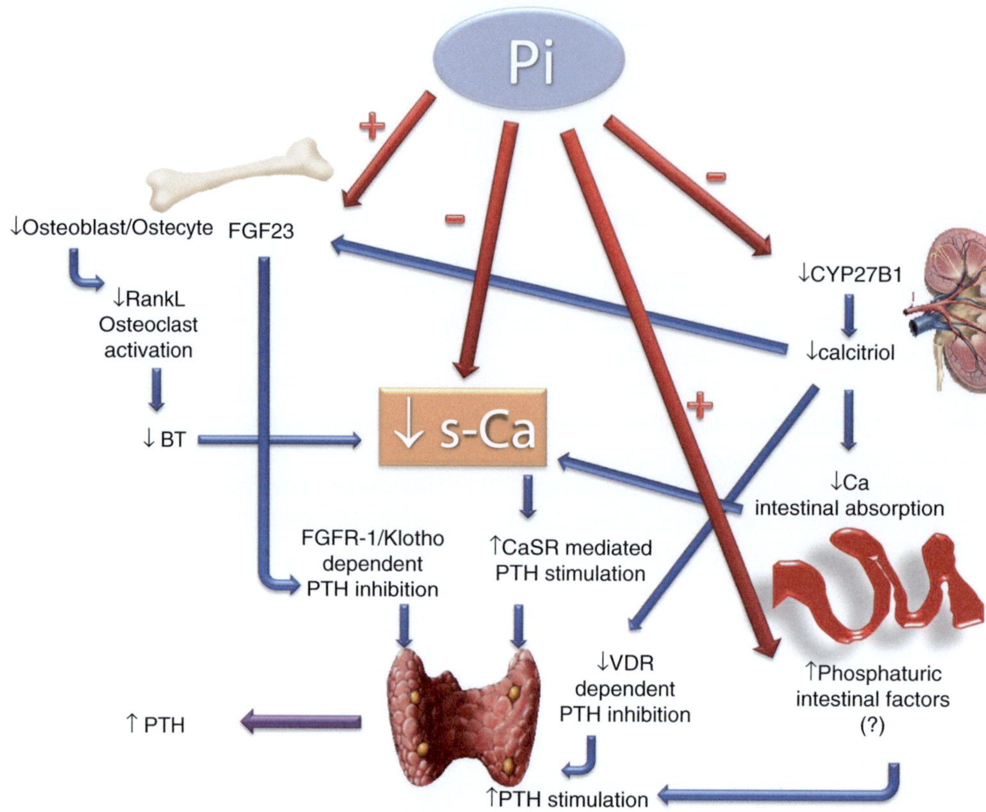

Fig. 6.2 Indirect mechanisms by which phosphorus controls PTH production. *Pi* inorganic phosphate, *RANK-L* ligand of receptor activator of nuclear factor kappa-B, *BT* bone turnover, *NP-Co 2a-2c* sodium-phosphate cotransporter 2a and 2c, *CYP27B1* cytochrome 27B1 (25-hydroxyvitamin D_3 1-alpha-hydroxylase), *VDR* vitamin D receptor, *CaSR* calcium-sensing receptor, *FGF23* fibroblast growth factor 23

[40, 41]. The reasons for these discrepancies are not clear. However, a common finding of all these studies was that any modification in dietary intake of phosphorus was always matched by a parallel change of its urinary excretion, independent of any significant change in the serum phosphorus and/or in phosphate-controlling hormone levels. This raised the question as to whether additional factors exist that might putatively influence phosphate balance.

There is some evidence that there is/are further, but not yet demonstrated, intestinal factor(s), directly modulated by dietary phosphate content, which might control phosphate excretion and possibly also PTH secretion. In fact, some experimental studies carried out in rats demonstrated that, while a force-feeding with a low-phosphate diet produces a rapid (within 15 min) reduction in both serum PTH and phosphorus levels [42], hypophosphatemia induced by an infusion of glucose did not yield any change in serum PTH. This putative additional mechanism opens the way to a string of unanswered questions on the possible existence of some mechanism(s) that are able to sense phosphorus in the different organs involved in phosphate control (Fig. 6.2).

6.2.5 Direct Effects of Phosphate on PTH

The possibility of a direct effect of phosphate on PTH secretion is strictly linked to this putative sensing mechanism for phosphate. The presence of mechanisms able to directly sense phosphate concentration has been recognized both in plants and in some bacteria. Plant metabolism is largely dependent on the phosphate availability in the soil, and a sensing mechanism, involving the activation of a microRNA (miR399), has been described that might induce signalling pathways able to increase the uptake of this mineral, followed by the activation of other metabolic processes [43].

In unicellular organisms (bacteria and yeast), a system directed at sensing extracellular phosphate levels has been described in great detail. This mechanism is characterized by a multicomponent system with some periplasmic membrane proteins (PstS, PstC, PstA, PstB associated with PhoU) sensing low phosphate concentration and triggering an increased flux of phosphate into the cell. The increased intracellular phosphate availability induces phosphorylation of PhoR, which in turn phosphorylates another Pho protein (PhoB), which ultimately acts as a transcription factor activating genes that encode for proteins involved in phosphate transport, in the production of alkaline phosphatase, or in the control of cyclin-dependent kinases [44–46].

There is also some experimental evidence that sensing mechanisms for phosphate also exist in multicellular organisms, at least in intestinal, renal tubular, and bone cells. In fact, intestinal, renal, osteoblast, or marrow stromal cells, cultured in low or high phosphate media, change the expression of some phosphate transport proteins and proteins with enzymatic control (alkaline phosphatase) or involved in regulating gene transcription (Runx2/Cbfa1, BMP4) [7, 47, 48]. However, the exact mechanisms underlying this sensor activity remain unknown.

Though the possibility that phosphate might be directly sensed by parathyroid cells, inducing direct control of PTH secretion and synthesis has long been suggested; the actual proof of this concept and the possible way by which it might work are far from having been convincingly demonstrated. One critical difficulty in achieving convincing results in this regard is the lack of a reliable cell line available for studying parathyroid function in culture. Another reason that makes it difficult to demonstrate a direct effect of phosphate on parathyroid cells is that most experimental conditions inducing a change in phosphate levels are invariably associated with changes also in calcium concentration, which is the most important factor in the control of PTH secretion and synthesis. Furthermore, most of the experimental models where this hypothesis have been challenged were those characterizing the secondary hyperparathyroidism of chronic kidney disease, and it is well recognized that in this condition, a number of confounding factors (changes in calcitriol, FGF23/Klotho levels, and in the levels of expression of CaSR and VDR) can muddy the interpretation of any result.

Seminal studies in animals demonstrated that a reduction of phosphate intake proportional to the GFR reduction prevents the development of SHP, independently of any major change in calcium or calcitriol levels [49, 50]. In a further study, the same group reported data that strongly suggested that phosphate per se, independently of any change in calcium or calcitriol, can directly stimulate PTH secretion. In fact, these authors reported that cultured rat parathyroid glands, when incubated with high medium phosphate (2.8 mmol/l), secreted much more PTH than glands incubated with low medium phosphate (0.2 mmol/l), with these changes being independent of changes in either calcitriol or calcium concentration [51]. Subsequent studies produced further data supporting a direct effect of phosphorus on PTH secretion [52–54] and suggested that this effect might be mostly posttranscriptional. In fact, these authors demonstrated that in parathyroid cells, some specific cytosolic proteins bound a 26-nucleotide sequence of PTH mRNA at its 3'-UTR, stabilizing this molecule and increasing PTH translation. In these experiments, hypocalcemia and hypophosphatemia increased and decreased, respectively, the capacity of these proteins to bind to PTH mRNA, with the final effect of hypocalcemia increasing and hypophosphatemia decreasing the PTH levels.

Another proposed mechanism for explaining the putative direct effect of phosphorus on parathyroid cells was that it was mediated by changes in the activity of phospholipase A_2 and its metabolic product, arachidonic acid (AA), which acts

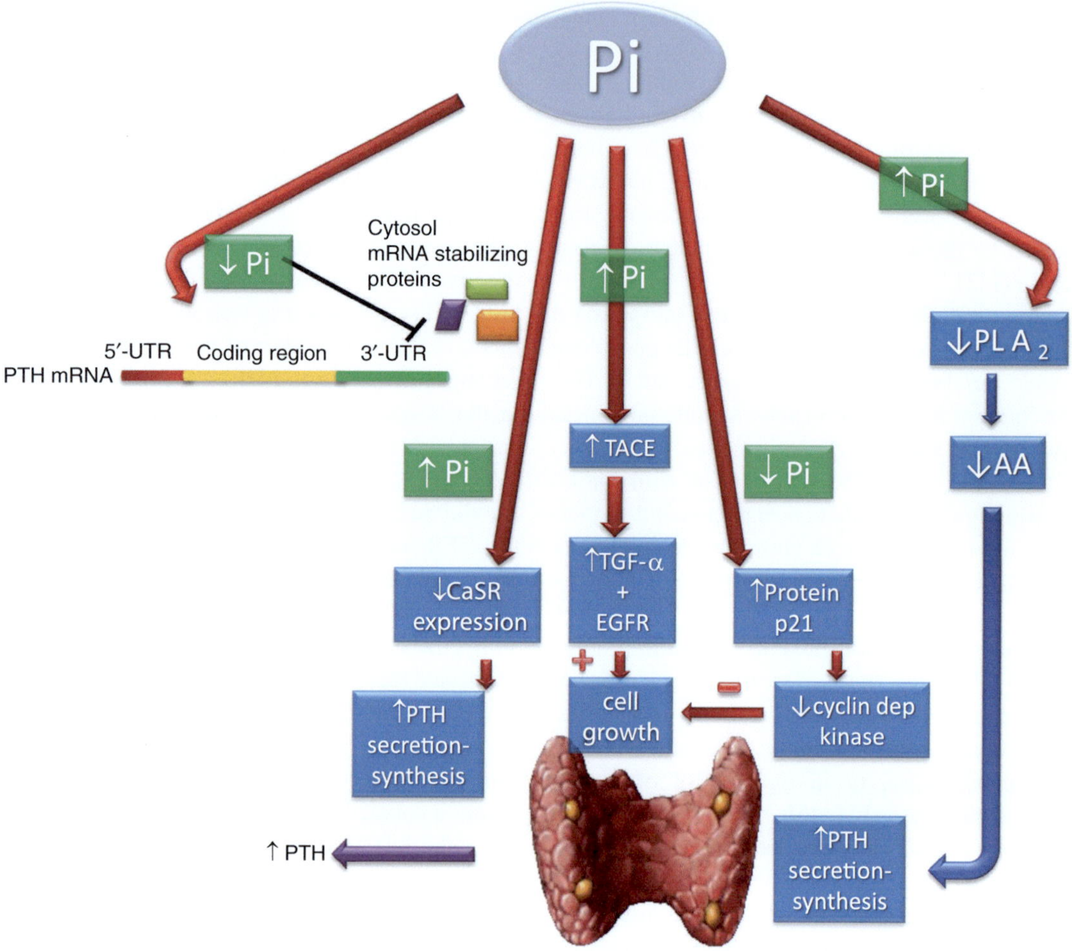

Fig. 6.3 Putative direct mechanisms by which phosphorus controls PTH production. *Pi* inorganic phosphate, *CaSR* calcium-sensing receptor, *TGF-α* transforming growth factor alpha, *TACE* TGF-α-converting enzyme, *EGFR* epidermal growth factor receptor, *PLA2* phospholipase A2, *AA* arachidonic acid

as an inhibitor of PTH. Phosphate was suggested to decrease the production of AA and, as a consequence of its inhibitory activity, leading to increased secretion of PTH. Other authors proposed that high phosphorus intake can induce increased secretion and synthesis of PTH, by reducing the expression of CaSR on parathyroid cells [55]. Finally, a series of studies, coming from the Slatopolsky group, suggested another potential direct mechanism of phosphate on parathyroid glands. Namely, Dusso and co-workers [56] demonstrated that dietary phosphate restriction counteracts the development of the parathyroid hyperplasia induced by uremia through the induction of the protein p21, an inhibitor of cyclin-dependent kinase, which mediated cell growth arrest; on the other hand, high phosphate intake induced transforming growth factor-α (TGF-α), which stimulates cell growth. Subsequently, Cozzolino et al. [57] clearly showed that a high dietary phosphate intake in chronic kidney disease (CKD) rats increases the expression of both TGF-α and the epidermal growth factor receptor (EGFR), which is essential for TGF-α signalling, in the parathyroid gland. Further studies from the same group [58] proposed that the increase of TGF-α was secondary to increased levels of TGF-α-converting enzyme (TACE), a metalloproteinase which plays an essential role in TGF-α signalling.

All these data strongly support the notion of the existence of some phosphate-mediated mechanism(s) that directly control PT cell activity and growth and suggest different models for explaining how these mechanisms function. However, while for all the factors that control PT cell function (calcium, vitamin D, FGF23), a defined receptor has been clearly defined (CaSR, VDR, Klotho-FGFR-1, respectively), no demonstration of an equivalent phosphate sensor exists so far (Fig. 6.3).

Conclusion

Phosphate plays many key biological roles in the body and its metabolism is under the control of a large number of factors. PTH is one of the main regulators of phosphate metabolism, and phosphate, in turn, has manifold mechanisms for controlling parathyroid function itself, by both indirect and direct pathways. The indirect mechanisms, through which phosphate control PTH, are far more well defined than the putative direct ones. However, it is worth stressing that most of the evidence comes from experimental models characterized by a reduced renal function, which makes the available data not easily generalizable to the normal condition. Furthermore, in the in vivo studies, given the complex interrelationship among all the components of mineral metabolism (calcium, vitamin D, FGF23/Klotho, etc.), it is not easy to dissect the effect of phosphate from those of all these other factors on PTH.

References

1. Kornberg A (1979) The enzymatic replication of DNA. CRC Crit Rev Biochem 7:23–43
2. Krebs EG, Beavo JA (1979) Phosphorylation-dephosphorylation of enzymes. Annu Rev Biochem 48:923–959
3. Lardy HA, Ferguson SM (1969) Oxidative phosphorylation in mitochondria. Annu Rev Biochem 38:991–1034
4. Bevington A, Brough D, Baker FE, Hattersley J, Walls J (1995) Metabolic acidosis is a potent stimulus for cellular inorganic phosphate generation in uraemia. Clin Sci (Lond) 88(4):405–412
5. Fleisch H (1980) Homeostasis of inorganic phosphate. In: Urist MR (ed) Fundamental and clinical bone physiology. Lippincott, Philadelphia
6. Lee DB, Walling MW, Brautbar N (1986) Intestinal phosphate absorption: influence of vitamin D and nonvitamin D factors. Am J Physiol Gastrointest Liver Physiol 250:G369–G373
7. Segawa H, Kaneko I, Yamanaka S, Ito M, Kuwahata M, Inoue Y, Kato S, Miyamoto K (2004) Intestinal Na-P(i) cotransporter adaptation to dietary P(i) content in vitamin D receptor null mice. Am J Physiol Renal Physiol 287:F39–F47
8. Berndt T, Kumar R (2009) Novel mechanisms in the regulation of phosphorus homeostasis. Physiology (Bethesda) 24:17–25
9. Hruska K, Slatopolsky E (1996) Disorders of phosphorous, calcium, and magnesium metabolism. In: Schrier R, Gottschalk C (eds) Diseases of the kidney. Little, Brown, and Company, London, pp 2477–2526
10. Kronenberg HM (2002) NPT2a—the key to phosphate homeostasis. N Engl J Med 347:1022–1024
11. Tanimura A, Yamada F, Saito A, Ito M, Kimura T, Anzai N, Horie D, Yamamoto H, Miyamoto K, Taketani Y, Takeda E (2011) Analysis of different complexes of type IIa sodium-dependent phosphate transporter in rat renal cortex using blue-native polyacrylamide gel ecletrophoresis. J Med Invest 58(1–2):140–147
12. Forster IC, Hernando N, Biber J, Murer H (2013) Phosphate transporters of the SLC20 and SLC34 families. Mol Aspects Med 34(2–3):386–395
13. Gattineni J, Alphonse P, Zhang Q, Mathews N, Bates CM, Baum M (2014) Regulation of renal phosphate transport by FGF23 is mediated by FGFR1 and FGFR4. Am J Physiol Renal Physiol 306(3):F351–F358
14. Dallas SL, Prideaux M, Bonewald LF (2013) The osteocyte: an endocrine cell … and more. Endocr Rev 34(5):658–690
15. Wesseling-Perry K, Jüppner H (2013) The osteocyte in CKD: new concepts regarding the role of FGF23 in mineral metabolism and systemic complications. Bone 54:222–229
16. DeLuca HF (2004) Overview of general physiologic features and functions of vitamin D. Am J Clin Nutr 80:1689S–1696S
17. Aurbach GD, Heath DA (1974) Parathyroid hormone and calcitonin regulation of renal function. Kidney Int 6(5):331–345
18. Berndt TJ, Schiavi S, Kumar R (2005) "Phosphatonins" and the regulation of phosphorus homeostasis. Am J Physiol Renal Physiol 289(6):F1170–F1182
19. Feng JQ, Clinkenbeard EL, Yuan B, White KE, Drezner MK (2013) Osteocyte regulation of phosphate homeostasis and bone mineralization underlies the pathophysiology of the heritable disorders of rickets and osteomalacia. Bone 54(2):213–221

20. Berndt T, Thomas LF, Craig TA, Sommer S, Li X, Bergstralh EJ, Kumar R (2007) Evidence for a signaling axis by which intestinal phosphate rapidly modulates renal phosphate reabsorption. Proc Natl Acad Sci U S A 104(26):11085–11090
21. Burnett SM, Gunawardene SC, Bringhurst FR et al (2006) Regulation of C-terminal and intact FGF-23 by dietary phosphate in men and women. J Bone Miner Res 21(8):1187–1196
22. Ferrari SL, Bonjour JP, Rizzoli R (2005) Fibroblast growth factor-23 relationship to dietary phosphate and renal phosphate handling in healthy young men. J Clin Endocrinol Metab 90:1519–1524
23. Baxter LA, DeLuca HF (1976) Stimulation of 25-hydroxy-vitamin D3-1α- hydroxylase by phosphate depletion. J Biol Chem 251:3158–3161
24. Bellido T, Saini V, Pajevic PD (2013) Effects of PTH on osteocyte function. Bone 54(2):250–257
25. Kaplan MA, Canterbury JM, Gavellas G, Jaffe D, Bourgoignie JJ, Reiss E, Bricker NS (1978) Interrelations between phosphorus, calcium, parathyroid hormone, and renal phosphate excretion in response to an oral phosphorus load in normal and uremic dogs. Kidney Int 14:207–214
26. Boyle IT, Gray RW, DeLuca HF (1971) Regulation by calcium of in vivo synthesis of 1,25-dihydroxycholecalciferol and 21,25-dihydroxycholecalciferol. Proc Natl Acad Sci U S A 68:2131–2134
27. Tanaka Y, DeLuca HF (1973) The control of 25-hydroxyvitamin D metabolism by inorganic phosphorus. Arch Biochem Biophys 154:566–574
28. Razzaque MS (2009) The FGF23–Klotho axis: endocrine regulation of phosphate homeostasis. Nat Rev Endocrinol 5:611–619
29. Saini RK, Kaneko I, Jurutka PW, Forster R, Hsieh A, Hsieh JC, Haussler MR, Whitfield GK (2013) 1,25-dihydroxyvitamin D(3) regulation of fibroblast growth factor-23 expression in bone cells: evidence for primary and secondary mechanisms modulated by leptin and interleukin-6. Calcif Tissue Int 92(4):339–353
30. Ben-Dov IZ, Galitzer H, Lavi-Moshayoff V, Goetz R, Kuro-o M, Mohammadi M, Sirkis R, Naveh-Many T, Silver J (2007) The parathyroid is a target organ for FGF23 in rats. J Clin Invest 117(12):4003–4008
31. Krajisnik T, Bjorklund P, Marsell R et al (2007) Fibroblast growth factor-23 regulates parathyroid hormone and 1alpha-hydroxylase expression in cultured bovine parathyroid cells. J Endocrinol 195:125–131
32. Komaba H, Fukagawa M (2010) FGF23–parathyroid interaction: implications in chronic kidney disease. Kidney Int 77:292–298
33. Silver J, Naveh-Many T, Mayer H et al (1986) Regulation by vitamin D metabolites of parathyroid hormone gene transcription in vivo in the rat. J Clin Invest 78:1296–1301
34. Silver J, Russell J, Sherwood LM (1985) Regulation by vitamin D metabolites of messenger ribonucleic acid for preproparathyroid hormone in isolated bovine parathyroid cells. Proc Natl Acad Sci U S A 82:4270–4273
35. Cantley LK, Russell J, Lettieri D, Sherwood LM (1985) 1,25-Dihydroxyvitamin D3 suppresses parathyroid hormone secretion from bovine parathyroid cells in tissue culture. Endocrinology 117:2114–2119
36. Russell J, Lettieri D, Sherwood LM (1986) Suppression by 1,25(OH)2D3 of transcription of the pre-proparathyroid hormone gene. Endocrinology 119:2864–2866
37. Brown AJ, Dusso A, Lopez-Hilker S, Lewis-Finch J, Grooms P, Slatopolsky E (1989) 1,25-(OH)2D receptors are decreased in parathyroid glands from chronically uremic dogs. Kidney Int 35:19–23
38. Perwad F, Azam N, Zhang MY, Yamashita T, Tenenhouse HS, Portale AA (2005) Dietary and serum phosphorus regulate fibroblast growth factor 23 expression and 1,25-dihydroxyvitamin D metabolism in mice. Endocrinology 146(12):5358–5364
39. Sommer S, Berndt T, Craig T, Kumar R (2007) The phosphatonins and the regulation of phosphate transport and vitamin D metabolism. J Steroid Biochem Mol Biol 103:497–503
40. Ito N, Fukumoto S, Takeuchi Y, Takeda S, Suzuki H, Yamashita T, Fujita T (2007) Effect of acute changes of serum phosphate on fibroblast growth factor (FGF)23 levels in humans. J Bone Miner Metab 25:419–422
41. Larsson T, Nisbeth U, Ljunggren O, Juppner H, Jonsson KB (2003) Circulating concentration of FGF-23 increases as renal function declines in patients with chronic kidney disease, but does not change in response to variation in phosphate intake in healthy volunteers. Kidney Int 64:2272–2279
42. Martin DR, Ritter CS, Slatopolsky E et al (2005) Acute regulation of parathyroid hormone by dietary phosphate. Am J Physiol Endocrinol Metab 289:E729–E734
43. Kuo HF, Chiou TJ (2011) The role of microRNAs in phosphorus deficiency signaling. Plant Physiol 156(3):1016–1024
44. Suzuki S, Ferjani A, Suzuki I, Murata N (2004) The SphS-SphR two component system is the exclusive sensor for the induction of gene expression in response to phosphate limitation in synechocystis. J Biol Chem 279(13):13234–13240
45. Mouillon JM, Persson BL (2006) New aspects on phosphate sensing and signaling in *Saccharomyces cerevisiae*. FEMS Yeast Res 6:171–176
46. Lamarche MG, Wanner BL, Crepin S, Harel J (2008) The phosphate regulon and bacterial virulence: a regulatory network connecting phosphate homeostasis and pathogenesis. FEMS Microbiol Rev 32:461–473
47. Markovich D, Verri T, Sorribas V, Forgo J, Biber J, Murer H (1995) Regulation of opossum kidney (OK) cell Na/Pi cotransport by Pi deprivation involves mRNA stability. Pflugers Arch 430:459–463
48. Fujita T, Izumo N, Fukuyama R, Meguro T, Nakamuta H, Kohno T, Koida M (2001) Phosphate provides an

extracellular signal that drives nuclear export of Runx2/Cbfa1 in bone cells. Biochem Biophys Res Commun 280:348–352
49. Slatopolsky E, Caglar S, Gradowska L et al (1972) On the prevention of secondary hyperparathyroidism in experimental chronic renal disease using 'proportional reduction' of dietary phosphorus intake. Kidney Int 2:147–151
50. Lopez-Hilker S, Dusso AS, Rapp NS et al (1990) Phosphorus restriction reverses hyperparathyroidism in uremia independent of changes in calcium and calcitriol. Am J Physiol 259:F432–F437
51. Almaden Y, Canalejo A, Hernandez A et al (1996) Direct effect of phosphorus on PTH secretion from whole rat parathyroid glands in vitro. J Bone Miner Res 11:970–976
52. Nielsen PK, Feldt-Rasmussen U, Olgaard K (1996) A direct effect in vitro of phosphate on PTH release from bovine parathyroid tissue slices but not from dispersed parathyroid cells. Nephrol Dial Transplant 11:1762–1768
53. Kilav R, Silver J, Naveh-Many T (1995) Parathyroid hormone gene expression in hypophosphatemic rats. J Clin Invest 96:327–333
54. Silver J, Naveh-Many T (2009) Phosphate and the parathyroid. Kidney Int 75:898–905
55. Ritter CS, Martin DR, Lu Y et al (2002) Reversal of secondary hyperparathyroidism by phosphate restriction restores parathyroid calcium-sensing receptor expression and function. J Bone Miner Res 17:2206–2213
56. Dusso AS, Pavlopoulos T, Naumovich L et al (2001) p21WAF1 and transforming growth factor-a mediate dietary phosphate regulation of parathyroid cell growth. Kidney Int 59:855–865
57. Cozzolino M, Lu Y, Sato T et al (2005) A critical role for enhanced TGF-a and EGFR expression in the initiation of parathyroid hyperplasia in experimental kidney disease. Am J Physiol Renal Physiol 289:F1096–F1102
58. Dusso A, Arcidiacono MV, Yang J et al (2010) Vitamin D inhibition of TACE and prevention of renal osteodystrophy and cardiovascular mortality. J Steroid Biochem Mol Biol 121:193–198

Role of Magnesium in Parathyroid Physiology

7

Oren Steen and Aliya Khan

7.1 Introduction

Magnesium (Mg^{2+}) is a divalent cation that is essential for numerous physiologic processes. By serving as a cofactor for various enzymes, it plays a vital role in energy metabolism as well as protein and nucleic acid synthesis [1]. Furthermore, Mg^{2+} is involved in maintenance of the electric potential of nervous tissues and cell membranes [1]. It is the fourth most abundant cation [2] and second most abundant intracellular cation in the body [3].

The average human body contains approximately 25 g of Mg^{2+} [1]. Ninety-nine percent of this is contained in the intracellular compartment, while 1 % is found in the extracellular fluid [1]. Approximately 90 % of the body's total Mg^{2+} content is contained in the osseous tissue and skeletal muscle [4]. Only 0.3 % is found in the serum, of which 30 % is protein bound [5], 10 % is complexed as salts (e.g., bicarbonate, citrate, phosphate, sulfate), and 60 % is present as free Mg^{2+} ions, the biologically active form [6, 7]. Unlike calcium (Ca^{2+}), however, serum Mg^{2+} is not routinely adjusted for albumin. Maintenance of serum ionized Mg^{2+} within a narrow range (0.44–0.59 mmol/L) [or total serum Mg^{2+} of 0.70–1.1 mmol/L] is dependent on the coordinated actions of the kidneys, gut, and bone [8, 9]. There is still much that remains to be established regarding the precise homeostatic mechanisms responsible for Mg^{2+} handling.

7.2 Magnesium Homeostasis

Mg^{2+} homeostasis is tightly regulated and involves the equilibrium between intestinal absorption and renal excretion of Mg^{2+} (Fig. 7.1) [5]. The average dietary intake of Mg^{2+} is 240–365 mg per day [1]. The majority of ingested Mg^{2+} is absorbed in the proximal small bowel, although absorption also occurs in the ileum and colon [10, 11]. Under normal conditions, 30–40 % of ingested Mg^{2+} is absorbed [12].

Ninety percent of Mg^{2+} absorption in the gut and reabsorption in the kidneys occurs passively via the paracellular route; the remaining 10 % occurs transcellularly, which is an energy-dependent process [5]. The main channel involved in active reabsorption of Mg^{2+} is transient receptor potential melastatin subtype 6 (TRPM6) [5].

The kidneys filter 80 % of total plasma Mg^{2+}, 95 % of which is reabsorbed under normal circumstances [13]. The proximal convoluted tubule (PCT) reabsorbs 15 % of filtered Mg^{2+}, while 70 % is reabsorbed in the thick ascending limb of the loop of Henle (TAL), and 10 % is reabsorbed in the distal convoluted tubule (DCT) [14].

O. Steen, MD, FRCPC • A. Khan, MD, FRCPC, FACP, FACE (✉)
Department of Medicine, McMaster University, 1280 Main St W, Hamilton, ON L8S 4L8, Canada
e-mail: oren.steen@medportal.ca; Aliya@mcmaster.ca

M.L. Brandi, E.M. Brown (eds.), *Hypoparathyroidism*,
DOI 10.1007/978-88-470-5376-2_7,

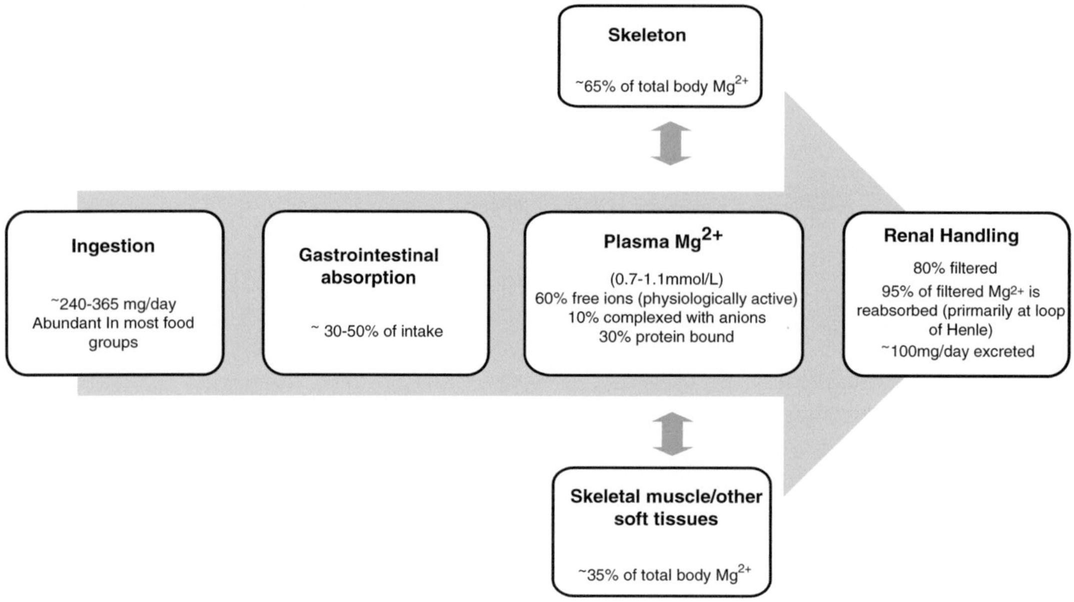

Fig. 7.1 Simplified scheme of magnesium homeostasis

This is in contrast to most other ions, where the PCT is the major site of reabsorption. In the TAL, Mg^{2+} is driven paracellularly by the positive transepithelial voltage via the tight junction proteins claudin-16 and claudin-19 [15, 16]. These proteins copolymerize within the plasma membrane together with the integral protein occludin to form the tight junctions [17]. Mutations in either claudin-16 or claudin-19 result in familial hypomagnesemia with hypercalciuria and nephrocalcinosis (FHHNC) [13]. The DCT determines the final urinary Mg^{2+} concentration, since the more distal nephron tends to be impermeable to Mg^{2+} [18]. This is mediated by the apical TRPM6 [13, 19, 20]. It involves active transcellular reabsorption, since the transepithelial voltage in this segment of the nephron is negative [13]. The basolateral mechanism involved in Mg^{2+} transport has yet to be elucidated, although CNNM2 (cyclin M2) may be a possible candidate [13]. Mg^{2+} absorption in the gut appears to be mediated by TRPM6 as well, located at the brush border membrane [21].

Serum Mg^{2+} is considered a poor marker of the body's Mg^{2+} stores [22–28]. More precise methods for estimating total body magnesium content have been developed (such as the erythrocyte Mg^{2+} concentration and the "magnesium tolerance test"); however, these are primarily used in research settings [27, 28]. Intracellular Mg^{2+} depletion may occur despite normal serum Mg^{2+} levels [29, 30]. It has therefore been postulated that intracellular Mg^{2+} content may be a valuable indicator of sufficiency and more important regulator of serum parathyroid hormone (PTH) levels [22–28].

7.3 Hypomagnesemia

Though often asymptomatic, hypomagnesemia (serum total Mg^{2+} <0.70 mmol/L) is a common electrolyte disturbance and may result in complications due to consequent hypocalcemia and hypokalemia [5]. Causes of hypomagnesemia can be classified into the following categories: decreased intake, decreased absorption, increased losses, and redistribution (Fig. 7.2) [5]. As magnesium is present in essentially all the food groups, it is very unusual for deficiency to occur on the basis of inadequate intake. The most common causes of decreased absorption include

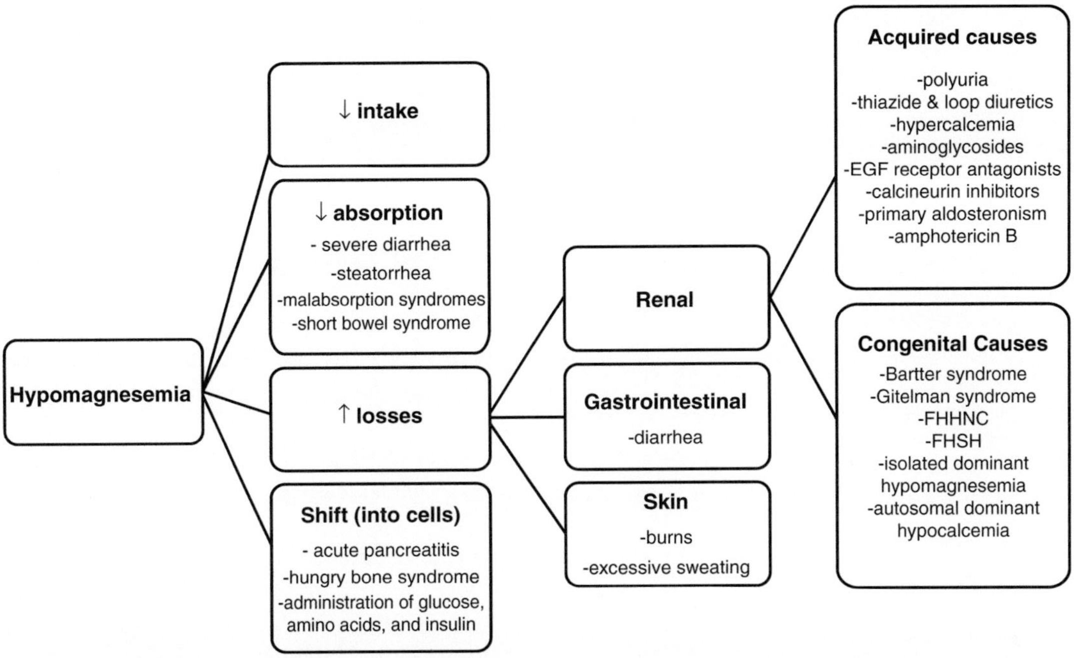

Fig. 7.2 Diagnostic approach to hypomagnesemia. *EGF* epidermal growth factor, *FHHNC* familial hypomagnesemia with hypercalciuria and nephrocalcinosis, *FHSH* familial hypomagnesemia with secondary hypocalcemia

severe diarrhea, steatorrhea, malabsorption syndromes, and short bowel syndrome [5]. Familial hypomagnesemia with secondary hypocalcemia (FHSH) is secondary to a mutation in the epithelial cation channel TRPM6 [5] and results in decreased intestinal Mg^{2+} absorption and increased renal Mg^{2+} excretion. FHSH results in severe hypomagnesemia (0.1–0.4 mmol/L), secondary hypocalcemia, altered neuromuscular excitability, muscle spasms, tetany, and seizures [13].

Proton pump inhibitors (PPIs) have recently been identified as a cause of drug-induced hypomagnesemia. Although the exact mechanism has yet to be elucidated, it is believed that chronic PPI use may cause severe hypomagnesemia via gastrointestinal Mg^{2+} loss [31, 32], possibly by inhibiting TRPM6-mediated active transport of Mg^{2+} as a result of intestinal pH alteration [33]. Causes of renal Mg^{2+} loss can be subdivided into increased flow (i.e., any cause of polyuria) and decreased tubular reabsorption. Medications can result in decreased tubular reabsorption and include diuretics, antibiotics, calcineurin inhibitors, and epiderrmal growth factor (EGF) receptor antagonists [5]. Downregulation of TRPM6 is the mechanism by which diuretics, calcineurin inhibitors, and EGF receptor antagonists may increase urinary Mg^{2+} losses [34–37]. As previously mentioned, FHHNC involves a mutation of claudin-16 or claudin-19. These proteins form cation-permeable channels and are responsible for Ca^{2+} and Mg^{2+} reabsorption in the PCT and TAL [8]. Inherited disorders predisposing to hypomagnesemia have been vital in furthering our understanding of Mg^{2+} transport, and these are outlined in Table 7.1.

7.4 Hypermagnesemia

Hypermagnesemia (serum Mg^{2+} >1.1 mmol/L) is considerably less common than hypomagnesemia. The most common cause of the former is renal insufficiency [38]. The kidneys are able to maintain Mg^{2+} balance by increasing its fractional excretion of Mg^{2+} until severe renal impairment develops (GFR < 30 mL/min) [38]. Other causes of impaired renal Mg^{2+} excretion

Table 7.1 Inherited disorders leading to disturbances in magnesium balance

Disorder	Mutation
Familial hypomagnesemia with hypercalciuria and nephrocalcinosis	Claudin 16 or 19
Isolated recessive hypomagnesemia	EGF receptor
Dominant hypomagnesemia	CNNM2
Familial hypomagnesemia with secondary hypocalcemia	TRPM6
Isolated dominant hypomagnesemia	γ-subunit of Na^+K^+ ATPase
Autosomal dominant hypomagnesemia	Kv1.1
Bartter syndrome	NKCC2 (type I), ROMK (type II), ClC-Kb (type III), barttin (type IV)
Gitelman syndrome	NCC
EAST (epilepsy, ataxia, sensorineural deafness, and renal tubulopathy) syndrome	Kir4.1

EAST epilepsy, ataxia, sensorineural deafness, and renal tubulopathy, *EGF* epidermal growth factor, *CNNM2* cyclin M2, *TRPM6* transient receptor potential melastatin subtype 6, Kv1.1 potassium voltage-gated channel subfamily A member 1, *NKCC2* sodium-potassium-chloride co-transporter 2, *ROMK* renal outer medullary potassium channel, ClC-Kb chloride channel, voltage-sensitive Kb, *NCC* sodium-chloride co-transporter, *Kir4.1* inward rectifying potassium channel 4.1

include lithium therapy (via unclear mechanisms) [38] and familial hypocalciuric hypercalcemia, the latter due to hyporesponsiveness of the calcium-sensing receptor (CaSR) to hypercalcemia [39]. Hypermagnesemia can also be caused by increased intake. The most common such scenario would be parenteral magnesium infusions during the treatment of preterm labor or preeclampsia/eclampsia [38]. Though uncommon, hypermagnesemia due to oral ingestion has been reported with antacids, laxative abuse, cathartics (used to treat overdoses), and accidental ingestion of Epsom salts [38, 40, 41]. There have also been reports of hypermagnesemia caused by Mg^{2+}-containing enemas and aspiration from near-drowning in the Dead Sea [42–45].

7.5 Relationship Between Mg^{2+} and PTH

The relationship between Mg^{2+} and parathyroid hormone (PTH) is rather complex. Serum Mg^{2+} levels are known to influence serum PTH levels, although the precise nature of this relationship has yielded conflicting results in the literature [22–26].

The CaSR is physiologically activated not only by Ca^{2+} but by Mg^{2+} as well; therefore, serum Mg^{2+} levels can influence PTH secretion [2] (see also Chap. 5). The release of PTH from the parathyroid gland partially depends on the intracellular signaling of cyclic AMP (cAMP), and Mg^{2+} plays an important role in adenylate cyclase activation [46]. Stimulation of the CaSR by Mg^{2+} results in transient intracellular Ca^{2+} elevations [47], stimulation of phospholipases C and A_2 [48], and inhibition of cellular cAMP generation, resulting in inhibition of PTH release [25, 26, 49]. These effects are mediated by the G_i and G_q subclasses of G-proteins [2]. Mg^{2+} is two to three times less potent than Ca^{2+} in activating phospholipase C by means of the CaSR [49]. Other mechanisms by which Mg^{2+} regulates PTH secretion are as yet unclear.

PTH plays a role in renal Mg^{2+} handling by increasing Mg^{2+} reabsorption in the DCT [2]. Although the precise mechanisms remain elusive, it is believed that protein kinase A, phospholipase C, and protein kinase C-mediated pathways play a role [2]. PTH has also been shown to increase Mg^{2+} reabsorption in the cortical TAL (cTAL) by increasing paracellular pathway permeability [50]. Activating mutations of the CaSR resulting in autosomal dominant hypocalcemic hypercalciuria lead to asymptomatic hypocalcemia, hypomagnesemia, and PTH levels in the lower-to-normal range [51]. Inactivating mutations of the CaSR result in familial hypocalciuric hypercalcemia (FHH) or its recessive form neonatal severe hyperparathyroidism [52]. Patients possessing this mutation manifest with hypercalcemia, hypermagnesemia, and normal-to-high PTH levels [8]. The biochemical features of these conditions suggest that the CaSR plays an important role in not only Ca^{2+} homeostasis but also Mg^{2+} homeostasis.

7.6 Effects of Hypomagnesemia on PTH

Hypomagnesemia may cause relative hypoparathyroidism, as severe hypomagnesemia interferes with PTH secretion [5]. Although mild decreases in Mg^{2+} levels to as low as 0.5 mmol/L result in stimulation of PTH secretion (as seen with hypocalcemia), more severe levels of hypomagnesemia actually inhibit PTH secretion [2]. This has been demonstrated at Mg^{2+} levels <0.4 mmol/L [53–55]. This phenomenon has been referred to as "paradoxical block of PTH secretion" [2]. The block of PTH secretion is believed to be related to the effect of intracellular Mg^{2+} depletion on the α-subunits of the heterotrimeric G-proteins associated with the CaSR [2]. Since these proteins contain a Mg^{2+}-binding site, Mg^{2+} deficiency can disinhibit the Gα subunits, thereby mimicking the effect of CaSR activation and consequently suppressing PTH secretion [2, 5]. As a result, hypocalcemia occurs due to decreased renal Ca^{2+} reabsorption [56]. The inhibition of PTH secretion tends to occur with prolonged, rather than acute, Mg^{2+} deficiency [57]. The resultant hypocalcemia can only be corrected by replacement of Mg^{2+} [57–60]. Serum PTH levels increase within minutes of Mg^{2+} administration [2]. Hence, it is believed that hypomagnesemia inhibits the secretion of PTH, rather than its biosynthesis [59–61].

Hypomagnesemia may also cause hypocalcemia by decreasing the responsiveness of target organs (renal tubules and bone in particular) to PTH activity [62–64]. PTH exerts its renal activity via activation of adenylate cyclase in the renal tubules, resulting in intracellular production of cAMP [65]. Free intracellular Mg^{2+} is a cofactor of adenylate cyclase; thus, when the intracellular concentration of ionized Mg^{2+} is reduced, resistance to the action of PTH may ensue as a result of the disturbed signal transduction pathway [4, 62–64]. There is also evidence implicating hypomagnesemia with inhibition of PTH binding to bone [66]. End-organ resistance, however, is felt to play less of a significant role in hypomagnesemia-induced hypocalcemia compared to inhibition of PTH release [53, 61].

7.7 Effects of Hypermagnesemia on PTH

Hypermagnesemia may cause hypocalcemia, although it is typically mild in severity and asymptomatic [67, 68]. Similar to hypomagnesemia, the cause of hypocalcemia in hypermagnesemia has been postulated to involve inhibition of PTH release [69], although some studies have actually found elevations of PTH in this setting [68].

Conclusion

Serum levels of Mg^{2+} and PTH depend on each other in an intricate manner. Disturbances in magnesium balance may lead to secondary hypocalcemia through its inhibitory effects on PTH secretion and activity. The effects of hypomagnesemia on calcium homeostasis depend on whether this process is acute versus chronic, as well as on the severity of the hypomagnesemia. Further research is required to improve our understanding of Mg^{2+} homeostasis, as well as the role of Mg^{2+} in parathyroid physiology.

References

1. Navarro-González JF, Mora-Fernández C, García-Pérez J (2009) Clinical implications of disordered magnesium homeostasis in chronic renal failure and dialysis. Semin Dial 22(1):37–44
2. Vetter T, Lohse M (2002) Magnesium and the parathyroid. Curr Opin Nephrol Hypertens 11:403–410
3. Wacker WE, Parisi AF (1968) Magnesium metabolism. N Engl J Med 278:772–776
4. Pironi L, Malucetti E, Guidetti M et al (2009) The complex relationship between magnesium and serum

parathyroid hormone: a study in patients with chronic intestinal failure. Magnes Res 22(1):37–43
5. Hoorn EJ, Zietse R (2013) Disorders of calcium and magnesium balance: a physiology-based approach. Pediatr Nephrol 28:1195–1206
6. Speich M, Bousquet B, Nicolas G (1981) Reference values for ionized, complexed, and protein-bound plasma magnesium in men and women. Clin Chem 27:246–248
7. Elin RJ (1988) Magnesium metabolism in health and disease. Dis Mon 34:165–218
8. Ferrè S, Hoenderop JG, Bindels RJ (2012) Sensing mechanisms involved in Ca2+ and Mg2+ homeostasis. Kidney Int 82:1157–1166
9. Glaudemans B, Knoers NV, Hoenderop JG, Bindels RJ (2010) New molecular players facilitating Mg(2+) reabsorption in the distal convoluted tubule. Kidney Int 77(1):17–22
10. Graham LA, Caesar JJ, Burgen AS (1960) Gastrointestinal absorption and excretion of Mg2+ in man. Metabolism 9:646–659
11. Brannan PG, Bergne-Marini P, Pak CY et al (1976) Magnesium absorption in the human small intestine. J Clin Invest 57:1412–1418
12. Quamme GA (2008) Recent developments in intestinal magnesium absorption. Curr Opin Gastroenterol 24:230–235
13. Ferrè S, Hoenderop JJ, Bindels RJ (2011) Role of the distal convoluted tubule in renal Mg2+ handling: molecular lessons from inherited hypomagnesemia. Magnes Res 24(3):S101–S108
14. Bindels RJ (2010) 2009 Homer W. Smith Award: minerals in motion: from new ion transporters to new concepts. J Am Soc Nephrol 21:1263–1269
15. Gunzel D, Yu AS (2009) Function and regulation of claudins in the thick ascending limb of Henle. Pflugers Arch 458:77–88
16. Hou J, Renigunta A, Gomes AS et al (2009) Claudin-16 and claudin-19 interaction is required for their assembly into tight junctions and for renal reabsorption of magnesium. Proc Natl Acad Sci U S A 106:15350–15355
17. Hou J, Goodenough DA (2010) Claudin-16 and claudin-19 function in the thick ascending limb. Curr Opin Nephrol Hypertens 19(5):483–488
18. Reilly RF, Ellison DH (2000) Mammalian distal tubule: physiology, pathophysiology, and molecular anatomy. Physiol Rev 80:277–313
19. Schlingmann KP, Weber S, Peters M et al (2002) Hypomagnesemia with secondary hypocalcemia is caused by mutations in TRPM6, a new member of the TRPM gene family. Nat Genet 31(2):166–170
20. Walder RY, Landau D, Meyer P et al (2002) Mutation of TRPM6 causes familial hypomagnesemia with secondary hypocalcemia. Nat Genet 31:171–174
21. Voets T, Nilius B, Hoefs S et al (2004) TRPM6 forms the Mg2+ influx channel involved in intestinal and renal Mg2+ absorption. J Biol Chem 279:19–25
22. Takahashi S, Okada K, Yanai M (1994) Magnesium and parathyroid hormone changes to magnesium-free dialysate in continuous ambulatory peritoneal dialysis patients. Perit Dial Int 14:75–78
23. Pletka P, Bernstein DS, Hampers CL et al (1974) Relationship between magnesium and secondary hyperparathyroidism during long-term hemodialysis. Metabolism 23:619–630
24. Parsons V, Papapoulos SE, Weston MJ et al (1980) The long-term effect of lowering dialysate magnesium on circulating parathyroid hormone in patients on regular haemodialysis therapy. Acta Endocrinol 93:455–460
25. Gonella M, Bonaguidi F, Buzzigoli G, Bartolini V, Mariani G (1981) On the effect of magnesium on the PTH secretion in uremic patients on maintenance hemodialysis. Nephron 27:40–42
26. Massry SG, Coburn JW, Kleeman CR (1970) Evidence for suppression of parathyroid gland activity by hypermagnesemia. J Clin Invest 49:1619–1629
27. Alfrey AC, Miller NL, Butkus D (1974) Evaluation of body magnesium stores. J Lab Clin Med 84:153–162
28. Rob PM, Dick K, Bley N et al (1999) Can one really measure magnesium deficiency using the short-term magnesium loading test? J Intern Med 246(4):373–378
29. Dynker T, Wester PO (1978) The relation between extra- and intracellular electrolytes in patients with hypokalemia and/or diuretic treatment. Acta Med Scand 204:269–282
30. Rob PM, Bley N, Dick K et al (1995) Magnesium deficiency after renal transplantion and cyclosporine treatment despite normal serum magnesium concentration detected by a modified magnesium-loading-test. Transplant Proc 27:3442–3443
31. Hess MW, Hoenderop JG, Bindels RJ, Drenth JP (2012) Systematic review: hypomagnesaemia induced by proton pump inhibition. Aliment Pharmacol Ther 36:405–413
32. Hoorn EJ, van der Hoek J, de Man RA et al (2010) A case series of proton pump inhibitor-induced hypomagnesemia. Am J Kidney Dis 56:112–116
33. Cundy T, Dissanayake A (2008) Severe hypomagnesaemia in long-term users of proton-pump inhibitors. Clin Endocrinol (Oxf) 69:338–341
34. Groenestege WT, Thébault S, van der Wijst J et al (2007) Impaired basolateral sorting of pro-EGF causes isolated recessive renal hypomagnesemia. J Clin Invest 117:2260–2267
35. Hoorn EJ, Walsh SB, McCormick JA et al (2011) The calcineurin inhibitor tacrolimus activates the renal sodium chloride cotransporter to cause hypertension. Nat Med 17:1304–1309
36. Nijenhuis T, Hoenderop JG, Bindels RJ (2004) Downregulation of Ca(2+) and Mg(2+) transport proteins in the kidney explains tacrolimus (FK506)-induced hypercalciuria and hypomagnesemia. J Am Soc Nephrol 15:549–557
37. Nijenhuis T, Vallon V, van der Kemp AW et al (2005) Enhanced passive Ca2+ reabsorption and reduced Mg2+ channel abundance explains thiazide-induced hypocalciuria and hypomagnesemia. J Clin Invest 115:1651–1658

38. Topf JM, Murray PT (2003) Hypomagnesemia and hypermagnesemia. Rev Endocr Metab Disord 4:195–206
39. Sutton RA, Domrongkitchaiporn S (1993) Abnormal renal magnesium handling. Miner Electrolyte Metab 19:232–240
40. Weber C, Santiago R (1989) Hypermagnesemia: a potential complication during treatment of theophylline intoxication with oral activated charcoal and magnesium-containing cathartics. Chest 95:56–59
41. Gren J, Woolf A (1989) Hypermagnesemia associated with catharsis in a salicylate-intoxicated patient with anorexia nervosa. Ann Emerg Med 18(2):200–203
42. Ashton MR, Sutton D, Nielsen M (1990) Severe magnesium toxicity after magnesium sulphate enema in a chronically constipated child. BMJ 300:541
43. Collinson PO, Burroughs AK (1986) Severe hypermagnesaemia due to magnesium sulphate enemas in patients with hepatic coma. Br Med J (Clin Res Ed) 293:1013–1014
44. Outerbridge EW, Papageorgiou A, Stern L (1973) Severe hypermagnesaemia due to magnesium sulphate enemas in patients with hepatic coma. JAMA 224:1392–1393
45. Porath A, Mosseri M, Harmon I et al (1989) Dead sea water poisoning. Ann Emerg Med 18(2):187–191
46. Grubbs RD, Maguire ME (1987) Magnesium as a regulatory cation: criteria and evaluation. Magnesium 6:113–127
47. Butters RR Jr, Chattopadhyay N, Nielsen P et al (1997) Cloning and characterization of a calcium-sensing receptor from the hypercalcemic New Zealand white rabbit reveals unaltered responsiveness to extracellular calcuim. J Bone Miner Res 12(4):568–579
48. Ruat M, Snowman AM, Hester LD, Snyder SH (1996) Cloned and expressed rat Ca2+-sensing receptor. J Biol Chem 271(11):5972–5975
49. Chang W, Pratt S, Chen TH et al (1998) Coupling of calcium receptors to inositol phosphate and cyclic AMP generation in mammalian cells and Xenopus laevis oocytes and immunodetection of receptor protein by region-specific antipeptide sera. J Bone Miner Res 13(4):570–580
50. Wittner M, Mandon B, Roinel N et al (1993) Hormonal stimulation of Ca2+ and Mg2+ transport in the cortical thick ascending limb of Henle's loop of the mouse: evidence for a change in the paracellular pathway permeability. Pflugers Arch 423:387–396
51. Pollak MR, Brown EM, Estep HL et al (1994) Autosomal dominant hypocalcaemia caused by a Ca(2+)-sensing receptor gene mutation. Nat Genet 8:303–307
52. Pollak MR, Brown EM, Chou YH et al (1993) Mutations in the human Ca(2+)-sensing receptor gene cause familial hypocalciuric hypercalcemia and neonatal severe hyperparathyroidism. Cell 75(7): 1297–1303
53. Duran MJ, Borst GC 3rd, Osburne RC, Eil C (1984) Concurrent renal hypomagnesemia and hypoparathyroidism with normal parathormone responsiveness. Am J Med 76(1):151–154
54. Anast CS, Mohs JM, Kaplan SL, Burns TW (1973) Magnesium, vitamin D, and parathyroid hormone. Lancet 1(7816):1389–1390
55. Mennes P, Rosenbaum R, Martin K, Slatopolsky E (1978) Hypomagnesemia and impaired parathyroid hormone secretion in chronic renal disease. Ann Intern Med 88(2):206–209
56. Quitterer U, Hoffmann M, Freichel M, Lohse MJ (2001) Paradoxical block of parathormone secretion is mediated by increased activity of G alpha subunits. J Biol Chem 276(9):6763–6769
57. Fatemi S, Ryzen E, Flores J et al (1991) Effect of experimental human magnesium depletion on parathyroid hormone secretion and 1,25-dihydroxyvitamin D metabolism. J Clin Endocrinol Metab 73:1067–1072
58. Stromme JH, Nesbakken R, Normann T et al (1969) Familial hypomagnesemia: biochemical, histological and hereditary aspects studied in two brothers. Acta Paediatr Scand 58(5):433–444
59. Rude RK, Oldham SB, Sharp CF Jr, Singer FR (1978) Parathyroid hormone secretion in magnesium deficiency. J Clin Endocrinol Metab 47(4):800–806
60. Anast CS, Mohs JM, Kaplan SL, Burns TW (1972) Evidence for parathyroid failure in magnesium deficiency. Science 177(4049):606–608
61. Suh SM, Tashijah AH Jr, Matsuo N et al (1973) Pathogenesis of hypocalcemia in primary hypomagnesemia: normal end-organ responsiveness to parathyroid hormone, impaired parathyroid gland function. J Clin Invest 52(1):153–160
62. Mune T, Yasuda K, Ishii M et al (1993) Tetany due to hypomagnesemia induced by cisplatin and doxorubicin treatment for synovial sarcoma. Intern Med 32:434–437
63. Mori S, Harada S, Okazaki R et al (1992) Hypomagnesemia with increased metabolism of parathyroid hormone and reduced responsiveness to calcitropic hormones. Intern Med 31:820–824
64. Mihara M, Kamikubo K, Hiramatsu K et al (1995) Renal refractoriness to phosphaturic action of parathyroid hormone in a patient with hypomagnesemia. Intern Med 34:666–669
65. Leicht E, Biro G (1992) Mechanisms of hypocalcaemia in the clinical form of severe magnesium deficit in the human. Magnes Res 5:37–44
66. Risco F, Traba ML, de la Piedra C (1995) Possible alterations of the *in vivo* 1,25(OH)2D3 synthesis and its tissue distribution in magnesium-deficient rats. Magnes Res 8:27–35
67. Clark B, Brown R (1992) Unsuspected morbid hypermagnesemia in elderly patients. Am J Nephrol 12:336–343
68. Cruikshank DP, Pitkin RM, Reynolds WA et al (1979) Effects of magnesium sulfate treatment on perinatal calcium metabolism. I. Maternal and fetal responses. Am J Obstet Gynecol 134:243–249
69. Cholst IN, Steinberg SF, Tropper PJ et al (1984) The influence of hypermagnesemia on serum calcium and parathyroid hormone levels in human subjects. N Engl J Med 310:1221–1225

The PTH/Vitamin D/FGF23 Axis

8

David Goltzman and Andrew C. Karaplis

8.1 Calcium and Phosphorus Balance

The skeleton, the gut, and the kidney each play a major role in assuring calcium (Ca^{++}) homeostasis. Overall, in a typical individual if 1,000 mg of Ca^{++} are ingested in the diet per day, approximately 200 mg will be absorbed across the intestinal epithelium and about 200 mg excreted.

The traditional model of transcellular Ca^{++} transport consists of influx through an apical calcium channel (TRPV6), which provides the rate-limiting step, diffusion through the cytosol, and active extrusion at the basolateral membrane by a plasma membrane ATPase (PMCA1b) [1]. Although entry of Ca^{++} has been reported to involve TRPV6, other Ca^{++} channels may also be involved. Ca^{++} binding proteins including calmodulin and calbindin-D9k (CaBP9k) may be important for fine-tuning Ca^{++} channel activity, and in the cytosol, calbindin 9 k may "buffer" and/or mediate the transit of intracellular absorbed Ca^{++} to the basolateral membrane. An additional modulator of transcellular Ca^{++} transport is a Ca ATPase, PMCA1b, encoded by *ATP2B1*, which is important for the extrusion of Ca^{++} at the basolateral membrane to complete the transcellular transport of this ion. Increasing evidence also suggests the importance of paracellular Ca^{++} transport in Ca^{++} absorption via tight junctions.

The skeleton, where approximately 1,000 mg of calcium is stored, is the major Ca^{++} reservoir in the body. Skeletal Ca^{++} is stored mainly in the form of hydroxyapatite crystals, the major inorganic component of the mineralized bone matrix. Ordinarily as a result of normal bone turnover, approximately 500 mg of Ca^{++} is resorbed from the bone per day and the equivalent amount is accreted. Approximately 10 g of Ca^{++} will be filtered daily through the kidney and most will be reabsorbed, with about 200 mg being excreted in the urine. The normal 24-h urine excretion of Ca^{++} may however vary between 100 and 300 mg per day (2.5–7.5 mmoles per day).

The average consumption of phosphorus (Pi) (i.e., about 1,000–1,500 mg) in a Western diet is similar to that of Ca^{++} (about 1,000 mg); however, about 70 % of phosphorus is absorbed daily compared with only about 20–30 % of Ca^{++}. Pi ingested through the diet is absorbed by the small intestine through sodium–phosphate cotransporters, as well as by sodium-independent diffusional absorption across intercellular spaces in the

D. Goltzman, MD (✉)
Calcium Research Laboratory and Department of Medicine, McGill University Health Centre, Royal Victoria Hospital, 687 Pine Ave West, Montreal, QC H3A1A1, Canada

Departments of Medicine and Physiology, McGill University, Montreal, QC, Canada
e-mail: david.goltzman@mcgill.ca

A.C. Karaplis
Division of Endocrinology, Department of Medicine and Lady Davis Institute for Medical Research, Jewish General Hospital, McGill University, Montreal, QC, Canada

M.L. Brandi, E.M. Brown (eds.), *Hypoparathyroidism*,
DOI 10.1007/978-88-470-5376-2_8, © Springer-Verlag Italia 2015

lumen. The skeleton also represents the largest reservoir in the body of Pi which is stored in an exact stoichiometry with Ca^{++} as hydroxyapatite crystals. The major control site for Pi homeostasis is the kidney where enhanced reabsorption can occur from the tubular lumen via sodium–phosphate cotransporters, NaPi2s (also known as SLC34As) that are expressed in the renal proximal convoluted tubule [2] (see also Chap. 6).

8.2 Hormones Regulating Ca^{++} and Pi Homeostasis

8.2.1 Parathyroid Hormone (PTH)

PTH secretion from the parathyroid gland and parathyroid cell proliferation are inhibited by serum Ca^{++}acting via a Ca^{++}-sensing receptor (CaSR) [3], and PTH gene transcription [4] and parathyroid cell proliferation [5, 6] may be inhibited by the active form of vitamin D, 1,25-dihydroxyvitamin D [1,25$(OH)_2$D], acting via the vitamin D receptor (VDR). Although the major glandular form of PTH is an 84 amino acid peptide, virtually all of the biological activity resides within its amino (NH_2)-terminal domain [7, 8]. Intracellular degradation of PTH(1–84) within the parathyroid cell, which is enhanced by high ambient calcium concentrations, provides a means of modulating the fraction of secreted hormone that is PTH(1–84) (see Chap. 4 for details). The NH_2-terminal domain interacts in target tissues with a classical G protein-coupled receptor (GPCR) termed the PTH/PTHrP receptor type 1, or PTHR1 [9, 10]. PTHR1 couples to several G protein subclasses, including Gs, Gq/11, and G12/13, resulting in the activation of many pathways, although the best studied are the adenylate cyclase (AC) and phospholipase C (PLC) pathways.

In the kidney, Ca^{++} reabsorption primarily occurs in the distal tubules and collecting ducts [11]. Ca^{++} ions cross the apical membrane from the tubular lumen via TRPV5, a highly selective Ca^{++}channel, and are then transported across the basolateral membrane into the blood system by the sodium/calcium exchanger 1 (NCX1) and a plasma membrane ATPase. PTH regulates the expression of TRPV5 and NCX1 [12] as well as their activity. The PTH regulation of at least TRPV5 appears to occur through protein kinase C (PKC) [13]. TRPV5 is also regulated through the PKC-signaling pathway by the Ca^{++} sensing receptor (CaSR) [14] (see also Chap. 5).

Pi reabsorption across the apical membrane of renal proximal tubules mainly occurs through two sodium-dependent phosphate cotransporters, NaPi-2a and NaPi-2c, that are exclusively expressed in the brush border membrane of the proximal tubules [15] (see also Chap. 6). PTH increases Pi excretion in the proximal tubule mainly by reducing the levels of these transporters. Thus, PTH binding to PTHR1 at either apical or basolateral membrane results in removal of the transporters from the brush border membrane via clathrin-coated pits [16, 17]. PTH activation of apical PTHR1 leads to phospholipase C (PLC)–protein kinase C (PKC) stimulation mediated by sodium–hydrogen exchanger regulatory factor (NHERF), whereas activation of basolateral PTHR1 utilizes the AC/protein kinase A (PKA) pathway [18]. Following endocytosis NaPi-2 is eventually transported to the lysosomes for degradation.

PTH also stimulates the conversion of 25-hydroxyvitamin D [25(OH)D] to 1,25$(OH)_2$D in the kidney by transcriptional activation of the gene encoding the 25-hydroxyvitamin D-1α hydroxylase [1(OH)ase] enzyme (CYP27B1) apparently by the AC/PKA pathway [19] (see also Chap. 11).

The bone undergoes constant remodeling in response to endocrine, autocrine/paracrine, and intracrine signals and bone mineral is maintained through a balance between bone formation and resorption. PTH binds to PTHR1 on cells of the osteoblastic lineage [20] including progenitor cells, osteoblasts, and osteocytes and can stimulate a variety of factors ultimately leading to increased proliferation of mesenchymal stem cells, such that these cells are committed into the osteoblast lineage, to enhance osteoblast differentiation and activity with new bone matrix production [21] and ultimately mineralization of bone tissue. However, osteoblastic cells also produce the TNF-related cytokine, receptor activator of nuclear factor κ-B (RANK) ligand (RANKL), a critical stimulator of osteoclast production and action, as well as the soluble RANKL decoy receptor, osteoprotegerin (OPG) [22]. PTH enhances production of RANKL and inhibits production of OPG leading to increased osteoclastogenesis and osteoclastic bone resorption [23]. Bone resorption leads to the release of Ca^{++} and Pi as a result of the degradation of hydroxyapatite, and it is the resorptive effect of PTH which plays a major role in mineral and particularly Ca^{++} homeostasis

8.2.2 Vitamin D

8.2.2.1 Metabolism of Vitamin D

Vitamin D can be obtained as vitamin D3 (cholecalciferol), via UV-light irradiation of a skin precursor, 7-dehydrocholesterol [24], or can be ingested from the diet as vitamin D3 or as the plant-derived sterol vitamin D2 (ergocalciferol). Vitamin D is then transported to the liver, bound to a plasma vitamin D binding protein (DBP) [25], where it is hydroxylated at the C-25 position of the side chain to produce 25(OH)D, the most abundant circulating form of vitamin D [26]. The final step in the activation to the hormonal form, $1,25(OH)_2D$, occurs mainly, but not exclusively, in the kidney via a tightly regulated 1α-hydroxylation reaction catalyzed by a mitochondrial enzyme 1(OH)ase, or CYP27B1 [27]. The renal *CYP27B1* gene is stimulated by PTH and hypocalcemia and inhibited by hyperphosphatemia thus sensing the need for mineral homeostasis. The renal *CYP27B1* gene is also product-inhibited by $1,25(OH)_2D$, which acts via a short negative feedback loop to limit its own production [28]. $1,25(OH)_2D$ circulates bound to vitamin D binding protein (DBP), to exert its endocrine actions in various target tissues. Extrarenal 1α-hydroxylation has also been described resulting in the production of $1,25(OH)_2D$, which can act locally in an intracrine mode [29]. The $1,25(OH)_2D$-mediated endocrine or intracrine signal may be terminated in all target cells via the catalytic action of CYP24A1, an enzyme that initiates the process of $1,25(OH)_2D$ catabolism [30]. The *CYP24A1* gene is transcriptionally activated by $1,25(OH)_2D$ [31]. This feed forward induction of $1,25(OH)_2D$ catabolism therefore prevents hypervitaminosis D.

8.2.2.2 Actions of Vitamin D

The active metabolite, $1,25(OH)_2D$, functions by initially binding to the VDR. The ligand-activated VDR interacts with the retinoid X receptor (RXR) to form a heterodimer that binds to vitamin D responsive elements in the region of genes directly regulated by $1,25(OH)_2D$ [32]. By recruiting complexes of either coactivators or corepressors, ligand-activated VDR–RXR modulates the transcription of genes encoding proteins that carry out the functions of vitamin D.

Under conditions of low dietary Ca^{++}, the $1,25(OH)_2D$/VDR system induces TRPV6 [33] and vitamin D-dependent calcium binding protein 9K (CaBPD9k) [34], to promote transcellular intestinal calcium absorption; the extrusion of calcium at the basolateral membrane is likely constitutive in part and is executed by the Ca ATPases PMCA1b/PMCA2c but could be amplified via induction of these molecules by $1,25(OH)_2D$. $1,25(OH)_2D$ also increases expression of claudins 2 and 12 to possibly promote paracellular Ca^{++} entry [35]. Ca^{++} absorption under normal dietary conditions postnatally does not absolutely require TRPV6 and CaBPD9K.

Pi is predominantly absorbed in the intestine via paracellular mechanisms. Although $1,25(OH)_2D$ may also increase intestinal Pi absorption by increasing expression of NaPi2b [36], because Pi is abundant in the diet, the Pi absorption effect of $1,25(OH)_2D$ may not be as profound as the effect on Ca^{++} transport. The $1,25(OH)_2D$/VDR system can promote bone resorption [37–39], suggesting that at least part of its skeletal action is catabolic. In support of this, the $1,25(OH)_2D$/VDR system enhances the expression of RANKL in osteoblastic cells to stimulate bone resorption through osteoclastogenesis [40]. Furthermore, osteoprotegerin, the soluble decoy receptor for RANKL that tempers its activity, is repressed by the $1,25(OH)_2D$/VDR system, thus amplifying the bioeffect of RANKL.

The action of the 1,25(OH)$_2$D/VDR system in the parathyroid gland to suppress PTH synthesis [4] effectively limits PTH-induced bone-resorbing activity and restricts PTH stimulation of renal CYP27B1 thereby preventing hypercalcemia.

8.2.3 Fibroblast Growth Factor 23 (FGF23)

Fibroblast growth factors (FGFs) are a large superfamily of peptides that act mainly as paracrine/autocrine substances to exert a broad range of biological functions in development and organogenesis. They act by binding and activation of FGF receptor (FGFR) tyrosine kinases [41, 42]. The unique FGF19 subfamily consists of FGF19, FGF21, and FGF23 [43] which act as hormones to regulate energy and mineral metabolism (see also Chap. 6).

FGF23 is a phosphaturic hormone produced and secreted by bone cells of the osteoblastic lineage, predominantly late osteoblasts and osteocytes, and was initially identified as the mediator of the human disorder autosomal dominant hypophosphatemic rickets (ADHR) [44]. FGF23 is synthesized as a 32 kDa, 251 amino acid protein, with a signal (leader) sequence of 24 amino acids, an NH$_2$-terminal FGF homology domain of 155 amino acids, and a unique sequence in its carboxyl (COOH)-domain of 72 amino acids [45]. The molecule can be cleaved between Arg^{179} and Ser^{180} by a subtilisin-like proprotein convertase, yet to be identified, to generate NH$_2$-terminal and COOH-terminal fragments; the entire sequence of the secreted 25–251 amino acid protein appears to be necessary for its biological action. O-glycosylation within the 162–228 region apparently reduces the susceptibility of the protein to proteolysis [46], and it is possible that the COOH-terminal fragment may compete with intact FGF23 for binding to its receptor complex and function as a competitive inhibitor [47]. ADHR patients carry missense mutations at the proteolytic cleavage site of FGF23 (176RXXR179), which confers resistance to inactivation by proteolytic cleavage [48]. As a result, ADHR patients exhibit increased blood levels of intact FGF23 and Pi-wasting phenotypes.

8.2.3.1 Regulation of FGF23 Production

Local Regulators of FGF23 Production

Local regulators produced in osteoblast/osteocytes appear to be important in modulating the release of FGF23. Mutations in the gene encoding the membrane protein PHEX (phosphate-regulating neutral endopeptidase with homology to endopeptidase on the X chromosome) [49–51] and in the gene encoding the SIBLING (small integrin-binding ligand interacting glycoproteins) protein DMP-1 (dentin matrix protein-1) [52, 53] increase FGF23 expression and induce renal Pi wasting in mice and humans. In one apparent mechanism [54], DMP-1 binds to the osteocyte through its integrin-binding domains and to PHEX through its ASARM domain [acidic serine aspartate-rich matrix extracellular phosphoglycoprotein (MEPE)-associated motif]. Binding of DMP-1 to PHEX inhibits production of active FGF-23, while disruption of either DMP-1 or PHEX releases this inhibition, and active FGF23 production is increased. Local relative levels of the mineralization regulators pyrophosphate and phosphate appear to be important modulators [55, 56] and disruption of bone mineralization may release low molecular weight FGFs from the bone matrix which may activate the osteocyte FGFR and stimulate transcription of the *FGF-23* gene [57]. Finally reduced iron and tissue hypoxemia have also been reported to stimulate FGF23 release [58].

Systemic Regulators

Extremely high FGF23 levels are observed in primary deficiency of the Klotho protein (genetic deletion or mutational hypomorph) [59, 60] and the much more common secondary Klotho deficiency in chronic kidney disease (CKD) [61, 62]. However, there are no in vitro data to date to support a direct effect of Klotho, either locally produced or circulating, on FGF23 production.

A critical regulator of FGF23 release is 1,25(OH)$_2$D, acting via the VDR [63] in part by directly upregulating gene expression [64] and in part by inhibiting PHEX and by inducing expression of the gene encoding ecto-nucleotide pyrophosphatase/phosphodiesterase (Enpp1), which could alter local levels of pyrophosphate and phosphate [65]. 1,25(OH)$_2$D may also increase the levels of the hypoxia-inducible transcription factor, HIF1A, which might mediate the effects of tissue hypoxemia on FGF23 release [66]. PTH has also been reported to stimulate synthesis and secretion of FGF23 through activation of the PTHR1 on osteocytes/osteoblasts [67, 68].

Increased Pi in the diet and increases in serum Pi can both increase FGF23 release by osteocytic cells; however, the mechanism is still unclear. Nevertheless, this observation has led to the concept that the skeleton therefore serves as a sensor of Pi levels in a manner analogous to the function of the parathyroid glands as a Ca^{++} sensor.

FGF23 is also regulated by serum Ca^{++} [69–71], as initially suggested by several lines of evidence. Thus, serum Ca^{++} levels are independently associated with FGF23 levels in dialysis patients, in transplant recipients, and in patients with primary hyperparathyroidism [3]. In patients with severe secondary and tertiary (persistent) hyperparathyroidism referred for parathyroidectomy, postoperative changes of FGF23 are limited and related to changes of Ca^{++} [72]. In patients with acute untreated hypoparathyroidism occurring after thyroidectomy [73], FGF23 levels are initially reduced when patients are hyperphosphatemic but still hypocalcemic. In contrast, in patients with chronic hypoparathryoidism and hyperphosphatemia who are normocalcemic on calcium and calcitriol treatment, serum FGF-23 levels are elevated [74]. Consequently treatment of hypocalcemia in patients with chronic hypoparathyroidism may raise FGF23 which will promote phosphaturia, but the serum levels of phosphorus concentrations may not completely normalize, possibly because the concerted phosphaturic actions of both PTH and FGF23 may be required.

In vitamin D receptor-null mice, dietary calcium supplementation significantly increases serum calcium levels, FGF23 messenger RNA abundance, and circulating FGF23 levels. In PTH-null mice that are hypocalcemic and hyperphosphatemic, FGF23 levels are reduced [75]. In wild-type mice and PTH-null mice, acute elevation of either serum Ca^{++} or Pi by intraperitoneal injection increased serum FGF23 levels. However, increases in serum Pi by chronic exposure to a high dietary Pi load were accompanied by severe hypocalcemia, which appeared to blunt stimulation of FGF23 release. Calcium-mediated increases in serum FGF23 required a threshold of at least normal serum Pi levels. Similarly, Pi-elicited increases in FGF23 were markedly blunted if serum Ca^{++} was less than normal. The best correlation between Ca^{++} and Pi and serum FGF23 was found between FGF23 and the $Ca^{++} \times$ Pi product.

8.2.3.2 Actions of FGF23

In addition to being regulated by serum Ca^{++} and Pi levels, FGF23 in turn acts to modulates serum Ca^{++} and Pi levels, thus ensuring that the $Ca^{++} \times$ Pi product remains within a physiological range. FGF23 is therefore not only a phosphoregulatory hormone but a dual calciophosphoregulatory hormone (Figs. 8.1 and 8.2).

FGF23 combines with an FGFR, as well as with Klotho, an obligate coreceptor, in order to transmit the signal of FGF23 to target organs [76]. Thus, Klotho protein forms constitutive binary complexes with FGFR1c, FGFR3c, and FGFR4 which increase the affinity of these FGFRs selectively to FGF23 [77] and produce a heterotrimeric complex which is required for FGF23 to activate downstream signaling molecules, including FGFR substrate-2α and mitogen-activated protein kinases (MAPKs) such as extracellular signal-regulated kinases (ERK1/2). Klotho is a type I membrane protein but may also be expressed, due to alternative RNA splicing, as a secreted form that lacks the transmembrane and intracellular domains; it may therefore also act as a humoral factor [78] and may also have β-glucuronidase activity [79]. A unique structural feature of the endocrine FGFs, including FGF23, is their lack of a heparin-binding domain that is conserved in all paracrine/autocrine FGFs [80].

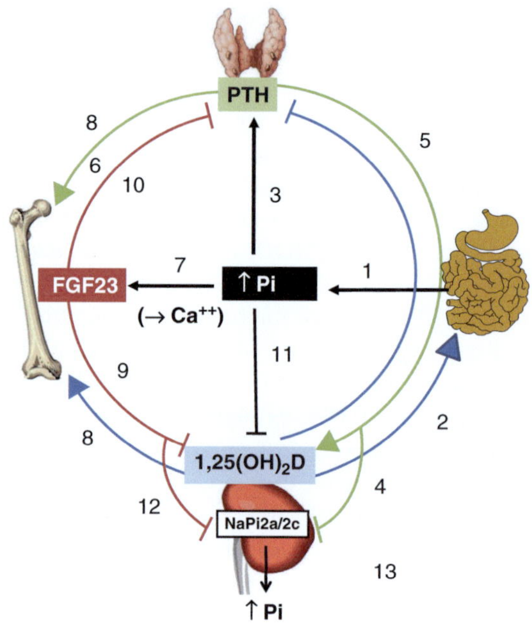

Fig. 8.1 Model of hormonal regulation of Pi (and Ca^{++}) homeostasis. In the presence of normal renal function, increased dietary phosphorus (Pi) (*1*) facilitated by $1{,}25(OH)_2D$ action on the intestine (*2*) may increase serum Pi levels. Increased Pi levels can directly or indirectly increase PTH secretion (*3*) leading to increased renal Pi excretion (*4*), but also to increased renal Ca^{++} retention, increased $1{,}25(OH)_2D$ production (*5*), and increased release of Ca^{++} and Pi from the bone (*6*). The increased serum Pi in the presence of a threshold level of serum Ca^{++} can now increase FGF23 secretion from the bone (*7*), as well as PTH (*8*), and $1{,}25(OH)_2D$ (*8*) which may also each increase FGF23 production. Secreted FGF23 can decrease $1{,}25(OH)_2D$ (*9*) thus reducing its capacity to further enhance intestinal Pi absorption and can also inhibit PTH (*10*) thus reducing stimulation of renal production of $1{,}25(OH)_2D$, Pi mobilization from the bone, and enhancement of Pi excretion. Serum Pi per se may also reduce $1{,}25(OH)_2D$ production (*11*). FGF23 can inhibit proximal tubular NaPi2a and 2c (*12*) and produce phosphaturia (*13*) in place of the suppressed PTH, thereby reducing the Pi (and $Ca^{++} \times$ Pi product). With prolonged elevation of FGF23 and suppression of $1{,}25(OH)_2D$, secondary hyperparathyroidism may occur to prevent significant hypocalcemia

This heparin-binding domain binds to heparan sulfate (HS) in the extracellular matrix, thereby imposing some restriction to the secretion of non-endocrine FGFs and increasing their local concentration to support their paracrine/autocrine mode of action; in addition, the HS-binding domain is essential for FGFR activation, forming a complex of HS, FGF, and FGFR [81]. Absence of the heparin-binding domain in endocrine FGFs may facilitate their release from sites of production and Klotho proteins substitute for HS in enhancing receptor binding by endocrine FGFs. Although FGFRs are quite ubiquitous, Klotho expression is relatively restricted and may confer tissue specificity for FGF23 action [42].

The major target for FGF23 is the kidney, where it acts to promote phosphate excretion and to decrease production and increase clearance of $1{,}25(OH)_2D$. FGF23 suppresses Pi reabsorption [28] by inhibiting NaPi2a and NaPi2c on the apical brush border membrane of proximal tubular cells [82]. Although all the FGF23 actions seem to occur in the proximal tubule, Klotho expression is higher in the distal tubules [83]. Because proximal tubules also express Klotho, albeit in lower quantities [84], FGF23 may signal directly in proximal tubules to regulate their function with a small number of FGFR–Klotho complexes. Alternatively FGF23 may act on distal convoluted tubules where Klotho is most abundantly expressed and initiate release of a paracrine factor(s) that acts on adjacent proximal tubules. It is currently unclear but unlikely that circulating Klotho can serve as a coreceptor for FGFR.

FGF23 also lowers blood levels of $1{,}25(OH)_2D$ by downregulating the expression of the *CYP27B1* gene [31] and by upregulating gene expression of 24-hydroxylase (CYP24A1), which converts $1{,}25(OH)_2D$ to inactive metabolites [85]. Thus, FGF23 suppresses synthesis and promotes degradation of the active hormonal form of vitamin D. By diminishing circulating $1{,}25(OH)_2D$ levels, FGF23 can therefore also indirectly reduce vitamin D-stimulated Pi (and Ca^{++}) absorption in the intestine (Figs. 8.1 and 8.2).

In addition, FGF23 acts directly on the parathyroid gland FGFRs (likely via FGFR1 and FGFR3) and Klotho [26, 27] in an ERK1/2-dependent manner [86, 87] to suppress PTH synthesis and secretion [86, 87]; FGF23 may also increase the levels of parathyroid CaSR and VDR to indirectly inhibit PTH gene expression, secretion, and cell proliferation via Ca^{++} and $1{,}25(OH)_2D$ respectively [88]. However, in most clinical and pathological situations associated with chronically increased circulating FGF23 concentrations,

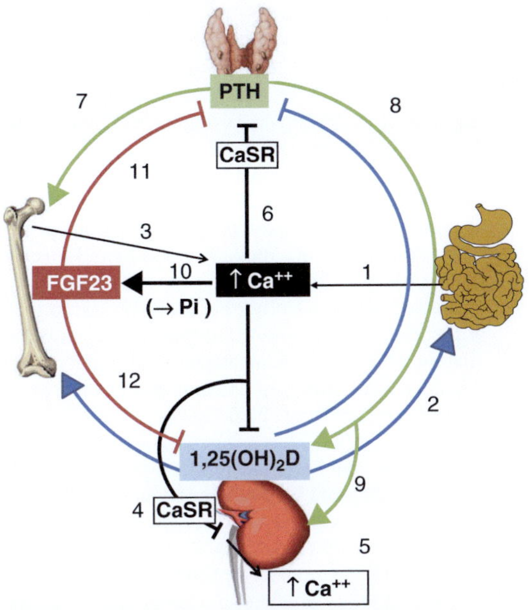

Fig. 8.2 Model of hormonal regulation of Ca^{++} (and Pi) homeostasis. Increased intestinal absorption of Ca^{++} (*1*) facilitated by increased $1,25(OH)_2D$ (*2*) or increased bone resorption (*3*) or both may result in increased serum calcium (Ca^{++}) (and Pi). Serum Ca^{++}, by stimulating the renal CaSR (*4*), can enhance Ca^{++} excretion (*5*). Serum Ca^{++}, acting via the CaSR in the parathyroid gland, can inhibit PTH secretion (*6*). Decreased PTH results in reduced mobilization of skeletal Ca^{++} (and Pi) (*7*), reduced renal production of $1,25(OH)_2D$ (*8*), and reduced renal Ca^{++} retention (*9*) but also in reduced renal Pi clearance. Elevated serum Ca^{++} in the presence of high normal or elevated serum Pi can then increase FGF23 secretion (*10*) which can further inhibit PTH secretion (*11*) and kidney-derived $1,25(OH)_2D$ (*12*) and normalize the $Ca^{++} \times Pi$ product

secondary hyperparathyroidism is present. In mice overexpressing FGF23 [89], secondary hyperparathyroidism also tends to occur, suggesting that the effects of FGF23 on Pi, Ca^{++}, and 1,25(OH)D metabolism, which act to stimulate PTH production, may overcome any direct inhibitory effects of FGF23 on PTH release. Downregulation of FGFRs and Klotho in the parathyroids that reduces sensitivity to FGF23 signaling has been reported as the underlying cause of FGF23 resistance in some [90, 91], but not all studies [92] of secondary hyperparathyroidism in uremia. In *Hyp* mice which have a loss of Phex function with resultant increased FGF23 and which are phenocopies of X-linked hypophosphatemic rickets (XLH) in man, secondary hyperparathyroidism occurs and deletion of the gene encoding PTH results in early lethality due to hypocalcemia [93]. Hyperparathyroidism, therefore, is an integral component in the pathophysiology of Hyp, and likely XLH, despite excess circulating FGF23 and may serve as a compensatory mechanism to prevent severe hypocalcemia in mice and perhaps in patients afflicted with the disorder.

8.3 Endocrine Regulation of Pi and Ca^{++} Metabolism

When renal function is normal, in the presence of an increased dietary load of Pi, serum Pi levels may increase, enhanced by $1,25(OH)_2D$ action on the intestine. This increased serum Pi may directly [94] or indirectly (by reducing serum Ca^{++} levels) increase PTH secretion and facilitate renal Pi excretion. However, PTH may also mobilize Pi (and Ca^{++}) from the bone and enhance $1,25(OH)_2D$ production with resultant increased intestinal absorption of Pi (and Ca^{++}). In the presence of a threshold level of serum Ca^{++}, Pi can then increase FGF23 secretion from the bone. $1,25(OH)_2D$ per se and PTH may also increase skeletal production of FGF23. Secreted FGF23 can then inhibit $1,25(OH)_2D$ production, thus reducing its capacity to further enhance intestinal Pi absorption, and can also inhibit PTH thus reducing its capacity to mobilize Pi from the bone, as well as to stimulate renal production of $1,25(OH)_2D$ and to promote Pi excretion. Pi per se may also inhibit $1,25(OH)_2D$ production. FGF23 can inhibit proximal tubular NaPi2a and 2c in the kidney and can produce phosphaturia in place of the suppressed PTH, restoring the serum Pi (and Ca^{++} x Pi product) to normal. With prolonged elevation of FGF23 and prolonged suppression of $1,25(OH)_2D$, secondary hyperparathyroidism may occur to prevent significant hypocalcemia (Fig. 8.1).

Increased serum Ca^{++} may arise from increased intestinal absorption of Ca^{++}, (with Pi) or from increased bone resorption (with Pi) or from both mechanisms. The increased serum Ca^{++} per se by stimulating the renal CaSR can enhance renal

Ca^{++} excretion. The increased serum Ca^{++} acting via the CaSR in the parathyroid gland can inhibit PTH secretion. Decreased PTH results in reduced bone resorption and concomitant mobilization of skeletal Ca^{++} (and Pi), decreased renal Ca^{++} retention, and reduced production of $1,25(OH)_2D$ in the kidney, but also in increased Pi retention. Elevated serum Ca^{++} in the presence of high normal or elevated serum Pi can also increase FGF23 secretion which can further inhibit PTH secretion, can inhibit kidney-derived $1,25(OH)_2D$, and replace the action of PTH in promoting phosphaturia (Fig. 8.2), thereby normalizing the $Ca^{++} \times Pi$ product.

These complex interrelationships underlie the exquisite controls that have evolved to maintain serum Pi and Ca^{++} levels within a defined range when one of these ions is dysregulated but also to ensure an appropriate ratio by maintaining a normal $Ca^{++} \times Pi$ product.

Conclusion

Initial studies on mineral ion regulation focussed on the Ca^{++}-regulating hormones PTH and $1,25(OH)_2D$, but these were also known to have profound effects on renal and intestinal handling of Pi. Analysis of genetic diseases of renal Pi wasting identified FGF23, as a primary hormonal regulator of Pi homeostasis. Subsequently FGF23 was found to participate in complex feedback loops with the classic Ca^{++}-regulating hormones, and all three hormones are now known to share interacting functions on regulating both Ca^{++} and Pi homeostasis.

References

1. Christakos S (2012) Recent advances in our understanding of 1,25-dihydroxyvitamin D(3) regulation of intestinal calcium absorption. Arch Biochem Biophys 523:73–76
2. Forster IC, Hernando N, Biber J, Murer H (2006) Proximal tubular handling of phosphate: A molecular perspective. Kidney Int 70:1548–1559
3. Brown EM (2013) Role of the calcium-sensing receptor in extracellular calcium homeostasis. Best Pract Res Clin Endocrinol Metab 27:333–343
4. Demay MB, Kiernan MS, DeLuca HF, Kronenberg HM (1992) Sequences in the human parathyroid hormone gene that bind the 1,25-dihydroxyvitamin D3 receptor and mediate transcriptional repression in response to 1,25-dihydroxyvitamin D3. Proc Natl Acad Sci U S A 89:8097–8101
5. Kremer R, Bolivar I, Goltzman D, Hendy GN (1989) Influenceofcalciumand1,25-dihydroxycholecalciferol on proliferation and proto-oncogene expression in primary cultures of bovine parathyroid cells. Endocrinology 125:935–941
6. Panda DK, Miao D, Bolivar I et al (2004) Inactivation of the 25-hydroxyvitamin D 1alpha-hydroxylase and vitamin D receptor demonstrates independent and interdependent effects of calcium and vitamin D on skeletal and mineral homeostasis. J Biol Chem 279:16754–16766
7. Tregear GW, Van Rietschoten J, Greene E et al (1973) Bovine parathyroid hormone: minimum chain length of synthetic peptide required for biological activity. Endocrinology 93:1349–1353
8. Goltzman D, Peytremann A, Callahan E et al (1975) Analysis of the requirements for parathyroid hormone action in renal membranes with the use of inhibiting analogues. J Biol Chem 250:3199–3203
9. Jüppner H, Abou-Samra AB, Freeman M et al (1991) A G protein-linked receptor for parathyroid hormone and parathyroid hormone-related peptide. Science 254:1024
10. Abou-Samra AB, Jüppner H, Force T et al (1992) Expression cloning of a common receptor for parathyroid hormone and parathyroid hormone-related peptide from rat osteoblast-like cells: a single receptor stimulates intracellular accumulation of both cAMP and inositol trisphosphates and increases intracellular free calcium. Proc Natl Acad Sci U S A 89:2732–2736
11. Lambers TT, Bindels RJ, Hoenderop JG (2006) Coordinated control of renal Ca2+ handling. Kidney Int 69:650–654
12. van Abel M, Hoenderop JG, van der Kemp AW et al (2005) Coordinated control of renal Ca(2+) transport proteins by parathyroid hormone. Kidney Int 68:1708–1721
13. Cha SK, Wu T, Huang CL (2008) Protein kinase C inhibits caveolae-mediated endocytosis of TRPV5. Am J Physiol Renal Physiol 294:F1212–F1221
14. Topala CN, Schoeber JP, Searchfield LE et al (2009) Activation of the Ca(2+)-sensing receptor stimulates the activity of the epithelial Ca(2+) channel TRPV5. Cell Calcium 45:331–339
15. Custer M, Lotscher M, Biber J et al (1994) Expression of Na-P(i) cotransport in rat kidney: localization by RT-PCR and immunohistochemistry. Am J Physiol 266:F767–F774
16. Bacic D, Lehir M, Biber J et al (2006) The renal Na+/phosphate cotransporter NaPi-IIa is internalized via the receptor-mediated endocytic route in response to parathyroid hormone. Kidney Int 69:495–503
17. Segawa H, Yamanaka S, Onitsuka A et al (2007) Parathyroid hormone-dependent endocytosis of renal

type IIc Na-Pi cotransporter. Am J Physiol Renal Physiol 292:F395–F403
18. Traebert M, Volkl H, Biber J et al (2000) Luminal and contraluminal action of 1–34 and 3–34 PTH peptides on renal type IIa Na-P(i) cotransporter. Am J Physiol Renal Physiol 278:F792–F798
19. Brenza HL, Kimmel-Jehan C, Jehan F et al (1998) Parathyroid hormone activation of the 25-hydroxyvitamin D3-1alpha-hydroxylase gene promoter. Proc Natl Acad Sci U S A 95:1387–1391
20. Rouleau MF, Mitchell J, Goltzman D (1990) Characterization of the major parathyroid hormone target cell in the endosteal metaphysis of rat long bones. J Bone Miner Res 5:1043–1053
21. Miao D, He B, Karaplis AC, Goltzman D (2002) Parathyroid hormone is essential for normal fetal bone formation. J Clin Invest 109:1173–1182
22. Boyle WJ, Simonet WS, Lacey DL (2003) Osteoclast differentiation and activation. Nature 423:337–342
23. Silva BC, Costa AG, Cusano NE et al (2011) Catabolic and anabolic actions of parathyroid hormone on the skeleton. J Endocrinol Invest 34:801–810
24. MacLaughlin JA, Anderson RR, Holick MF (1982) Spectral character of sunlight modulates photosynthesis of previtamin D3 and its photoisomers in human skin. Science 216:1001–1003
25. Dastani Z, Berger C, Langsetmo L et al (2014) In healthy adults, biological activity of vitamin D, as assessed by serum PTH, is largely independent of DBP concentrations. J Bone Miner Res 29:494–499
26. Zhu JG, Ochalek JT, Kaufmann M et al (2013) CYP2R1 is a major, but not exclusive, contributor to 25-hydroxyvitamin D production in vivo. Proc Natl Acad Sci U S A 110:15650–15655
27. Jones G, Strugnell SA, DeLuca HF (1998) Current understanding of the molecular actions of vitamin D. Physiol Rev 78:1193–1231
28. Murayama A, Takeyama K, Kitanaka S et al (1999) Positive and negative regulations of the renal 25-hydroxyvitamin D3 1alpha-hydroxylase gene by parathyroid hormone, calcitonin, and 1alpha,25(OH)2D3 in intact animals. Endocrinology 140:2224–2231
29. Liu P, Stenger S, Li H et al (2006) Toll-like receptor triggering of a vitamin D-mediated human antimicrobial response. Science 311:1770–1773
30. St-Arnaud R (2010) CYP24A1-deficient mice as a tool to uncover a biological activity for vitamin D metabolites hydroxylated at position 24. J Steroid Biochem Mol Biol 121:254–256
31. Shimada T, Hasegawa H, Yamazaki Y et al (2004) FGF-23 is a potent regulator of vitamin D metabolism and phosphate homeostasis. J Bone Miner Res 19:429–435
32. Pike JW, Meyer MB (2014) Fundamentals of vitamin D hormone-regulated gene expression. J Steroid Biochem Mol Biol 144PA:5–11. pii: S0960-0760(13)00234-3. doi:10.1016/j.jsbmb.2013.11.004. PMID: 24239506
33. Meyer MB, Watanuki M, Kim S et al (2006) The human transient receptor potential vanilloid type 6 distal promoter contains multiple vitamin D receptor binding sites that mediate activation by 1,25-dihydroxyvitamin D3 in intestinal cells. Mol Endocrinol 20:1447–1461
34. Fleet JC, Wood RJ (1994) Identification of calbindin D-9 k mRNA and its regulation by 1,25-dihydroxyvitamin D3 in Caco-2 cells. Arch Biochem Biophys 308:171–174
35. Christakos S, Dhawan P, Ajibade D et al (2010) Mechanisms involved in vitamin D mediated intestinal calcium absorption and in non-classical actions of vitamin D. J Steroid Biochem Mol Biol 121:183–187
36. Haussler MR, Whitfield GK, Kaneko I et al (2013) Molecular mechanisms of vitamin D action. Calcif Tissue Int 92:77–98
37. Suda T, Takahashi N, Martin TJ (1992) Modulation of osteoclast differentiation. Endocr Rev 3:66–80
38. Miao D, He B, Lanske B et al (2004) Skeletal abnormalities in Pth-null mice are influenced by dietary calcium. Endocrinology 145:2046–2053
39. Tanaka H, Seino Y (2004) Direct action of 1,25-dihydroxyvitamin D on bone: VDRKO bone shows excessive bone formation in normal mineral condition. J Steroid Biochem Mol Biol 89–90:343–345
40. Kim S, Yamazaki M, Zella LA et al (2006) Activation of receptor activator of NF-kappaB ligand gene expression by 1,25-dihydroxyvitamin D3 is mediated through multiple long-range enhancers. Mol Cell Biol 26:6469–6486
41. Eswarakumar VP, Lax I, Schlessinger J (2005) Cellular signaling by fibroblast growth factor receptors. Cytokine Growth Factor Rev 16:139–149
42. Mohammadi M, Olsen SK, Ibrahimi OA (2005) Structural basis for fibroblast growth factor receptor activation. Cytokine Growth Factor Rev 16:107–137
43. Beenken A, Mohammadi M (2012) The structural biology of the FGF19 subfamily. Adv Exp Med Biol 728:1–24
44. White KE, Evans WE, O'Riordan JLH et al (2000) Autosomal dominant hypophosphataemic rickets is associated with mutations in FGF23. Nat Genet 26:345–348
45. Yamashita T (2005) Structural and biochemical properties of fibroblast growth factor 23. Ther Apher Dial 9:313–318
46. Kato K, Jeanneau C, Tarp MA et al (2006) Polypeptide GalNAc-transferase T3 and familial tumoral calcinosis. Secretion of fibroblast growth factor 23 requires O-glycosylation. J Biol Chem 281:18370–18377
47. Goetz R, Nakada Y, Hu MC et al (2010) Isolated C-terminal tail of FGF23 alleviates hypophosphatemia by inhibiting FGF23-FGFR-Klotho complex formation. Proc Natl Acad Sci U S A 107:407–412. PubMed: 19966287
48. Bai XY, Miao D, Goltzman D, Karaplis AC (2003) The autosomal dominant hypophosphatemic rickets R176Q mutation in fibroblast growth factor 23 resists proteolytic cleavage and enhances in vivo biological potency. J Biol Chem 278(11):9843–9849
49. Francis F, Hennig S, Korn B et al (1995) A gene (PEX) with homologies to endopeptidases is mutated

in patients with X-linked hypophosphatemic rickets. Nat Genet 11:130–136
50. Yamazaki Y, Okazaki R, Shibata M et al (2002) Increased circulatory level of biologically active full-length FGF-23 in patients with hypophosphatemic rickets/osteomalacia. J Clin Endocrinol Metab 87:4957–4960
51. Weber TJ, Liu S, Quarles LD (2003) Serum FGF23 levels in normal and disordered phosphorus homeostasis. J Bone Miner Res 18:1227–1234
52. Feng JQ, Ward LM, Liu S et al (2006) Loss of DMP1 causes rickets and osteomalacia and identifies a role for osteocytes in mineral metabolism. Nat Genet 38:1310–1315
53. Turan S, Aydin C, Bereket A, Akcay T, Güran T, Yaralioglu BA, Bastepe M, Jüppner H (2010) Identification of a novel dentin matrix protein-1 (DMP-1) mutation and dental anomalies in a kindred with autosomal recessive hypophosphatemia. Bone 46:402–409
54. Rowe PS (2012) The chicken or the egg: PHEX, FGF23 and SIBLINGs unscrambled. Cell Biochem Funct 30:355–375
55. Huitema LF, Apschner A, Logister I et al (2012) Entpd5 is essential for skeletal mineralization and regulates phosphate homeostasis in zebrafish. Proc Natl Acad Sci U S A 109:21372–21377
56. Mackenzie NC, Zhu D, Milne EM et al (2012) Altered bone development and an increase in FGF-23 expression in Enpp1−/− mice. PLoS One 7(2012)
57. Wohrle S, Bonny O, Beluch N et al (2011) FGF receptors control vitamin D and phosphate homeostasis by mediating renal FGF-23 signaling and regulating FGF-23 expression in bone. J Bone Miner Res 26:2486–2497
58. Wolf M, Koch TA, Bregman DB (2013) Effects of iron deficiency anemia and its treatment on fibroblast growth factor 23 and phosphate homeostasis in women. J Bone Miner Res 28:1793–1803
59. Kuro-o M, Matsumura Y, Aizawa H et al (1997) Mutation of the mouse klotho gene leads to a syndrome resembling ageing. Nature 390:45–51
60. Ichikawa S, Imel EA, Kreiter ML et al (2007) A homozygous missense mutation in human KLOTHO causes severe tumoral calcinosis. J Clin Invest 117:2684–2691
61. Hu MC, Shi M, Zhang J et al (2011) Klotho deficiency causes vascular calcification in chronic kidney disease. J Am Soc Nephrol 22:124–136
62. Shimamura Y, Hamada K, Inoue K et al (2012) Serum levels of soluble secreted α-Klotho are decreased in the early stages of chronic kidney disease, making it a probable novel biomarker for early diagnosis. Clin Exp Nephrol 16:722–729
63. Saito H, Maeda A, Ohtomo S et al (2005) Circulating FGF-23 is regulated by 1α,25-dihydroxyvitamin D3 and phosphorus in vivo. J Biol Chem 280:2543–2549
64. Kolek OI, Hines ER, Jones MD et al (2005) 1{alpha},25-Dihydroxyvitamin D3 upregulates FGF23gene expression in bone: the final link in a renal-gastrointestinal-skeletal axis that controls phosphate transport. Am J Physiol Gastrointest Liver Physiol 289(6):G1036–G1042
65. Turner AG, Hanrath MA, Morris HA et al (2014) The local production of 1,25(OH)2D3 promotes osteoblast and osteocyte maturation. J Steroid Biochem Mol Biol 144:114–118. doi:10.1016/j.jsbmb.2013.10.003. pii: S0960-0760(13)00196-9
66. Ormsby RT, Findlay DM, Kogawa M et al (2014) Analysis of vitamin D metabolism gene expression in human bone: Evidence for autocrine control of bone remodelling. J Steroid Biochem Mol Biol 144:110–113. doi:10.1016/j.jsbmb.2013.09.016. pii: S0960-0760(13)00190-8
67. Rhee Y, Bivi N, Farrow E et al (2011) Parathyroid hormone receptor signaling in osteocytes increases the expression of fibroblast growth factor-23 in vitro and in vivo. Bone 49:636–643
68. Lavi-Moshayoff V, Wasserman G, Meir T et al (2010) PTH increases FGF23 gene expression and mediates the high-FGF23 levels of experimental kidney failure: a bone parathyroid feedback loop. Am J Physiol Renal Physiol 299:F882–F889
69. Lopez I, Rodriguez-Ortiz ME, Almaden Y et al (2011) Direct and indirect effects of parathyroid hormone on circulating levels of fibroblast growth factor 23 in vivo. Kidney Int 80:475–482
70. Rodriguez-Ortiz ME, Lopez I, Munoz-Castaneda JR et al (2012) Calcium deficiency reduces circulating levels of FGF23. J Am Soc Nephrol 23:1190–1197
71. Haussler MR, Whitfield GK, Kaneko I et al (2011) The role of vitamin D in the FGF23, klotho, and phosphate bone-kidney endocrine axis. Rev Endocr Metab Disord 13:57–69
72. Sato T, Tominaga Y, Ueki T et al (2004) Total parathyroidectomy reduces elevated circulating fibroblast growth factor 23 in advanced secondary hyperparathyroidism. Am J Kidney Dis 44:481–487
73. Yamashita H, Yamazaki Y, Hasegawa H et al (2007) Fibroblast growth factor-23 (FGF23) in patients with transient hypoparathyroidism: its important role in serum phosphate regulation. Endocr J 54(3):465–470
74. Gupta A, Winer K, Econs MJ et al (2004) FGF-23 is elevated by chronic hyperphosphatemia. J Clin Endocrinol Metab 89:4489–4492
75. Quinn SJ, Thomsen AR, Pang JL et al (2013) Interactions between calcium and phosphorus in the regulation of the production of fibroblast growth factor 23 in vivo. Am J Physiol Endocrinol Metab 304(3):E310–E320
76. Juppner H, Wolf M (2012) αKlotho: FGF23 coreceptor and FGF23-regulating hormone. J Clin Invest 122:4336–4339
77. Kurosu H, Ogawa Y, Miyoshi M et al (2006) Regulation of fibroblast growth factor-23 signaling by klotho. J Biol Chem 281:6120–6123
78. Kurosu H, Yamamoto M, Clark JD et al (2005) Suppression of aging in mice by the hormone Klotho. Science 309:1829–1833
79. Tohyama O, Imura A, Iwano A et al (2004) Klotho is a novel β-glucuronidase capable of hydrolyzing steroid β-glucuronides. J Biol Chem 279:9777–9784

80. Goetz R, Beenken A, Ibrahimi OA et al (2007) Molecular insights into the Klotho dependent, endocrine mode of action of FGF19 subfamily members. Mol Cell Biol 27:3417–3428
81. Urakawa I, Yamazaki Y, Shimada T et al (2006) Klotho converts canonical FGF receptor into a specific receptor for FGF23. Nature 444:770–774
82. Murer H, Hernando N, Forster I, Biber J (2003) Regulation of Na/Pi transporter in the proximal tubule. Annu Rev Physiol 65:531–542
83. Farrow EG, Davis SI, Summers LJ, White KE (2009) Initial FGF23-mediated signaling occurs in the distal convoluted tubule. J Am Soc Nephrol 20:955–960
84. Hu MC, Shi M, Zhang J et al (2010) Klotho: a novel phosphaturic substance acting as an autocrine enzyme in the renal proximal tubule. FASEB J 24:3438–3450
85. Shimada T, Kakitani M, Yamazaki Y et al (2004) Targeted ablation of Fgf23 demonstrates an essential physiological role of FGF23 in phosphate and vitamin D metabolism. J Clin Invest 113(4):561–568
86. Ben-Dov IZ, Galitzer H, Lavi-Moshayoff V et al (2007) The parathyroid is a target organ for FGF23 in rats. J Clin Invest 117:4003–4008
87. Krajisnik T, Bjorklund P, Marsell R et al (2007) Fibroblast growth factor-23 regulates parathyroid hormone and 1α-hydroxylase expression in cultured bovine parathyroid cells. J Endocrinol 195:125–131
88. Canalejo R, Canalejo A, Martinez-Moreno JM et al (2010) FGF23 fails to inhibit uremic parathyroid glands. J Am Soc Nephrol 21:1125–1135
89. Bai X, Miao D, Li J et al (2004) Transgenic mice overexpressing human fibroblast growth factor 23 (R176Q) delineate a putative role for parathyroid hormone in renal phosphate wasting disorders. Endocrinology 145(11):5269–5279
90. Galitzer H, Ben Dov IZ, Silver J, Naveh-Many T (2010) Parathyroid cell resistance to fibroblast growth factor 23 in secondary hyperparathyroidism of chronic kidney disease. Kidney Int 77:211–218
91. Komaba H, Goto S, Fujii H et al (2010) Depressed expression of Klotho and FGF receptor 1 in hyperplastic parathyroid glands from uremic patients. Kidney Int 77:232–238
92. Hofman-Bang J, Martuseviciene G, Santini MA et al (2010) Increased parathyroid expression of klotho in uremic rats. Kidney Int 78:1119–1127
93. Bai X, Miao D, Goltzman D, Karaplis AC (2007) Early lethality in Hyp mice with targeted deletion of Pth gene. Endocrinology 148(10):497
94. Slatopolsky E, Finch J, Denda M et al (1996) Phosphorus restriction prevents parathyroid gland growth-high phosphorus directly stimulates PTH secretion in vitro. J Clin Invest 97:2534–2540

The PTH Receptorsome and Transduction Pathways

9

Thomas J. Gardella

9.1 Introduction

The parathyroid hormone receptor type 1 (PTHR1) mediates the biological actions of parathyroid hormone (PTH) in target cells of the bone and kidney. The agonist-bound PTHR1 couples efficiently to several signal transduction pathways, including the cAMP/PKA and PLC/IP3/PKC pathways, and activation of signal transduction systems in target cells results in the deployment of a number of downstream effector responses, such as an increase in the production of RANK ligand in osteoblasts and a suppression of surface expression of the sodium-dependent phosphate co-transporter type 2A in renal proximal tubular cells. Such effector responses to PTHR1 signaling ultimately result in the finely tuned changes blood calcium (Ca) and inorganic phosphate (Pi) levels that serve to keep these mineral ions within a remarkably narrow range of their ideal set-point concentrations. The PTHR1 also serves as the receptor for PTH-related proteins (PTHrP) and thus plays a completely distinct role by mediating the paracrine control of primordial cell proliferation and differentiation in a number of developing tissues, such as the skeleton. Understanding the molecular mechanisms by which the PTHR1 engages its ligands, PTH and PTHrP, and mediates cellular signal transduction responses is an important goal to achieve in order to be able to fully explain and control systems of calcium homeostasis and tissue growth development in normal and diseases states.

9.2 The PTHR1: Background

The PTHR1 utilizes the seven transmembrane domain protein architecture used by all *G protein-coupled receptors* (GPCRs) (Fig. 9.1). As occurs in each GPCR upon activation by a cognate external stimulating agent, the PTHR1, when acted upon by a PTH agonist ligand, undergoes a series of conformational changes which result in coupling of the PTHR1 to cytoplasmic effector proteins, most prominently Gαs-containing heterotrimeric G proteins and hence activation of downstream intracellular signaling cascades. The activation of Gαs thus results in an increase in the activity of membrane-bound adenylyl cyclase, and the resulting increase in intracellular cAMP activates protein kinase A (PKA), which, in turn, phosphorylates and hence activates other downstream signaling proteins, such as the cAMP response element-binding protein transcription factor CREB. The activated PTHR1 can also couple to several other signaling cascades, including the Gαq/phospholipase C (PLC)/inositol trisphosphate (IP_3)/intracellular Ca/protein kinase C

T.J. Gardella, PhD
Endocrine Unit, Massachusetts General Hospital, Harvard Medical School, Thier 10, 50 Blossom Street, Boston, MA 02114, USA
e-mail: Gardella@Helix.MGH.Harvard.edu

M.L. Brandi, E.M. Brown (eds.), *Hypoparathyroidism*,
DOI 10.1007/978-88-470-5376-2_9, © Springer-Verlag Italia 2015

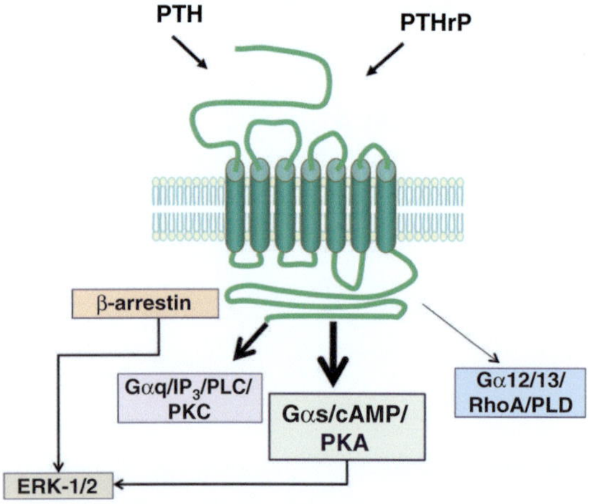

Fig. 9.1 The PTH/PTHrP receptor (PTHR1). The PTHR1 mediates the actions of two peptide ligands, parathyroid hormone and PTH-related protein; the receptor displays the seven-membrane-spanning helical domain protein architecture characteristic of a GPCR. The PTHR1 couples to a variety of signal transduction pathways, including most prominently the Gαs/cAMP/PKA pathway, the Gαq/PLC/PKC pathway, the Gα12/13/RhoA/PLD pathway, and the ERK-1/2-MAP-kinase pathway via indirect, Gαs-dependent and β-arrestin-dependent mechanisms

(PKC) pathway [1], the Gα12/13/phospholipase D (PLD)/RhoA pathway [2], and the mitogen-activated protein kinase (MAPK) signaling cascades, such as extracellular signal-regulated kinase, ERK1/2. The last of these can occur through G protein-dependent mechanisms or G protein-independent, β-arrestin-dependent mechanisms [3].

9.2.1 Parathyroid Hormone: Ligand Determinants of Biological Activity

While endogenous PTH in humans is an 84 amino acid polypeptide, synthetic peptides comprised of the first 34 amino acids are fully active for most if not all PTHR1-mediated biological responses. The key determinants of hormone signaling and receptor binding reside within the N-terminal and C-terminal portions of the PTH(1–34) fragment, respectively. Critical ligand determinants of signaling include the first two amino acids, particularly valine-2, and N-terminal truncated peptides such as PTH(7–34) function as effective PTH antagonist [4]. The PTH(15–34) peptide represents the shortest-length peptide fragment that retains at least some capacity to bind to the PTHR1 [5]. Short amino-terminal PTH fragments generally lack detectable activity, but analogs such as "M"-PTH(1–14) that are optimized with a set of six substitutions, collectively called the "M" modifications, exhibit signaling potencies in the low-nanomolar range, similar to that of PTH(1–34) [6].

Native PTHrP is 141 amino acids in length, with splice variants terminating at positions 139 or 173 also detectable [7]. Here again the (1–34) peptide exhibits full activity on the PTHR1. The PTHrP peptide shares strongest homology with PTH in the N-terminal region, in which eight of the first 13 amino acid residues are identical, but then diverges considerably. The PTHrP(15–34) peptide nevertheless competes with PTH for the same or an overlapping binding site on the receptor. These findings suggested that the principal binding domains of PTH and PTHrP adopt similar conformations when binding to the receptor, likely an amphipathic α-helix, which indeed was confirmed by the recent x-ray crystallographic analyses of each ligand fragment in complex with the cognate portion of the receptor [8].

9.2.2 The PTHR1: Structural and Functional Properties

While the cloning of the PTHR1 cDNA in 1991 revealed the seven transmembrane domain motif of the GPCR class, there was no direct homology with most GPCRs identified at the time, such as the β_2-adrenergic receptor [9]. It was realized, however, that the PTHR1, together with several other recently identified receptors, formed a distinct GPCR subgroup, thenceforth called the family B GPCRs. This family B GPCR subgroup is comprised of about 15 distinct receptors that bind peptide hormones, including, in addition to PTH, calcitonin, secretin, glucagon, and corticotrophin-releasing factor (CRF) [10]. The sequence homologies between the related family B GPCRs, though low, at ~35 % identity, predict similarities in their structures and modes of action, as also suggested by the finding that each binds a single-chain polypeptide ligand of 30–40 amino acids.

The human PTHR1 is comprised of 593 amino acids. The mature receptor contains a relatively large amino-terminal extracellular domain (ECD) of ~160 amino acids, a transmembrane domain (TMD) region containing the seven-membrane-spanning helices and interconnecting loops (ICLs and ECLs), and a carboxy-terminal tail of about 130 amino acids. The PTHR1 ECD contains a segment of 44 amino acids encoded by exon E2 that is not found in other family B receptors and which can be deleted without an effect on function. Several hallmark features are well conserved in the family B GPCRs, including six extracellular cysteines in the ECD. These conserved residues presumably help to maintain a protein fold that is used by each of the family B GPCRs and defines a common mechanism of action.

9.2.3 Mechanism of Binding: The Two-Site Model

The mode of ligand binding and activation used by the PTHR1 has been approached using biochemical and mutational methods employing mutant receptors and altered ligand analogs. Such studies led to the so-called two-site model of binding for the PTH/PTHR1 interaction. By this model, the C-terminal (15–34) portion of PTH(1–34) first docks to the ECD region of the receptor to establish initial affinity interactions, and then the N-terminal (1–14) portion of the ligand engages the TMD region to induce the conformational changes involved in receptor activation and G protein coupling [11]. The two-site model that emerged from the early mutation-based functional studies gained strong support from more direct photoaffinity cross-linking studies, which mapped specific sites of physical proximity between residues in the ligand and receptor. Examples of key intermolecular proximities thus assigned were those between Lys13 in the ligand and Arg186 at the extracellular of TMD helix 1, between Val2 in the ligand and Met425 at the extracellular end of TMD helix 6, and between Trp23 in the ligand and Thre33/Gln37 in the receptor's ECD [12, 13]. The general two-site model is well validated for the PTHR1 and is likely used by other members of the family B GPCR subgroup [14]. Recent use of direct structural approaches provides further support for this general mode of interaction.

9.3 The PTHR1 ECD Region: Affinity Interactions

The complete three-dimensional structure of the intact PTHR1 has not yet been determined; however, such data are now available for the isolated ECD region. Pioszak and Xu thus determined the x-ray crystal structure of the PTHR1 ECD in complex with the (12–34) fragment of PTH as well as the (15–34) fragment of PTHrP [8]. Similar approaches have been used to acquire structures of the ECD regions of several other family B GPCRs [15]. The common protein fold is defined by a network of three intramolecular disulfide bonds that braces the structure, a central core comprised of two pairs of antiparallel beta-strands and a prominent loop, and a long N-terminal flanking α-helix (Fig. 9.2). A prominent groove lies along the center of the structure and this groove serves as the binding site for the C-terminal ligand domain, which is indeed bound as an amphipathic α-helix. Extensive contacts are made between Trp23, Leu24, and Leu28 that form the hydrophobic surface of the PTH ligand helix and hydrophobic residues that line the groove of the ECD. The observed mode of binding directly confirms prior PTH structure-activity studies which predicted an important role for the hydrophobic face of an amphipathic α-helix in this region of PTH in binding to the receptor [8, 18]. The binding mode for PTHrP shows nearly complete overlap with that used by PTH, as predicted by the previous binding data; however, a slight bend in the PTHrP helix at position-27 leads to a divergence in contacts made by the C-terminal portions of the two ligand helices. Such differences in binding could potentially contribute to any difference in action of the two ligands, as seen in some cell- and clinical-based studies [19–21]. The structures suggest how the C-terminal domain of the ligand binds so as to

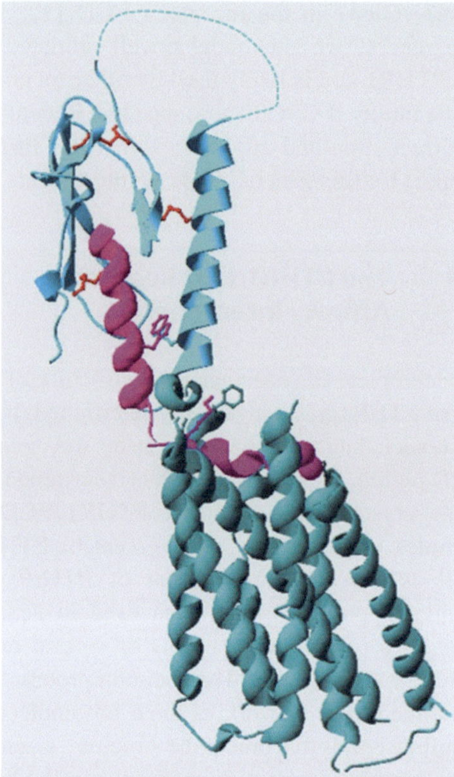

Fig. 9.2 Molecular model of the PTH(1–34)·PTHR1 complex: Shown is a plausible model of PTH(1–34) (*magenta*) bound to the PTHR1 (*blue–green*), developed based on the x-ray crystal structures of the ECD region of the PTHR1 in complex with PTH(15–34) [16], and for the TMD region of the related CRFR1 [17]. The protein backbones are displayed in ribbon format, with the side chains displayed for a few selected residues, including Val2, Lys13, and Trp23 in PTH and Gln37 and Phe184 in the PTHR1. The three disulfide bonds in the ECD are shown in *red*, and the segment encoded by exon E2, absent in the crystal structure, is represented by a *dashed line*

present the N-terminal portion of the ligand in optimal position for interaction with the TMD portion of the receptor.

9.4 The PTHR1 Transmembrane Domain Region: Signaling Interactions

Recent x-ray crystal structures have been determined for the TMD regions of two related family B GPCRs, the CRF receptor-1 and the glucagon receptor (GlucR) [17]. These breakthrough structures followed on the heels of the crystallographic structures obtained by Kobilka and colleagues for several family A GPCRs, including the β_2-adrenergic receptor [22]. The family B TMD structures exhibit a heptahelical bundle that superimposes fairly well with that of the family A adrenergic receptors; however, the opening at the extracellular surface is wider in the family B receptors than in the family A receptors, which likely reflects the larger size of the peptide ligand bound by the family B receptors versus the small catecholamine ligands bound by the adrenergic receptors. An extensive mutational and cross-linking analysis of the predicted ligand-binding pocket of the CRFR1 TMD indeed reveals as many as 35 distinct ligand-contact sites, dispersed among the extracellular loops and the extracellular ends of the TM helices [23]. The structures of the family B TMD regions were obtained in the presence of a bound small molecule antagonist ligand, rather than a bound peptide agonist. Current data suggest, however, that the N-terminal portion of PTH adopts at least some helical conformation when it is bound to the TMD region [24]. Determining the binding site in the receptor for Val2 in PTH ligands is a particularly crucial goal, given the importance of this residue for inducing receptor activation. Cross-linking and mutational studies have identified several receptor residues likely to be involved [12, 25, 26]. These residues include Ser370, Met425, and Gln440 located at the extracellular ends of TMD helices 5, 6, and 7, respectively. How these residues contribute to the receptor activation processes remains to be elucidated.

9.5 The PTHR1 Cytoplasmic Surface: Interaction with Effectors and Signal Regulators

In the structural models, the intracellular ends of the TM helices of the family B receptors are seen to spatially align closely with those of the family A GPCRs. This finding likely reflects an overlap in the repertoire of effector and regulatory proteins with which the different GPCRs interact on their cytoplasmic surfaces. A key component or the activation mechanism is likely to be an outward movement of some of the TM helices at their intracellular ends, such that access to cytoplasmic G proteins and signal regulating proteins is increased. Evidence for dynamic movement of the intracellular ends of TM3 and TM6 upon agonist-induced PTHR1 activation is provided by prior zinc-chelation-based mutational studies [27]. The highly conserved His223 in TM2, Thr410 in TM6, and Arg458 in TM7 were identified as the sites of PTHR1 mutations in patients with Jansen's chondrodysplasia, and the identified mutations result in high levels of ligand-independent (constitutive) PTHR1 signaling, which fully explains the disease phenotype [28]. These three residues are each located at the intracellular base of a TM helix and can be seen in the crystal structures to participate in a network of interhelical interactions that likely plays a key role in the receptor activation process [17].

Several residues on the cytoplasmic surface of the PTHR1 have been identified by mutational methods as determinants of G protein coupling. Thus, Lys388 in ICL3 is a determinant of Gαs and Gαq interaction [29], whereas Lys319 in ICL2 is a selective determinant of Gαq interaction, as its mutation to Glu reduces coupling to the PLC/IP3 pathway but not the cAMP/PKA pathway [1]. This mutation along with three neighboring mutations, comprise the "DSEL" clustered mutation, which has been engineered into the PTHR1 gene in mice by "knock-in" approaches for the purpose of assessing the relative roles of PTH-mediated cAMP/PKA versus PLC/PKC signaling in vivo [1].

As for most GPCRs, termination of signaling at the PTHR1 involves phosphorylation of the receptor's C-terminal tail by G protein receptor kinases, followed by β -arrestin recruitment and receptor internalization. A cluster of seven serine residues in the mid-region of the cytoplasmic tail are the sites of ligand-induced receptor phosphorylation and thus serve to regulate the interaction of activated PTHR1 with arrestin proteins and thus mediate receptor desensitization and subsequent receptor internalization [30–32]. Arrestins also mediate interaction of GPCRs with clathrin components of internalization vesicles and can also mediate signaling through the ERK-1/2 MAP-kinase pathway, as shown for the PTHR1 [3, 33]. The PTHR cytoplasmic tail also mediates interaction with the sodium-hydrogen-regulating factor (NHERF) family of proteins, which occurs via a PDZ domain-based interaction involving the last five residues of the receptor's C-terminal tail [34, 35]. Interaction with NHERF proteins regulate the docking of the receptor to the actin cytoskeleton via the EZRIN adaptor protein and thereby modulate intracellular trafficking and recycling of the ligand-activated PTHR1 [36].

9.6 Novel Mechanisms of Prolonged Signaling at the PTHR1

Recent studies suggest that the PTHR1 can form surprisingly stable complexes with certain PTH ligand analogs and can thus mediate persistent cAMP signaling responses [20, 37]. The mechanism appears to involve binding of the ligand to a unique high-affinity PTHR1 conformation, which can couple persistently to Gαs, even when following internalization of the complex to endosomal vesicles [38]. Moreover, the PTH analogs that bind efficiently to the R^0 conformation not only mediate markedly prolonged cAMP responses in cells but also induce markedly prolonged hypercalcemic and hypophosphatemic responses when injected into animals [32, 39]. One particularly long-acting analog, called LA-PTH and consisting of a unique M-PTH(1–14)/PTHrP(15–36) hybrid structure, can induce elevations of serum calcium in mice that persist for nearly 24 h following a single subcutaneous injection, which contrasts markedly with PTH(1–34) which raises calcium for only 2–4 h [32] (Fig. 9.3). The prolonged responses of these analogs in vivo cannot be explained by a prolonged half-life in the circulation, but rather by their stable binding to the PTHR1 in bone and kidney target cells. This class of R^0-selective PTH analogs is thus of interest as a potential new mode of therapy for patients with hypoparathyroidism [40] (see also Chaps. 30 and 31).

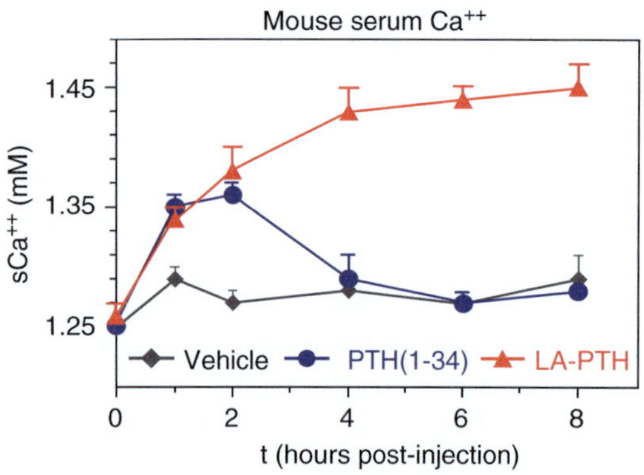

Fig. 9.3 A long-acting PTH analog: Shown is the capacity of a long-acting PTH analog, called LA-PTH, to stimulate prolonged increases in blood-ionized calcium in mice following a single injection. Mice were injected (S.C.) with either vehicle, PTH(1–34) at 50 nmol/kg or LA-PTH at 10 nmol/kg), and blood Ca^{++} was measured at times indicated (Adapted from Maeda et al. [32])

References

1. Guo J, Liu M, Yang D et al (2010) Phospholipase C signaling via the parathyroid hormone (PTH)/PTH-related peptide receptor is essential for normal bone responses to PTH. Endocrinology 151:3502–13
2. Singh AT, Gilchrist A, Voyno-Yasenetskaya T et al (2005) G{alpha}12/G{alpha}13 subunits of heterotrimeric G proteins mediate parathyroid hormone activation of phospholipase D in UMR-106 osteoblastic cells. Endocrinology 146:2171–5
3. Luttrell LM (2014) Minireview: more than just a hammer: ligand 'bias' and pharmaceutical discovery. Mol Endocrinol 28(3):281–94. me.20131314
4. Goldman ME, McKee RL, Caulfield MP et al (1988) A new highly potent parathyroid hormone antagonist: [d-Trp12,Tyr34]bPTH(7-34)NH2. Endocrinology 123:2597–9
5. Nussbaum SR, Rosenblatt M, Potts JT Jr (1980) Parathyroid hormone/renal receptor interactions: demonstration of two receptor-binding domains. J Biol Chem 255:10183–7
6. Shimizu N, Dean T, Khatri A et al (2004) Amino-terminal parathyroid hormone fragment analogs containing alpha, alpha-dialkyl amino acids at positions 1 and 3. J Bone Mineral Res 19:2078–86
7. Suva LJ, Winslow GA, Wettenhall RE et al (1987) A parathyroid hormone-related protein implicated in malignant hypercalcemia: cloning and expression. Science 237:893–6
8. Pioszak AA, Parker NR, Gardella TJ et al (2009) Structural basis for parathyroid hormone-related protein binding to the parathyroid hormone receptor and design of conformation-selective peptides. J Biol Chem 284:28382–91
9. Jüppner H, Abou-Samra A-B, Freeman M et al (1991) A G protein-linked receptor for parathyroid hormone and parathyroid hormone-related peptide. Science 254:1024–6
10. Barwell J, Gingell JJ, Watkins HA et al (2012) Calcitonin and calcitonin receptor-like receptors: common themes with family B GPCRs? Br J Pharmacol 166:51–65
11. Hoare S, Usdin T (2001) Molecular mechanisms of ligand-recognition by parathyroid hormone 1 (PTH1) and PTH2 receptors. Curr Pharm Des 7:689–713
12. Gensure RC, Gardella TJ, Juppner H (2005) Parathyroid hormone and parathyroid hormone-related peptide, and their receptors. Biochem Biophys Res Commun 328:666–78
13. Wittelsberger A, Corich M, Thomas BE et al (2006) The mid-region of parathyroid hormone (1–34) serves as a functional docking domain in receptor activation. Biochemistry 45:2027–34
14. Dong M, Koole C, Wootten D et al (2014) Structural and functional insights into the juxtamembranous amino-terminal tail and extracellular loop regions of class B GPCRs. Br J Pharmacol 171:1085–101
15. Pioszak AA, Parker NR, Suino-Powell K et al (2008) Molecular recognition of corticotropin-releasing factor by its G protein-coupled receptor CRFR1. J Biol Chem 283:32900–12
16. Pioszak AA, Xu HE (2008) Molecular recognition of parathyroid hormone by its G protein-coupled receptor. Proc Natl Acad Sci U S A 105:5034–9
17. Hollenstein K, de Graaf C, Bortolato A et al (2014) Insights into the structure of class B GPCRs. Trends Pharmacol Sci 35:1
18. Dean T, Khatri A, Potetinova Z et al (2006) Role of amino acid side chains in the (17–31) domain of parathyroid hormone in binding to the PTH receptor. J Biol Chem 281:32485–95
19. Dean T, Vilardaga JP, Potts JT Jr et al (2008) Altered selectivity of parathyroid hormone (PTH) and PTH-related protein (PTHrP) for distinct conformations of the PTH/PTHrP receptor. Mol Endocrinol 22:156–66
20. Ferrandon S, Feinstein TN, Castro M et al (2009) Sustained cyclic amp production by parathyroid hormone receptor endocytosis. Nat Chem Biol 5:734–42
21. Horwitz MJ, Tedesco MB, Sereika SM et al (2003) Direct comparison of sustained infusion of human parathyroid hormone-related protein-(1–36) [hPTHrP-(1–36)] versus hPTH-(1–34) on serum calcium, plasma 1,25-dihydroxyvitamin D concentrations, and fractional calcium excretion in healthy human volunteers. J Clin Endocrinol Metab 88:1603–9
22. Granier S, Kobilka B (2012) A new era of GPCR structural and chemical biology. Nat Chem Biol 8:670–3
23. Coin I, Katritch V, Sun T et al (2013) Genetically encoded chemical probes in cells reveal the binding path of urocortin-1 to CRF class B GPCR. Cell 155:1258–69
24. Tsomaia N, Pellegrini M, Hyde K et al (2004) Toward parathyroid hormone minimization: conformational studies of cyclic PTH(1–14) analogues. Biochemistry 43:690–9

25. Behar V, Bisello A, Bitan G et al (2000) Photoaffinity cross-linking identifies differences in the interactions of an agonist and an antagonist with the parathyroid hormone/parathyroid hormone-related protein receptor. J Biol Chem 275:9–17
26. Gensure R, Carter P, Petroni B et al (2001) Identification of determinants of inverse agonism in a constitutively active parathyroid hormone/parathyroid hormone related peptide receptor by photoaffinity cross linking and mutational analysis. J Biol Chem 276:42692–9
27. Vilardaga J, Frank M, Krasel C et al (2001) Differential conformational requirements for activation of G proteins and regulatory proteins, arrestin and GRK in the G protein coupled receptor for parathyroid hormone (PTH)/PTH-related protein. J Biol Chem 31:31
28. Calvi L, Schipani E (2000) The PTH/PTHrP receptor in jansen's metaphyseal chondrodysplasia. J Endocrinol Invest 23:545–54
29. Huang Z, Chen Y, Pratt S et al (1996) The N-terminal region of the third intracellular loop of the parathyroid hormone (PTH)/PTH-related peptide receptor is critical for coupling to camp and inositol phosphate/Ca++ signal transduction pathways. J Biol Chem 271:33382–9
30. Malecz N, Bambino T, Bencsik M et al (1998) Identification of phosphorylation sites in the G protein-coupled receptor for parathyroid hormone. Receptor phosphorylation is not required for agonist-induced internalization. Mol Endocrinol 12:1846–56
31. Rey A, Manen D, Rizzoli R et al (2006) Proline-rich motifs in the parathyroid hormone (PTH)/PTH-related protein receptor C-terminus mediate scaffolding of c-src with beta-arrestin2 for erk1/2 activation. J Biol Chem 281:38181–8
32. Maeda A, Okazaki M, Baron DM et al (2013) Critical role of parathyroid hormone (PTH) receptor-1 phosphorylation in regulating acute responses to PTH. Proc Natl Acad Sci U S A 110:5864–9
33. Sneddon WB, Friedman PA (2007) Beta-arrestin-dependent parathyroid hormone-stimulated extracellular signal-regulated kinase activation and parathyroid hormone type 1 receptor internalization. Endocrinology 148:4073–9
34. Mahon M, Donowitz M, Yun C et al (2002) Na(+)/H(+) exchanger regulatory factor 2 directs parathyroid hormone 1 receptor signalling. Nature 417:858–61
35. Mamonova T, Kurnikova M, Friedman PA (2012) Structural basis for NHERF1 pdz domain binding. Biochemistry 51:3110–20
36. Ardura JA, Wang B, Watkins SC et al (2011) Dynamic Na-H+ exchanger regulatory factor-1 association and dissociation regulate parathyroid hormone receptor trafficking at membrane microdomains. J Biol Chem 286:35020–9
37. Dean T, Linglart A, Mahon MJ et al (2006) Mechanisms of ligand binding to the PTH/PTHrP receptor: selectivity of a modified pth(1–15) radioligand for g{alpha}s-coupled receptor conformations. Mol Endocrinol 20:931–42
38. Vilardaga JP, Gardella TJ, Wehbi VL et al (2012) Non-canonical signaling of the PTH receptor. Trends Pharmacol Sci 33:423–31
39. Okazaki M, Ferrandon S, Vilardaga JP et al (2008) Prolonged signaling at the parathyroid hormone receptor by peptide ligands targeted to a specific receptor conformation. Proc Natl Acad Sci U S A 105:16525–30
40. Shimizu M, Ichikawa F, Noda H et al (2008) A new long-acting PTH/PTHrP hybrid analog that binds to a distinct PTHR conformation has superior efficacy in a rat model of hypoparathyroidism. J Bone Mineral Res 23(S1):S128 (Abstract)

G Protein Gsα and *GNAS* Imprinting

10

Murat Bastepe

10.1 The α-Subunit of the Stimulatory Heterotrimeric G Protein

Heterotrimeric guanine nucleotide binding proteins transduce signals from a wide range of endogenous molecules into different cellular actions and, thus, mediate a plethora of biological processes. Expressed in nearly all tissues and cells, the α-subunit of the stimulatory G protein (Gsα) is essential for the cellular actions of many hormones, neurotransmitters, and autocrine/paracrine factors. Accordingly, complete loss of Gsα, as shown in mouse models, results in embryonic lethality [1–3]. Furthermore, inactivating or activating mutations within the gene encoding Gsα (*GNAS*) are responsible for various human diseases, including Albright's hereditary osteodystrophy, progressive osseous heteroplasia, pseudohypoparathyroidism (PHP) type Ia, PHP-Ic, pseudopseudohypoparathyroidism, different endocrine and non-endocrine tumors, and McCune-Albright syndrome (see Chaps. 32, 33, 34, and 35 for additional details) [4–11]. In addition, imprinting abnormalities of *GNAS* are found in patients with PHP-Ib [12–14] (see also Chaps. 32, 33, 34, and 35).

M. Bastepe
Endocrine Unit, Department of Medicine, Massachusetts General Hospital, Harvard Medical School, 50 Blossom St. Thier 10, Boston, MA 02114, USA
e-mail: bastepe@helix.mgh.harvard.edu

Similar to the α-subunits of other heterotrimeric G proteins, agonist activation of a specific cell-surface receptor results in a GDP-GTP exchange on Gsα, causing its dissociation from G$\beta\gamma$ subunits [15] (Fig. 10.1). GTP-bound, free Gsα can directly activate several different effectors. These effectors include Src tyrosine kinase [16] and certain Ca channels [17, 18]; however, the most extensively investigated effector molecule stimulated by Gsα is adenylyl cyclase. This enzyme catalyzes the synthesis of the ubiquitous second messenger cyclic AMP (cAMP), which activates intracellular targets including protein kinase A (the cAMP-dependent protein kinase) and the exchange proteins directly activated by cAMP [19, 20] (Fig. 10.1).

Different molecular mechanisms tightly regulate the cAMP signaling pathway, and some of those mechanisms function at the level of Gsα. An important regulatory mechanism is the intrinsic GTP hydrolase (GTPase) activity of Gsα, which converts the active GTP-bound Gsα to its inactive GDP-bound state (Fig. 10.1). GDP-bound Gsα has a much higher affinity for G$\beta\gamma$, resulting in the re-formation of the heterotrimer and, thereby, preventing further effector stimulation [15]. Some amino acid residues, such as Arg^{201} and Gln^{227}, are particularly critical for the GTPase activity, and modifications of these residues, such as ADP-ribosylation of Arg^{201} that can be induced by cholera toxin or mutations at either residue, lead to inhibition of the GTPase activity and, therefore, constitutive Gsα signaling [8, 21–23]. Gsα-mediated signaling is also regulated by

M.L. Brandi, E.M. Brown (eds.), *Hypoparathyroidism*,
DOI 10.1007/978-88-470-5376-2_10, © Springer-Verlag Italia 2015

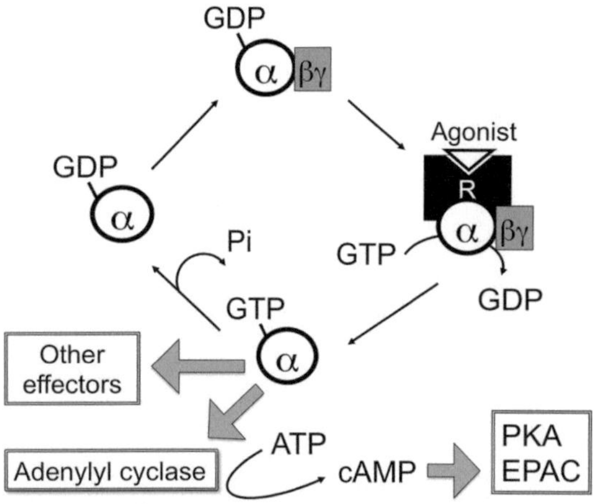

Fig. 10.1 Activation-inactivation cycle of the heterotrimeric stimulatory G protein. Upon binding of an agonist to its Gsα-coupled receptor (*R*), the GDP molecule bound to the α-subunit is replaced with a GTP molecule. The GTP-bound form of the α-subunit dissociates from βγ subunits and, thereby, stimulates its downstream effectors including adenylyl cyclase. The intrinsic GTP hydrolase activity of the α-subunit converts the GTP into GDP, resulting in the reassembly of the heterotrimer and, thereby, termination of effector stimulation. Adenylyl cyclase catalyzes the conversion of ATP into cAMP, which stimulates protein kinase A (*PKA*) and the exchange proteins directly activated by cAMP (*EPAC*)

activation-induced subcellular redistribution of Gsα to the cytosol from the plasma membrane [24]. Gsα is palmitoylated at its N-terminus [25], and the plasma membrane avidity of activated Gsα protein is reduced due to its depalmitoylation and dissociation from Gβγ subunits [26–29]. Another mechanism regulating Gsα-induced signaling entails changes in protein turnover. Upon activation, the rate of Gsα degradation increases through a mechanism that appears to be independent of the plasma membrane localization of this protein [30].

The Gsα transcript has a long and a short variant (Gsα-L and Gsα-S) formed as a result of alternative splicing of exon 3 [31–33]. In addition, each of these Gsα variants either includes or excludes a CAG trinucleotide (encoding serine) at the start of exon 4. The functional significance of this additional serine residue is unknown. On the other hand, several lines of biochemical evidence suggest that Gsα-L may be slightly more efficient than Gsα-S in transducing receptor signals, although it remains to be determined whether this difference is biologically significant [34–40]. Another Gsα variant, termed Gsα-N1, has been identified in brain [41]. Owing to the use of a novel exon in intron 3 that comprises an in-frame termination codon, Gsα-N1 is truncated in the C-terminus and lacks the portion encoded by exons 4–13.

10.2 The *GNAS* Complex Locus

The human *GNAS* locus maps to the telomeric end of the long arm of chromosome 20 (20q13.2-20q13.3) [42–44], while its mouse ortholog is located in the distal region of chromosome 2 [45, 46]. Human Gsα is encoded by *GNAS* exons 1–13 [32]. *GNAS* in humans and mice appear structurally and functionally similar to one another, although the mouse Gsα protein is encoded by 12 rather than 13 exons [47]. In addition to the exons encoding Gsα, *GNAS* includes several additional exons and promoters located upstream of exon 1, thus leading to multiple additional transcripts (Fig. 10.2). These include the neuroendocrine secretory protein 55 (NESP55), the extra-large variant of Gsα (XLαs), the A/B transcript (also known 1A or 1′), and the *GNAS* antisense transcript (GNAS-AS1).

In humans, the NESP55 protein is encoded by a single exon, while in mice, the open reading frame consists of two separate exons; however, in both species, these exons splice onto Gsα exons 2–13, which comprise the 3′-untranslated region (Fig. 10.2) [48, 49]. The promoter and the exon(s) encoding NESP55 are located in a differentially methylated region (DMR), and the expression takes place from the unmethylated maternal allele [48, 49]. NESP55 is a chromogranin-like protein expressed in neuroendocrine tissues, peripheral and central nervous system, and some endocrine tissues [50–53]. The knockout of NESP55 protein in mice leads to a mild phenotype characterized by increased reactivity to novel environments [54]; however, no discernible phenotype has yet been attributed to the loss of NESP55 protein in humans, as the clinical findings of patients with PHP-Ib who show gain of NESP55 DMR methylation appear indistinguishable from the clinical

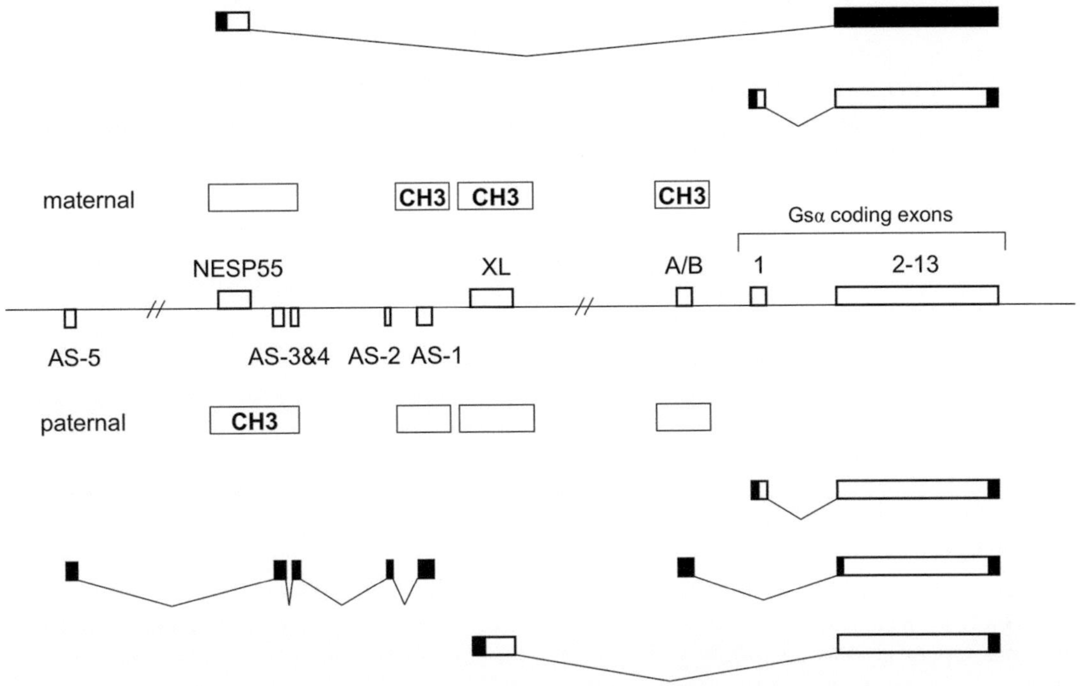

Fig. 10.2 The *GNAS* complex locus yields multiple imprinted sense and antisense transcripts. Exons 1–13 encode Gsα, which is biallelically expressed in most tissues; however, paternal Gsα allele is silenced in a small number of tissues, including the renal proximal tubule. From differentially methylated promoters arise several other transcripts, including the maternally expressed NESP55 and the paternally expressed XLαs. Both of these transcripts use individual first exons that splice onto exons 2–13. In addition, the paternal *GNAS* allele gives rise to a transcript termed A/B (also referred to as 1A or 1′). A noncoding antisense transcript is also derived from the paternal *GNAS* allele (GNAS-AS1). Open rectangles and rectangles filled with CH3 show non-methylated and methylated DMRs, respectively. The distance between GNAS-AS1 exon 5 and exon NESP55 is ~19 kb and the distance between exon A/B and exon XL ~35 kb. Boxes and connecting lines depict exons and introns, respectively; splicing patterns are indicated by broken lines; filled boxes, untranslated sequences; only the major splicing patterns are depicted

findings of those who have normal NESP55 DMR methylation [55]. On the other hand, NESP55 transcription is critical for *GNAS* imprinting (see below).

XLαs is expressed exclusively from the paternal allele, consistent with methylation of its maternal promoter [49, 56, 57] (Fig. 10.2); however, variable biallelic expression of XLαs has been demonstrated in clonal bone stromal cells [58]. XLαs mRNA is expressed abundantly in the brain, cerebellum, and neuroendocrine tissues, but its expression can be detected at many other tissues [56, 59–63]. The N-terminal portion of XLαs, which is encoded by exon XL, consists of multiple repetitive amino acid motifs. Exon XL splices onto *GNAS* exons 2–13, which, unlike in NESP55, belong to the open reading frame. Thus, XLαs protein is partially identical to Gsα protein and comprises most of the functional domains of the latter [49, 56]. Accordingly, XLαs can mimic Gsα in vitro and in vivo [64–69]. Multiple XLαs variants and alternative translation products of XLαs mRNA have been described [47, 56, 60, 70–72], and disruption of exon XL leads to a severe phenotype in mice, including poor adaptation to feeding and defective glucose and energy metabolism [61, 73]. Based on additional XLαs knockout mouse models, some of those phenotypes reflect XLαs deficiency, while others are caused by the loss of other products that utilize exon XL [74–77]. Some studies in mice also suggest that XLαs plays a role in bone and mineral metabolism [76–78], and alterations of XLαs activity/level are implicated in several human disorders, such as the intrauterine growth retardation observed in pseudopseudohypoparathyroidism [79]. Functional properties and in vivo roles of XLαs and its variants are reviewed elsewhere [80, 81].

Located ~2.5 kb upstream of the Gsα promoter is the maternally methylated, paternally active promoter of the A/B transcript, which is broadly expressed [82–84] (Fig. 10.2). The first exon of the A/B transcript also splices onto Gsα exons 2–13 [83]. Exon A/B does not contain an in-frame translation initiation codon, but translation can be initiated by an in-frame AUG located in exon 2, leading to an N-terminally truncated Gsα variant that localizes to the plasma membrane [84]. Thus, the A/B protein can interact with adenylyl cyclase and may, therefore, exert a dominant negative effect on Gsα. Indeed, a recent study has provided evidence supporting this possibility [85]. On the other hand, the A/B transcript exerts important actions at the transcriptional level (see below).

A promoter located immediately upstream of the XLαs promoter drives the expression of GNAS-AS1 transcript, which extends past the exon(s) encoding NESP55 [86, 87] (Fig. 10.2). The structural features of this transcript indicate that it is noncoding. The promoter of the GNAS-AS1 resides in a DMR and is active exclusively on the paternal allele [57, 86–88]; however, the antisense transcript shows biallelic expression in the adrenal and testes [57]. Similar to many noncoding RNAs in the genome, GNAS-AS1 has regulatory actions on *GNAS* expression (see below).

In contrast to the promoters of these recently described *GNAS* products, which are located within DMRs and show activity on the non-methylated allele, the promoter of Gsα is not methylated and, in most tissues, Gsα expression is biallelic [48, 56, 89]. However, the paternal Gsα promoter is repressed in a small number of tissues, including renal proximal tubule, thyroid, pituitary, gonads, and certain parts of the brain [1, 90–95]. The loss of paternal Gsα silencing in mice causes a phenotype consistent with increased PTH sensitivity at the renal proximal tubule, demonstrating the significance of this epigenetic event at least in this tissue [96]. The tissue-specific monoallelic (maternal) expression of Gsα is also critical in determining the phenotype in certain diseases caused by *GNAS* mutations, such as PHP and growth hormone-secreting pituitary adenomas [93, 97, 98].

10.3 Control of Imprinting at the *GNAS* Complex Locus

Many familial PHP-Ib cases have isolated loss of methylation at the A/B DMR [12, 14]. These patients either have deletions within the neighboring *STX16* locus – overlapping at exon 4 or in one case a ~19-kb deletion removing exon NESP55 and a large upstream genomic region [99–101]. Thus, the deleted regions are predicted to comprise critical cis-acting elements necessary for the establishment or maintenance of A/B methylation (Fig. 10.3). Note that the loss of one *STX16* allele is not predicted to contribute to the phenotype, as this gene is biallelically expressed and the ablation *Stx16* exons 4–6 in mice does not alter *Gnas* imprinting or cause PTH resistance [100, 102]. A few familial PHP-Ib cases show broad epigenetic alterations at the *GNAS* locus, including loss of A/B methylation (for additional details on PHP1b, see Chaps. 32, 33, and 34) [14, 103, 104]. These cases have deletions affecting the NESP55 DMR, with the shortest region of overlap including exons 3 and 4 of the GNAS-AS1 transcript [103, 104]. Thus, this genomic region is also likely to include a cis-acting element controlling all maternal *GNAS* imprints. In addition, important novel insights have been gained from generation and the analysis of mice in which the Nesp55 transcript is prematurely terminated through insertion of a polyadenylation cassette immediately downstream of exons encoding this protein [105]. Maternal inheritance of this defect leads to loss of A/B methylation and, in some animals, additional loss of methylation at the DMR comprising the promoter of GNAS-AS1 (Nespas in mice) and exon XL [105]. Thus, NESP55 transcription is essential for the establishment of maternal *GNAS* imprints, and it is thus possible

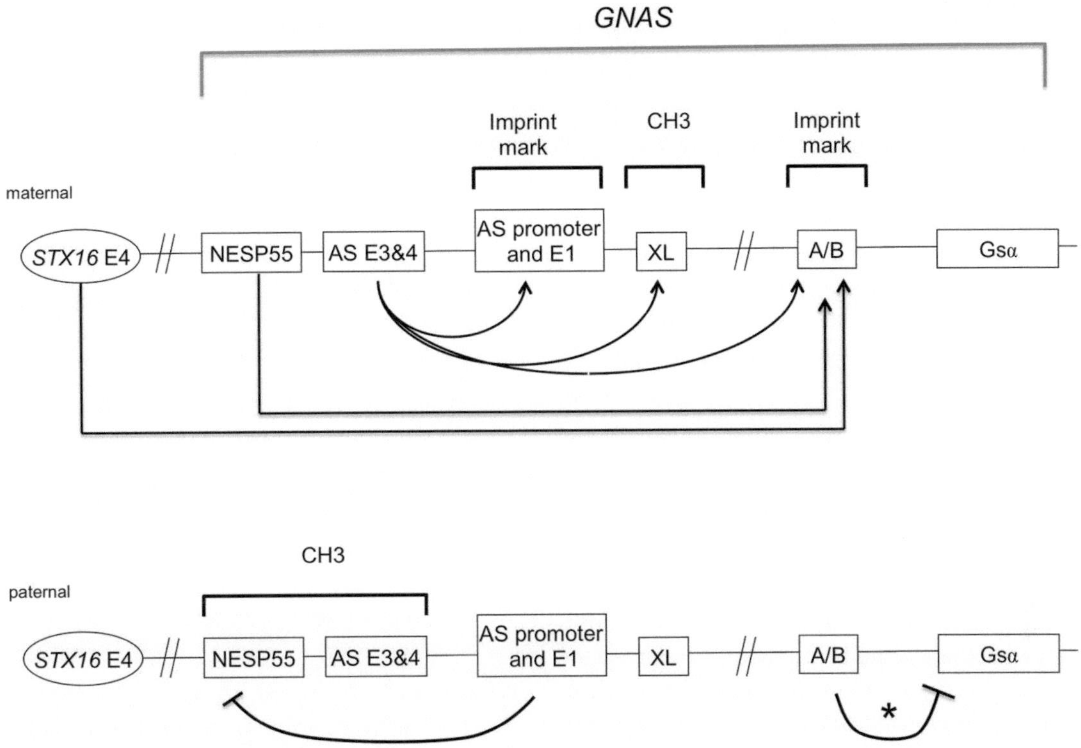

Fig. 10.3 Regulation of gene expression from the *GNAS* locus. The maternal *GNAS* allele comprises two female germ-line imprint marks, one at the GNAS-AS1 promoter and the other at exon A/B. Deletions identified in PHP-Ib cases point to *STX16* exon 4, the NESP55 promoter and exon, and the region comprising exons 3 and 4 of GNAS-AS1 as important cis-acting elements regulating *GNAS* imprinting. *Arrows* indicate the regulatory effects revealed by these deletions. Findings from different mouse models indicate that the GNAS-AS1 transcript silences Nesp55 expression in cis and that the paternal exon A/B region and/or the A/B transcript silences Gsα expression in cis; the latter occurs in a tissue-specific manner (*)

that the maternal deletions affecting the NESP55 DMR in PHP-Ib patients result in loss of maternal *GNAS* methylation because of disrupting NESP55 transcription.

The study of the mouse *Gnas* locus has revealed two female germ-line imprint marks, one at the exon A/B (1A in mouse) DMR [83] and the other at the promoter of GNAS-AS1 (Nespas in mice) [106]. Ablation of the homologous non-methylated chromosomal region alters expression of different imprinted *Gnas* transcripts. Ablation of the entire paternal A/B DMR derepresses the Gsα transcript, in cis, in those tissues where paternal Gsα expression is normally silenced [96, 107]. Deletion of the paternal GNAS-AS1 DMR leads to derepression of the paternal Nesp55 transcription [108, 109]. In addition, loss of all maternal *Gnas* methylation in mice, including A/B and GNAS-AS1 (as in some familial PHP-Ib cases), which results from ablation of the Nesp55 DMR, is associated with diminished Gsα expression in kidney [110]. Thus, the A/B and the GNAS-AS1 germ-line imprints are necessary for maternal expression of Gsα in a tissue-specific manner and of NESP55, respectively.

10.4 Regulation of Allelic Gsα Silencing

Data from mouse models indicate that the paternal Gsα silencing is developmentally regulated in some tissues. For example, the paternal Gsα allele is silenced in brown adipose tissue at birth, whereas the silencing mechanisms in the renal proximal tubule operate only after the early postnatal period [90, 107]. The loss of A/B methylation observed in patients with PHP-Ib [12–14] and the findings in mice with the ablation of paternal exon A/B [96, 107] clearly indicate that the A/B DMR and/or the A/B transcript plays an important role in the paternal silencing of Gsα. However, the mechanisms governing this important epigenetic event remain poorly defined. One possible mechanism involves binding of a trans-acting factor to the non-methylated exon A/B region. This could be a tissue-specific repressor that acts directly on the Gsα promoter. Alternatively, a tissue-specific or universal insulator could bind to this region and block the action of an upstream universal or a tissue-specific enhancer, respectively. Another possibility is that there could be competition between the promoters of A/B and Gsα transcripts for common regulatory elements (promoter competition) or transcriptional interference on the paternal Gsα promoter due to the upstream A/B transcription. Data from mice in which the entire paternal exon A/B DMR is ablated do not argue for or against these possibilities, because both the A/B promoter and the A/B transcript are deleted [96, 107]. However, consistent with mechanisms involving promoter competition or transcriptional interference, a recent study measured higher A/B levels in tissues in which the paternal Gsα allele is silenced than in those in which Gsα expression is biallelic [77]. In fact, paternal Gsα expression is derepressed in mice in which the A/B transcript is prematurely terminated through insertion of a polyadenylation cassette downstream of exon A/B, thus supporting the mechanism involving transcriptional interference [77]. The control mice, however, in which the polyadenylation cassette is inserted in the reverse orientation, also show derepressed Gsα expression, thus making it difficult to entirely rule out any mechanisms involving binding of a trans-acting factor in this region [77]. Future studies are necessary to elucidate the mechanisms governing the tissue-specific paternal silencing of Gsα.

References

1. Yu S, Yu D, Lee E et al (1998) Variable and tissue-specific hormone resistance in heterotrimeric G_s protein a-subunit (G_sa) knockout mice is due to tissue-specific imprinting of the G_sa gene. Proc Natl Acad Sci U S A 95:8715–8720
2. Germain-Lee EL, Schwindinger W, Crane JL et al (2005) A mouse model of Albright hereditary osteodystrophy generated by targeted disruption of exon 1 of the Gnas gene. Endocrinology 146: 4697–4709
3. Chen M, Gavrilova O, Liu J et al (2005) Alternative Gnas gene products have opposite effects on glucose and lipid metabolism. Proc Natl Acad Sci U S A 102:7386–7391
4. Weinstein LS, Gejman PV, Friedman E et al (1990) Mutations of the Gs alpha-subunit gene in Albright hereditary osteodystrophy detected by denaturing gradient gel electrophoresis. Proc Natl Acad Sci U S A 87:8287–8290
5. Patten JL, Johns DR, Valle D et al (1990) Mutation in the gene encoding the stimulatory G protein of adenylate cyclase in Albright's hereditary osteodystrophy. N Engl J Med 322:1412–1419
6. Eddy MC, De Beur SM, Yandow SM et al (2000) Deficiency of the alpha-subunit of the stimulatory G protein and severe extraskeletal ossification. J Bone Miner Res 15:2074–2083
7. Linglart A, Carel JC, Garabedian M et al (2002) GNAS1 lesions in pseudohypoparathyroidism Ia and Ic: genotype phenotype relationship and evidence of the maternal transmission of the hormonal resistance. J Clin Endocrinol Metab 87:189–197
8. Landis CA, Masters SB, Spada A et al (1989) GTPase inhibiting mutations activate the alpha chain of Gs and stimulate adenylyl cyclase in human pituitary tumours. Nature 340:692–696
9. Lyons J, Landis CA, Harsh G et al (1990) Two G protein oncogenes in human endocrine tumors. Science 249:655–659
10. Weinstein LS, Shenker A, Gejman PV et al (1991) Activating mutations of the stimulatory G protein in the McCune-Albright syndrome. N Engl J Med 325:1688–1695
11. Schwindinger WF, Francomano CA, Levine MA (1992) Identification of a mutation in the gene encoding the alpha subunit of the stimulatory G protein of adenylyl cyclase in McCune-Albright syndrome. Proc Natl Acad Sci U S A 89:5152–5156
12. Liu J, Litman D, Rosenberg M et al (2000) A GNAS1 imprinting defect in pseudohypoparathyroidism type IB. J Clin Invest 106:1167–1174
13. Bastepe M, Lane AH, Jüppner H (2001) Paternal uniparental isodisomy of chromosome 20q (patUPD20q) – and the resulting changes in *GNAS1* methylation – as a plausible cause of pseudohypoparathyroidism. Am J Hum Genet 68:1283–1289

14. Bastepe M, Pincus JE, Sugimoto T et al (2001) Positional dissociation between the genetic mutation responsible for pseudohypoparathyroidism type Ib and the associated methylation defect at exon A/B: evidence for a long-range regulatory element within the imprinted *GNAS1* locus. Hum Mol Genet 10:1231–1241
15. Bourne HR, Sanders DA, McCormick F (1991) The GTPase superfamily: conserved structure and molecular mechanism. Nature 349:117–127
16. Ma YC, Huang J, Ali S et al (2000) Src tyrosine kinase is a novel direct effector of G proteins. Cell 102:635–646
17. Yatani A, Imoto Y, Codina J et al (1988) The stimulatory G protein of adenylyl cyclase, Gs, also stimulates dihydropyridine-sensitive Ca2+ channels. Evidence for direct regulation independent of phosphorylation by cAMP-dependent protein kinase or stimulation by a dihydropyridine agonist. J Biol Chem 263:9887–9895
18. Mattera R, Graziano MP, Yatani A et al (1989) Splice variants of the alpha subunit of the G protein Gs activate both adenylyl cyclase and calcium channels. Science 243:804–807
19. Taylor SS, Buechler JA, Yonemoto W (1990) cAMP-dependent protein kinase: framework for a diverse family of regulatory enzymes. Annu Rev Biochem 59:971–1005
20. Bos JL (2006) Epac proteins: multi-purpose cAMP targets. Trends Biochem Sci 31:680–686
21. Sondek J, Lambright DG, Noel JP et al (1994) GTPase mechanism of Gproteins from the 1.7-A crystal structure of transducin alpha-GDP-AIF-4. Nature 372:276–279
22. Coleman DE, Berghuis AM, Lee E et al (1994) Structures of active conformations of Gi alpha 1 and the mechanism of GTP hydrolysis. Science 265:1405–1412
23. Graziano MP, Gilman AG (1989) Synthesis in Escherichia coli of GTPase-deficient mutants of Gs alpha. J Biol Chem 264:15475–15482
24. Wedegaertner PB, Bourne HR, von Zastrow M (1996) Activation-induced subcellular redistribution of Gs alpha. Mol Biol Cell 7:1225–1233
25. Linder ME, Middleton P, Hepler JR et al (1993) Lipid modifications of G proteins: alpha subunits are palmitoylated. Proc Natl Acad Sci U S A 90:3675–3679
26. Wedegaertner PB, Bourne HR (1994) Activation and depalmitoylation of Gs alpha. Cell 77:1063–1070
27. Huang C, Duncan JA, Gilman AG et al (1999) Persistent membrane association of activated and depalmitoylated G protein alpha subunits. Proc Natl Acad Sci U S A 96:412–417
28. Iiri T, Backlund PS Jr, Jones TL et al (1996) Reciprocal regulation of Gs alpha by palmitate and the beta gamma subunit. Proc Natl Acad Sci U S A 93:14592–14597
29. Evanko DS, Thiyagarajan MM, Siderovski DP et al (2001) Gbeta gamma isoforms selectively rescue plasma membrane localization and palmitoylation of mutant Galphas and Galphaq. J Biol Chem 276:23945–23953
30. Levis MJ, Bourne HR (1992) Activation of the alpha subunit of Gs in intact cells alters its abundance, rate of degradation, and membrane avidity. J Cell Biol 119:1297–1307
31. Bray P, Carter A, Simons C et al (1986) Human cDNA clones for four species of G alpha s signal transduction protein. Proc Natl Acad Sci U S A 83:8893–8897
32. Kozasa T, Itoh H, Tsukamoto T et al (1988) Isolation and characterization of the human Gsα gene. Proc Natl Acad Sci U S A 85:2081–2085
33. Robishaw JD, Smigel MD, Gilman AG (1986) Molecular basis for two forms of the G protein that stimulates adenylate cyclase. J Biol Chem 261:9587–9590
34. Sternweis PC, Northup JK, Smigel MD et al (1981) The regulatory component of adenylate cyclase. Purification and properties. J Biol Chem 256:11517–11526
35. Walseth TF, Zhang HJ, Olson LK et al (1989) Increase in Gs and cyclic AMP generation in HIT cells. Evidence that the 45-kDa alpha-subunit of Gs has greater functional activity than the 52-kDa alpha-subunit. J Biol Chem 264:21106–21111
36. Graziano MP, Freissmuth M, Gilman AG (1989) Expression of Gs alpha in Escherichia coli. Purification and properties of two forms of the protein. J Biol Chem 264:409–418
37. Seifert R, Wenzel-Seifert K, Lee TW et al (1998) Different effects of Gsalpha splice variants on beta2-adrenoreceptor-mediated signaling. The beta2-adrenoreceptor coupled to the long splice variant of Gsalpha has properties of a constitutively active receptor. J Biol Chem 273:5109–5116
38. Kvapil P, Novotny J, Svoboda P et al (1994) The short and long forms of the alpha subunit of the stimulatory guanine-nucleotide-binding protein are unequally redistributed during (-)-isoproterenol-mediated desensitization of intact S49 lymphoma cells. Eur J Biochem 226:193–199
39. el Jamali A, Rachdaoui N, Jacquemin C et al (1996) Long-term effect of forskolin on the activation of adenylyl cyclase in astrocytes. J Neurochem 67:2532–2539
40. Bourgeois C, Duc-Goiran P, Robert B et al (1996) G protein expression in human fetoplacental vascularization. Functional evidence for Gs alpha and Gi alpha subunits. J Mol Cell Cardiol 28: 1009–1021
41. Crawford JA, Mutchler KJ, Sullivan BE et al (1993) Neural expression of a novel alternatively spliced and polyadenylated Gs alpha transcript. J Biol Chem 268:9879–9885
42. Gejman PV, Weinstein LS, Martinez M et al (1991) Genetic mapping of the Gs-a subunit gene (GNAS1) to the distal long arm of chromosome 20 using a

polymorphism detected by denaturing gradient gel electrophoresis. Genomics 9:782–783
43. Rao VV, Schnittger S, Hansmann I (1991) G protein Gs alpha (GNAS 1), the probable candidate gene for Albright hereditary osteodystrophy, is assigned to human chromosome 20q12-q13.2. Genomics 10:257–261
44. Levine MA, Modi WS, O'Brien SJ (1991) Mapping of the gene encoding the alpha subunit of the stimulatory G protein of adenylyl cyclase (GNAS1) to 20q13.2–q13.3 in human by in situ hybridization. Genomics 11:478–479
45. Blatt C, Eversole-Cire P, Cohn VH et al (1988) Chromosomal localization of genes encoding guanine nucleotide-binding protein subunits in mouse and human. Proc Natl Acad Sci U S A 85:7642–7646
46. Peters J, Beechey CV, Ball ST et al (1994) Mapping studies of the distal imprinting region of mouse chromosome 2. Genet Res 63:169–174
47. Abramowitz J, Grenet D, Birnbaumer M et al (2004) XLalphas, the extra-long form of the alpha-subunit of the Gs G protein, is significantly longer than suspected, and so is its companion Alex. Proc Natl Acad Sci U S A 101:8366–8371
48. Hayward BE, Moran V, Strain L et al (1998) Bidirectional imprinting of a single gene: GNAS1 encodes maternally, paternally, and biallelically derived proteins. Proc Natl Acad Sci U S A 95:15475–15480
49. Peters J, Wroe SF, Wells CA et al (1999) A cluster of oppositely imprinted transcripts at the Gnas locus in the distal imprinting region of mouse chromosome 2. Proc Natl Acad Sci U S A 96:3830–3835
50. Ischia R, Lovisetti-Scamihorn P, Hogue-Angeletti R et al (1997) Molecular cloning and characterization of NESP55, a novel chromogranin-like precursor of a peptide with 5-HT1B receptor antagonist activity. J Biol Chem 272:11657–11662
51. Lovisetti-Scamihorn P, Fischer-Colbrie R, Leitner B et al (1999) Relative amounts and molecular forms of NESP55 in various bovine tissues. Brain Res 829:99–106
52. Weiss U, Ischia R, Eder S et al (2000) Neuroendocrine secretory protein 55 (NESP55): alternative splicing onto transcripts of the GNAS gene and posttranslational processing of a maternally expressed protein. Neuroendocrinology 71:177–186
53. Bauer R, Weiss C, Marksteiner J et al (1999) The new chromogranin-like protein NESP55 is preferentially localized in adrenaline-synthesizing cells of the bovine and rat adrenal medulla. Neurosci Lett 263:13–16
54. Plagge A, Isles AR, Gordon E et al (2005) Imprinted Nesp55 influences behavioral reactivity to novel environments. Mol Cell Biol 25:3019–3026
55. Linglart A, Bastepe M, Jüppner H (2007) Similar clinical and laboratory findings in patients with symptomatic autosomal dominant and sporadic pseudohypoparathyroidism type Ib despite different epigenetic changes at the GNAS locus. Clin Endocrinol (Oxf) 67:822–831
56. Hayward B, Kamiya M, Strain L et al (1998) The human GNAS1 gene is imprinted and encodes distinct paternally and biallelically expressed G proteins. Proc Natl Acad Sci U S A 95:10038–10043
57. Li T, Vu TH, Zeng ZL et al (2000) Tissue-specific expression of antisense and sense transcripts at the imprinted Gnas locus. Genomics 69:295–304
58. Michienzi S, Cherman N, Holmbeck K et al (2007) GNAS transcripts in skeletal progenitors: evidence for random asymmetric allelic expression of Gs{alpha}. Hum Mol Genet 16:1921–1930
59. Kehlenbach RH, Matthey J, Huttner WB (1994) XLas is a new type of G protein (Erratum in Nature 1995 375:253). Nature 372:804–809
60. Pasolli H, Klemke M, Kehlenbach R et al (2000) Characterization of the extra-large G protein alpha-subunit XLalphas. I. Tissue distribution and subcellular localization. J Biol Chem 275:33622–33632
61. Plagge A, Gordon E, Dean W et al (2004) The imprinted signaling protein XLalphas is required for postnatal adaptation to feeding. Nat Genet 36:818–826
62. Pasolli H, Huttner W (2001) Expression of the extra-large G protein alpha-subunit XLalphas in neuroepithelial cells and young neurons during development of the rat nervous system. Neurosci Lett 301:119–122
63. Krechowec SO, Burton KL, Newlaczyl AU et al (2012) Postnatal changes in the expression pattern of the imprinted signalling protein XLalphas underlie the changing phenotype of deficient mice. PLoS One 7:e29753
64. Klemke M, Pasolli H, Kehlenbach R et al (2000) Characterization of the extra-large G protein alpha-subunit XLalphas. II. Signal transduction properties. J Biol Chem 275:33633–33640
65. Bastepe M, Gunes Y, Perez-Villamil B et al (2002) Receptor-mediated adenylyl cyclase activation through XLalphas, the extra-large variant of the stimulatory G protein alpha-subunit. Mol Endocrinol 16:1912–1919
66. Linglart A, Mahon MJ, Kerachian MA et al (2006) Coding GNAS mutations leading to hormone resistance impair in vitro agonist- and cholera toxin-induced adenosine cyclic 3′,5′-monophosphate formation mediated by human XLas. Endocrinology 147:2253–2262
67. Liu Z, Segawa H, Aydin C et al (2011) Transgenic overexpression of the extra-large Gs{alpha} variant XL{alpha}s enhances Gs{alpha}-mediated responses in the mouse renal proximal tubule in vivo. Endocrinology 152:1222–1233
68. Mariot V, Wu JY, Aydin C et al (2011) Potent constitutive cyclic AMP-generating activity of XLalphas implicates this imprinted GNAS product in the pathogenesis of McCune-Albright syndrome and fibrous dysplasia of bone. Bone 48:312–320
69. Liu Z, Turan S, Wehbi VL et al (2011) Extra-long Galphas variant XLalphas protein escapes

activation-induced subcellular redistribution and is able to provide sustained signaling. J Biol Chem 286:38558–38569
70. Aydin C, Aytan N, Mahon MJ et al (2009) Extralarge XLαs (XXLαs), a variant of stimulatory G protein alpha-subunit (Gsα), is a distinct, membrane-anchored *GNAS* product that can mimic Gsα. Endocrinology 150:3567–3575
71. Klemke M, Kehlenbach RH, Huttner WB (2001) Two overlapping reading frames in a single exon encode interacting proteins–a novel way of gene usage. EMBO J 20:3849–3860
72. Freson K, Jaeken J, Van Helvoirt M et al (2003) Functional polymorphisms in the paternally expressed XLalphas and its cofactor ALEX decrease their mutual interaction and enhance receptor-mediated cAMP formation. Hum Mol Genet 12:1121–1130
73. Xie T, Plagge A, Gavrilova O et al (2006) The alternative stimulatory G protein alpha-subunit XLalphas is a critical regulator of energy and glucose metabolism and sympathetic nerve activity in adult mice. J Biol Chem 281:18989–18999
74. Skinner J, Cattanach B, Peters J (2002) The imprinted oedematous-small mutation on mouse chromosome 2 identifies new roles for Gnas and Gnasxl in development. Genomics 80:373
75. Kelly ML, Moir L, Jones L et al (2009) A missense mutation in the non-neural G-protein alpha-subunit isoforms modulates susceptibility to obesity. Int J Obes (Lond) 33:507–518
76. Cheeseman MT, Vowell K, Hough TA et al (2012) A mouse model for osseous heteroplasia. PLoS One 7:e51835
77. Eaton SA, Williamson CM, Ball ST et al (2012) New mutations at the imprinted Gnas cluster show gene dosage effects of Gsalpha in postnatal growth and implicate XLalphas in bone and fat metabolism but not in suckling. Mol Cell Biol 32:1017–1029
78. Fernandez-Rebollo E, Maeda A, Reyes M et al (2012) Loss of XLalphas (extra-large alphas) imprinting results in early postnatal hypoglycemia and lethality in a mouse model of pseudohypoparathyroidism Ib. Proc Natl Acad Sci U S A 109:6638–6643
79. Richard N, Molin A, Coudray N et al (2013) Paternal GNAS mutations lead to severe intrauterine growth retardation (IUGR) and provide evidence for a role of XLalphas in fetal development. J Clin Endocrinol Metab 98:E1549–E1556
80. Plagge A, Kelsey G, Germain-Lee EL (2008) Physiological functions of the imprinted Gnas locus and its protein variants Galpha(s) and XLalpha(s) in human and mouse. J Endocrinol 196:193–214
81. Turan S, Bastepe M (2013) The GNAS complex locus and human diseases associated with loss-of-function mutations or epimutations within this imprinted gene. Horm Res Paediatr 80:229–241
82. Swaroop A, Agarwal N, Gruen JR et al (1991) Differential expression of novel Gs alpha signal transduction protein cDNA species. Nucleic Acids Res 19:4725–4729
83. Liu J, Yu S, Litman D et al (2000) Identification of a methylation imprint mark within the mouse Gnas locus. Mol Cell Biol 20:5808–5817
84. Ishikawa Y, Bianchi C, Nadal-Ginard B et al (1990) Alternative promoter and 5′ exon generate a novel G_sa mRNA. J Biol Chem 265:8458–8462
85. Puzhko S, Goodyer CG, Mohammad AK et al (2011) Parathyroid hormone signaling via Galphas is selectively inhibited by an NH(2) -terminally truncated Galphas: implications for pseudohypoparathyroidism. J Bone Miner Res 26:2473–2485
86. Hayward B, Bonthron D (2000) An imprinted antisense transcript at the human GNAS1 locus. Hum Mol Genet 9:835–841
87. Wroe SF, Kelsey G, Skinner JA et al (2000) An imprinted transcript, antisense to Nesp, adds complexity to the cluster of imprinted genes at the mouse Gnas locus. Proc Natl Acad Sci U S A 97:3342–3346
88. Li T, Vu TH, Ulaner GA et al (2004) Activating and silencing histone modifications form independent allelic switch regions in the imprinted Gnas gene. Hum Mol Genet 13:741–750
89. Campbell R, Gosden CM, Bonthron DT (1994) Parental origin of transcription from the human GNAS1 gene. J Med Genet 31:607–614
90. Turan S, Fernandez-Rebollo E, Aydin C et al (2013) Postnatal establishment of allelic Galphas silencing as a plausible explanation for delayed onset of parathyroid hormone-resistance due to heterozygous Galphas disruption. J Bone Miner Res. doi:10.1002/jbmr.2070
91. Liu J, Erlichman B, Weinstein LS (2003) The stimulatory G protein a-subunit Gsa is imprinted in human thyroid glands: implications for thyroid function in pseudohypoparathyroidism types 1A and 1B. J Clin Endocrinol Metabol 88:4336–4341
92. Germain-Lee EL, Ding CL, Deng Z et al (2002) Paternal imprinting of Galpha(s) in the human thyroid as the basis of TSH resistance in pseudohypoparathyroidism type 1a. Biochem Biophys Res Commun 296:67–72
93. Hayward B, Barlier A, Korbonits M et al (2001) Imprinting of the G(s)alpha gene GNAS1 in the pathogenesis of acromegaly. J Clin Invest 107:R31–R36
94. Mantovani G, Ballare E, Giammona E et al (2002) The gsalpha gene: predominant maternal origin of transcription in human thyroid gland and gonads. J Clin Endocrinol Metab 87:4736–4740
95. Chen M, Wang J, Dickerson KE et al (2009) Central nervous system imprinting of the G protein G(s) alpha and its role in metabolic regulation. Cell Metab 9:548–555
96. Liu J, Chen M, Deng C et al (2005) Identification of the control region for tissue-specific imprinting of the stimulatory G protein alpha-subunit. Proc Natl Acad Sci U S A 102:5513–5518

97. Davies AJ, Hughes HE (1993) Imprinting in Albright's hereditary osteodystrophy. J Med Genet 30:101–103
98. Jüppner H, Schipani E, Bastepe M et al (1998) The gene responsible for pseudohypoparathyroidism type Ib is paternally imprinted and maps in four unrelated kindreds to chromosome 20q13.3. Proc Natl Acad Sci U S A 95:11798–11803
99. Bastepe M, Fröhlich LF, Hendy GN et al (2003) Autosomal dominant pseudohypoparathyroidism type Ib is associated with a heterozygous microdeletion that likely disrupts a putative imprinting control element of GNAS. J Clin Invest 112:1255–1263
100. Linglart A, Gensure RC, Olney RC et al (2005) A novel STX16 deletion in autosomal dominant pseudohypoparathyroidism type Ib redefines the boundaries of a cis-acting imprinting control element of GNAS. Am J Hum Genet 76:804–814
101. Richard N, Abeguile G, Coudray N et al (2012) A new deletion ablating NESP55 causes loss of maternal imprint of A/B GNAS and autosomal dominant pseudohypoparathyroidism type Ib. J Clin Endocrinol Metab 97:E863–E867
102. Fröhlich LF, Bastepe M, Ozturk D et al (2007) Lack of Gnas epigenetic changes and pseudohypoparathyroidism type Ib in mice with targeted disruption of syntaxin-16. Endocrinology 148:2925–2935
103. Bastepe M, Fröhlich LF, Linglart A et al (2005) Deletion of the NESP55 differentially methylated region causes loss of maternal GNAS imprints and pseudohypoparathyroidism type-Ib. Nat Genet 37:25–37
104. Chillambhi S, Turan S, Hwang D-Y et al (2008) Deletion of the GNAS antisense transcript results in parent-of-origin specific GNAS imprinting defects and phenotypes including PTH-resistance (Abstract No. 1052). In: 30th annual meeting of The American Society of Bone and Mineral Research, Montreal
105. Chotalia M, Smallwood SA, Ruf N et al (2009) Transcription is required for establishment of germline methylation marks at imprinted genes. Genes Dev 23:105–117
106. Coombes C, Arnaud P, Gordon E et al (2003) Epigenetic properties and identification of an imprint mark in the Nesp-Gnasxl domain of the mouse Gnas imprinted locus. Mol Cell Biol 23:5475–5488
107. Williamson CM, Ball ST, Nottingham WT et al (2004) A cis-acting control region is required exclusively for the tissue-specific imprinting of Gnas. Nat Genet 36:894–899
108. Williamson CM, Turner MD, Ball ST et al (2006) Identification of an imprinting control region affecting the expression of all transcripts in the Gnas cluster. Nat Genet 38:350–355
109. Williamson CM, Ball ST, Dawson C et al (2011) Uncoupling antisense-mediated silencing and DNA methylation in the imprinted Gnas cluster. PLoS Genet 7:e1001347
110. Fröhlich LF, Mrakovcic M, Steinborn R et al (2010) Targeted deletion of the Nesp55 DMR defines another Gnas imprinting control region and provides a mouse model of autosomal dominant PHP-Ib. Proc Natl Acad Sci U S A 107:9275–9280

Parathyroid Hormone Actions on Bone and Kidney

11

Paola Divieti Pajevic, Marc N. Wein, and Henry M. Kronenberg

11.1 Introduction

The major actions of parathyroid hormone (PTH) to regulate calcium and phosphorus homeostasis involve actions of PTH on bone and kidney. In this chapter, we will consider actions on these target organs individually, but we emphasize that these actions cannot be viewed separately. The $1,25(OH)_2vitaminD_3$ made by the proximal tubule, and possibly by bone cells, in response to PTH synergizes with PTH in stimulating bone resorption. FGF23, whose synthesis is stimulated in osteocytes by $1,25(OH)_2vitaminD_3$ and perhaps directly by PTH itself [1], acts along with PTH to decrease phosphate reabsorption in the proximal tubule. Nevertheless, it is useful to consider the actions of PTH on bone and kidney separately, not just because the topics are too large to consider in an integrated fashion, but also because PTH's actions on these two organs appear to be quite different in their mechanisms at the cellular level. In the kidney, PTH acts primarily to regulate calcium, phosphorus, and vitamin D homeostasis without significantly changing the cellular composition and anatomy of the organ itself. In striking contrast, in bone, PTH not only regulates calcium, phosphorus, and vitamin D metabolism, but profoundly affects the cellular composition, anatomy, and function of bone as an organ. Here we will first consider the actions of PTH on bone and then on the kidney.

P.D. Pajevic • M.N. Wein • H.M. Kronenberg (✉)
Endocrine Unit, Massachusetts General Hospital, Harvard Medical School,
50 Blossom St., Thier 1101,
Boston, MA 02114, USA
e-mail: hkronenberg@mgh.harvard.edu

11.2 PTH Actions on Bone

PTH modulates bone turnover and calcium and phosphorous homeostasis by binding to the PTH/PTH related-peptide (PTHrP) type 1 receptor (PTHR1), a G-protein-coupled receptor highly expressed in bone and kidney and in a variety of other tissues in which it likely mediates the local paracrine effects of PTHrP [2] (see also Chaps. 9 and 10). In bone, the receptor is expressed on chondrocytes and on cells of the osteoblastic lineage, which include, among others, osteoprogenitors, mature osteoblasts, and osteocytes [3]. A detailed description and analysis of hormone-receptor interactions and the subsequent intracellular signaling can be found in Chap. 9.

11.2.1 Actions of PTHR1 on Growth Plate Chondrocytes

During bone growth, activation of the PTHR1 is required to slow the differentiation of chondrocytes. Studies on genetically modified animals lacking the hormones (PTH-KO and PTHrP-KO)

M.L. Brandi, E.M. Brown (eds.), *Hypoparathyroidism*,
DOI 10.1007/978-88-470-5376-2_11, © Springer-Verlag Italia 2015

[4, 5] or the receptor (PTHR1-KO animals) [6] have demonstrated that, during embryonic development, PTHrP plays a critical role in skeletal development. Mice lacking either PTHrP or the PTHR1 die at birth and display a similar phenotype, which includes short bones and a domed skull that is a phenocopy of the human lethal form of dwarfism known as Blomstrand chondrodysplasia, in which homozygous inactivating mutations of the PTHR1 are found [7, 8] (see also Chaps. 23 and 36). The more severe phenotype of the PTHR1 versus PTHrP knockout mouse suggests that PTH may contribute modestly to the activation of the PTHR1 in the growth plate, as well. In the PTH knockout mouse at birth, chondrocyte differentiation is normal, with modest expansion of the hypertrophic zone, decreased vascularization of the chondro-osseous junction, and decreased mineralization of the cartilage matrix [5]. Thus, PTHrP, not PTH, is the major activator of the PTHR1 on growth plate chondrocytes.

11.2.2 Actions of PTHR1 on Cells of the Osteoblast Lineage

11.2.2.1 Osteoblastic Cell Types Targeted

The PTHR1 receptor is expressed on cells of the osteoblastic lineage; however, the effects of PTH on bone cells at varying stages of differentiation have been challenging to study. Most such studies were first performed on osteoblastic cells in primary culture. Early studies from Bellows et al. demonstrated that PTH had no detectable effects on early progenitors in vitro and exerted most of its actions on late-stage progenitors, osteoblasts, and osteocytes [9]. These earlier studies were further supported by work of Huang et al. [10], who reported that PTH significantly regulated both receptor activator of NF-kb ligand (RANKL) and osteoprotegerin (OPG) in bone marrow osteoblasts with a maximal sensitivity at later stages of their differentiation (28 days in culture), demonstrating that receptor expression and activity increased with cell differentiation. Further in vitro studies have suggested that the actions of PTH might vary, depending on the state of differentiation of the PTH target cell. For example, Isogai et al. [11] noted that PTH added to confluent, differentiated primary osteoblastic cells suppressed indices of osteoblast differentiation. In contrast, when these investigators added PTH to osteoblastic cells growing at low cell density, the levels of alkaline phosphatase activity in these cells increased. This same group [12] noted that, when PTH was added to differentiated osteoblastic cells, the effects on differentiation changed, depending on whether exposure to PTH was continuous (suppression of differentiation, as just noted above) or only 6 h of every 48 h period (stimulation of differentiation in this condition). These studies show that PTH has some direct actions that require only cells of the osteoblast lineage and that these actions, at least in vitro, can depend crucially on the stage of differentiation of the cells and the duration of exposure to the hormone.

More precise analysis of the actions of PTH on osteoblastic cells at varying stages of differentiation has become possible with the tools provided by genetic manipulation of mice. In recent years, a series of papers has identified early cells of the osteoblast lineage in vivo in genetically manipulated mice. In one such study, Mendez-Ferrer et al. [13] noted that pericytic cells in the bone marrow of mice, marked with expression of green fluorescent protein (GFP) driven by a nestin gene promoter, have properties in vivo of multipotential, self-renewing cells that can become osteoblasts, a finding ascertained through the use of a lineage tracking strategy. Strikingly, they showed that these nestin-GFP+cells contain immunologically defined PTH receptors. Further, after intermittent PTH administration, the numbers of the nestin-GFP+cells increased. Purified nestin-GFP+cells responded to PTH in vitro. Further studies will be needed to clarify the roles of the PTHR1 on early cells of the osteoblast lineage in vivo, but their existence raises the possibility that some of the effects of PTH on precursor cells may reflect direct actions of PTH on these cells.

PTHR1 expression has been noted for years on osteoblasts on the bone surfaces and on osteocytes as well. When a constitutively active PTHR1 was expressed in osteoblasts using a collagen I(α1) promoter [14], that activity led to a dramatic increase in trabecular bone and to an accumulation of stromal cells in the marrow, perhaps resembling those changes seen in severe primary hyperparathyroidism. The increase in osteoblast number was accompanied both by an increase in proliferation of these cells and by a decrease in their rate of apoptosis. Indices of bone resorption were also increased. When the same constitutively active PTHR1 was activated only in osteocytes through the use of the dentin matrix protein-1 (DMP-1) promoter [15], an increase in trabecular bone mass was also seen along with an increase in resorption. This increase in bone mass was accompanied by a suppression of osteoblastic apoptosis. Sclerostin expression in osteocytes (see below) was suppressed and the increase in bone mass was attenuated by deletion of LRP5, a co-receptor for wnt signaling. Thus, it was suggested that increased wnt signaling in cells of the osteoblast lineage is part of the mechanism whereby activation of the PTHR1 in osteocytes leads to increased bone mass. The apparently much more striking increase in stromal cells (perhaps osteoblast precursors) in the mouse in which the PTHR1 was activated by the collagen I(α 1) promoter suggests that this accumulation requires direct activation of the receptor in cells less differentiated than osteocytes.

These two mouse models differ in another way, as well. Cells of the osteoblast lineage provide part of a niche environment that fosters hematopoiesis. While the mouse with activation of the PTHR1 in osteoblasts exhibits an increase in hematopoietic stem cells [16], the mouse with activation of the PTHR1 only in osteocytes does not [17]. This result suggests that osteoblasts and other less mature cells in the lineage may be more important PTH targets for the effects of PTH on hematopoietic stem cells.

Studies of mice with knockout of the PTHR1 suggest that the endogenous PTHR1 has properties predicted by the findings in the transgenic mice with an activated PTHR1. Ablation of the receptor in osteocytes (and some osteoblasts), by using the 10Kb DMP-1 promoter fragment to drive Cre recombinase to target the floxed PTHR1 [18], leads to an increase in bone mass that may reflect a low bone turnover state. Further, the bone mass and indices of bone formation failed to increase when PTH was administered by intermittent daily injection, a regimen that increases bone mass in normal mice. The failure of sclerostin expression to fall after PTH administration may contribute to this blunted response in bone formation to PTH. In contrast, ablation of the receptor in mature osteoblasts (and consequently in osteocytes also), using the osteocalcin (Oc) promoter to drive Cre expression, or in osteoprogenitors (using the osterix promoter to drive Cre expression) causes profound osteopenia (PDP, unpublished data), suggesting an essential role for PTHR1 signaling in earlier cells of the osteoblast lineage.

PTH also affects the function of inactive bone lining cells on the surfaces of bone. Recent lineage-tracing experiments [19] confirm earlier suggestions [20, 21] that PTH administration leads to conversion of inactive bone lining cells to active osteoblasts, synthesizing large amounts of collagen I. The mechanism for this activation and whether this mechanism involves direct actions on the lining cells are unknown.

11.2.2.2 Cellular Mechanisms of PTH Action on Cells of the Osteoblast Lineage

This summary of the cell types in bone targeted by PTH indicates the complexity of PTH's actions on cells of the osteoblast lineage. These actions are likely to be a mixture of direct, cell autonomous actions caused by activation of PTHR1 on the target cell, as well as indirect actions through autocrine and paracrine communication between various cell types with PTHR1s. The complexity of PTH's actions is further complicated by the substantial differences in the effects of PTH, depending upon whether it is administered by continuous infusion or by intermittent injection, typically subcutaneously. While both modes of administration lead to an increase in bone formation and bone resorption [22], the continuous administration of PTH leads to a fall in bone mass, while the intermittent administration of PTH leads to an increase in bone mass. Presumably, the extent of resorption exceeds that of formation when PTH is administered continuously and the rate of formation exceeds that of resorption after intermittent PTH administration.

When PTH is administered continuously to experimental animals, stromal cells increase in number in the bone marrow. In one such model in the rat, Lotinun et al. [23] noted that, when tritiated thymidine was used to mark proliferating cells, a large fraction of the fibroblastic marrow cells incorporated thymidine, while almost no osteoblasts did so. A week after stopping the PTH infusion and thymidine administration, however, the fibroblastic stroma disappeared, and many of the osteoblasts exhibited incorporation of the previously administered thymidine, suggesting that some of the labeled stromal cells had become osteoblasts. This observation is consistent with the in vitro findings noted earlier [24] that continuous exposure to PTH in vitro blocks differentiation of precursors into mature osteoblasts. These findings may correlate with the observation that prolonged exposure of osteoblastic cells to PTH leads to degradation of Runx2, a key transcription factor for driving differentiation of osteoblasts [25].

Several cellular mechanisms have been demonstrated that may contribute to the increase in bone formation when PTH is administered intermittently. PTH suppresses apoptosis of osteoblasts in vivo and in vitro and this action is predicted to increase the number of osteoblasts on the bone surface [26]. Further, as noted earlier, PTH can activate previously dormant lining cells, thereby increasing the number of active osteoblasts on the bone surface [14, 20, 21]. PTH may also increase the number of osteoblast precursors [27], though this is a difficult issue to study, because currently the only way to identify and quantitate osteoblast precursors is by counting colonies of putative precursors from the marrow after plating cells in vitro [28].

11.2.2.3 Biochemical Mediators and Pathways That Regulate PTH Action on Cells of the Osteoblast Lineage

Though PTH may exert some actions through yet uncloned receptors for PTH (1–84) [29], the known actions of PTH appear to be mediated by the cloned PTH/PTHrP receptor [30]. This receptor activates heterotrimeric G proteins and arrestin in ways that lead to activation of intracellular signaling cascades [31]. Most of the actions of PTH in bone appear to involve predominantly activation of Gs and consequent activation of adenylate cyclase. Studies of mice with mutations in the PTHR1 that do not allow stimulation of phospholipase C in response to receptor activation [32], for example, show that such mice have fairly normal bones in adulthood, though, when treated with prolonged infusions of PTH, these mice fail to mount a marrow stromal proliferative response. When PTH is added to a cultured osteoblastic cell line or to primary osteoblasts, the levels of a large number of messenger RNAs change [33]. How to relate these direct actions of PTH to specific physiologic functions in intact bone remains a major challenge.

PTH acts to increase bone resorption by increasing the expression of RANK ligand (RANKL) and decreasing the expression of OPG in cells of the osteoblast lineage [34]. Since increased activity of RANKL, by itself, can increase osteoblast number [35], the stimulation of RANKL expression and bone resorption by PTH probably contributes to the increase in bone formation caused by PTH administration. The release of TGF-β from bone matrix by osteoclastic action may signal to osteoblast precursors to come to the bone surface [36], and signals from osteoclasts themselves may increase bone formation [37].

PTH's activation of several signaling programs probably contributes to PTH's stimulation of bone formation. PTH increases the activity of the canonical wnt pathway by a variety of mechanisms. PTH decreases the expression of the wnt antagonists, dkk1 [38] and sclerostin [39, 40]. Genetic evidence in mice suggests that the suppression of sclerostin expression contributes to the increase in bone formation after PTH administration [41]. Further, PTHR1s bind directly to the wnt co-receptor, LRP6 [42] and to disheveled [43], a key mediator of canonical wnt signaling. Moreover, through activation of protein kinase A, PTH leads to the activation of β-catenin by phosphorylation. β-catenin is the key transcriptional mediator of canonical wnt signaling [38]. Thus, PTH activation of wnt signaling by multiple cell-autonomous and non-cell-autonomous pathways may contribute to stimulation of bone formation.

The interactions between PTH and IGF I signaling are important for PTH's ability to increase bone formation, as well. PTH increases IGF I production by cells of the osteoblast lineage [44, 45], and knockout of the IGF receptor leads to blunted bone formation after intermittent administration of PTH [46]. PTH also increases the production of FGF2 in osteoblastic cells, and the action of PTH to increase bone formation is blunted in the FGF2 knockout mouse [47].

11.3 PTH Actions in Kidney

Here we will review the renal actions of PTH that have clear physiologic importance. In the kidney, PTH exerts its effects on mineral ion metabolism via receptors expressed by epithelial cells in the proximal and distal tubules. In the proximal convoluted tubule, PTH has two principal actions: (1) increasing active vitamin D ($1,25(OH)_2D_3$) synthesis and (2) decreasing phosphate reabsorption. In the distal nephron, PTH stimulates calcium reabsorption.

Through these three actions (and in conjunction with the intimately related hormones, $1,25(OH)_2D_3$, FGF-23, and calcitonin), PTH plays a crucial role in maintaining normal mineral ion homeostasis (see also Chaps. 6, 7, and 8). In addition to these three major actions of PTH in the nephron, chronic hyperparathyroidism leads to a mild hyperchloremic metabolic acidosis via promoting renal bicarbonate wasting and increasing chloride efflux [48]. Herein we will focus on the molecular mechanisms underlying these three major renal PTH actions with respect to mineral ion homeostasis. Potential non-epithelial renal effects of PTH and PTHrP [49], such as modulation of glomerular filtration rate and vascular tone, are less well established and beyond the scope of this chapter.

11.3.1 Control of $1,25(OH)_2D_3$ Synthesis by *Cyp27B1* Expression

PTH control of $1,25(OH)_2D_3$ synthesis was first described in experimental animals over four decades ago. In seminal studies, parathyroidectomized rats were unable to convert $25(OH)D_3$ to $1,25(OH)_2D_3$, while purified parathyroid extract containing PTH restored normal $1,25(OH)_2D_3$ production [50]. Subsequent in vitro studies using cultured proximal tubule cells demonstrated that this activity of PTH requires new RNA synthesis [51]. Signaling downstream of PTHR1 activates multiple second messenger pathways, including the Gsα-linked adenylate cyclase-cAMP-protein kinase A (PKA) pathway and the $G_{q/11}$-linked phospholipase C (PLC)-protein kinase C (PKC) pathway [52, 53]. Although inhibitor-based studies have suggested that the PKC activity participates in PTH-induced $1,25(OH)_2D_3$ synthesis [54], multiple laboratories have described a cAMP/PKA pathway in cultured proximal tubule cells [51, 55]. More recently, experiments using mice expressing a PTH receptor that activates adenylate cyclase normally but cannot activate PLC/PKC [56] showed that PTH-dependent increases in $1,25(OH)_2D_3$ levels in vivo do not require PKC signaling by PTHR1 [57].

Cyp27B1 encodes the 1α-hydroxylase responsible for conversion of $25(OH)_2D_3$ into $1,25(OH)_2D_3$. Cloning of the *Cyp27B1* gene [58–60] allowed subsequent analysis of its proximal promoter. In kidney cell lines, PTH increases the activity of the murine and human *Cyp27B1* proximal promoters [61, 62]. This same promoter region linked to a reporter gene in transgenic mice is activated in vivo by secondary hyperparathyroidism due to vitamin D or dietary calcium deficiencies [63]. While a complete understanding of the transcription factors mediating PTH-dependent *Cyp27B1* promoter activation is lacking, overexpression and mutagenesis studies have suggested stimulatory roles for NR4A2 (Nurr1, whose expression itself is induced by PTH), Sp1, and NF-Y and an inhibitory role for C/EBPß [64, 65].

$1,25(OH)_2D_3$ itself negatively regulates *Cyp27B1* expression as part of a negative feedback loop. At the molecular level, this has been mapped to vitamin D receptor (VDR) binding to the proximal *Cyp27B1* promoter [66]. This inhibitory role of $1,25(OH)_2D_3$/VDR has been linked to the basic helix loop helix (bHLH) transcription factor, VDIR. Interestingly, VDIR is a PKA substrate. When phosphorylated, VDIR shows decreased VDR and transcriptional corepressor binding and increased association with transcriptional coactivators [67]. Additional in vivo studies are required to confirm the potential roles for all these transcription factors in controlling renal PTH-mediated *Cyp27B1* expression. Finally, although extrarenal *Cyp27B1* expression is well documented [68], we currently do not understand the molecular determinants of tissue-specific expression and why this gene's expression in some tissue types is not regulated by PTH or other cAMP-inducing agents [69].

11.3.2 Control of Phosphate Handling in the Proximal Tubule

A second, distinct action of PTH in the proximal tubule is to increase urinary phosphate excretion. As a phosphaturic factor, PTH's action synergizes with those of fibroblast growth factor-23 (FGF-23), the other known circulating phosphatonin [48] (see also Chap. 6). A complete review of FGF-23-dependent physiology is beyond the scope of this chapter. The rate of phosphate reabsorption in the proximal tubule is mainly controlled by the number of type II sodium/phosphate cotransporters (Npt2a and Npt2c) found on the brush-border membrane [70]. Both PTH and FGF-23 inhibit renal phosphate reabsorption in the proximal tubule by reducing surface levels and activity of these cotransporters [71].

The signaling pathways leading from the PTH receptor to Npt2a downregulation have been the object of intense investigation over the past 15 years [72]. A major advance came in 2002 when two groups identified the sodium/hydrogen exchanger regulatory factor-1 (NHERF-1) as an important regulator of apical membrane levels of Npt2a. Mice lacking this Npt2a-binding scaffolding protein have renal phosphate wasting and ineffective membrane targeting of Npt2a [73, 74]. Furthermore, mutations in NHERF1 have been described in humans with low tubular phosphate reabsorption, nephrolithiasis, and bone demineralization [75].

Proximal tubule cells from mice lacking NHERF-1 show resistance to the inhibitory effects of PTH on phosphate transport [76]. While Npt2a itself is not a phosphoprotein regulated by PTH, NHERF-1 is phosphorylated at serine 77 downstream of PTH receptor signaling. Phosphorylation at this site decreases NHERF-1/Npt2a interaction and leads to Npt2a-internalization [77]. NHERF-1 and NHERF-2 constitutively bind to the PTH receptor and may dictate downstream signaling by coupling to a PKA versus a PKC pathway [78]. More recently, the cytoskeleton-associated protein ezrin has been implicated in dynamic regulation of NHERF-1/Npt2a interactions by PTH [79, 80].

The relative contribution to PTH-mediated PKA versus PKC activation with respect to Npt2a-mediated phosphate reabsorption has been tested in vivo using two different experimental systems. First, PTH analogs defective in PKC activation were competent to cause phosphaturia in the acute setting [81]. Second, phosphaturic responses to PTH were investigated in mice expressing PTH receptors that cannot activate PKC. Interestingly, while acute hypophosphatemia due to PTH infusion was comparable in control and PTH receptor mutant mice, prolonged hypophosphatemia required PTH receptors that can activate PKC [57]. Future studies will be required to completely delineate the mechanism whereby PKC signaling stimulated by PTHR1 is required for prolonged (but not acute) PTH-mediated phosphaturia. In addition, an area ripe for future investigation is the interplay between PTH- and FGF-23-mediated effects on phosphate handling in the proximal tubule.

11.3.3 Control of Calcium Reabsorption in the Distal Tubule

Parathyroid hormone promotes calcium reabsorption in the distal convoluted tubule (DCT) [82]. Conversely, hypoparathyroidism can cause hypercalciuria and nephrolithiasis. Classical studies using isolated rabbit kidney connecting tubules demonstrated that PTH (and cAMP) rapidly increases cytosolic intracellular calcium concentrations only when calcium was present in luminal fluid [83]. Pharmacologic inhibitor studies indicated that cAMP-dependent kinase activation is required for this effect [84]. An understanding of the precise pathways by which PTH controls transcellular calcium fluxes in the DCT required identification of the apical calcium channel TRPV5 (transient receptor potential vanilloid 5) [85]. Once calcium enters the cell via the TRPV5 channel, it is chaperoned across the cell via association with carrier proteins (most notably calbindin D28K and calmodulin) and ultimately extruded into the bloodstream via the Na^+/Ca^{2+} exchanger (NCX1) and the plasma membrane Ca^{2+}-ATPase (PMCA1b) [86].

At least three lines of evidence indicate that TRPV5-mediated transcellular calcium transport is regulated by PTH. First, PTH tonically controls expression levels of TRPV5, as evidenced by reduced TRPV5 mRNA after parathyroidectomy [87]. Second, PTH increases cell surface TRPV5 levels by reducing caveolae-mediated TRPV5 endocytosis in a mechanism that appears to be sensitive to PKC inhibitors [88]. Third, the TRPV5 channel itself is a direct PKA substrate at threonine 709. This phosphorylation event increases the probability of TRPV5 channel opening [89]. TRPV5 channel activity is negatively modulated by calmodulin binding to its intracellular domain; interestingly, calmodulin binding is diminished by PTH-mediated threonine 709 phosphorylation [89]. At this point, the relative contributions of PKA versus PKC signaling with respect to PTH-mediated transcellular calcium reabsorption in the DCT in vivo have not been reported. In addition, the importance of TRPV5 threonine 709 phosphorylation in vivo remains to be established.

In addition to PTH-mediated regulation of TRPV5 activity by the mechanisms outlined above, parathyroidectomy in rodents reduces expression of calbindin D28, NCX1, and PMCA1b in the DCT [87]. Therefore, it is likely that PTH coordinates a program of gene expression in DCT cells necessary for optimal transcellular calcium reabsorption.

Conclusions

Renal actions of parathyroid hormone are predominantly due to expression of PTH receptors on epithelial targets in the proximal and distal nephron. As detailed above, PTH coordinates a response in the kidney including increased $1{,}25(OH)_2D_3$ synthesis, phosphaturia, and increased calcium reabsorption. While much is known about molecular mechanisms underlying these physiologic effects, we have also pointed out some of the many outstanding questions for future study. In addition, the precise mechanisms underlying failure of these pathways in the setting of renal insufficiency remain unknown. As such, the future of this field remains exciting for discoveries of major physiological and clinical significance.

References

1. Rhee Y, Bivi N, Farrow E, Lezcano V, Plotkin LI, White KE et al (2011) Parathyroid hormone receptor signaling in osteocytes increases the expression of fibroblast growth factor-23 in vitro and in vivo. Bone 49(4):636–643. doi:10.1016/j.bone.2011.06.025. Epub 2011/07/06. doi: S8756-3282(11)01066-0 [pii] PubMed PMID: 21726676; PubMed Central PMCID: PMC3167030
2. Juppner H, Abou-Samra AB, Freeman M, Kong XF, Schipani E, Richards J et al (1991) A G protein-linked receptor for parathyroid hormone and parathyroid hormone-related peptide. Science 254(5034):1024–1026
3. Fermor B, Skerry TM (1995) PTH/PTHrP receptor expression on osteoblasts and osteocytes but not resorbing bone surfaces in growing rats. J Bone Miner Res 10(12):1935–1943. PubMed PMID: 8619374
4. Karaplis AC, Luz A, Glowacki J, Bronson RT, Tybulewicz VL, Kronenberg HM et al (1994) Lethal skeletal dysplasia from targeted disruption of the parathyroid hormone-related peptide gene. Genes Dev 8(3):277–289. Epub 1994/02/01. PubMed PMID: 8314082
5. Miao D, He B, Lanske B, Bai XY, Tong XK, Hendy GN et al (2004) Skeletal abnormalities in Pth-null mice are influenced by dietary calcium. Endocrinology

145(4):2046–2053. doi:10.1210/en.2003-1097. Epub 2004/01/01. en.2003-1097 [pii]. PubMed PMID: 14701672
6. Lanske B, Karaplis AC, Lee K, Luz A, Vortkamp A, Pirro A et al (1996) PTH/PTHrP receptor in early development and Indian hedgehog-regulated bone growth [see comments]. Science 273(5275): 663–666
7. Jobert AS, Zhang P, Couvineau A, Bonaventure J, Roume J, Le Merrer M et al (1998) Absence of functional receptors for parathyroid hormone and parathyroid hormone-related peptide in Blomstrand chondrodysplasia. J Clin Invest 102(1):34–40. doi:10.1172/JCI2918. Epub 1998/07/03. PubMed PMID: 9649554; PubMed Central PMCID: PMC509062
8. Zhang P, Jobert AS, Couvineau A, Silve C (1998) A homozygous inactivating mutation in the parathyroid hormone/parathyroid hormone-related peptide receptor causing Blomstrand chondrodysplasia. J Clin Endocrinol Metab 83(9):3365–3368. doi:10.1210/jcem.83.9.5243. Epub 1998/09/24. PubMed PMID: 9745456
9. Bellows CG, Ishida H, Aubin JE, Heersche JN (1990) Parathyroid hormone reversibly suppresses the differentiation of osteoprogenitor cells into functional osteoblasts. Endocrinology 127(6):3111–3116. doi:10.1210/endo-127-6-3111. Epub 1990/12/01. PubMed PMID: 2174346
10. Huang JC, Sakata T, Pfleger LL, Bencsik M, Halloran BP, Bikle DD et al (2004) PTH differentially regulates expression of RANKL and OPG. J Bone Miner Res 19(2):235–244. doi:10.1359/JBMR.0301226. Epub 2004/02/19. PubMed PMID: 14969393
11. Isogai Y et al (1996) Parathyroid hormone regulates osteoblast differentiation positively or negatively depending on the differentiation stages. J Bone Miner Res 11:1384–1393. PMID: 888983
12. Ishizuya et al (1997) Parathyroid hormone exerts disparate effects on osteoblast differentiation depending on exposure time in rat osteoblastic cells. J Clin Invest 99:2961–2970. PMID: 9185520
13. Mendez-Ferrer S, Michurina TV, Ferraro F, Mazloom AR, Macarthur BD, Lira SA et al (2010) Mesenchymal and haematopoietic stem cells form a unique bone marrow niche. Nature 466(7308):829–834. PubMed PMID: 20703299
14. Schipani E, Kruse K, Juppner H (1995) A constitutively active mutant PTH-PTHrP receptor in Jansen-type metaphyseal chondrodysplasia. Science 268(5207):98–100. PubMed PMID: 7701349
15. O'Brien CA, Plotkin LI, Galli C, Goellner JJ, Gortazar AR, Allen MR et al (2008) Control of bone mass and remodeling by PTH receptor signaling in osteocytes. PLoS One 3(8):e2942. doi:10.1371/journal.pone.0002942. Epub 2008/08/14. PubMed PMID: 18698360; PubMed Central PMCID: PMC2491588
16. Calvi LM, Adams GB, Weibrecht KW, Weber JM, Olson DP, Knight MC et al (2003) Osteoblastic cells regulate the haematopoietic stem cell niche. Nature 425(6960):841–846. doi:10.1038/nature02040. Epub 2003/10/24. nature02040 [pii]. PubMed PMID: 14574413
17. Calvi LM, Bromberg O, Rhee Y, Weber JM, Smith JN, Basil MJ et al (2012) Osteoblastic expansion induced by parathyroid hormone receptor signaling in murine osteocytes is not sufficient to increase hematopoietic stem cells. Blood 119(11):2489–2499. doi:10.1182/blood-2011-06-360933. Epub 2012/01/21. doi: blood-2011-06-360933 [pii]. PubMed PMID: 22262765; PubMed Central PMCID: PMC3311272
18. Saini V, Marengi DJ, Barry KJ, Fulzele KS, Heiden E, Liu X et al (2013) Parathyroid hormone (PTH)/PTH-related peptide type 1 receptor (PPR) signaling in osteocytes regulates anabolic and catabolic skeletal responses to PTH. J Biol Chem. doi:10.1074/jbc.M112.441360. Epub 2013/06/05. doi: M112.441360 [pii] PubMed PMID: 23729679
19. Kim SW, Pajevic PD, Selig M, Barry KJ, Yang JY, Shin CS et al (2012) Intermittent PTH administration converts quiescent lining cells to active osteoblasts. J Bone Miner Res. doi:10.1002/jbmr.1665. Epub 2012/05/25. PubMed PMID: 22623172
20. Dobnig H, Turner RT (1995) Evidence that intermittent treatment with parathyroid hormone increases bone formation in adult rats by activation of bone lining cells. Endocrinology 136(8):3632–3638
21. Leaffer D, Sweeney M, Kellerman LA, Avnur Z, Krstenansky JL, Vickery BH et al (1995) Modulation of osteogenic cell ultrastructure by RS-23581, an analog of human parathyroid hormone (PTH)-related peptide-(1–34), and bovine PTH-(1–34). Endocrinology 136(8):3624–3631
22. Tam CS, Heersche JNM, Murray TM, Parsons JA (1982) Parathyroid hormone stimulates the bone apposition rate independently of its resorptive action: differential effects of intermittent and continuous administration. Endocrinology 110:506–512
23. Lotinun S, Sibonga JD, Turner RT (2005) Evidence that the cells responsible for marrow fibrosis in a rat model for hyperparathyroidism are preosteoblasts. Endocrinology 146(9):4074–4081. PubMed PMID: 15947001
24. Ishizuya T, Yokose S, Hori M, Noda T, Suda T, Yoshiki S et al (1997) Parathyroid hormone exerts disparate effects on osteoblast differentiation depending on exposure time in rat osteoblastic cells. J Clin Invest 99(12):2961–2970. doi:10.1172/JCI119491. Epub 1997/06/15. PubMed PMID: 9185520; PubMed Central PMCID: PMC508148
25. Bellido T, Ali AA, Plotkin LI, Fu Q, Gubrij I, Roberson PK et al (2003) Proteasomal degradation of Runx2 shortens parathyroid hormone-induced anti-apoptotic signaling in osteoblasts. A putative explanation for why intermittent administration is needed for bone anabolism. J Biol Chem 278(50):50259–50272. PubMed PMID: 14523023
26. Jilka RL, Weinstein RS, Bellido T, Roberson P, Parfitt AM, Manolagas SC (1999) Increased bone formation by prevention of osteoblast apoptosis with parathyroid hormone. [see comments.]. J Clin Invest 104(4):439–446

27. Nishida S, Yamaguchi A, Tanizawa T, Endo N, Mashiba T, Uchiyama Y et al (1994) Increased bone formation by intermittent parathyroid hormone administration is due to the stimulation of proliferation and differentiation of osteoprogenitor cells in bone marrow. Bone 15:717–723
28. Robey PG, Kuznetsov SA, Riminucci M, Bianco P (2014) Bone marrow stromal cell assays: in vitro and in vivo. Methods Mol Biol 1130:279–293. doi:10.1007/978-1-62703-989-5_21. PubMed PMID: 24482181
29. Murray TM, Rao LG, Divieti P, Bringhurst FR (2005) Parathyroid hormone secretion and action: evidence for discrete receptors for the carboxyl-terminal region and related biological actions of carboxyl- terminal ligands. Endocr Rev 26:78–113. PubMed PMID: 15546922
30. Potts JT, Gardella TJ (2007) Progress, paradox, and potential: parathyroid hormone research over five decades. Ann N Y Acad Sci 1117:196–208. doi:10.1196/annals.1402.088. PubMed PMID: 18056044
31. Vilardaga JP, Gardella TJ, Wehbi VL, Feinstein TN (2012) Non-canonical signaling of the PTH receptor. Trends Pharmacol Sci 33(8):423–431. doi:10.1016/j.tips.2012.05.004. PubMed PMID: 22709554; PubMed Central PMCID: PMC3428041
32. Guo J, Liu M, Yang D, Bouxsein ML, Thomas CC, Schipani E et al (2010) Phospholipase C signaling via the parathyroid hormone (PTH)/PTH-related peptide receptor is essential for normal bone responses to PTH. Endocrinology 151(8):3502–3513. PubMed PMID: 20501677
33. Qin L, Qiu P, Wang L, Li X, Swarthout JT, Soteropoulos P et al (2003) Gene expression profiles and transcription factors involved in parathyroid hormone signaling in osteoblasts revealed by microarray and bioinformatics. J Biol Chem 278(22):19723–19731. PubMed PMID: 12644456
34. Horwood NJ, Elliott J, Martin TJ, Gillespie MT (1998) Osteotropic agents regulate the expression of osteoclast differentiation factor and osteoprotegerin in osteoblastic stromal cells. Endocrinology 139(11):4743–4746
35. Jilka RL, O'Brien CA, Bartell SM, Weinstein RS, Manolagas SC (2010) Continuous elevation of PTH increases the number of osteoblasts via both osteoclast-dependent and -independent mechanisms. J Bone Miner Res 25(11):2427–2437. doi:10.1002/jbmr.145. PubMed PMID: 20533302; PubMed Central PMCID: PMC3179285
36. Tang Y, Wu X, Lei W, Pang L, Wan C, Shi Z et al (2009) TGF-beta1-induced migration of bone mesenchymal stem cells couples bone resorption with formation. Nat Med 15(7):757–765. PubMed PMID: 19584867
37. Takeshita S, Fumoto T, Matsuoka K, Park KA, Aburatani H, Kato S et al (2013) Osteoclast-secreted CTHRC1 in the coupling of bone resorption to formation. J Clin Invest 123(9):3914–3924. doi:10.1172/JCI69493. PubMed PMID: 23908115; PubMed Central PMCID: PMC3754269
38. Guo J, Liu M, Yang D, Bouxsein ML, Saito H, Galvin RJ et al (2010) Suppression of Wnt signaling by Dkk1 attenuates PTH-mediated stromal cell response and new bone formation. Cell Metab 11(2):161–171. PubMed PMID: 20142103
39. Bellido T, Ali AA, Gubrij I, Plotkin LI, Fu Q, O'Brien CA et al (2005) Chronic elevation of parathyroid hormone in mice reduces expression of sclerostin by osteocytes: a novel mechanism for hormonal control of osteoblastogenesis. Endocrinology 146(11):4577–4583. PubMed PMID: 16081646
40. Keller H, Kneissel M (2005) SOST is a target gene for PTH in bone. Bone 37(2):148–158. PubMed PMID: 15946907
41. Kramer I, Loots GG, Studer A, Keller H, Kneissel M (2010) Parathyroid hormone (PTH)-induced bone gain is blunted in SOST overexpressing and deficient mice. J Bone Miner Res 25(2):178–189. doi:10.1359/jbmr.090730. PubMed PMID: 19594304; PubMed Central PMCID: PMC3153379
42. Wan M, Yang C, Li J, Wu X, Yuan H, Ma H et al (2008) Parathyroid hormone signaling through low-density lipoprotein-related protein 6. Genes Dev 22(21):2968–2979. PubMed PMID: 18981475
43. Romero G, Sneddon WB, Yang Y, Wheeler D, Blair HC, Friedman PA (2010) Parathyroid hormone receptor directly interacts with dishevelled to regulate beta-Catenin signaling and osteoclastogenesis. J Biol Chem 285(19):14756–14763. doi:10.1074/jbc.M110.102970. PubMed PMID: 20212039; PubMed Central PMCID: PMC2863183
44. Canalis E, Centrella M, Burch W, McCarthy TL (1989) Insulin-like growth factor I mediates selective anabolic effects of parathyroid hormone in bone cultures. J Clin Invest 83(1):60–65. PubMed PMID: 2910920
45. Pfeilschifter J, Laukhuf F, Muller-Beckmann B, Blum W, Pfister T, Ziegler R (1995) Parathyroid hormone increases the concentration of insulin-like growth factor-I and transforming growth factor beta 1 in rat bone. J Clin Invest 96:767–774
46. Wang Y, Nishida S, Boudignon BM, Burghardt A, Elalieh HZ, Hamilton MM et al (2007) IGF-I receptor is required for the anabolic actions of parathyroid hormone on bone. J Bone Miner Res 22(9):1329–1337. PubMed PMID: 17539737
47. Hurley MM, Okada Y, Xiao L, Tanaka Y, Ito M, Okimoto N et al (2006) Impaired bone anabolic response to parathyroid hormone in Fgf2–/– and Fgf2+/– mice. Biochem Biophys Res Commun 341(4):989–994. PubMed PMID: 16455048
48. Bergwitz C, Juppner H (2010) Regulation of phosphate homeostasis by PTH, vitamin D, and FGF23. Annu Rev Med 61:91–104
49. Esbrit P et al (2001) Parathyroid hormone-related protein as a renal regulating factor. From vessels to glomeruli and tubular epithelium. Am J Nephrol 21(3):179–184
50. Garabedian M et al (1972) Control of 25-hydroxycholecalciferol metabolism by parathyroid glands. Proc Natl Acad Sci U S A 69(7):1673–1676

51. Korkor AB et al (1987) Evidence that stimulation of 1,25(OH)2D3 production in primary cultures of mouse kidney cells by cyclic AMP requires new protein synthesis. J Bone Miner Res 2(6):517–524
52. Gensure RC, Gardella TJ, Juppner H (2005) Parathyroid hormone and parathyroid hormone-related peptide, and their receptors. Biochem Biophys Res Commun 328(3):666–678
53. Bringhurst FR et al (1993) Cloned, stably expressed parathyroid hormone (PTH)/PTH-related peptide receptors activate multiple messenger signals and biological responses in LLC-PK1 kidney cells. Endocrinology 132(5):2090–2098
54. Janulis M, Tembe V, Favus MJ (1992) Role of protein kinase C in parathyroid hormone stimulation of renal 1,25-dihydroxyvitamin D3 secretion. J Clin Invest 90(6):2278–2283
55. Henry HL (1985) Parathyroid hormone modulation of 25-hydroxyvitamin D3 metabolism by cultured chick kidney cells is mimicked and enhanced by forskolin. Endocrinology 116(2):503–510
56. Guo J et al (2002) The PTH/PTHrP receptor can delay chondrocyte hypertrophy in vivo without activating phospholipase C. Dev Cell 3(2):183–194
57. Guo J et al (2013) Activation of a non-cAMP/PKA signaling pathway downstream of the PTH/PTHrP receptor is essential for a sustained hypophosphatemic response to PTH infusion in male mice. Endocrinology 154(5):1680–1689
58. Shinki T et al (1997) Cloning and expression of rat 25-hydroxyvitamin D3-1alpha-hydroxylase cDNA. Proc Natl Acad Sci U S A 94(24):12920–12925
59. Takeyama K et al (1997) 25-Hydroxyvitamin D3 1alpha-hydroxylase and vitamin D synthesis. Science 277(5333):1827–1830
60. St-Arnaud R et al (1997) The 25-hydroxyvitamin D 1-alpha-hydroxylase gene maps to the pseudovitamin D-deficiency rickets (PDDR) disease locus. J Bone Miner Res 12(10):1552–1559
61. Brenza HL et al (1998) Parathyroid hormone activation of the 25-hydroxyvitamin D3-1alpha-hydroxylase gene promoter. Proc Natl Acad Sci U S A 95(4):1387–1391
62. Murayama A et al (1998) The promoter of the human 25-hydroxyvitamin D3 1 alpha-hydroxylase gene confers positive and negative responsiveness to PTH, calcitonin, and 1 alpha,25(OH)2D3. Biochem Biophys Res Commun 249(1):11–16
63. Hendrix I et al (2005) Response of the 5′-flanking region of the human 25-hydroxyvitamin D 1alpha-hydroxylase gene to physiological stimuli using a transgenic mouse model. J Mol Endocrinol 34(1):237–245
64. Zierold C, Nehring JA, DeLuca HF (2007) Nuclear receptor 4A2 and C/EBPbeta regulate the parathyroid hormone-mediated transcriptional regulation of the 25-hydroxyvitamin D3-1alpha-hydroxylase. Arch Biochem Biophys 460(2):233–239
65. Gao XH et al (2002) Basal and parathyroid hormone induced expression of the human 25-hydroxyvitamin D 1alpha-hydroxylase gene promoter in kidney AOK-B50 cells: role of Sp1, Ets and CCAAT box protein binding sites. Int J Biochem Cell Biol 34(8):921–930
66. Murayama A et al (1999) Positive and negative regulations of the renal 25-hydroxyvitamin D3 1alpha-hydroxylase gene by parathyroid hormone, calcitonin, and 1alpha,25(OH)2D3 in intact animals. Endocrinology 140(5):2224–2231
67. Murayama A et al (2004) Transrepression by a liganded nuclear receptor via a bHLH activator through co-regulator switching. EMBO J 23(7):1598–1608
68. Anderson PH et al (2008) Co-expression of CYP27B1 enzyme with the 1.5 kb CYP27B1 promoter-luciferase transgene in the mouse. Mol Cell Endocrinol 285(1–2):1–9
69. Young MV et al (2004) The prostate 25-hydroxyvitamin D-1 alpha-hydroxylase is not influenced by parathyroid hormone and calcium: implications for prostate cancer chemoprevention by vitamin D. Carcinogenesis 25(6):967–971
70. Biber J et al (1996) Renal Na/Pi-cotransporters. Kidney Int 49(4):981–985
71. Pfister MF et al (1998) Parathyroid hormone leads to the lysosomal degradation of the renal type II Na/Pi cotransporter. Proc Natl Acad Sci U S A 95(4):1909–1914
72. Weinman EJ, Lederer ED (2012) PTH-mediated inhibition of the renal transport of phosphate. Exp Cell Res 318(9):1027–1032
73. Shenolikar S et al (2002) Targeted disruption of the mouse NHERF-1 gene promotes internalization of proximal tubule sodium-phosphate cotransporter type IIa and renal phosphate wasting. Proc Natl Acad Sci U S A 99(17):11470–11475
74. Hernando N et al (2002) PDZ-domain interactions and apical expression of type IIa Na/P(i) cotransporters. Proc Natl Acad Sci U S A 99(18):11957–11962
75. Karim Z et al (2008) NHERF1 mutations and responsiveness of renal parathyroid hormone. N Engl J Med 359(11):1128–1135
76. Cunningham R et al (2006) Adenoviral expression of NHERF-1 in NHERF-1 null mouse renal proximal tubule cells restores Npt2a regulation by low phosphate media and parathyroid hormone. Am J Physiol Renal Physiol 291(4):F896–F901
77. Weinman EJ et al (2007) Parathyroid hormone inhibits renal phosphate transport by phosphorylation of serine 77 of sodium-hydrogen exchanger regulatory factor-1. J Clin Invest 117(11):3412–3420
78. Mahon MJ et al (2002) Na(+)/H(+) exchanger regulatory factor 2 directs parathyroid hormone 1 receptor signalling. Nature 417(6891):858–861
79. Wang B et al (2012) Ezrin-anchored protein kinase A coordinates phosphorylation-dependent disassembly of a NHERF1 ternary complex to regulate hormone-sensitive phosphate transport. J Biol Chem 287(29):24148–24163
80. Guo J et al (2012) Fluorescent ligand-directed co-localization of the parathyroid hormone 1 receptor with the brush-border scaffold complex of the proxi-

mal tubule reveals hormone-dependent changes in ezrin immunoreactivity consistent with inactivation. Biochim Biophys Acta 1823(12):2243–2253
81. Nagai S et al (2011) Acute down-regulation of sodium-dependent phosphate transporter NPT2a involves predominantly the cAMP/PKA pathway as revealed by signaling-selective parathyroid hormone analogs. J Biol Chem 286(2):1618–1626
82. Lau K, Bourdeau JE (1995) Parathyroid hormone action in calcium transport in the distal nephron. Curr Opin Nephrol Hypertens 4(1):55–63
83. Bourdeau JE, Lau K (1989) Effects of parathyroid hormone on cytosolic free calcium concentration in individual rabbit connecting tubules. J Clin Invest 83(2):373–379
84. Lau K, Bourdeau JE (1989) Evidence for cAMP-dependent protein kinase in mediating the parathyroid hormone-stimulated rise in cytosolic free calcium in rabbit connecting tubules. J Biol Chem 264(7):4028–4032
85. Hoenderop JG et al (2003) Renal Ca2+ wasting, hyperabsorption, and reduced bone thickness in mice lacking TRPV5. J Clin Invest 112(12):1906–1914
86. Hoenderop JG, Nilius B, Bindels RJ (2002) Molecular mechanism of active Ca2+ reabsorption in the distal nephron. Annu Rev Physiol 64:529–549
87. van Abel M et al (2005) Coordinated control of renal Ca(2+) transport proteins by parathyroid hormone. Kidney Int 68(4):1708–1721
88. Cha SK, Wu T, Huang CL (2008) Protein kinase C inhibits caveolae-mediated endocytosis of TRPV5. Am J Physiol Renal Physiol 294(5):F1212–F1221
89. de Groot T et al (2009) Parathyroid hormone activates TRPV5 via PKA-dependent phosphorylation. J Am Soc Nephrol 20(8):1693–1704

PTH and PTHrP: Nonclassical Targets

12

Luisella Cianferotti

12.1 Introduction

In the last 25 years, it has become evident that the PTH/PTHrP/PTH receptorsome system is also present in organs not classically related to mineral homeostasis. Indeed, there is accumulating evidence that PTH itself displays multiple effects, which can contribute to the complexity of symptoms of diseases caused by PTH excess or deficiency [1]. In parallel, PTHrP, initially identified as the circulating factor responsible for the development of malignant hypercalcemia in the setting of paraneoplastic syndromes [2], had been found to be expressed in multiple developing and differentiating adult organs and tissues, both in physiology and pathology [3, 4]. In physiology, besides the conditions of pregnancy and lactation, in which PTHrP produced in the breast and uterus, respectively, circulates thus exerting proper endocrine functions (i.e., transplacental transfer of calcium to the rapidly mineralizing skeleton and milk production), PTHrP acts where it is produced and is not detected in the circulation in physiologic states. In particular, as demonstrated in mice devoid of PTHrP or PTH/PTHrP receptor, this molecule displays key paracrine roles in cartilage, bone, and mammary gland development and maintenance; in tooth eruption; and in endocrine pancreas, hair follicle, and smooth muscle physiology [5–8]. PTHrP is a growth factor in several tissues. This implies a potential key role in regeneration, but it can also serve as a tumor-promoting factor.

Human PTH and PTHrP are encoded by different, yet related, genes (*PTH* and *PTHLH*) with different chromosomal locations and likely derive from duplication of a common ancestral sequence (also see Chap. 3). Alternative splicing variants has been found for the *PTHLH* gene, so that the PTHrP group is made of at least three main isoforms of different length (1–139, 1–141, and 1–173), whose transcription is regulated by three different promoter regions (Fig. 12.1). They share a similar N-terminal structure with PTH (first 34 amino acids), so that they can bind and activate the same receptor, although PTH is a more potent agonist than PTHrP [9] (refer to Chap. 2 for further details). The fact that PTHrP can be processed into fragments by posttranslational modifications (proteolysis) further highlights the complexity of this molecule, opening new fields of research (Fig. 12.1).

PTHrP mRNA itself harbors two different transcription initiation sites, so it can be translated into two different principal forms of PTHrP. The first retains a signal peptide which drives PTHrP to the exocytic pathway (endoplasmic reticulum), while the second disrupts the signal peptide and directs PTHrP towards the cytoplasm and, then, the nucleus because of a

L. Cianferotti, MD, PhD
Unit of Bone and Mineral Metabolism, Department of Surgery and Translational Medicine, University of Florence, Viale Pieraccini, 6, Florence 50139, Italy
e-mail: luisella.cianferotti@unifi.it

M.L. Brandi, E.M. Brown (eds.), *Hypoparathyroidism*,
DOI 10.1007/978-88-470-5376-2_12, © Springer-Verlag Italia 2015

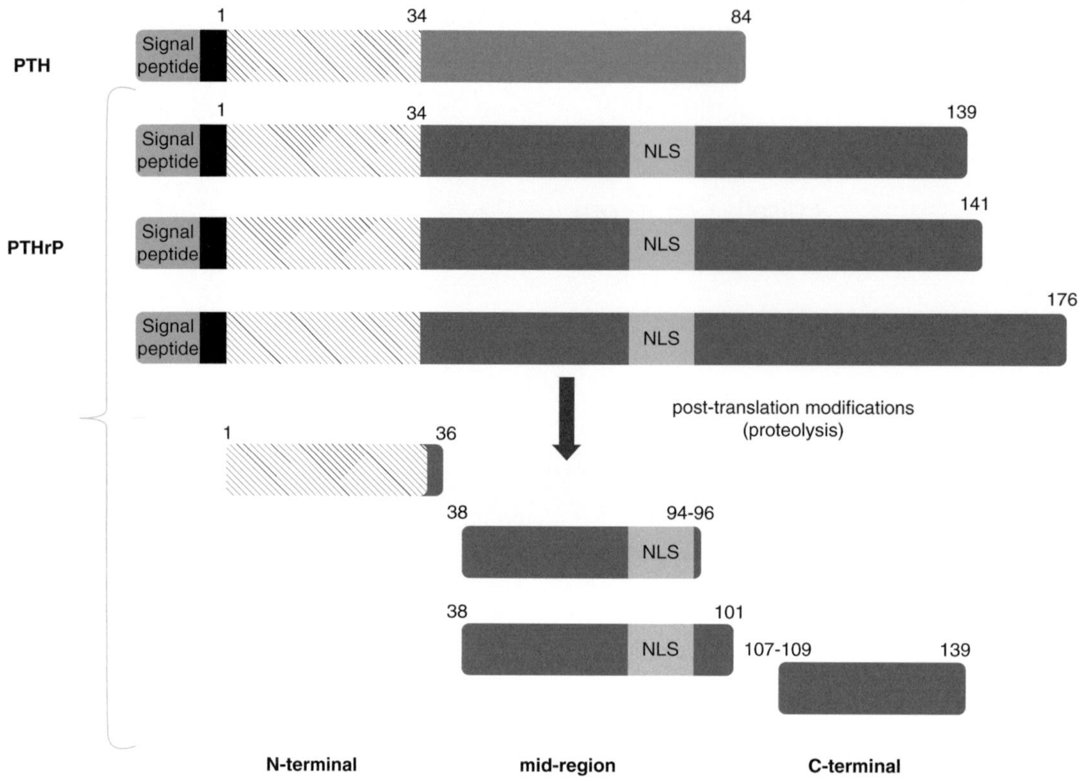

Fig. 12.1 Parathyroid hormone-related peptide (PTHrP): main isoforms resulting from different transcripts (whole forms) and posttranslational processing (fragments). N-terminal fragments (PTHrP1-36) share homology with PTH in the first 34 amino acids (*dashed area*), are secreted, and display PTH-like actions and growth regulatory activities. Mid-region fragments (PTHrP38-94/95/96/101) harbor the nuclear localization signal (NLS, amino acids 84–93), are retained in the cell, trafficking between the cytoplasm and the nucleus, and thus act intracrinally and regulating calcium transport and cell proliferation. C-terminal fragments (PTHrP107/109-11/139, with PTHrP107-111 also referred to as osteostatin), inhibit osteoclast function and stimulate osteoblast proliferation

classic nuclear localization signal (NLS) retained in the C-terminal domain [10, 11]. The form of PTHrP lacking the signal peptide, also referred to as nuclear PTHrP, does not act as a secretory protein [12]. Its intracrine functions are not completely understood and characterized, although it seems to be able to bind to RNA or to the nucleolus thus regulating gene transcription and translation. In some systems where cell proliferation is dysregulated (i.e., colon, breast, and prostate cancer), the pro-apoptotic functions of nuclear PTHrP seem opposite to the pro-apoptotic actions of secreted PTH [13, 14]. It is reasonable to infer that the balance of PTHrP isoforms would be important in determining its overall mitogenic or anti-mitogenic effects.

While the classical calciotropic actions of the PTH/PTHrP system are mediated by circulating PTH, the nonclassical effects are mainly determined by local actions of PTHrP. The locally produced PTHrP acts mainly nearby, thus exerting intracrine-autocrine-paracrine actions. Moreover, its constitutive production is elicited during pathologic conditions. Hence, in general, in the absence of malignant hypercalcemia, nonclassical actions of the PTH/PTHrP/PTH receptorsome system are sustained by circulating PTH (endocrine actions) and locally synthesized PTHrP (autocrine/paracrine actions) and are mediated by parathyroid hormone receptor 1 (PTH1R), which can be activated both by PTH and PTHrP (also see Chap. 3) [15] (Fig. 12.2). In contrast to PTH,

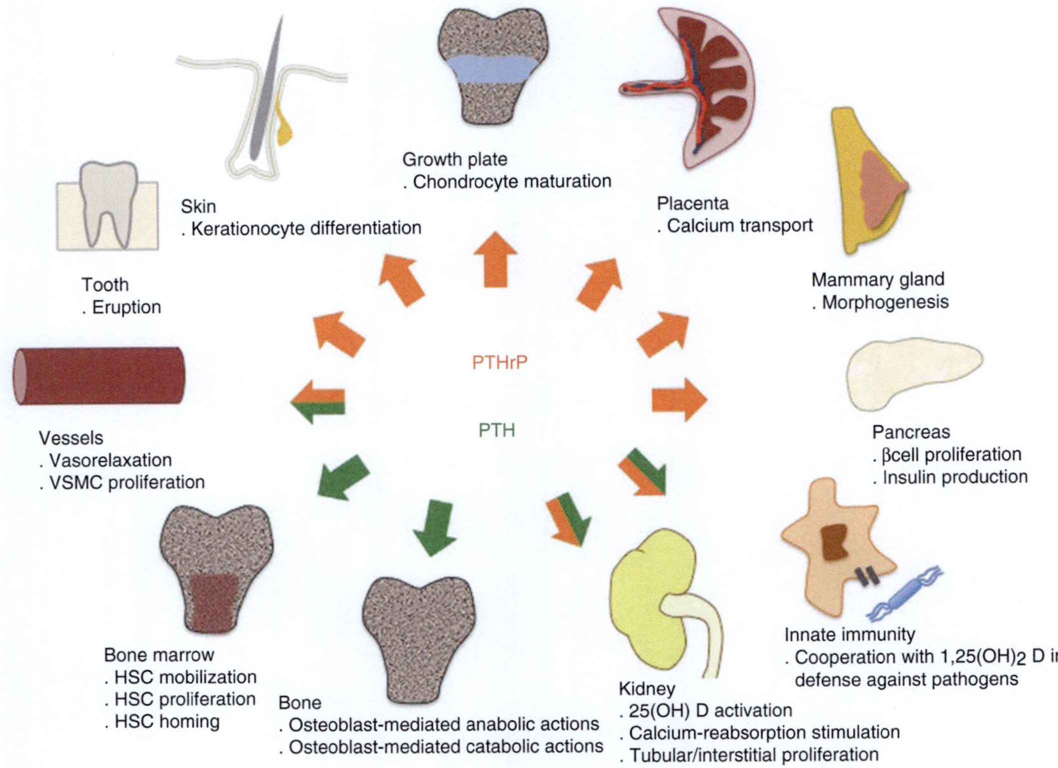

Fig. 12.2 Pleiotropic actions of endocrine PTH (pointed by the *green arrow*), paracrine/autocrine PTHrP (pointed by the *orange arrow*) or both (*dual color arrow*): effects of development of ectodermal and mesenchymal tissues (i.e., bone, cartilage, teeth, mammary gland) and key effects in adult organ physiology

PTHrP is not capable of activating parathyroid hormone receptor 2 (PTH2R), mainly expressed in the nervous system [16], the natural ligand of which is the tuberoinfundibular peptide of 39 residues (TIP39), which displays a modest similarity to PTH and PTHrP but a different 3D structure [17].

In this chapter the main physiological endocrine, paracrine, and autocrine effects of PTH and PTHrP through PTH1R in adult tissues not classically related to the maintenance of mineral and skeletal homeostasis will be described, mostly focusing on recently described effects in the vascular/cardiovascular system, in the hematopoietic system, and in the cell cycle, with the majority of the studies dealing with in vitro and animal studies and fewer translational approaches.

12.2 Actions on Vascular and Cardiovascular Targets

A century ago, pioneering experiments in animal models showed that the infusion of parathyroid extracts had the ability to increase blood flow in several organs and lower systemic blood pressure. Later on, the direct vasorelaxant activity of parathyroid extracts and N-terminal fragments of intact PTH was demonstrated [18]. These vasorelaxant actions of PTH were demonstrated to be mediated by the activation of PTH1R expressed on the vascular smooth muscle myocytes mainly belonging to resistance vessels and to occur independently of PTH-related changes in mineral homeostasis. Intravenous delivery PTH1R itself in rats was shown to reduce blood pressure and directly modulate renin-angiotensin system [19]. Similarly, PTH1R overexpression in smooth muscle cells increased the vasodilatory response to acute saline volume expansion and could counteract the vasoconstrictor effect of angiotensin II [20]. While it appeared clear that the vasorelaxant properties of PTH at nanomolar concentration could counteract the action of vasoactive agents by means of an independent mode of action, these results could not be applied to physiological conditions. Indeed, PTH is normally present in the range of picomolar concentration in the serum. Moreover, conditions where PTH is constitutively high (i.e., primary hyperparathyroidism) can be associated with the development of hypertension.

It then became evident that PTHrP, the molecule responsible for the paraneoplastic syndrome of humoral hypercalcemia of malignancy, is also a vasoactive peptide produced in the endothelium and in the vascular smooth muscle cells (VSMC), and it could be responsible, at least in part, for the autocrine/paracrine activation of the PTHR1 in the cardiovascular system [21]. When injected systemically, PTHrP displays vasodilator and hypotensive properties through nitric oxide-dependent and nitric oxide-independent mechanisms [22, 23]. In humans, infusion of PTHrP at a dose not causing hypotension leads to an important increase in renal blood flow [24]. Recent studies designed to establish the physiologic roles of endogenous PTHrP in the cardiovascular system have demonstrated that selective transient inhibition of endogenous PTHrP in adult mice does affect renal hemodynamics and renin release [25].

The PTHrP/PTHR1 system is also important in vessel development, since knockdown of either PTHrP or PTH1R in zebra fish leads to a aortic coarctation due to altered notch signaling, as it is observed in cases of Blomstrand's chondrodysplasia [26] (refer to Chap. 36 for further details).

Besides vasoactive properties, PTHrP regulates VSMC proliferation, thus potentially playing a key role in vasculoproliferative diseases. Mechanical stretch and vasoconstrictors are able to induce PTHrP in the vascular wall [27]. In vitro experiments had shown that overexpression of PTHrP leads to VSMC proliferation, while overexpression of PTHrP devoid of NLS decreases VSMC number and might prevent in vivo neointimal hyperplasia [11, 28]. Indeed, PTHrP is upregulated in VSMC and triggers neointimal development after vascular injury or a therapeutic procedure such as angioplasty [29]. The regulation of the VSMC cell cycle is due, at least in part, to an induction of the phosphorylation retinoblastoma protein and c-myc-mediated downregulation of the cell-cycle inhibitor p27 [30–32]. Conversely, secreted PTHrP or PTHrP devoid of

NLS is a potent inhibitor of VSMC proliferation, upregulating p27 via PTHR1 signaling triggering and cAMP formation through protein kinase A activation [33–35]. However, secreted PTHrP is capable of downregulating PTH1R, thus contributing to enhances VSMC proliferation [36]. Moreover, effectors of PTH1R signaling, such as NHERF1, which is also upregulated after vascular injury or angioplasty, may inhibit the receptor-mediated antiproliferative actions of secreted PTHrP, thus representing local potential pharmaceutical targets to avoid restenosis after angioplasty procedures [37]. These studies further highlight the key roles of the PTHrP/PTHR1 system in vascular remodeling.

Additional evidence highlights a new role of PTHrP as a proinflammatory cytokine and pro-atherogenic factor. Indeed, PTHrP overexpression in atherosclerotic plaques has been proposed as a mechanism for their instability [38]. Moreover, PTHrP expression in atheromata is regulated by LDL and can be modulated by statins [39]. On the other hand, it has been recently shown that PTHrP could also inhibit apoptosis of coronary endothelial cells by inducing antiapoptotic genes such as bcl-2 and metalloproteinase (TIMP-1) [40]. Thus, also a downregulation of PTHrP could trigger atherosclerosis, since endothelial cell apoptosis is considered a key initial event for the development of atherosclerosis. Ex vivo and in vitro experiments on arteries obtained from patients on chronic hemodialysis and human aortic muscle cells, respectively, point towards a possible role of PTHrP in modulating vascular calcification in hemodialysis patients because of possible interactions with known key players in the process of Runx2 and BMP-2 co-expressed in the vascular wall [41]. Indeed, previous studies have shown that PTHrP secreted in a paracrine manner by the vasculature may limit VSMC calcification [42], but it is known that increased PTH in kidney failure is linked to the development of uremic vasculopathy.

In the last years, along with the use of PTH and its analogs in the treatment of osteoporosis, new attention has been paid to the vasorelaxant and pro- or anti-atherogenic properties of these molecules. In vitro experiments on umbilical vein endothelial cells, expressing PTH1R, have shown that full-length PTH administered in picomolar concentrations, thus mimicking physiological and pathophysiological conditions, is capable of inducing the expression of nitric oxide synthase via c-AMP, providing an explanation for the vasodilation observed after acute treatment with PTH [43]. A recent ex vivo experiment on femoral principal nutrient arteries has further demonstrated that PTH induces vasodilatation in bone arteries through an increase in endothelium-dependent nitric oxide production, which is mediated, at least in part, by VEGF signaling [44]. These effects could provide an explanation for the bone anabolism elicited upon PTH administration in vivo. While intermittent PTH increases microvessel size and prevents the decrease in bone perfusion induced by ovariectomy in mice, continuous PTH produce the opposite effect [45]. Further translational studies are necessary to better characterize the effect on the vasculature of intermittent administration of PTH and PTH analogs versus the possible adverse effects observed in pathophysiologic conditions characterized by persistently elevated PTH levels, where the increase in aldosterone contributes to increased cardiovascular risk [46]. Indeed, full-length PTH is also able of inducing atherosclerotic parameters such as the receptor for advanced glycation end products, the proinflammatory IL-6, and the vascular endothelial growth factor via activation of protein kinase pathways, accounting for a possible PTH-mediated endothelial dysfunction potentially responsible for chronic vascular lesions in the presence of long-standing hyperparathyroidism [47, 48]. It is likely that exposure to high levels of PTH and other circulating factors such as FGF23 might trigger the progression of vascular calcification, as highlighted in recent studies [49].

It had been known for some time that ventricular cardiomyocytes can respond to PTH and PTHrP, which seemed to display opposite and antagonistic effects at a molecular level [50, 51]. Indeed, recombinant full-length PTH(1–84) could induce creatine kinase, which is responsible for cardiomyocytes hypertrophy, while synthetic PTHrP was not able to stimulate creatine kinase and, indeed, antagonized PTH(1–84) actions when administered simultaneously [51]. Several groups have demonstrated that PTHrP is locally produced in the heart (by coronary endothelial cells and cardiomyocytes), especially under hypoxic stress conditions and/or congestive heart failure, and, like PTH, is able to induce vasodilatation, to increase directly heart rate and contractile function [52–56]. In particular, the release of PTHrP from the coronary endothelial cells seems to be dependent, at least in part, on mechanical forces or strains applied by blood flow to the vessel walls [57]. These effects are mediated, at least in part, by a PTH1R-independent mechanisms, such as endothelial cells hyperpolarization, since PTHrP was still able to dilate the vessels in the presence of cAMP/protein kinase antagonists [58]. In vivo studies demonstrated that PTHrP was indeed detectable in the serum of patients with congestive heart failure and its levels were significantly directly correlated with ventricular ejection fraction and heart end-diastolic and end-systolic sizes [59]. The release of PTHrP by cardiomyocytes in conditions of ischemia-reperfusion seems to be estrogen dependent, indicating that PTHrP could improve cardiac performance during postischemic conditions to a greater extent in females than in males [60]. The reduction in the activity of nitric oxide synthase observed in conditions of estrogen deficiency such as menopause or nicotine excess has been shown to downregulate PTH1R, suggesting that PTHrP signaling can be hampered in these conditions [61, 62]. As demonstrated by the aberrant PTHrP action in the myocardium of aged spontaneously hypertensive rats as compared to normotensive animals, the protective effects of PTHrP might be lost in pathologic states and during aging, contributing to the reduction in the ischemic tolerance that usually characterizes these conditions [63].

Although persistently, markedly elevated levels of PTH have been linked to increased cardiovascular risk and mortality [64], PTH levels are positively associated with cardiac function in patients with congestive heart failure. It has been recently shown that cardiomyocytes exposed to low (i.e., picomolar) concentrations of PTH in a nonacute situation (i.e., after 24 h) respond better to electric stimulation in terms of cellular shortening taken as an index of inotropic responsiveness. This suggests that the modestly elevated PTH levels in heart failure can be interpreted as an adaptive response. Nonetheless, further in vivo and translational studies are necessary to assess whether PTH and PTH analogs might improve heart contractility [65].

12.3 Proliferative/Mitogenic Effects of the PTH/PTHrP System

After the first identification of PTHrP as the responsible circulating factor for humoral hypercalcemia of malignancy, it has become evident that PTHrP may primarily act as a growth factor in several developing and adult organs and tissues, and its expression can be increased in proliferative states such as in various types of cancer, where it also regulates angiogenesis, invasiveness, and the progression towards a metastatic phenotype [66, 67].

During the last years several studies have examined whether the increased expression of PTHrP in certain types of tumors is to be considered an epiphenomenon or an index of poor prognosis. In prostate cancer cells, nuclear PTHrP protects cells from anoikis, a form of apoptosis in cells detaching from extracellular matrix, potentially leading cancer cells to acquire metastatic capacity, as indirectly confirmed by the decreased number of skeletal metastatic lesions in mice injected with PTHrP-knockdown prostate cancer cells relative to the ones injected with control cancer cells [14]. Neutralizing antibodies against PTHrP blunt the bone marrow-mediated angiogenic and growth-promoting properties of circulating PTHrP [68]. It has been recently shown that prostate cancer cells overexpressing PTHrP have the ability to promote epithelial-to-mesenchymal transition (i.e., a critical process for cancer invasiveness and metastasis) in vitro and bone metastasis in vivo when injected in nude mice [69]. Several pathways, such as canonical Wnt signaling, which are involved in cancer growth and invasiveness, have been identified as being activated by PTHrP, and they represent a potential target for anticancer drug development [67, 70–74]. Other factors, such as 1,25(OH)$_2$ vitamin D and specific microRNAs, might modulate PTHrP pro-oncogenic and pro-metastatic effects in cancer. Indeed, 1,25(OH)$_2$ vitamin D has been shown to inhibit PTHrP transcription and translation and to promote PTHrP intracellular degradation in prostate cancer cells [75]. MicroRNA33a has been shown to act as a potent tumor suppressor in that it decreases PTHrP expression in lung cancer cells and inhibits PTHrP-induced osteoclastogenesis in vitro [76].

In a mouse model of breast cancer, ablation of PTHrP delayed the initiation and progression of primary and metastatic lesions through modulation of cellular proliferation and angiogenesis [77]. Indeed, in breast cancer patients, expression of PTHrP in the primary lesion was significantly related to an overall decreased survival per se and is associated with an increased risk of bone metastasis if a positive lymph node is also present at baseline [78].

PTHrP plays a key role also in the development of primary bone lesions. In primary stromal cell cultures obtained from patients with giant cell tumor of bone, the addition of a PTHrP neutralizing antibody inhibited cell proliferation and induced apoptosis, suggesting that PTHrP is an autocrine/paracrine inhibitory factor of programmed cell death [79]. Thus, neutralizing PTHrP by means of specific antisera could serve as an anticancer therapy, as suggested by in vitro experiments where anti-PTHrP neutralizing antibodies induced caspase-mediated cell-cycle-mediated apoptosis and modulating cell adhesion, migration, and invasion of giant cell tumor stromal cells [80, 81].

Given the anabolic properties of full-length PTH and its analogs and the pro-proliferative properties of PTHrP, carcinogenicity studies were undertaken to assess whether these substances could induce bone neoplasms in rats. In this animal model, long-term (i.e., 2 years) treatment with both PTH(1–84) at doses >10 μg/kg/day and PTH(1–34) at doses >4.5 μg/kg/day produced osteosarcomas [82, 83]. Given these results, a post marketing 15-year-long surveillance study has assessed the incidence of osteosarcoma in patients with severe osteoporosis treated with PTH(1–34) for up to 2 years but has failed to detect any association between this treatment and the occurrence of osteosarcoma in humans [84], underlying the safety of this drug at the currently employed doses in humans.

12.4 PTH and Regeneration: Actions on Hematopoietic and Endothelial Stem Cells, Nervous System, and the Kidney

The anabolic effects of PTH and PTH analogs have been widely characterized for their effects in mineral and skeletal metabolism, and, as a result of these studies, today PTH analogs have been widely used in severe osteoporosis to enhance osteoblast function (refer to Chaps. 30 and 31 for further details). Nonetheless, there is experimental evidence that other cellular systems (e.g., the hematopoietic stem cell niche) can be both directly and/or indirectly influenced by these compounds, which can then be possibly employed in regenerative medicine. These effects are likely to be mediated by modulation of the osteoblastic and/or vascular niches, defined and restricted microenvironments within the bone marrow where stem cells reside, self-renew, differentiate, and then migrate into the periphery. Even if endothelial cells do not secrete PTH, they are able to produce PTHrP and express PTH1R, thus establishing autocrine/paracrine positive feedback loops.

Preliminary experiments in mice with constitutively active PTH1R in osteoblasts demonstrated an expansion of the pool of hematopoietic stem cells (HSCs) [85]. Since HSCs do not express the PTH1R, this effect was likely to be indirectly mediated by osteoblast actions on HSC niche within the bone marrow [85]. Indeed, PTH produces a cAMP-mediated expansion of the HSC pool and favors the engraftment of transplanted bone marrow. In adult mice, a 5-week treatment with PTH(1–34), administered intraperitoneally at a dose of 80 μg/kg/day followed by a standard HSC mobilization procedure with G-CSF, increased significantly the mobilization of HSCs into the peripheral blood, as measured by the number of colony forming units (CFUs) detected in the peripheral circulation after such a treatment compared to mice treated with PTH or G-CSF alone [86]. The proportion of mobilized mature and progenitor elements in the periphery was not altered by PTH treatment, as compared to treatment with G-CSF alone. Moreover, the administration of PTH for 11 days protected the animals from the reduction in HSC pool in the bone marrow and then in the peripheral circulation that is induced by chemotherapy (cyclophosphamide) and combined pretreatment with G-CSF. PTH administered after bone marrow transplantation led to an increase in the number of HSCs in the bone marrow by expanding the exogenous stem cells received from the donors [86]. Additional studies in healthy mice have further shown that the PTH-dependent mobilization of hematopoietic stem cells occurs without a concurrent depletion of bone marrow, and that this is likely due to the direct actions of PTH on osteoblasts [87]. The mechanism by which PTH induces the expansion of the HSCs involves PTH/PTH1R signaling in T cells and the resulting production of Wnt10b, since mice devoid of PTH1R specifically in T-lymphocytes fail to show any of the above described effects of PTH on HSCs number [88]. These results have highlighted the potential of PTH(1–34) as a stem cell therapy in humans, especially to increase the number of HSCs in particular conditions, such as umbilical cord blood transplantation in adults, whose use is limited by the fact that the available number of HSCs is generally low. Thus, during the last several years, this area of research has become translational, and the results of clinical studies are now available. The effects of a prolonged PTH(1–34) treatment on circulating HSCs have been assessed in postmenopausal women at high risk for fracture receiving this treatment for 2 years [89]. The administration of PTH(1–34) produced an early significant increase in circulating CD34/CD45 positive HSCs (40 % ± 14 % at 3 months, $p = 0.004$ versus baseline) and early transitional B cells, as assessed by flow cytometry, in the absence of any change in blood count profile. This increase persisted up to 18 months and then HSCs number returned to baseline at +24 months [89]. Although the origin of the modulation of HSCs could not be explored (increased proliferation of bone marrow HSCs or augmented peripheral HSC mobilization), this clinical study highlights for the first time the importance of PTH in regulating HSCs in vivo in humans.

The capacity of PTH to induce the mobilization of stem cells has paved the way for new possible applications of this hormone in the field of the early treatment of ischemic disorders. Pioneering experiments in dogs long before the bone anabolic properties of PTH were fully revealed demonstrated that administration of PTH(1–34), as repeated infusions of PTH every 30 min, to dogs in whom acute ischemic injury of the myocardium was induced resulted in a reduction of the infarct size. This was interpreted as a result of the vasoactive relaxant properties of PTH and the increased oxygen supply in the damaged area [90]. Years later, after the evidence that PTH may indirectly act on stem and endothelial cells, in order to further assess the potential and mechanisms of PTH in regeneration after vascular injury, additional studies were undertaken in this field. In mice with experimentally induced myocardial infarction by coronary artery ligation, treatment with PTH(1–34) at a dose of 80 μg/kg/day for 14 days produced an increased number of CD34/CD45 positive progenitor cells recruited within the injured myocardium, an increased level of VEGF mRNA expression leading to enhanced neovascularization, as demonstrated in the sacrificed animals [91]. In living mice, the treatment with PTH improved myocardial function and survival, demonstrating that intermittent PTH can be successfully employed in ischemic cardiomyopathy [91]. Later studies in mice with knockout of granulocyte-colony stimulating factor (G-CSF) have demonstrated that, while the PTH-induced mobilization of stem cells is indeed G-CSF-dependent, the enhanced homing of stem cells in the injured myocardium and the positive outcomes of cardiac function and survival rely entirely on PTH and are due to an increased expression of stromal cell-derived factor-1 (SDF-1) in the injured myocardium [92].

The vascular regenerative properties of PTH have been recently tested also in a major neurovascular disorder such as stroke. PTHrP and the PTH1R are widely expressed in the brain [93]. PTHrP itself is upregulated after an injury and can have a role in maintaining the dedifferentiation state for a proper nerve regeneration [94–96]. Given the actions of PTH on stem cells, a recent study has investigated whether PTH(1–34) could mobilize endogenous stem cells/progenitors cells early after the induction of focal ischemic stroke in adult mice. Indeed, the administration of PTH(1–34) at a dose of 80 μg/kg/day for 6 days induced an increase in circulating CD-34/fetal liver kinase-1 positive endothelial progenitors and increased the expression of trophic and regenerative factors and promoted neuroblast migration in the damaged tissue leading to enhanced angiogenesis. This was reflected by an improved sensorimotor functional recovery compared to control mice [97].

Further studies on the use of PTH in a translational setting are needed to establish its possible use in regenerative medicine, both after bone marrow transplantation or after ischemic injury.

The PTH1R and PTHrP are abundantly expressed in the kidney, both in the renal parenchyma and the vasculature. Locally produced PTHrP has been shown to regulate renal blood flow and glomerular filtration rate and the proliferation of glomerular mesangial cells and tubular epithelial cells [98]. As originally demonstrated in experimental models of nephropathy, PTHrP is transiently upregulated in the kidney in the very first stages of acute renal failure, obstructive and diabetic nephropathy, and behaved unexpectedly as a proinflammatory agent and not as a regenerative factor [99–104]. Interestingly, the upregulation of PTHrP in the damaged kidney seemed to be mediated, at least in part, by angiotensin II, since pretreatment with angiotensin II blockers prevented the increase in the expression of PTHrP after the administration of nephrotoxins, suggesting one mechanism by which these drugs offer protective effects in renal disease [103, 105, 106]. To dissect the effects of PTHrP in the kidney, mice overexpressing this growth factor in the renal proximal tubule cells were generated. In this murine model, overexpression of PTHrP in the renal proximal tubule failed to offer any protection against ischemia-induced renal failure, which was initially explained by the concomitant reduction in PTH1R expression due to renal injury [107]. Moreover, in a model of folic acid-induced nephropathy, mice overexpressing PTHrP in the proximal nephron displayed a massive tubulointerstitial fibrosis leading to sustained impairment of renal function in the long term as compared to control littermates [108]. This occurred mainly through an inhibition of apoptosis of interstitial fibroblasts, a key mechanisms in the development of renal fibrogenesis in damaged kidneys [108], and through cooperation with VEGF and other growth factors such as TGF-beta and RGF in inducing epithelial-mesenchymal transition of the renal tubuloepithelium [109, 110]. In diabetic nephropathy, hypertrophy of mesangial cells is a key early event. In ex vivo experiments on kidneys from diabetic patients, increased tubular and glomerular immunostaining for PTHrP was observed and strongly associated with hypertrophy as estimated by cell protein content and observed in vitro upon administration of PTHrP(1–36) in primary cultures of human mesangial cells cultured with high glucose compared with controls [111, 112]. Recent in vitro studies on murine mesangial cells devoid of PTHrP and transfected with wild-type PTHrP, PTHrP lacking its signal peptide, or PTHrP lacking the NLS have further dissected the mechanisms and the pathways involved in PTHrP effects in this cellular system [113]. Transfection of cells with PTHrP lacking the signal peptide displayed an increased proliferation rate, with minimal changes in apoptosis, while transfection with PTHrP lacking the NLS protected cells from apoptosis. Thus, PTHrP acts as a mitogenic factor in an autocrine way and as an antiapoptotic agent acting in an intracrine manner [113].

Besides the reduction in apoptosis of interstitial mesangial cells, overexpression of PTHrP has been shown to inhibit apoptosis of tubular epithelial cells via a Runx2-dependent mechanism, which also has been demonstrated to be implicated in the epithelial-to-mesenchymal transition in some metastatic tumors. Indeed, increased expression of Runx2, osteopontin, and Bcl-2 was demonstrated in the proximal kidney in association with the antiapoptotic effects [114]. These lines of evidence point towards a protective role of PTHrP in enhancing cell survival in renal tubules after renal damage and an additional pro-regenerative effect in this system. Further studies need to be performed to assess whether administration of PTHrP or PTH/PTH analogs in the long term could modify the outcome of kidney injury.

12.5 Other Nonclassical Actions of the PTH/PTHrP/PTH1R System: Effects on Endocrine Pancreas, Adrenals, and Innate Immune System

Endocrine organs nonclassically related to the maintenance of calcium and skeletal homeostasis can constitute the target of the PTH/PTHrP system. Pancreatic β-cells express the PTH1R [115]. Preliminary studies demonstrated local production of PTHrP in rat and human endocrine pancreas (e.g., in all four cell types, α, β, δ, and pancreatic polypeptide cells), both in normal and neoplastic islets of Langerhans (insulinomas), suggesting a putative role in pancreatic physiology and pathophysiology [116]. Mitogenic effects on pancreatic islet cells were shown in vitro by N-terminal PTHrP peptides capable of binding to the PTH1R [117]. From a functional point of view, transgenic mice overexpressing PTHrP selectively in the β-cell under control of the rat insulin promoter displayed an increased islet mass apparently and unexpectedly not the result of either enhanced proliferation or cellular hyperplasia and hyperinsulinemia with consequent hypoglycemia [118, 119]. The fact that the β-cells overexpressing PTHrP were resistant to the cytotoxic, diabetogenic effect of streptozotocin suggested that the main mechanism by which PTHrP-induced pancreatic islet hyperplasia was inhibition of β-cell death, i.e., by apoptosis [120]. In vitro experiments in which PTHrP was overexpressed confirmed the ability of PTHrP to induce growth and increase insulin secretion in well-differentiated β-cell lines and further demonstrated that these effects occurred through activation of protein kinase pathways with the increased cAMP production of [121] and activation of MAP kinase-specific phosphatase 1 downregulating c-jun NH2-terminal kinase (JNK) [122]. Additional experiments have demonstrated that PTHrP(1–36) induces specific cell cycle activators (cyclin-dependent kinase 2 and cyclin E) [123]. These actions are reflected in vivo in mice, in which the systemic administration of PTHrP(1–36) enhances β-cell proliferation [124]. Even if the role of PTHrP in the pancreas is fully not clear under physiologic conditions, all these observations support the concept that PTHrP could act as an autocrine/paracrine growth factor within the endocrine pancreas regulating islet mass and function, suggesting a potential use of PTHrP in type 1 diabetes and islet transplantation. Nonetheless, pancreatic PTHrP can be released into the circulation, potentially acting as an endocrine factor. An in vivo study demonstrated that serum PTHrP levels in patients with type 2 diabetes were detectable and higher relative to controls and increased in parallel to insulin in response to a glucose load [125]. Indeed, PTHrP and insulin are co-packaged into secretory vesicles and are released together upon the same stimuli. Additional in vivo and translational studies are necessary to explore the endocrine role of pancreatic PTHrP and its putative role in the modulation of insulin resistance.

PTH(1–84), PTH(1–34), and PTHrP stimulate aldosterone secretion in vitro acting on the PTH1R expressed in the zona glomerulosa cells and activating both the adenylate cyclase/protein kinase A and phospholipase C (PKC) pathways [126–129]. In addition, PTHrP expression has been documented in adrenocortical tumors, where it acts as a growth factor [130]. In primary hyperparathyroidism, the increased morbidity and mortality resulting from cardiovascular diseases have been attributed, at least in part, to the increase in serum aldosterone, which is directly linked to a greater risk for atherosclerotic cardiovascular disease [131]. After parathyroidectomy, aldosterone levels return to normal, and this is linked to a parallel improvement in cardiovascular outcomes (i.e., decrease in blood pressure and in cardiovascular morbidity) in most studies [46, 132]. Indeed, after parathyroidectomy a decrease in aldosterone levels in parallel with improved cardiovascular outcomes is observed. PTH and aldosterone are closely related in a complex interplay. Aldosterone per se modulates PTH secretion acting directly on the mineralocorticoid receptor expressed on the parathyroid cell and indirectly through induction of hypocalcemia and hypomagnesemia owing to the mineralocorticoid receptor-mediated hypercalciuric effect [46]. Secondary hyperparathyroidism induced by aldosterone excess, such as in primary hyperaldosteronism, is blunted by adrenalectomy or MR blockade.

Recent studies have ascribed a new role to PTH and PTHrP in innate immunity [133].

Cathelicidin, an antimicrobial peptide produced in cells belonging to the innate immune system, such as macrophages, is also produced in keratinocytes in response to pathogens. Active vitamin D is able to enhance cathelicidin production in macrophages in response to pathogens such as *Mycobacterium tuberculosis* [134]. Active vitamin D can induce the expression of the PTH1R in keratinocytes. On the other hand, PTH can enhance $1{,}25(OH)_2$ vitamin D production in this cellular system, thus establishing a positive feedback loop. In keratinocytes, PTH and PTHrP cooperate with active vitamin D in increasing cathelicidin expression in response to pathogens such as group A *Streptococcus*, likely through modulation of epigenetic mechanisms (i.e., DNA methylation). In this view, the development of secondary hyperparathyroidism in response to vitamin D deficiency might compensate for the decreased vitamin D and its role in innate immune system, thereby participating in the protection against infections [134]. These studies further expand the known pleiotropic actions of PTH/PTHrP, adding new unpredicted roles to this complex hormonal system.

Conclusion

Novel, formerly unexpected effects in organs nonclassically related to mineral homeostasis are today attributed to PTH and PTHrP. These molecules act primarily through the specific receptor, PTH1R, expressed in many tissues. While PTH exerts its actions systemically, PTHrP is an autocrine/paracrine/intracrine factor, with distinct abilities depending on whether it is retained in the cell or secreted into the extracellular milieu. Additional studies are needed to assess the various properties of nuclear and secreted PTHrP in different tissues (e.g., cardiovascular system, cancer, and chronic kidney disease), especially in light of its potential for targeted treatments of a number of diseases.

Further translational studies are needed in order to assess specific extraskeletal effects of PTH and PTH analogs. Besides their known anabolic actions on the osteoblast, the proven effects of these peptides on HSC open new possibilities for the application of these molecules to the field of regenerative medicine.

References

1. Bro S, Olgaard K (1997) Effect of excess PTH on nonclassical target organs. Am J Kidney Dis 30:606–620
2. Suva LJ, Winslow GA, Wettenhall RE et al (1987) A parathyroid hormone-related protein implicated in malignant hypercalcemia: cloning and expression. Science 237:893–896
3. Asa SL, Henderson J, Goltzman D et al (1990) Parathyroid hormone-like peptide in normal and neoplastic human endocrine tissues. J Clin Endocrinol Metab 71:1112–1118
4. Wysolmerski JJ (2012) Parathyroid hormone-related protein: an update. J Clin Endocrinol Metab 97:2947–2956
5. Philbrick WM, Wysolmerski JJ, Galbraith S et al (1996) Defining the roles of parathyroid hormone-related protein in normal physiology. Physiol Rev 76:127–173
6. McCauley LK, Martin TJ (2012) Twenty-five years of PTHrP progress: from cancer hormone to multifunctional cytokine. J Bone Miner Res 27:1231–1239
7. El-Hashash AH, Esbrit P, Kimber SJ (2005) PTHrP promotes murine secondary trophoblast giant cell differentiation through induction of endocycle,

upregulation of giant-cell-promoting transcription factors and suppression of other trophoblast cell types. Differentiation 73:154–174
8. Hiremath M, Wysolmerski J (2013) Parathyroid hormone-related protein specifies the mammary mesenchyme and regulates embryonic mammary development. J Mammary Gland Biol Neoplasia 18:171–177
9. Gensure RC, Gardella TJ, Jüppner H (2005) Parathyroid hormone and parathyroid hormone-related peptide, and their receptors. Biochem Biophys Res Commun 328:666–678
10. Nguyen MT, Karaplis AC (1998) The nucleus: a target site for parathyroid hormone-related peptide (PTHrP) action. J Cell Biochem 70:193–199
11. de Miguel F, Fiaschi-Taesch N, López-Talavera JC et al (2001) The C-terminal region of PTHrP, in addition to the nuclear localization signal, is essential for the intracrine stimulation of proliferation in vascular smooth muscle cells. Endocrinology 142:4096–4105
12. Stewart AF, Fiaschi-Taesch NM (2003) Minireview: parathyroid hormone-related protein as an intracrine factor—trafficking mechanisms and functional consequences. Endocrinology 144:407–411
13. Shen X, Qian L, Falzon M (2004) PTH-related protein enhances MCF-7 breast cancer cell adhesion, migration, and invasion via an intracrine pathway. Exp Cell Res 294:420–433
14. Park SI, McCauley LK (2012) Nuclear localization of parathyroid hormone-related peptide confers resistance to anoikis in prostate cancer cells. Endocr Relat Cancer 19:243–254
15. Clemens TL, Cormier S, Eichinger A et al (2001) Parathyroid hormone-related protein and its receptors: nuclear functions and roles in the renal and cardiovascular systems, the placental trophoblasts and the pancreatic islets. Br J Pharmacol 134:1113–1136
16. Dobolyi A, Dimitrov E, Palkovits M et al (2012) The neuroendocrine functions of the parathyroid hormone 2 receptor. Front Endocrinol (Lausanne) 3:121. doi:10.3389/fendo.2012.00121
17. John MR, Arai M, Rubin DA et al (2002) Identification and characterization of the murine and human gene encoding the tuberoinfundibular peptide of 39 residues. Endocrinology 143:1047–1057
18. Mok LL, Nickols GA, Thompson JC et al (1989) Parathyroid hormone as a smooth muscle relaxant. Endocr Rev 10:420–436
19. Fritsch S, Lindner V, Welsch S et al (2004) Intravenous delivery of PTH/PTHrP type 1 receptor cDNA to rats decreases heart rate, blood pressure, renal tone, renin angiotensin system, and stress-induced cardiovascular responses. J Am Soc Nephrol 15:2588–2600
20. Noonan WT, Qian J, Stuart WD et al (2003) Altered renal hemodynamics in mice overexpressing the parathyroid hormone (PTH)/PTH-related peptide type 1 receptor in smooth muscle. Endocrinology 144:4931–4938
21. Jiang B, Morimoto S, Yang J et al (1998) Expression of parathyroid hormone/parathyroid hormone-related protein receptor in vascular endothelial cells. J Cardiovasc Pharmacol 31:S142–S144
22. Sutliff RL, Weber CS, Qian J et al (1999) Vasorelaxant properties of parathyroid hormone-related protein in the mouse: evidence for endothelium involvement independent of nitric oxide formation. Endocrinology 140:2077–2083
23. Kalinowski L, Dobrucki LW, Malinski T (2001) Nitric oxide as a second messenger in parathyroid hormone-related protein signaling. J Endocrinol 170:433–440
24. Wolzt M, Schmetterer L, Dorner G et al (1997) Hemodynamic effects of parathyroid hormone-related peptide-(1–34) in humans. J Clin Endocrinol Metab 82:2548–2551
25. Raison D, Coquard C, Hochane M et al (2013) Knockdown of parathyroid hormone related protein in smooth muscle cells alters renal hemodynamics but not blood pressure. Am J Physiol Renal Physiol 305:F333–F342
26. Gray C, Bratt D, Lees J et al (2013) Loss of function of parathyroid hormone receptor 1 induces Notch-dependent aortic defects during zebrafish vascular development. Arterioscler Thromb Vasc Biol 33:1257–1263
27. Schordan E, Welsch S, Rothhut S et al (2004) Role of parathyroid hormone-related protein in the regulation of stretch-induced renal vascular smooth muscle cell proliferation. J Am Soc Nephrol 15:3016–3025
28. Massfelder T, Dann P, Wu TL et al (1997) Opposing mitogenic and anti-mitogenic actions of parathyroid hormone-related protein in vascular smooth muscle cells: a critical role for nuclear targeting. Proc Natl Acad Sci U S A 94:13630–13635
29. Ishikawa M, Akishita M, Kozaki K et al (2000) Expression of parathyroid hormone-related protein in human and experimental atherosclerotic lesions: functional role in arterial intimal thickening. Atherosclerosis 152:97–105
30. Fiaschi-Taesch N, Takane KK, Masters S et al (2004) Parathyroid-hormone-related protein as a regulator of pRb and the cell cycle in arterial smooth muscle. Circulation 110:177–185
31. Fiaschi-Taesch N, Sicari BM, Ubriani K et al (2006) Cellular mechanism through which parathyroid hormone-related protein induces proliferation in arterial smooth muscle cells: definition of an arterial smooth muscle PTHrP/p27kip1 pathway. Circ Res 99:933–942
32. Sicari BM, Troxell R, Salim F et al (2012) c-myc and skp2 coordinate p27 degradation, vascular smooth muscle proliferation, and neointima formation induced by the parathyroid hormone-related protein. Endocrinology 153:861–872

33. Stuart WD, Maeda S, Khera P et al (2000) Parathyroid hormone-related protein induces G1 phase growth arrest of vascular smooth muscle cells. Am J Physiol Endocrinol Metab 279:E60–E67
34. Bakre MM, Zhu Y, Yin H, Burton DW et al (2002) Parathyroid hormone-related peptide is a naturally occurring, protein kinase A-dependent angiogenesis inhibitor. Nat Med 8:995–1003
35. Fiaschi-Taesch N, Sicari B, Ubriani K et al (2009) Mutant parathyroid hormone-related protein, devoid of the nuclear localization signal, markedly inhibits arterial smooth muscle cell cycle and neointima formation by coordinate up-regulation of p15Ink4b and p27kip1. Endocrinology 150:1429–1439
36. Song GJ, Fiaschi-Taesch N, Bisello A (2009) Endogenous parathyroid hormone-related protein regulates the expression of PTH type 1 receptor and proliferation of vascular smooth muscle cells. Mol Endocrinol 23:1681–1690
37. Song GJ, Barrick S, Leslie KL et al (2010) EBP50 inhibits the anti-mitogenic action of the parathyroid hormone type 1 receptor in vascular smooth muscle cells. J Mol Cell Cardiol 49:1012–1021
38. Martín-Ventura JL, Ortego M, Esbrit P et al (2003) Possible role of parathyroid hormone-related protein as a proinflammatory cytokine in atherosclerosis. Stroke 34:1783–1789
39. Martin-Ventura JL, Blanco-Colio LM, Aparicio C et al (2008) LDL induces parathyroid hormone-related protein expression in vascular smooth muscle cells: modulation by simvastatin. Atherosclerosis 198:264–271
40. Conzelmann C, Krasteva G, Weber K et al (2009) Parathyroid hormone-related protein (PTHrP)-dependent regulation of bcl-2 and tissue inhibitor of metalloproteinase (TIMP)-1 in coronary endothelial cells. Cell Physiol Biochem 24:493–502
41. Liu F, Fu P, Fan W et al (2012) Involvement of parathyroid hormone-related protein in vascular calcification of chronic haemodialysis patients. Nephrology (Carlton) 17:552–560
42. Jono S, Nishizawa Y, Shioi A et al (1997) Parathyroid hormone-related peptide as a local regulator of vascular calcification. Its inhibitory action on in vitro calcification by bovine vascular smooth muscle cells. Arterioscler Thromb Vasc Biol 17:1135–1142
43. Rashid G, Bernheim J, Green J et al (2007) Parathyroid hormone stimulates the endothelial nitric oxide synthase through protein kinase A and C pathways. Nephrol Dial Transplant 22:2831–2837
44. Prisby R, Menezes T, Campbell J (2013) Vasodilation to PTH (1–84) in bone arteries is dependent upon the vascular endothelium and is mediated partially via VEGF signaling. Bone 54:68–75
45. Roche B, Vanden-Bossche A, Malaval L et al (2014) Parathyroid hormone 1–84 targets bone vascular structure and perfusion in mice: impacts of its administration regimen and of ovariectomy. J Bone Miner Res. doi:10.1002/jbmr.2191
46. Tomaschitz A, Ritz E, Pieske B et al (2014) Aldosterone and parathyroid hormone interactions as mediators of metabolic and cardiovascular disease. Metabolism 63:20–31
47. Rashid G, Bernheim J, Green J et al (2007) Parathyroid hormone stimulates endothelial expression of atherosclerotic parameters through protein kinase pathways. Am J Physiol Renal Physiol 292:F1215–F1218
48. Rashid G, Bernheim J, Green J et al (2008) Parathyroid hormone stimulates the endothelial expression of vascular endothelial growth factor. Eur J Clin Invest 38:798–803
49. Jean G, Bresson E, Lorriaux C et al (2012) Increased levels of serum parathyroid hormone and fibroblast growth factor-23 are the main factors associated with the progression of vascular calcification in long-hour hemodialysis patients. Nephron Clin Pract 120:c132–c138
50. Bui TD, Shallal A, Malik AN et al (1993) Parathyroid hormone-related peptide gene expression in human fetal and adult heart. Cardiovasc Res 127:1204–1208
51. Schlüter KD, Wingender E, Tegge W et al (1996) Parathyroid hormone-related protein antagonizes the action of parathyroid hormone on adult cardiomyocytes. J Biol Chem 271:3074–3078
52. Crass MF III, Pang PKT (1980) Parathyroid hormone: a coronary artery vasodilator. Science 207:1087–1089
53. Deftos LJ, Burton DW, Brandt DW (1993) Parathyroid hormone-like protein (PLP) is a secretory product of atrial myocytes. J Clin Invest 92:727–735
54. Schlüter K, Katzer C, Frischkopf K et al (2000) Expression, release, and biological activity of parathyroid hormone-related peptide from coronary endothelial cells. Circ Res 86:946–951
55. Jansen J, Gres P, Umschlag C et al (2003) Parathyroid hormone-related peptide improves contractile function of stunned myocardium in rats and pigs. Am J Physiol Heart Circ Physiol 284:H49–H55
56. Monego G, Arena V, Pasquini S et al (2009) Ischemic injury activates PTHrP and PTH1R expression in human ventricular cardiomyocytes. Basic Res Cardiol 104:427–434
57. Degenhardt H, Jansen J, Schulz R et al (2002) Mechanosensitive release of parathyroid hormone-related peptide from coronary endothelial cells. Am J Physiol Heart Circ Physiol 283:H1489–H1496
58. Abdallah Y, Ross G, Dolf A et al (2006) N-terminal parathyroid hormone-related peptide hyperpolarizes endothelial cells and causes a reduction of the coronary resistance of the rat heart via endothelial hyperpolarization. Peptides 27:2927–2934
59. Ogino K, Ogura K, Kinugasa Y et al (2002) Parathyroid hormone-related protein is produced in the myocardium and increased in patients with

congestive heart failure. J Clin Endocrinol Metab 87:4722–4727
60. Grohe C, van Eickels M, Wenzel S et al (2004) Sex-specific differences in ventricular expression and function of parathyroid hormone-related peptide. Cardiovasc Res 61:307–316
61. Schreckenberg R, Wenzel S, da Costa Rebelo RM et al (2009) Cell-specific effects of nitric oxide deficiency on parathyroid hormone-related peptide (PTHrP) responsiveness and PTH1 receptor expression in cardiovascular cells. Endocrinology 150:3735–3741
62. Röthig A, Schreckenberg R, Weber K et al (2012) Effects of nicotine on PTHrP and PTHrP receptor expression in rat coronary endothelial cells. Cell Physiol Biochem 29:485–492
63. Ross G, Schlüter KD (2005) Cardiac-specific effects of parathyroid hormone-related peptide: modification by aging and hypertension. Cardiovasc Res 66:334–344
64. Grandi NC, Breitling LP, Hahmann H et al (2011) Serum parathyroid hormone and risk of adverse outcomes in patients with stable coronary heart disease. Heart 97:1215–1221
65. Tastan I, Schreckenberg R, Mufti S et al (2009) Parathyroid hormone improves contractile performance of adult rat ventricular cardiomyocytes at low concentrations in a non-acute way. Cardiovasc Res 82:77–83
66. Akino K, Ohtsuru A, Kanda K et al (2000) Parathyroid hormone-related peptide is a potent tumor angiogenic factor. Endocrinology 141:4313–4316
67. Kremer R, Li J, Camirand A, Karaplis AC (2011) Parathyroid hormone related protein (PTHrP) in tumor progression. Adv Exp Med Biol 720:145–160
68. Park SI, Lee C, Sadler WD et al (2013) Parathyroid hormone-related protein drives a CD11b+Gr1+ cell-mediated positive feedback loop to support prostate cancer growth. Cancer Res 73:6574–6583
69. Ongkeko WM, Burton D, Kiang A et al (2014) Parathyroid hormone related-protein promotes epithelial-to-mesenchymal transition in prostate cancer. PLoS One 9(1):e85803
70. Downs TM, Burton DW, Araiza FL et al (2011) PTHrP stimulates prostate cancer cell growth and upregulates aldo-keto reductase 1C3. Cancer Lett 306:52–59
71. Danilin S, Sourbier C, Thomas L et al (2009) von Hippel-Lindau tumor suppressor gene-dependent mRNA stabilization of the survival factor parathyroid hormone-related protein in human renal cell carcinoma by the RNA-binding protein HuR. Carcinogenesis 30:387–396
72. Mula RV, Bhatia V, Falzon M (2010) PTHrP promotes colon cancer cell migration and invasion in an integrin α6β4-dependent manner through activation of Rac1. Cancer Lett 298:119–127
73. Zhang H, Yu C, Dai J et al (2013) Parathyroid hormone-related protein inhibits DKK1 expression through c-Jun-mediated inhibition of β-catenin activation of the DKK1 promoter in prostate cancer. Oncogene. doi:10.1038/onc.2013.203
74. Bhatia V, Mula RV, Falzon M (2013) Parathyroid hormone-related protein regulates integrin α6 and β4 levels via transcriptional and post-translational pathways. Exp Cell Res 319:1419–1430
75. Bhatia V, Mula RV, Falzon M (2011) 1,25-Dihydroxyvitamin D(3) regulates PTHrP expression via transcriptional, post-transcriptional and post-translational pathways. Mol Cell Endocrinol 342:32–40
76. Kuo PL, Liao SH, Hung JY et al (2013) MicroRNA-33a functions as a bone metastasis suppressor in lung cancer by targeting parathyroid hormone related protein. Biochim Biophys Acta 1830:3756–3766
77. Li J, Karaplis AC, Huang DC et al (2011) PTHrP drives breast tumor initiation, progression, and metastasis in mice and is a potential therapy target. J Clin Invest 121:4655–4669
78. Takagaki K, Takashima T, Onoda N et al (2012) Parathyroid hormone-related protein expression, in combination with nodal status, predicts bone metastasis and prognosis of breast cancer patients. Exp Ther Med 3:963–968
79. Mak IW, Cowan RW, Turcotte RE et al (2011) PTHrP induces autocrine/paracrine proliferation of bone tumor cells through inhibition of apoptosis. PLoS One 6:e19975. doi:10.1371/journal.pone.0019975
80. Mak IW, Turcotte RE, Ghert M (2012) Transcriptomic and proteomic analyses in bone tumor cells: deciphering parathyroid hormone-related protein regulation of the cell cycle and apoptosis. J Bone Miner Res 27:1976–1991
81. Mak IW, Turcotte RE, Ghert M (2013) Parathyroid hormone-related protein (PTHrP) modulates adhesion, migration and invasion in bone tumor cells. Bone 55:198–207
82. Jolette J, Wilker CE, Smith SY et al (2006) Defining a noncarcinogenic dose of recombinant human parathyroid hormone 1–84 in a 2-year study in Fischer 344 rats. Toxicol Pathol 34:929–940
83. Watanabe A, Yoneyama S, Nakajima M et al (2012) Osteosarcoma in Sprague–Dawley rats after long-term treatment with teriparatide (human parathyroid hormone (1–34)). J Toxicol Sci 37:617–629
84. Andrews EB, Gilsenan AW, Midkiff K et al (2012) The US postmarketing surveillance study of adult osteosarcoma and teriparatide: study design and findings from the first 7 years. J Bone Miner Res 27:2429–2437
85. Calvi LM, Adams GB, Weibrecht KW et al (2003) Osteoblastic cells regulate the haematopoietic stem cell niche. Nature 425:841–846
86. Adams GB, Martin RP, Alley IR et al (2007) Therapeutic targeting of a stem cell niche. Nat Biotechnol 25:238–243
87. Brunner S, Zaruba MM, Huber B et al (2008) Parathyroid hormone effectively induces mobilization

of progenitor cells without depletion of bone marrow. Exp Hematol 36:1157–1166
88. Li JY, Adams J, Calvi LM et al (2012) PTH expands short-term murine hemopoietic stem cells through T cells. Blood 120:4352–4362
89. Yu EW, Kumbhani R, Siwila-Sackman E et al (2014) Teriparatide (PTH 1–34) treatment increases peripheral hematopoietic stem cells in postmenopausal women. J Bone Miner Res. doi:10.1002/jbmr.2171
90. Feola M, Crass MF 3rd (1986) Parathyroid hormone reduces acute ischemic injury of the myocardium. Surg Gynecol Obstet 163:523–530
91. Zaruba MM, Huber BC, Brunner S et al (2008) Parathyroid hormone treatment after myocardial infarction promotes cardiac repair by enhanced neovascularization and cell survival. Cardiovasc Res 77:722–731
92. Brunner S, Weinberger T, Huber BC et al (2012) The cardioprotective effects of parathyroid hormone are independent of endogenous granulocyte-colony stimulating factor release. Cardiovasc Res 93:330–339
93. Weir EC, Brines ML, Ikeda K et al (1990) Parathyroid hormone-related peptide gene is expressed in the mammalian central nervous system. Proc Natl Acad Sci U S A 87:108–112
94. Chatterjee O, Nakchbandi IA, Philbrick WM et al (2002) Endogenous parathyroid hormone-related protein functions as a neuroprotective agent. Brain Res 930:58–66
95. Funk JL, Trout CR, Wei H et al (2001) Parathyroid hormone-related protein (PTHrP) induction in reactive astrocytes following brain injury: a possible mediator of CNS inflammation. Brain Res 915:195–209
96. Macica CM, Liang G, Lankford KL et al (2006) Induction of parathyroid hormone-related peptide following peripheral nerve injury: role as a modulator of Schwann cell phenotype. Glia 53:637–648
97. Wang LL, Chen D, Lee J et al (2014) Mobilization of endogenous bone marrow derived endothelial progenitor cells and therapeutic potential of parathyroid hormone after ischemic stroke in mice. PLoS One 9(2):e87284. doi:10.1371/journal.pone.0087284
98. Esbrit P, Santos S, Ortega A et al (2001) Parathyroid hormone-related protein as a renal regulating factor. From vessels to glomeruli and tubular epithelium. Am J Nephrol 21:179–184
99. Soifer NE, Van Why SK, Ganz MB et al (1993) Expression of parathyroid hormone-related protein in the rat glomerulus and tubule during recovery from renal ischemia. J Clin Invest 92:2850–2857
100. Largo R, Gómez-Garre D, Santos S et al (1999) Renal expression of parathyroid hormone-related protein (PTHrP) and PTH/PTHrP receptor in a rat model of tubulointerstitial damage. Kidney Int 55:82–90
101. Santos S, Bosch RJ, Ortega A et al (2001) Up-regulation of parathyroid hormone-related protein in folic acid-induced acute renal failure. Kidney Int 60:982–995
102. Lorenzo O, Ruiz-Ortega M, Esbrit P et al (2002) Angiotensin II increases parathyroid hormone-related protein (PTHrP) and the type 1 PTH/PTHrP receptor in the kidney. J Am Soc Nephrol 13:1595–1607
103. Izquierdo A, López-Luna P, Ortega A et al (2006) The parathyroid hormone-related protein system and diabetic nephropathy outcome in streptozotocin-induced diabetes. Kidney Int 69:2171–2177
104. Rámila D, Ardura JA, Esteban V et al (2008) Parathyroid hormone-related protein promotes inflammation in the kidney with an obstructed ureter. Kidney Int 73:835–847
105. Ortega A, Rámila D, Izquierdo A et al (2005) Role of the renin-angiotensin system on the parathyroid hormone-related protein overexpression induced by nephrotoxic acute renal failure in the rat. J Am Soc Nephrol 16:939–949
106. Bosch RJ, Ortega A, Izquierdo A et al (2011) A transgenic mouse model for studying the role of the parathyroid hormone-related protein system in renal injury. J Biomed Biotechnol 2011:290874. doi:10.1155/2011/290874
107. Fiaschi-Taesch NM, Santos S, Reddy V et al (2004) Prevention of acute ischemic renal failure by targeted delivery of growth factors to the proximal tubule in transgenic mice: the efficacy of parathyroid hormone-related protein and hepatocyte growth factor. J Am Soc Nephrol 15:112–125
108. Ortega A, Rámila D, Ardura JA et al (2006) Role of parathyroid hormone-related protein in tubulointerstitial apoptosis and fibrosis after folic acid-induced nephrotoxicity. J Am Soc Nephrol 17:1594–1603
109. Ardura JA, Berruguete R, Rámila D et al (2008) Parathyroid hormone-related protein interacts with vascular endothelial growth factor to promote fibrogenesis in the obstructed mouse kidney. Am J Physiol Renal Physiol 295:F415–F425
110. Ardura JA, Rayego-Mateos S, Rámila D et al (2010) Parathyroid hormone-related protein promotes epithelial-mesenchymal transition. J Am Soc Nephrol 21:237–248
111. Ortega A, Romero M, Izquierdo A et al (2012) Parathyroid hormone-related protein is a hypertrophy factor for human mesangial cells: implications for diabetic nephropathy. J Cell Physiol 227:1980–1987
112. Romero M, Ortega A, Olea N et al (2013) Novel role of parathyroid hormone-related protein in the pathophysiology of the diabetic kidney: evidence from experimental and human diabetic nephropathy. J Diabetes Res 2013:162846. doi:10.1155/2013/162846
113. Hochane M, Raison D, Coquard C et al (2013) Parathyroid hormone-related protein is a mitogenic and a survival factor of mesangial cells from male mice: role of intracrine and paracrine pathways. Endocrinology 154:853–864

114. Ardura JA, Sanz AB, Ortiz A et al (2013) Parathyroid hormone-related protein protects renal tubuloepithelial cells from apoptosis by activating transcription factor Runx2. Kidney Int 83:825–834
115. Fujinaka Y, Sipula D, Garcia-Ocaña A et al (2004) Characterization of mice doubly transgenic for parathyroid hormone-related protein and murine placental lactogen: a novel role for placental lactogen in pancreatic beta-cell survival. Diabetes 53:3120–3130
116. Drucker DJ, Asa SL, Henderson J et al (1989) The parathyroid hormone-like peptide gene is expressed in the normal and neoplastic human endocrine pancreas. Mol Endocrinol 3:1589–1595
117. Villanueva-Peñacarrillo ML, Cancelas J, de Miguel F et al (1999) Parathyroid hormone-related peptide stimulates DNA synthesis and insulin secretion in pancreatic islets. J Endocrinol 163:403–408
118. Vasavada RC, Cavaliere C, D'Ercole AJ et al (1996) Overexpression of parathyroid hormone-related protein in the pancreatic islets of transgenic mice causes islet hyperplasia, hyperinsulinemia, and hypoglycemia. J Biol Chem 271:1200–1208
119. Porter SE, Sorenson RL, Dann P et al (1998) Progressive pancreatic islet hyperplasia in the islet-targeted, parathyroid hormone-related protein-overexpressing mouse. Endocrinology 139:3743–3751
120. Cebrian A, García-Ocaña A, Takane KK et al (2002) Overexpression of parathyroid hormone-related protein inhibits pancreatic beta-cell death in vivo and in vitro. Diabetes 51:3003–3013
121. Sawada Y, Zhang B, Okajima F et al (2001) PTHrP increases pancreatic beta-cell-specific functions in well-differentiated cells. Mol Cell Endocrinol 182:265–275
122. Zhang B, Hosaka M, Sawada Y et al (2003) Parathyroid hormone-related protein induces insulin expression through activation of MAP kinase-specific phosphatase-1 that dephosphorylates c-Jun NH2-terminal kinase in pancreatic beta-cells. Diabetes 52:2720–2730
123. Guthalu Kondegowda N, Joshi-Gokhale S, Harb G et al (2010) Parathyroid hormone-related protein enhances human ß-cell proliferation and function with associated induction of cyclin-dependent kinase 2 and cyclin E expression. Diabetes 59:3131–3138
124. Williams K, Abanquah D, Joshi-Gokhale S et al (2011) Systemic and acute administration of parathyroid hormone-related peptide(1–36) stimulates endogenous beta cell proliferation while preserving function in adult mice. Diabetologia 54:2867–2877
125. Shor R, Halabe A, Aberbuh E et al (2006) PTHrP and insulin levels following oral glucose and calcium administration. Eur J Intern Med 17:408–411
126. Rosenberg J, Pines M, Hurwitz S (1987) Response of adrenal cells to parathyroid hormone stimulation. J Endocrinol 112:431–437
127. Isales CM, Barrett PQ, Brines M et al (1991) Parathyroid hormone modulates angiotensin II induced aldosterone secretion from the adrenal glomerulosa cell. Endocrinology 129:489–495
128. Olgaard K, Lewin E, Bro S et al (1994) Enhancement of the stimulatory effect of calcium on aldosterone secretion by parathyroid hormone. Miner Electrolyte Metab 20:309–314
129. Mazzocchi G, Aragona F, Malendowicz LK et al (2001) PTH and PTH-related peptide enhance steroid secretion from human adrenocortical cells. Am J Physiol Endocrinol Metab 280:E209–E213
130. Rizk-Rabin M, Assie G, Rene-Corail F et al (2008) Differential expression of parathyroid hormone-related protein in adrenocortical tumors: autocrine/paracrine effects on the growth and signaling pathways in H295R cells. Cancer Epidemiol Biomarkers Prev 17:2275–2285
131. Gennari C, Nami R, Gonnelli S (1995) Hypertension and primary hyperparathyroidism: the role of adrenergic and renin–angiotensin–aldosterone systems. Miner Electrolyte Metab 21:77–81
132. Kovacs L, Goth MI, Szabolcs I et al (1998) The effect of surgical treatment on secondary hyperaldosteronism and relative hyperinsulinemia in primary hyperparathyroidism. Eur J Endocrinol 138:543–547
133. Muehleisen B, Bikle DD, Aguilera C et al (2012) PTH/PTHrP and vitamin D control antimicrobial peptide expression and susceptibility to bacterial skin infection. Sci Transl Med 4:135ra66. doi:10.1126/scitranslmed.3003759
134. Liu PT, Stenger S, Li H et al (2006) Toll-like receptor triggering of a vitamin D-mediated human antimicrobial response. Science 311:1770–1773

13 In Vitro Cellular Models of Parathyroid Cells

Ana Rita Gomez, Sergio Fabbri, and Maria Luisa Brandi

13.1 Introduction

The parathyroid gland is an endocrine organ formed by four small glands, which have the important role of maintaining serum calcium levels within a narrow physiological range, through the calcium-sensing receptor (CaSR) [1] (see also Chap. 5). It is mainly formed by two types of cells, named parathyroid chief cells and parathyroid oxyphil cells. Parathyroid chief cells are the most abundant cells of this gland, whereas parathyroid oxyphil cells are present in lower numbers [2]. Parathyroid chief cells secrete parathyroid hormone (PTH) in response to acute variations in the concentration of extracellular calcium ($Ca^{2+}{}_{o}$), which is sensed by the CaSR, located in the surface of these cells [1]. The function of parathyroid oxyphil cells is still unknown. In the interior of the gland, chief cells and oxyphil cells are organized in small islets delimited by connective tissue. Blood vessels, endothelial cells, and a few adipose cells are also present in the interior of the gland, while more adipose tissue is present in the exterior of the gland. The glands are surrounded by connective tissue in the form of a capsule that provides support to the parathyroid tissue and separation from other organs [2].

A.R. Gomez (✉) • S. Fabbri • M.L. Brandi
Department of Surgery and Translational Medicine, University of Florence, Florence, FI, Italy
e-mail: marialuisa.brandi@unifi.it

In vitro cell cultures of parathyroid cells have long represented an issue of key importance for several researchers who have aimed their studies at the physiology and pathologies of this endocrine gland. Diverse pathologies can affect the parathyroid gland, such as parathyroid hyperplasia, parathyroid adenomas, parathyroid carcinomas, hypoparathyroidism, and hyperparathyroidism [3–6]. In order to learn more about these pathologies, as well as the physiological functions of this gland, it is highly desirable to develop adequate in vitro parathyroid cell models. However, in vitro parathyroid cell culture has proved to be challenging, presenting several difficulties to be overcome. Bovine parathyroid glands and human parathyroid glands have been extensively used to develop in vitro cell models of parathyroid cells. Human pathologic parathyroid glands, such as parathyroid adenomas and hyperplastic parathyroid glands from patients with secondary hyperparathyroidism due to chronic kidney disease (CKD) constitute the models most commonly used for the development of parathyroid cell systems. On the other hand, normal human parathyroid glands have been used less commonly to establish parathyroid cell cultures, because they have proved to be more difficult to grow in culture due to their very low proliferative activity [7]. Similarly, reports of parathyroid cell cultures derived from parathyroid carcinomas are scarce in literature given the very low frequency of parathyroid cancer [5].

M.L. Brandi, E.M. Brown (eds.), *Hypoparathyroidism*,
DOI 10.1007/978-88-470-5376-2_13, © Springer-Verlag Italia 2015

Most of the in vitro parathyroid cell models described in the literature are primary cultures, which are only viable for a short period of time, while there are only a few reports regarding the achievement of long-term viable parathyroid cell cultures or cell lines. The term "primary culture" refers to proliferating cells isolated from a tissue until confluence and first subculture, while the term "primary cell line" or "cell line" represents a primary culture that has been subcultured. Primary cell lines are finite, i.e., are only viable and able to proliferate for a limited amount of time, since the cells lose the ability to proliferate and became senescent after a given number of passages. In contrast, continuous cell lines are immortal with an infinite ability to proliferate [8]. However, the only report present in the literature regarding a continuous parathyroid cell line is a clonal rat cell line named PT-r [9].

13.2 In Vitro Primary Cultures of Bovine Parathyroid Cells

Primary cultures of bovine parathyroid cells were among the first in vitro parathyroid cell models used to study the physiological functions of the parathyroid, such as the regulated secretion of *PTH* in response to alterations in $Ca^{2+}{}_o$ [10–12]. However, primary cultures of bovine parathyroid cells have been shown to quickly lose responsiveness to $Ca^{2+}{}_o$ [12–14]. Freshly dispersed bovine parathyroid cells are able to sense changes in $Ca^{2+}{}_o$ and inhibit PTH secretion at high $Ca^{2+}{}_o$ levels. However, the ability to respond to $Ca^{2+}{}_o$ levels and suppress PTH secretion begins to decrease starting the first day in culture, and complete lack of response to changes in $Ca^{2+}{}_o$ has been shown to occur over a time period as short as 6 days of in vitro cell culture. The rapid loss of sensitivity to $Ca^{2+}{}_o$ and the absence of modulation of PTH secretion were attributed to the concomitant loss of CaSR at mRNA and protein level, which was reported to decrease rapidly in bovine parathyroid cells cultured in a monolayer [15, 16]. Several explanations were proposed for the decrease in CaSR mRNA and protein levels; however, the responsible mechanisms are still not fully understood. Deficiency in 1,25-dihydroxyvitamin D_3 ($1,25(OH)_2D_3$), the active form of vitamin D, was one of the proposed reasons, since $1,25(OH)_2D_3$ was shown to increase CaSR mRNA levels [17]. However, the addition of $1,25(OH)_2D_3$ to primary bovine parathyroid cell cultures did not prevent the loss of CaSR expression [15]. Changes in concentrations of $Ca^{2+}{}_o$ were also proposed to contribute to decreasing CaSR levels, but variations in $Ca^{2+}{}_o$ concentrations did not prevent the decrease in expression of the receptor or increase CaSR expression [15, 16]. Some researchers reported that changes in serum concentrations had no effect in preventing a decrease in CaSR expression [15], while others stated that reduced concentrations of serum or its replacement by bovine serum albumin (BSA) led to a delay in the decrease of CaSR expression levels [18]. Since parathyroid cells in vivo present a very slow proliferation rate [19], it was also proposed that in vitro parathyroid cell culture could stimulate cell proliferation [20] and lead to loss of CaSR mRNA levels [15]. Furthermore, high levels of $Ca^{2+}{}_o$ have been shown to inhibit parathyroid cell proliferation [21], and this effect is thought to be CaSR mediated [22]. However, the incubation of primary cultures of bovine parathyroid cells with high $Ca^{2+}{}_o$ was shown to be ineffective in decreasing cell proliferation [13]. The lack of effect of $Ca^{2+}{}_o$ in parathyroid cell proliferation was associated with decreased receptor expression in bovine primary cell cultures.

Another important obstacle that arises in primary cultures of bovine parathyroid cells is the overgrowth of fibroblasts in the culture that ultimately may lead to the complete loss of parathyroid cells in culture [12, 13]. In spite of all the difficulties associated with primary cultures of bovine parathyroid cells, i.e., such a rapid loss of parathyroid function, loss of sensitivity to changes in $Ca^{2+}{}_o$, and contamination of the culture with fibroblasts, some researchers have been able to maintain functional cultures of bovine parathyroid cells for long periods of time. Brandi et al. [23] maintained bovine parathyroid cells in culture for an impressive time period of 140 doublings. The cells maintained secretion of PTH during the first 30 passages, which showed that this parathyroid cell model was able to maintain one of the most important characteristics of the parathyroid gland. However, this primary cell line showed some limitations, such as becoming rapidly senescent, inability to be cloned [9], and, similar to other reports in the literature [12–14], decreased sensitivity to $Ca^{2+}{}_{o}$.

13.3 In Vitro Primary Cultures of Human Parathyroid Cells

Normal parathyroid cells have a very slow rate of proliferation [7] and in vitro primary cell cultures derived from normal human parathyroid cells are almost impossible to obtain [24]. Therefore, primary cultures of human parathyroid glands have mainly been derived from parathyroid adenomas and hyperplastic glands from patients with secondary hyperparathyroidism, which were shown to have higher proliferative activity than normal glands [7] and have led to the establishment of some long-term primary parathyroid cell lines [25–27]. The CaSR was shown to be decreased both at mRNA and protein expression levels in parathyroid adenomas [28], parathyroid carcinomas [29], and hyperplastic parathyroid glands from patients with secondary hyperparathyroidism [30]. In addition, the CaSR present in parathyroid tumors demonstrated decreased sensitivity to $Ca^{2+}{}_{o}$, with a higher concentration of $Ca^{2+}{}_{o}$ required to produce half-maximal inhibition of PTH secretion [28, 30]. The culture of human parathyroid cells presents several difficulties similar to what was observed in bovine parathyroid cell cultures. The overgrowth of fibroblasts in the culture is perhaps the most important problem encountered with human parathyroid cells, together with the slow proliferation exhibited by these cells. The growth rate of fibroblasts exceeds that of parathyroid cells, leading to the eventual dominance of the culture by fibroblasts. In addition to the use of pathological parathyroid glands, culture medium and technical skills applied in the preparation of human parathyroid cell cultures seem to have a high importance in the successful development of a functional viable cell culture. It could be that the initial amount of connective tissue present in the parathyroid gland may also be an important factor, since an increased amount of connective tissue may possibly lead to an increased number of fibroblasts in the culture (Figs. 13.1 and 13.2).

Primary cultures of human parathyroid cells have been used to study parathyroid cell function and proliferation, as well as the effects of several biological molecules involved in parathyroid downstream functions. For example, the effects of $Ca^{2+}{}_{o}$ and calcimimetics, allosteric modulators of the CaSR, on PTH secretion were studied using in vitro cell cultures derived from parathyroid adenomas and hyperplastic glands from patients suffering from secondary

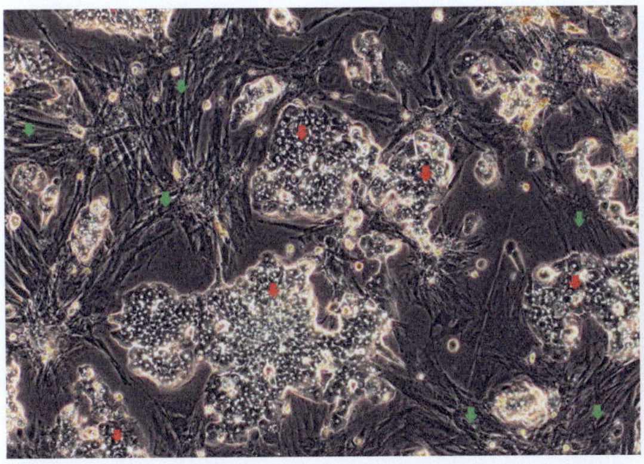

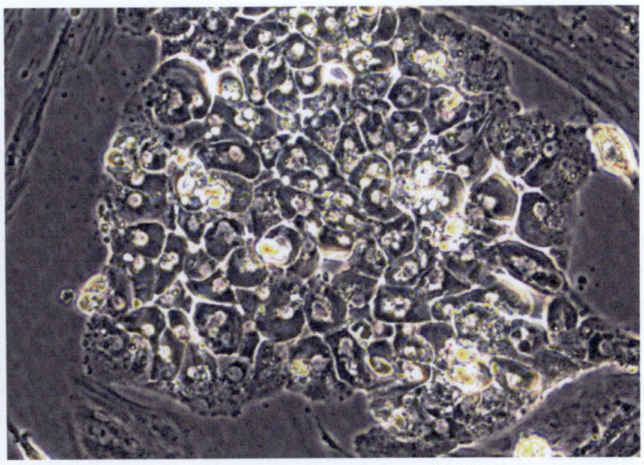

Figs. 13.1 and 13.2 Primary culture of human parathyroid cells derived from a parathyroid adenoma. In Fig. 13.1 are visible parathyroid cells organized in islets (*red arrows*), which are surrounded by cells with a fibroblastic shape, probably derived from connective tissue of the gland (*green arrows*). In Fig. 13.2 an islet of parathyroid cells can be seen in detail. The nuclei and the polygonal shape of the parathyroid cells are clearly visible in the figure. Human parathyroid cells were cultured for about 1 week after appropriate treatment of the parathyroid tissue. Images were acquired in an Axiovert 200 M inverted microscope (Zeiss, Oberkochen, Germany) in phase contrast with a magnification of 10× and 40×

hyperparathyroidism [28, 31]. Furthermore, the effects of calcium, calcitriol, and phosphate on parathyroid cell proliferation [32, 33], as well as the downstream signaling pathways of the CaSR, such as the mitogen-activated protein kinase (MAPK) signaling pathway [34], modifications in intracellular calcium ($Ca^{2+}{}_{i}$) release, and changes in intracellular cAMP levels were studied in human parathyroid cell cultures [28].

13.4 Long-Term In Vitro Cell Models of Human Parathyroid Cells

Several researchers have tried to establish in vitro models of parathyroid cells. However, only a few reports of successful long-term parathyroid cell cultures with maintained proliferation and functional parathyroid activity have been described in literature. The works of Liu et al., Roussanne et al., and Björklund et al. with human parathyroid cells drew attention from the others involved in such studies and will be discussed further below.

Roussanne et al. [25] were the first to establish a long-term culture of human parathyroid cells. The culture was developed from hyperplastic parathyroid glands derived from patients with secondary hyperparathyroidism, and the cells maintained functional activity for a period of time as long as 5 months. This parathyroid cell system showed secretion of PTH and adequate inhibition of this hormone in response to increased concentrations of $Ca^{2+}{}_{o}$. The ability to modulate PTH secretion in response to changes in $Ca^{2+}{}_{o}$ was maintained until the fifth passage, even though the concentration of PTH secreted decreased with each passage. This primary parathyroid cell line also maintained expression of the CaSR, both at mRNA and protein levels, which constitutes the probable explanation for the presence of PTH-regulated response to changes in $Ca^{2+}{}_{o}$. The success of this in vitro cell model was attributed to several factors, such as the different types of cells present in the culture; the organization of cell population in aggregates, described as clusters, instead of a monolayer; and the low cell proliferation. The majority of cells were of epithelial origin and a low percentage of cells of endothelial origin (1 %). Therefore, the mixed cell population could have contributed to the maintenance of active parathyroid cell function. Other contributing factors could also be the contact between adjacent cells, provided by cell aggregates, and the development of architecture more similar to the parathyroid gland in vivo.

Liu et al. have established a human primary parathyroid cell line derived from parathyroid adenomas with maintained parathyroid functional activity for 2 months. The cell culture medium used contained a low concentration of calcium and did not contain serum, which proved to be effective in altering the proliferation of fibroblasts. Furthermore, cells showed ability to proliferate until confluency and sense changes in $Ca^{2+}{}_{o}$ levels during 2 months in culture, as was shown by the ability to secrete PTH and to modulate the release of $Ca^{2+}{}_{i}$ in response to changes in $Ca^{2+}{}_{o}$ concentration. However, after subculture, the cells ceased to proliferate and lost the capacity to modulate PTH and $Ca^{2+}{}_{i}$ in response to variations in $Ca^{2+}{}_{o}$, showing sustained levels of $Ca^{2+}{}_{i}$ at high $Ca^{2+}{}_{o}$. Interestingly, simultaneous culture of parathyroid cells in serum-enriched medium showed expression of PTH protein only for 10 days, absence of proliferation from the third day in culture, and increased growth of fibroblasts, which were the predominant cells in culture at day 10 [26]. Therefore, cell culture medium applied in this in vitro cell culture could have been a key point for the establishment of this cell system.

More recently, another research group successfully established a long-term primary parathyroid cell culture. The cell line was named sHPT-1 and was derived from hyperplastic parathyroid glands from patients with secondary hyperparathyroidism [27]. The sHPT-1 cell line exhibited accumulation of non-phosphorylated stabilized β-catenin, a feature previously observed in other parathyroid tumors [27], and the role of β-catenin was studied in this cell line. The cells were maintained viable, with active proliferation and expression of PTH protein for more than 45 days. The cells were cultured in a growth medium enriched with serum, which may contradict the poor results obtained by other groups when culturing primary human parathyroid cells in serum-rich medium. The culture protocol utilized by the authors presents two particular characteristics: the use of lithium chloride for the first four passages and the culture of cells in suspension [27]. Consequently, perhaps these two features have also contributed to the success of this cell culture.

13.5 In Vitro Cell Models of Continuous Parathyroid Cell Lines

In the light of all the difficulties associated with in vitro parathyroid cell cultures, and in particular the difficulty encountered in cloning bovine parathyroid cells, a group of researchers tried to use rat parathyroid glands in an attempt to establish a long-term functional parathyroid cell line [9]. The authors have cloned epithelial cells derived from hyperplastic parathyroid glands from Sprague–Dawley rats due to a diet with low calcium and high phosphorus levels. The authors were successful in developing a clonal epithelial parathyroid cell line, named PT-r, which showed features similar to parathyroid chief cells. The PT-r cell line exhibited secretion of parathyroid-hormone-related peptide (PTHrP), expression of PTHrP at mRNA level, presence of cell proliferation (doubling time of 20 h), and modulation of cell proliferation and Ca^{2+}_{i} with Ca^{2+}_{o}. Furthermore, the cell line exhibits characteristics of an immortal cell line, with continuous cell proliferation [9, 35–37]. The authors have attributed the success of the PT-r cell line to the prevention of overgrowth of fibroblasts through selective cell isolation and the use of hyperplastic parathyroid tissue from rodents, enriched in epithelial cells [9] (Figs. 13.3, 13.4, and 13.5).

However, initially expression of *PTH* mRNA and protein was not found in the PT-r cell line [35]. More recently, expression of PTH mRNA was found to be present in the PT-r cell line [38]. In addition, this cell line was used to study parathyroid cell proliferation [39], the role of 25OHD in PTH gene transcription in hypovitaminosis D [38], and the regulation of PTH gene transcription by transcription factors [40].

13.6 Three-Dimensional In Vitro Parathyroid Cell Models

In vitro cell cultures of parathyroid cells were not the only in vitro systems used to study the parathyroid gland. Researchers have also employed intact parathyroid glands [41], tissue slices of parathyroid

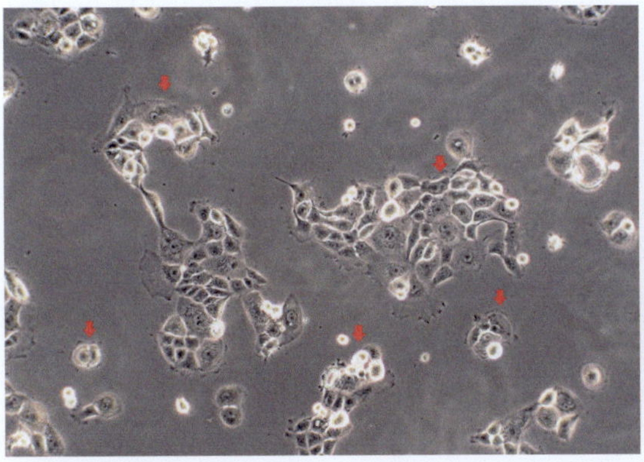

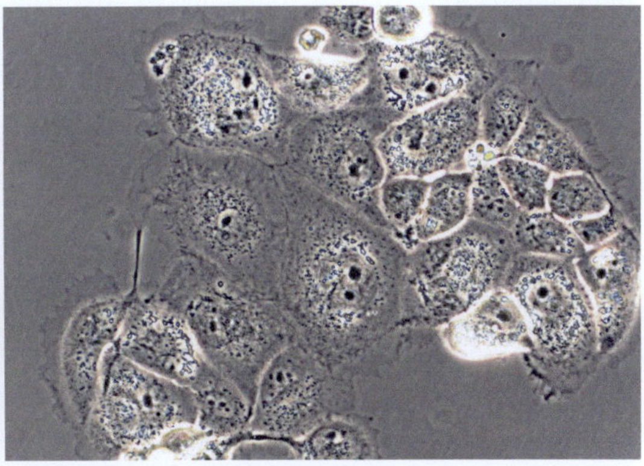

Figs. 13.3 and 13.4 Continuous culture of clonal epithelial parathyroid cells from the PT-r cell line. In Fig. 13.3 larger and smaller islets of PT-r cells (*red arrows*) are visible, illustrating the typical type of proliferation of these cells, which grow in the form of islets. In Fig. 13.4 an islet of PT-r cells is seen in the image and the morphology of the cells is seen in detail, particularly the nuclei and the polygonal shape of the cells. Images were acquired in an Axiovert 200 M inverted microscope (Zeiss, Oberkochen, Germany) in phase contrast with a magnification of 10× and 40×

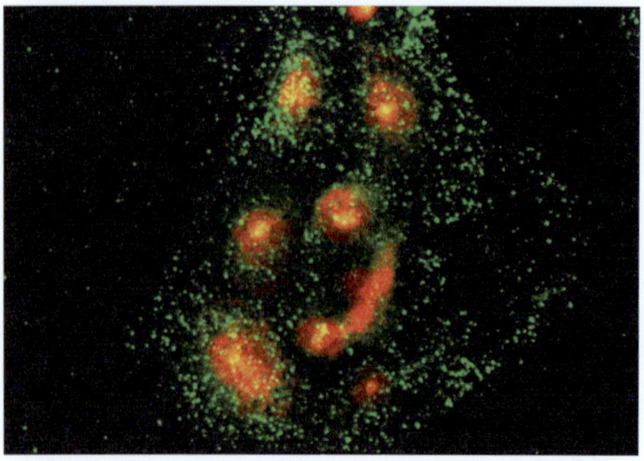

Fig. 13.5 Epithelial parathyroid cells from the PT-r cell line stably overexpressing the parathyroid hormone gene. In the image, the parathyroid hormone is visible in *green* and the nuclei in *red*. Image was acquired in laser scanner confocal microscopy (LSCM) using a LSM510META microscope (Zeiss, Oberkochen, Germany) equipped with Ar/ML458/477/488/514, HeNe543, and HeNe633 laser lines

glands [42], and explants [43] to study this endocrine organ. However, even these in vitro cell systems showed calcium-sensitive PTH secretion for short periods of time. In the view of this and the numerous difficulties discussed here in establishing long-term functional in vitro parathyroid cell systems, some research groups have also attempted to use three-dimensional in vitro cell culture models to culture parathyroid cells. The advantage of using such models lies in maintaining in cell culture a structural architecture and environment more similar to the parathyroid gland in vivo. The first parathyroid cell model of this kind reported in literature was that of Ridgeway et al. [44], who developed a model of bovine parathyroid cells organized in multicellular aggregates, named organoids. The cells in the organoids presented a morphology similar to fresh parathyroid tissue and were able to sense alterations in $Ca^{2+}{}_{o}$ concentration and respond with adequate PTH secretion for 2 weeks in culture; after this time PTH release suffered a rapid decrease. In spite of the short time of maintained parathyroid function, this model constituted an advance in parathyroid cell culture since it allowed the secretion of similar amounts of PTH by the organoids, which was not possible to obtain with parathyroid explants or tissue. Subsequently, Roussanne et al. [25] established long-term parathyroid cell cultures of human parathyroid cells derived from patients with secondary hyperparathyroidism. Cells were aggregated in clusters and were kept viable for 5 months with secretion of PTH in response to $Ca^{2+}{}_{o}$. The structure of parathyroid cells obtained in this cell model, with features more similar to the parathyroid gland in vivo and the possible cell-to-cell interactions due to the close proximity of the cells in the clusters, constitute two of the reasons claimed to explain the success obtained. Picariello et al. [45] have developed a model of human parathyroid cells encapsulated in membranes of alginate–polylysine–alginate with the aim of using this three-dimensional model of parathyroid cells in the treatment of hypoparathyroidism. The main advantage of microencapsulation consists in the protection of parathyroid cells from immunological rejection, which is the main risk of parathyroid transplantation [46]. Microencapsulated parathyroid cells were derived from two different pathological tissues, human parathyroid adenomas and hyperplastic parathyroid glands. The microencapsulated cells were maintained viable for 3 months and showed parathyroid cell growth, secretion of PTH, and sensitivity to changes in $Ca^{2+}{}_{o}$. Interestingly, primary cultures of parathyroid cells in monolayer showed decreased PTH secretion after 20 days in culture, while microencapsulated parathyroid cells stably secreted PTH for 3 months. Another example of a three-dimensional cell culture system was developed by Ritter et al. [47]. This model used bovine parathyroid cells cultured in type I collagen matrix, which coalesced to form a cellular mass named a pseudogland. Pseudoglands maintained sensitivity to $Ca^{2+}{}_{o}$ and regulated PTH secretion for 3 weeks, even if these two features were absent in the first week in collagen culture. The set point for calcium was equivalent to the values seen in freshly dispersed parathyroid cells, while CaSR mRNA expression was decreased. More recently, another three-dimensional cell model of human parathyroid cells derived from patients suffering from secondary hyperparathyroidism due to CKD was reported in the literature. In this model, the cells were cultured in nonadherent plates and formed compact masses of parathyroid cells that were separated by acellular substrate, an architecture named spheroids. The cells organized in spheroids did not show overgrowth of fibroblasts, maintained CaSR expression, and maintained responsiveness to changes in Ca^{2+}_{o} during 2–3 months [48]. The tissue-like structural models of bovine parathyroid cells described above are still cell models that maintain parathyroid function for a short period of time, but the organization of parathyroid cells in a three-dimensional architecture appears to maintain viable parathyroid function for a period of time longer than usually observed in primary cultures of bovine parathyroid cells in monolayers [15, 16]. The same is seen in three-dimensional cell cultures of human parathyroid cells, where the development of an environment more similar to that of parathyroid glands in vivo is possibly an advantage to maintaining long-term cell viability and sensitivity to Ca^{2+}_{o}. In addition, the culture of parathyroid cells in a three-dimensional cell system appears to be less prone to the overgrowth with fibroblasts usually seen in primary cultures of bovine and human parathyroid cells. It can be concluded that culture of parathyroid cells in three-dimensional systems can possibly constitute a useful alternative to partially overcome the several difficulties associated with primary cultures of parathyroid cells in monolayer.

13.7 Differentiation of Parathyroid Cells from Stem Cells

Recently, researchers have also focused on attempting to differentiate parathyroid-like cells from stem cells, as well as to investigate whether stem cells are present in parathyroid gland tissue. Ignatoski et al. have tried to differentiate human embryonic stem cells (hESCs) into parathyroid-like cells. H1 hESCs were differentiated into cells expressing parathyroid cell markers such as glial cells missing-2 (GCM2), CaSR, CXCR4, and PTH. In addition, differentiated cells exhibited PTH secretion and, at the same time, did not show production of other hormones characteristic of other endocrine organs that develop from the same pouch common to the parathyroid gland (pharyngeal endoderm), such as cells secreting thyroid-stimulating hormone (TSH), thyroxine (T4), and calcitonin, supporting the evidence that the stem cells differentiated into parathyroid-like cells [49]. Previously, Bingham et al. had already differentiated hESCs into cells with a parathyroid phenotype, using the BG01-hES cell line [50]. The development of parathyroid-like cells from hESCs could represent an alternative treatment for hypoparathyroidism. In the future, hESCs could possibly be collected from a patient suffering from hypoparathyroidism, differentiated in vitro into parathyroid-like cells, and implanted in the patient as an autograft [49, 50]. In addition, parathyroid-like cells derived from stem cells could possibly represent an adequate in vitro cell model to study parathyroid gland function.

Conversely, other researchers have tried to determine if the human parathyroid gland possesses stem cells. Stem cells have been found to be present in several differentiated tissues such as bone marrow, adipose tissue, umbilical cord blood, and others [51]. This supports the hypothesis that the parathyroid gland may contain stem cells. In this regard, Shih et al. attempted to isolate stem cells from the human parathyroid gland. The obtained cells presented a phenotype similar to mesenchymal stem cells, with the presence of markers such as CD73 and CD105, telomerase activity, and expression of CaSR gene and were differentiated into osteoblasts, chondrocytes, and adipocytes. However, the isolated cells were not able to secrete *PTH* in response to changes in Ca^{2+}_{o} [52]. Consequently, although parathyroid-like cells differentiated from parathyroid stem cells could be a good in vitro parathyroid cell model, further research is needed to determine if the human parathyroid gland possesses stem cells and if so, whether these cells can be differentiated into functional parathyroid-like cells.

References

1. Brown EM, MacLeod RJ (2001) Extracellular calcium sensing and extracellular calcium signaling. Physiol Rev 81:239–297
2. Zhang S-X (1999) An atlas of histology. Springer, New York

3. Bilezikian JP (2012) Primary hyperparathyroidism. Endocr Pract 18:781–790
4. Jamal SA, Miller PD (2013) Secondary and tertiary hyperparathyroidism. J Clin Densitom 16:64–68
5. Wei CH, Harari A (2012) Parathyroid carcinomas: update and guidelines for management. Curr Treat Options Oncol 13:11–23
6. De Sanctis V, Soliman A, Fiscina B (2012) Hypoparathyroidism: from diagnosis to treatment. Curr Opin Endocrinol Diabetes Obes 19:435–442
7. Yamaguchi S, Yachiku S, Morikawa M (1997) Analysis of proliferative activity of the parathyroid glands using proliferating cell nuclear antigen in patients with hyperparathyroidism. J Clin Endocrinol Metab 82:2681–2688
8. Matter JP, Roberts PE (1998) Introduction to cell and tissue culture: theory and technique. Plenum Press, New York
9. Sakaguchi K, Santora A, Zimering M, Curcio F, Aurbach GD, Brandi ML (1987) Functional epithelial cell line clones from rat parathyroid glands. Proc Natl Acad Sci U S A 84:3269–3273
10. Brown EM, Hurwitz S, Aurbach GD (1976) Preparation of viable isolated bovine parathyroid cells. Endocrinology 99:1582–1588
11. Nygren P, Gylfe E, Larsson R, Johansson H, Juhlin C, Klareskog L, Akerström G, Rastad J (1988) Modulation of the Ca2+-sensing function of parathyroid cells in vitro and in hyperparathyroidism. Biochim Biophys Acta 968:253–260
12. MacGregor RR, Sarras MP Jr, Houle A, Cohn DV (1983) Primary monolayer cell culture of bovine parathyroids: effects of calcium, isoproterenol and growth factors. Mol Cell Endocrinol 30:313–328
13. LeBoff MS, Rennke HG, Brown EM (1983) Abnormal regulation of parathyroid cell secretion in primary cultures of bovine parathyroid cells. Endocrinology 113:277–284
14. LeBoff MS, Shoback DM, Brown EM, Thatcher J, Leombruno R, Beaudoin D, Henry M, Wilson R, Pallota J, Marynick S, Stock J, Leight G (1985) Regulation of parathyroid hormone release and cytosolic calcium by extracellular calcium in dispersed and cultured bovine and pathological human parathyroid cells. J Clin Invest 75:49–57
15. Brown AJ, Zhong M, Ritter C, Brown EM, Slatopolsky E (1995) Loss of calcium responsiveness in cultured bovine parathyroid cells is associated with decreased calcium receptor expression. Biochem Biophys Res Commun 212:861–867
16. Mithal A, Kifor O, Kifor I, Vassilev P, Butters R, Krapcho K, Simin R, Fuller F, Herbert SC, Brown EM (1995) The reduced responsiveness of cultured bovine parathyroid cells to extracellular Ca^{2+} is associated with marked reduction in the expression of extracellular Ca^{2+}-sensing receptor messenger ribonucleic acid and protein. Endocrinology 136:3087–3092
17. Brown AJ, Zhong M, Finch J, Ritter C, McCracken R, Morrisey J, Slatopolsky E (1996) Rat calcium-sensing receptor is regulated by vitamin D but not by calcium. Am J Physiol 270:F454–F460
18. Nygren P, Larsson R, Johansson H, Gylfe E, Rastad J, Akerström G (1988) Inhibition of cell growth retains differentiated function of bovine parathyroid cells in monolayer culture. Bone Miner 4:123–132
19. Wang Q, Palnitkar S, Parfitt AM (1997) The basal rate of cell proliferation in normal human parathyroid tissue: implications for the pathogenesis of hyperparathyroidism. Clin Endocrinol 46:343–349
20. Kremer R, Bolivar I, Goltzman D, Hendy GN (1986) Influence of calcium and 1,25-dihydroxycholecalciferol on proliferation and proto-oncogene expression in primary cultures of bovine parathyroid cells. Endocrinology 125:935–941
21. Lee MJ, Roth SI (1975) Effect of calcium and magnesium on deoxyribonucleic acid synthesis in rat parathyroid glands in vitro. Lab Invest 33:72–79
22. Yano S, Sugimoto T, Tsukamoto T, Chihara K, Kobayashi A, Kitazawa S, Maeda S, Kitazawa R (2000) Association of decreased calcium-sensing receptor expression with proliferation of parathyroid cells in secondary hyperparathyroidism. Kidney Int 58:1980–1986
23. Brandi ML, Fitzpatrick LA, Coon HG, Aurbach GD (1986) Bovine parathyroid cells: cultures maintained for more than 140 population doublings. Proc Natl Acad Sci U S A 83:1709–1713
24. Björklund P, Hellman P (2012) Culture of parathyroid cells. Methods Mol Biol 806:43–53
25. Roussane M-C, Gogusev J, Hory B, Duchambon P, Souberbielle JC, Nabarra B, Pierrat D, Sarfait E, Drüke T, Bourdeau A (1998) Persistence of Ca^{2+}-sensing receptor expression in functionally active, long-term human parathyroid cell cultures. J Bone Miner Res 13:354–362
26. Liu W, Ridefelt P, Åkerström G, Hellman P (2001) Differentiation of human parathyroid cells in culture. J Endocrinol 168:417–425
27. Björklund P, Åkerström G, Westin G (2007) Activated β-catenin in the novel human parathyroid tumor cell line sHPT-1. Biochem Biophys Res Commun 352:532–536
28. Corbetta S, Mantovani G, Lania A, Borgato S, Vicentini L, Beretta E, Faglia G, Di Blasio AM, Spada A (2000) Calcium-sensing receptor expression and signaling in human parathyroid adenomas and primary hyperplasia. Clin Endocrinol 52:339–348
29. Haven CJ, Puijenbroek MV, Karperien M, Fleuren GJ, Morreau H (2004) Differential expression of the calcium sensing receptor and combined loss of chromosomes 1q and 11q in parathyroid carcinoma. J Pathol 202:86–94
30. Gogusev J, Duchambon P, Hory B, Giovannini M, Goureau Y, Sarfati E, Drueke T (1997) Depressed expression of calcium receptor in parathyroid gland tissue of patients with hyperparathyroidism. Kidney Int 51:328–336
31. Kawata T, Imanishi Y, Kobayashi K, Onoda N, Okuno S, Takemoto K, Komo T, Tahara H, Wada M, Nagano

N, Ishimura E, Miki T, Ishikawa T, Inaba M, Nishizawa Y (2006) Direct in vitro evidence of the suppressive effect of cinacalcet HCL on parathyroid hormone secretion in human parathyroid cells with pathologically reduced calcium-sensing receptor levels. J Bone Miner Metab 24:300–306
32. Roussanne M-C, Lieberherr M, Souberbielle JC, Sarfati E, Drueke T, Bourdeau A (2001) Human parathyroid cell proliferation in response to calcium, NPS R-467, calcitriol and phosphate. Eur J Clin Invest 31:610–616
33. Almaden Y, Felsenfeld AJ, Rodriguez M, Canadillas S, Luque F, Bas A, Bravo J, Torregrosa V, Palma A, Ramos B, Sanchez C, Martin-Malo A, Canalejo A (2003) Proliferation in hyperplastic human and normal rat parathyroid glands: role of phosphate, calcitriol and gender. Kidney Int 64:2311–2317
34. Corbetta S, Lania A, Filopanti M, Vicentini L, Ballaré E, Spada A (2002) Mitogen-activated protein kinase in human normal and tumoral parathyroids cells. J Clin Endocrinol Metab 87:2201–2205
35. Ikeda K, Weir EC, Sakaguchi K, Burtis WJ, Zimering M, Mangin M, Dreyer BE, Brandi ML, Aurbach GD, Broadus AE (1989) Clonal rat parathyroid cell line expresses a parathyroid hormone-related peptide but not parathyroid hormone itself. Biochem Biophys Res Commun 162:108–115
36. Sakaguchi K, Ikeda K, Curcio F, Aurbach GD, Brandi ML (1990) Subclones of rat parathyroid cell line (PT-r): regulation of growth and production of parathyroid hormone-related peptide (PTHRP). J Bone Miner Res 5:863–869
37. Sakaguchi K (1994) Autocrine and paracrine functions of parathyroid tissue. In: Bilezikian JP, Levine MA, Marcus R (eds) The parathyroids. Raven Press, New York, pp 93–105
38. Kawahara M, Iwasaki Y, Sakaguchi K, Taguchi T, Nishiyama M, Nigawara T, Tsugita M, Kambayashi M, Suda T, Hashimoto K (2008) Predominant role of 25OHD in the negative regulation of PTH expression: clinical relevance for hypovitaminosis D. Life Sci 82:677–683
39. Bianchi S, Fabiani S, Muratori M, Arnold A, Sakaguchi K, Miki T, Brandi ML (1994) Calcium modulates the cyclin D1 expression in a rat parathyroid cell line. Biochem Biophys Res Commun 204:691–700
40. Kawahara M, Iwasaki Y, Sakaguchi K, Taguchi T, Nishiyama M, Nigawara T, Kambayashi M, Sawada T, Jing X, Miyajima M, Terada Y, Hashimoto K, Suda T (2010) Involvement of GCMB in the transcriptional regulation of the human parathyroid hormone gene in a parathyroid-derived cell line PT-r. Effects of calcium and 1,25(OH_2)D_3. Bone 47:534–541
41. Raisz LG (1963) Regulation by calcium of parathyroid growth and secretion in vitro. Nature 197:1115–1116
42. Sherwood LM, Hermann I, Basset CA (1970) Parathyroid hormone secretion in vitro: regulation by calcium and magnesium ions. Nature 225:1056–1058
43. MacGregor RR, Hamilton JW, Cohn DV (1975) The bypass of tissue hormone stores during the secretion of newly synthesized parathyroid hormone. Endocrinology 97:178–188
44. Ridgeway RD, Hamilton JW, MacGregor R (1986) Characteristics of bovine parathyroid cell organoids in culture. In Vitro Cell Dev Biol 22:91–99
45. Picariello L, Benvenuti S, Recenti R, Formigli L, Falcheti A, Morelli A, Masi L, Tonelli F, Cicchi P, Brandi ML (2001) Microencapsulation of human parathyroid cells: an "*in vitro*" study. J Surg Res 96:81–89
46. Lim F, Sun AM (1980) Microencapsulated islets as bioartificial endocrine pancreas. Science 210: 908–910
47. Ritter CS, Slatopolsky E, Santoro S, Brown AJ (2004) Parathyroid cells cultured in collagen matrix retain calcium responsiveness: importance of three-dimensional tissue architecture. J Bone Miner Res 19:491–498
48. Kanai G, Kakuta T, Sawada K, Yokoyama TA, Tanaka R, Saito A (2009) Suppression of parathyroid hormone production *in vitro* and *in vivo* by RNA interference. Kidney Int 75:490–498
49. Ignatoski KMW, Bingham EL, Frome LK, Doherty GM (2010) Differentiation of precursors into parathyroid-like cells for treatment of hypoparathyroidism. Surgery 148:1186–1190
50. Bingham EL, Cheng S-P, Ignatoski KMW, Doherty GM (2009) Differentiation of hES cells to a parathyroid-like phenotype. Stem Cells Dev 18: 1071–1080
51. Odabas S, Elçin AE, Elçin YM (2014) Isolation and characterization of mesenchymal stem cells. Methods Mol Biol 1109:47–63
52. Shih Y-RV, Kuo TK, Yang A-H, Lee OK, Lee C-H (2009) Isolation and characterization of stem cells from the human parathyroid gland. Cell Prolif 42:461–470

Part II

Conditions of Hypoparathyroidism

14 Epidemiology of Hypoparathyroidism

Bart L. Clarke

14.1 Introduction

Hypoparathyroidism is a rare disorder diagnosed by the presence of low serum calcium and low or inappropriately low-normal serum parathyroid hormone. This condition may be acquired or inherited (see Table 14.1). The acquired form is most often due to the removal of, or damage to, the parathyroid glands or their blood supply at the time of neck surgery for thyroid disease, head and neck cancer, or parathyroid disease. Postsurgical hypoparathyroidism explains about 75 % of acquired cases. The next most common cause in adults is thought to be autoimmune disease, either affecting only the parathyroid glands, or multiple other endocrine organs. Remaining cases are due to a variety of rare infiltrative disorders, metastatic disease, iron or copper overload, ionizing radiation exposure, or rare genetic disorders.

This chapter describes the known epidemiology of hypoparathyroidism. Current estimates of the prevalence, incidence, risk factors, details regarding hospitalization for complications, medical costs of this disorder, and estimates of the morbidity and mortality of this disorder are reviewed.

Financial Support
None
The author states that he has no conflicts of interest.

B.L. Clarke, MD
Division of Endocrinology, Diabetes, Metabolism, and Nutrition, Department of Internal Medicine, Mayo Clinic College of Medicine, East 18-A, 200 First Street SW, Rochester, MN 55905, USA
e-mail: clarke.bart@mayo.edu

14.2 Prevalence

There are only a few estimates of the prevalence of hypoparathyroidism in the published literature. A recent retrospective study that analyzed a large US claims database gave an estimated prevalence of 65,389 insured individuals with hypoparathyroidism for more than 6 months in 2008 [1]. This prevalence estimate was extrapolated to 78,000 total insured and uninsured individuals. The database used in this retrospective study included fully adjudicated medical and pharmaceutical claims from all insured members, and contained longitudinal data for nearly 77 million patients from 75 health plans in the USA. The estimate of prevalence was obtained by calculating the number of diagnoses of hypoparathyroidism over a 12-month period using different methods. The first method was diagnosis based, and the second was surgical based. This insurance database was not cross-validated with other databases, although it has been used in the epidemiologic assessment of other diseases.

Underbjerg et al. [2] identified patients with hospital discharge diagnosis of postsurgical hypoparathyroidism through the Danish National Patient Registry. A national prescription database

M.L. Brandi, E.M. Brown (eds.), *Hypoparathyroidism*,
DOI 10.1007/978-88-470-5376-2_14, © Springer-Verlag Italia 2015

Table 14.1 Classification of hypoparathyroid disorders

Destruction or removal of parathyroid tissue with inadequate secretory reserve
Postsurgical hypoparathyroidism
Autoimmune hypoparathyroidism
Deposition of heavy metals in the parathyroid tissue
Radiation-induced destruction of the parathyroid tissue
Metastatic infiltration of the parathyroid glands
Reversible impairment of PTH secretion or PTH action with intact underlying secretory action
Severe magnesium depletion
Hypermagnesemia
Constitutively active CaSR
Genetic disorders of PTH biosynthesis and parathyroid gland development
PTH gene mutations
Mutations or deletions in transcription factors and other regulators of the development of the parathyroid glands
Mutations in mitochondrial DNA

Adapted from Table 1, Bilezikian et al. [13], 2318. Used with permission
Abbreviations: *PTH* parathyroid hormone, *CaSR* calcium-sensing receptor, *DNA* deoxyribonucleic acid

was used to confirm that these patients were treated with calcium and active vitamin D supplementation. All diagnoses were confirmed by review of individual patient hospital records. For each patient with postsurgical hypoparathyroidism where surgery was done for nonmalignant disease between 1988 and 2012, 3 age- (±2 years) and gender-matched controls were selected from the general Danish population. The prevalence estimate of postsurgical hypoparathyroidism was 22 per 100,000 person-years. A total of 688 patients were identified who had undergone neck surgery for benign disease since 1988, with subsequent diagnosis of hypocalcaemia and inappropriately low parathyroid hormone levels that necessitated treatment with calcium and/or active vitamin D supplementation for more than 6 months. The average age at diagnosis was 49 years (range, 17–87 years), with 88 % women. Sixteen percent of patients had neck surgery prior to the operation that caused hypoparathyroidism. Compared to controls, patients with postsurgical hypoparathyroidism had a 3.7-fold increased risk of renal complications (HR 3.67, 95 % CI 2.41–5.59) and 3.8-fold increased risk of hospitalization due to seizures (HR 3.82; 95 % CI, 2.15–6.79), whereas risk of cardiac arrhythmias (HR 1.11; 95 % CI, 0.79–1.57) or cardiovascular disease or death (HR 0.89, 95 % CI, 0.73–1.09) were not increased. The study concluded that, while risk of convulsions and renal disease is increased, mortality and risk of cardiovascular diseases or arrhythmias were not increased in patients with postsurgical hypoparathyroidism.

In a preliminary report, the longitudinal population-based Rochester Epidemiology Project medical records–linkage resources were used to identify all persons residing in Olmsted County, Minnesota, in 2009 with any diagnosis of hypoparathyroidism assigned by a health care provider since 1945 [3]. Detailed medical records were reviewed to confirm the diagnosis of hypoparathyroidism and assign an etiology. Subjects were then assigned 2 age- and sex-matched controls per confirmed case, and all medical diagnoses from 2006 to 2008 were then evaluated to compare cases with controls for the percent of cases with any diagnosis in each chapter and subchapter of the International Classification of Diseases, Version 9, Clinical Modification (ICD-9-CM). There were 54 confirmed cases, giving a prevalence estimate of 37 per 100,000 person-years, which translates into approximately 115,000 patients in the USA having hypoparathyroidism of any cause. Of these, 71 % were female, with a mean age for affected individuals of 58 ± 20 years. Hypoparathyroidism was caused by neck surgery in 78 % of cases, and due to recognized secondary causes in 9 %, familial disorders in 7 %, and no identified cause in 6 %. Cases were more likely than controls ($p<0.05$) to have 1 or more diagnosis within 7 of 17 chapters and 15 subchapters. These population-based data on confirmed hypoparathyroidism prevalence and case characteristics revealed that, compared to unaffected controls, persons with hypoparathyroidism exhibited a substantial burden of comorbid disease across multiple disease categories.

The Osteoporotic Fractures in Men (MrOS) study and Dallas Heart Study (DHS) were used to identify asymptomatic subjects with normocalcemic hypoparathyroidism [4]. Normocalcemic hypoparathyroidism is defined as occurring in patients with normal serum calcium and serum PTH decreased below the normal range. Cross-sectional data obtained from these studies showed that of 2,364 men in MrOS, 26 had normocalcemic hypoparathyroidism, for a prevalence of 1.1 %. Baseline data from the DHS showed that of 3,450 men and women, 68 had normocalcemic hypoparathyroidism, for a prevalence of 1.9 %. Follow-up data from these patients over 8 years showed that none developed overt hypoparathyroidism and that only two (0.09 %) had persistent normocalcemic hypoparathyroidism. The lack of persistence of the biochemical changes seen in most patients in this study may mean that the hypoparathyroidism was secondary to unidentified factors that resolved during follow-up or that laboratory measurement of parathyroid hormone (PTH) was variable, with some values incidentally found to be low.

Because hypoparathyroidism is a rare disorder, large population-based studies will be required to determine the true prevalence of this condition in each country.

14.3 Incidence

14.3.1 Postsurgical Hypoparathyroidism

Acquired hypoparathyroidism is typically due to removal or irreversible damage to the parathyroid glands, sometimes due to damage to their blood supply, during various types of neck surgery [5] (Table 14.1) (see also Chap. 22). The rate of postsurgical hypoparathyroidism depends on the center, the type of intervention, and surgical expertise. Larger studies report total rates of 5.4–8.8 %, although most cases are transient (4.9–7.3 %) [1, 6, 7]. Smaller series report an incidence of temporary hypoparathyroidism of 25.4–83 % [8–12].

Definitions of permanent postsurgical hypoparathyroidism vary, but the definition most generally accepted is insufficient parathyroid hormone to maintain normocalcemia with adequate daily intake of calcium and vitamin D longer than 6 months after surgery [5, 13]. Permanent hypoparathyroidism occurs less frequently than temporary hypoparathyroidism, with estimates ranging from 0.12 to 4.6 % [6–11].

The development of permanent hypoparathyroidism depends on a variety of risk factors. The risk is greater when more than one parathyroid gland is inadvertently removed during thyroidectomy [6, 12], when the serum calcium level is ≤8.0 mg/dL (≤2 mmol/L) 1 week after surgery, or when serum phosphorus is ≥4 mg/dL on oral calcium supplementation [6].

The frequency of postsurgical hypoparathyroidism also depends on the experience of the surgeon. One study showed that 32.8 % of cases performed by surgical residents were documented to have transient postoperative hypoparathyroidism, compared to only 19.4 % when surgeries were performed by an experienced endocrine surgeon [9]. Permanent hypoparathyroidism was reported to be more frequent during the earliest period of a surgeon's practice during retrospective review of total thyroidectomy cases for thyroid cancer performed by one surgeon [8, 14].

The type of diagnosis of thyroid disease also bears on the risk of postoperative hypoparathyroidism. Advanced thyroid cancer, Graves' disease, and other types of hyperthyroidism are associated with higher rates of postoperative hypoparathyroidism compared to small thyroid cancers or benign euthyroid disease. More extensive surgery also significantly increases the incidence of permanent hypoparathyroidism, with greater risk associated with total thyroidectomy, repeat thyroid surgery, and thyroid surgery with central compartment or more extensive neck dissection [12, 14].

Serum parathyroid hormone or calcium after surgery may also be predictors of postoperative hypoparathyroidism. Several studies have shown that patients who have intact PTH levels below the lower limit of normal, or serum calcium levels ≤8.0 mg/dL (2.0 mmol/L) are at greater risk of developing long-term hypoparathyroidism [9, 15]. One series of 170 postoperative patients showed that measuring serum intact PTH 1 day after total thyroidectomy, in combination with measuring the serum calcium level on the second day after surgery, predicted the development of hypoparathyroidism with high sensitivity, specificity, and positive predictive value. The highest sensitivity of predicting postoperative hypoparathyroidism was 97.7 % with measurement of intact PTH 1 day after surgery, with the best specificity 96.1 % with measurement of serum calcium 1 day after surgery. When both serum intact PTH and calcium were analyzed using a combined approach, the greatest predictive value was when intact PTH values were less than 15 pg/mL measured 24 h after surgery, and serum calcium values ≤7.6 mg/dL (1.9 mmol/L) measured 48 h after surgery. This combined approach resulted in a sensitivity of 96.3 %, and specificity of 96.1 %, with positive predictive value (PPV) of 86.0 %, and negative predictive value (NPV) of 99.0 % [9].

Parathyroid injury may be caused by inadvertent removal of the parathyroid glands, tying off blood vessels supplying the glands, or destruction of tissue due to intracapsular bleeding [16]. In order to prevent the development of permanent hypoparathyroidism, parathyroid autotransplantation is often recommended where these injuries are suspected to have occurred. Autotransplantation has been shown to predict transient postoperative hypoparathyroidism because of the time that engrafted parathyroid tissue needs to regain its function, but the risk of permanent hypoparathyroidism after autotransplantation is generally low [8, 16–18]. One study demonstrated that the risk of transient hypoparathyroidism increased when a greater number of parathyroid glands were autotransplanted, from 9.8 % if glands were not autotransplanted to 11.9, 15.1, and 31.4 % if 1,2, or 3 glands were autotransplanted, respectively ($p<0.05$) [17]. The risk of permanent hypoparathyroidism decreased with a greater number of glands autotransplanted, with risk 0.98 % for one gland autotransplanted, and 0.77, 0.97, and 0 % for two to four glands autotransplanted, respectively ($p=\text{NS}$).

14.3.2 Autoimmune Hypoparathyroidism

Autoimmune hypoparathyroidism is currently recognized as the second most common cause of adult hypoparathyroidism. Autoimmune isolated hypoparathyroidism occurs sporadically, with a low remission rate of 3.8 % [19]. Autoimmune hypoparathyroidism also occurs in combination with other autoimmune endocrine disorders as part of an autoimmune polyglandular syndrome type 1 (APS-1), otherwise known as autoimmune polyendocrinopathy-candidiasis-ectodermal dystrophy (APECED) [20]. This disorder is associated with hypoparathyroidism, Addison's disease, and candidiasis, and at least two other conditions, including insulin-dependent diabetes mellitus, primary hypogonadism, autoimmune thyroid disease, pernicious anemia, chronic active hepatitis, steatorrhea, alopecia, or vitiligo. At least 80 % of APS-1 patients have hypoparathyroidism, which may be the only manifestation of the disorder. APS-1 is most often an autosomal recessive disorder caused by mutations in the autoimmune regulator (*AIRE*) gene, but autosomal dominant versions have been reported. The *AIRE* gene product is a zinc-finger transcription factor located in the thymus gland and lymph nodes that is essential in mediation of central tolerance by the thymus [21]. In contrast to other immune conditions, this disorder is monogenic, and not associated with the major histocompatibility complex, and does not have a genotype-phenotype correlation [22].

The majority of patients with APS-1 are identified in childhood or adolescence, but must be followed long term because the other conditions associated with the syndrome only emerge gradually. APS-1 is estimated to occur in 1 per 1,000,000 person-years, but is much more common in three genetically distinct populations. It occurs in Finns at a frequency of 1:25,000, and 1:14,500 in Sardinians, and 1:9,000 in Iranian Jews [23].

NACHT leucine-rich repeat protein 5 (NALP5) is an intracellular signaling molecule expressed in parathyroid glands that is thought to be a parathyroid cell-specific autoantigen in APS-1 patients with hypoparathyroidism. No patients without APS-1 have been shown to have antibodies to NALP5 to date [24]. The extracellular domain of the calcium-sensing receptor (CaSR) may also be an autoantigen in some patients with autoimmune hypoparathyroidism. Activating antibodies to the extracellular domain of the receptor have been reported in both APS-1 and acquired hypoparathyroidism [25–27]. These studies suggest that, even though the majority of patients with APS-1 do not have autoantibodies to the CaSR, a subset of patients may exist with hypoparathyroidism due to functional suppression of parathyroid gland activity, rather than irreversible destruction of the parathyroid glands [28, 29].

The CaSR is a G protein-coupled receptor (GPCR) of the same family (family 3 or C) as GPCRs sensing glutamate, gamma-aminobutyric acid (GABA), odorants, sweet taste, and pheromones [30]. This family of GPCRs has large amino terminal extracellular domains, with 612 amino acids in the human CaSR, and seven membrane-spanning helices characteristic of the superfamily of GPCRs. The heavily glycosylated CaSR resides on the cell membrane as a disulfide-linked dimer. The extracellular domain contains important determinants for binding calcium, although additional calcium binding sites are found within the 7 membrane-spanning domain, since a receptor lacking the extracellular domain still responds to extracellular calcium. The CaSR functions to inhibit parathyroid cell proliferation, PTH secretion, and PTH gene expression, to stimulate calcitonin secretion, and to directly inhibit renal tubular calcium reabsorption [31]. Other less well-documented actions include stimulating proliferation, chemotaxis, and differentiation of osteoblasts, mineralization of newly formed bone by osteoblasts, and inhibition of osteoclast differentiation and activity [32].

One early study showed anti-parathyroid gland antibodies in 38 % of 75 patients with idiopathic hypoparathyroidism, 26 % of 92 patients with idiopathic Addison's disease, 12 % of 49 patients with Hashimoto thyroiditis, and 6 % of 245 normal control patients [33]. Later studies demonstrated that some anti-parathyroid gland antibodies are specific for mitochondrial or endomysial antigens. Li et al. reported that sera from 20 % of 25 patients with autoimmune hypoparathyroidism, idiopathic hypoparathyroidism, or APS-1 had CaSR autoantibodies [34]. Patients with short-duration autoimmune hypoparathyroidism of less than 5 years were shown to be more likely to have CaSR autoantibodies, whereas CaSR autoantibodies were not found in 22 healthy control patients and 50 patients with autoimmune disorders without hypoparathyroidism. It is possible that CaSR autoantibodies play a causal role in the development of hypoparathyroidism, but also possible that they are simply markers of tissue injury [35]. Another report of two patients with activating CaSR autoantibodies showed that their antibodies inhibited PTH secretion by dispersed cells from parathyroid adenomas, suggesting that hypoparathyroidism in these two cases resulted from inhibition of PTH release mediated by the autoantibodies via the CaSR, and not permanent parathyroid gland damage [36].

14.3.3 Excess Accumulation of Iron and Copper

Patients rarely develop hypoparathyroidism due to parathyroid gland storage of excessive iron deposits, resulting from either repeated transfusions in thalassemia or increased intestinal iron absorption in hemochromatosis [37]. Vogiatzi et al. showed that subclinical hypoparathyroidism and hypercalciuria were fairly common in patients with various forms of thalassemia in North America [38]. The prevalence of hypoparathyroidism in β-thalassemia major patients treated with multiple transfusions in the United Arab Emirates was recently estimated to be 10.5 % [39]. A similar study in northwest Saudi Arabia showed the prevalence of hypoparathyroidism was 11.1 % [40]. No recent studies have been published on the incidence of hypoparathyroidism in hemochromatosis.

Excessive copper deposition in the parathyroid glands may cause hypoparathyroidism in Wilson's disease. The estimated prevalence of hypoparathyroidism in Wilson's disease is 1:50,000 to 1:100,000 [41].

14.3.4 Magnesium Deficiency or Excess

Hypoparathyroidism developing due to magnesium deficiency [42] because of malabsorption, alcoholism, or poor nutrition may be reversible. Proton pump inhibitor therapy may cause hypocalcemia associated with hypoparathyroidism [43]. Magnesium excess due to infusion of magnesium during preterm labor may also cause hypoparathyroidism, thought due to magnesium-mediated inhibition of PTH secretion [44]. The hypoparathyroidism associated with hypermagnesemia is also typically reversible. Estimates of the incidence of hypoparathyroidism due to magnesium deficiency or excess have not been published.

14.3.5 Ionizing Radiation

Hypoparathyroidism may rarely be acquired after iodine-131 therapy is given for thyroid overactivity or thyroid cancer [45]. By extension, external beam radiation therapy could also theoretically cause hypoparathyroidism.

14.3.6 Metastatic Disease

Hypoparathyroidism may occur in occasional patients with metastatic disease spreading to the parathyroid glands, when significant gland destruction occurs [46].

14.3.7 Genetic Causes

Genetic forms of isolated hypoparathyroidism are rare (see Table 14.2) (see also Chaps. 16, 17, 18, 19, 20, and 21). The incidence of the genetic causes of hypoparathyroidism is currently unknown, except for DiGeorge syndrome. Familial isolated hypoparathyroidism may occur with autosomal dominant, autosomal recessive, or X-linked recessive inheritance. Autosomal forms of hypoparathyroidism may be caused by mutations in the genes that synthesize the CaSR, PTH, and GCMB (glial cells missing homologue B) [47–51]. For most cases of idiopathic hypoparathyroidism, the genetic mutation remains unknown, but multiple unknown genes are suspected to cause isolated hypoparathyroidism.

Table 14.2 Classification of congenital hypoparathyroid disorders with genetic characterization

Disorder	Gene defect/ chromosome locus
Isolated hypoparathyroidism	
Autosomal recessive	*PTH*/11p15
	GCMB/6p24.2
Autosomal dominant	*PTH*/11p15
	CaSR/3q21/1
	GCMB/6p24.2
X-linked	*SOX3*/Xq26–27
Hypoparathyroidism with additional features	
Autoimmune polyglandular syndrome type 1	*AIRE*/21q22.3
DiGeorge syndrome	*TBX1*/22q11
Hypoparathyroidism-retardation-dysmorphism syndrome	*TBCE*/1q42–43
Hypoparathyroidism-deafness-renal dysplasia syndrome	*GATA3*/10p13–14
Mitochondrial disorders associated with hypoparathyroidism	
Kearns-Sayre syndrome	Mitochondrial genome
Mitochondrial encephalopathy, lactic acidosis, and stroke-like episodes	Mitochondrial genome
Mitochondrial trifunctional protein deficiency syndrome	
Several other forms	Unknown

Adapted from Table 1 in Shoback [5]. Used with permission

Autosomal dominant hypocalcemia type 1 due to mutations in the *CaSR* gene that result in constitutive activation of the protein may be among the most common nonsurgical causes of hypoparathyroidism [52]. Affected individuals have biochemical values similar to patients with idiopathic or postoperative hypoparathyroidism, typically with rare symptoms due to mild hypocalcemia and mildly decreased or inappropriately low PTH levels, with relatively increased urinary calcium excretion due to constitutive activation of the CaSR in the renal tubule. Families with autosomal dominant hypocalcemia usually have wide variability in the severity of hypocalcemia within the family. The diagnosis is confirmed definitively by sequencing the proband's *CaSR* gene, along with the *CaSR* genes from unaffected family members. Many point mutations in the *CaSR* gene have been reported in patients with this condition that cause hyperactivity of the

CaSR in the presence of decreased or low-normal extracellular calcium. It may be difficult to know the functional effect of new CaSR mutations until they are assessed in vitro, because the *CaSR* gene has had multiple presumably benign polymorphisms reported. Autosomal dominant hypocalcemia type 2 has been reported to occur due to activating mutations in the G protein subunit α11 [53], whereas autosomal dominant hypocalcemia type 3 due to activating mutations in the adaptor protein 2 sigma subunit (AP2S1) has not yet been identified in patients [54]. The type 2 form of autosomal dominant hypocalcemia is much less common than type 1.

Other known genetic causes of isolated hypoparathyroidism include mutations in the *PTH* and *GCMB* (glial cells missing homologue B) genes. In most patients with genetic hypoparathyroidism, the mutation has not yet been identified. Mutations in the *PTH* gene that lead to altered processing of the pre-pro-PTH protein and/or to altered mRNA translation may be due to autosomal recessive or dominant inheritance [55, 56]. Homozygous mutations in the pre-pro-PTH gene may cause very low or undetectable PTH levels. Patients with autosomal dominant isolated hypoparathyroidism may have a single thymine/cytosine base substitution in exon 2, codon 18, with the resultant mutant PTH having a dominant-negative effect that leads to absent or very inefficient translocation of the nascent wild-type and mutant PTH molecules across the endoplasmic reticulum and to apoptosis of parathyroid cells [57, 58].

The *GCMB* gene is expressed mainly in parathyroid cells [59]. Mutations in this gene lead to lack of normal development of parathyroid glands, leading to hypoparathyroidism. These mutations do not affect the ability of cells to respond to PTH.

X-linked recessive hypoparathyroidism has been reported in two related kindreds in the state of Missouri in the USA [60, 61]. Male infants are affected by seizures due to hypocalcemia, with the mutation identified on chromosome Xq26–27 [62]. Genetic material from chromosome 2p25.3 is inserted into the Xq27.1 region, causing a positional effect on possibly regulatory elements controlling *SOX3* gene transcription, resulting in impaired parathyroid gland development [63].

DiGeorge syndrome is believed to occur in 1:4,000–5,000 live births [64], with complete expression associated with asymptomatic hypocalcemia due to hypoparathyroidism in 60 % of cases, thymic aplasia or hypoplasia with immunodeficiency, congenital heart defects, cleft palate, dysmorphic facies, and renal abnormalities with impaired renal function. DiGeorge syndrome is associated with variable phenotypes due to defects that occur during early embryologic development. DiGeorge syndrome most commonly develops due to new mutations, but autosomal dominant inheritance may occur. Molecular studies show that 70–80 % of cases of DiGeorge syndrome are due to a hemizygous microdeletion within the chromosomal region 22q11.21–q11.23 [62]. The *TBX1* gene within this region has been shown to carry inactivating point mutations in some DiGeorge syndrome patients. Other patients with DiGeorge syndrome-like features have been shown to have deletions in chromosome 10p13, 17p13, and 18q21. Deletions within the chromosome 22q11 region may cause the conotruncal anomaly facies and velocardiofacial syndrome. Hypocalcemia due to hypoparathyroidism is found in up to 20 % of cases with the velocardiofacial syndrome. The CATCH-22 deletion of chromosome 22q11 is associated with abnormal facies, thymic hypoplasia, cleft palate, and hypocalcemia.

Hypoparathyroidism-retardation-dysmorphism (HRD) syndrome is a rare form of autosomal recessive hypoparathyroidism incorporating the Sanjad-Sakati and the Kenny-Caffey syndromes [65, 66]. Mutations within the *TCBE* gene on chromosome 1q42–43 are associated with alterations in microtubule assembly in affected tissues. The Sanjad-Sakati syndrome is characterized by parathyroid dysgenesis, short stature, mental retardation, microphthalmia, microcephaly, small hands and feet, and abnormal teeth in individuals of mostly Arab descent. The Kenny-Caffey syndrome is characterized by hypoparathyroidism, dwarfism, medullary stenosis of the long bones, and eye abnormalities.

The hypoparathyroidism-deafness-renal dysplasia (HDR) syndrome is due to autosomal dominant mutations or deletions in the *GATA3* gene on chromosome 10p14-10pter. These lead to haploinsufficiency of the GATA3 transcription factor, a protein critical for normal parathyroid, kidney, and otic vesicle development [67]. This syndrome was first reported in a kindred in 1992 [68]. Affected subjects have asymptomatic hypocalcemia with undetectable or inappropriately normal serum PTH, and normal response to PTH.

Hypoparathyroidism is associated with mitochondrial dysfunction in three disorders: the Kearns-Sayre syndrome, the MELAS (mitochondrial encephalopathy, lactic acidosis, and stroke-like episodes) syndrome, and the mitochondrial trifunctional protein deficiency syndrome. Point mutations, deletions, rearrangements, and duplication of maternally inherited mitochondrial DNA have been described in these disorders [69].

14.4 Hospitalization

The population-based study by Leibson et al. [70] quantitated overall cost of medical care for patients with hypoparathyroidism in Olmsted County, Minnesota. Unfortunately, the study was not able to quantify the individual costs related to, or the frequency of utilization of, outpatient clinics, hospital, emergency department, or pharmacy. No other studies to date have addressed the frequency of hospitalization of patients with hypoparathyroidism relative to normal controls, but it is assumed that hospitalization for complications of hypoparathyroidism, such as bronchospasm, laryngospasm, seizures, or cardiac dysrhythmias is increased.

14.5 Cost

The population-based longitudinal medical records-linkage resources of the Rochester Epidemiology Project in Rochester, Minnesota, were also used to assess the cost of caring for patients with hypoparathyroidism [70]. All persons residing in Olmsted County in 2009 with any diagnosis of hypoparathyroidism ever assigned by a health care provider since 1945 were identified, and their detailed medical records reviewed to confirm their diagnosis of hypoparathyroidism and assign the most likely cause. Two age- and sex-matched controls were assigned per confirmed case, and follow-up censored for every case/control set member at the shortest follow-up for each member. Since 1987, Rochester Epidemiology Project resources have included provider-linked line item billing data for essentially all medical services and procedures received by residents of Olmsted County, Minnesota, with the ability to assign nationally standardized wage- and inflation-adjusted dollar estimates. Data on outpatient prescription costs are not included in these estimates. Using these resources, all medical care costs for each year 2006 through 2008 were obtained for cases and controls for 2009 estimated dollar costs. Results of cost comparisons between cases and controls showed that average medical care for each patient with hypoparathyroidism cost about three times that of each control. These population-based data on medical care costs of patients with confirmed hypoparathyroidism reveal that, although a relatively rare condition, the burden of costs associated with hypoparathyroidism is substantial and consistent. Additional investigation is needed to elucidate the source of excess costs for cases compared to controls.

14.6 Morbidity

In view of the fact that there are currently no formal guidelines, management of hypoparathyroidism is based on experience and clinical judgment [5]. The primary goals of management of chronic hypoparathyroidism include maintaining serum total calcium in the low-normal range, serum phosphorus in the high-normal range, 24-h urine calcium less than 300 mg (7.5 mmol), and the calcium x phosphate product less than 55 mg^2/dL^2 (4.4 $mmol^2/L^2$) [71–73].

The currently accepted standard treatment for hypoparathyroidism consists of supplementation with calcium and vitamin D, active vitamin D metabolites, or vitamin D analogs, but does not include hormone replacement therapy with PTH [5]. Complications of treatment of chronic hypoparathyroidism result from both inadequate treatment and overtreatment. The rates of complications from hypoparathyroidism, or the management of hypoparathyroidism, however, are difficult to estimate given the lack of large natural history studies.

14.6.1 Hypocalcemia

Suboptimal treatment of hypoparathyroidism with inadequate doses of calcium or vitamin D in the diet or with supplements may cause symptomatic hypocalcemia (see also Chaps. 15, 30, and 31). One study projected that 33 % of patients with chronic hypoparathyroidism required at least one emergency department visit or hospital admission each year [74]. Of hospital or emergency department visits, 62 % were due to symptomatic hypocalcemia. Seizures may occur in up to 15 % of patients with hypoparathyroidism each year. Dilated cardiomyopathy may also rarely occur due to prolonged or frequent hypocalcemia in affected individuals.

14.6.2 Hypercalcemia and Hypercalciuria

Given that patients with hypoparathyroidism require relatively high doses of calcium and vitamin D and its analogues to maintain serum calcium levels close to the normal range, hypercalcemia is a relatively frequent development in patients with hypoparathyroidism [13]. The replacement regimen with calcium and active vitamin D required for hypoparathyroidism can lead to hypercalciuria, because lack of circulating PTH reduces renal calcium reabsorption [75]. Prolonged or significant hypercalciuria may lead to nephrolithiasis, nephrocalcinosis, or renal insufficiency [75, 76].

The rate of nephrolithiasis reported in patients with hypoparathyroidism differs depending on the number of subjects studied. One cross-sectional study of 25 patients with postsurgical hypoparathyroidism showed that 23 % had 24-h urine calcium excretion greater than 320 mg, whereas 8 % had asymptomatic nephrolithiasis noted on renal ultrasound. All patients had normal renal function [76]. Another cross-sectional study of 33 patients with hypoparathyroidism of diverse etiologies showed that 15 % had a history of nephrolithiasis [77]. A larger retrospective cohort of 120 patients reported nephrolithiasis and nephrocalcinosis in 31 % of patients, most of whom were asymptomatic. The rate of chronic kidney disease stage 3 or higher was 2- to 17-fold greater than in age-matched controls [74].

Winer et al. have reported a higher rate of renal complications in patients participating in Clinical Research Center studies at the National Institutes of Health [78–80] (see also Chap. 30). In a short-term randomized controlled trial comparing therapies for hypoparathyroidism [78], evidence of renal insufficiency was reported in 80 % of patients ($n = 10$). Four of the subjects had radiographic evidence of nephrocalcinosis, and two suffered from recurrent nephrolithiasis. In a separate cohort of 17 patients, 8 patients (47 %) had evidence of nephrocalcinosis by renal computerized tomography scan, and 14 patients (80 %) had renal insufficiency [79]. In another randomized controlled trial comparing therapies for hypoparathyroidism over a longer period, 40 % of 27 patients had nephrocalcinosis, and two-thirds had creatinine clearance values below the normal range [80].

14.6.3 Neuropsychological State

Patients with chronic hypoparathyroidism who are treated with standard doses of calcium and vitamin D report suffering from significant impairment in their neuropsychological state. Psychometric evaluation performed in a cross-sectional controlled study of 25 unselected women treated for postsurgical hypoparathyroidism for 6.4 ± 8.0 years with calcium and vitamin D (or analogs) and 25 controls matched for sex, age, and time since surgery reported a variety of abnormalities [73]. Three validated questionnaires were used, including the revised version Symptom Checklist-90-R (SCL-90-R), the von Zerssen Symptom List (B-L Zerssen), and the short form of the Giessen Complaint List (GBB-24). The higher the score or subscale score in any of the three psychometrical instruments, the greater the impairment of well-being as assessed by the respective questionnaire. Compared with controls, hypoparathyroid patients in this study had significantly higher global complaint scores in the SCL-90-R ($P=0.020$), B-L Zerssen ($P=0.002$), and GBB-24 ($P=0.036$) instruments, with predominant increases in the subscale scores for anxiety, phobic anxiety, and their physical equivalents.

Aggarwal et al. [81] showed that significantly more patients with idiopathic hypoparathyroidism showed neuropsychological dysfunction than controls [32.3 % (95 % CI: 20.9–45.3) vs. 5.7 % (95 % CI: 1.6–14.0), $P<0.001$]. Neurological signs were present in 35.5 % patients (extrapyramidal: 16.1 %; cerebellar: 20.9 %). Volume of basal ganglia calcifications and number of sites with intracranial calcifications including the cerebellum and dentate nucleus were comparable in patients with and without neuropsychological, extrapyramidal, or cerebellar dysfunctions. Cognitive dysfunction score was lower by 1.7 points in men than in women ($P=0.02$), and increased by 0.21 and 5.5 for each year increase in duration of hypoparathyroidism ($P=0.001$), and each unit increase in serum calcium × phosphorus product ($P=0.01$), respectively. These scores improved by 0.27 for every 1.0 mg/dL increase in serum calcium ($P=0.001$). The study concluded that neuropsychological dysfunction was present in up to a third of patients with idiopathic hypoparathyroidism and that dysfunction correlated with duration of illness, female gender, serum calcium, and calcium × phosphate product during follow-up, but not with intracranial calcification. Neuropsychological dysfunction may affect daily functions, safety, and drug compliance.

Hadker et al. [82] evaluated symptoms of patients with hypoparathyroidism aged 18 years or older who were diagnosed 6 months or more previously using an Internet-based self-reported questionnaire. The study population ($N=374$) included 85 % women with mean age 49 years. Surgery of the thyroid, parathyroid, or neck for cancer was the cause of hypoparathyroidism in 43 %. Mean disease duration was 13 years, and moderate or severe disease reported by 79 %. Patients reported visiting an average of 6 different specialists or physicians before and after their diagnosis. More than ten symptoms were experienced by 72 % of patients in the preceding 12 months, despite standard symptomatic management with calcium and active vitamin D supplementation. Symptoms were experienced for an average 13 h each day. Comorbidities were experienced by 69 % of patients. Disease-associated hospital stays or emergency department visits were required by 79 % of patients. Fifty-six percent of subjects strongly agreed that they felt unprepared to manage their condition at diagnosis, 60 % revealed that controlling their hypoparathyroidism was harder than expected, and 75 % were concerned about long-term complications of their current medications. Forty-five percent reported significant interference from hypoparathyroidism in their daily lives. The study concluded that patients with hypoparathyroidism have a substantial multidimensional burden of illness, experiencing comorbidities, acute episodes of hypocalcemia, and a nearly continuous presence of symptoms despite standard symptomatic management.

14.6.4 Basal Ganglia Calcification

Basal ganglia calcification is a well-known complication of hypoparathyroidism [83], but it is not clear why the basal ganglia, among other intracranial tissues, should be subjected to this. In patients with long-standing hypoparathyroidism, with duration of disease longer than 8–10 years, basal ganglia calcification, or even more diffuse brain calcification, may occur [84–87]. While this condition is generally asymptomatic, in some cases an association with cognitive dysfunction [87] and even with organic mood disorder [88, 89] may be seen.

In the general population, basal ganglia calcification prevalence estimates are not well established, but have been reported to be low at 2–12.5 % [90, 91]. Reported rates of basal ganglia calcification in hypoparathyroidism vary, from 12 % in a cohort of 33 patients [77] to 36 % of 25 patients with CaSR mutations [92]. In one cohort of mostly postsurgical hypoparathyroidism cases, 52 % of 31 patients showed basal ganglia calcification on head computerized tomography scan [74]. In contrast, in a cohort of 145 patients with idiopathic hypoparathyroidism, all of whom had head computerized tomography scans, 74 % had basal ganglia calcification, and this correlated with the duration of hypocalcemia, choroid plexus calcification, seizures, and cataracts [93].

Familial idiopathic basal ganglia calcification has been shown to be caused by a mutation in a type III sodium-phosphate transporter leading to impaired cellular uptake of inorganic phosphate [94]. This finding suggests that increased extracellular phosphate in the setting of chronic hyperphosphatemia may contribute to basal ganglia calcification in hypoparathyroidism. Serum phosphorus has been found to be quantitatively higher in patients with basal ganglia calcification compared to those without [74].

14.6.5 Cataracts

The presence of cataracts have long been associated with both postsurgical (55 %) [73] and idiopathic hypoparathyroidism (41–51 %) [93, 95]. Patients with cataracts tend to have a longer duration of hypoparathyroidism than those without (7.5 ± 11.0 vs. 4.8 ± 4.0 years, $P = 0.49$) and tend to be older (53.6 ± 15.3 vs. 43.2 ± 11.5 years, $P = 0.11$) [73].

Patients with idiopathic hypoparathyroidism with intracranial calcification had a higher frequency of occurrence of cataracts when compared with those without calcification (19/39, 48.7 % vs. 2/12, 17.7 %, $P = 0.048$) [95]. Mean duration of illness was greater in patients with intracranial calcification or cataracts as compared to patients without these complications (9.0 ± 9.5 years vs. 2.4 ± 4.1 years, $P = 0.002$; and 11.6 ± 10.4 years vs. 4.4 ± 6.4 years, $P = 0.01$, respectively). Linear regression analysis of the data in models where age of onset of symptoms, duration of illness, and serum calcium levels were considered independent variables showed that duration of illness alone explained the variation in the frequency of occurrence of cataract or basal ganglia calcification. However, duration of the illness could explain the variation in intracranial calcification and cataracts in only 15–16 % of patients. These findings suggest a role for other factors in causation of cataracts in hypoparathyroidism.

14.6.6 Skeletal Disease

In the absence of PTH, bone remodeling is markedly reduced [96–98]. Chronically low bone turnover in patients with hypoparathyroidism typically leads to bone mass that is higher than in age- and sex-matched controls [99–104]. One study evaluated percutaneous iliac crest bone biopsies after double labeling with tetracycline from 33 subjects with hypoparathyroidism and 33 age- and sex-matched control subjects with no known metabolic bone disease and assessed histomorphometry for both static and dynamic structural skeletal parameters. Subjects with hypoparathyroidism had greater cancellous bone volume, trabecular width, and cortical width than controls. Dynamic skeletal indices, including mineralizing surface and bone formation rate, were profoundly decreased in the hypoparathyroid patients [77].

14.7 Mortality

While patients with hypoparathyroidism likely have increased mortality due to the effects of chronic hypocalcemia, intermittent hypercalcemia, significant hypercalciuria, and multiple comorbidities, few studies have yet quantified overall or cause-specific mortality due to hypoparathyroidism. Underbjerg et al. [2] were unable to show increased cardiovascular mortality in the Danish population.

Conclusion

The understanding of the epidemiology of hypoparathyroidism remains incomplete due to the fact that it is a rare condition, with only a few recent studies able to quantitate the incidence of the disorder and associated risk factors, prevalence, cost, hospitalization, morbidity, and mortality. Further large population-based studies are required to provide the missing information necessary to complete our understanding of this disorder. Previous studies have estimated the incidence and risk factors for postsurgical hypoparathyroidism, with less information available on autoimmune hypoparathyroidism and other less common nonsurgical causes. The genetic causes for hypoparathyroidism are quite rare, and in many cases limited to a few reported kindreds or individuals. Future studies will address these issues and further clarify the epidemiology of hypoparathyroidism.

References

1. Powers J, Joy K, Ruscio A, Lagast H (2013) Prevalence and incidence of hypoparathyroidism in the USA using a large claims database. J Bone Miner Res 28:2570–2576
2. Underbjerg L, Sikjaer T, Mosekilde L, Rejnmark L (2013) Cardiovascular and renal complications to postsurgical hypoparathyroidism: a Danish nationwide controlled historic follow-up study. J Bone Miner Res 28:2277–2285
3. Clarke BL, Leibson C, Emerson J, Ransom JE, Lagast H (2011) Co-morbid-medical conditions associated with prevalent hypoparathyroidism: a population-based study. J Bone Miner Res 26:S182 (Abstract SA1070)
4. Cusano NE, Maalouf NM, Wang PY, Zhang C, Cremers SC, Haney EM, Bauer DC, Orwoll ES, Bilezikian JP (2013) Normocalcemic hyperparathyroidism and hypoparathyroidism in two community-based non-referral populations. J Clin Endocrinol Metab 98:2734–2741
5. Shoback D (2008) Clinical practice. Hypoparathyroidism. N Engl J Med 359:391–400
6. Pattou F, Combemale F, Fabre S et al (1998) Hypocalcemia following thyroid surgery: incidence and prediction of outcome. World J Surg 22: 718–724
7. Thomusch O, Machens A, Sekulla C, Ukkat J, Brauckhoff M, Dralle H (2003) The impact of surgical technique on postoperative hypoparathyroidism in bilateral thyroid surgery: a multivariate analysis of 5846 consecutive patients. Surgery 133:180–185
8. Paek SH, Lee YM, Min SY, Kim SW, Chung KW, Youn YK (2013) Risk factors of hypoparathyroidism following total thyroidectomy for thyroid cancer. World J Surg 37:94–101
9. Asari R, Passler C, Kaczirek K, Scheuba C, Niederle B (2008) Hypoparathyroidism after total thyroidectomy: a prospective study. Arch Surg 143:132–137
10. Page C, Strunski V (2007) Parathyroid risk in total thyroidectomy for bilateral, benign, multinodular goitre: report of 351 surgical cases. J Laryngol Otol 121:237–241
11. Pereira JA, Jimeno J, Miquel J, Iglesias M, Munné A, Sancho JJ, Sitges-Serra A (2005) Nodal yield, morbidity, and recurrence after central neck dissection for papillary thyroid carcinoma. Surgery 138:1095–1101
12. Wingert DJ, Friesen SR, Iliopoulos JI, Pierce GE, Thomas JH, Hermreck AS (1986) Post-thyroidectomy hypocalcemia. Incidence and risk factors. Am J Surg 152:606–610
13. Bilezikian JP, Khan A, Potts JT Jr, Brandi ML, Clarke BL, Shoback D, Juppner H, D'Amour P, Fox J, Rejnmark L, Mosekilde L, Rubin MR, Dempster D, Gafni R, Collins MT, Sliney J, Sanders J (2011) Hypoparathyroidism in the adult: epidemiology, diagnosis, pathophysiology, target-organ involvement, treatment, and challenges for future research. J Bone Miner Res 26:2317–2337
14. Toniato A, Boschin IM, Piotto A, Pelizzo M, Sartori P (2008) Complications in thyroid surgery for carcinoma: one institution's surgical experience. World J Surg 32:572–575
15. Lindblom P, Westerdahl J, Bergenfelz A (2002) Low parathyroid hormone levels after thyroid surgery: a feasible predictor of hypocalcemia. Surgery 131:515–520
16. Shaha AR, Burnett C, Jaffe BM (1991) Parathyroid autotransplantation during thyroid surgery. J Surg Oncol 46:21–24
17. Palazzo FF, Sywak MS, Sidhu SB, Barraclough BH, Delbridge LW (2005) Parathyroid autotransplantation during total thyroidectomy-does the number of glands transplanted affect outcome? World J Surg 29:629–631
18. Zedenius J, Wadstrom C, Delbridge L (1999) Routine autotransplantation of at least one parathyroid gland during total thyroidectomy may reduce permanent hypoparathyroidism to zero. Aust N Z J Surg 69:794–797
19. Goswami R, Goel S, Tomar N, Gupta N, Lumb V, Sharma YD (2010) Prevalence of clinical remission in patients with sporadic idiopathic hypoparathyroidism. Clin Endocrinol (Oxf) 72:328–333
20. Ahonen P, Myllarniemi S, Sipila I, Perheentupa J (1990) Clinical variation of autoimmune polyendocrinopathy-candidiasis-ectodermal dystro-

phy (APECED) in a series of 68 patients. N Engl J Med 322:1829–1836
21. Bjorses P, Halonen M, Palvimo JJ, Kolmer M, Aaltonen J, Ellonen P, Perheentupa J, Ulmanen I, Peltonen L (2000) Mutations in the AIRE gene: effects on subcellular location and transactivation function of the autoimmune polyendocrinopathy-candidiasis-ectodermal dystrophy protein. Am J Hum Genet 66:378–392
22. Su MA, Giang K, Zumer K, Jiang H, Oven I, Rinn JL, Devoss JJ, Johannes KP, Lu W, Gardner J, Chang A, Bubulya P, Chang HY, Peterlin BM, Anderson MS (2008) Mechanisms of autoimmunity syndrome in mice caused by dominant mutation in Aire. J Clin Invest 358:1018–1028
23. Lankisch TO, Jaeckel E, Strassburg CP (2009) The autoimmune polyendocrinopathy-candidiasis-ectodermal dystrophy or autoimmune polyglandular syndrome type 1. Semin Liver Dis 29:307–314
24. Alimohammadi M, Bjorklund P, Hallgren A, Pöntynen N, Szinnai G, Shikama N, Keller MP, Ekwall O, Kinkel SA, Husebye ES, Gustafsson J, Rorsman F, Peltonen L, Betterle C, Perheentupa J, Akerström G, Westin G, Scott HS, Holländer GA, Kämpe O (2008) Autoimmune polyendocrine syndrome type 1 and NALP5, a parathyroid autoantigen. N Engl J Med 358:1018–1028
25. Hendy GN, Guarnieri V, Canaff L (2009) Calcium-sensing receptor and associated diseases. Prog Mol Biol Transl Sci 89:31–95
26. Gavalas NG, Kemp EH, Krohn KJE, Brown EM, Watson PF, Weetman AP (2007) The calcium-sensing receptor is a target of autoantibodies in patients with autoimmune polyendocrine syndrome type 1. J Clin Endocrinol Metab 92:2107–2114
27. Husebye ES, Perhentupa J, Rautemaa R, Kampe O (2009) Clinical manifestations and management of patients with autoimmune polyendocrine syndrome type 1. J Intern Med 265:514–529
28. Kemp EH, Gavalas NG, Krohn KJE, Brown EM, Watson PF, Weetman AP (2009) Activating autoantibodies against the calcium-sensing receptor detected in two patients with autoimmune polyendocrine syndrome type 1. J Clin Endocrinol Metab 94:4749–4756
29. Tomar N, Gupta N, Goswami R (2013) Calcium-sensing receptor autoantibodies and idiopathic hypoparathyroidism. J Clin Endocrinol Metab 98:3884–3891
30. Brauner-Osborne H, Wellendorph P, Jensen AA (2007) Structure, pharmacology and therapeutic prospects of family C G-protein coupled receptors. Curr Drug Targets 8:169–184
31. Hauache OM (2001) Extracellular calcium-sensing receptor: structural and functional features and association with diseases. Braz J Med Biol Res 34:577–584
32. Brown EM, MacLeod RJ (2001) Extracellular calcium sensing and extracellular calcium signaling. Physiol Rev 81:239–297
33. Blizzard RM, Chee D, Davis W (1966) The incidence of parathyroid and other antibodies in the sera of patients with idiopathic hypoparathyroidism. Clin Exp Immunol 1:119–128
34. Li Y, Song YH, Rais N, Connor E, Schatz D, Muir A, Maclaren N (1996) Autoantibodies to the extracellular domain of the calcium sensing receptor in patients with acquired hypoparathyroidism. J Clin Invest 97:910–914
35. Brown EM (2009) Anti-parathyroid and anti-calcium sensing receptor antibodies in autoimmune hypoparathyroidism. Endocrinol Metab Clin North Am 38:437–445
36. Kifor O, McElduff A, LeBoff MS, Moore FD Jr, Butters R, Gao P, Cantor TL, Kifor I, Brown EM (2004) Activating antibodies to the calcium-sensing receptor in two patients with autoimmune hypoparathyroidism. J Clin Endocrinol Metab 89:548–556
37. Toumba M, Sergis A, Kanaris C, Skordis N (2007) Endocrine complications in patients with thalassemia major. Pediatr Endocrinol Rev 5:642–648
38. Vogiatzi MG, Macklin EA, Trachtenberg FL, Fung EB, Cheung AM, Vichinsky E, Olivieri N, Kirby M, Kwiatkowski JL, Cunningham M, Holm IA, Fleisher M, Grady RW, Peterson CM, Giardina PJ, Thalassemia Clinical Research Network (2009) Differences in the prevalence of growth, endocrine and vitamin D abnormalities among the various thalassaemia syndromes in North America. Br J Haematol 146:546–556
39. Belhoul KM, Bakir ML, Kadhim AM, Dewedar HE, Eldin MS, Alkhaja FA (2013) Prevalence of iron overload complications among patients with b-thalassemia major treated at Dubai Thalassemia Centre. Ann Saudi Med 33:18–21
40. Habeb AM, Al-Hawsawi ZM, Morsy MM, Al-Harbi AM, Osilan AS, Al-Magamsi MS, Zolaly MA (2013) Endocrinopathies in beta-thalassemia major. Prevalence, risk factors, and age at diagnosis in Northwest Saudi Arabia. Saudi Med J 34:67–73
41. Carpenter TO, Carnes DL Jr, Anast CS (1983) Hypoparathyroidism in Wilson's disease. N Engl J Med 309:873–877
42. Tong GM, Rude RK (2005) Magnesium deficiency in critical illness. J Intensive Care Med 20:3–17
43. Milman S, Epstein EJ (2011) Proton pump inhibitor-induced hypocalcemic seizure in a patient with hypoparathyroidism. Endocr Pract 17:104–107
44. Cholst IN, Steinberg SF, Tropper PJ, Fox HE, Segre GV, Bilezikian JP (1984) The influence of hypermagnesemia on serum calcium and parathyroid hormone levels in human subjects. N Engl J Med 310:1221–1225
45. Pauwels EK, Smit JW, Slats A, Bourguignon M, Overbeek F (2000) Health effects of therapeutic use of 131I in hyperthyroidism. Q J Nucl Med 44:333–339
46. Goddard CJ (1990) Symptomatic hypocalcaemia associated with metastatic invasion of the parathyroid glands. Br J Hosp Med 43:72

47. Thakker RV (1996) Molecular basis of PTH underexpression. In: Principles of bone biology. Academic, New York, pp 837–851
48. Bilezikian JP, Thakker RV (1998) Hypoparathyroidism. Curr Opin Endocrinol Diabetes 4:427–432
49. Mirczuk SM, Bowl MR, Nesbit MA, Cranston T, Fratter C, Allgrove J, Brain C, Thakker RV (2010) A missense glial cells missing homolog B (GCMB) mutation, Asn502His, causes autosomal dominant hypoparathyroidism. J Clin Endocrinol Metab 95:3512–3516
50. Thakker RV (2001) Genetic developments in hypoparathyroidism. Lancet 357:974–976
51. Thakker RV, Juppner H (2001) Genetic disorders of calcium homeostasis caused by abnormal regulation of parathyroid hormone secretion or responsiveness. In: DeGroot LJ, Jameson JL (eds) Endocrinology, 4th edn. WB Saunders Company, Philadelphia, pp 1062–1074
52. Pollak MR, Brown EM, Estep HL, McLaine PN, Kifor O, Park J, Hebert SC, Seidman CE, Seidman JG (1994) Autosomal dominant hypocalcemia caused by a calcium-sensing receptor mutation. Nat Genet 8:303–307
53. Nesbit MA, Hannan FM, Howles SA, Babinsky VN, Head RA, Cranston T, Rust N, Hobbs MR, Heath H 3rd, Thakker RV (2013) Mutations affecting G-protein subunit α11 in hypercalcemia and hypocalcemia. N Engl J Med 368:2476–2486
54. Rogers A, Nesbit MA, Hannan FM, Howles SA, Gorvin CM, Cranston T, Allgrove J, Bevan JS, Bano G, Brain C, Datta V, Grossman AB, Hodgson SV, Izatt L, Millar-Jones L, Pearce SH, Robertson L, Selby PL, Shine B, Snape K, Warner J, Thakker RV (2014) Mutational analysis of the adaptor protein 2 sigma subunit (AP2S1) gene: search for autosomal dominant hypocalcemia type 3 (ADH3). J Clin Endocrinol Metab 99:E1300–E1305
55. Sunthornthepvarakul T, Churesigaew S, Ngowngarmratana S (1999) A novel mutation of the signal peptide of the pre-pro-parathyroid hormone gene associated with autosomal recessive familial isolated hypoparathyroidism. J Clin Endocrinol Metab 84:3792–3796
56. Parkinson DB, Thakker RV (1992) A donor splice site mutation in the parathyroid hormone gene is associated with autosomal recessive hypoparathyroidism. Nat Genet 1:149–153
57. Arnold A, Horst SA, Gardella TJ, Baba H, Levine MA, Kronenberg HM (1990) Mutations in the signal peptide encoding region of preproparathyroid hormone gene in isolated hypoparathyroidism. J Clin Invest 86:0184–1087
58. Datta R, Waheed A, Shah GN, Sly WS (2007) Signal sequence mutation in autosomal dominant form of hypoparathyroidism induces apoptosis that is corrected by a chemical chaperone. Proc Natl Acad Sci U S A 104:19989–19994
59. Gunther T, Chen ZF, Kim J, Priemel M, Rueger JM, Amling M, Moseley JM, Martin TJ, Anderson DJ, Karsenty G (2000) Genetic ablation of parathyroid glands reveals another source of parathyroid hormone. Nature 406:199–203
60. Whyte MP, Weldon VV (1981) Idiopathic hypoparathyroidism presenting with seizures during infancy: X-linked recessive inheritance in a large Missouri kindred. J Pediatr 99:608–611
61. Mumm S, Whyte MP, Thakker RV, Buetow KH, Schlessinger D (1997) mtDNA analysis shows common ancestry in two kindreds with X-linked recessive hypoparathyroidism and reveals a heteroplasmic silent mutation. Am J Hum Genet 60:153–159
62. Thakker RV, Davies KE, Whyte MP, Wooding C, Riordan JL (1990) Mapping the gene causing X-linked recessive idiopathic hypoparathyroidism to Xq26-Xq27 by linkage studies. J Clin Invest 86:40–45
63. Bowl MR, Nesbit MA, Harding B, Levy E, Jefferson A, Volpi E, Rizzoti K, Lovell-Badge R, Schlessinger D, Whyte MP, Thakker RV (2005) An interstitial deletion-insertion involving chromosome 2p25.3 and Xq27.1, near SOX3, causes X-linked recessive hypoparathyroidism. J Clin Invest 115:2822–2833
64. Kobrynski LJ, Sullivan KE (2007) Velocardiofacial syndrome, DiGeorge syndrome: the chromosome 22q11.2 deletion syndrome. Lancet 370:1443–1452
65. Parvari R, Diaz GA, Hershkovitz E (2007) Parathyroid development and the role of tubulin chaperone E. Horm Res 358:12–21
66. Sanjad SA, Sakati NA, Abu-Osba YK, Kaddoura R, Nilner RDG (1991) A new syndrome of congenital hypoparathyroidism, severe growth failure, and dysmorphic features. Arch Dis Child 66:193–196
67. Ali A, Christie PT, Grigorieva IV et al (2007) Functional characterization of GATA3 mutations causing the hypoparathyroidism-deafness-renal (HDR) dysplasia syndrome: insight into mechanisms of DNA binding by the GATA3 transcription factor. Hum Mol Genet 16:265–275
68. Bilous RW, Murty G, Parkinson DB et al (1992) Brief report: autosomal dominant familial hypoparathyroidism, sensorineural deafness and renal dysplasia. N Engl J Med 327:1069–1074
69. Thakker RV (2004) Genetics of endocrine and metabolic disorders: parathyroid. Rev Endocr Metab Disord 5:37–51
70. Leibson C, Clarke BL, Ransom JE, Lagast H (2011) Medical care costs for persons with and without prevalent hypoparathyroidism: a population-based study. J Bone Miner Res 26:S183 (Abstract SA1071)
71. Maeda SS, Fortes EM, Oliveira UM, Borba VC, Lazaretti-Castro M (2006) Hypoparathyroidism and pseudohypoparathyroidism. Arq Bras Endocrinol Metabol 50:664–673
72. Noordzij M, Voormolen NMC, Boeschoten EW, Dekker FW, Bos WJ, Krediet RT, Korevaar JC, NECOSAD Study Group (2009) Disordered mineral

metabolism is not a risk factor for loss of residual renal function in dialysis patients. Nephrol Dial Transplant 24:1580–1587
73. Arlt W, Fremerey C, Callies F, Reincke M, Schneider P, Timmermann W, Allolio B (2002) Well-being, mood and calcium homeostasis in patients with hypoparathyroidism receiving standard treatment with calcium and vitamin D. Eur J Endocrinol 146:215–222
74. Mitchell DM, Regan S, Cooley MR, Lauter KB, Vrla MC, Becker CB, Burnett-Bowie SM, Mannstadt M (2012) Long-term follow-up of patients with hypoparathyroidism. J Clin Endocrinol Metab 97:4507–4514
75. Santos F, Chan JC (1986) Idiopathic hypoparathyroidism: a case study on the interactions between exogenous parathyroid hormone infusion and 1,25-dihydroxyvitamin D. Pediatrics 78:1139–1141
76. Weber G, Cazzuffi MA, Frisone F, de Angelis M, Pasolini D, Tomaselli V, Chiumello G (1988) Nephrocalcinosis in children and adolescents: sonographic evaluation during long-term treatment with1,25-dihydroxycholecalciferol. Child Nephrol Urol 9:273–276
77. Rubin MR, Dempster DW, Zhou H, Shane E, Nickolas T, Sliney J Jr, Silverberg SJ, Bilezikian JP (2008) Dynamic and structural properties of the skeleton in hypoparathyroidism. J Bone Miner Res 23:2018–2024
78. Winer KK, Yanovski JA, Cutler GB Jr (1996) Synthetic human parathyroid hormone 1-34 vs calcitriol and calcium in the treatment of hypoparathyroidism. JAMA 276:631–636
79. Winer KK, Yanovski JA, Sarani B, Cutler GB Jr (1998) A randomized, cross-over trial of once-daily versus twice-daily parathyroid hormone 1-34 in treatment of hypoparathyroidism. J Clin Endocrinol Metab 83:3480–3486
80. Winer KK, Ko CW, Reynolds JC, Dowdy K, Keil M, Peterson D, Gerber LH, McGarvey C, Cutler GB Jr (2003) Long-term treatment of hypoparathyroidism: a randomized controlled study comparing parathyroid hormone (1–34) versus calcitriol and calcium. J Clin Endocrinol Metab 88:4214–4220
81. Aggarwal S, Kailash S, Sagar R, Tripathi M, Sreenivas V, Sharma R, Gupta N, Goswami R (2013) Neuropsychological dysfunction in idiopathic hypoparathyroidism and its relationship with intracranial calcification and serum total calcium. Eur J Endocrinol 168:895–903
82. Hadker N, Egan J, Sanders J, Lagast H, Clarke BL (2014) Understanding the burden of illness associated with hypoparathyroidism reported among patients in the PARADOX study. Endocr Pract 20:671–679
83. Eaton LM, Camp JD, Love JG (1939) Symmetric cerebral calcification, particularly of the basal ganglia, demonstrable roentgenographically. Arch Neurol Psychiatry 41:921–942
84. Posen S, Clifton-Bligh P, Cromer T (1979) Computerized tomography of the brain in surgical hypoparathyroidism. Ann Intern Med 91:415–417
85. Forman MB, Sandler MP, Danziger A, Kalk WJ (1980) Basal ganglia calcification in postoperative hypoparathyroidism. Clin Endocrinol (Oxf) 12:385–390
86. Illum F, Dupont E (1985) Prevalences of CT-detected calcification in the basal ganglia in idiopathic hypoparathyroidism and pseudohypoparathyroidism. Neuroradiology 27:32–37
87. Kowdley KV, Coull BM, Orwoll ES (1999) Cognitive impairment and intracranial calcification in chronic hypoparathyroidism. Am J Med Sci 317:273–277
88. Trautner RJ, Cummings JL, Read SL, Benson DF (1988) Idiopathic basal ganglia calcification and organic mood disorder. Am J Psychiatry 145:350–353
89. Lopez-Villegas D, Kulisevsky J, Deus J, Junque C, Pujol J, Guardia E et al (1996) Neuropsychological alterations in patients with computed tomography-detected basal ganglia calcification. Arch Neurol 53:251–256
90. Fenelon G, Gray F, Paillard F, Thibierge M, Mahieux F, Guillani A (1993) A prospective study of patients with CT-detected pallidal calcifications. J Neurol Neurosurg Psychiatry 56:622–625
91. Gomille T, Meyer RA, Falkai P, Gaebel W, Konigshausen T, Christ F (2001) Prevalence and clinical significance of computerized tomography verified idiopathic calcinosis of the basal ganglia. Radiologe 41:205–210 (German)
92. Raue F, Pichl J, Dorr HG, Schnabel D, Heidemann P, Hammersen G, Jaursch-Hancke C, Santen R, Schofl C, Wabitsch M, Haag C, Schulze E, Frank-Raue K (2001) Activating mutations in the calcium-sensing receptor: genetic and clinical spectrum in 25 patients with autosomal dominant hypocalcaemia: a German survey. Clin Endocrinol (Oxf) 75:760–765
93. Goswami R, Sharma R, Sreenivas V, Gupta N, Ganapathy A, Das S (2012) Prevalence and progression of basal ganglia calcification and its pathogenic mechanism in patients with idiopathic hypoparathyroidism. Clin Endocrinol (Oxf) 77: 200–206
94. Wang C, Li Y, Shi L, Ren J, Patti M, Wang T, de Oliveira JR, Sobrido MJ, Quintans B, Baquero M, Cui X, Zhang XY, Wang L, Xu H, Wang J, Yao J, Dai X, Liu J, Zhang L, Ma H, Gao Y, Ma X, Feng S, Liu M, Wang QK, Forster IC, Zhang X, Liu JY (2012) Mutations in SLC20A2 link familial idiopathic basal ganglia calcification with phosphate homeostasis. Nat Genet 44:254–256
95. Goswami R, Brown EM, Kochupillai N, Gupta N, Rani R, Kifor O, Chattopadhyay N (2004) Prevalence of calcium sensing receptor autoantibodies in patients with sporadic idiopathic hypoparathyroidism. Eur J Endocrinol 150:9–18

96. Langdahl BL, Mortensen L, Vesterby A, Eriksen EF, Charles P (1996) Bone histomorphometry in hypoparathyroid patients treated with vitamin D. Bone 18:103–108
97. Kruse K, Kracht U, Wohlfart K, Kruse U (1989) Biochemical markers of bone turnover, intact serum parathyroid hormone, and renal calcium excretion in patients with pseudohypoparathyroidism and hypoparathyroidism before and during vitamin D treatment. Eur J Pediatr 148:535–539
98. Mizunashi K, Furukawa Y, Miura R, Yumita S, Sohn HE, Yoshinaga K (1988) Effects of active vitamin D3 and parathyroid hormone on the serum osteocalcin in idiopathic hypoparathyroidism and pseudohypoparathyroidism. J Clin Invest 82:861–865
99. Abugassa S, Nordenstrom J, Eriksson S, Sjoden G (1993) Bone mineral density in patients with chronic hypoparathyroidism. J Clin Endocrinol Metab 76:1617–1621
100. Fujiyama K, Kiriyama T, Ito M, Nakata K, Yamashita S, Yokoyama N, Nagataki S (1995) Attenuation of postmenopausal high turnover bone loss in patients with hypoparathyroidism. J Clin Endocrinol Metab 80:2135–2138
101. Seeman E, Wahner HW, Offord KP, Kumar R, Johnson WJ, Riggs BL (1982) Differential effects of endocrine dysfunction on the axial and the appendicular skeleton. J Clin Invest 69:1302–1309
102. Touliatos JS, Sebes JI, Hinton A, McCommon D, Karas JG, Palmieri GM (1995) Hypoparathyroidism counteracts risk factors for osteoporosis. Am J Med Sci 310:56–60
103. Sikjaer T, Rejnmark L, Thomsen JS, Tietze A, Bruel A, Andersen G, Mosekilde L (2012) Changes in 3-dimensional bone structure indices in hypoparathyroid patients treated with PTH (1–84): a randomized controlled study. J Bone Miner Res 24:781–788
104. Takamura Y, Miyauchi A, Yabuta T, Kihara M, Ito Y, Miya A (2013) Attenuation of postmenopausal bone loss in patients with transient hypoparathyroidism after total thyroidectomy. World J Surg 37:2860–2865

Clinical Presentation of Hypoparathyroidism

15

Amber L. Wheeler and Dolores M. Shoback

15.1 Introduction

Hypocalcemia, defined as low serum levels of albumin-corrected total calcium or of ionized calcium, is a common clinical occurrence and has many potential causes. In general, hypocalcemia results from inadequate parathyroid hormone (PTH) secretion or receptor activation, an insufficient supply of vitamin D metabolites or activity of the vitamin D receptor, or abnormal magnesium metabolism [1]. The presence of severe systemic illness such as pancreatitis, sepsis, and shock as well as a myriad of congenital and acquired disorders should also be considered when establishing the etiology.

Hypocalcemia occurs in hypoparathyroidism because levels of PTH secretion are inadequate to mobilize calcium from the bone, reabsorb calcium from the distal nephron, and stimulate renal 1α-hydroxylase activity; as a result, insufficient 1,25-dihydroxyvitamin D (1,25(OH)$_2$vitamin D) is generated for efficient intestinal absorption of calcium. Ultimately, the duration, severity, and rate of development of hypocalcemia as well as the age of onset determine the clinical presentation. This can range from the acute onset of neuromuscular irritability, seizures, and altered mental status to no symptomatology in incidentally discovered or chronic cases. In this chapter, we review the clinical presentations of hypocalcemia secondary to hypoparathyroidism and other features of the disorder.

15.2 Clinical Presentation of Hypocalcemia

15.2.1 In Adult Patients

Hypocalcemia can present dramatically with tetany (symptoms due to intermittent tonic spasms of voluntary muscles), seizures, altered mental status and sensorium, congestive heart failure, or stridor (Table 15.1). Neuromuscular symptoms are typically the most prominent and include muscle cramping, twitching, and spasms; circumoral and acral numbness and paresthesias; laryngospasm and or bronchospasm; and seizures that can manifest as generalized, focal, or petit mal. Many

A.L. Wheeler, MD
Department of Medicine, University of California, San Francisco, San Francisco, CA USA

Metabolism and Endocrinology Section, San Francisco Department of Veterans Affairs Medical Center, San Francisco, CA, USA

D.M. Shoback, MD (✉)
Department of Medicine, University of California, San Francisco, San Francisco, CA, USA

Metabolism and Endocrinology Section, San Francisco Department of Veterans Affairs Medical Center, San Francisco, CA, USA

Endocrine Research Unit, San Francisco Department of Veteran Affairs Medical Center, 4150 Clement Street, 111 N, San Francisco, CA 94121, USA
e-mail: dolores.shoback@ucsf.edu

M.L. Brandi, E.M. Brown (eds.), *Hypoparathyroidism*,
DOI 10.1007/978-88-470-5376-2_15, © Springer-Verlag Italia 2015

Table 15.1 Signs and symptoms of hypocalcemia

Paresthesias of the fingers, toes, and circumoral region
Increased neuromuscular irritability
Tetany
Muscle cramping and twitching
Muscle weakness
Abdominal cramping
Laryngospasm
Bronchospasm
Central nervous system involvement
Seizures
Altered mental status
Impaired memory and concentration
Papilledema
Pseudotumor cerebri
Personality disturbances
Extrapyramidal disorders
Chvostek's sign
Trousseau's sign
Cardiac involvement
Prolonged QT interval
QRS and ST segment changes that may mimic the changes of myocardial infarction
Ventricular arrhythmias
Congestive heart failure
Cataracts
Abnormal dentition (enamel hypoplasia)

patients report that muscular symptoms interfere with exertion and exercise, such that these activities are limited because of painful cramping and stiffness of the involved muscle groups [2]. Cardiac function may be affected, manifested by a prolonged QT interval corrected for heart rate (QTc) on electrocardiogram and, in rare cases, depressed systolic function and congestive heart failure. If the disturbance is chronic, patients with extraordinarily low levels of ionized calcium may be asymptomatic. Other complications include premature cataracts, pseudotumor cerebri, and calcifications of the basal ganglia. Rarely, calcifications of the basal ganglia can cause extrapyramidal neurological dysfunction [3].

15.2.2 In Pediatric Patients

Similar to adults, hypocalcemia in pediatric patients can present acutely with symptoms that include generalized or focal seizures, depressed consciousness, tachycardia, and stridor due to laryngospasm. However, the diagnosis can be challenging since the clinical manifestations of hypocalcemia in pediatric patients are often milder or nonspecific. Younger children can present with a range of signs and symptoms including neuromuscular irritability, attention deficit disorders, poor academic performance, mental retardation, and poor dental development (e.g., enamel hypoplasia), while older children can also exhibit psychological and behavioral problems including depression and sleep disturbances [4].

The diagnosis of hypocalcemia is frequently made or suspected during periods of accelerated growth when there are increasing calcium demands for bone accrual, for instance, during infancy or puberty. The pediatric patient with mild or persistent hypocalcemia may become symptomatic or display nonspecific symptoms of hypocalcemia, including changes in growth velocity, delayed intellectual development, or behavioral problems in school [4]. These findings may trigger a referral to a pediatrician who subsequently uncovers hypocalcemia as the cause.

In contrast to hypoparathyroidism in adulthood, which is acquired usually due to damage incurred during thyroid, parathyroid, or neck surgery, hypoparathyroidism in childhood is mostly due to a congenital disorder, even if symptoms did not manifest during the neonatal period. Mutations may be present in genes involved in parathyroid gland development; in the *PTH* gene itself that affects the intracellular processing and/or secretion of the hormone; and in the extracellular calcium-sensing receptor (CaSR) that controls responsiveness to changes in the extracellular calcium concentration. Mutations in molecules involved in PTH receptor signal transduction cause resistance to PTH or pseudohypoparathyroidism [5]. The most frequently identified disorder of parathyroid gland development is the DiGeorge syndrome. The genetic abnormalities that cause the DiGeorge syndrome result in congenital hypoplasia or agenesis of the parathyroid gland and thymus and affect structures derived from the third and fourth pharyngeal pouches. Anomalies are seen in association with microdeletions of chromosome 22q11.2. The DiGeorge syndrome includes hypoparathyroidism, T-cell defects caused by a partial or absent thymus, and conotruncal heart defects (tetralogy of Fallot, truncus arteriosus) or aortic arch abnormalities. Cleft palate and facial dysmorphism may also occur [6]. Other genetic defects that result in disrupted parathyroid gland development include mutations in *GCM2* or *GATA3*. Altered PTH production in response to hypocalcemia that leads to hypoparathyroidism includes PTH mutations that affect intracellular processing, secretion, and functional properties of the hormone. Also, reduced PTH secretion may be due to altered extracellular calcium-sensing and CaSR-induced signal transduction [7]. Individuals with pseudohypoparathyroidism, caused by loss of function of the Gsα protein (type 1), may not present with clinical manifestations of hypocalcemia until later in childhood.

15.2.3 In Neonates

Neonatal hypocalcemia is a potentially life-threatening condition, with reported prevalence varying by gestational age, maternal and infant comorbidities, and perinatal factors [8]. Although sometimes clinically asymptomatic, it may present with signs of neuromuscular irritability ranging from myoclonic jerks to seizures, apnea, cyanosis, or arrhythmias including tachycardia, atrioventricular heart block in premature babies [9], and severe bradycardia. Hypocalcemia in this age group is commonly differentiated by the time of onset [10].

Early neonatal hypocalcemia occurs during the first 3 days of life and is seen in premature babies, infants of diabetic mothers, and asphyxiated infants [5]. The premature infant has an exaggerated postnatal depression in circulating calcium, such that total calcium levels drop below 7.0 mg/dl, but the proportional drop in ionized calcium is less [11]. Inadequate PTH secretion may contribute to early neonatal hypocalcemia in premature infants; a delay in the phosphaturic action of PTH and resultant hyperphosphatemia may further decrease serum calcium [5].

Late neonatal hypocalcemia most often presents clinically as tetany or seizures between 5 and 10 days of life, and the differential diagnosis includes transient hypoparathyroidism, transient PTH resistance, DiGeorge syndrome, maternal vitamin D deficiency, malabsorption, intake of formula high in phosphorus content, and hypomagnesemia [10–12].

Maternal hyperparathyroidism or hypercalcemia may result in neonatal hypocalcemia. Infants who are exposed in utero to increased calcium delivery can have suppressed parathyroid function and responsiveness [3]. As a result, normal calcium levels are not maintained postnatally because of persistent parathyroid gland suppression in the baby. The resultant hypocalcemia can occur during the first few weeks of life but may persist up to 1 year of age [5].

15.3 Approach to Hypocalcemia

15.3.1 Biochemistry

Total serum calcium includes a fraction of approximately 50 % that is free or ionized and another fraction approximately 45–50 % that is protein-bound, primarily to albumin. There is a small percentage (5–10 %) of the total calcium that is complexed to other anions in the circulation. Thus, the total serum calcium level should be "corrected" if hypoalbuminemia exists [13]. The total calcium may be adjusted as follows to correct for calcium binding to albumin:

$$\text{Corrected total calcium} = \text{measured total calcium} + 0.8\left(4.0 - \text{serum albumin}\right),$$

where calcium is measured in milligrams per deciliter and albumin is measured in grams per deciliter. Hypocalcemia associated with hypoparathyroidism can be differentiated from other causes of hypocalcemia by routine laboratory tests [2]. Serum calcium is low due to lack of PTH-mediated bone resorption and urinary calcium reabsorption. Serum phosphate is increased because of impaired renal clearance. Serum $1,25(OH)_2$ vitamin D is low because of the lack of PTH-elicited stimulation of the renal 25(OH)vitamin D 1α-hydroxylase. Consequently, 1,25$(OH)_2$vitamin D-mediated intestinal calcium absorption is markedly decreased, further exacerbating the hypocalcemia. The phosphate-regulating hormone FGF23 (fibroblast growth factor 23) which can lower serum phosphate tends to be elevated in hypoparathyroidism, but it is unable to exert its full actions on the proximal tubule to increase phosphate excretion in the absence of PTH. It has also become clear that calcium stimulates FGF23 and that the stimulation of FGF23 by phosphate requires an extracellular calcium level of ~8 mg/dL in rats and mice [14, 15]. Therefore, the effects of mediators of FGF23 production in vivo likely depend on the calcemic status of the patient, which can vary over a wide range in patients with hypoparathyroidism. The relative deficiency of FGF23 in the severely hypocalcemic patient with hypoparathyroidism may contribute to the accompanying hyperphosphatemia since neither PTH nor FGF23 can exert a robust phosphaturic action. Moreover, correction of hypocalcemia during treatment of hypoparathyroidism may contribute to the reciprocal drop in serum phosphorus concentration by stimulating FGF23 production.

In the laboratory workup of hypocalcemic patients, PTH levels, measured by sensitive intact PTH assays, are usually low or undetectable. They may be in some cases within the normal range, depending on the assay used, but such levels are inappropriately normal. This reflects the fact that some of the capacity to secrete hormone is still present in many hypoparathyroid patients, but it is insufficient to meet physiologic needs. Patients with pseudohypoparathyroidism have laboratory profiles that resemble those of patients with hypoparathyroidism (i.e., low calcium and high phosphate levels), but they have elevated PTH levels [1].

Serum levels of magnesium should be measured to rule out a deficiency that could contribute to reduced serum calcium levels. It may be difficult to rule out hypomagnesemia completely as the cause of or a contributor to hypocalcemia because the serum magnesium level may be normal, even when intracellular magnesium stores are reduced.

Measurement of 25(OH)vitamin D levels is essential to rule out vitamin D deficiency as a contributor to or cause of hypocalcemia. In classic vitamin D deficiency, intact PTH levels are elevated, and serum phosphate levels are low or in the low-normal range. In contrast, serum phosphate levels are high in hypoparathyroidism. Measurement of 1,25$(OH)_2$ vitamin D levels is generally not necessary in the initial evaluation of patients with hypocalcemia, but may be needed later to assess 1,25$(OH)_2$vitamin D production if the PTH levels are found to be elevated.

Renal function should also be assessed by the measurement of serum creatinine and the estimation of glomerular filtration rates by standard formula. A 24-h urine collection for formal creatinine clearance can be helpful and can be added to the assessment of calcium and or magnesium excretion. The latter analytes are often helpful in the differential diagnosis of hypoparathyroidism [16]. Patients with activating mutations of the CaSR often have an elevated 24-h urinary calcium excretion, compared to patients with postsurgical hypoparathyroidism. This results from the fact that the activating mutation of the CaSR in the kidney leads to increased renal excretion of calcium because the mutant receptor is providing the signal of excessive levels of serum calcium erroneously to the kidney. This can be a helpful initial tip to the diagnosis if the patient does the urine collection when hypocalcemic.

15.3.2 History and Family History

A thorough history is very important in the workup of hypocalcemic patients. This includes surgical history, history of neck radiation or systemic illness, and family history. A history of neck surgery suggests that parathyroid function may have been compromised during surgery. A family history of hypocalcemia suggests a genetic cause. The presence of autoimmune disorders or of candidiasis should prompt an evaluation for autoimmune polyendocrine syndrome type 1 (APS-1). APS-1 has a variable clinical presentation, but the classic triad is mucocutaneous candidiasis, adrenal insufficiency, and hypoparathyroidism. Any two of these three conditions are sufficient to establish the diagnosis. Most patients with APS-1 present during childhood with very early onset candidiasis, but presentations later in life of the other disease manifestations have been seen. Other features seen in APS-1 include hypogonadism, type 1 diabetes mellitus, hypothyroidism, vitiligo, alopecia, keratoconjunctivitis, hepatitis, pernicious anemia, and malabsorption [17, 18]. Immunodeficiency and other congenital defects point to the DiGeorge syndrome, which occurs in 1 in 3,000–4,000 live births [6, 19]. History of symptoms such as cramping, tetany, twitching, and seizures should be elicited in the patient and family members.

A personal or family history of deafness, renal anomalies, and hypocalcemia should trigger an evaluation for the hypoparathyroidism, deafness, and renal dysplasia (HDR) syndrome, an autosomal dominant disorder caused by mutations in GATA3, a zinc finger transcription factor. Clinically, the HDR syndrome is characterized by low plasma calcium concentrations with low-normal to undetectable intact PTH levels. The sensorineural hearing loss is bilateral and is most pronounced at higher frequencies. Hearing loss is typically moderate to severe and present at birth. Numerous renal anomalies have been observed with variable penetrance, including renal dysplasia, hypoplasia, aplasia, and vesicoureteral reflux [20]. These anomalies are often only detected on renal imaging, but affected patients often show some mild degrees of renal dysfunction with mildly elevated serum creatinine levels.

15.3.3 Physical Exam

Recognition of the signs and symptoms of hypocalcemia can be a valuable tool when diagnosing hypocalcemia and monitoring therapeutic responses. On physical examination, assessment of Trousseau's and Chvostek's signs, which are indicative of neuromuscular irritability, should be done. Chvostek's sign is elicited by tapping the cheek (2 cm anterior to the earlobe below the zygomatic process) over the path of the facial nerve. A positive sign is ipsilateral twitching of the upper lip. Of note, a degree of twitching may be seen in 10–15 % of normal individuals. Trousseau's sign is elicited by inflating a sphygmomanometer placed on the upper arm to 20 mmHg above the systolic blood pressure for 3 min. A positive sign is the occurrence of a painful carpal spasm. Table 15.1 summarizes the signs and symptoms of hypocalcemia.

A history of muscle cramps, commonly involving the legs and feet, is also common. These muscle spasms may progress to carpopedal spasm with tetany, laryngospasm and stridor, or bronchospasm. The skin should be examined carefully for a neck scar (which suggests a postsurgical cause of hypocalcemia); for candidiasis and vitiligo (which are suggestive of APS-1); and for generalized bronzing (which are suggestive of hemochromatosis) and signs of liver disease (which are seen in hemochromatosis or Wilson's disease). Features such as growth failure, congenital anomalies, hearing loss, or retardation suggest the possibility of genetic disease.

15.4 Causes of Hypocalcemia Secondary to Parathyroid-Related Disorders

Assessing the intact PTH level is essential to establishing the etiology of hypocalcemia. It is important to ascertain whether the hypocalcemia is associated with an undetectable or inappropriately low serum PTH concentration (hypoparathyroidism) or is associated with an appropriate compensatory increase in PTH. This evaluation can guide the appropriate approach the various causes of hypocalcemia secondary to PTH-related disorders that are outlined in Table 15.2.

15.4.1 Absence of the Parathyroid Glands or of PTH

Postsurgical hypoparathyroidism (see also Chap. 22) is the most common acquired cause of hypoparathyroidism in adults [1, 21–23]. Surgery on the thyroid, parathyroid glands, and adjacent neck structures or neck dissection surgery for malignancy may lead to acute or chronic

Table 15.2 Causes of hypocalcemia secondary to PTH-related disorders

Absence of the parathyroid glands or of PTH
Postsurgical hypoparathyroidism
Autoimmune
Isolated
Autoimmune polyglandular syndrome type I
Idiopathic
Congenital/genetic
DiGeorge syndrome
X-linked
Autosomal dominant or recessive
Kenny–Caffey syndrome
Mitochondrial disorders
PTH gene mutations
Mutations of GATA3 and GCMB genes
Destruction of the parathyroid glands due to infiltrative disorders or other acquired causes
Infiltrative disorders
Hemochromatosis
Wilson's disease
Metastases
Thalassemia due to iron overload from chronic transfusions
Other Acquired Causes
Radioactive iodine destruction following thyroid ablation
Impaired secretion of PTH
Activating mutations of the CaSR and G α11
Secondary to maternal hyperparathyroidism or hypercalcemia
Hypomagnesemia
Target organ resistance
Hypomagnesemia
Pseudohypoparathyroidism
Type 1
Type 2

hypoparathyroidism. Postoperative hypoparathyroidism usually is due to inadvertent or unavoidable removal of or damage to the parathyroid glands and/or their blood supply. While transient hypoparathyroidism after neck surgery is relatively common, often called "stunning" of the glands, chronic partial hypoparathyroidism is less common, and chronic complete hypoparathyroidism is relatively rare. Most patients with postoperative hypoparathyroidism recover parathyroid gland function within several weeks to months after surgery and thus do not develop permanent disease [23].

Autoimmune hypoparathyroidism (see also Chap. 17) may be isolated or part of APS-1 which is due to loss-of-function mutations in *AIRE*, a transcription factor present in thymus and lymph nodes and critical for mediating central tolerance by the thymus [17, 18].

Familial occurrences of hypoparathyroidism with autosomal dominant, autosomal recessive, or X-linked recessive modes of inheritance have been established and will be discussed in the subsequent chapters. Autosomal forms of hypoparathyroidism are caused by mutations in the genes encoding PTH, GATA3, and GCMB (glial cells missing homologue B) [7] and molecules in the pathway leading to calcium ion sensing by the parathyroid glands [24]. Genes in this pathway include the CaSR and the G protein α subunit (Gα11) that couples the CaSR to intracellular signaling pathways [25, 26]. X-linked hypoparathyroidism is due to a deletion and insertion involving genetic material from chromosomes 2p25.3 and Xq27.1, causing a position effect on possible regulatory elements controlling transcription of SOX3, a transcription factor thought to be expressed in the developing parathyroid glands [27].

DiGeorge syndrome (see also Chap. 18) is due to loss of function of genes on chromosome 22q11, most notably *TBX1*, a transcription factor responsible for regulating expression of other transcription and growth factors important in development of thymus and parathyroid glands; parathyroid and thymic defects are caused by abnormal development in the third and fourth branchial pouches [6, 7].

Syndromes of hypoparathyroidism, growth and mental retardation, and dysmorphism including Kenny–Caffey and Sanjad–Sakati syndromes (see also Chap. 20) are due to mutations in tubulin-folding cofactor E (*TBCE*) causing loss of function and probably altered microtubule assembly in affected tissues [28]. The Kenny–Caffey syndrome presents with short stature, osteosclerosis, cortical bone thickening, calcification of basal ganglia, ocular abnormalities, and hypoparathyroidism that is probably due to agenesis of the glands [29], while the Sanjad–Sakati syndrome presents with parathyroid aplasia, growth failure, ocular malformations, microencephaly, and retardation [30]. These syndromes are exceedingly rare.

Mitochondrial disorders with hypoparathyroidism are due to deletions of varying size, mutations, rearrangements, and duplications in the mitochondrial genome (see also Chap. 21). These syndromes include the Kearns–Sayre syndrome; mitochondrial encephalomyopathy, lactic acidosis, and stroke-like episodes (MELAS); and mitochondrial trifunctional protein deficiency (MTPDS). The Kearns–Sayre syndrome is characterized by progressive external ophthalmoplegia and pigmentary retinopathy before age 20. It is often associated with heart block or cardiomyopathy [31]. In MELAS, the constellation of symptoms typically occurs in childhood [31]. A varying degree of proximal myopathy and insulin-dependent diabetes and hypoparathyroidism can be seen in both Kearns–Sayre syndrome and MELAS. MTPDS is a disorder of fatty-acid oxidation associated with peripheral neuropathy, retinopathy, acute fatty liver in pregnant women who carry an affected fetus, and hypoparathyroidism [32].

In addition to postsurgical hypoparathyroidism, there are other less common acquired causes of hypoparathyroidism (see also Chap. 24). They include infiltrative disorders with excessive accumulation of iron in the parathyroid glands owing to thalassemia (post multiple transfusions) or untreated hemochromatosis or of copper in Wilson's disease [33–36]. Acquired hypoparathyroidism has been reported to occur very rarely after iodine-131 therapy or due to metastatic infiltration of the parathyroid glands [1].

15.4.2 Impaired Secretion of PTH

Activating mutations of the CaSR are most commonly caused by mutations in the gene itself, but rare cases are due to production of antibodies that stimulate the CaSR and suppress PTH secretion. These activating mutations appear to be one of the most common causes of hypoparathyroidism. Activating CaSR mutations cause a left-shifted set point for PTH secretion, defined as the extracellular calcium level required for half-maximal suppression of secretion, causing inappropriately normal or low PTH levels even at low serum calcium levels [37–41]. Once thought to cause mild, asymptomatic hypocalcemia that did not require any intervention, it is now clear that, depending on the mutation and its effects on the parathyroid cells, the disorder can present in the neonatal period with refractory hypocalcemia and severe symptomatology such as seizures. Two recent reports [25, 26] demonstrate that activating mutations in the α subunit (Gα11) that mediate the coupling of the CaSR to its intracellular signaling partners in parathyroid cells can produce suppression of PTH secretion, inherited in an autosomal dominant manner in families and present as hypoparathyroidism. Thus, defects in CaSR function and or CaSR signaling can cause the hypoparathyroid phenotype.

Hypomagnesemia can be seen in chronic conditions that include alcoholism, malnutrition, pancreatitis, malabsorption, diarrhea, or diabetes; chronic drug intake (e.g., diuretics, cisplatinum, aminoglycoside antibiotics, amphotericin B, and cyclosporine); metabolic acidosis; and renal disorders leading to magnesium wasting (chronic pyelonephritis, post-obstructive nephropathy, renal tubular acidosis, primary renal magnesium wasting, and diuretic stage of acute tubular necrosis). During chronic and severe magnesium depletion, PTH secretion is impaired, and most patients will have PTH concentrations that are undetectable or inappropriately normal for the degree of hypocalcemia. The basis for the defects in PTH secretion and in PTH end-organ resistance in patients with hypomagnesemia is unclear [42].

15.4.3 Target Organ Resistance

Several clinical disorders characterized by PTH end-organ resistance have been described and are collectively referred to as pseudohypoparathyroidism (see also Chaps. 32, 33, 34, and 35). PTH resistance is defined as hypocalcemia and hyperphosphatemia in the presence of high plasma PTH levels, after excluding chronic renal failure or magnesium and or vitamin D deficiency states [3].

Pseudohypoparathyroidism type 1a is characterized by short stature and other skeletal abnormalities, known collectively as Albright's hereditary osteodystrophy (AHO), as well as hypocalcemia and high serum concentrations of parathyroid hormone. It is caused by heterozygous inactivating mutations in the α subunit of G_s (GNAS) and is inherited as an autosomal dominant trait with maternal transmission of the biochemical phenotype. Many patients with pseudohypoparathyroidism type 1a have resistance not only to PTH but also to thyrotropin and less frequently, gonadotropin resistance resulting in hypogonadism [21, 43]. Pseudohypoparathyroidism type 1b has the same biochemical features as pseudohypoparathyroidism type 1a but has selective resistance to PTH and not to other hormone receptors. Patients with pseudohypoparathyroidism 1b do not manifest AHO [21, 43]. Pseudohypoparathyroidism type 2 has the same biochemical profile as type 1a or 1b but is less common. However, pseudohypoparathyroidism type 2 lacks a clear genetic or familial basis, and the mechanism of PTH resistance is unknown [1].

15.5 Imaging in Hypoparathyroidism

Although no specific imaging studies are used to diagnose hypoparathyroidism or distinguish hypoparathyroidism from other forms of hypocalcemia, computed tomography (CT) of the brain, which may be done for other reasons, may show calcifications in the basal ganglia and other locations in the brain [44]. These deposits do not usually cause neurological problems. Generally, such calcifications are found in patients who have been treated for extended periods of time with calcium salts and activated vitamin D metabolites (e.g., calcitriol) that tend to raise the calcium × phosphate product. It has been noted, however, that brain calcifications can be seen at the time of diagnosis, before treatment has been initiated. This is particularly the case in patients with activating mutations in the CaSR.

Renal imaging by ultrasound or CT can be helpful in the chronic care of patients with hypoparathyroidism to monitor nephrocalcinosis and nephrolithiasis. At the time of diagnosis and before the initiation of therapy, one rarely sees these pathologic findings in the kidneys. There is one exception, however, and that is hypoparathyroidism due to activating CaSR mutations. In that situation, nephrocalcinosis may be seen even in young children—many severely affected—and even before long-term treatment is initiated with calcium salts and activated vitamin D metabolites. Renal anomalies may also be noted in patients with the HDR syndrome. Patients with pseudohypoparathyroidism type 1a may have subcutaneous or soft tissue calcifications as part of the features of Albright's osteodystrophy. When these subdermal and soft tissue lesions are imaged, diffuse calcifications may be appreciated on plain X-rays.

Patients with chronic hypoparathyroidism usually do not have bone mineral density (BMD) or T-scores by dual energy X-ray absorptiometry (DXA) indicative of osteopenia or osteoporosis. In fact, bone mass, by DXA, is generally substantially higher in hypoparathyroidism than in age and sex-matched controls [45–48]. It has been suggested that chronic treatment with calcium salts and vitamin D and its metabolites contribute to the preservation of bone mass. Low levels of bone turnover, assessed by circulating biochemical markers and by dynamic histomorphometry, have been shown in hypoparathyroidism [49]. This is likely to be the other key reason for the higher BMD in these patients because when PTH is replaced and turnover increases, BMD declines modestly verified by biochemical markers [50].

Conclusions

The signs and symptoms of hypoparathyroidism are manifestations of hypocalcemia including increased neuromuscular irritability, paresthesias, muscle spasm and cramping, and central nervous system involvement (Table 15.1). The initial evaluation of a patient with hypocalcemia should include a detailed family history that might suggest a genetic cause, and relevant medical history, including surgical history and known history of autoimmune disease. Laboratory testing should include measurements of serum total and ionized calcium, albumin, phosphorus, magnesium, and intact PTH levels. When determining the etiology of hypocalcemia, it is important to determine whether it is associated with a detectable or inappropriately low PTH or with an appropriate compensatory increase in PTH. The causes of hypocalcemia secondary to PTH-related disorders are outlined in Table 15.2; however, there is much to uncover regarding the etiology and clinical phenotypes of many of the disorders associated with hypoparathyroidism. Imaging studies, including CT, ultrasound, and DXA, can add further diagnostic value but cannot be used at this time to diagnose hypoparathyroidism or rule out other etiologies of hypocalcemia. Recognition of hypocalcemia is paramount to establish its etiology in order to initiate the appropriate evaluation and treatment (discussed in subsequent chapters) especially since acute symptomatic hypocalcemia can be medical emergencies necessitating rapid, life-saving interventions.

References

1. Shoback D (2008) Hypoparathyroidism. N Engl J Med 359:391–403
2. Bringhurst FR, Demay MB, Kronenberg KM (2011) Hormones and disorders of mineral metabolism. In: Melmed S, Polonsky KS, Larsen PR, Kronenberg HM (eds) Williams textbook of endocrinology, 12th edn. Elsevier/Saunders, Philadelphia, pp 1237–1304
3. Rubin MR, Levine M (2013) Hypoparathyroidism and pseudohypoparathyroidism. In: Rosen CJ (ed) Primer on the metabolic bone diseases and disorders of mineral metabolism, 8th edn. Wiley-Blackwell, Ames, pp 579–589
4. Lienhardt-Roussie A, Linglart A (2012) Hypoparathyroidism in children. In: Licata AA, Lerma EV (eds) Diseases of the parathyroid glands. Springer, New York, pp 299–310
5. Carpenter TO (2013) Disorders of mineral metabolism in childhood. In: Rosen CJ (ed) Primer on the metabolic bone diseases and disorders of mineral metabolism, 8th edn. Wiley-Blackwell, Ames, pp 651–658
6. Webber SA, Hatchwell E, Barber JC, Daubeney PE, Crolla JA, Salmon AP, Keeton BR, Temple IK, Dennis NR (1996) Importance of microdeletions of chromosomal region 22q11 as a cause of selected malformations of the ventricular outflow tracts and aortic arch: a three-year prospective study. J Pediatr 129(1):26–32
7. Thakker RV (2004) Genetics of endocrine and metabolic disorders: parathyroid. Rev Endocr Metab Disord 5:37–51
8. Tsang RC, Light IJ, Sutherland JM, Kleinman LI (1973) Possible pathogenetic factors in neonatal hypocalcemia of prematurity. The role of gestation, hyperphosphatemia, hypomagnesemia, urinary calcium loss, and parathormone responsiveness. J Pediatr 82(3):423–429
9. Stefanaki E, Koropuli M, Stefanaki S, Tsilimigaki A (2005) Atrioventricular block in preterm infants caused by hypocalcemia: a case report and review of the literature. Eur J Obstet Gynecol Reprod Biol 120(1):115–116
10. Thomas TC, Smith JM, White PC, Adhikari S (2012) Transient neonatal hypocalcemia: presentation and outcomes. Pediatrics. doi:10.1542/peds.2011-2659
11. Hsu SC, Levine MA (2004) Perinatal calcium metabolism: physiology and pathophysiology. Semin Neonatol 9(1):23–36
12. Gittleman IF, Pincus JB (1951) Influence of diet on the occurrence of hyperphosphatemia and hypocalcemia in the newborn infant. Pediatrics 8(6):778–787
13. Fukugawa M, Kurokawa K (2002) Calcium homeostasis and imbalance. Nephron 92(suppl 1):41–45
14. Rodriguez-Ortiz ME, Lopez I, Munoz-Castaneda JR, Martinez-Moreno JM, Ramirez AP, Pineda C, Canalejo A, Jaeger P, Aguilera-Tejero E, Rodriguez M, Felsenfeld A, Almaden Y (2012) Calcium deficiency reduces circulating levels of FGF23. J Am Soc Nephrol 23:1190–1197
15. Quinn SJ, Thomsen AR, Pang JL, Kantham L, Brauner-Osborne H, Pollak M, Goltzman D, Brown EM (2013) Interactions between calcium and phosphorus in the regulation of the production of fibroblast growth factor 23 in vivo. Am J Physiol Endocrinol Metab 304(3):E310–E320
16. Al-Azem H, Khan AA (2012) Hypoparathyroidism. Best Pract Res Clin Endocrinol Metab 26(4):517–522
17. Dittmar M, Kahaly GJ (2003) Polyglandular autoimmune syndromes: immunogenetics and long-term follow-up. J Clin Endocrinol Metab 88:2983–2992
18. Eisenbarth GS, Gottlieb PA (2004) Autoimmune polyendocrine syndromes. N Engl J Med 350:2068–2079
19. Kobrynski LJ, Sullivan KE (2007) Velocardiofacial syndrome, DiGeorge syndrome: the chromosome 22q11.2 deletion syndromes. Lancet 370:1443–1452
20. Zahirieh A, Nesbit MA, Ali A, Wang K, He N, Stangou M, Bamichas G, Sombolos K, Thakker RV, Pei Y (2005) Functional analysis of a novel GATA3 mutation in a family with the hypoparathyroidism, deafness, and renal dysplasia (HDR) syndrome. J Clin Endocrinol Metab 90(4):2445–2450
21. Marx SJ (2000) Hyperparathyroid and hypoparathyroid disorders. N Engl J Med 343:1863–1875
22. Mannstadt M, Clarke BL, Vokes T, Brandi ML, Ranganath L, Fraser WD, Lakatos PL, Bajnok L, Garceau R, Mosekilde L, Lagast H, Shoback D, Bilezikian JP (2013) Efficacy and safety of recombinant human parathyroid hormone (1–84) in hypoparathyroidism (REPLACE): a double-blind,

placebo-controlled, randomised phase 3 study. Lancet Diabetes Endocrinol 1(4):275–283

23. Bilezikian JP, Khan A, Potts JT Jr, Brandi ML, Clarke BL, Shoback D, Jüppner H, D'Amour P, Fox J, Rejnmark L, Mosekilde L, Rubin MR, Dempster D, Gafni R, Collins MT, Sliney J, Sanders J (2011) Hypoparathyroidism in the adult: epidemiology, diagnosis, pathophysiology, target-organ involvement, treatment, and challenges for future research. J Bone Miner Res 26(10):2317–2337
24. Baron J, Winer KK, Yanovski JA, Cunningham AW, Laue L, Zimmerman D, Cutler GB Jr (1996) Mutations in the Ca2+-sensing receptor gene cause autosomal dominant and sporadic hypoparathyroidism. Hum Mol Genet 5(5):601–606
25. Nesbit MA, Hannan FM, Howles SA, Babinsky VN, Head RA, Cranston T, Rust N, Hobbs MR, Heath H 3rd, Thakker RV (2013) Mutations affecting G-protein subunit α11 in hypercalcemia and hypocalcemia. N Engl J Med 368(26):2476–2486
26. Mannstadt M, Harris M, Bravenboer B, Chitturi S, Dreijerink KM, Lambright DG, Lim ET, Daly MJ, Gabriel S, Jüppner H (2013) Germline mutations affecting Gα11 in hypoparathyroidism. N Engl J Med 368(26):2532–2534
27. Bowl MR, Nesbit MA, Harding B, Levy E, Jefferson A, Volpi E, Rizzoti K, Lovell-Badge R, Schlessinger D, Whyte MP, Thakker RV (2005) An interstitial deletion-insertion involving chromosomes 2p25.3 and Xq27.1, near SOX3, causes X-linked recessive hypoparathyroidism. J Clin Invest 115:2822–2831
28. Parvari R, Diaz GA, Hershkovitz E (2007) Parathyroid development and the role of tubulin chaperone E. Horm Res 67:12–21
29. Parvari R, Hershkovitz E, Grossman N, Gorodischer R, Loeys B, Zecic A, Mortier G, Gregory S, Sharony R, Kambouris M, Sakati N, Meyer BF, Al Aqeel AI, Al Humaidan AK, Al Zanhrani F, Al Swaid A, Al Othman J, Diaz GA, Weiner R, Khan KT, Gordon R, Gelb BD, HRD/Autosomal Recessive Kenny-Caffey Syndrome Consortium (2002) Mutation of TBCE causes hypoparathyroidism-retardation-dysmorphism and autosomal recessive Kenny-Caffey syndrome. Nat Genet 32:448–452
30. Sanjad SA, Sakati NA, Abu-Osba YK, Kaddoura R, Milner RDG (1991) A new syndrome of congenital hypoparathyroidism, severe growth failure, and dysmorphic features. Arch Dis Child 66:193–196
31. Moraes CT, DiMauro S, Zeviani M, Lombes A, Shanske S, Miranda AF, Nakase H, Bonilla E, Werneck LC, Servidei S et al (1989) Mitochondrial DNA deletions in progressive external ophthalmoplegia and Kearns-Sayre syndrome. N Engl J Med 320(20):1293–1299
32. Thakker RV, Bringhurst FR, Juppner H (2013) Calcium regulation calcium homeostasis and genetic disorders. In: Potts JT, Jameson JL, De Groot L (eds) Endocrinology adult and pediatric: the parathyroid gland and bone metabolism. Elsevier/Saunders, Philadelphia, p 1148
33. Angelopoulos NG, Goula A, Rombopoulos G, Kaltzidou V, Katounda E, Kaltsas D, Tolis G (2006) Hypoparathyroidism in transfusion-dependent patients with β-thalassemia. J Bone Miner Metab 24:138–145
34. Toumba M, Sergis A, Kanaris C, Skordis N (2007) Endocrine complications in patients with thalassaemia major. Pediatr Endocrinol Rev 5:642–648
35. de Seze S, Solnica J, Mitrovic D, Miravet L, Dorfmann H (1972) Joint and bone disorders and hypoparathyroidism in hemochromatosis. Semin Arthritis Rheum 2:71–94
36. Carpenter TO, Carnes DL Jr, Anast CS (1983) Hypoparathyroidism in Wilson's disease. N Engl J Med 309:873–877
37. Brown EM (2007) Clinical lessons from the calcium-sensing receptor. Nat Clin Pract Endocrinol Metab 3:122–133
38. Egbuna OI, Brown EM (2008) Hypercalcaemic and hypocalcaemic conditions due to calcium-sensing receptor mutations. Best Pract Res Clin Rheumatol 22:129–148
39. Yamamoto M, Akatsu T, Nagase T, Ogata E (2000) Comparison of hypocalcemic hypercalciuria between patients with idiopathic hypoparathyroidism and those with gain-of-function mutations in the calcium-sensing receptor: is it possible to differentiate the two disorders? J Clin Endocrinol Metab 85:4583–4591
40. Lienhardt A, Bai M, Lagarde JP, Rigaud M, Zhang Z, Jiang Y, Kottler ML, Brown EM, Garabédian M (2001) Activating mutations of the calcium sensing receptor: management of hypocalcemia. J Clin Endocrinol Metab 86:5313–5323
41. Kifor O, McElduff A, LeBoff MS, Moore FD Jr, Butters R, Gao P, Cantor TL, Kifor I, Brown EM (2004) Activating antibodies to the calcium-sensing receptor in two patients with autoimmune hypoparathyroidism. J Clin Endocrinol Metab 89:548–556
42. Tong GM, Rude RK (2005) Magnesium deficiency in critical illness. J Intensive Care Med 20:3–17
43. Bastepe M (2008) The GNAS locus and pseudohypoparathyroidism. Adv Exp Med Biol 626:27–40
44. Fukunaga M, Otsuka N, Ono S, Kajihara Y, Nishishita S, Nakano Y, Yamamoto I, Torizuka K, Morita R (1987) Computed tomography of basal ganglia calcifications in pseudo- and idiopathic hypoparathyroidism. Radiat Med 5:187–190
45. Abugassa S, Nordenström J, Eriksson S, Sjödén G (1993) Bone mineral density in patients with chronic hypoparathyroidism. J Clin Endocrinol Metab 76:1617–1621
46. Seeman E, Wahner HW, Offord KP, Kumar R, Johnson WJ, Riggs BL (1982) Differential effects of endocrine dysfunction on the axial and the appendicular skeleton. J Clin Invest 69:1302–1309
47. Fujiyama K, Kiriyama T, Ito M, Nakata K, Yamashita S, Yokoyama N, Nagataki S (1995) Attenuation of

postmenopausal high turnover bone loss in patients with hypoparathyroidism. J Clin Endocrinol Metab 80:2135–2138
48. Touliatos JS, Sebes JI, Hinton A, McCommon D, Karas JG, Palmieri GM (1995) Hypoparathyroidism counteracts risk factors for osteoporosis. Am J Med Sci 310:56–60
49. Rubin MR, Bilezikian JP (2010) Hypoparathyroidism: clinical features, skeletal microstructure and parathyroid hormone replacement. Arq Bras Endocrinol Metabol 54(2):220–226
50. Cusano NE, Rubin MR, Sliney J Jr, Bilezikian JP (2012) Mini-review: new therapeutic options in hypoparathyroidism. Endocrine 41(3):410–414

Familial Isolated Hypoparathyroidism

16

Geoffrey N. Hendy and David E.C. Cole

16.1 Introduction

This chapter reviews those disorders that are classified strictly as isolated hypoparathyroidism due to single gene mutations [1]. The hypoparathyroidism is characterized clinically by hypocalcemia and hyperphosphatemia and low or low-normal 1,25-dihydroxyvitamin D [1,25(OH)$_2$D] [2]. It occurs when parathyroid hormone (PTH) secreted from the parathyroid glands is either absent or insufficient to maintain normal extracellular fluid calcium concentrations. The Mendelian inheritance of FIH can be autosomal dominant, autosomal recessive, or X-linked [3] (Table 16.1).

In 8 families with 23 affected individuals fulfilling strict criteria for familial isolated hypoparathyroidism (FIH, OMIM# 146200), Ahn et al [4] noted autosomal dominant inheritance in five and autosomal recessive inheritance in three. In one dominant and two recessive kindreds, the pedigrees were consistent with X-linked inheritance. The presence of large deletions, insertions, or rearrangements of the *PTH* gene was excluded in all cases. In four families, linkage with the *PTH* locus (11p15.2) was excluded by restriction fragment length polymorphism (RFLP) analysis, whereas in two families the hypoparathyroidism was concordant with *PTH* gene markers.

16.2 PTH

The preproPTH mRNA, translated on rough endoplasmic reticulum (ER) polyribosomes, encodes an amino-terminal 25-amino-acid signal or prepeptide that directs the nascent chain into the ER where it is removed cotranslationally by the signalase enzyme, followed by a six–amino acid propeptide that is removed by the proprotein convertases furin and PC7 after trafficking of the prohormone to the Golgi apparatus [5, 6]. The mature 84–amino acid PTH molecule is packaged into secretory vesicles and is released from the parathyroid endocrine cells following a decrease in blood calcium monitored by the plasma membrane G-protein-coupled calcium-sensing receptor (CaSR). The human *PTH* gene (OMIM# 168450) has three exons. Exon

Work from our laboratories was supported by the Canadian Institutes of Health Research (to G.N.H.) and the Dairy Farmers of Canada (to D.E.C.C.).

G.N. Hendy (✉)
Departments of Medicine, Physiology, and Human Genetics (G.N.H.), McGill University, Montreal, QC H3A 1A1, Canada

Calcium Research Laboratory and Hormones and Cancer Research Unit, Royal Victoria Hospital, Montreal, QC H3A 1A1, Canada
e-mail: geoffrey.hendy@mcgill.ca

D.E.C. Cole
Departments of Laboratory Medicine and Pathobiology, Medicine, and Genetics (D.E.C.C.), Sunnybrook Health Sciences Centre, University of Toronto, Toronto, ON MSG 1L5, Canada
e-mail: davidec.cole@utoronto.ca

M.L. Brandi, E.M. Brown (eds.), *Hypoparathyroidism*,
DOI 10.1007/978-88-470-5376-2_16, © Springer-Verlag Italia 2015

Table 16.1 Disorders causing familial isolated hypoparathyroidism

Disorder, [a]OMIM#	Chromosomal locus	Gene name, OMIM#	Mode of inheritance	Pathophysiology
FIH, 146200	11p15.2	Parathyroid hormone, *PTH*, 168450	AD, AR	Defective PTH synthesis
ADH1, 601198	3q13.3-21	Calcium-sensing receptor, *CASR*, 601199	AD	Enhanced Ca^{2+} sensing, with excess suppression of parathyroid response
ADH2, 615361	19p13.3	Guanine nucleotide-binding protein, α-11 *GNA11*, 139313	AD	Enhanced parathyroid CaSR signaling and excess suppression of parathyroid response
FIH, 146200	6p24.2	Glial cells missing-2, *GCM2*, 603716	AD, AR	Defective parathyroid gland development
X-linked, 307700	Xq27.1	del(X)(q27.1)inv ins(X;2)(q27.1;p25.3)	XLR	Breakpoint on Xq27.1 67-kb downstream of SOX3 OMIM# 313430 SRY-Box 3. Positional effect on SOX3 thought to result in defective parathyroid gland development

[a]*Abbreviations*: *OMIM* Online Mendelian Inheritance in Man, *FIH* familial isolated hypoparathyroidism, *AD* autosomal dominant, *AR* autosomal recessive, *XLR* X-linked recessive, *ADH1* autosomal dominant hypocalcemia type 1, *ADH2* autosomal dominant hypocalcemia type 2

1 encodes the 5′-untranslated region, exon 2 encodes the signal peptide and part of the propeptide, while exon 3 encodes the propeptide cleavage site and the mature PTH molecule [7, 8].

In a few instances of autosomal dominant disease, a mutation in the *PTH* gene has been identified. In one family presenting with reduced hormone production and chronic hypocalcemia, a missense mutation (p.Cys18Arg) in the signal sequence of the preproPTH precursor was identified [9] and the mutant shown to be defective in vitro in processing of preproPTH to proPTH [10]. Patients had one normal gene copy, leaving the autosomal dominant mode of inheritance unexplained. Further studies in transfected cells showed that the mutant protein accumulated abnormally in the ER, promoting ER stress and apoptosis [11]. The intracellular accumulation of the mutant hormone was corrected by a chemical chaperone, 4-phenylbutyric acid, with decrease in expression of ER stress markers and protection from cell death. The same mutation was found in an unrelated individual with idiopathic hypoparathyroidism [12]. A hypoparathyroid patient was reported to be heterozygous for a Met1_Asp6 deletion mutation [13]. The c.2T>C mutation in the methionine initiation codon predicted abnormal initiation of translation of the mutant prepro-PTH mRNA starting at the Met +7 codon. How the mutant protein might function in an autosomal dominant fashion in this patient is unclear. Previous functional studies of an engineered Met1_Asp6 mutant found that loss of the first 6 amino acids did not impair cleavage and translocation of the mutant preproPTH in a cell-free translation system, nor did it inhibit export of processed PTH from intact secretory cells in culture [14].

In three siblings of a consanguineous family with autosomal recessive hypoparathyroidism, a donor splice site mutation (c.86+1G>C) at the exon2/intron 2 junction of the *PTH* gene was identified [15]. The mutation leads to exon skipping and loss of exon 2 containing the initiation codon and signal sequence of preproPTH mRNA. In another family, a novel mutation in the signal sequence (p.Ser23Pro) segregated with affected status [16]. This mutation may prevent proper cleavage of the signal peptide during processing of the nascent protein. In a girl with isolated hypoparathyroidism, a homozygous p. Ser23X signal sequence mutation was found predicting a truncated inactive PTH peptide [17]. However, the circulating PTH level was not undetectable suggesting some translational readthrough of the mutant PTH mRNA.

16.3 CaSR

The human *CASR* gene maps to 3q13.3-21, and the CaSR protein encoded by exons 2–7 is a member of group 3 or C of the GPCR superfamily [18]. After removal of the signal peptide, the mature CaSR has a ~600–amino acid extracellular domain comprised of a bilobed Venus-flytrap-like domain connected by a cysteine-rich region and a peptide linker to the 250-amino-acid transmembrane domain, followed by an intracellular COOH-terminal tail of 216 amino acids [19]. The CaSR is predominantly expressed in the parathyroid gland and the nephron. Binding of the receptor at the plasma membrane by extracellular calcium couples it to Gi, Gq, and G11 proteins that activate signaling pathways, modulating PTH secretion and renal mineral ion handling [20].

Altered activity or function of the CaSR contributes to a variety of human disorders [21–23]. Loss-of-function mutations underlie familial hypocalciuric hypercalcemia type 1 (FHH1; heterozygous mutations), neonatal (severe) hyperparathyroidism (heterozygous, homozygous), and primary hyperparathyroidism (heterozygous, homozygous). Gain-of-function mutations cause autosomal dominant hypocalcemia type 1 (ADH1; heterozygous) and sometimes Bartter syndrome subtype V (heterozygous) [24, 25]. Loss-of-function mutations in the *GNA11* and *AP2S1* genes whose products mediate signaling and endocytic activity of the CaSR underlie FHH2 (heterozygous) and FHH3 (heterozygous), respectively [26–28]. Mutations of the *AP2S1* gene do not appear to be involved in familial hypoparathyroidism [29]. Gain-of-function mutations in *GNA11* cause autosomal dominant hypocalcemia type 2 (ADH2, heterozygous) [26, 30] (see also Chap. 5). Phenocopies due to inactivating and activating CaSR antibodies are well known and can cause autoimmune hypocalciuric hypercalcemia (AHH) and autoimmune hypoparathyroidism (AH), respectively [31–34].

16.3.1 ADH1

ADH1 (OMIM# 146200) due to *CASR* gain-of-function mutation may be associated with relative hypercalciuria [35–37] (see also Chap. 5). The hypocalcemia is mild and asymptomatic in some but, when more pronounced, may be associated with neuromuscular symptoms such as paresthesias, carpopedal spasm, and seizures [35, 36]. Serum phosphate concentrations are elevated or in the upper-normal range, and the serum magnesium concentrations are low or low-normal [35]. ADH1 patients have serum PTH levels that are also in the low-normal range and may not be overtly hypoparathyroid.

In most ADH1 patients, the relative hypercalciuria is accompanied by urinary calcium to creatinine (Ca/Cr) ratios that are within or above the upper reference limit, whereas most patients with untreated hypoparathyroidism of other etiologies have urinary Ca/Cr ratios below or within the lower half of the reference range [37]. Treatment of ADH1 patients with active vitamin D preparations may exacerbate or induce hypercalciuria and result in nephrocalcinosis, nephrolithiasis, and renal impairment [35, 38]. Both promoters of the *CASR* gene have vitamin D response elements that mediate upregulation of CaSR expression by active vitamin D [39].

Some ADH patients receiving vitamin D preparations can develop polyuria and polydipsia. This may reflect the role of the apical renal collecting duct CaSR in sensing the luminal calcium concentration and signaling the suppression of vasopressin-dependent aquaporin-2-expression and/or insertion in the apical membrane that promotes water reabsorption, actions that would be accentuated by an activating CaSR mutation, particularly during treatment [40].

To minimize the adverse effects of vitamin D in ADH1, it is important to distinguish ADH1 from other causes of hypoparathyroidism. When possible, vitamin D therapy may be reserved for those with symptomatic ADH1 to avoid the long-term consequences of nephrocalcinosis, nephrolithiasis, and renal impairment [36]. Therapeutic alternatives include recombinant human PTH that raises serum calcium levels, and helps prevent symptomatic hypocalcemic episodes. It may lower urinary calcium excretion to some extent but may not completely mitigate the risk of nephrocalcinosis [41–43]. The combination of a thiazide diuretic with calcitriol may alleviate the hypocalcemia in some but not all patients and lower the urinary calcium excretion relative to those receiving calcitriol alone [36, 44].

Thus, *CASR* mutation analysis can be key in diagnosing ADH1 [45]. Thus far, about one hundred different *CASR* mutations have been identified in ADH1 affected individuals, and of these, 95 % are heterozygous missense substitutions [CASRdb – calcium-sensing receptor database, www.casrdb.mcgill.ca/; 21, 22]. In vitro expression studies demonstrate a gain of function of the mutant receptors with a leftward shift of the extracellular calcium-response curve and a reduction of the EC_{50} for extracellular calcium concentration relative to wild-type CaSR [46]. It has been suggested that the relative in vitro functional activity of mutant CaSRs also correlates with the serum magnesium level, renal magnesium handling, and PTH levels in patients with ADH1 [47].

The CaSR, after entry into the ER and removal of the signal peptide, is synthesized as a constitutive dimer and functions as such at the plasma membrane [48, 49]. Studies of ADH1-associated mutations have provided insight into critical structural regions of the receptor. Activating mutations cluster within the second peptide loop of the extracellular domain (residues 116–131) that contributes in part to the interface between individual protomers in the CaSR dimer. It may be that these mutations enhance the sensitivity of the CaSR to extracellular calcium by promoting conformational changes that facilitate receptor activation [50]. A further clustering of ADH1-associated mutations is found in a region (residues 820–836) that encompasses the sixth and seventh transmembrane helices and the intervening third extracellular loop. Residues in this region are critical for maintaining the CaSR in an inactive conformation and also for the binding and activity of allosteric modulators such as calcimimetics and calcilytics [50].

In the future, calcilytics (synthetic allosteric inhibitors of the CaSR) may provide a novel treatment for patients with ADH1. In vitro analyses demonstrated distinct effects of the calcilytic NPS-2143 on different ADH1 CaSR mutations, with some being responsive and others less so or unresponsive [51]. However, under in vitro conditions that mimicked the heterozygous in vivo situation in patients, the calcilytic corrected the overactivity of all the mutants [51]. In a separate study, NPS-2143 suppressed the enhanced activity of gain-of-function CaSR mutants without altering cell-surface expression levels [52].

16.3.2 Bartter Syndrome Subtype V

Although Bartter syndrome subtype V is represented by only a handful of cases with heterozygous severe activating mutations in the *CASR*, they provide special insight into the functioning of the CaSR in the thick ascending limb of the nephron [53–55]. Bartter syndrome encompasses a heterogeneous group of electrolyte homeostasis disorders, the common features of which are hypokalemic alkalosis, hyperreninemia, and hyperaldosteronism. Bartter syndrome subtypes I–IV are autosomal recessive disorders due to inactivating mutations in the following ion transporters or channels active in the thick ascending limb of the loop of Henle: type I, the sodium-potassium-chloride cotransporter (NKCC2); type II, the outwardly rectifying potassium channel (ROMK); type III, the voltage-gated chloride channel (CLC-Kb); type IV, Barttin, a β subunit that is required for trafficking of CLC-Ka and CLC-Kb. Patients with the autosomal dominant Bartter syndrome subtype V have, in addition to the classic features of the syndrome, hypocalcemia and may exhibit neuromuscular manifestations, seizures, and basal ganglia calcifications.

NKCC2 and ROMK in the apical membrane (luminal side) of the thick ascending limb of the loop of Henle generate a transepithelial electrochemical gradient that drives passive paracellular transport of Na^+, Mg^{2+}, and Ca^{2+} from the lumen to blood [25]. The CaSR is situated in the basolateral membrane (antiluminal side) and when activated increases 20-hydroxyeicosatetraenoic acid and decreases cAMP concentrations, both of which would inhibit ROMK and NKCC2 activities. Thus, severe activating mutations of the CaSR lead to the salt wasting of Bartter syndrome in addition to the hypercalciuric hypocalcemia of ADH1.

16.3.3 ADH2

Heterozygous gain-of-function missense mutations of *GNA11* have been identified in ADH patients without detectable *CASR*-activating mutations [26, 30]. The *GNA11*-activating mutations increase the sensitivity of the parathyroid gland and renal tubule to the extracellular calcium concentration. Autosomal dominant hypocalcemia and hypoparathyroidism due to *CASR* mutations and *GNAS11* are now designated as ADH type1 and type 2, respectively. The human Gα11 protein (a Gq family member) has 359 amino acids with an α-helical domain in the NH_2-terminal region, a GTPase domain in the COOH-terminal region, and three switch regions (SR1-3) in the middle portion that change conformation based on whether GTP or GDP is bound [56]. The R60C, R181Q, S211W, and F341L mutations found in hypocalcemic individuals are predicted by 3D modeling to disrupt the normal Gα11 protein structure. Moreover, cells stably expressing the CaSR and transfected with the mutants exhibit increased sensitivity to changes in extracellular calcium [26, 30].

16.4 GCM2

Glial cells missing (GCM) acts in *Drosophila* as a binary switch between glial cell and neuronal development. In mammals, there are two orthologs GCM1/GCMA and GCM2/GCMB important for parathyroid and placental development, respectively [57–59] (see also Chap. 2). The *GCM2* gene (OMIM# 603716) localizes to chromosome 6p24.2 and its five exons encode a transcription factor of 509 amino acids. It is expressed in the PTH-secreting cells of the developing parathyroid glands, is critical for their development in terrestrial vertebrates, and continues to be expressed in the adult [60–63]. From NH_2 to COOH termini of the GCM2 protein, there is a DNA-binding domain, transactivation domain 1, an inhibitory domain, and transactivation domain 2.

In a few cases, homozygous or heterozygous inactivating mutations in the *GCM2* gene have been implicated in FIH inherited in an autosomal recessive or dominant manner, respectively. Autosomal recessive mutations include gene deletion, missense, stop, and frameshift: specifically intragenic deletion of exons 1–4, R39X, R47L, G63S, R110W, Y136X, I298fsX307, and T370M [64–70]. Autosomal dominant mutations include C106R, N502H, c.1389delT, and c.1399delC [71–74]. In vitro functional studies of some of these mutants have demonstrated loss of GCM response element binding and/or transcriptional activity in the case of recessive mutations, as well as an ability of dominant mutants to inhibit activity of wild-type GCM2 when the two are cotransfected into cells [71–73].

The prevalence of genetic defects affecting GCM2 function is not high in FIH, and a recent study investigating 20 unrelated cases with this disorder (10 familial and 10 sporadic) identified several polymorphic variants, but failed to identify actual GCM2 mutations segregating with the disease and/or leading to loss of function [75].

Absent or reduced levels of parathyroid GCM2, either in *gcm2* knockout mice or in cultured human parathyroid cells treated with GCM2 siRNA, correlate with maintenance of expression of the parathyroid early differentiation marker, CaSR [60, 63, 76]. GCM2 transactivates the *CASR* gene via GCM response elements in promoters P1 and P2 [72, 77]. This provides the mechanistic link for the association between GCM2 and CaSR and the development of the evolutionarily related parathyroid glands (in terrestrial vertebrates) and gills (in fish) [62].

The transcriptional activator, v-maf musculoaponeurotic fibrosarcoma oncogene homologue B (MafB), is expressed in developing and mature parathyroid glands [78]. MafB acts downstream of GCM2 and is necessary for proper localization of the developing parathyroid glands between the thymus and the thyroid. Although GCM2 alone does not stimulate the *PTH* promoter, as it does the *CASR* promoter, it associates with MafB to synergistically activate PTH expression [78].

Haploinsufficiency of the dual zinc-finger transcription factor, GATA3, results in the congenital hypoparathyroidism-deafness-renal dysplasia (HDR) syndrome. *Gata3* knockout mouse embryos lack *Gcm2* expression and have gross defects in the third and fourth pharyngeal pouches including absent parathyroid-thymus primordia [79]. GATA3 transactivates the *GCM2* gene by binding specifically to a double-GATA-motif within the *GCM2* promoter. Thus, GATA3 and GCM2 are part of a critical transcriptional cascade in the parathyroid morphogenesis pathway [80].

16.5 Xq27.1

In two multigenerational families with X-linked recessive hypoparathyroidism exhibiting neonatal onset of hypocalcemia and parathyroid agenesis, the trait was mapped to a 906-kb region on distal Xq27 that contains three genes including SOX3, but no intragenic mutations were found (OMIM# 307700) [81]. A deletion-insertion [del(X) (q27.1) inv ins (X;2)(q27.1;p25.3)] mutation was identified that was speculated to exert a positional effect on SOX3 expression and affect embryonic development of the parathyroid glands.

Conclusion

Studies of familial isolated hypoparathyroidism have highlighted the critical roles played by several genes in parathyroid gland development and function.

References

1. Brandi ML (2011) Genetics of hypoparathyroidism and pseudohypoparathyroidism. J Endocrinol Invest 34(Supp l7):27–34
2. Shoback D (2008) Clinical practice. Hypoparathyroidism. N Engl J Med 359:391–403

3. McKusick V (2001) Online Mendelian inheritance in man (OMIM). Johns Hopkins University Press, Baltimore
4. Ahn TG, Antonarakis SE, Kronenberg HM, Igarashi T, Levine MA (1986) Familial isolated hypoparathyroidism: a molecular genetic analysis of 8 families with 23 affected persons. Medicine (Baltimore) 65:73–81
5. Hendy GN, Bennett HPJ, Gibbs BF et al (1995) Proparathyroid hormone (ProPTH) is preferentially cleaved to parathyroid hormone (PTH) by the prohormone convertase furin: a mass spectrometric analysis. J Biol Chem 270:9517–9525
6. Canaff L, Bennett HPJ, Hou Y, Seidah NG, Hendy GN (1999) Proparathyroid hormone processing by the proprotein convertase PC7: comparison with furin and assessment of modulation of parathyroid convertase mRNA levels by calcium and 1,25-dihydroxyvitamin D_3. Endocrinology 140:3633–3641
7. Hendy GN, Kronenberg HM, Potts JT Jr, Rich A (1981) Nucleotide sequence of cloned cDNAs encoding human preproparathyroid hormone. Proc Natl Acad Sci U S A 78:7365–7369
8. Vasicek TJ, McDevitt BE, Freeman MW et al (1983) Nucleotide sequence of the human parathyroid hormone gene. Proc Natl Acad Sci U S A 80:2127–2131
9. Arnold A, Horst SA, Gardella TJ et al (1990) Mutation of the signal peptide-encoding region of the preproparathyroid hormone gene in familial isolated hypoparathyroidism. J Clin Invest 86:1084–1087
10. Karaplis AC, Lim SK, Baba H, Arnold A, Kronenberg HM (1995) Inefficient membrane targeting, translocation, and proteolytic processing by signal peptidase of a mutant preproparathyroid hormone protein. J Biol Chem 270:1629–1635
11. Datta R, Waheed A, Shah GN, Sly WS (2007) Signal sequence mutation in autosomal dominant form of hypoparathyroidism induces apoptosis that is corrected by a chemical chaperone. Proc Natl Acad Sci U S A 104:19989–19994
12. Suprasdngsin C, Wattanachal A, Preeysombat C (2001) Parathyroid hormone gene mutation in patient with hypoparathyroidism. Madhidol University Research Abstracts, 28 Abstract 164
13. Tomar N, Gupta N, Goswami R (2013) Calcium-sensing receptor autoantibodies and idiopathic hypoparathyroidism. J Clin Endocrinol Metab 98:3884–3891
14. Freeman MW, Wiren KM, Rapoport A et al (1987) Consequences of amino-terminal deletions of preproparathyroid hormone signal sequence. Mol Endocrinol 1:628–638
15. Parkinson DB, Thakker RV (1992) A donor splice site mutation in the parathyroid hormone gene is associated with autosomal recessive hypoparathyroidism. Nat Genet 1:149–152
16. Sunthornthepvarakul T, Churesigaew S, Ngowngarmratana S (1999) A novel mutation of the signal peptide of the preproparathyroid hormone gene associated with autosomal recessive familial isolated hypoparathyroidism. J Clin Endocrinol Metab 84:3792–3796
17. Ertl D-A, Stary S, Streubel B, Raimann A, Haeusler G (2012) A novel homozygous mutation in the parathyroid hormone gene (*PTH*) in a girl with isolated hypoparathyroidism. Bone 51:629–632
18. Hendy GN, Canaff L, Cole DEC (2013) The *CASR* gene: alternative splicing and transcriptional control, and calcium-sensing receptor (CaSR) protein: structure and ligand binding sites. Best Pract Res Clin Endocrinol Metab 27:285–301
19. Brown EM (2013) Role of the extracellular calcium-sensing receptor in extracellular calcium homeostasis. Best Pract Res Clin Endocrinol Metab 27:333–343
20. Conigrave AD, Ward DT (2013) Calcium-sensing receptor (CaSR); pharmacological properties and signalling pathways. Best Pract Res Clin Endocrinol Metab 27:315–331
21. Hendy GN, Guarnieri V, Canaff L (2009) Calcium-sensing receptor and associated diseases. Prog Mol Biol Transl Sci 89C:31–95
22. Hannan FM, Nesbit MA, Zhang C et al (2012) Identification of 70 calcium-sensing receptor mutations in hyper- and hypo-calcaemic patients: evidence for clustering of extracellular domain mutations at calcium-binding sites. Hum Mol Genet 21:2768–2778
23. Hannan FM, Thakker RV (2013) Calcium-sensing receptor (CaSR) mutations and disorders of calcium, electrolyte and water metabolism. Best Pract Res Clin Endocrinol Metab 27:359–371
24. Pollak MR, Brown EM, Estep HL et al (1994) Autosomal dominant hypocalcemia caused by a Ca^{2+}-sensing receptor mutation. Nat Genet 8:303–307
25. Hebert SC (2003) Bartter syndrome. Curr Opin Nephrol Hypertens 12:527–532
26. Nesbit MA, Hannan FM, Howles SA et al (2013) Mutations affecting G-protein subunit α11 in hypercalcemia and hypocalcemia. N Engl J Med 368:2476–2486
27. Nesbit MA, Hannan FM, Howles SA et al (2013) Mutations in *AP2S1* cause familial hypocalciuric hypercalcemia type 3. Nat Genet 45:93–97
28. Hendy GN, Cole DEC (2013) Ruling in a suspect: the role of *AP2S1* mutations in familial hypocalciuric hypercalcemia type 3. J Clin Endocrinol Metab 98:4666–4669
29. Lambert A-S, Grybek V, Francou B et al (2014) Analysis of *AP2S1*, a calcium-sensing receptor regulator, in familial and sporadic isolated hypoparathyroidism. J Clin Endocrinol Metab 99(3):E469–E473
30. Mannstadt M, Harris M, Bravenboer B et al (2013) Germline mutations affecting Gα11 in hypoparathyroidism. N Engl J Med 368:2532–2534
31. Kifor O, McElduff A, LeBoff MS et al (2004) Activating antibodies in the calcium-sensing receptor in two patients with autoimmune hypoparathyroidism. J Clin Endocrinol Metab 89:548–556
32. Goswami R, Brown EM, Kochupillai N et al (2004) Prevalence of calcium sensing receptor autoantibodies in patients with sporadic idiopathic hypoparathyroidism. Eur J Endocrinol 150:9–18
33. Brown EM (2009) Anti-parathyroid and anti-calcium-sensing receptor antibodies in autoimmune

hypoparathyroidism. Endocrinol Metab Clin North Am 38:437–445
34. Kemp EH, Gavalas NG, Krohn KJ et al (2009) Activating autoantibodies against the calcium-sensing receptor detected in two patients with autoimmune polyendocrine syndrome type 1. J Clin Endocrinol Metab 94:4749–4756
35. Pearce SHS, Williamson C, Kifor O et al (1996) A familial syndrome of hypocalcemia with hypercalciuria. N Engl J Med 335:1115–1122
36. Raue F, Pichl J, Dorr HG et al (2011) Activating mutations in the calcium-sensing receptor: genetic and clinical spectrum in 25 patients with autosomal dominant hypocalcaemia – a German survey. Clin Endocrinol (Oxf) 75:760–765
37. Yamamoto M, Akatsu T, Nagase T et al (2000) Comparison of hypocalcemic hypercalciuria between patients with idiopathic hypoparathyroidism and those with gain-of-function mutations in the calcium-sensing receptor: is it possible to differentiate the two disorders? J Clin Endocrinol Metab 85:4583–4591
38. Lienhardt A, Bai M, Lagarde JP et al (2001) Activating mutations of the calcium-sensing receptor: management of hypocalcemia. J Clin Endocrinol Metab 86:5313–5323
39. Canaff L, Hendy GN (2002) Human calcium-sensing receptor gene: vitamin D response elements in promoters P1 and P2 confer transcriptional responsiveness to 1,25-dihydroxyvitamin D. J Biol Chem 277:30337–30350
40. Sands JM, Naruse M, Baum M et al (1997) Apical extracellular calcium/polyvalent cation-sensing receptor regulates vasopressin-elicited water permeability in rat kidney inner medullary collecting duct. J Clin Invest 99:1399–1405
41. Mittelman SD, Hendy GN, Fefferman RA et al (2006) A hypocalcemic child with a novel activating mutation of the calcium-sensing receptor gene: successful treatment with recombinant human parathyroid hormone. J Clin Endocrinol Metab 91:2474–2479
42. Sanda S, Schlingmann KP, Newfield RS (2008) Autosomal dominant hypoparathyroidism with severe hypomagnesemia and hypocalcemia, successfully treated with recombinant PTH and continuous subcutaneous magnesium infusion. J Pediatr Endocrinol Metab 21:385–391
43. Theman TA, Collins MT, Dempster DW et al (2009) PTH(1–34) replacement therapy in a child with hypoparathyroidism caused by a sporadic calcium receptor mutation. J Bone Miner Res 24:964–973
44. Sato K, Hasegawa Y, Nakae J et al (2002) Hydrochlorothiazide effectively reduces urinary calcium excretion in two Japanese patients with gain-of-function mutations of the calcium-sensing receptor gene. J Clin Endocrinol Metab 87:3068–3073
45. Cole DEC, Yun FHJ, Wong BYL et al (2009) Calcium-sensing receptor mutations and denaturing high performance liquid chromatography. J Mol Endocrinol 42:1–9
46. Hendy GN, Minutti C, Canaff L et al (2003) Recurrent familial hypocalcemia due to germline mosaicism for an activating mutation of the calcium-sensing receptor gene. J Clin Endocrinol Metab 88:3674–3681
47. Kinoshita Y, Hori M, Taguchi M, Watanabe S, Fukumoto S (2014) Functional activities of mutant calcium-sensing receptors determine clinical presentations in patients with autosomal dominant hypocalcemia. J Clin Endocrinol Metab 99(3):E363–E368
48. Pidasheva S, Canaff L, Simonds WF, Marx SJ, Hendy GN (2005) Impaired cotranslational processing of the calcium-sensing receptor due to missense mutations in familial hypocalciuric hypercalcemia. Hum Mol Genet 14:1679–1690
49. Pidasheva S, Grant M, Canaff L et al (2006) The calcium-sensing receptor (CASR) dimerizes in the endoplasmic reticulum: biochemical and biophysical characterization of novel CASR mutations causing familial hypocalciuric hypercalcemia. Hum Mol Genet 15:2200–2209
50. Hu J, Spiegel AM (2007) Structure and function of the human calcium-sensing receptor: insights from natural and engineered mutations and allosteric modulators. J Cell Mol Med 11:908–922
51. Letz S, Rus R, Haag C et al (2010) Novel activating mutations of the calcium-sensing receptor: the calcilytic NPS-2143 mitigates excessive signal transduction of mutant receptors. J Clin Endocrinol Metab 95:E229–E233
52. Nakamura A, Hotsubo T, Kobayashi K et al (2013) Loss-of-function and gain-of-function mutations of calcium-sensing receptor: functional analysis and the effect of allosteric modulators NPS R-568 and NPS 2143. J Clin Endocrinol Metab 98:E1692–E1701
53. Vargas-Poussou R, Huang C, Hulin P et al (2002) Functional characterization of a calcium-sensing receptor mutation in severe autosomal dominant hypocalcemia with a Bartter-like syndrome. J Am Soc Nephrol 13:2259–2266
54. Watanabe S, Fukumoto S, Chang H et al (2002) Association between activating mutations of calcium-sensing receptor and Bartter's syndrome. Lancet 360:692–694
55. Vezzoli G, Arcdiacono T, Paloschi V et al (2006) Autosomal dominant hypocalcemia with mild type 5 Bartter syndrome. J Am Soc Nephrol 19:525–528
56. Mizuno N, Itoh H (2009) Functions and regulatory mechanisms of Gq-signaling pathways. Neurosignals 17:42–54
57. Kim J, Jones BW, Zock C et al (1998) Isolation and characterization of mammalian homologs of the *Drosophila* gene *glial cells missing*. Proc Natl Acad Sci U S A 95:12364–12369
58. Kanemura Y, Hiraga S, Arita N, Ohnishi T, Izumoto S, Mori K, Matsumura H, Yamasaki M, Fushiki S, Yoshimine T (1999) Isolation and expression analysis of a novel human homologue of the Drosophila glial cells missing (gcm) gene. FEBS Lett 442:151–156

59. Kammerer M, Pirola B, Giglio A, Giangrande A (1999) GCMB, a second human homolog of the fly glide/gcm gene. Cytogenet Cell Genet 84:43–47
60. Gunther T, Chen ZF, Kim J, Priemel M, Rueger JM, Amling M, Moseley JM, Martin TJ, Anderson DJ, Karsenty G (2000) Genetic ablation of parathyroid glands reveals another source of parathyroid hormone. Nature 406:199–203
61. Maret A, Bourdeau I, Ding C et al (2004) Expression of GCMB by intrathymic parathyroid hormone-secreting adenomas indicates their parathyroid cell origin. J Clin Endocrinol Metab 89:8–12
62. Okabe M, Graham A (2004) The origin of the parathyroid gland. Proc Natl Acad Sci U S A 101:17716–17719
63. Liu Z, Yu S, Manley NR (2007) GCM2 is required for the differentiation and survival of parathyroid precursor cells in the parathyroid/thymus primordia. Dev Biol 305:333–346
64. Ding C, Buckingham B, Levine MA (2001) Familial isolated hypoparathyroidism caused by a mutation in the gene for the transcription factor GCMB. J Clin Invest 108:1215–1220
65. Baumber L, Tufarelli C, Patel S et al (2005) Identification of a novel mutation disrupting the DNA binding activity of GCM2 in autosomal recessive familial isolated hypoparathyroidism. J Med Genet 42:443–448
66. Thomee C, Schubert SW, Parma J et al (2005) GCMB mutation in familial isolated hypoparathyroidism with residual secretion of parathyroid hormone. J Clin Endocrinol Metab 90:2487–2492
67. Canaff L, Tham EB, Human D, Chanoine JP, Hendy GN (2007) Homozygous glial cell missing-2 (GCM-2) gene mutation in familial isolated hypoparathyroidism. Bone 40S2:Abs.180Th. p S285
68. Bowl MR, Mirczuk SM, Grigorieva IV et al (2010) Identification and characterization of novel parathyroid-specific transcription factor Glial Cells Missing Homolog B (GCMB) mutations in eight families with autosomal recessive hypoparathyroidism. Hum Mol Genet 19:2028–2038
69. Tomar N, Bora H, Singh R et al (2010) Presence and significance of a R110W mutation in the DNA-binding domain of *GCM2* gene in patients with isolated hypoparathyroidism and their family members. Eur J Endocrinol 162:407–421
70. Doyle D, Kirwin SM, Sol-Church K, Levine MA (2012) A novel mutation in the *GCM2* gene causing severe congenital isolated hypoparathyroidism. J Pediatr Endocrinol Metab 25:741–746
71. Mannstadt M, Bertrand G, Muresan M et al (2008) Dominant-negative GCMB mutations cause an autosomal dominant form of hypoparathyroidism. J Clin Endocrinol Metab 93:3568–3576
72. Canaff L, Zhou X, Mosesova I, Cole DEC, Hendy GN (2009) Glial cells missing-2 (GCM2) transactivates the calcium-sensing receptor gene: effect of a dominant-negative GCM2 mutant associated with autosomal dominant hypoparathyroidism. Hum Mutat 30:85–92
73. Mirczuk SM, Bowl MR, Nesbit MA et al (2010) A missense glial cells missing homolog B (GCMB) mutation, Asn502His, causes autosomal dominant hypoparathyroidism. J Clin Endocrinol Metab 95:3512–3516
74. Yi H-S, Eom YS, Park IB et al (2012) Identification and characterization of C106R, a novel mutation in the DNA-binding domain of GCMB, in a family with autosomal-dominant hypoparathyroidism. Clin Endocrinol (Oxf) 76:625–633
75. Maret A, Ding C, Kornfield SL, Levine MA (2008) Analysis of the GCM2 gene in isolated hypoparathyroidism: a molecular and biochemical study. J Clin Endocrinol Metab 93:1426–1432
76. Mizobuchi M, Ritter CS, Krits I et al (2009) Calcium-sensing receptor expression is regulated by glial cells missing-2 in human parathyroid cells. J Bone Miner Res 24:1173–1179
77. Canaff L, Zhou X, Cole DEC, Hendy GN (2008) Glial cells missing-2 (GCM2), the regulator of parathyroid cell fate, transactivates the calcium-sensing receptor gene (*CASR*): identification of GCM-response elements in *CASR* promoters P1 and P2. J Bone Miner Res 23S1:Abs. M187. p S429
78. Kamitani-Kawamoto A, Hamada M, Moriguchi T et al (2011) MafB interacts with Gcm2 and regulates parathyroid development. J Bone Miner Res 26:2463–2472
79. Grigorieva IV, Mirczuk S, Gaynor KU et al (2010) Gata3-deficient mice develop parathyroid abnormalities due to dysregulation of the parathyroid-specific transcription factor Gcm2. J Clin Invest 120:2144–2155
80. Grigorieva IV, Thakker RV (2011) Transcription factors in parathyroid development: lessons from hypoparathyroid disorders. Ann N Y Acad Sci 1237:24–38
81. Bowl MR, Nesbit MA, Harding B et al (2005) An interstitial deletion-insertion involving chromosomes 2p25.3 and Xq27.1, near SOX3, causes X-linked recessive hypoparathyroidism. J Clin Invest 115:2822–2831

Autoimmune Hypoparathyroidism 17

E. Helen Kemp and Anthony P. Weetman

17.1 Introduction

The earliest suggestion of a possible autoimmune basis for parathyroid disease came from observations on patients with autoimmune thyroid disease, the archetype of autoimmune disorders, combined with Addison's disease: this clustering was first recognized by Schmidt in 1926 [1] (see also Chaps. 14 and 15). The association between these two conditions implied that other diseases occurring with them might also be autoimmune, and several cases were already known of Addison's disease accompanied by idiopathic hypoparathyroidism when thyroid autoimmunity was discovered in 1953 [2]. The first attempt to determine whether autoantibodies to parathyroid tissue could be detected in such patients used an indirect immunofluorescence assay with selected human parathyroid adenoma tissue as substrate; autoantibodies were found in 38 % of 74 patients with idiopathic hypoparathyroidism, 26 % of 92 patients with idiopathic Addison's disease, 12 % of 49 patients with Hashimoto's thyroiditis, and 6 % of 245 controls [3]. Patients with hypoparathyroidism also had a higher than expected frequency of adrenal, thyroid, or gastric parietal cell autoantibodies, and parathyroid hormone was excluded as a target of the parathyroid autoantibodies. The specificity of the autoantibodies was confirmed by the inability of any tissue extracts, other than parathyroid, to absorb out the reactivity.

Further work on the distribution of autoantibodies in affected families identified a subset of patients with early-onset Addison's disease and hypoparathyroidism that could be distinguished from Schmidt's syndrome, with a possible autosomal recessive inheritance [4]. This syndrome was initially called autoimmune polyendocrinopathy-candidiasis-ectodermal dystrophy, with chronic mucocutaneous candidiasis added as a third major disease component and manifesting in early childhood (Fig. 17.1), but it is now usually referred to as autoimmune polyendocrine syndrome type 1 (APS1; OMIM 240300) [5].

APS1 has become well characterized clinically [6, 7], and a variety of features distinguish it from the much more common Schmidt's syndrome, which is now termed APS type 2 (Table 17.1). It is more prevalent in isolated populations, including Finland, Sardinia, and Iranian Jews. APS1 is known to be due to mutations in the autoimmune regulator (*AIRE*) gene which leads to deficient negative selection of autoreactive T cells in the fetal thymus and an aberrant thymic microenvironment in which autoreactive T cells are more likely to occur [8]. Although the original study distinguishing different types of polyendocrine autoimmunity [5] and others [9]

E.H. Kemp • A.P. Weetman (✉)
Department of Human Metabolism,
Faculty of Medicine, Dentistry and Health,
University of Sheffield, The Medical School,
Beech Hill Road, Sheffield South Yorkshire
S10 2RX, UK
e-mail: e.h.kemp@sheffield.ac.uk; a.p.weetman@sheffield.ac.uk

M.L. Brandi, E.M. Brown (eds.), *Hypoparathyroidism*,
DOI 10.1007/978-88-470-5376-2_17,

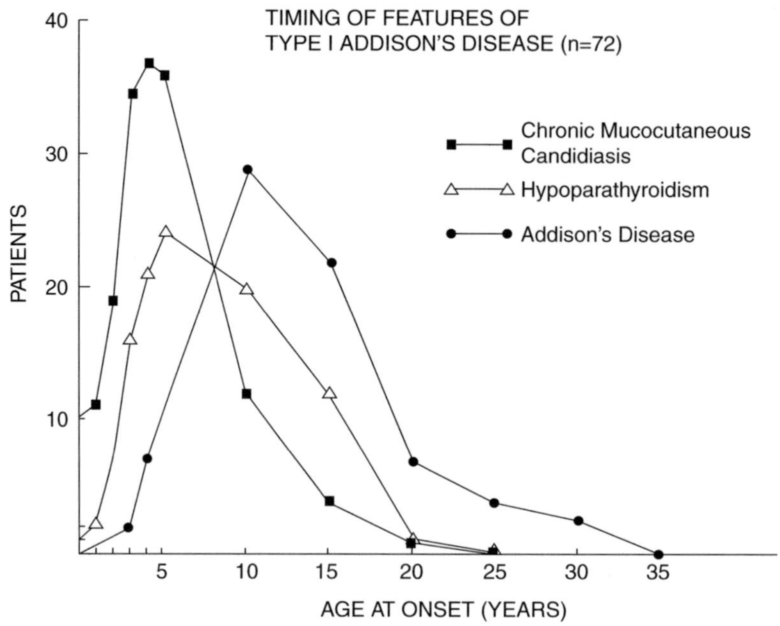

Fig. 17.1 The ages of onset of the three major components of autoimmune polyendocrine syndrome type 1 (From Neufeld et al. [5]. With permission from Lippincott Williams & Wilkins)

Table 17.1 Key features of autoimmune polyendocrine syndromes (APS) types 1 and 2

Characteristic	APS1	APS2
Prevalence	Rare	Common
Age at onset	Childhood	Adulthood
Genetics	Autosomal recessive: mutations in *AIRE* gene	Polygenic: associated with HLA DR3-DQB1*0201
Interferon autoantibodies	Present	Absent
Major components	Mucocutaneous candidiasis (100 %)	Autoimmune thyroid disease (70 %)
	Hypoparathyroidism (80 %)[a]	Type 1 diabetes mellitus (60 %)
	Addison's disease (70 %)	Addison's disease (50 %)[b]
Significant minor components[c]	Alopecia (30 %)	Vitiligo (15 %)
	Type 1 diabetes mellitus (15 %)	Alopecia (5 %)
	Vitiligo (10 %)	Pernicious anemia (5 %)
	Pernicious anemia (10 %)	Celiac disease (3 %)
Non-autoimmune manifestations	Enamel hypoplasia (90 %)	None
	Dental, nail, and tympanic membrane dystrophies (60 %)	
	Keratopathy (50 %)	
	Asplenia (10 %)	

[a]Females, 95 %; males, 70 %. This is the only disease component of APS1 that shows a sex difference

[b]An alternative classification system has been proposed in which Addison's disease is an essential (100 % present) component of APS2; those family clusters associated with autoimmune thyroid disease have been termed APS3, and clusters of organ-specific autoimmunity without either component, APS4 [7]

[c]Many other autoimmune disorders occur in these syndromes; only the key disorders are listed

identified hypoparathyroidism as occurring uniquely in the context of APS1 and not APS2, idiopathic hypoparathyroidism is a well-known but rare entity, and autoimmunity now appears to be a cause for some of these cases too. Nonetheless, a recent study of such sporadic cases of hypoparathyroidism surprisingly failed to find any increase in associated thyroid autoimmunity, which is the commonest of the organ-specific autoimmune disorders [10]. However, this was a small series of only 87 patients, and it is probable that the patients are heterogeneous, with some having an autoimmune basis for their disorder (only these would be expected to show an increase in the prevalence of other autoimmune conditions like thyroid disease) and others with a non-autoimmune basis.

17.2 Histological and T Cell Studies

As well as the finding of autoantibodies, the presence of autoimmunity is suggested by the establishment of animal models replicating the human disease, the presence of a characteristic lymphocytic infiltration in the target tissue, and the demonstration of T cell reactivity to putative tissue antigens [11]. Lymphocytic infiltration of the parathyroid glands, accompanied by hypoparathyroidism, has been induced in rats and dogs by immunization with parathyroid tissue; autoantibodies against parathyroid tissue were also detectable in these animals [12]. Immunization with homologous thyroid tissue induced both thyroiditis and parathyroiditis in chickens, presumably through contamination of the thyroid extract with parathyroid tissue; thymectomy reduced the severity of disease, indicating an important role for T cells in pathogenesis [13]. Immunization of rabbits with bovine or rat parathyroid homogenate produced species- and parathyroid tissue-specific autoantibodies, and passive immunization of rats with the serum from the rabbits immunized with rat parathyroid tissue induced a severe lymphocytic infiltration of the parathyroid glands in a fifth of the animals, but hypoparathyroidism was not observed [14].

Sporadic case reports have appeared of histological changes in cases of idiopathic hypoparathyroidism which include lymphocytic and plasma cell infiltration (Fig. 17.2); in one case this was accompanied by a positive leukocyte migration inhibition test and hypergammaglobulinemia [15, 16]. The appearance described contrasts with the finding of a lesser degree of lymphocytic

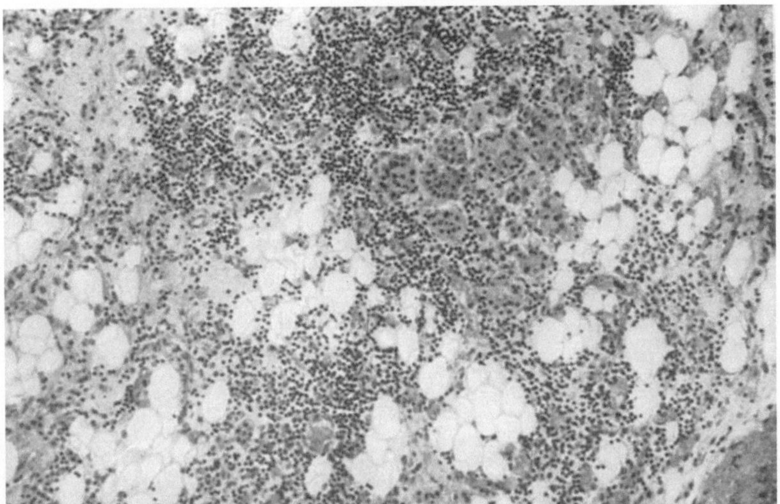

Fig. 17.2 Lymphocytic infiltration in the parathyroid in autoimmune hypoparathyroidism (From Van de Casseye and Gepts [16]. With permission from Springer)

infiltration in around 6 % of autopsy cases of unselected patients, associated with generalized inflammatory conditions or venous congestion, and is also different to the much rarer parathyroiditis associated with parathyroid hyperplasia and hyperparathyroidism [17]. T cell reactivity to parathyroid tissue antigens has not been convincingly demonstrated so far, although in a study of eight patients with idiopathic hypoparathyroidism, there was an increase in circulating activated T cells compared to controls [18].

17.3 Parathyroid Autoantibodies

The first detection of parathyroid autoantibodies employed indirect immunofluorescence microscopy techniques which are prone to artifacts and difficult to quantitate [3]. These difficulties are illustrated by the attempt to replicate the original description of parathyroid autoantibodies, in which only one of nine sera from patients with idiopathic hypoparathyroidism reacted with parathyroid oxyphil cells and with chief cells of the parathyroid (using a human adenoma as substrate), but whether other sera contained low-titer antibody to chief cells could not be determined [19]. The positive serum came from a patient with APS1.

In an attempt to refine the assay, cultures of bovine parathyroid cells were used in immunofluorescence or cytotoxicity assays [20]. Seven sera were studied from patients with idiopathic hypoparathyroidism and all were positive by immunofluorescence against these cells; pooled normal sera were negative. Moreover, all seven serum samples lysed the parathyroid cells in the presence of complement, unlike control sera. Although preincubation with parathyroid cells absorbed out the activity, so did preincubation with adrenal cells, raising some concerns over tissue specificity of these autoantibodies.

The same group subsequently reported that six of these hypoparathyroid sera reacted with bovine endothelial cells by flow cytometry and indirect immunofluorescence microscopy, and by adsorption experiments it was shown that the reactivity to parathyroid cells was due to endothelial cell cross-reaction [21]. Although the authors postulated that parathyroid-specific endothelial rather than epithelial cells may therefore be the target of autoimmunity, this would be at odds with the pathogenesis of other known endocrinopathies. Further indications of the difficulties of these techniques came from another study in which sera from 5 of 32 patients with idiopathic hypoparathyroidism and 1 of 50 controls gave strong reactivity against parathyroid oxyphil cells and weak reactivity against chief cells, but this reactivity was exactly replicated by reactivity with a variety of other mitochondria-rich tissues and was absorbed out with mitochondria [22].

Better evidence in support of parathyroid cell autoimmunity came from studies using dispersed human parathyroid cells, with immunofluorescence microscopy and inhibition of parathyroid hormone (PTH) secretion being measured as indices of antibody binding [23]. Eight out of 23 sera from patients with idiopathic hypoparathyroidism reacted by immunofluorescence; two of the positive sera came from patients who probably had APS1, but the others were adults, two of whom also had thyroid or adrenal autoantibodies. None of the sera reacted with bovine parathyroid cells. Three of these eight positive sera inhibited PTH action, and in one, sequential samples showed a diminution in antibody levels with spontaneous amelioration of disease. These findings clearly suggested an etiological role for specific parathyroid autoantibodies in some patients with hypoparathyroidism, but further characterization of the parathyroid autoantigen(s) was not attempted for nearly a decade.

17.4 Calcium-Sensing Receptor Autoantibodies

Most organ-specific autoantigens are either intracellular enzymes or extracellular receptors; well-known examples of the latter include the thyroid-stimulating hormone (TSH) receptor in Graves' disease and the acetylcholine receptor in myasthenia gravis. The calcium-sensing receptor (CaSR) was cloned in 1993 and identified to have a central role in control of PTH secretion in response to extracellular calcium concentration [24, 25] (see also Chap. 5). The relative restriction of this receptor to the PTH-producing chief cells and renal tubular cells led its identification as a candidate parathyroid autoantigen by Li and colleagues [26] who performed immunoblotting experiments which revealed that of 25 sera from patients with hypoparathyroidism (17 with APS1 and 8 with adult-onset disease plus thyroid autoimmunity), five reacted with a parathyroid antigen of the appropriate size. Further immunoblotting experiments showed similar reactivity using HEK293 cells transfected with the CaSR, and 14 (56 %) of the sera immunoprecipitated the extracellular domain of receptor in radioimmunoprecipitation assays (Table 17.2), whereas there was no reactivity with the intracellular domain. Control sera were uniformly negative, as were experiments using non-transfected HEK293 cells. With regard to the patients who were positive for CaSR autoantibodies, six (35 %) had APS1 and eight (100 %) had adult-onset disease, and there was significantly less chance of detecting positive autoantibodies in the patients if hypoparathyroidism had been present for more than 5 years. No effect of the CaSR autoantibodies on intracellular calcium levels could be identified in preliminary experiments.

Initial attempts aimed at confirming these results were unsuccessful [27]. In a series of 90 APS1 patients (16 of whom were studied within a year of diagnosis), 19 % had autoantibodies that recognized canine parathyroid tissue in indirect immunofluorescence assays, and this species difference was taken to exclude mitochondrial cross-reactivity. By immunoblotting, multiple bands of reactivity to human parathyroid tissue were found in 37 % of APS1 sera, although there was no difference found when this reactivity was compared with the control sera. Using a radioimmunoprecipitation technique, 12 % of APS1 sera were positive for autoantibodies to the extracellular domain of the CaSR, but so were 4 % of control sera. There was no apparent difference between the prevalence of CaSR autoantibodies in APS1 patients with and those without hyperparathyroidism (Table 17.2). No autoantibodies could be detected to PTH. A further paper from some of the same authors reported an absence of CaSR autoantibodies in APS1 patients when the radioimmunoprecipitation assay was used (Table 17.2), but it is unclear whether these cases overlapped with those previously reported [28].

In contrast, other studies were able to confirm the presence of CaSR autoantibodies in hypoparathyroidism. By immunoblotting with human parathyroid tissue extracts containing the CaSR, 49 % of 51 patients with idiopathic hypoparathyroidism (mean disease duration of 7 years) were positive for autoantibodies, and although a surprisingly high prevalence of CaSR autoantibodies was also noted in controls (13 %), this difference was significant (Table 17.2) [29]. In a further study of 147 idiopathic hypoparathyroid patients, between 16 and 25 % were positive for CaSR autoantibodies depending on the assay used (Table 17.2) [30]. Controls were positive at a frequency of 0.5–14 %. In addition, using a recombinant extracellular domain of the CaSR in immunoblotting, 29 % of 17 sera from patients with idiopathic hypoparathyroidism were positive for autoantibodies, compared with one patient from eight (13 %) with APS1 and one patient from six (17 %) with APS2 (Table 17.2) [31]. The mean duration of hypoparathyroidism was 4.3 years in this study. Controls were negative, reactivity could be absorbed out by recombinant receptor, and this assay was found to be more sensitive than radioimmunoprecipitation.

The first detection of functional activity of CaSR autoantibodies came from a study of four patients with hypocalciuric hypercalcemia who also had celiac disease or thyroid autoantibodies, indicating a possible autoimmune origin for their calcium disorder [32]. Parathyroid autoantibodies could be detected in these patients by immunofluorescence, CaSR autoantibodies were detected by immunoprecipitation (Table 17.2), and functional effects on the receptor were demonstrated by showing that patient IgG could increase PTH release from cultured human parathyroid cells. This receptor blocking activity could be inhibited specifically by preabsorption with CaSR-transfected HEK293 cells. In two cases, affinity-purified autoantibodies inhibited CaSR-mediated accumulation of inositol phosphates and activation of MAP kinase activity in response to high

Table 17.2 Prevalence of CaSR antibodies in idiopathic and APS1-associated hypoparathyroidism

Method of CaSR antibody detection	Number of participants with CaSR antibodies (%)		*P* value[b]	*P* value[c]	Reference
Radioimmunoprecipitation with in vitro translated CaSR	APS1 with HP[a]	6/17 (35)	0.004	–	[26]
	HP	8/8 (100)	<0.0001		
	Controls	0/22 (0)			
Radioimmunoprecipitation with in vitro translated CaSR	APS1 with HP	6/50 (12)	0.044	1.000	[27]
	APS1 no HP	1/10 (10)	0.370		
	Controls	8/192 (4)			
Radioimmunoprecipitation with in vitro translated CaSR	APS1 with HP	0/73 (0)	–	–	[28]
	APS1 no HP	0/17 (0)	–		
	Controls	0/100 (0)			
Immunoblotting of CaSR expressed in *Escherichia coli*	APS1 with HP	1/8 (13)	0.200	–	[31]
	HP	5/17 (29)	0.003	–	
	APS2 with HP	1/6 (17)	0.162	0.182	
	APS2 no HP	0/27 (0)	–		
	Controls	0/32 (0)			
Immunoblotting of human parathyroid tissue containing CaSR	HP	25/51 (49)	0.0002	–	[23]
	Controls	6/45 (13)			
Immunoprecipitation of CaSR expressed in HEK293 cells	HP	2/2 (100)	0.048	–	[32]
	Controls	0/5 (0)			
Immunoprecipitation of CaSR expressed in HEK293 cells	APS1 with HP	12/13 (92)	<0.0001	0.143	[34]
	APS1 no HP	0/1 (0)	–		
	Controls	0/28 (0)			
Immunoblotting of CaSR expressed in HEK293 cells	HP	36/147 (25)	0.011	–	[30]
	Controls	27/199 (14)			

[a]HP, hypoparathyroidism
[b]Patients compared with controls
[c]APS1 or APS2 patients with hypoparathyroidism compared to those without hypoparathyroidism

extracellular calcium signaling. Thereafter, two unusual patients with idiopathic hypoparathyroidism were described in whom CaSR-stimulating autoantibodies could be detected by similar methods; one had Graves' disease but intact parathyroid glands were observed at neck surgery (and confirmed histologically), and the other had Addison's disease and spontaneously remitting hypoparathyroidism, thus indicating in both cases that parathyroid tissue had not been destroyed by any autoimmune process [33]. It therefore seemed likely that the hypoparathyroidism in these patients was the direct result of the CaSR autoantibodies affecting the function of the receptor, akin to the effect of thyroid-stimulating autoantibodies in Graves' disease.

Taken together, these observations support the existence of CaSR autoantibodies in a moderate proportion of patients with idiopathic hypoparathyroidism, and in rare cases these autoantibodies may have an etiological role in activating the receptor, thereby inhibiting PTH secretion. However, the relatively low frequency or even apparent absence of these autoantibodies in APS1 [26–28] begged the question of whether CaSR autoantibodies are associated more with idiopathic hypoparathyroidism than the hypoparathyroidism that occurs in APS1. This possibility would be compatible with other features of APS1; for instance, antibody markers of type 1 diabetes mellitus have different sensitivity and specificity in APS1 compared to idiopathic diabetes, possibly reflecting a different pathogenic basis [28]. However, using an immunoprecipitation technique with CaSR expressed in HEK293 cells, 12 out of 14 (86 %) APS1 sera were found to be positive for CaSR autoantibodies; controls were negative, but 7 % of 20 Graves' disease sera were also positive (Table 17.2) [34]. Activity could be absorbed out specifically with CaSR in the form of CaSR-transfected HEK293 cell extract. By contrast, the same sera were negative when the previously described radioimmunoprecipitation assay for CaSR autoantibodies [26–28] was employed, indicating that the type of assay is critical to the results. Overall, the prevalence of CaSR autoantibodies was significantly increased in APS1 patients compared with controls. Autoantibodies against the CaSR were absent in the only APS1 patient who was not hypoparathyroid and available for study [34].

Subsequent studies using phage-display technology and ELISAs revealed that the major autoepitope for CaSR autoantibodies in this group of APS1 patients was localized between amino acids 41 and 69 in the extracellular domain of the receptor, with antibody reactivity demonstrated in all 12 individuals [35]. Minor epitopes were located between amino acids 114 and 126 and between 171 and 195 with antibody responses detected in five out of 12 (42 %) and four out of 12 (33 %) APS1 patients, respectively. Furthermore, IgG purified from two of the patients with CaSR autoantibodies increased both Ca^{2+}-dependent MAP kinase phosphorylation and inositol phosphate accumulation (Fig. 17.3) in HEK293 cells expressing the CaSR, indicating a stimulatory action upon the receptor [36]. Both patients had hypoparathyroidism. An important implication of this finding was that although the majority of APS1 patients do not have detectable CaSR-stimulating autoantibodies, there may be a small but significant minority of patients in whom the hypoparathyroid state is the result of functional suppression of the parathyroid glands rather than their irreversible destruction. Other CaSR antibody binding sites include those between amino acids 214 and 236, 344 and 358, and 374 and 391, as reported for idiopathic hypoparathyroid patients [30, 33].

Overall, several aspects of CaSR autoantibodies in idiopathic hypoparathyroidism and APS1 remain to be addressed. For example, the variability of their prevalence in patients with idiopathic hypoparathyroidism or APS1 (Table 17.2) could be due to differences in the type of assay employed as currently there is no gold standard test for CaSR autoantibodies. Particularly, the source of the CaSR antigen used can have a significant effect upon the frequency of antibody positives [27, 34]. In addition, the study design including the size and origin of the patient and control groups and the criteria for positivity may influence the reported prevalence of CaSR autoantibodies. Secondly, although they may just be markers for parathyroid gland damage caused by

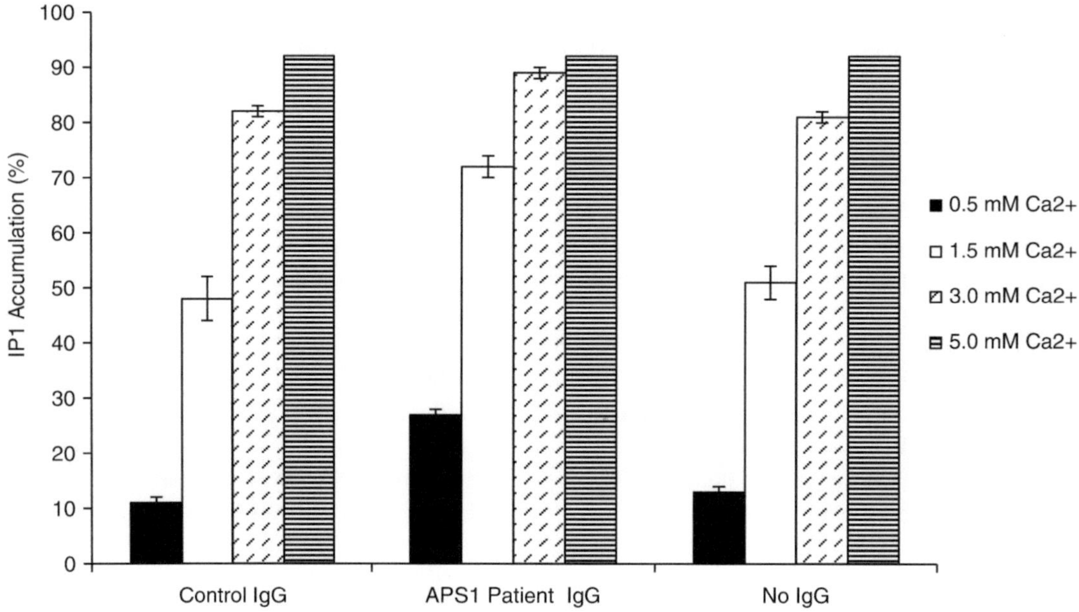

Fig. 17.3 Effect of APS1 patient and control IgG on the response of the CaSR to Ca^{2+} stimulation by measuring inositol-1-phosphate (IP1) accumulation. Changes in IP1 accumulation were measured in response to treatment with 0.5, 1.5, 3.0, and 5.0 mM Ca^{2+} in HEK293-CaSR cells preincubated with either an IgG sample from either an APS1 patient or a control subject or with no IgG. Only the patient IgG sample significantly increased the levels of IP1 accumulation at Ca^{2+} concentrations of 0.5, 1.5, and 3.0 mM when compared with cells treated with the control IgG sample or no IgG (*P* values <0.0001)

cytotoxic T cells [18], the functional relevance of CaSR autoantibodies in the development of hypoparathyroidism requires further exploration. So far, only a minority of patients have been identified with receptor-activating autoantibodies that could potentially reduce PTH secretion [33, 36]. However, CaSR autoantibodies could also impair parathyroid cells through complement-fixation or antibody-dependent cellular cytotoxicity [20]. Finally, the question as to the relationship between the presence of CaSR autoantibodies and clinical hypoparathyroidism in APS1 requires clarification.

17.5 NALP5 Autoantibodies

The absence of CaSR autoantibodies in APS1 patients in some reports prompted the search for other parathyroid autoantigens in these patients, who are well known to have a plethora of tissue autoantibodies [28]. Immunoscreening of a human parathyroid cDNA library with the serum of a patient with APS1 and severe hypoparathyroidism initially revealed reactivity to NALP5 (NACHT leucine-rich-repeat protein 5) [37]. Subsequent assay with a sequential immunoprecipitation assay showed that NALP5-specific autoantibodies were detectable in 41 % of 87 patients with APS1 but were absent in all of the 11 APS1 patients without hypoparathyroidism and in patients with other autoimmune endocrine disorders and in controls. Immunostaining of human parathyroid glands with serum samples from patients with APS1 specifically identified parathyroid chief cells rather than oxyphilic cells: there was no staining of parathyroid tissue by APS1 sera without NALP5 autoantibodies, and absorption of positive sera with recombinant NALP5 blocked the parathyroid staining. Hypoparathyroidism in APS1 is more common in women than men, and the authors noted that NALP5 expression is found in both the parathyroid glands and the ovaries in women, so the increased amount of the autoantigen in women might explain this sex difference. NALP5 autoantibodies were absent in 20 patients with idiopathic hypoparathyroidism.

This association of NALP5 autoantibodies with APS1 has been confirmed in a study of 145 patients with idiopathic hypoparathyroidism, only one of whom was positive by immunoprecipitation assay for NALP5 autoantibodies [38]. Further genetic and serological study revealed that this patient had occult APS1. Nonetheless, this study also identified two patients with APS1 who had hypoparathyroidism, and both were negative for NALP5 autoantibodies. Their disease duration was 6 and 8 years. NALP5 autoantibodies have also been detected in a 64-year-old patient with acquired APS1 and hypoparathyroidism as a result of a thymoma [39] and in an APS1 patient with chronic hypoparathyroidism [40].

The apparent specificity of NALP5 autoantibodies for the hypoparathyroidism of APS1 in these studies is striking and suggestive of a specific type of autoimmune pathogenesis in this syndrome, although it is notable that less than half of all patients were positive, possibly due to disease duration. It must also be emphasized that autoimmune destruction in organ-specific autoimmune disorders like Hashimoto's thyroiditis and type 1 diabetes mellitus is mediated by T cells and autoantibodies have only a limited role in pathogenesis. This distinction might be particularly relevant when considering a role for an intracellular autoantigen such as NALP5. Regarding how NALP5 might be involved, the NALP family of molecules are components of the "inflammasome" involved in particle sensing and activation of the innate immune system, and so it is possible that autoimmunity to NALP5 could lead to an enhanced inflammatory response in parathyroid tissue [41].

17.6 Immunogenetics

Most autoimmune diseases are associated with particular human leukocyte antigen (HLA) specificities and with a number of other gene polymorphisms which regulate various aspects of immune function; sharing of these genetic associations accounts in large part for the clustering of different autoimmune diseases within individuals and within families [42, 43]. Due to its rarity, there is a paucity of information relating to the immunogenetic basis of idiopathic hypoparathyroidism. *AIRE* gene polymorphisms have been reported as being associated with the disease, but these are rare, dependent upon the size and origin of the study population, and beg the question as to whether these are really APS1 patients [40, 44]. A strong association has been reported of *HLA-A*26:01* with idiopathic hypoparathyroidism, suggesting an important role in its pathogenesis for major histocompatibility complex (MHC) class I-mediated presentation of autoantigenic peptides to $CD8^+$ cytotoxic T cells [45]. It is also noteworthy that a critical role for disordered CD8 cell regulation has been suggested to occur in APS1 [46], and this association therefore provides further circumstantial evidence for an autoimmune etiology for idiopathic hypoparathyroidism. In another study, MHC class II *HLA-DRB1*01* and *HLA-DRB1*09* alleles were more frequent in idiopathic hypoparathyroidism patients than in controls, and this too supports an autoimmune basis for the disease [29]. So far other immunogenetic associations, including polymorphisms in *CTLA-4* and *PTPN22*, which are associated with susceptibility to several autoimmune diseases, have not been linked to the development of idiopathic hypoparathyroidism [44, 47]. Here, small sample size and disease heterogeneity may be masking modest associations.

Conclusion

Overall, there is considerable evidence for an autoimmune etiology for hypoparathyroidism both as a sporadic disease and as part of APS1. Most work has been done to identify parathyroid autoantibodies in these patients, but these do not appear to be either specific or sensitive as diagnostic tools. Future investigations should aim to identify whether any novel parathyroid antigens exist besides CaSR and NALP5 and clarify the pathogenic role played by autoantibodies against the CaSR and any newly discovered autoantigens in the pathogenesis of hypoparathyroidism. However, apart from rare cases with functional CaSR autoantibodies, parathyroid autoantibodies seem unlikely to have a primary role in producing disease. Because mutations in the *AIRE* gene cause the development of APS1 following a failure to establish self-tolerance against peripheral tissue-specific antigens in developing T lymphocytes [48, 49], a cellular immune response against known parathyroid autoantigens such as NALP5 and the CaSR must be present in such patients. By analogy with other autoimmune endocrinopathies, T cells also are likely to play the major role in parathyroid destruction in sporadic hypoparathyroidism, and so it will also be essential to characterize the parathyroid autoantigens and epitopes that are targeted by T cells.

References

1. Schmidt MB (1926) Eine biglandulare Erkrankung (Nebennieren und Schilddrüse) bei Morbus Addisonii. Verh Dtsch Ges Pathol 21:212–221
2. Leifer E, Hollander W Jr (1953) Idiopathic hypoparathyroidism and chronic adrenal insufficiency: a case report. J Clin Endocrinol Metab 13:1264–1269
3. Blizzard RM, Chee D, Davis W (1966) The incidence of parathyroid and other antibodies in the sera of patients with idiopathic hypoparathyroidism. Clin Exp Immunol 1:119–128
4. Spinner MW, Blizzard RM, Gibbs J, Abbey H, Childs B (1969) Familial distributions of organ specific antibodies in the blood of patients with Addison's disease and hypoparathyroidism and their relatives. Clin Exp Immunol 5:461–468
5. Neufeld M, MacLaren NK, Blizzard RM (1981) Two types of autoimmune Addison's disease associated with different polyglandular autoimmune (PGA) syndromes. Medicine 60:355–362
6. Perheentupa J (2006) Autoimmune polyendocrinopathy-candidiasis-ectodermal dystrophy. J Clin Endocrinol Metab 91:2843–2850
7. Meloni A, Willcox N, Meager A, Atzeni M, Wolff AS, Husebye ES, Furcas M, Rosatelli MC, Cao A, Congia M (2012) Autoimmune polyendocrine syndrome type 1: an extensive longitudinal study in Sardinian patients. J Clin Endocrinol Metab 97: 1114–1124
8. Kisand K, Peterson P (2011) Autoimmune polyendocrinopathy candidiasis ectodermal dystrophy: known and novel aspects of the syndrome. Ann N Y Acad Sci 1246:77–91
9. Betterle C, Dal Pra C, Mantero F, Zanchetta R (2002) Autoimmune adrenal insufficiency and autoimmune polyendocrine syndromes: autoantibodies, autoantigens, and their applicability in diagnosis and disease. Endocr Rev 23:327–364
10. Goswami R, Marwaha RK, Goswami D, Gupta N, Ray D, Tomar N, Singh S (2006) Prevalence of thyroid autoimmunity in sporadic idiopathic hypoparathyroidism in comparison to type 1 diabetes and premature ovarian failure. J Clin Endocrinol Metab 91:4256–4259
11. Rose NR, Bona C (1993) Defining criteria for autoimmune diseases (Witebsky's postulates revisited). Immunol Today 14:426–430
12. Lupuleseu A, Potorac E, Pop A, Heitmanek C, Merculiev E, Chisiu N, Oprisan R, Neacsu C (1968) Experimental investigations on immunology of the parathyroid gland. Immunology 14:475–482
13. Jancovic BD, Isvaneski M, Ljuhjana P, Mitrovic K (1965) Experimental allergic thyroiditis (and parathyroiditis) in neonatally thymectomized and bursectomized chickens. Int Arch Allergy 26:18–33
14. Altenähr E, Jenke W (1974) Experimental parathyroiditis in the rat by passive immunisation. Virchows Arch A Pathol Anat Histol 363:333–342
15. Kössling FK, Emmrich P (1971) Demonstration eines Falles von Kindlichem Morbus Addison mit Hypoparathyreoidismus. Verh Dtsch Ges Pathol 55:155–160
16. Van de Casseye M, Gepts W (1973) Primary (autoimmune?) parathyroiditis. Virchows Arch A Pathol Anat 361:257–261
17. Talat N, Diaz-Cano S, Schulte KM (2011) Inflammatory diseases of the parathyroid gland. Histopathology 59:897–908
18. Wortsman J, McConnachie P, Baker JR Jr, Mallette LE (1992) T-lymphocyte activation in adult-onset idiopathic hypoparathyroidism. Am J Med 92:352–356
19. Irvine WJ, Scarth L (1969) Antibody to the oxyphil cells of the human parathyroid in idiopathic hypoparathyroidism. Clin Exp Immunol 4:505–510
20. Brandi ML, Aurbach GD, Fattorossi A, Quarto R, Marx SJ, Fitzpatrick LA (1986) Antibodies cytotoxic to bovine parathyroid cells in autoimmune hypoparathyroidism. Proc Natl Acad Sci U S A 83:8366–8369
21. Fattorossi A, Aurbach GD, Sakaguchi K, Cama A, Marx SJ, Streeten EA, Fitzpatrick LA, Brandi ML (1988) Anti-endothelial cell antibodies: detection and characterization in sera from patients with autoimmune hypoparathyroidism. Proc Natl Acad Sci U S A 85:4015–4019
22. Betterle C, Caretto A, Zeviani M, Pedini B, Salviati C (1985) Demonstration and characterization of anti-human mitochondria autoantibodies in idiopathic hypoparathyroidism and in other conditions. Clin Exp Immunol 62:353–360
23. Posillico JT, Wortsman J, Srikanta S, Eisenbarth GS, Mallette LE, Brown EM (1986) Parathyroid cell sur-

face autoantibodies that inhibit parathyroid hormone secretion from dispersed human parathyroid cells. J Bone Miner Res 1:475–483
24. Brown EM, Gamba G, Riccardi D, Lombardi M, Butters R, Kifor O, Sun A, Hediger MA, Lytton J, Hebert SC (1993) Cloning and characterization of an extracellular Ca(2+)-sensing receptor from bovine parathyroid. Nature 366:575–580
25. Brown EM, Pollak M, Hebert SC (1995) Sensing of extracellular Ca2+ by parathyroid and kidney cells: cloning and characterization of an extracellular Ca(2+)-sensing receptor. Am J Kidney Dis 25:506–513
26. Li Y, Song YH, Rais N, Connor E, Schatz D, Muir A, Maclaren N (1996) Autoantibodies to the extracellular domain of the calcium sensing receptor in patients with acquired hypoparathyroidism. J Clin Invest 97:910–914
27. Gylling M, Kääriäinen E, Väisänen R, Kerosuo L, Solin ML, Halme L, Saari S, Halonen M, Kämpe O, Perheentupa J, Miettinen A (2003) The hypoparathyroidism of autoimmune polyendocrinopathy-candidiasis-ectodermal dystrophy protective effect of male sex. J Clin Endocrinol Metab 88:4602–4608
28. Söderbergh A, Myhre AG, Ekwall O, Gebre-Medhin G, Hedstrand H, Landgren E, Miettinen A, Eskelin P, Halonen M, Tuomi T, Gustafsson J, Husebye ES, Perheentupa J, Gylling M, Manns MP, Rorsman F, Kämpe O, Nilsson T (2004) Prevalence and clinical associations of 10 defined autoantibodies in autoimmune polyendocrine syndrome type I. J Clin Endocrinol Metab 89:557–562
29. Goswami R, Brown EM, Kochupillai N, Gupta N, Rani R, Kifor O, Chattopadhyay N (2004) Prevalence of calcium sensing receptor autoantibodies in patients with sporadic idiopathic hypoparathyroidism. Eur J Endocrinol 150:9–18
30. Tomar N, Gupta N, Goswami R (2013) Calcium-sensing receptor autoantibodies in idiopathic hypoparathyroidism. J Clin Endocrinol Metab 98:3884–3891
31. Mayer A, Ploix C, Orgiazzi J, Desbos A, Moreira A, Vidal H, Monier JC, Bienvenu J, Fabien N (2004) Calcium-sensing receptor autoantibodies are relevant markers of acquired hypoparathyroidism. J Clin Endocrinol Metab 89:4484–4488
32. Kifor O, Moore FD Jr, Delaney M, Garber J, Hendy GN, Butters R, Gao P, Cantor TL, Kifor I, Brown EM, Wysolmerski J (2003) A syndrome of hypocalciuric hypercalcemia caused by autoantibodies directed at the calcium-sensing receptor. J Clin Endocrinol Metab 88:60–72
33. Kifor O, McElduff A, LeBoff MS, Moore FD Jr, Butters R, Gao P, Cantor TL, Kifor I, Brown EM (2004) Activating antibodies to the calcium-sensing receptor in two patients with autoimmune hypoparathyroidism. J Clin Endocrinol Metab 89:548–556
34. Gavalas NG, Kemp EH, Krohn KJ, Brown EM, Watson PF, Weetman AP (2007) The calcium-sensing receptor is a target of autoantibodies in patients with autoimmune polyendocrine syndrome type 1. J Clin Endocrinol Metab 92:2107–2114
35. Kemp EH, Gavalas NG, Akhtar S, Krohn KJE, Pallais JC, Brown EM, Watson PF, Weetman AP (2010) Mapping of human autoantibody binding sites on the calcium-sensing receptor. J Bone Miner Res 25:132–140
36. Kemp EH, Gavalas NG, Krohn KJE, Brown EM, Watson PF, Weetman AP (2009) Activating autoantibodies against the calcium-sensing receptor in two patients with autoimmune polyendocrine syndrome type 1. J Clin Endocrinol Metab 94:4749–4756
37. Alimohammadi M, Björklund P, Hallgren A, Pöntynen N, Szinnai G, Shikama N, Keller MP, Ekwall O, Kinkel SA, Husebye ES, Gustafsson J, Rorsman F, Peltonen L, Betterle C, Perheentupa J, Akerström G, Westin G, Scott HS, Holländer GA, Kämpe O (2008) Autoimmune polyendocrine syndrome type 1 and NALP5, a parathyroid autoantigen. N Engl J Med 358:1018–1028
38. Tomar N, Kaushal E, Das M, Gupta N, Betterle C, Goswami R (2012) Prevalence and significance of NALP5 autoantibodies in patients with idiopathic hypoparathyroidism. J Clin Endocrinol Metab 97:1219–1226
39. Cheng MH, Fan U, Grewal N, Barnes M, Mehta A, Taylor S, Husebye ES, Murphy EJ, Anderson MS (2010) Acquired autoimmune polyglandular syndrome, thymoma, and an AIRE defect. N Engl J Med 362:764–766
40. Cervato S, Morlin L, Albergoni MP, Masiero S, Greggio N, Meossi C, Chen S, del Pilar Larosa M, Furmaniak J, Rees Smith B, Alimohammadi M, Kämpe O, Valenzise M, Betterle C (2010) AIRE gene mutations and autoantibodies to interferon omega in patients with chronic hypoparathyroidism without APECED. Clin Endocrinol (Oxf) 73:630–636
41. Eisenbarth G (2008) Do NALP5 antibodies correlate with hypoparathyroidism in patients with APS-1? Nat Clin Pract Endocrinol Metab 4:544–545
42. Boelaert K, Newby PR, Simmonds MJ, Holder RL, Carr-Smith JD, Heward JM, Manji N, Allahabadia A, Armitage M, Chatterjee KV, Lazarus JH, Pearce SH, Vaidya B, Gough SC, Franklyn JA (2010) Prevalence and relative risk of other autoimmune diseases in subjects with autoimmune thyroid disease. Am J Med 123:183.e1–9
43. Weetman AP (2011) Diseases associated with thyroid autoimmunity: explanations for the expanding spectrum. Clin Endocrinol (Oxf) 74:411–418
44. Goswami R, Gupta N, Ray D, Rani R, Tomar N, Sarin R, Vupputuri MR (2005) Polymorphisms at +49A/G and CT60 sites in the 3′ UTR of the CTLA-4 gene and APECED-related AIRE gene mutations analysis in sporadic idiopathic hypoparathyroidism. Int J Immunogenet 32:393–400
45. Goswami R, Singh A, Gupta N, Indian Genome Variation Consortium, Rani R (2012) Presence of

strong association of the major histocompatibility complex (MHC) class I allele *HLA-A*26:01* with idiopathic hypoparathyroidism. J Clin Endocrinol Metab 97:E1820–E1824
46. Laakso SM, Kekäläinen E, Rossi LH, Laurinolli TT, Mannerström H, Heikkilä N, Lehtoviita A, Perheentupa J, Jarva H, Arstila TP (2011) IL-7 dysregulation and loss of CD8+ T cell homeostasis in the monogenic human disease autoimmune polyendocrinopathy-candidiasis-ectodermal dystrophy. J Immunol 187: 2023–2030
47. Ray D, Tomar N, Gupta N, Goswami R (2006) Protein tyrosine phosphatase non-receptor type 22 (PTPN22) gene R620W variant and sporadic idiopathic hypoparathyroidism in Asian Indians. Int J Immunogenet 33:237–240
48. Nagamine K, Peterson P, Scott HS, Kudoh J, Minoshima S, Heino M, Krohn KJ, Lalioti MD, Mullis PE, Antonarakis SE, Kawasaki K, Asakawa S, Ito F, Shimizu N (1997) Positional cloning of the APECED gene. Nat Genet 17:393–398
49. Zuklys S, Balciunaite G, Agarwal A, Fasler-Kan E, Palmer E, Holländer GA (2000) Normal thymic architecture and negative selection are associated with Aire expression, the gene defective in the autoimmune-polyendocrinopathy-candidiasis-ectodermal dystrophy (APECED). J Immunol 165:1976–1983

DiGeorge Syndrome

18

Marina Tarsitano, Andrea Vitale,
and Francesco Tarsitano

18.1 Introduction

DiGeorge syndrome (DGS), described in 1968 by the pediatric endocrinologist Angelo DiGeorge, is a genetic disorder (see also Chaps. 14 and 15). The term "genomic disorders" refers to those diseases that are caused by chromosomal rearrangements involving large regions of one to several megabase pairs in size [1]. The worldwide incidence is estimated at 1/2,000–1/4,000 live births [2]. DiGeorge or velocardiofacial syndrome, caused in over 90 % of cases by the deletion of a small piece of chromosome 22, is also known as the deletion 22q11.2 syndrome (Del22) and results in the poor development of several body systems, with resultant symptoms that vary greatly between individuals but commonly include a primary immunodeficiency, often, but not always, characterized by deficiency in cellular (T-cell) immunity, characteristic facies, congenital heart disease, and hypocalcemia. Endocrinopathies are common in patients with a 22q11.2 deletion. Hypoparathyroidism, which results in hypocalcemia, was the first endocrine disturbance documented in DGS. A wider phenotype is recognized today, including congenital heart defects, abnormal facies, lack of resistance to infection, and cognitive, behavioral, and psychiatric problems.

Most features show variable expressivity and penetrance due to genetic modifiers, chance association, or environmental interactions. Somatic mosaicism or postzygotic second hit have been hypothesized as potential mechanisms underlying such phenotypic discordance. The deletion in chromosome 22 was first identified by cytogenetic methods and was later confirmed by molecular approaches, including FISH and haplotype analysis with genetic markers.

M. Tarsitano, PhD (✉)
Department of Genetics, Biochemical Laboratory,
Via Aversano 35, Salerno 84124, Italy
e-mail: marinatarsitano@hotmail.com

A. Vitale, PhD Student
Department for Exercise Science and Research,
Parthenope University, Naples 80133, Italy
e-mail: andrevi@hotmail.it

F. Tarsitano, MD, PhD Student
Obstetrics and Gynecology Department,
A.O.U. Federico II,
Via S.Pansini 5, Napoli 80131, Italy
e-mail: fratarsitano@hotmail.it

18.2 Clinical Features

Signs and symptoms of DGS can vary significantly in type and severity. This variation depends on what body systems are affected and how severe the defects are. Some signs and symptoms may be apparent at birth; others may not appear until later in infancy or early childhood or not appear for a lifetime. This is called phenotypic variability. The clinical manifestations leading to the diagnosis in the first 2 years of life are frequently congenital heart disease, defects in the palate, mild

M.L. Brandi, E.M. Brown (eds.), *Hypoparathyroidism*,
DOI 10.1007/978-88-470-5376-2_18, © Springer-Verlag Italia 2015

abnormalities in facial features, and/or convulsions due to neonatal hypocalcemia. After 2 years of age, the manifestations that can give rise to suspicion of the disease include recurrent infections, delay in psychomotor developmental and/or speech, hypothyroidism, hypoparathyroidism, and changes in behavior [3–5].

Patients with DGS may have any or all of the following.

18.2.1 Unusual Facial Appearance

The facies of children with 22q11.2DS commonly exhibit the following characteristics: underdeveloped chin; small mouth; eyes with heavy eyelids (narrow palpebral fissures); ears that are rotated backward with, in some cases, defective upper portions of their ear lobes; nasal bossing; hypertelorism; micrognathia, high arched palate; and periorbital fullness. In infancy, micrognathia may be present [6]. These facial characteristics may diminish with age, vary greatly from person to person, and may not be prominent in many patients [7].

18.2.2 Heart Defects/Congenital Heart Defect (CHD)

These include a variety of cardiovascular defects. These usually involve the aorta and the part of the heart from which the aorta develops [8]. A variety of cardiac malformations are seen, especially affecting the outflow tract. These include tetralogy of Fallot, type B interrupted aortic arch, truncus arteriosus, right aortic arch, and aberrant right subclavian artery [9].

In most cases, when present together with hypocalcemia, the onset of symptoms in affected infants and the discovery of a heart murmur on a routine physical exam may lead to the diagnosis of the condition. Affected individuals may show signs of heart failure, or they may have low oxygen saturation of their arterial blood and appear "blue" or cyanotic. Associated with these forms of heart disease, specific cardiovascular anatomic variants have been described in patients with 22q11.2DS that require accurate diagnosis and may influence the surgical treatment of these patients. In some patients, in contrast, heart defects may be very mild or absent [10].

18.2.3 Parathyroid Gland Abnormalities

These glands may be underdeveloped in patients with DGS, causing hypoparathyroidism. The parathyroids are small glands found in the front of the neck, generally close to (and usually posterior to) the thyroid gland, hence the name "parathyroid." They function to control the normal metabolism and blood levels of calcium. People with DGS may have trouble maintaining normal levels of calcium, and this may cause seizures (convulsions). In some cases, the parathyroid abnormality is not present at all, relatively mild, or only a problem during times of stress such as severe illness or surgery. The parathyroid defect often becomes less severe over time [11].

18.2.4 Hypocalcemia

It is considered one of the cardinal features of DiGeorge syndrome. This symptom is related to hypoparathyroidism due to absence or underdevelopment of parathyroid glands, which leads to low blood calcium levels [12]. In the immediate neonatal period, hypoparathyroidism can present with symptoms of hypocalcemia, including seizures, tremors, or tetany. These are due to abrupt discontinuation of the active transport of calcium from mother to fetus at birth. The calcium level usually improves over the first year of life because of parathyroid gland hypertrophy and dietary calcium intake. More commonly, however, hypocalcemia is transient, and as dietary calcium intake increases, the remaining parathyroid activity supplies sufficient PTH to meet metabolic demands. Therefore, in children with severe parathyroid hypoplasia, hypocalcemia is persistent. Bastian et al. reported 18 patients with at least two of four features of DGS (typical facial features, characteristic cardiac lesion, hypocalcemia within the first month of life, and absent thymus), 13 (72 %) of them had hypocalcemia [13]. Muller et al., using similar criteria, reported 16 patients with DGS, 11 (69 %) of them had hypocalcemia. These results confirm that in patients with confirmed DGS, the prevalence of hypocalcemia may be as high as 70 % [14].

18.2.5 Thymus Gland Abnormalities

As part of the fetal developmental defect, the thymus gland may be affected, and development of the cellular (T-cell) immune system may be impaired. The thymus is normally located in the upper area of the front of the thoracic cavity behind the breastbone (sternum). The thymus begins its development high in the neck during the first 3 months of fetal development. As the thymus matures and enlarges, it drops down into the chest to its ultimate location under the sternum and in front of the heart [15].

DGS is a primary immunodeficiency disease caused by abnormal migration and development of certain cells and tissues during development [16]. The thymus controls the development and maturation of one kind of lymphocyte, the T-lymphocyte. Patients with a small thymus produce fewer T-lymphocytes than someone with a normally sized thymus [17]. T-lymphocytes are essential for resistance to certain viral and fungal infections. Some T-lymphocytes, the cytotoxic T-lymphocytes, directly kill cells infected with viruses and some other pathogens.

T-lymphocytes also help B-lymphocytes to develop into plasma cells and produce immunoglobulins or antibodies. Patients with DGS may have poor T-cell production compared to their peers, and as a result, they may have an increased susceptibility to viral, fungal, and bacterial infections [11].

As with the other defects in DGS, the T-lymphocyte defect varies from patient to patient. In addition, small or mild deficiencies may disappear with time [18]. In a very small number of patients with DGS, the thymus is completely absent, so the number of T-cells is severely low. These patients require prompt medical attention since they are severely immunocompromised. The majority of patients with DGS have less severe or mild deficiencies.

18.2.6 Autoimmunity

The immunological alterations of 22q11.2DS may predispose to the onset of autoimmune manifestations [19]. Autoimmune disease occurs when the immune system inappropriately attacks its own body. The most common autoimmune diseases in DGS are idiopathic thrombocytopenia purpura (antibodies against platelets) [20], autoimmune hemolytic anemia (antibodies against red blood cells) [21], and autoimmune disease of the thyroid gland [22]. Further, it may include juvenile rheumatoid arthritis [23], pancytopenia [24], autoimmune diabetes [25], vitiligo [26], and hepatitis [27].

18.2.7 Miscellaneous Clinical Features

Patients with DGS may occasionally have a variety of other developmental abnormalities including cleft palate, poor function of the palate, delayed acquisition of speech, and difficulty in feeding and swallowing. In addition, some patients have learning disabilities, behavioral problems, psychiatric disorders, and hyperactivity. For example, schizophrenia occurs at a higher rate in patients with DGS compared to the rate in the general population [5].

Salient features can be summarized using the mnemonic CATCH-22 to describe DiGeorge syndrome, with the 22 to remind one that the chromosomal abnormality is found on chromosome 22, as in the following: *C*ardiac abnormality (especially tetralogy of Fallot), *A*bnormal facies, *T*hymic aplasia, *C*left palate, and *H*ypocalcemia/*H*ypoparathyroidism [28].

18.3 Diagnosis

The diagnosis of DGS is made on the basis of signs and symptoms that are present at birth, or develop soon after birth, including the typical facial features that, if you look carefully, are always present. Further, the dysmorphic facial appearance in an individual with a major outflow tract defect of the heart or a history of recurrent infection should raise suspicion. In infancy, hypocalcemia is a characteristic feature although this may be intermittent and has a tendency to resolve during the first year. Immunological assessment relies on chest radiography to detect a thymic shadow. In some children, all of the classical features are present and the diagnosis of DGS is made very early. In other people, all of the different organs and tissues may not be affected, and the organs and tissues that are involved may be impaired to different degrees so that the presentation is more subtle and the diagnosis is not made until later on in life when a delay speech development, feeding problems, or autoimmune disease(s) are noted. In the past, the diagnosis of DGS was usually made when all the characteristic findings described above were present without obtaining a confirmatory genetic test. Unfortunately, this caused many mild cases to be missed. In recent years, the genetic test has been more widely used [29, 30].

The clinical suspicion must be confirmed on a blood sample with genetic testing that demonstrate the microdeletion in the region 22q11.2: FISH (fluorescent in situ hybridization) analysis, multiplex ligation-dependent probe amplification (MLPA), and/or array-comparative genomic hybridization (array-CGH) [2]. In FISH one DNA probe from the 22q11.2 chromosomal region is used at a time. In MLPA testing, various probes for selected regions of 22q11.2 are used to identify microdeletions [31]. However, any findings should be further confirmed by other techniques (FISH or array-CGH) (Fig. 18.1). Array-CGH is a technique of molecular karyotyping and can identify the extent of the microdeletion in the region of chromosome 22q11.2 characterizing any missing genes. In case of a positive genetic test in a child, it is advisable to perform the same examination in the parents to evaluate the potential familial involvement of the disease. Some cases of DGS have defects in other chromosomes, notably a deletion in chromosomal region 10p14 [32]. They may have variant deletions of DGS that may be detectable on a research basis only or with other more advanced clinical testing methods.

18.4 Genetics

The syndrome is caused by hemizygous deletion found on the long arm of one of the pair of chromosomes 22, at a location designated q11.2. Of 83 % of patients with a detectable deletion, 90 % had a similar 3 Mb deletion. Another 7 % of the patients with deletions had the same proximal breakpoint as those with the 3 Mb deletion but had an additional nested distal deletion endpoint resulting in a 1.5 Mb deletion [33] (Fig. 18.2).

Deletions of other sizes have also been identified in the interval in a small subset of patients [34, 35]. The deletion results from nonallelic homologous recombination, occurring during meiosis, and is mediated by low-copy repeats (LCR) on chromosome 22 [34]. The deletion of the 3 Mb region on 22q11.2 includes about 30 genes, whereas the deletion of the 1.5 Mb region contains 24 genes. No correlation between the severity of the phenotype and the different size of the deletions has been documented [33]. However, while several studies have analyzed the phenotypic variability of the syndrome, extensive and conclusive intergenerational and intrafamilial comparisons have not yet been reported. DGS can be inherited, but this is the case in the minority of newly diagnosed individuals. Only 5–10 % have inherited the 22q11.2 deletion from a parent with an autosomal dominant pattern, following standard Mendelian inheritance, so an individual carrying the deletion 22q11.2 has a 50 % (one in two) chance of passing it on to their offspring. Conversely, 90–95 % of cases have a de novo (new to the family) deletion of 22q11.2.

Recent studies have shown that the recombination rate in the 22q11.2 region in females was about 1.6–1.7 times greater than that for males, suggesting that for this region in the genome, enhanced meiotic recombination rates, as well as other 22q11.2-specific features, could be responsible for the observed excess in maternal origin [36].

To determine the molecular basis of DGS, several studies have used techniques of mouse genetics. The complete sequence of chromosome 22 [37] and orthologous regions in the mouse have provided the tools for such efforts [37, 38]. For the identification of candidate genes for

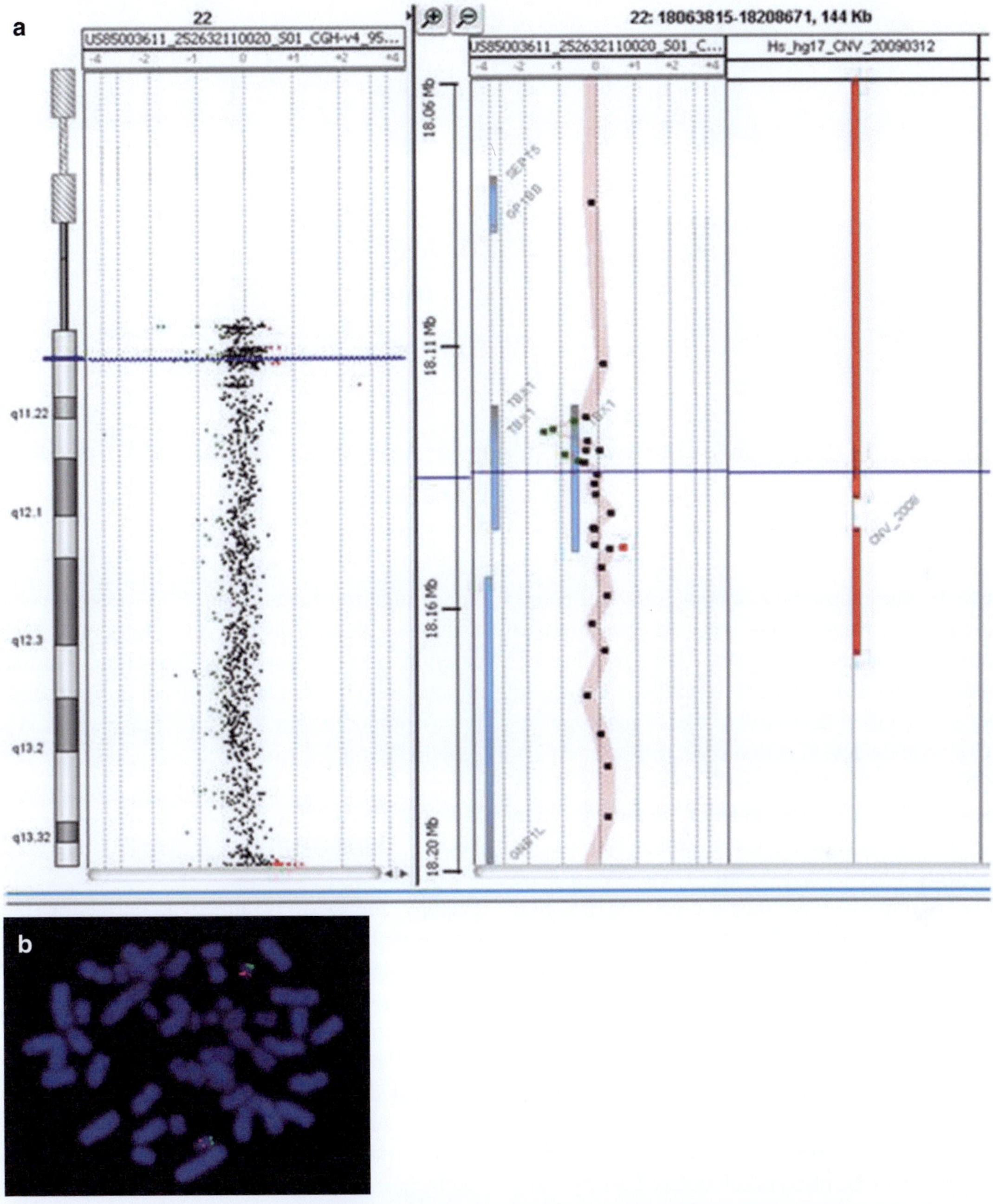

Fig. 18.1 Array-CGH identified a de novo microdeletion in 22q11.21 (arr 22q11.21 (19746363–19747209) × 1 dn) in a case (**a**), which had gone undetected by fluorescence in situ hybridization (FISH) with the DiGeorge TBX1/22q13.3 combination probe (**b**) [62]

DGS, nested deletions and duplications of the orthologous region on MMU16 were generated [39–42]. Mice that harbor a large 1.5 Mb deletion, containing 124 genes and mimicking the nested 1.5 Mb deletion in humans, had reduced viability, conotruncal heart defects, and hypoparathyroidism.

Many of the tissues and structures affected in patients with DGS derive, during embryonic development, from the pharyngeal arches [43],

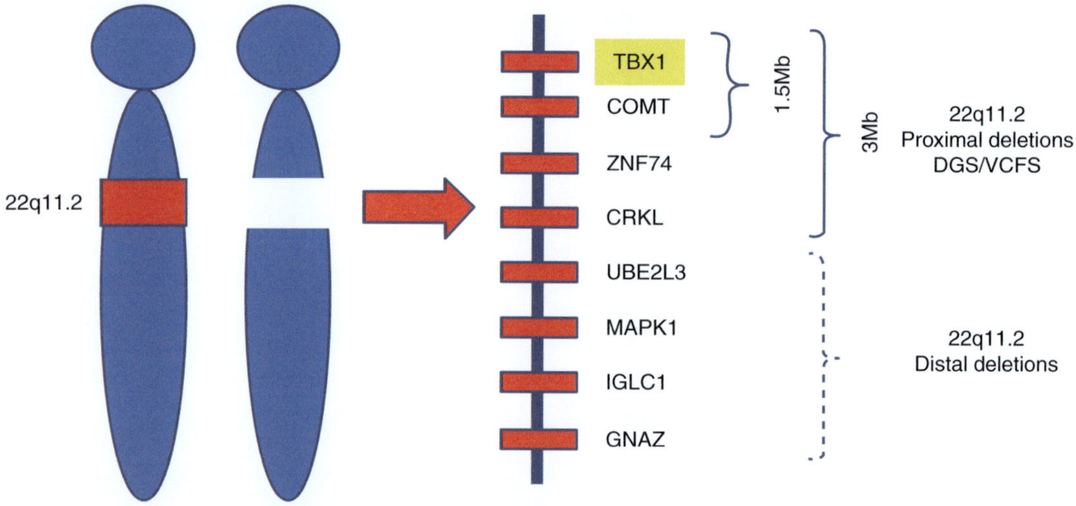

Fig. 18.2 Schematic diagram showing the 22q11.2 deletions and some of the genes included in this region

which are conserved among all vertebrate organisms. Neural-crest cells migrate from a position adjacent to the neural tube and participate in the formation of both the pharyngeal arches and their derivatives. It has been hypothesized that defects in neural-crest cells are responsible for the characteristic features of DGS [44–47]. Neural-crest ablation generates mice with malformations that are similar to those in patients with DGS [48–50]. Therefore, a gene that is important for neural-crest cell function would be a candidate for DGS.

A BAC that harbored four human genes, GP1Bb, PNUTL1, TBX1, and WDR14, provided complete rescue in most mice [42], suggesting that one of these four genes is responsible for the defects. One of the four genes in the BAC, TBX1, a member of the T-box-containing family of transcription-factor genes, is highly expressed in the pharyngeal arches during mouse embryonic development [51]. Tbx1 hemizygotes had mild cardiovascular defects but did not show reduced viability, whereas homozygotes had more severe defects. Tbx1 homozygosity was perinatally lethal, with thymus and parathyroid gland aplasia and major ear malformations. Homozygotes also showed cleft palate and truncus arteriosus, a more severe conotruncal heart defect than that shown in the heterozygotes. Mice, haploinsufficient for TBX1, share several features with humans carrying the homologous deletion and, in particular, structural cardiac anomalies [51–54]. On the basis of these studies, it was proposed that TBX1 in humans is a key gene in the etiology of DGS.

18.5 Treatment

Therapy for DGS is aimed at correcting the defects in the affected organs or tissues. The therapeutic approach varies according to the clinical manifestations of individual patients and depends on the nature of the various defects and their severity. The type and timing of cardiac treatment is evaluated based on the congenital heart abnormalities, and cardiac surgery is often required to improve the function of the heart [55]. Surgery can be performed before any immune defects are corrected. Indeed, it is important that the immune problems are identified early as special precautions are required regarding blood transfusion and immunization with live vaccines. The administration of vaccines that consist of purified proteins (tetanus, diphtheria, pertussis, hepatitis B, *Haemophilus influenzae*, influenza, pneumococcal) is recommended in all subjects; in fact, they are not harmful and can induce an antibody response [56]. For patients with an immunological phenotype with a severity similar to that of patients with severe immunodeficiency, however a rare event, the only experimental treatment showing promise is the transplantation of allogeneic [57], postnatal thymus tissue or alternatively the transplantation of hematopoietic cells from healthy donors. Indeed, thymus transplantation can be used to address the absence of the thymus in the rare, so-called "complete" DGS [58]. Immunologic care for patients with DGS includes monitoring the overall immune system including the numbers and function of T-lymphocytes.

Treatment of severe symptomatic hypocalcemia requires prompt administration of parenteral calcium, 10–15 mg/kg elemental calcium, infused slowly to avoid cardiac dysfunction [59].

Asymptomatic hypocalcemia may be treated with oral calcium supplements, 75–100 mg/kg/day elemental calcium. Maintenance therapy is usually accomplished with 1,25-dihydroxy vitamin D, with or without calcium supplementation [60].

Finally, the key issue is the early intervention speech therapy and psychomotor physiotherapy to limit the difficulty of articulation and language delay and motor learning. It is important that a speech-language pathologist specializing in the evaluation of this assessment should participate in the patient's care within the first year of life. Many children need speech therapy to learn how to properly articulate sounds [61]. The correction of cleft palate and renal velopharyngeal can be performed by different specialists such as plastic surgeons, maxillofacial surgeons, and pediatric surgeons. The physiotherapy offered should be directed at the antigravity extensor muscles (especially the hamstring and paraspinal crural) in order to restore the physiological curves lost as a result of the disease.

18.6 Expectations for Patients with DiGeorge Syndrome

DGS is a multisystem disorder and early diagnosis is important, and optimal management of patients with DGS requires a multidisciplinary approach to management. Each child, regardless of age, should have an echocardiogram, renal ultrasound scan, lymphocyte count and functional assessment, and screening for hypocalcemia; their parents should have their karyotype checked. Each family should be offered a review by a clinical geneticist. Affected children should be under regular medical review from a community pediatrician or general pediatrician who would be best placed to monitor developmental and behavioral aspects of this condition and to coordinate the wide-ranging medical care that these patients need. With regard to the risks for developing hypoparathyroidism, it is suggested that families with DGS be aware of the symptoms that might occur with hypocalcemia. Serum calcium determination should be considered at the time of diagnosis of DGS, when symptoms of hypocalcemia occur, prior to surgery, and during pregnancy. Finally, genetic investigation for DGS should be considered in patients with idiopathic hypoparathyroidism.

Acknowledgments We are grateful to the Dr. M. Chen who agreed for the publication of his array-CGH illustration in this chapter.

References

1. Lupski JR (1998) Genomic disorders: structural features of the genome can lead to DNA rearrangements and human disease traits. Trends Genet 14(10):417–422
2. Schwinger E, Devriendt K, Rauch A, Philip N (2010) Clinical utility gene card for: DiGeorge syndrome, velocardiofacial syndrome, Shprintzen syndrome, chromosome 22q11.2 deletion syndrome (22q11.2, TBX1). Eur J Hum Genet 18(9). doi:10.1038/ejhg.2010.5
3. Debbané M, Glaser B, David MK et al (2006) Psychotic symptoms in children and adolescents with 22q11.2 deletion syndrome: neuropsychological and behavioral implications. Schizophr Res 84(2–3):187–193
4. Bassett AS, Chow EW, AbdelMalik P et al (2003) The schizophrenia phenotype in 22q11 deletion syndrome. Am J Psychiatry 160(9):1580–1586
5. Horowitz A, Shifman S, Rivlin N et al (2005) A survey of the 22q11 microdeletion in a large cohort of schizophrenia patients. Schizophr Res 73(2–3):263–267
6. Lammer EJ, Chen DT, Hoar RM et al (1985) Retinoic acid embryopathy. N Engl J Med 313:837–841

7. Huang RY, Shapiro N (2000) Structural airway anomalies in patients with DiGeorge syndrome: a current review. Am J Otolaryngol 21(5):326–330
8. Digilio MC, Angioni A, De Santis M, Lombardo A et al (2003) Spectrum of clinical variability in familial deletion 22q11.2: from full manifestation to extremely mild clinical anomalies. Clin Genet 63(4):308–313
9. Digilio MC, Marino B, Capolino R, Dallapiccola B (2005) Clinical manifestations of Deletion 22q11.2 syndrome (DiGeorge/Velo-Cardio-Facial syndrome). Images Paediatr Cardiol 7(2):23–34
10. Cirillo E, Giardino G, Gallo V et al (2014) Intergenerational and intrafamilial phenotypic variability in 22q11.2 deletion syndrome subjects. BMC Med Genet 15:1
11. Barrett DJ, Ammann AJ, Wara DW et al (1981) Clinical and immunologic spectrum of the DiGeorge syndrome. J Clin Lab Immunol 6(1):1–6
12. Garabdian M (1999) Hypocalcemia and chromosome 22q11 microdeletion. Genet Couns 10:389–394
13. Müller W, Peter HH, Wilken M et al (1988) The DiGeorge syndrome: I. Clinical evaluation and course of partial and complete forms of the syndrome. Eur J Pediatr 147:496–502
14. Gordon J, Manley NR (2011) Mechanisms of thymus organogenesis and morphogenesis. Development 138(18):3865–3878
15. Bastian J, Law S, Vogler L et al (1989) Prediction of persistent immunodeficiency in the DiGeorge anomaly. J Pediatr 115(3):391–396
16. Jawad FA, Mcdonald-Mcginn DM, Zackai E et al (2001) Immunologic features of chromosome 22q11.2 deletion syndrome (DiGeorge Syndrome/Velocardiofacial syndrome). J Pediatr 139:715–723
17. Junker AK, Driscoll DA (1995) Humoral immunity in DiGeorge syndrome. J Pediatr 127(2):231–237
18. Muller W, Peter HH, Kallfelz HC et al (1989) The DiGeorge sequence. II. Immunological findings in partial and complete forms of the disorder. Eur J Pediatr 19:96–103
19. Gennery AR, Barge D, O'Sullivan JJ et al (2002) Antibody deficiency and autoimmunity in 22q11.2 deletion syndrome. Arch Dis Child 86(6):422–425
20. Hernández-Nieto L, Yamazaki-Nakashimada MA, Lieberman-Hernández E et al (2011) Autoimmune thrombocytopenic purpura in partial DiGeorge syndrome: case presentation. J Pediatr Hematol Oncol 33(6):465–466
21. Sakamoto O et al (2004) Refractory autoimmune hemolytic anemia in a patient with chromosome 22q11.2 deletion syndrome. Pediatr Int 46:612–614
22. Choi JH, Shin YL, Kim GH et al (2005) Endocrine manifestations of chromosome 22q11.2 microdeletion syndrome. Horm Res 63(6):294–299
23. Flato B, Aasland A, Vinje O et al (1998) Outcome and predictive factors in juvenile rheumatoid arthritis and juvenile spondyloarthropathy. J Rheumatol 25:366–375
24. Bruno B, Barbier C, Lambilliotte A et al (2002) Auto-immune pancytopenia in a child with DiGeorge syndrome. Eur J Pediatr 161(7):390–392
25. Elder DA, Kaiser-Rogers K, Aylosworth AS et al (2001) Type I diabetes mellitus in a patient with chromosome 22q11.2 deletion syndrome. Am J Med Genet 101:17–19
26. Niepomniszcze H, Amad RH (2001) Skin disorders and thyroid diseases. J Endocrinol Invest 24(8):628–638
27. Pinchas-Hamiel O, Mandel M, Engelberg S et al (1994) Immune hemolytic anaemia, thrombocytopenia and liver disease in a patient with DiGeorge syndrome. Isr J Med Sci 30:530–532
28. Burn J (1999) Closing time for CATCH22. J Med Genet 36(10):737–738
29. Driscoll DA, Salvin J, Sellinger B et al (1993) Prevalence of 22q11 microdeletions in DiGeorge and velocardiofacial syndromes: implications for genetic counselling and prenatal diagnosis. J Med Genet 30:813–817
30. Botto LD, May K, Fernhoff PM et al (2003) A population-based study of the 22q11.2 deletion: phenotype, incidence, and contribution to major birth defects in the population. Pediatrics 112:101–107
31. Jalali GR, Vorstman JA, Errami A et al (2008) Detailed analysis of 22q11.2 with a high density MLPA probe set. Hum Mutat 29(3):433–440
32. Henwood J, Pickard C, Leek JP et al (2001) A region of homozygosity within 22q11.2 associated with congenital heart disease: recessive DiGeorge/velocardiofacial syndrome? J Med Genet 38:533–536
33. Carlson C, Sirotkin H, Pandita R et al (1997) Molecular definition of 22q11 deletions in151 velo-cardio-facial syndrome patients. Am J Hum Genet 61:620–629
34. Edelmann L, Pandita RK, Morrow BE (1999) Low-copy repeats mediate the common 3-Mb deletion in patients with velo-cardio-facial syndrome. Am J Hum Genet 64:1076–1086
35. Shaikh TH, Kurahashi H, Saitta SC et al (2000) Chromosome 22-specific low copy repeats and the 22q11.2 deletion syndrome: genomic organization and deletion endpoint analysis. Hum Mol Genet 9:489–501
36. Delio M, Guo T, McDonald-McGinn DM et al (2013) Enhanced maternal origin of the 22q11.2 deletion in velocardiofacial and DiGeorge syndromes. Am J Hum Genet 92:439–447
37. Dunham I, Shimizu N, Roe BA (1999) The DNA sequence of human chromosome 22. Nature 402(6761):489–495
38. Lund J, Roe B, Chen F et al (1999) Sequence-ready physical map of the mouse chromosome 16 region with conserved synteny to the human velocardiofacial syndrome region on 22q11.2. Mamm Genome 10(5):438–443
39. Puech A, Saint-Jore B, Funke B et al (1997) Comparative mapping of the human 22q11 chromosomal region and the orthologous region in mice reveals complex changes in gene organization. Proc Natl Acad Sci U S A 94(26):14608–14613

40. Kimber WL, Hsieh P, Hirotsune S et al (1999) Deletion of 150 kb in the minimal DiGeorge/velocardiofacial syndrome critical region in mouse. Hum Mol Genet 8(12):2229–2237
41. Lindsay EA, Botta A, Jurecic V et al (1999) Congenital heart disease in mice deficient for the DiGeorge syndrome region. Nature 401(6751):379–383
42. Merscher S, Funke B, Epstein JA et al (2001) TBX1 is responsible for cardiovascular defects in velo-cardio-facial/DiGeorge syndrome. Cell 104(4): 619–629
43. Kirby ML, Waldo KL (1995) Neural crest and cardiovascular patterning. Circ Res 77(2):211–215
44. Kirby ML, Bockman DE (1984) Neural crest and normal development: a new perspective. Anat Rec 209(1):1–6
45. Bockman DE, Kirby ML (1985) Neural crest interactions in the development of the immune system. J Immunol 135(2 Suppl):766s–768s
46. Phillips MT, Kirby ML, Forbes G (1987) Analysis of cranial neural crest distribution in the developing heart using quail-chick chimeras. Circ Res 60(1):27–30
47. Couly GF, Le Douarin NM (1987) Mapping of the early neural primordium in quail-chick chimeras. II. The prosencephalic neural plate and neural folds: implications for the genesis of cephalic human congenital abnormalities. Dev Biol 120:198–214
48. Bockman DE, Kirby ML (1984) Dependence of thymus development on derivatives of the neural crest. Science 223:498–500
49. Kirby ML, Gale TF, Stewart DE (1983) Neural crest cells contribute to normal aorticopulmonary septation. Science 220(4601):1059–1061
50. Bockman DE, Redmond ME, Waldo K et al (1987) Effect of neural crest ablation on development of the heart and arch arteries in the chick. Am J Anat 180(4):332–341
51. Chapman DL, Garvey N, Hancock S (1996) Expression of the T-box family genes, Tbx1-Tbx5, during early mouse development. Dev Dyn 206(4):379–390
52. Jerome LA, Papaioannou VE (2001) DiGeorge syndrome phenotype in mice mutant for the T-box gene, Tbx1. Nat Genet 27(3):286–291
53. Lindsay EA, Vitelli F, Su H (2001) Tbx1 haploinsufficiency in the DiGeorge syndrome region causes aortic arch defects in mice. Nature 410(6824):97–101
54. Zweier C, Sticht H, Aydin-Yaylagül I (2007) Human TBX1 missense mutations cause gain of function resulting in the same phenotype as 22q11.2 deletions. Am J Hum Genet 80(3):510–517
55. Formigari R, Michielon G, Digilio MC et al (2009) Genetic syndromes and congenital heart defects: how is surgical management affected? Eur J Cardiothorac Surg 35(4):606–614
56. Azzari C, Gambineri E, Resti M (2005) Safety and immunogenicity of measles-mumps-rubella vaccine in children with congenital immunodeficiency (DiGeorge syndrome). Vaccine 23(14): 1668–1671
57. Markert ML, Boeck A, Hale LP et al (1999) Transplantation of thymus tissue in complete DiGeorge syndrome. N Engl J Med 341(16):1180–1189
58. Markert ML, Devlin BH, Alexieff MJ (2007) Review of 54 patients with complete DiGeorge anomaly enrolled in protocols for thymus transplantation: outcome of 44 consecutive transplants. Blood 109(10):4539–4547
59. Tohme JF, Bilezekian JP (1993) Hypocalcemic emergencies. Endocrinol Metab Clin North Am 22:363–375
60. Weinzimer SA (2001) Endocrine aspects of the 22q11.2 deletion syndrome. Genet Med 3(1):19–22
61. Ruda JM, Krakovitz P, Rose AS (2012) A review of the evaluation and management of velopharyngeal insufficiency in children. Otolaryngol Clin North Am 45(3):653–669
62. Chen M et al (2014) Microdeletions/duplications involving TBX1 gene in fetuses with conotruncal heart defects which are negative for 22q11.2 deletion on fluorescence in-situ hybridization. Ultrasound Obstet Gynecol 43(4):396–403

19 Hypoparathyroidism, Deafness, and Renal Anomaly Syndrome

M. Andrew Nesbit

19.1 Introduction

Heterozygous mutations of *GATA3*, which encodes a dual zinc-finger transcription factor leading to haploinsufficiency, cause the autosomal dominant hypoparathyroidism, deafness, and renal dysplasia (HDR) syndrome. The HDR phenotype is consistent with the expression pattern of *GATA3* during embryogenesis (see also Chap. 2). The spectrum of HDR-associated GATA3 mutations comprises complex chromosomal translocations, whole gene loss, missense, nonsense, frameshifting intragenic insertions and deletions, in-frame deletion, and splice site mutations. Analysis of the effects of key missense mutations has revealed DNA- and protein-binding structure-function relationships of the GATA3 molecule. There is variability of the HDR phenotype with no apparent correlation with the underlying genetic defect, suggesting the influence of genetic modifiers or epigenetic modification. Clinical description of an increasing number of HDR syndrome patients is revealing roles for GATA3 in tissues beyond the original triad. Mouse models have demonstrated the important roles of GATA3 in the embryonic development of the parathyroids, inner ear, and kidney and in parathyroid cell proliferation in the adult in response to hypocalcemia.

M.A. Nesbit
Centre for Biomedical Sciences, University of Ulster, Cromore Road, Coleraine, County Londonderry BT51 4JS, Northern Ireland, UK
e-mail: andrew.nesbit@ulster.ac.uk

19.2 Description of the Syndrome and Gene Discovery

The combined occurrence of familial hypoparathyroidism, nerve deafness, and nephrosis (OMIM #146255) was first described in 1977 by Barakat [1] in two brothers with steroid-resistant progressive renal failure resulting in death at the ages of 5 and 8 years. Postmortem investigations revealed that the parathyroid glands were absent in one child and hypoplastic in the other. Similar findings were described in male twins from another family [1]. In 1991, Shaw et al. [2] described four cases of autosomal recessive hypoparathyroidism with renal insufficiency, developmental delay, and lack of auditory responses. The children were the products of consanguineous marriages in three related Asian families, and all died within 15 months after birth.

Inheritance of hypoparathyroidism, deafness, and renal dysplasia (HDR) as an autosomal dominant trait was first reported in one family by Bilous in 1992 [3]. The patients, two brothers and the daughters of one of the brothers, had asymptomatic hypocalcemia with undetectable or inappropriately normal serum concentrations of PTH. Mutation of the *PTH* gene was excluded, and PTH infusion resulted

M.L. Brandi, E.M. Brown (eds.), *Hypoparathyroidism*,
DOI 10.1007/978-88-470-5376-2_19, © Springer-Verlag Italia 2015

in normal brisk increases in plasma cAMP, indicating unaffected sensitivity of the PTH receptor cAMP signaling pathway. The patients also had bilateral, symmetrical, sensorineural nonprogressive deafness involving all frequencies. The renal abnormalities consisted mainly of bilateral cystic or dysplastic kidneys with abnormally compressed glomeruli and tubules leading, in some patients, to renal impairment. Cytogenetic analysis of this family revealed no detectable abnormalities [4].

The gene responsible for the HDR syndrome was identified as a result of the identification of a subset of individuals with DiGeorge syndrome without a microdeletion or translocation involving chromosome 22q11 that is present in the majority of DiGeorge patients. In this small group of patients, there is evidence of deletion or aberration in chromosome 10p. First described by Elliot [5], more than 50 patients with a "DiGeorge-like" phenotype and a partial deletion of chromosome 10p have been reported in the literature (reviewed [6]). Analysis of the deletions in these patients allowed the delineation of two nonoverlapping regions on chromosome 10p that contribute separately to this phenotype. Terminal 10p deletions (10p14–10pter) are associated with hypoparathyroidism, sensorineural deafness, and renal anomalies (HDR), whereas interstitial deletions (10p13–14) are associated with heart defects and immunodeficiency [6–10]. Construction of a 1.2 megabase pair (Mbp) yeast artificial chromosome (YAC), bacterial artificial chromosome (BAC), and P1-derived artificial chromosome (PAC) contig of the breakpoint region facilitated a molecular analysis by fluorescent in situ hybridization (FISH) of the breakpoint region in defining a 900kbp deletion in a patient who had a complex reciprocal, insertional translocation of chromosomes 10p and 8q [4]. An analysis in members of the family ascertained by Bilous [3], using polymorphic microsatellite markers from the 10p15 region, revealed a single allele at D10S1779 in the affected individuals, suggesting the absence of this marker on the other chromosome [4]. The presence of a microdeletion in the affected family members was confirmed by FISH analysis using PACs and BACs spanning the D10S1779 locus. The combined results from the patient with the reciprocal insertional translocation and the original HDR family delineated a critical HDR deletion region about 200 kbp in extent. A search for candidate genes within this region identified GATA3, which has an expression pattern that includes the parathyroid, the embryonic kidney, and the inner ear [11, 12]. FISH analysis using a cosmid clone containing the entire coding region of the GATA3 gene showed that one GATA3 allele was deleted both in the translocation-deletion patient and in the affected family members from the original HDR family. Mutation analysis in HDR patients without cytogenetic abnormalities revealed GATA3 mutations that were shown to result in loss of GATA3 function and consequent haploinsufficiency, confirming the involvement of GATA3 in the human HDR syndrome [4].

In addition to the cases of HDR associated with whole gene loss caused by chromosomal-scale deletions and insertions, 46 mutations of GATA3 have been described in 84 individuals (Table 19.1) [4, 8, 13–39]. These consist of 12 missense mutations, six nonsense mutations, seven frameshifting intragenic insertions, 16 frameshifting intragenic deletions, one in-frame intragenic deletion, and four splice site mutations. The majority of these mutations are unique to one family, but five are recurrent: c.404_405insC [14, 16–18], c.431delG [14, 19], c.431_432insG [20, 21], c.829C>T [4, 14], and c.1099C>T [25, 28]. Examination of the exome variant server database reveals that the HDR-associated GATA3 mutations are not seen in over 13,000 alleles studied and that 11 different missense polymorphisms of unknown functional significance are observed at frequencies of less than 0.1 % [40].

GATA3 belongs to a family of six related zinc-finger transcription factors that are involved in vertebrate embryonic development [41–43]. The mammalian GATA factor family can be subdivided into two families based on their structures and expression patterns [44, 45]. Thus, GATA1, GATA2, and GATA3 are expressed in the hematopoietic cell lineages in which they control differentiation and development of the erythroid, megakaryocyte, hematopoietic stem cell, and

Table 19.1 Clinical findings in reported HDR probands and relatives

a	Mutation (cDNA)[b]	Mutation(protein)[c]	Proband/ relative	Sex [d]	Hc/Hp[e]	Age[f]	D[g]	Age[f]	Renal anomalies	Age[f]	Other phenotypes	Ref.
12	c.35_36delGC	p.Ser12Thr fsX40	Proband	M	Hc/Hp	3 weeks	B		Hypoplasia (R)	1 month	Hypomagnesemia	[13]
22	c.64C>T	p.Gln22X	Proband	M	Hc/Hp	5 years	B	2 years	None			[14]
			Mother	F		Adult	B	Adult	MPGN[h]/RI[i]			
36	c.108_109delGG	p.Met36IlefsX16	Proband	M	Hc/Hp	14 years	B	3 years	Hypoplasia (R), VUR[j] (R/L)	In utero	BGC[n], BPP[o], cataract	[15]
135	c.404delC	p.Pro135ArgfsX60	Proband	F	Hc/Hp	31 years	B	13 years	Normal			[14]
136	c.404_405insC	p.Ala136GlyfsX168	Proband	F	Hc/LN	11 years	Y	<11 years	Normal		Hemimegalencephaly	[16]
			Father	M	Hc/ND	40 years	Y	40 years	Normal			
			Proband	F	Hc/Hp	mid-20s	B	4 years	Right kidney cyst			[17]
			Proband	F	Hc/Hp	8 months	B	3 years	Dysplasia, VUR	8 months		[14]
			Father	M	Hc/Hp	25 years	B	7 years	Aplasia, dysplasia, ERF[k]	25 years		
			Sister	F	Hc/Hp	8 years	B	8 years	VUR	10 months		
			Proband	F	Hc/Hp	10 years	B	13 years	Renal hypoplasia	9 years		[18]
144	c.431delG	p.Gly144AlafsX51	Proband	M	Hc/Hp	4 years	B	6 years	Aplasia	4 years		[14]
			Proband	F	Normal	25 years	B	8 years	Agenesis (R)	24 years	Uterus didelphys, septate vagina	[19]
			Mother	F	Normal	58 years	B	6 years	Nonfunctional kidney (R), ERF	28 years	Left choanal stenosis, septate uterus	
145	c.431_432insG	p.His145ProfsX159	Proband	M	Hc/Hp	24 years	B	20 years	Normal			[20]
			Proband	F	Hc/Hp	1 year	B	8 years	Calcification (R)	1 year	Epilepsy	[21]
156	c.465_513del	p.Thr156ArgfsX23	Proband	M	Hc/Hp		B	8 years	Cysts (R), dysplasia (L)	Infancy		[4, 17]
160	c.478delG	p.Asp160ThrfsX35	Proband	M	Hc/Hp	50 years	B	Childhood	Normal			[22]
			Son	M	Normal	23 years	B	10	Normal			
			Son	M	Normal/ Hp	18 years	B	17	Normal			
164	c.490delG	p.Asp164ThrfsX31	Proband	F	Hc/Hp	Neonate	B	10 months	Normal			[14]
173	c.517delG	p.Ala173ProfsX22	Proband	F	Hc/Hp		B		Dysplasia			[23]
198	c.593C>A	p.Ser198X	Proband	M	Hc/LN	58 years	B	Childhood	Normal		BPP, ICC[p]	[24]

(continued)

Table 19.1 (continued)

a	Mutation (cDNA)[b]	Mutation(protein)[c]	Proband/ relative	Sex [d]	Hc/Hp[e]	Age[f]	D[g]	Age[f]	Renal anomalies	Age[f]	Other phenotypes	Ref.
202	c.603delC	p.Arg202ValfsX4	Proband	F	Hc/LN	3 years	B	3 years	Sepsis, VUR, cysts	3 years		[25]
			Father	M	Hc/LN	<40 years	B, L>R	<30 years	VUR			
228	c.682G>T	p.Glu228X	Proband	F	Hc/LN	20 years	B	4	Hypoplasia, ERF			[25]
			Mother	F	Hc/ normal	45 years	B	<25 years	Normal			
234	c.700T>C, c.708_709insC	p.Phe234Leu, Ser237GlnfsX67	Proband	F	Hc/ND	13 years	B	6 years	Pelvicalyceal deformity L/R	27 years		[18]
237	c.708delC	p.Ser237AlafsX29	Proband	F	Hc/Hp	35 years	L>R	3 years	Cysts, hypoplasia	36 years		[14]
			Daughter	F	Hc/ normal	14 years	B	2.5 years	Normal			
			Son	M	Hc/Hp	7 years	B	Birth	Normal			
	c.708_709insC	p.Ser237GlnfsX67	Proband	F	Hc/Hp	Adult	B	Adult	Agenesis (L)	Adult	Diffuse goiter	[26]
			Daughter	F	Hc/Hp	Adult	B	Adult	Hypoplasia (L)	Adult	Diffuse goiter	
247	c.737_738insG	p.Phe247LeufsX57	Proband	M	Hc/LN	17 months	Normal		Right pelvic duplication	17 months		[18]
248	c.742G>T	p.Gly248X	Proband	F	Hc/Hp	2 days	B	5 years	Renal cysts			[14]
262	c.784A>G	p.Arg262Gly	Proband	F	Hc/ND	2 years	B	11 years	Normal			[21]
			Mother	F	Hc/ND	27 years	B	Childhood	ND			
272	c.815C>T	p.Thr272Ile	Proband	M	Hc/Hp	Neonate	B	8 months	Pelvic kidney	Neonate	DPD[q], hyperopia, astigmatism, FDF	[27]
275	c.823T>A	p.Trp275Arg	Proband	M	Hc/Hp	39 years	B	Childhood	Hypoplasia (L)	36 years		[28]
			Mother	F	Hc/LN	72 years	B	Childhood	Hypoplasia (L)	67 years		
	c.824G>T	p.Trp275Leu	Proband	F	Hc/Hp	3 years	B	11 years	Renal dysfunction			[18]
276	c.827G>C	p.Arg276Pro	Proband	F	Hc/LN	22 years	B	Childhood	Agenesis (R), ERF	22 years		[29]
			Mother	F	Hc/LN	45 years	B	Childhood	Normal			
			Sister	F	Hc/LN	14 years	B	Childhood	Normal			
277	c.828C>T	p.Arg277X	Proband	F	Hc/Hp		B		Normal			[4]
			Son	M	Hc/Hp		B		Normal			
			Proband	F	Hc/Hp	33 years	B	Childhood	Hypoplasia, ERF			[14]

[a]	Mutation (cDNA)[b]	Mutation(protein)[c]	Proband/ relative	Sex [d]	Hc/Hp[e]	Age[f]	D[g]	Age[f]	Renal anomalies	Age[f]	Other phenotypes	Ref.
295	c.883_886delAACG	p.Asn295AspfsX60	Proband	M	Hc/Hp	6 years	B	5 years	Hypoplasia, sepsis	10 years		[14]
			Sister	F	Hc/LN	1 year	B	7 years	Hypoplasia, sepsis, ERF	1 year		
			Mother	F	Hc/LN	Adult	B	38 years	Hypoplasia, sepsis, ERF	Adult		
300	c.900_901insAA, c.901_902insCCT	p.Leu301AsnfsX56	Proband	F	Hc/Hp	25 years	B	27 years	Aplasia (R)	27 years		[28]
			Father	M	Hc/Hp	57 years	B, L>R	Childhood	Normal			
308	c.924+2 T>GCTTACTTCCC	p.Glu260ValfsX44	Proband	F	Hc/LN	33 years	B	Childhood	Normal			[22]
			Daughter	F	Hc/LN	16 years	B	2 years	Hypoplasia, VUR, ERF	5 years		
316	c.946_957del	p.(Thr316_Ala319) del	Proband	M	Hc/Hp		B		Dysplasia			[4]
318	c.952T>C	p.Cys318Arg	Proband	M	Hc/ND	11 years	B	8 years	Sepsis, VUR, hypoplasia, ERF	<3 years		[25]
	c.952T>A	p.Cys318Ser	Proband	M	Hc/Hp	12 years	B	Childhood	Normal		Strabismus	[21]
			Father	M	Hc/ND		B		Hydronephrosis			
			Sister	F	LN/LN	Childhood	B	Childhood	Multicystic kidney (R), ectopic ureter		Vaginal atresia	
	c.951delC	p.Cys318ValfsX38	Proband	M	Hc/Hp	17 months	B	3 years	ERF, NI[l], TIN[m]			[30]
320	c.960C>A	p.Asn320Lys	Proband	F	Hc/Hp	27 years	B	4 years	Cysts	23 years		[25]
			Mother	F	Hc/LN	51 years	B	24 years	Normal			
321	c.963G>C	p.Cys321Ser	Proband	M	Hc/LN	5 days	B	Neonate	Hypoplasia (R)			[31]
			Mother	F			B				Bicornuate uterus	
342	c.1025G>A		Proband	F	Hc/Hp	28 years	B	Infancy	Normal		Vagina/uterus agenesis, ovary cyst	[32]
348	c.1043T>G	p.Leu348Arg	Proband	M	Hc/Hp	14 years	R>L	Neonate	Normal		Long QT interval	[14]
350	c1050+1 G>C		Proband	M	Hc/Hp	17 months	B	12 months	Hypoplasia (L)	17 months		[18]

(continued)

Table 19.1 (continued)

[a]	Mutation (cDNA)[b]	Mutation(protein)[c]	Proband/ relative	Sex [d]	Hc/Hp[e]	Age[f]	D[g]	Age[f]	Renal anomalies	Age[f]	Other phenotypes	Ref.
351	c.1051-1G>T	p.Ile351ThrfsX18	Proband	F	Hc/LN	1 year	B, R>L	5 years	Agenesis			[25]
			Proband	F	Hc/Hp	1 month	B	5 months	Hypoplasia (L)	5 months	Uterus didelphys	[21]
	c.1051-2 A>G	p.Ile351ThrfsX18	Proband	F	Hc/LN	2 months	B	2 months	Dysplasia, VUR, sepsis	1 year		[14]
353	c.1059A>T	p.Arg353Ser	Proband	F	Hc/Hp	32 years	B	Childhood	Normal		BGC, ICC	[22]
			Daughter	F	Normal	10 years	B	5 years	ND			
			Son	M	LN/LN	7 years	B	4 years	ND			
355	c.1063_1064delCT	p.Leu355AspfsX15	Proband	F	Hc/ND	4 years	B	<8 months	Agenesis			[25]
	c.1064delT	p.Leu355ArgfsX1	Proband	M	Hc/Hp	8 months	B	1 year	Hypoplasia	8 months		[14]
	c.1063delC	p.Leu355X	Proband	F	Hc/Hp	1 month	B	<3 years	Normal		Epilepsy	[21]
367	c.1099C>T	p.Arg367X	Proband	F	Hc/Hp	3 years	ND		Proteinuria, hematuria	20 years		[28]
			Daughter	F	LN/ normal		Normal		Normal			
			Proband	M	Hc/Hp	13 years	B, R>L	<1 year	Normal			[25]
			Proband	M	Hc/Hp	10 years	B	3 years	Normal		ICC	[33]
			Sister	F	Hc/Hp	5 years	B	5 years	Normal		ICC	
401	c.1200_1201delCA	p.Met401ValfsX104	Proband	F	Hc/Hp	3 months	B	1 year	Aplasia (R), dysplasia (L)	9 years	Diabetes mellitus	[34]
408	c.1221_1222insC	p.Ser408LeufsX100	Proband	F	Hc/Hp	1 month	B	Neonate	Hypoplasia	In utero		[14]
	D10S1751-D10S1779		Proband	F	Hc/Hp		B		Aplasia		Uterus bicornis	[4]
	D10S1779		Proband	M	Hc/Hp		B	Adult	Hypoplasia, pelvis cyst (R)			[3, 4]
			Brother	M	Hc/ normal		B	1 year	Renal pelvis cyst (L)			
			Niece	F	Hc/ normal		B	Neonate				
			Niece	F	Normal		B	5 years	Dysplasia			
	D10S226-10pter		Proband	F	Hc/Hp	6 days	B	5 months	Pelvicalyceal deformity L/R			[28]

[a]	Mutation (cDNA)[b]	Mutation(protein)[c]	Proband/ relative	Sex [d]	Hc/Hp[e]	Age[f]	D[g]	Age[f]	Renal anomalies	Age[f]	Other phenotypes	Ref.
	D10S226-10pter		Proband	M	Hc/Hp	1 months	B	3 years	Renal scar (L), VUR(R/L)			[28]
	D10S547-10pter		Proband	F	Hc/Hp	2.5 years	B	2.7 years	Agenesis (R), VUR (L), pelvicalyceal deformity (L)	2.7 years	Hypertelorism, FDF[r], clinodactyly	[28]
	D10S465-D10S189		Proband	M	Hc/Hp	1.4 years	B	7 years	Pelvicalyceal deformity R/L		BGC	[8, 28]
	10p12.1-pter		Proband	F	Normal		B	15 months	Hypodysplasia, VUR (R/L)	15 months	FDF, DPD	[35]
	10p14-10p15.3		Proband	M	Hc/Hp	2 months	ND		Normal		BGC, FDF, DD[(s)]	[36]
	10p14		Proband	M	Normal		B	4.7 years	Normal		FDF, skeletal anomalies, cryptorchidism, DD	[37]

[a]Codon
[b]GenBank accession number NM_001002295.1
[c]NP_001002295.1
[d]*M* male, *F* female
[e]*Hc*/*Hp* *h*ypo*c*alcemia/*h*ypo*p*arathyroidism, *LN* low normal, *ND* not determined
[f]Age at presentation or diagnosis
[g]*D* deafness, *B* bilateral, *R* right, *L* left
[h]Mesangioproliferative glomerulonephritis
[i]Renal insufficiency
[j]Vesicoureteral reflux
[k]End-stage renal failure
[l]Nephrocalcinosis
[m]Tubulointerstitial nephritis
[n]Basal ganglia calcification
[o]Bilateral palpebral ptosis
[p]Intracranial calcification
[q]Delayed psychomotor development
[r]Facial dysmorphic features
[s]Developmental delays

T-cell lineages, whereas the structurally related proteins GATA4, GATA5, and GATA6 are expressed in overlapping patterns in the heart, gut, urogenital system, and smooth muscle cell lineages [44, 45].

19.3 Structure-Function Relationships of HDR-Associated GATA3 Mutations

GATA3 is a 444-amino-acid protein which has two C4-type (Cys-*X*2-Cys-X17-Cys-*X*2-Cys (where X represents any amino acid residue)) zinc-finger DNA-binding domains that bind to the consensus motif 5′-(A/T)GATA(A/G)-3′ [46] – the C-terminal finger (ZnF2) is essential for DNA binding, whereas the N-terminal finger (ZnF1), which has a preference for GATC motifs [47], stabilizes this DNA binding and interacts with other zinc-finger proteins, such as the multi-type zinc-finger proteins friends of GATA (FOG) [48–50].

The majority (>70 %) of HDR-associated mutations that do not involve whole gene loss are predicted to result in truncated forms of the GATA3 protein. Most result in the loss of both ZnF1 And ZnF2 or only the loss of ZnF2, while one nonsense mutation results in the loss of a stretch of basic amino acids immediately C-terminal to ZnF2 [25, 28, 33]. Less commonly C-terminal mutations result in a long C-terminal extension [14, 34]. All but one of the 12 missense mutations described to date are clustered within either ZnF1 or ZnF2, emphasizing the importance of these domains to GATA3 function (Fig. 19.1a).

The functional consequences of HDR-associated mutations have been investigated by a variety of methods. An assessment of the subcellular localization of GATA3 mutants using GFP-tagged GATA3 constructs revealed that mutations that resulted in protein truncation before ZnF1 did not accumulate in the nucleus [25], findings that are consistent with the nuclear localization signal (NLS) for GATA3 being contained within residues 249–311 that encompass ZnF1 [53, 54]. Interestingly, mutation of the first zinc-chelating cysteine residue of ZnF1, Cys264Arg (equivalent to the HDR-associated ZnF2 mutation Cys318Arg [25]), has no effect upon nuclear localization of the mutant GATA3, suggesting that the GATA3 NLS is intrinsic to the amino acid sequence rather than to either the tertiary structure of ZnF1 or its ability to bind DNA [54]. Electrophoretic mobility shift assays have demonstrated that mutations within GATA3 ZnF2 or C-terminally adjacent basic amino acids result in a loss of DNA binding (Fig. 19.1b), whereas mutation of particular residues within ZnF1 altered DNA-binding affinity [14, 25, 27]. Yeast two-hybrid and GST pull-down assays revealed that other ZnF1 mutations lead to a loss of interaction with FOG2 ZnFs [25, 27]. These findings are consistent with the three-dimensional model of GATA3 ZnF1, which has separate DNA- and protein-binding surfaces [14, 25, 27, 51, 55] (Fig. 19.1c). Finally, luciferase reporter assays have been used to demonstrate the functional consequences of mutations upon the transactivation ability of GATA3 [27, 31].

Thus, it may be useful for an understanding of the structure-function relationships of GATA3 to divide the HDR-associated GATA3 mutations into three broad classes, which depend upon their functional consequences with respect to alterations in DNA binding and interactions with cofactors such as FOG2. The first class comprises the majority of mutations which result in truncated or deleted forms of GATA3 lacking ZnF2, loss of the important C-terminal basic tail, or missense mutations in ZnF2 resulting in a loss of DNA binding [4, 13–16, 18, 19, 21–26, 28, 30–34] (Fig. 19.1b). The second class is defined by a loss of DNA-binding affinity and is represented by one missense mutation, Arg276Pro, in ZnF1 [29]. This mutant GATA3 binds to DNA but with a reduced affinity such that it rapidly dissociates from the bound DNA in EMSA dissociation assays. The crystallographic study of GATA3 zinc fingers bound to DNA confirms that Arg276 is involved in binding DNA and reveals more specifically that hydrogen bonds between its guanidinium group and guanine lead to specific recognition of the base-pair identity of the GATA motif [51] (Fig. 19.1c). Interestingly, the crystal structure reveals that, in addition to binding GATA or GATC motifs in palindromic GATA sites, ZnF1 residues may also stabilize binding of GATA3 to DNA by interaction with the stretch of

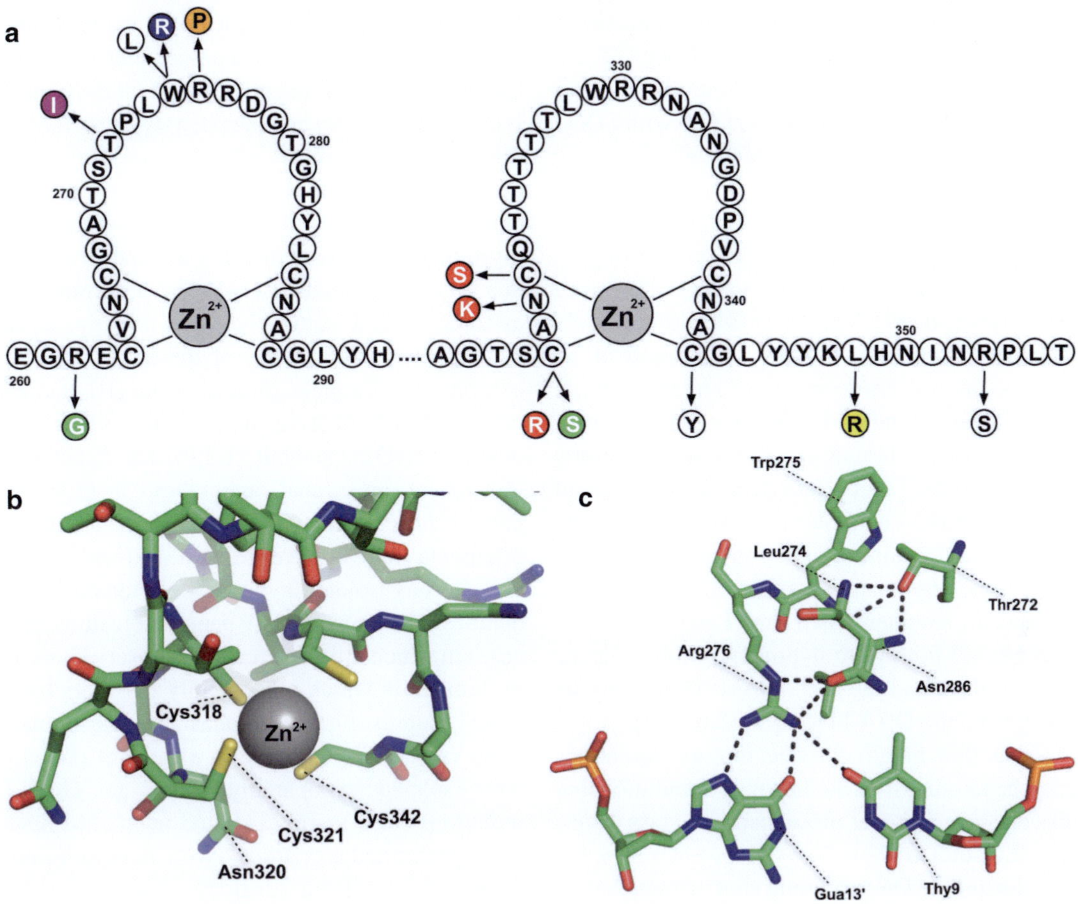

Fig. 19.1 (**a**) Schematic representation of the clustering of HDR-associated *GATA3* missense mutations in the zinc-finger (ZnF) domains and their functional consequences. The ZnF domains consist of a C-X$_2$-C-X$_{17}$-C-X2-C consensus sequence, where X represents any amino acid. The zinc ion coordinates with four cysteine residues. The location of the missense mutations is indicated by arrows. Missense mutations which have been shown to affect DNA binding are shown in red (C318R, N320K, C321S [25, 31]), to affect DNA-binding stability in orange (R276P [29]), to affect interaction with FOG2 in blue (W275R [25, 28]), to affect DNA binding and interaction with FOG2 in purple (T272I [27]), to reduce transactivation activity in green (R262G, C318S [21]), and to affect neither DNA binding nor stability of binding in yellow (L348R [14]). Mutations in which functionality has not been assessed are left uncolored (W275L, C318S, C342Y, R353S [18, 21, 22, 32]). (**b**) Three-dimensional model of the human GATA3 ZnF2 showing clustering of HDR-associated missense GATA3 mutations around the zinc ion based on a structure of GATA3 bound to a 20 mer palindromic GATA site, AATGTC*CATCTGATA*AGACG (Protein Data Bank 4HCA [51]). Mutation of the cysteine residues (Cys318, Cys321, and Cys342) directly results in loss of coordination of the zinc ion, while mutation of Asn320 will likely result in structural changes affecting its binding. (**c**) Three-dimensional model of the human GATA3 ZnF1 showing important hydrogen bonds between residues and DNA. The residues Thr272, Trp275, and Arg276, missense mutations of which cause HDR, all lie in close proximity to one another. However, while Arg276 forms hydrogen bonds with the DNA and is involved in specific recognition of the GATA motif, Trp275 faces away from the DNA and can interact with FOG2 [25]. Loss of hydrogen bonds between Arg276 and Asn286 and between Thr272 and Asn286 and Leu274 will result in changes in the structure affecting DNA binding and/or protein interaction [27, 29]. Hydrogen bonds are shown as black dashed lines [52]

basic residues C-terminal to ZnF2 [51]. An Arg262Gly mutation may also belong to this class. It has been shown to significantly diminish the transactivation function of GATA3 [21] and to form a hydrogen bond with DNA that would be lost on mutation to glycine [51]. The third class of mutation is characterized by normal DNA binding and affinity and is represented by three HDR-associated missense mutations, Trp275Arg, Trp275Leu, and Leu348Arg [14, 18, 25, 28]. The Trp275Arg GATA3 mutant has been shown to result in a loss of interaction with FOG2 zinc fingers 1, 5, and 8 [25]. Such mutations may interfere with the long-range control of gene expression that GATA and FOG proteins mediate by facilitating chromosome looping, bringing distant enhancers and promoters into close proximity [56, 57]. The Leu348 residue lies six residues C-terminal to ZnF2 at the end of an α-helix linking ZnF2 and the C-terminal basic domain. ZnF2 binds to DNA in the major groove, whereas the C-terminal basic domain inserts deeply into the DNA minor groove [51, 58], and mutation from a nonpolar leucine residue to the larger polar arginine residue may have been predicted to significantly affect DNA binding. Indeed, the residues which flank Leu348, Lys347, and His349 both form hydrogen bonds with the bound DNA, and the Leu348 side chain may lie away from the DNA-binding surface [51].

That this broad classification of mutations may be too simplistic is shown by a mutation that has characteristics of more than one class. Thus, mutation of Thr272 to Leu causes not a loss, but a reduction in DNA binding that is nonetheless stable, a loss of interaction with FOG2 zinc fingers, and a reduction in transactivation that is consistent with the degree of the reduction in DNA binding [27]. It should be noted that GATA3 mutations are likely to have effects beyond the interactions with FOG2 as GATA3 has also been shown to interact with other proteins including GATA1, GATA2 [59, 60], smad3 [61], sp1 [62], EKLF [62], RBTN2 [63], menin [64], MTA-2 [65], IRX5 [66], and BRCA1 [67]. GATA2 [68], smad3 [69], and RBTN2 [69] are expressed in the kidney, whereas SP1 and menin are expressed in both the kidney and parathyroids [70–72].

19.4 Genotype-Phenotype Relationships of HDR-Associated GATA3 Mutations

An examination of the spectrum of HDR-associated GATA3 mutations shows that there is both great intra- and interfamilial variability of the HDR phenotype with each proband and family generally having its own unique mutation, and there appears to be no correlation with the underlying genetic defect and the phenotypic variation. Studies have demonstrated that more than 90 % of patients with at least two of the defining clinical features of HDR syndrome—hypoparathyroidism, deafness, or renal anomalies— harbor a GATA3 mutation [14]. The penetrance of all three clinical features has been investigated in 87 patients (Table 19.1)—of these 54 (62.1 %) exhibit the complete clinical triad (HDR), 25 patients (28.7 %) have hypoparathyroidism and deafness (HD), 1 patient (1.2 %) has hypoparathyroidism and renal anomalies, 3 (3.4 %) patients have deafness and renal anomalies (DR), 3 patients (3.4 %) have isolated deafness, and 1 (1.2 %) has isolated hypoparathyroidism. Thus, deafness is the most highly penetrant feature. The cases in which only one feature is observed occur in families in which other family members have other features further emphasizing the phenotypic variability. However, in a study of patients with isolated hypoparathyroidism, no *GATA3* mutations were found [14], and no studies have been performed in patients with nonsyndromic deafness or isolated renal dysplasia given the large number of alternative genes which are known to cause these disorders.

Some of those HDR patients without a GATA3 mutation of the coding region and that appear to be cytogenetically normal, who nonetheless appear indistinguishable phenotypically from those with a GATA3 mutation [4, 14], may harbor mutations in the regulatory sequences flanking the GATA3 gene. These regulatory elements are located at substantial distances both upstream and downstream of the gene [73–75]. Alternatively, these patients may represent genetic heterogeneity in HDR with mutations in other genes.

The variability in the penetrance of GATA3 mutation-associated characteristics is exemplified by two unrelated families from Japan and Britain who had an identical Arg367X mutation [25, 28]. Thus, the Japanese patients had hypoparathyroidism and renal abnormalities but no deafness [28], whereas the British patient had hypoparathyroidism and deafness but no renal abnormalities [25]. Furthermore, even within families with patients harboring identical GATA3 mutations, there appears to be a variable expression of the triad [4, 14, 22, 25, 28]

The severity of the HDR phenotype is variable, with the greatest variation reported for the development of kidney defects. Some GATA3 mutations, such as Cys318Arg [14] which disrupts zinc ion coordination and causes a complete loss of GATA3 function, are associated with abnormalities of the parathyroids, kidneys, and hearing. However, a similar severity of effect is also found in patients with a GATA3 Thr272Ile mutation, which retains approximately 30 % of the wild-type function [27]. These findings suggest that there is a critical threshold of GATA3 essential for parathyroid, otic, and renal development. Indeed, a study of the effects of GATA3 dosage on metanephrogenesis [75] demonstrated that at least 70 % of diploid GATA3 levels are required to restore renal development in mice. Thus, any residual level of activity that mutant GATA3 molecules retain is unlikely to be sufficient to reach this threshold and prevent hypoparathyroidism, deafness, or renal deficiency.

There is a wide variability in the presentation of hypoparathyroidism in HDR patients, ranging from asymptomatic hypocalcemia to paresthesias, muscular aching, and tetany, with hypocalcemia ranging from low to normal and serum PTH levels ranging from undetectable to slightly elevated. The clinical feature of early-onset sensorineural deafness is the most completely penetrant aspect of the HDR syndrome. The deafness is, most often, bilateral and symmetrical, involves all frequencies, and is more severe at higher frequencies, with the severity ranging from moderate to severe (40–105 dB), necessitating the use of hearing aids [4, 17]. The renal phenotype displayed in patients with HDR syndrome shows the greatest variation, even in the same family, with no detectable anomalies in some [4, 14, 16, 20–22, 24, 25, 28, 29, 33, 36, 37] and renal agenesis, renal hypoplasia, renal dysplasia, multicystic kidneys, or vesicoureteric reflux in others often leading to end-stage renal failure. Additional renal phenotypes observed include hypomagnesemia [13] and nephrocalcinosis with or without detected distal renal tubular acidosis [21, 30, 76, 77].

19.5 Additional Phenotypic Characteristics of HDR Syndrome Patients

With the description of a growing number of HDR syndrome patients, a number of phenotypic characteristics in addition to hypoparathyroidism, deafness, or renal anomalies have been described in a small but significant proportion of patients (Table 19.1). These include malformations in the female genital tract [4, 19, 21, 31, 32], basal ganglia calcification [8, 15, 22, 28, 36], intracranial calcification [22, 24, 33], moderate mental retardation with or without recurrent cerebral infarction [8, 29], seizures or epilepsy [16, 21], polycystic ovaries [29, 32], pyloric stenosis [28], diffuse goiter [26], and diabetes mellitus [34],

19.6 The Gata3 Heterozygous Knockout Mouse as a Model of Human HDR

Homozygous disruption of *Gata3* in mice causes severe deformities in the brain and spinal cord, fetal liver hematopoiesis with a total block of T-cell differentiation, and massive internal bleeding, resulting in mid-gestation embryonic lethality (E11.5–E12.5) [42]. Partial pharmacological rescue of mutant mice (to E16.5) by feeding catechol intermediates to the pregnant dams reveals that embryonic lethality is partially due to a noradrenaline deficiency of the sympathetic nervous system [78]. These rescued embryos reveal later-onset defects that more closely mimic the human pathology with cephalic neural crest abnormalities, thymic hypoplasia, renal hypoplasia, a failure to form the metanephros, and an aberrant elongation of the nephric duct [79, 80].

Although heterozygous *Gata3* knockout mice were initially reported to be normal with a normal life span and fertility [42], careful reexamination of these mice revealed the presence of hypoparathyroidism and sensorineural deafness [81, 82]. *Gata3*-null mice lack a parathyroid-thymus primordium, and in heterozygous *Gata3* knockout mice, it is smaller [81]. Although normocalcemic, when challenged with a low-calcium/low-vitamin D diet, the parathyroids do not enlarge or increase cellular proliferation rate in response to the induced hypocalcemia, resulting in an inadequate increase in plasma PTH [81], consistent with the observed hypocalcemia that occurs in HDR patients. The role of GATA3 in parathyroid development and function appears to be maintenance of the differentiation and survival of parathyroid and thymus progenitor cells, at least in part by the transcriptional regulation of *GCMB* (*Gcm2* in mouse) by GATA3 [81].

The hearing loss in *Gata3* heterozygous mutant mice is peripheral and is associated with cochlear abnormalities, which consist of a significant progressive morphological degeneration that starts with the outer hair cells at the apex and ultimately affects all the inner hair cells, pillar cells, and nerve fibers [82, 83]. These studies have shown that hearing loss in *Gata3* heterozygous mutant mice is detectable in the early postnatal period with outer hair cells of the cochlea showing early signs of cell degeneration in affected mice as young as 1–2 months and progressing through adulthood [82, 83].

19.7 GATA3 as a Determinant of Serum Calcium Levels

Apart from *CASR*, the genes associated with the determination of serum calcium concentration are largely unknown. The potential involvement of GATA3 was suggested by a recent genome-wide association study (GWAS) of 39,400 individuals which identified six loci, in addition to *CASR*, that are in association with serum calcium [84]. One of these, rs10491003, lies within a long noncoding RNA upstream of GATA3 and may influence the expression of GATA3. In the tibia, *CASR* was markedly upregulated in response to a low-calcium diet, as was *GATA3*. In parathyroid, alterations in GATA3 expression have been demonstrated to affect parathyroid cell proliferation in response to hypocalcemia and thus the maximum PTH secretion capacity of the gland [81]. GATA3 has been shown to regulate *GCMB* expression [81] that, in turn, may regulate expression of *CASR* in the parathyroid [85].

Conclusion

In conclusion, GATA3 haploinsufficiency causes the HDR syndrome. Mutations of GATA3 have not been found in a small, but significant, proportion of HDR patients, suggesting the involvement of other genes or distant enhancer elements in the etiology of the disease. The advent of next-generation sequencing of HDR patients will undoubtedly help to resolve this. Description of further HDR patients will allow further definition of the spectrum of developmental abnormalities associated with this syndrome and additional dissection of the structure-function relationships of the GATA3 transcription factor.

References

1. Barakat AY, D'Albora JB, Martin MM et al (1977) Familial nephrosis, nerve deafness, and hypoparathyroidism. J Pediatr 91:61–64
2. Shaw NJ, Haigh D, Lealmann GT et al (1991) Autosomal recessive hypoparathyroidism with renal insufficiency and developmental delay. Arch Dis Child 66:1191–1194
3. Bilous RW, Murty G, Parkinson DB et al (1992) Brief report: autosomal dominant familial hypoparathyroidism, sensorineural deafness, and renal dysplasia. N Engl J Med 327:1069–1074
4. Van Esch H, Groenen P, Nesbit MA et al (2000) GATA3 haplo-insufficiency causes human HDR syndrome. Nature 406:419–422
5. Elliott D, Thomas GH, Condron CJ et al (1970) C-group chromosome abnormality (? 10p-). Occurrence in a child with multiple malformations. Am J Dis Child 119:72–73
6. Lindstrand A, Malmgren H, Verri A et al (2010) Molecular and clinical characterization of patients with overlapping 10p deletions. Am J Med Genet A 152A:1233–1243
7. Dasouki M, Jurecic V, Phillips JA 3rd et al (1997) DiGeorge anomaly and chromosome 10p deletions: one or two loci? Am J Med Genet 73:72–75
8. Fujimoto S, Yokochi K, Morikawa H et al (1999) Recurrent cerebral infarctions and del(10)(p14p15.1) de novo in HDR (hypoparathyroidism, sensorineural deafness, renal dysplasia) syndrome. Am J Med Genet 86:427–429
9. Gottlieb S, Driscoll DA, Punnett HH et al (1998) Characterization of 10p deletions suggests two nonoverlapping regions contribute to the DiGeorge syndrome phenotype. Am J Hum Genet 62:495–498
10. Lichtner P, Konig R, Hasegawa T et al (2000) An HDR (hypoparathyroidism, deafness, renal dysplasia)

syndrome locus maps distal to the DiGeorge syndrome region on 10p13/14. J Med Genet 37:33–37
11. Labastie MC, Catala M, Gregoire JM et al (1995) The GATA-3 gene is expressed during human kidney embryogenesis. Kidney Int 47:1597–1603
12. Debacker C, Catala M, Labastie MC (1999) Embryonic expression of the human GATA-3 gene. Mech Dev 85:183–187
13. Al-Shibli A, Al AI, Willems PJ (2011) Novel DNA mutation in the GATA3 gene in an Emirati boy with HDR syndrome and hypomagnesemia. Pediatr Nephrol 26:1167–1170
14. Ali A, Christie PT, Grigorieva IV et al (2007) Functional characterization of GATA3 mutations causing the hypoparathyroidism-deafness-renal (HDR) dysplasia syndrome: insight into mechanisms of DNA binding by the GATA3 transcription factor. Hum Mol Genet 16:265–275
15. Ferraris S, Del Monaco AG, Garelli E et al (2009) HDR syndrome: a novel "de novo" mutation in GATA3 gene. Am J Med Genet A 149A:770–775
16. Adachi M, Tachibana K, Asakura Y et al (2006) A novel mutation in the GATA3 gene in a family with HDR syndrome (Hypoparathyroidism, sensorineural Deafness and Renal anomaly syndrome). J Pediatr Endocrinol Metab 19:87–92
17. van Looij MA, Meijers-Heijboer H, Beetz R et al (2006) Characteristics of hearing loss in HDR (hypoparathyroidism, sensorineural deafness, renal dysplasia) syndrome. Audiol Neurootol 11:373–379
18. Fukami M, Muroya K, Miyake T et al (2011) GATA3 abnormalities in six patients with HDR syndrome. Endocr J 58:117–121
19. Hernandez AM, Villamar M, Rosello L et al (2007) Novel mutation in the gene encoding the GATA3 transcription factor in a Spanish familial case of hypoparathyroidism, deafness, and renal dysplasia (HDR) syndrome with female genital tract malformations. Am J Med Genet A 143:757–762
20. Saito T, Fukumoto S, Ito N et al (2009) A novel mutation in the GATA3 gene of a Japanese patient with PTH-deficient hypoparathyroidism. J Bone Miner Metab 27:386–389
21. Nakamura A, Fujiwara F, Hasegawa Y et al (2011) Molecular analysis of the GATA3 gene in five Japanese patients with HDR syndrome. Endocr J 58:123–130
22. Chiu WY, Chen HW, Chao HW et al (2006) Identification of three novel mutations in the GATA3 gene responsible for familial hypoparathyroidism and deafness in the Chinese population. J Clin Endocrinol Metab 91:4587–4592
23. Kobayashi H, Kasahara M, Hino M et al (2006) A novel heterozygous deletion frameshift mutation of GATA3 in a Japanese kindred with the hypoparathyroidism, deafness and renal dysplasia syndrome. J Endocrinol Invest 29:851–853
24. Nanba K, Usui T, Nakamura M et al (2013) A novel GATA3 nonsense mutation in a newly diagnosed adult patient of hypoparathyroidism, deafness, and renal dysplasia (HDR) syndrome. Endocr Pract 19:e17–e20
25. Nesbit MA, Bowl MR, Harding B et al (2004) Characterization of GATA3 mutations in the hypoparathyroidism, deafness, and renal dysplasia (HDR) syndrome. J Biol Chem 279:22624–22634
26. Mino Y, Kuwahara T, Mannami T et al (2005) Identification of a novel insertion mutation in GATA3 with HDR syndrome. Clin Exp Nephrol 9:58–61
27. Gaynor KU, Grigorieva IV, Nesbit MA et al (2009) A missense GATA3 mutation, Thr272Ile, causes the hypoparathyroidism, deafness, and renal dysplasia syndrome. J Clin Endocrinol Metab 94:3897–3904
28. Muroya K, Hasegawa T, Ito Y et al (2001) GATA3 abnormalities and the phenotypic spectrum of HDR syndrome. J Med Genet 38:374–380
29. Zahirieh A, Nesbit MA, Ali A et al (2005) Functional analysis of a novel GATA3 mutation in a family with the hypoparathyroidism, deafness, and renal dysplasia syndrome. J Clin Endocrinol Metab 90:2445–2450
30. Chenouard A, Isidor B, Allain-Launay E et al (2013) Renal phenotypic variability in HDR syndrome: glomerular nephropathy as a novel finding. Eur J Pediatr 172:107–110
31. Ohta M, Eguchi-Ishimae M, Ohshima M et al (2011) Novel dominant-negative mutant of GATA3 in HDR syndrome. J Mol Med (Berl) 89:43–50
32. Moldovan O, Carvalho R, Jorge Z et al (2011) A new case of HDR syndrome with severe female genital tract malformation: comment on "Novel mutation in the gene encoding the GATA3 transcription factor in a Spanish familial case of hypoparathyroidism, deafness, and renal dysplasia (HDR) syndrome with female genital tract malformations" by Hernandez et al. Am J Med Genet A 155A:2329–2330
33. Sun Y, Xia W, Xing X et al (2009) Germinal mosaicism of GATA3 in a family with HDR syndrome. Am J Med Genet A 149A:776–778
34. Muroya K, Mochizuki T, Fukami M et al (2010) Diabetes mellitus in a Japanese girl with HDR syndrome and GATA3 mutation. Endocr J 57:171–174
35. Benetti E, Murer L, Bordugo A et al (2009) 10p12.1 deletion: HDR phenotype without DGS2 features. Exp Mol Pathol 86:74–76
36. Verri A, Maraschio P, Devriendt K et al (2004) Chromosome 10p deletion in a patient with hypoparathyroidism, severe mental retardation, autism and basal ganglia calcifications. Ann Genet 47:281–287
37. Melis D, Genesio R, Boemio P et al (2012) Clinical description of a patient carrying the smallest reported deletion involving 10p14 region. Am J Med Genet A 158A:832–835
38. Bernardini L, Sinibaldi L, Capalbo A et al (2009) HDR (Hypoparathyroidism, Deafness, Renal dysplasia) syndrome associated to GATA3 gene duplication. Clin Genet 76:117–119
39. Hayashi Y, Suwa T, Inuzuka T (2013) Intracranial calcification in a patient with HDR syndrome and a GATA3 mutation. Intern Med 52:161–162

40. Exome Variant Server, NHLBI Exome Sequencing Project (ESP), Seattle, Washington, USA, http://evs.gs.washington.edu/EVS/
41. Kuo CT, Morrisey EE, Anandappa R et al (1997) GATA4 transcription factor is required for ventral morphogenesis and heart tube formation. Genes Dev 11:1048–1060
42. Pandolfi PP, Roth ME, Karis A et al (1995) Targeted disruption of the GATA3 gene causes severe abnormalities in the nervous system and in fetal liver haematopoiesis. Nat Genet 11:40–44
43. Simon MC (1995) Gotta have GATA. Nat Genet 11:9–11
44. Weiss MJ, Orkin SH (1995) GATA transcription factors: key regulators of hematopoiesis. Exp Hematol 23:99–107
45. Molkentin JD (2000) The zinc finger-containing transcription factors GATA-4, -5, and -6. Ubiquitously expressed regulators of tissue-specific gene expression. J Biol Chem 275:38949–38952
46. Orkin SH (1992) GATA-binding transcription factors in hematopoietic cells. Blood 80:575–581
47. Pedone PV, Omichinski JG, Nony P et al (1997) The N-terminal fingers of chicken GATA-2 and GATA-3 are independent sequence-specific DNA binding domains. EMBO J 16:2874–2882
48. Svensson EC, Tufts RL, Polk CE et al (1999) Molecular cloning of FOG-2: a modulator of transcription factor GATA-4 in cardiomyocytes. Proc Natl Acad Sci U S A 96:956–961
49. Tevosian SG, Deconinck AE, Cantor AB et al (1999) FOG-2: A novel GATA-family cofactor related to multitype zinc-finger proteins Friend of GATA-1 and U-shaped. Proc Natl Acad Sci U S A 96:950–955
50. Chlon TM, Crispino JD (2012) Combinatorial regulation of tissue specification by GATA and FOG factors. Development 139:3905–3916
51. Chen Y, Bates DL, Dey R et al (2012) DNA binding by GATA transcription factor suggests mechanisms of DNA looping and long-range gene regulation. Cell Rep 2:1197–1206
52. DeLano WL(2007) MacPyMOL: a PyMOL-based molecular graphics application for MacOS X. DeLano Scientific LLC, Palo Alto, CA, USA
53. Yang Z, Gu L, Romeo PH et al (1994) Human GATA-3 trans-activation, DNA-binding, and nuclear localization activities are organized into distinct structural domains. Mol Cell Biol 14:2201–2212
54. Gaynor KU, Grigorieva IV, Allen MD et al (2013) GATA3 mutations found in breast cancers may be associated with aberrant nuclear localization, reduced transactivation and cell invasiveness. Horm Cancer 4:123–139
55. Liew CK, Simpson RJ, Kwan AH et al (2005) Zinc fingers as protein recognition motifs: structural basis for the GATA-1/friend of GATA interaction. Proc Natl Acad Sci U S A 102:583–588
56. Spilianakis CG, Lalioti MD, Town T et al (2005) Interchromosomal associations between alternatively expressed loci. Nature 435:637–645
57. Vakoc CR, Letting DL, Gheldof N et al (2005) Proximity among distant regulatory elements at the beta-globin locus requires GATA-1 and FOG-1. Mol Cell 17:453–462
58. Omichinski JG, Clore GM, Schaad O et al (1993) NMR structure of a specific DNA complex of Zn-containing DNA binding domain of GATA-1. Science 261:438–446
59. Crossley M, Merika M, Orkin SH (1995) Self-association of the erythroid transcription factor GATA-1 mediated by its zinc finger domains. Mol Cell Biol 15:2448–2456
60. Yang HY, Evans T (1995) Homotypic interactions of chicken GATA-1 can mediate transcriptional activation. Mol Cell Biol 15:1353–1363
61. Blokzijl A, ten Dijke P, Ibanez CF (2002) Physical and functional interaction between GATA-3 and Smad3 allows TGF-beta regulation of GATA target genes. Curr Biol 12:35–45
62. Merika M, Orkin SH (1995) Functional synergy and physical interactions of the erythroid transcription factor GATA-1 with the Kruppel family proteins Sp1 and EKLF. Mol Cell Biol 15:2437–2447
63. Osada H, Grutz G, Axelson H et al (1995) Association of erythroid transcription factors: complexes involving the LIM protein RBTN2 and the zinc-finger protein GATA1. Proc Natl Acad Sci U S A 92:9585–9589
64. Nakata Y, Brignier AC, Jin S et al (2010) c-Myb, Menin, GATA-3, and MLL form a dynamic transcription complex that plays a pivotal role in human T helper type 2 cell development. Blood 116:1280–1290
65. Hwang SS, Lee S, Lee W et al (2010) GATA-binding protein-3 regulates T helper type 2 cytokine and ifng loci through interaction with metastasis-associated protein 2. Immunology 131:50–58
66. Bonnard C, Strobl AC, Shboul M et al (2012) Mutations in IRX5 impair craniofacial development and germ cell migration via SDF1. Nat Genet 44:709–713
67. Tkocz D, Crawford NT, Buckley NE et al (2012) BRCA1 and GATA3 corepress FOXC1 to inhibit the pathogenesis of basal-like breast cancers. Oncogene 31:3667–3678
68. Uchida S, Matsumura Y, Rai T et al (1997) Regulation of aquaporin-2 gene transcription by GATA-3. off. Biochem Biophys Res Commun 232:65–68
69. Oxburgh L, Robertson EJ (2002) Dynamic regulation of Smad expression during mesenchyme to epithelium transition in the metanephric kidney. Mech Dev 112:207–211
70. Cohen HT, Bossone SA, Zhu G et al (1997) Sp1 is a critical regulator of the Wilms' tumor-1 gene. J Biol Chem 272:2901–2913
71. Alimov AP, Langub MC, Malluche HH et al (2003) Sp3/Sp1 in the parathyroid gland: identification of an Sp1 deoxyribonucleic acid element in the parathyroid hormone promoter. Endocrinology 144:3138–3147
72. Lemmens I, Van de Ven WJ, Kas K et al (1997) Identification of the multiple endocrine neoplasia type

1 (MEN1) gene. The European Consortium on MEN1. Hum Mol Genet 6:1177–1183
73. Lakshmanan G, Lieuw KH, Grosveld F et al (1998) Partial rescue of GATA-3 by yeast artificial chromosome transgenes. Dev Biol 204:451–463
74. Lakshmanan G, Lieuw KH, Lim KC et al (1999) Localization of distant urogenital system-, central nervous system-, and endocardium-specific transcriptional regulatory elements in the GATA-3 locus. Mol Cell Biol 19:1558–1568
75. Hasegawa SL, Moriguchi T, Rao A et al (2007) Dosage-dependent rescue of definitive nephrogenesis by a distant Gata3 enhancer. Dev Biol 301:568–577
76. Kato Y, Wada N, Numata A et al (2007) Case of hypoparathyroidism, deafness and renal dysplasia (HDR) syndrome associated with nephrocalcinosis and distal renal tubular acidosis. Int J Urol 14:440–442
77. Taslipinar A, Kebapcilar L, Kutlu M et al (2008) HDR syndrome (hypoparathyroidism, sensorineural deafness and renal disease) accompanied by renal tubular acidosis and endocrine abnormalities. Intern Med 47:1003–1007
78. Lim KC, Lakshmanan G, Crawford SE et al (2000) Gata3 loss leads to embryonic lethality due to noradrenaline deficiency of the sympathetic nervous system. Nat Genet 25:209–212
79. Grote D, Souabni A, Busslinger M et al (2006) Pax 2/8-regulated Gata 3 expression is necessary for morphogenesis and guidance of the nephric duct in the developing kidney. Development 133:53–61
80. Grote D, Boualia SK, Souabni A et al (2008) Gata3 acts downstream of beta-catenin signaling to prevent ectopic metanephric kidney induction. PLoS Genet 4:e1000316
81. Grigorieva IV, Mirczuk S, Gaynor KU et al (2010) Gata3-deficient mice develop parathyroid abnormalities due to dysregulation of the parathyroid-specific transcription factor Gcm2. J Clin Invest 120:2144–2155
82. van der Wees J, van Looij MA, de Ruiter MM et al (2004) Hearing loss following Gata3 haploinsufficiency is caused by cochlear disorder. Neurobiol Dis 16:169–178
83. van Looij MA, van der Burg H, van der Giessen RS et al (2005) GATA3 haploinsufficiency causes a rapid deterioration of distortion product otoacoustic emissions (DPOAEs) in mice. Neurobiol Dis 20:890–897
84. O'Seaghdha CM, Wu H, Yang Q et al (2013) Meta-analysis of genome-wide association studies identifies six new Loci for serum calcium concentrations. PLoS Genet 9:e1003796
85. Canaff L, Zhou X, Mosesova I et al (2009) Glial cells missing-2 (GCM2) transactivates the calcium-sensing receptor gene: effect of a dominant-negative GCM2 mutant associated with autosomal dominant hypoparathyroidism. Hum Mutat 30:85–92

20 Hypoparathyroidism, Dwarfism, Medullary Stenosis of Long Bones, and Eye Abnormalities (Kenny-Caffey Syndrome) and Hypoparathyroidism, Retardation, and Dysmorphism (Sanjad-Sakati) Syndrome

Eli Hershkovitz and Ruti Parvari

20.1 Introduction

In 1966, Kenny and Linarelli described a mother and son who had severe short stature, thin long bones with narrow diaphyses, and episodes of hypocalcemia [1]. In 1967, Caffey described the radiographic features of the same patients [2]. The condition has since been known as Kenny-Caffey syndrome (KCS [MIM127000]) (see also Chap. 15). The inheritance of this disorder is autosomal dominant, and it was recently recognized to be allelic (i.e., caused by mutations in the same gene) to a lethal disorder, osteocraniostenosis (OCS [MIM 602361]), characterized by gracile bones with thin diaphyses, premature closure of basal cranial sutures, and microphthalmia [3, 4]. Hypocalcemia due to hypoparathyroidism has been reported among patients with OCS who survived the perinatal period.

E. Hershkovitz
Pediatric Endocrinology Unit, Soroka Medical Center and Faculty of Health Sciences, Ben Gurion University of the Negev,
Ben Gurion, Beer Sheva 84105, Israel

R. Parvari (✉)
Shraga Segal Department of Microbiology, Immunology and Genetics, Faculty of Health Sciences, Ben Gurion University of the Negev and National Institute of Biotechnology Negev,
Ben Gurion, Beer Sheva 84105, Israel
e-mail: ruthi@bgu.ac.il

The syndrome of hypoparathyroidism, retardation (of growth and mental development) with dysmorphic features, HRD syndrome, also known as Sanjad-Sakati syndrome has been described by Sanjad and Sakati in 1988 in an abstract followed by a detailed report 3 years later [5]. This syndrome has been described mostly in Arab patients and is inherited by the autosomal recessive mode. HRD syndrome shares several important clinical features with KCS, a fact that has caused some confusion in the literature. This syndrome has been classified by some authors as autosomal recessive KCS or KCS type 1 in contrast to KCS type 2, the autosomal dominant form.

The identification of the causative mutations for KCS/OCS and the HRD syndrome has clearly confirmed that KCS/OCS and the HRD syndrome are separate clinical and genetic disorders. KCS/OCS is caused by heterozygous mutations in the *FAM111A* gene, while HRD syndrome is caused by homozygous or compound heterozygous mutations in the *TBCE* gene as will be subsequently described.

The following clinical description of the clinical picture of KCS/OCS in this chapter will rely upon genetically diagnosed patients or sporadic patients from non-consanguineous, non-Arab families. Most of the currently described HRD patients were of Arab origin. Many were born to consanguineous families and carry a common single homozygous mutation in the TBCE gene.

M.L. Brandi, E.M. Brown (eds.), *Hypoparathyroidism*,
DOI 10.1007/978-88-470-5376-2_20, © Springer-Verlag Italia 2015

20.2 HRD/Sanjad-Sakati Syndrome

20.2.1 Epidemiology

Most genetically diagnosed HRD patients have been of Middle Eastern (Arab) origin. In Saudi Arabia, estimated incidence varies from 1:40,000 to 1:100,000 live births. In Kuwait, the estimated incidence of the syndrome is 7–18 per 100,000 live births [6]. Based on the number of new cases and the total live births over the past 10 years, we estimated the incidence of HRD syndrome at 1 per 10,000 live births among the Bedouin in southern Israel (unpublished data).

20.2.2 Clinical Phenotype

The early literature on the HRD syndrome has been reviewed previously [7].

Growth retardation is seen in most of the patients. Both prenatal and postnatal growths are impaired [6–9]. In a recent study, all of the reported children suffered from intrauterine growth restriction (IUGR) with a resultant low birth weight and short birth length. Mean birth weight was 2,100 ± 200 g (−2.2 ± 0.25 SDS) in boys and 1970 ± 450 g (−2.6 ± 0.7 SDS) in girls. Mean birth length was 44.7 ± 3.3 cm (−5.1 ± 1.27 SDS) in boys and 44.6 ± 2.75 cm (−4.7 ± 1.7 SDS) in girls. Analysis of growth in those patients by the infancy-childhood-puberty (ICP) growth model revealed that during the first year of life, linear growth followed a path of growth that, although very short, coincided with the first component (I) of the ICP model. However, further decrease in linear growth was observed during the second year of life. Growth analysis of the path of growth by the ICP model disclosed a markedly delayed appearance of the childhood component that normally occurs between 6 and 12 months of age, when the infancy component markedly decelerates. In HRD patients, the appearance of the C component occurred at the age of 17.6 ± 5.6 months in boys and 19.7 ± 6 months in girls. The latest available growth measurements, expressed as weight and height SDs, in boys were −13.1 ± 3.8 and −8.7 ± 1, respectively. In girls, the latest available weight and height SDs were −16.6 ± 4.4 and −9.5 ± 2.4, respectively. BMI SDs or weight for length SDSs (in patients younger than 3 years) was below −2 in almost all the patients [10].

Global developmental delay is a universal feature of the syndrome. Although, many patients have moderate to severe mental retardation, some had mild to moderate mental retardation [7–9]. Speech skills have been reported as variable. Some patients' speech improved after attending speech therapy [9]. Several characteristic dysmorphic features have been described in patients with HRD syndrome (Fig. 20.1). Microcephaly, deep-set eyes, external ear malformations, depressed nasal bridge, thin upper lip, hooked small nose, micrognathia, and small hands and feet are consistent features of the syndrome. Prominent forehead, microphthalmia, and long philtrum have been reported as well [7–9]. Cryptorchidism and micropenis have been reported in some of the male patients. No fertility has been reported in patients with HRD. Dental abnormalities include microdontia and oligodontia, delayed teeth eruption, enamel hypoplasia, and severely carious teeth [7, 11, 12].

HRD patients display a variety of ocular findings, including microphthalmia, microcornea, keratitis, errors of refraction, strabismus, and retinal vascular tortuosity [13, 14]. Seizures due to hypocalcemia may appear as early as in the neonatal period and are a common feature of the syndrome [7, 9]. Significant neurological disabilities are rare [9].

The patients are susceptible to severe infections including life-threatening pneumococcal infections especially during infancy. The syndrome carries a high risk for mortality with a reported rate ranging between 25 and 55 % during infancy and early childhood. Recurrent infections and hypocalcemic seizures are the main reported causes of death in some infants [7, 10]. Chronic intestinal pseudo-obstruction has been also implicated as a cause of mortality in a child with the HRD syndrome [15]. Patients have been described as late as in their third decade of life.

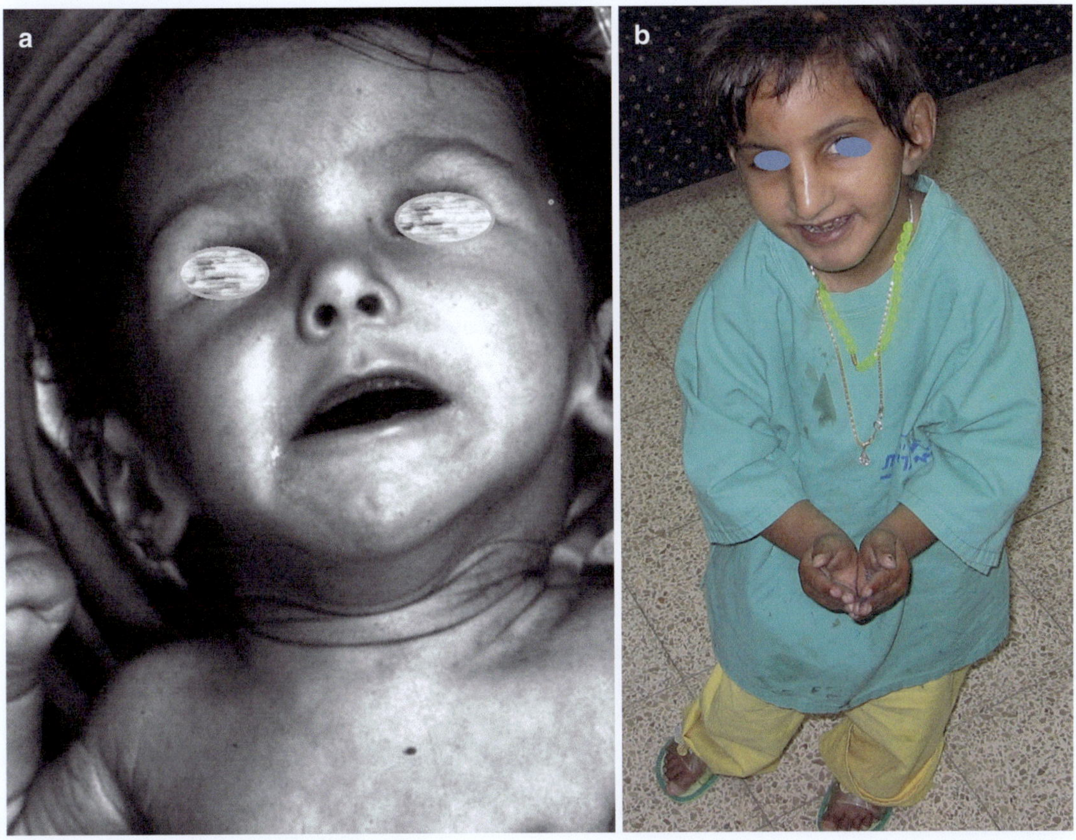

Fig. 20.1 HRD syndrome. (**a**) Facial dysmorphism. Note prominent forehead, small deep-set eyes, and depressed nasal bridge (Reproduced with permission from Hershkovitz et al. [7]) (**b**) A 14 years old patient with severe growth retardation and lack of pubertal signs

20.2.3 Biochemical and Radiological Findings

Hypocalcemia and hyperphosphatemia due to congenital permanent hypoparathyroidism are the hallmarks of HRD syndrome. Serum parathyroid hormone (PTH) levels are undetectable to very low in most of the patients [5, 6, 8, 10, 16–20]. Surprisingly, high PTH levels have been reported in two patients [17, 18]. Postmortem examinations of HRD patients have been seldom reported, but the absence of the parathyroid glands had been documented in one of the author's patient. Increased liver transaminases have been found in some patients without progression to chronic liver disease [17].

Partial growth hormone deficiency (GH<10 ng/ml) has been found following stimulation tests in several patients [10, 17, 21–23]. Low serum IGF-I concentrations were found in all patients investigated in two studies [10, 22]. Normal immunoglobulin levels were found in all but one of the patients tested [7]. Normal T-cell responses to mitogens were observed in about ten patients studied [5, 16, 23], but reduced numbers of all T-cells subclasses were found in five patients reported by Richardson and Kirk 1990 [17]. Chemotactic migration, random migration, and phagocytosis of PMN from HRD patients were significantly lower than in PMN from healthy controls. Functional hyposplenism has been demonstrated in most of the studied patients [10].

Delayed bone age and osteopenia are common findings [17, 23], while medullary stenosis of the long bones, a common finding in the KCS syndrome, is infrequently observed in HRD patients

[8, 9]. Cranial MRI showed severe hypoplasia of the anterior pituitary and corpus callosum, with decreased white matter bulk in one study [22].

20.2.4 Diagnosis

The clinical signs of severe growth and mental retardation, typical dysmorphism, and congenital hypoparathyroidism are highly suggestive of the HRD syndrome, especially in Arab patients. The HRD syndrome should be differentiated from the KCS (see Table 20.1), the CHARGE association (coloboma, heart anomaly, choanal atresia, retardation, genital and ear anomalies) and DiGeorge's syndrome.

20.2.5 Therapy

Early recognition and therapy of hypocalcemia is important. Humanized milk formulas containing low phosphorous, supplements with calcium salts, and administration of vitamin D analogs are effective in keeping serum calcium at the required low normal range. Hypercalciuria, which may cause nephrocalcinosis, should be treated by thiazides. Recombinant human PTH(1–34) may offer an advantage in the treatment of these patients, but it has not been tried on HRD patients. Since the recognition of the susceptibility of patients with HRD to serious bacterial infections, we recommend on daily prophylactic antibiotic therapy and prudent vaccination against pneumococci.

Table 20.1 Comparison between HRD/Sanjad-Sakati and Kenny-Caffey syndromes

Feature	HRD/Sanjad-Sakati syndrome	Kenny-Caffey syndrome
Mode of inheritance	Autosomal recessive	Autosomal dominant
Ethnicity	Mostly Arabs	Pan ethnic
Growth	IUGR in most patients	IUGR in less than 50 %
	Extreme short stature	Short stature
Mental development	Mental retardation	Normal mentality in most of the patients
Dysmorphic features	Microcephaly	Relative macrocephaly
	Deep-set eyes	Delayed closure of the anterior fontanel
	Micrognathia	Deep-set eyes
	Dental anomalies	Micrognathia
	Small hands and feet	Dental anomalies
Hypoparathyroidism	Universal	Common with variable age of onset
	Neonatal or early infantile onset	
Other clinical complications	Susceptibility to severe bacterial infections	None in most patients
	High mortality	Fertility has been reported in females only.
	Probably infertile	
Laboratory findings	Hyposplenism	Hypomagnesemia
	Impaired PMN chemotaxis and phagocytosis	Impaired T-cell function
	Normal cell-mediated immunity	
	Growth hormone deficiency	
Radiologic findings	Severe hypoplasia of anterior pituitary and corpus callosum.	Medullary stenosis and cortical thickening of the long bones
Molecular abnormality	Homozygous mutation in the tubulin-specific chaperone E gene (TBCE)	Heterozygous mutation in the FAM111A gene

20.3 KCS/OCS

KCS is a rare disorder, and sporadic cases have been reported from various parts of the world in different ethnic groups [1, 24–30].

20.3.1 Clinical Phenotype

Larsen, reviewing 20 previously described patients, has reported short stature in most of them. The attained adult height was between 121 and 155 cm, in contrast to the extreme dwarfism observed in adult HRD patients (around 100 cm). Intrauterine growth restriction was observed in less than half of the patients unlike the 95–97 % presence of IUGR in HRD syndrome [26].

Most of the patients exhibited delayed closure of the anterior fontanel [26, 30]. The patients had typical facial appearance, including prominent forehead, deep-set eyes, beaked nose, thin upper lip, micrognathia, and external ear abnormalities [26, 28–30]. Microcephaly is a common feature in the HRD syndrome, while normal head circumference or even macrocephaly characterizes patients with KCS [12, 27, 28].

Most reported KCS patients had normal mental development [12, 26, 29, 30]. Ophthalmic abnormalities are common in patients with KCS. The various findings include reduced visual acuity; hyperopia; myopia; small corneal diameter; elevated, blurred margin of the optic nerve; tortuous blood vessels; glaucoma; and strabismus [12, 26, 29, 30]. Defective dentition, accompanied by oligodontia and severe carries is very common in the KCS [12, 26, 29].

Mother-to-son transmission was reported in the original description of the disorder [1] and some other women with KCS were reported to have unaffected children [26]. No cases of paternity were reported among males with the KCS. Micropenis, hypospadias, and small testes have been reported in some patients [26, 28, 29]. Except for hypocalcemia most of the patients with KCS do not suffer from life-threatening complications.

20.3.2 Biochemical and Radiological Findings

Hypoparathyroidism is often found among KCS patients but is not a universal phenomenon as it is in the HRD syndrome. The age of onset of hypocalcemia is variable, ranging from the neonatal period to adulthood [12, 26]. Interestingly, some patients have hypomagnesemia [30]. Absence of the parathyroid glands has been reported in a patient with the KCS [31]. Medullary stenosis with cortical thickening of the long bones is a hallmark feature of KCS, while infrequently described in HRD patients [2, 26, 27, 29, 30]. Delayed bone age is a common finding [26].

20.3.3 Osteocraniostenosis (OCS)

Osteocraniostenosis was delineated by Verloes et al. [3]. It has been recently recognized as allelic to the KCS. Osteocraniostenosis was lethal in the few reported cases in the neonatal period, but survival to 21 months of age has been reported [29, 32]. It is characterized by IUGR; spleen hypoplasia or aplasia; a striking bone dysplasia consisting of thin ribs and long, thin, straight, or curved tubular diaphyses, flared metaphyses, hypoplastic distal phalanges, and drumstick metacarpals; marked cranial hypomineralization, leading to wide fontanels and cloverleaf head shape; and intrauterine fractures [3, 4]. The facial appearance was variable, from mild anomaly to a striking combination of midface hypoplasia, short, upturned nose, short philtrum, and inverted V-shaped mouth. Hypocalcemia and hypoparathyroidism are recognized in surviving neonates [29].

20.4 Molecular Genetics and Pathogenesis of HRD and KCS/OCS Syndromes

20.4.1 HRD

20.4.1.1 Tubulin Folding and Assembly

Microtubules are polymerized from α/β-tubulin heterodimers [33]. Newly synthesized α-and β-tubulin polypeptides undergo a sequence of folding steps catalyzed by chaperones. Initially, the tubulins are associated with the hexameric prefoldin complex that passes them on to the cytosolic chaperonin complex [34], and they are then further processed by tubulin-folding cofactors (the standard now seems to be TBCA-E or just cofactor) [35, 36]. In vitro folding assays suggest that in mammals α-tubulin binds to cofactors B and E, whereas β-tubulin binds to cofactors A and D. α-tubulin/TBCE and β-tubulin/TBCD are bound by TBCC, forming a super complex from which α/β-tubulin heterodimers are released by GTP hydrolysis of β-tubulin. The small G-protein Arl2 appears to play a regulatory role, binding to and sequestering cofactor D [37]. Budding yeast mutants lacking tubulin cofactor homologs have only conditional effects and are normally not lethal ([38] and references therein). In contrast, null mutations in fission yeast *TFC* genes cause abnormal cell shapes and mostly result in lethality [39–43]. Genetic analysis of tubulin cofactor function in fission yeast has led to a different model of tubulin folding: an essential pathway of α-tubulin folding involves, successively, cofactors B, E, and D, with the Arl2 homolog acting upstream of D, whereas a nonessential pathway of β-tubulin folding involves cofactor A passing β-tubulin on to cofactor D to associate with α-tubulin [40–42]. Results in the plant *Arabidopsis* suggest that cofactors C–E and Arl2 are stringently required for microtubule formation, similar to the requirements for in vitro assays using purified mammalian cofactors [44]. *PFI*, the ortholog of the vertebrate *Tbce* in *Arabidopsis*, is necessary for continuous microtubule organization, mitotic division, and cytokinesis but do not mediate cell cycle progression [45, 46]. Vesicle trafficking to the division plane during cytokinesis but not to the cell surface during interphase was impaired [45].

Coincident with the discovery that mutations in *TBCE* cause recessive HRD, a missense mutation in murine *Tbce* (W527G) inherited in homozygosity was described in a mouse model of peripheral motor neuropathy, *pmn* [47, 48]. The W527G mutation destabilized the chaperone, resulting in diminished protein levels [48]. The original *pmn* mice had low birth weight, decreased brain size, and hypogonadism, reminiscent of the human trait, but no hypoparathyroidism was noted and no report of continued low weight or size [49]. Since mice that lack parathyroid glands have PTH serum levels identical to those of wild-type mice, as do parathyroidectomized wild-type animals, are viable and fertile and have only a mildly abnormal bone phenotype [50], it is possible that the parathyroid defect has been overlooked. Although many embryonic cell lines enabling creation of a mouse in which *TBCE* is deleted are available [51], there are no reports of such a mouse model. In agreement with no existence of a mouse null for *TBCE Drosophila, tbce* nulls are embryonic lethal, requiring tissue-specific knockdown for the study of the effects of absence of TBCE. Tissue-specific knockdown and overexpression of *tbce* in neuromusculature resulted in disrupted and increased microtubules, respectively. Alterations in TBCE expression also affected neuromuscular synapses [52]. No other phenotype was observed in *Drosophila*.

As in mice and *Drosophila*, complete absence of TBCE function was not reported in human. The common homozygous mutation: a deletion of four amino acid deletion (del52-55) leaves tubulin GAP-enhancing activities (unpublished results), while the cryptic out-of-frame translational initiation caused by a heterozygous mutation of the TBCE gene, rescues tubulin formation in a compound heterozygous HRD patient carrying a second nonsense mutation [53].

Our studies on the effect of a homozygous four amino acid deletion of TBCE(del52-55) in patients' cells demonstrated that lymphoblastoid cells showed aberrant microtubule polarity and the microtubules arrays are not centered on centrosomes in disease cells. This effect was more pronounced in fibroblasts than in keratinocytes. Thus, the cellular phenotype may be tissue specific despite ubiquitous transcriptional *TBCE* expression. Organization of the Golgi complex in patients' fibroblasts was diffuse and surrounded the nucleus, in contrast to its compact and localization near one side of the nucleus in control cells. The distribution of late endosomes which, like the Golgi complex, is microtubule dependent revealed an abnormally diffuse pattern in the whole cell in contrast to the predominantly perinuclear pattern observed in control cells [54].

The specific absence of parathyroid glands, with accompanying normal development of the thyroid and other branchial pouch derivatives, is an intriguing and unexpected aspect of a derangement in tubulin physiology [55]. The interplay between the known factors involved in the parathyroid development and the TBCE and/or microtubule cytoskeleton remains to be elucidated by future studies.

20.4.2 KCS/OC

The mutation causing the dominant form of the KCS2 was recently identified by the power of exome sequencing. Interestingly, the same missense mutation, R569H in the gene *FAM111A* (NM_001142519.1), occurring de novo was detected in heterozygosity in five patients studied by a Swiss group [29] and in four Japanese patients [30]. *FAM111A* encodes a previously uncharacterized protein consisting of 611 amino acids. The carboxy-terminal half of the protein has homology to trypsin-like peptidases, and the catalytic triad specific to such peptidases is conserved [56], but its possible proteolytic activity was not studied. Similarly to *TBCE*, the transcriptional expression of *FAM111A* is ubiquitous according to the human protein atlas [57]. It is expressed in the parathyroid gland and bone, but the expression levels are similar to those in other tissues. A recent report showed that FAM111A functions as a host range restriction factor and is required for viral replication and gene expression by specifically interacting with Simian Virus 40 large T antigen (LT) [56]. In addition, *FAM111A* mRNA and protein levels have been shown to be regulated in a cell cycle-dependent manner with the lowest expression during the G0 or quiescent phase and peak expression during the G2/M phase [56]. Another recent report revealed that variants in the region including *FAM111A* and *FAM111B* were associated with prostate cancer [58]. However, the clinical course of disease in KCS2 patients revealed neither increased viral infections nor carcinogenesis up to early adulthood. Again, in similarity to *TBCE*, the de novo mutation (R569H) would not significantly affect the function of FAM111A as suggested by in silico analyses. Additionally, the mutant *FAM111A* mRNA was expressed similarly to the wild type in peripheral blood cells. This raises the question of how this mutation causes KCS2. One hypothesis is that this mutation does not cause loss of function of the protein but rather modulates its peptidase activity for a particular target peptide in a mutant-specific way. Another possibility is that FAM111A functions with some physiological partner(s) and the disease occurs as a result of specific modulation of this putative network. In agreement with the suggestion that the amino acid changed by the mutation interacts with other partners is exposed on the protein surface as indicated by molecular modeling. Since other LT-interacting proteins, such as RB, p53, FBXW7, and CDC73, are involved in gene transcription and are bona fide tumor suppressors, FAM111A is localized in the nucleus and its expression is cell-cycle dependent; it was suggested that FAM111A might be involved in the regulation of gene transcription [29]. KCS1 and KCS2 share distinctive phenotypic features.

FAM111A is important for skeletal development, the dysmorphic features, and primary hypoparathyroidism but not for intrauterine growth and mental development.

The autosomal recessive Kenny-Caffey syndrome [59] (AR-KCS; MIM244460), HRD [16] (MIM241410), or Sanjad-Sakati syndrome (SSS) [5] is caused by mutations in the tubulin-specific chaperone E gene, *TBCE* (Fig. 20.2) [54]. Presently, the only known function for TBCE is to serve as a chaperone of α-tubulin.

Conclusion

The partly phenotypic overlap between KCS and HRD syndromes might indicate a functional relationship between FAM111A and TBCE. It is tempting to speculate that the two proteins might be interlinked in a common regulatory pathway. It could also be that one of the candidate partner proteins of FAM111A

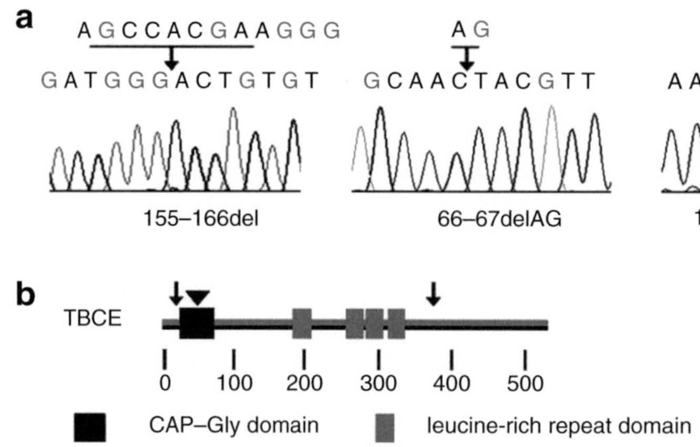

Fig. 20.2 *TBCE* mutation analysis. (**a**) Sequence traces from affected individuals. All affected Middle Eastern subjects evaluated were homozygous with respect to the 155–166del mutation, and a Belgian individual with HRD was compoundly heterozygous with respect to 66–67delAG and 1113 T→A (mutation positions are relative to +1 at the initiation ATG). The mutated sequences are shown above the electropherograms, with the positions of deletion or point mutations indicated by arrows and wild-type sequences given above the arrows. (**b**) Mutations in TBCE relative to the CAP-Gly and leucine-rich repeat domains. Positions of truncating mutations (Val23fs48X, Cys371X) are indicated by arrows, and the in-frame deletion (del52–55) is indicated by an arrowhead. Approximate amino acid positions are indicated below the cartoon (Reproduced with permission from Hershkovitz et al. [7])

is TBCE. The identification of mutations in these two genes as causative of KCS and HRD syndromes provides novel tools for the study of the pathophysiological mechanisms of these pathologies.

References

1. Kenny FM, Linarelli L (1966) Dwarfism and cortical thickening of tubular bones: transient hypocalcemia in a mother and son. Am J Dis Child 111:201–207
2. Caffey JP (1967) Congenital stenosis of medullary spaces in tubular bones and calvaria in two proportionate dwarfs, mother and son, coupled with transitory hypocalcemic tetany. Am J Roentgenol Radium Ther Nucl Med 100:1–11
3. Verloes A, Narcy F, Grattagliano B et al (1994) Osteocraniostenosis. J Med Genet 31:772–778
4. Elliott AM, Wilcox WR, Spear GS et al (2006) Osteocraniostenosis–hypomineralized skull with gracile long bones and splenic hypoplasia. Four new cases with distinctive chondro-osseous morphology. Am J Med Genet 140A:1553–1563
5. Sanjad SA, Sakati NA, Abu Obsa YK et al (1991) A new syndrome of congenital hypoparathyroidism severe growth failure and dysmorphic features. Arch Dis Child 66:193–196
6. Alawadi SA, Azab AS, Bastaki L et al (2009) Sanjad-Sakati syndrome/Kenny-Caffey syndrome type 1: a study of 21 cases in Kuwait. East Mediterr Health J 15:345–349
7. Hershkovitz E, Parvari R, Diaz GA et al (2004) Hypoparathyroidism, retardation and dysmorphism (HRD) syndrome – a review. J Pediatr Endocrinol Metab 17:1583–1590
8. Albaramki J, Akl K, Al-Muhtaseb A et al (2012) Sanjad Sakati syndrome: a case series from Jordan. East Mediterr Health J 18:527–531
9. Elhassanien AF, Alghaiaty HAA (2013) Neurological manifestations in children with Sanjad–Sakati syndrome. Int J Gen Med 6:393–398
10. Hershkovitz E, Rozin I, Limony Y et al (2007) Hypoparathyroidism, retardation, and dysmorphism syndrome: impaired early growth and increased susceptibility to severe infections due to hyposplenism and impaired polymorphonuclear cell functions. Pediatr Res 62:505–509
11. Al Malik MI (2004) The dentofacial features of Sanjad–Sakati syndrome: a case report. Int J Paediatr Dent 14:136–140
12. Moussaid Y, Griffiths D, Richard B et al (2012) Oral manifestations of patients with Kennye-Caffey syndrome. Eur J Med Genet 55:441–445
13. Al Dhoyan N, Al Hemidan AI, Ozand PT (2006) Ophthalmic manifestations of Sanjad-Sakati syndrome. Ophthalmic Genet 27:83–87
14. Khan AO, Al-Assiri A, Al-Mesfer S (2007) Ophthalmic features of hypoparathyroidism-retardation-dysmorphism. J AAPOS 11:288–290
15. Pal K, Moammar H, Mitra DK (2010) Visceral myopathy causing chronic intestinal pseudoobstruction and intestinal failure in a child with Sanjad-Sakati syndrome. J Pediatr Surg 45:430–434

16. Hershkovitz E, Shalitin S, Levy J et al (1995) The new syndrome of congenital hypoparathyroidism, growth retardation, and developmental delay. A report of six patients. Isr J Med Sci 31:293–297
17. Richardson RJ, Kirk JMW (1990) Short stature, mental retardation and hypoparathyroidism: a new syndrome. Arch Dis Child 65:1113–1117
18. Khan KTS, Uma R, Usha R et al (1997) Kenny–Caffey syndrome in six Bedouin sibships: autosomal recessive inheritance is confirmed. Am J Med Genet 69:126–132
19. Kamalesh P (2010) Sanjad – Sakati syndrome in a neonate. Indian Pediatr 47:443–444
20. Rafique B, Al-Yaarubi S (2010) Sanjad-Sakati syndrome in Omani children. Oman Med J 25:227–229
21. Marsden D, Nyhan WL, Sakati NO (1994) Syndrome of hypoparathyroidism with growth hormone deficiency and multiple minor anomalies. Am J Med Genet 52:334–338
22. Padidela R, Kelberman D, Press M et al (2009) Mutation in the TBCE gene is associated with hypoparathyroidism-retardation-dysmorphism syndrome featuring pituitary hormone deficiencies and hypoplasia of the anterior pituitary and the corpus callosum. J Clin Endocrinol Metab 94:2686–2691
23. Sabry MA, Farag TI, Shaltout AA et al (1999) Kenny-Caffey syndrome: an Arab variant? Clin Genet 55:44–49
24. Frech RS, McAlister WH (1968) Medullary stenosis of the tubular bones associated with hypocalcemic convulsions and short stature. Radiology 91:457–461
25. Majewski F, Rosendahl W, Ranke M et al (1981) The Kenny syndrome, a rare type of growth deficiency with tubular stenosis, transient hypoparathyroidism and anomalies of refraction. Eur J Pediatr 136:21–30
26. Larsen JL, Kivlin J, Odell WD (1985) Unusual cause of short stature. Am J Med 78:1025–1032
27. Bergada I, Schiffrin A, Abu Srair H et al (1988) Kenny syndrome: description of additional abnormalities and molecular studies. Hum Genet 80:39–42
28. Hoffman WH, Kovacs K et al (1998) Kenny-Caffey syndrome and microorchidism. Am J Med Genet 80:107–111
29. Unger S, Gorna MW, Le Béchec A et al (2013) FAM111A mutations result in hypoparathyroidism and impaired skeletal development. Am J Hum Genet 92:990–995
30. Isojima T, Doi K, Mitsui J et al (2014) A recurrent *de novo FAM111A* mutation causes Kenny-Caffey syndrome type 2. J Bone Miner Res 29:992–998
31. Boynton JR, Pheasant TR, Johnson BL et al (1979) Ocular findings in Kenny's syndrome. Arch Ophthalmol 97:896–900
32. Verloes A, Garel C, Robertson S et al (2005) Gracile bones, periostal appositions, hypomineralization of the cranial vault, and mental retardation in brothers: milder variant of osteocraniostenosis or new syndrome? Am J Med Genet A 137:199–203
33. Nogales E (2000) Structural insight into microtubule function. Annu Rev Biochem 69:277–302
34. Leroux MR, Hartl FU (2000) Protein folding: versatility of the cytosolic chaperonin TRiC/CCT. Curr Biol 10:R260–R264
35. Tian G, Huang Y, Rommelaere H et al (1996) Pathway leading to correctly folded beta-tubulin. Cell 86:287–296
36. Tian G, Lewis SA, Feierbach B et al (1997) Tubulin subunits exist in an activated conformational state generated and maintained by protein cofactors. J Cell Biol 138:821–832
37. Bhamidipati A, Lewis SA, Cowan NJ (2000) ADP ribosylation factor-like protein 2 (Arl2) regulates the interaction of tubulin-folding cofactor D with native tubulin. J Cell Biol 149:1087–1096
38. Fleming JA, Vega LR, Solomon F (2000) Function of tubulin binding proteins in vivo. Genetics 156:69–80
39. Radcliffe PA, Toda T (2000) Characterisation of fission yeast alp11 mutants defines three functional domains within tubulin-folding cofactor B. Mol Gen Genet 263:752–760
40. Radcliffe PA, Garcia MA, Toda T (2000) The cofactor-dependent pathways for alpha- and beta-tubulins in microtubule biogenesis are functionally different in fission yeast. Genetics 156:93–103
41. Radcliffe PA, Vardy L, Toda T (2000) A conserved small GTP-binding protein Alp41 is essential for the cofactor-dependent biogenesis of microtubules in fission yeast. FEBS Lett 468:84–88
42. Radcliffe PA, Hirata D, Vardy L, Toda T (1999) Functional dissection and hierarchy of tubulin-folding cofactor homologues in fission yeast. Mol Biol Cell 10:2987–3001
43. Grishchuk EL, McIntosh JR (1999) Sto1p, a fission yeast protein similar to tubulin folding cofactor E, plays an essential role in mitotic microtubule assembly. J Cell Sci 112:1979–1988
44. Tian G, Bhamidipati A, Cowan NJ, Lewis SA (1999) Tubulin folding cofactors as GTPase-activating proteins. GTP hydrolysis and the assembly of the /ß-tubulin heterodimer. J Biol Chem 274:24054–24058
45. Steinborn K, Maulbetsch C, Priester B et al (2002) The *Arabidopsis* PILZ group genes encode tubulin-folding cofactor orthologs required for cell division but not cell growth. Genes Dev 16:959–971
46. Mayer U, Herzog U, Berger F et al (1999) Mutations in the pilz group genes disrupt the microtubule cytoskeleton and uncouple cell cycle progression from cell division in Arabidopsis embryo and endosperm. Eur J Cell Biol 78:100–108
47. Bommel H, Xie G, Rossoll W et al (2002) Missense mutation in the tubulin-specific chaperone E (Tbce) gene in the mouse mutant progressive motor neuronopathy, a model of human motoneuron disease. J Cell Biol 159:563–569
48. Martin N, Jaubert J, Gounon P et al (2002) A missense mutation in Tbce causes progressive motor neuronopathy in mice. Nat Genet 32:443–447
49. Schmalbruch H, Jensen HJ, Bjaerg M et al (1991) A new mouse mutant with progressive motor neuronopathy. J Neuropathol Exp Neurol 50:192–204

50. Gunther T, Chen ZF, Kim J et al (2000) Genetic ablation of parathyroid glands reveals another source of parathyroid hormone. Nature 406:199–203
51. http://www.informatics.jax.org/searches/allele_report.cgi?_Marker_key=54410
52. Jin S, Pan L, Liu Z et al (2009) Drosophila Tubulin-specific chaperone E functions at neuromuscular synapses and is required for microtubule network formation. Development 136:1571–1581
53. Tian G, Huang MC, Parvari R et al (2006) Cryptic out-of-frame translational initiation of TBCE rescues tubulin formation in compound heterozygous HRD. Proc Natl Acad Sci U S A 103:13491–13496
54. Parvari R, Hershkovitz E, Grossman N et al (2002) Mutation of TBCE causes hypoparathyroidism-retardation-dysmorphism and autosomal recessive Kenny-Caffey syndrome. Nat Genet 32:448–452
55. Parvari R, Diaz GA, Hershkovitz E (2007) Parathyroid development and the role of tubulin chaperone E. Horm Res 67:12–21
56. Fine DA, Rozenblatt-Rosen O, Padi M et al (2012) Identification of FAM111A as an SV40 host range restriction and adenovirus helper factor. PLoS Pathog 8(10):e1002949
57. http://www.proteinatlas.org/ENSG00000166801/normal
58. Akamatsu S, Takata R, Haiman CA et al (2012) Common variants at 11q12, 10q26 and 3p11.2 are associated with prostate cancer susceptibility in Japanese. Nat Genet 44:426–429
59. Tahseen K, Khan S, Uma R et al (1997) Kenny-Caffey syndrome in six Bedouin sibships: autosomal recessive inheritance is confirmed. Am J Med Genet 69:126–132

Hypoparathyroidism in Mitochondrial Disorders

21

Daniele Orsucci, Gabriele Siciliano, and Michelangelo Mancuso

21.1 Introduction

The fundamental task of the mitochondrion is the generation of energy as ATP (adenosine triphosphate), by means of the respiratory chain or electron transport chain. This pathway is under control of both the nuclear and mitochondrial (mtDNA) genomes [1]. Mitochondrial diseases are a group of disorders caused by impairment of the respiratory chain. The genetic classification of these diseases distinguishes the disorders due to defects in mtDNA from those due to defects in nuclear DNA [2] (Table 21.1). The estimated prevalence of mitochondrial disorders is 1–2 in 10,000 [3]. They are, therefore, one of the most common genetic metabolic diseases.

The clinical features may be multisystem, with possible involvement of visual and auditory pathways, heart, central nervous system, and skeletal muscle. The "red flags" are myopathy with exercise intolerance, eyelid ptosis, ophthalmoparesis, axonal multifocal neuropathy, sensorineural hearing loss, pigmentary retinopathy, optic neuropathy, diabetes mellitus, hypertrophic cardiomyopathy, migraine-like headache, and short stature [4] (Table 21.2). Furthermore, even if only encountered rarely, mitochondrial diseases may present with hypoparathyroidism (see also Chap. 15).

Because of the frequent multisystem involvement caused by mitochondrial disorders, a wide range of medical specialists (including endocrinologists, pediatricians, internists, general practitioners, cardiologists, audiologists, ophthalmologists, neurologists) may first encounter these patients; therefore, an acute "clinical awareness" about this diagnosis is essential in order to initiate a correct diagnostic workup.

21.2 Mitochondrial Genetics

Mitochondrial diseases related to abnormalities in nuclear DNA are inherited according to the Mendelian rules. They are caused by mutations in structural components or ancillary proteins of the respiratory chain, by defects of the intergenomic signaling (associated to mitochondrial DNA (mtDNA) depletion or multiple deletions) and, rarely, by mutations in coenzyme Q10 biosynthetic genes (see Table 21.1).

MtDNA-related diseases are inherited according to the rules of mitochondrial genetics (maternal inheritance, heteroplasmy and the threshold effect, mitotic segregation). Each cell contains multiple copies of mtDNA. Heteroplasmy refers to the coexistence of two populations of mtDNA, normal and mutated. Mutated mtDNA in a given tissue has to reach a minimum critical number before oxidative metabolism is impaired severely

D. Orsucci • G. Siciliano • M. Mancuso (✉)
Neurological Clinic, University of Pisa,
Via Roma 67, Pisa, Italy
e-mail: orsuccid@gmail.com; gsicilia@neuro.med.unipi.it; mancusomichelangelo@gmail.com

M.L. Brandi, E.M. Brown (eds.), *Hypoparathyroidism*,
DOI 10.1007/978-88-470-5376-2_21, © Springer-Verlag Italia 2015

Table 21.1 Genetic classification of mitochondrial diseases (selected phenotypes)

Disorders of mitochondrial genome (mtDNA)	Diseases caused by nuclear gene mutations
Sporadic rearrangements	*Structural proteins of the respiratory chain*
Kearns-Sayre syndrome	Leigh syndrome
Pearson syndrome	Paraganglioma, pheochromocytomas
PEO	*Assembling factors of the respiratory chain*
Sporadic point mutations	Leigh syndrome
PEO	*Defects of mtDNA stability and integrity*
MELAS	Autosomal PEO
Maternal-inherited point mutations	MNGIE
MELAS	mtDNA depletion syndromes
MERRF	*Coenzyme Q10 deficiency*
PEO	*Defects of mitochondrial fission or fusion*
Leber hereditary optic neuropathy	Dominant optic atrophy
NARP	
Leigh syndrome	

MELAS mitochondrial encephalomyopathy with lactic acidosis and stroke-like episodes, *MERRF* myoclonic epilepsy with ragged-red fibers, *MNGIE* mitochondrial neurogastrointestinal encephalomyopathy, *NARP* neuropathy, ataxia, and retinitis pigmentosa, *PEO* progressive external ophthalmoplegia

enough to cause dysfunction (threshold effect) [1]. The pathogenic threshold varies from tissue to tissue according to the relative dependence of each tissue on oxidative metabolism [1]. Nervous and endocrine tissues have a highly active metabolism and are therefore frequently involved in mitochondrial diseases. MELAS (mitochondrial encephalomyopathy with lactic acidosis and stroke-like episodes) and MERRF (myoclonic epilepsy with ragged-red fibers) syndromes are two typical examples of diseases due to mtDNA point mutations.

Furthermore, the sporadic occurrence of a mitochondrial disease, such as progressive external ophthalmoplegia (PEO), Kearns-Sayre syndrome (ophthalmoplegia associated to pigmentary retinopathy and cardiac conduction block), and Pearson syndrome (pediatric refractory sideroblastic anemia associated with pancreatic insufficiency), is suggestive of a single, sporadic, mtDNA deletion [2].

Therefore, every inheritance pattern (maternal inheritance, autosomal dominant, autosomal recessive, sporadic) can be seen in mitochondrial diseases, and a negative family history does not rule out this potential diagnosis. Furthermore, the phenotypic heterogeneity between the affected family members may be remarkable, and gene-environment interactions could have a role as well. Drugs must also be considered; the medications with potential mitochondrion-toxic actions (i.e., aminoglycosides, valproic acid, metformin, linezolid, etc.) have been reviewed elsewhere [5].

Table 21.2 Selected clinical features of mitochondrial diseases

Nervous system	Migraine, myoclonus, cognitive impairment, stroke-like episodes, seizures, leukoencephalopathy, ataxia, dystonia, parkinsonism, tremor, psychiatric involvement, sensorineural hearing loss, neuropathy
Skeletal muscle	Weakness, ophthalmoparesis, eyelid ptosis, exercise intolerance, myoglobinuria, respiratory impairment, hypotonia
Visual system	Pigmentary retinopathy, cataract, optic neuropathy
Digestive system	Malabsorption, intestinal pseudo-obstruction
Kidney	Tubulopathy, Fanconi syndrome
Metabolic/endocrine system	Lactic acidosis, multiple lipomatosis, short stature, diabetes mellitus, hypothyroidism, hypoparathyroidism
Heart	Cardiomyopathy, conduction system defects, Wolff-Parkinson-White syndrome

21.3 Diagnostic Approach

The diagnostic process should start from patient and family history and complete physical and neurologic examination [6]. Diagnosis of mitochondrial disease requires a complex approach: measurements of serum lactate at rest and after exercise, electromyography, muscle histology and enzymology, and molecular analysis [4]. Serum creatine kinase (CK) levels are usually normal or moderately elevated. Unfortunately, a really reliable biomarker is not available yet, and muscle biopsy is still needed in the majority of patients with suspected mitochondrial disease [4].

21.4 Mitochondrial Hypoparathyroidism

Possible metabolic/endocrine disturbances in mitochondrial disorders include (but are not limited to) diabetes mellitus, hypothyroidism, hypogonadism, short stature, lactic acidosis, and multiple lipomatosis. Hypoparathyroidism has been occasionally reported in mitochondrial patients since the early years of mitochondrial medicine [7]. Most of the reported patients are children or teenagers (range 1–18 years) with complicated forms of mitochondrial disorder and multisystem involvement [7–24]. Frequently associated features include renal disease, especially tubulopathy [7, 15, 16, 19–21, 24]; lactic acidosis [7, 8, 10, 19]; diabetes [9, 10, 19, 25]; psychomotor retardation [7, 10, 11, 20, 24]; and hearing loss [7–11, 18, 19, 23, 24].

From a genetic perspective, the most typical molecular alteration is a large, single deletion of the mtDNA. Therefore, most cases are sporadic (as discussed above, single large mtDNA rearrangements are not inherited). A single deletion was first reported in an 11-year-old boy with hypoparathyroidism and combined features of Kearns-Sayre syndrome and MELAS (ptosis, progressive external ophthalmoplegia, pigmentary retinopathy, recurrent vomiting, and cerebral infarcts with lactic acidosis) [8]. Subsequently, it was observed in a girl who presented with painful carpopedal spasms due to hypoparathyroidism at the age of 4 years, followed by truncal and limb ataxia, spastic paraparesis, muscle weakness and wasting, pigmentary retinal degeneration, sensorineural hearing loss, hirsutism, anemia, diabetes mellitus, and exocrine pancreatic dysfunction [9].

Many other case reports have further confirmed the specific association between primary hypoparathyroidism and single mtDNA deletions (and/or duplications), e.g., a 26-month-old child with growth retardation, tubulopathy, and episodic encephalopathy [20]; a 5-year-old boy with myopathy, Addison's disease, and Fanconi syndrome [21]; a 5-year-old patient with sideroblastic anemia, failure to thrive, chronic diarrhea, and lactic acidosis [13]; an 8-year-old boy with Kearns-Sayre syndrome and atrophic gastritis with pernicious anemia [22]; an 11-year-old boy with short stature, bilateral ptosis, sensorineural hearing loss, muscle weakness, and growth hormone deficiency [18]; a 12-year-old patient with incomplete Kearns-Sayre syndrome [14]; a 17-year-old girl with Kearns-Sayre syndrome and insulin-dependent diabetes mellitus [10]; an 18-year-old girl with Kearns-Sayre syndrome and tubulopathy [15]; and an 18-year-old male patient with Kearns-Sayre syndrome, short stature, sensorineural hearing impairment, cerebellar ataxia, diabetes mellitus, hyperaldosteronism, and severe tubulopathy with complete Fanconi syndrome [19].

In four children with hypoparathyroidism and deafness as initial major manifestations of Kearns-Sayre syndrome, a unique pattern of mitochondrial DNA rearrangements was observed [11]. Hypocalcemic tetany caused by PTH deficiency started between age 6 and 13 years and was well controlled by small amounts of 1.25-$(OH)_2$-cholecalciferol. Rearranged mitochondrial genomes were present in blood cells of all patients and consisted of partially duplicated and deleted molecules. The deletions were localized between the origins of replication of heavy and light strands and encompassed at least eight polypeptide-encoding genes and six tRNA genes [11]. Sequence analysis revealed imperfect direct repeats present in all rearrangements flanking the break points. The duplicated population accounted for 25–53 % of the mitochondrial genome and was predominant to the deleted

DNA (5–30 %) in all cases. The proportions of the mutant populations (30–75 %) correlated with the age at onset of the disease, suggesting that, unlike heteroplasmic deletions, such "pleioplasmic" rearrangements may escape selection in rapidly dividing cells, distribute widely over many tissues, and thus cause multisystem involvement [11]. Hypoparathyroidism and deafness could represent the result of one or more altered signaling pathway caused by selective ATP deficiency [11], but further studies are still needed.

The respiratory chain cofactor coenzyme Q10 was reported to stabilize calcium levels within normal range in two pediatric patients suffering from hypoparathyroidism and Kearns-Sayre syndrome [23]. Unfortunately, the mechanism of action was unclear, and subsequent studies confirming this isolated observation are not available.

Other genetic causes of mitochondrial hypoparathyroidism are rarer. A mtDNA heteroplasmic point mutation (A3252G), in the mtDNA gene tRNA leucine (UUR), has been reported in a patient who also had dementia, pigmentary retinopathy, and diabetes mellitus [25]. Primary hypoparathyroidism was also observed in two children with the A3243G "MELAS" mutation with psychomotor retardation and diabetes mellitus [26], and in a 54-year-old woman with diabetes mellitus, hearing loss, and myopathy due to the same mutation [27] still in the tRNA leucine (UUR) gene.

Of note, even if it is not strictly a mitochondrial respiratory chain disorder, mitochondrial trifunctional protein deficiency has been associated with hypoparathyroidism [17, 28]. Mitochondrial trifunctional protein deficiency is a rare disorder of fatty-acid oxidation which may show characteristic features such as peripheral neuropathy, pigmentary retinopathy, and acute fatty liver degeneration in pregnant women with an affected fetus [28].

Conclusion

In some instances, primary mtDNA genetic abnormalities can directly cause endocrine dysfunction, including primary hypoparathyroidism. Mitochondrial hypoparathyroidism typically shows a pediatric onset, and features of multisystem involvement are commonly associated. Given that the most frequent genetic abnormality is the presence of a sporadic mtDNA rearrangement (followed by point mutations in the mtDNA tRNA leucine (UUR) gene), a negative family history does not rule out this diagnosis. It is important to consider the possibility of a mitochondrial disease especially in children/teenagers with primary hypoparathyroidism associated to other manifestations of mitochondrial dysfunction (e.g., eyelid ptosis, external ophthalmoplegia, myopathy with exercise intolerance and/or muscle weakness, pigmentary retinopathy, heart conduction block, ataxia, sideroblastic anemia, diabetes mellitus, psychomotor retardation, hearing loss, nephropathy, lactic acidosis). In most cases, muscle biopsy is needed for a correct diagnosis, showing signs of mitochondrial dysfunction (ragged-red fibers, COX-negative fibers) and representing the first-choice tissue for mtDNA analyses. Further studies are needed in order to study the prevalence of mitochondrial disorders among patients with idiopathic primary hypoparathyroidism, and conversely of hypoparathyroidism among mitochondrial patients. Endocrinologists and pediatricians should consider a possible diagnosis of mitochondrial disease (especially when other suggestive clinical features are associated), in order to start a correct diagnostic workup. Metabolic/endocrine screening may be warranted in patients with mitochondrial diseases, since it can guide the molecular diagnosis and because specific treatments are frequently needed. Large, multicenter studies are strongly needed to better characterize the clinical picture and natural history of these diseases.

References

1. DiMauro S, Schon EA (2003) Mitochondrial respiratory-chain diseases. N Engl J Med 348(26):2656–2668. doi:10.1056/NEJMra022567348/26/2656 [pii]
2. Filosto M, Mancuso M (2007) Mitochondrial diseases: a nosological update. Acta Neurol Scand 115(4):211–221. doi:10.1111/j.1600-0404.2006.00777.x, ANE777 [pii]

3. Schaefer AM, McFarland R, Blakely EL, He L, Whittaker RG, Taylor RW, Chinnery PF, Turnbull DM (2008) Prevalence of mitochondrial DNA disease in adults. Ann Neurol 63(1):35–39. doi:10.1002/ana.21217
4. Mancuso M, Orsucci D, Coppede F, Nesti C, Choub A, Siciliano G (2009) Diagnostic approach to mitochondrial disorders: the need for a reliable biomarker. Curr Mol Med 9(9):1095–1107. doi:CMM#03 [pii]
5. Mancuso M, Orsucci D, Filosto M, Simoncini C, Siciliano G (2012) Drugs and mitochondrial diseases: 40 queries and answers. Expert Opin Pharmacother. doi:10.1517/14656566.2012.657177
6. DiMauro S, Tay S, Mancuso M (2004) Mitochondrial encephalomyopathies: diagnostic approach. Ann N Y Acad Sci 1011:217–231
7. Kitano A, Nishiyama S, Miike T, Hattori S, Ohtani Y, Matsuda I (1986) Mitochondrial cytopathy with lactic acidosis, carnitine deficiency and DeToni-Fanconi-Debre syndrome. Brain Dev 8(3):289–295
8. Zupanc ML, Moraes CT, Shanske S, Langman CB, Ciafaloni E, DiMauro S (1991) Deletion of mitochondrial DNA in patients with combined features of Kearns-Sayre and MELAS syndromes. Ann Neurol 29(6):680–683. doi:10.1002/ana.410290619
9. Tulinius MH, Oldfors A, Holme E, Larsson NG, Houshmand M, Fahleson P, Sigstrom L, Kristiansson B (1995) Atypical presentation of multisystem disorders in two girls with mitochondrial DNA deletions. Eur J Pediatr 154(1):35–42
10. Isotani H, Fukumoto Y, Kawamura H, Furukawa K, Ohsawa N, Goto Y, Nishino I, Nonaka I (1996) Hypoparathyroidism and insulin-dependent diabetes mellitus in a patient with Kearns-Sayre syndrome harbouring a mitochondrial DNA deletion. Clin Endocrinol (Oxf) 45(5):637–641
11. Wilichowski E, Gruters A, Kruse K, Rating D, Beetz R, Korenke GC, Ernst BP, Christen HJ, Hanefeld F (1997) Hypoparathyroidism and deafness associated with pleioplasmic large scale rearrangements of the mitochondrial DNA: a clinical and molecular genetic study of four children with Kearns-Sayre syndrome. Pediatr Res 41(2):193–200. doi:10.1203/00006450-199704001-01165
12. Shankar RR, Haider A, Garvey WT, Freidenberg GR (1997) Multiple endocrinopathies in an infant with fatal neurodegenerative disease. Am J Med Genet 69(3):271–279. doi:10.1002/(SICI)1096-8628(19970331)69:3<271::AID-AJMG11>3.0.CO;2-O [pii]
13. Seneca S, De Meirleir L, De Schepper J, Balduck N, Jochmans K, Liebaers I, Lissens W (1997) Pearson marrow pancreas syndrome: a molecular study and clinical management. Clin Genet 51(5):338–342
14. Tengan CH, Kiyomoto BH, Rocha MS, Tavares VL, Gabbai AA, Moraes CT (1998) Mitochondrial encephalomyopathy and hypoparathyroidism associated with a duplication and a deletion of mitochondrial deoxyribonucleic acid. J Clin Endocrinol Metab 83(1):125–129. doi:10.1210/jcem.83.1.4497
15. Katsanos KH, Elisaf M, Bairaktari E, Tsianos EV (2001) Severe hypomagnesemia and hypoparathyroidism in Kearns-Sayre syndrome. Am J Nephrol 21(2):150–153. doi:46239 [pii] 46239
16. Lee YS, Yap HK, Barshop BA, Rajalingam S, Loke KY (2001) Mitochondrial tubulopathy: the many faces of mitochondrial disorders. Pediatr Nephrol 16(9):710–712. doi:10.1007/s0046710160710
17. Labarthe F, Benoist JF, Brivet M, Vianey-Saban C, Despert F, de Baulny HO (2006) Partial hypoparathyroidism associated with mitochondrial trifunctional protein deficiency. Eur J Pediatr 165(6):389–391. doi:10.1007/s00431-005-0052-5
18. Cassandrini D, Savasta S, Bozzola M, Tessa A, Pedemonte M, Assereto S, Stringara S, Minetti C, Santorelli FM, Bruno C (2006) Mitochondrial DNA deletion in a child with mitochondrial encephalomyopathy, growth hormone deficiency, and hypoparathyroidism. J Child Neurol 21(11):983–985
19. Mihai CM, Catrinoiu D, Toringhibel M, Stoicescu RM, Hancu A (2009) De Toni-Debre-Fanconi syndrome in a patient with Kearns-Sayre syndrome: a case report. J Med Case Rep 3:101. doi:10.1186/1752-1947-3-101
20. Chae JH, Lim BC, Cheong HI, Hwang YS, Kim KJ, Hwang H (2010) A single large-scale deletion of mtDNA in a child with recurrent encephalopathy and tubulopathy. J Neurol Sci 292(1–2):104–106. doi:10.1016/j.jns.2010.02.006, S0022-510X(10)00063-8 [pii]
21. Tzoufi M, Makis A, Chaliasos N, Nakou I, Siomou E, Tsatsoulis A, Zikou A, Argyropoulou M, Bonnefont JP, Siamopoulou A (2013) A rare case report of simultaneous presentation of myopathy, Addison's disease, primary hypoparathyroidism, and Fanconi syndrome in a child diagnosed with Kearns-Sayre syndrome. Eur J Pediatr 172(4):557–561. doi:10.1007/s00431-012-1798-1
22. Abramowicz MJ, Cochaux P, Cohen LH, Vamos E (1996) Pernicious anaemia and hypoparathyroidism in a patient with Kearns-Sayre syndrome with mitochondrial DNA duplication. J Inherit Metab Dis 19(2):109–111
23. Papadimitriou A, Hadjigeorgiou GM, Divari R, Papagalanis N, Comi G, Bresolin N (1996) The influence of Coenzyme Q10 on total serum calcium concentration in two patients with Kearns-Sayre Syndrome and hypoparathyroidism. Neuromuscul Disord 6(1):49–53
24. Hameed R, Raafat F, Ramani P, Gray G, Roper HP, Milford DV (2001) Mitochondrial cytopathy presenting with focal segmental glomerulosclerosis, hypoparathyroidism, sensorineural deafness, and progressive neurological disease. Postgrad Med J 77(910):523–526

25. Morten KJ, Cooper JM, Brown GK, Lake BD, Pike D, Poulton J (1993) A new point mutation associated with mitochondrial encephalomyopathy. Hum Mol Genet 2(12):2081–2087
26. Shigemoto M, Yoshimasa Y, Yamamoto Y, Hayashi T, Suga J, Inoue G, Okamoto M, Jingami H, Tsuda K, Yamamoto T, Yagura T, Oishi M, Tsujii S, Kuzuya H, Nakao K (1998) Clinical manifestations due to a point mutation of the mitochondrial tRNAleu(UUR) gene in five families with diabetes mellitus. Intern Med 37(3):265–272
27. Tanaka K, Takada Y, Matsunaka T, Yuyama S, Fujino S, Maguchi M, Yamashita S, Yuba I (2000) Diabetes mellitus, deafness, muscle weakness and hypocalcemia in a patient with an A3243G mutation of the mitochondrial DNA. Intern Med 39(3):249–252
28. Dionisi-Vici C, Garavaglia B, Burlina AB, Bertini E, Saponara I, Sabetta G, Taroni F (1996) Hypoparathyroidism in mitochondrial trifunctional protein deficiency. J Pediatr 129(1):159–162. doi:S0022-3476(96)70206-8 [pii]

Postoperative Hypoparathyroidism

22

Francesco Tonelli and Francesco Giudici

22.1 Introduction

Postsurgical hypoparathyroidism is the most common cause of acquired hypoparathyroidism. Surgery on the thyroid, parathyroid glands, adjacent neck structures (esophagus, larynx), or lymph node neck dissection may lead to hypocalcemia and/or hypoparathyroidism. This complication is more likely to occur in patients who have undergone repetitive neck operations or extensive cervical surgery.

Hypoparathyroidism is not a synonymous with hypocalcemia. Hypocalcemia is a multifactorial phenomenon that may occur also in the presence of normally functioning parathyroid glands. Approximately 50 % of total serum calcium is protein bound, principally (80 %) to albumin. Therefore, a decrease of serum calcium can occur after surgery owing to a lowering of the serum albumin level due to postoperative water retention, hemodilution, or transcapillary leak of albumin into the extravascular space. However, in this case the serum ionized or free calcium remains normal. This type of hypocalcemia is termed *spurious*.

F. Tonelli (✉) • F. Giudici
Department of Surgery and Translational Medicine, University of Florence, Florence, Italy
e-mail: francesco.tonelli@unifi.it

22.2 Etiopathogenesis of Postoperative Hypocalcemia

The main cause of postoperative hypocalcemia is insufficient secretion of PTH to maintain normocalcemia. The causes are (1) removal of parathyroid glands or damage to them or their blood supply and (2) functional deficiency of PTH secretion.

1. Damage of the blood supply to the parathyroid glands can explain the majority of postoperative hypoparathyroidism. The parathyroid arteries are tenuous terminal vessels that must be preserved during surgery. Most often, the blood supply to the parathyroid glands arises from the inferior thyroid artery. The vascular supply of the parathyroid glands also includes small vessels arising from the adjacent thyroid gland. These small bridging vessels can be divided without compromising the viability of the parathyroid glands as long as the parathyroid arteries are preserved, whereas the blood supply from the medial thyroidal site is not usually sufficient to maintain normal parathyroid vascularity [1]. The identification of the parathyroid glands must always be assured before ligation of the thyroidal arteries, taking care to preserve their integrity and to ligate the branches of the inferior thyroidal artery near the thyroid capsule distal to the emergence of the parathyroid vessels. However, even a meticulous dissection of the parathyroid

M.L. Brandi, E.M. Brown (eds.), *Hypoparathyroidism*,
DOI 10.1007/978-88-470-5376-2_22, © Springer-Verlag Italia 2015

glands, which at the end of the surgery seem perfectly preserved, can be associated with temporary hypoparathyroidism as can be documented by low levels of PTH postoperatively. It can be difficult to ascertain if the vascularity of the parathyroid glands is really preserved. Observation of change of color of the parathyroid glands, use of the knife test, and Doppler flow imaging have been proposed as ways of determining the viability, but may not be reliable or practical [2]. A possible explanation could be that surgical dissection induces vascular spasm with temporary ischemia. The resultant *stunning* usually lasts for a few weeks. Also hypothermia or dehydration due to the climatic variations of the operating theater has been referred as causal in the block of PTH secretion [1].

2. Functional suppression of the parathyroid glands is typically observed after excision of a parathyroid adenoma. Hyperparathyroidism causes increase in bone resorption and negative calcium balance with a consequent parathyroid suppression due to the feedback inhibition by hypercalcemia of PTH synthesis and secretion. Hypoparathyroidism can also be induced by an excess or a depletion of magnesium. Hypermagnesemia activates the extracellular calcium-sensing receptors and suppresses PTH secretion, while hypomagnesemia reduces PTH secretion [3]. Moreover, hypomagnesemia favors the calcium excretion in the kidney for a reduced competition with calcium at the transporter sites of the renal tubules [4]. In the last situation, hypocalcemia can be observed. Therefore, a latent functional hypoparathyroidism due to low circulating levels of magnesium can be revealed after parathyroidectomy.

Temporary causes of postoperative hypocalcemia independent of parathyroid damage or parathyroid suppression are the following: The first is *hungry bone syndrome*. This cause of hypocalcemia occurs frequently after parathyroidectomy for primary hyperparathyroidism (PHPT) and has been proposed to contribute to postoperative hypocalcemia in cases of hyperthyroidism after thyroidectomy. Both these endocrinopathies are accompanied by a negative calcium balance, hypercalciuria, and osteodystrophy (due to an increased osteoclastic activity and an impaired mineralization). Parathyroidectomy and thyroidectomy can acutely reverse this osteodystrophy with avid deposition of calcium and phosphorus in bone and consequently the rapid development of temporary hypocalcemia and hypophosphatemia. The serum levels of PTH are normal or high. Symptoms of hypocalcemia can last for several weeks until adequate mineralization of bone has been achieved. However, it seems unlikely that postoperative hypocalcemia after thyroidectomy for hyperthyroidism is due to hungry bone syndrome, since acute postoperative hypocalcemia is also observed in patients in whom osteodystrophy has been corrected by antithyroid drugs before surgery or in patients in whom surgery has not immediately decreased the circulating levels of thyroidal hormones [5]. The second is decreased renal reabsorption of calcium. It was hypothesized that this could happen owing to the release of calcitonin from thyroidal C cell reserve during thyroidectomy. Calcitonin favors an increase in renal excretion of calcium by decreasing its renal absorption and also inhibiting osteoclastic activity. However, the increase of calcitonin in relationship to thyroidectomy has not been universally proven, and the role of calcitonin in contributing to post-thyroidectomy hypocalcemia remains controversial [1]. The third is a low level of vitamin D. Vitamin D has an important role in calcium homeostasis and promotes bone formation by both indirect (i.e., increasing intestinal calcium absorption) and direct mechanisms. 25-OH vitamin D is activated by 25-hydroxyvitamin D 1-alpha-hydroxylase in the kidney, a process that is stimulated by PTH. The activated vitamin D increases gut absorption of dietary calcium and promotes calcium reabsorption in the kidney and bone resorption at high doses. Therefore, it has been hypothesized that the preoperative 25-OH vitamin D level could influence postoperative hypocalcemia. Vitamin D deficiency has been frequently observed in elderly patients and can be responsible for reduced intestinal calcium absorption. A significant difference in post-thyroidectomy hypocalcemia has been found between patients with vitamin D levels more than 20 ng/ml and those with levels of less than 10 ng/ml [6]. A serum 25-OH vitamin D level less than 15 ng/ml increases the risk of post-thyroidectomy hypocalcemia by 28-fold according to Erbil et al. [7]. However, other recent studies did not show any significant association between the preoperative values of 25-OH vitamin D and postoperative hypocalcemia [8, 9].

Thus, postoperative hypocalcemia is a multifactorial process that is not always due to the presence of impaired secretion of PTH. True postoperative hypoparathyroidism occurs only when the low serum level of calcium ion is accompanied and caused by a low serum level of PTH.

22.3 Symptomatology

Usually hypoparathyroidism has an acute phase that compare immediately after surgery with a latency of hours or days (usually one or two). The clinical presentation of hypocalcemia can be variable and sometimes aspecific. The first symptomatology is usually bland but progressive, characterized by circumoral tingling and numbness, tingling in the hands and feet, muscle cramps, carpopedal cramps, and tetany. The increased neuromuscular excitability that accompanies these symptoms can be tested by assessing for Chvostek's and Trousseau's signs. The most dramatic manifestations of hypocalcemia are tetany, laryngospasm, bronchospasm, and seizure. Fatigue and mental changes including confusion, depression, irritability, psychosis, and congestive heart failure may be observed especially when hypocalcemia becomes chronic. Most of the time in presence of hypocalcemia and hypoparathyroidism symptoms are present, but can also happen that patients do not present the typical symptoms in presence of these biochemical deficiencies or that aspecific neurological symptoms can be interpreted as due to hypocalcemia, but the value of calcemia is normal [10].

Postoperative hypoparathyroidism can be *transient* when hypocalcemia, low PTH, and symptoms requiring supplementation with calcium and active vitamin D last less than 6 months or *persistent* when they are present for more than 6 months. Other authors advice to wait at least 1 year for indicating persistent hypoparathyroidism [11]. However, a persistent postoperative hypoparathyroidism can disappear also more than 6–12 months after surgery [12, 13]. Recently, a recovery of parathyroid function in four patients with undetectable PTH assay in whom PTH1–84 was administered was documented. These patients had previous neck operation for hyperparathyroidism or for Graves's disease and presented hypoparathyroidism for at least 8 years [14]. A potential role for exogenous PTH influencing parathyroid cell function by increasing VEGF has been hypothesized.

22.4 Thyroid Surgery

Thyroid surgery is accompanied by frequent occurrence of hypocalcemia and hypoparathyroidism. The reported incidence is highly variable in the literature in relationship to the type of thyroidal pathology, to the choice of the surgical procedure, to the modality of the assessment of the hypoparathyroidism, and to the use of autotransplantation of parathyroid tissue [15–30]. The majority of the studies referred in the literature are retrospective, not following a strict methodological assessment of hypoparathyroidism. Some prospective studies made on a large number of patients have been recently produced and can be used to look for more detailed information on the epidemiology of postoperative hypoparathyroidism and the risk factors responsible for this occurrence (Table 22.1).

22.4.1 Potential Risk Factors for Postoperative Hypocalcemia and Hypoparathyroidism

Age

Even if a clear age cutoff has not clearly emerged, the majority of the studies observed a significant lower postoperative calcemia in patients over

Table 22.1 Prospective studies on the incidence of hypocalcemia (hypoparathyroidism) after thyroidectomy

Author (year)	Surgical skill	PTS N°	Type of disease	Type of surgery	Diagnosis of hypocalcemia	Auto graft %	Hypoparathyroidism % Transient	Persistent
Asari (2008) [10]	Expert	170	All	T.T.	<1.9 mmol/L	50	24.1	1.2
Lombardi (2006) [19]	Expert/resident	199	All	T.T.	<8 mg/dL	NA	38	0.9
Toniato (2008) [21]	Expert	160	All	T.T.	<8.5 mg/dL	NA	35.6	5.6
Barczynski (2008) [22]	Expert	170	All	T.T.	<2 mmol/L	Elective	22.4	0
		170	All	T.T.	<2 mmol/L	Selective	11.2	0
Di Fabio (2006) [23]	NA	89	All	T.T.	<8 mg/dL	29	33	3.4
Salinger (2013) [27]	Single surgeon	111	All	T.T.	Ca^{2+} < 4.4 mg/dL	14.4	25.2	0
Cayo (2012) [25]	Expert	143	All	T.T.	<8.1 mg/dL or symptoms	44	18	0.7
Thomusch (2003) [16]	Multicenter	5,846	All	All	Need for calcium and vitamin D	2	7.3	1.5
Rayes (2013) [28]	Multicenter	100	Nontoxic goiter	B.S.T D.T.	PTH < 12 pg/ml	5	6	0
Julian (2013) [29]	Expert	70	All	T.T.	<2.1 mmol/L	NA	62.9	7.1
Gonzales-Botas (2013) [30]	Expert	254	All	T.T.	<8 mg/dL (corrected for protein)	NA	29.1	4.7
De Andrade Sousa (2012) [26]	Multicenter	333	All	ALL	Ca^{2+} < 1.12 mmol/L	NA	40.8	4.2
Testini (2007) [20]	Expert	160	All	T.T. (67 %)	<8.5 mg/dL or symptoms	49.4 Yes	17.7	0
				LOB. (23.3 %)		50.6 No	48.1	2.5
Erbil (2009) [7]	NA	200	Nontoxic goiter	T.T.(26 %)	<8 mg/dL	8	13.5	0
				N.T.T. (74 %)			2.5	0

Rios-Zabbudo (2004) [17]	Expert	301	Nontoxic goiter	T.T.	<7.5 mg/dL or <8.5 mg/dL with symptoms	1	9.6	0.6
Pappalardo (1998) [15]	NA	141	Nontoxic goiter	T.T. (51 %) N.T.T. (49 %)	<2.1 mmol/L	NA	26.2	2.1
Colak (2004) [18]	NA	200	Nontoxic goiter	T.T. (52.5 %) N.T.T. (47.5 %)	Symptoms	NA	10.5	0.5

DT Dunhill thyroidectomy, *LOB* lobectomy, *B.S.T.* bilateral subtotal thyroidectomy, *N.T.T.* near total thyroidectomy, *T.T* total thyroidectomy, *NA* not available

50 years of age in comparison to those less than 50 years of age [7, 24, 26]. This has been correlated to the reduction of exchangeable calcium in the elderly and the lower levels of 1,25-OH vitamin D and therefore a decrease in intestinal absorption of calcium. Another explanation of the higher risk of postoperative hypoparathyroidism in the elderly could be the difficulties in the identification of the parathyroid glands that have an increasing fat deposition and can be confused with the fatty areolar tissue located adjacent to the thyroid gland and therefore damaged or accidentally removed [1].

Gender

Even if some authors [31, 32] are of the opinion that women have a higher incidence of postoperative hypocalcemia, the majority retain that gender does not interfere with calcium homeostasis after thyroidectomy.

Extent of Thyroidal Resection

The extent of thyroidal resection is the main risk factor for postoperative hypoparathyroidism. Both retrospective and prospective studies evaluating the incidence of postoperative hypocalcemia showed a significantly higher incidence of hypocalcemia after total thyroidectomy in comparison to more limited thyroidectomy. Total thyroidectomy is also considered the strongest independent risk factor for postoperative hypoparathyroidism in several multivariate analyses [16, 33]. Transient hypocalcemia can be observed even after thyroidal unilateral lobectomy, but in this setting the permanent hypoparathyroidism is exceptional [34]. When a bilateral subtotal resection is performed, the risk of transient hypocalcemia increases, but persistent hypoparathyroidism is exceptional (0–0.4 %) [28, 35]. Total lobectomy associated to contralateral subtotal resection (the Dunhill procedure) increases only lightly of both transient and persistent hypoparathyroidism (2 and 0.5 %, respectively) [28]. Completion thyroidectomy for recurrence of benign goiter or for remnant thyroid tissue in case of unsuspected thyroid carcinoma, which has been removed by a procedure less than total thyroidectomy, seems not to be accompanied by a higher risk of transient or persistent hypoparathyroidism than a primary total thyroidectomy [36].

Thyroidectomy for Thyroid Cancer

The majority of the surgeons did not observe differences of postoperative hypoparathyroidism after total thyroidectomy without central neck compartment or lateral neck lymph node dissection between patients operated for a differentiated thyroidal cancer or a benign euthyroidal goiter [10, 30]. However, lymph node dissection of the central neck compartment is accompanied by a higher risk of postoperative hypoparathyroidism in several surgical experiences [37–39].

Thyroidal Volume

Massive goiter and goiter with an intrathoracic component can be associated to longer operative time and major risk of damage of the parathyroid glands [17, 33].

Hyperthyroidism

A significant higher incidence of both transient and persistent hypoparathyroidism is observed in patients operated for hyperthyroidism or Graves's disease. Hyperthyroidism is considered an independent predictor of postoperative hypocalcemia when the risk factors are analyzed with a multivariate logistic regression analysis [17, 33]. A significant correlation between the decrease in serum magnesium in the first 48 h after surgery and development of persistent hypocalcemia has been observed in patients operated for Graves's disease [40]. The mechanism of this acute postoperative hypomagnesemia remains obscure. Perioperatively, serum magnesium should be assessed in order to diagnose or prevent this type of hypoparathyroidism. A meta-analysis of 35 clinical studies on surgery for Graves's disease reported no significant difference of persistent hypocalcemia between subtotal and total thyroidectomies (1.0 and 0.9 %, respectively) [41]. A prospective longitudinal cohort study of 149 patients undergoing surgery for Graves's disease and operated with total thyroidectomy, Dunhill procedure, or subtotal thyroidectomy by two consultant endocrine surgeons has been recently published: transient hypocalcemia was observed in 84.9 % of the total thyroidectomies, in 65.2 % of Dunhill procedure, and in 37 % of the subtotal thyroidectomy. Persistent hypocalcemia was observed in 7.5 % of the total thyroidectomies, in

1.3 % of the subtotal thyroidectomy, and in none of the Dunhill procedure [42].

Ligature of the Main Trunk of the Inferior Thyroidal Artery

It is a controversial point if the ligation of the main trunk of the inferior thyroidal artery can increase the incidence of postoperative hypoparathyroidism. Some randomized trials comparing the central ligation with ligation of the peripheral ramus near the thyroidal capsule have shown no differences with regard to postoperative hypoparathyroidism. However, a multivariate analysis of a very large number of patients operated on showed that the bilateral central ligation of the inferior thyroidal artery is an independent risk factor for both transient and persistent hypoparathyroidism. In particular, the risk of persistent hypoparathyroidism increases five times with bilateral central ligation but two times with peripheral ligation of the inferior thyroidal artery [16].

Lack of Identification of the Parathyroid Glands and the Number of Preserved Parathyroid Glands

There is no consensus on the recommendation to identify the parathyroid glands in order to preserve them. Some authors discourage identification because of the risk of injuring their blood supply due to parathyroid manipulation [43]. Furthermore, no correlation between the number of parathyroid identified during surgery and the rate of transient or persistent hypoparathyroidism has been found [30]. However, in the opinion of other surgeons, the careful dissection of the thyroid along the thyroidal capsule allows the identification of the parathyroid glands and most of the time the preservation of their vascularization with ligation of the terminal branches of the inferior thyroid artery close to the thyroid parenchyma [1, 2, 17]. The identification of the parathyroid glands allows also the assessment of their viability at the end of surgery and the necessity to transplant the nonviable parathyroid tissue. It has been also observed that the identification of none or only one parathyroid gland increases four times the risk of persistent hypoparathyroidism [16]. It is not always possible to preserve one or more parathyroid glands at the end of surgery. This happens especially if parathyroid glands are located in the anterior portion of the thyroid gland or within the thyroid gland. There is a relationship between the number of preserved parathyroid glands and the occurrence of transient or persistent hypoparathyroidism. In the opinion of some surgeons at least two parathyroid glands should be preserved to avoid postoperative hypocalcemia and overall a persistent hypoparathyroidism [44–46]. However, it seems that preserving only one parathyroid gland with a good blood supply is sufficient for avoiding a persistent hypoparathyroidism [44].

Autografting of the Parathyroid Gland

Autotransplantation of the parathyroid glands is considered an important tool for avoiding persistent hypoparathyroidism. Since 1909, Halsted showed the prevention of tetany with autotransplantation of the parathyroid glands in an experimental animal [47]. The transplanted tissue spontaneously revascularizes and reinnervates. This process is presumably favored by the local production of growth factors [48]. The most employed technique makes 10–20 little cubes (1 mm^3) of the parathyroid gland and inserts 2–3 pieces into individual muscle pockets. The sternocleidomastoid muscle is used most frequently because of convenience. The sites of the autotransplant are marked with stitches of nonadsorbable material or with clips to allow finding the transplanted tissue in case it becomes hyperfunctioning. This occurrence can be observed especially in the parathyroid autotransplantation performed after a total parathyroidectomy for primary hyperparathyroidism due to MEN1 or MEN2 syndrome. Exceptionally, hyperfunction of the autotransplant has been observed also after autotransplantation of histological normal parathyroid tissue several years after a period of hypoparathyroidism consequent to thyroidectomy [49]. Even if it has been shown from 1975 that autografting of normal or hyperfunctioning parathyroid tissue can secrete PTH, this procedure is not commonly adopted during thyroidectomy in the case of accidental or deliberate excision of the parathyroid glands. The success rate of parathyroid autotransplantation is very high ranging from 75 to 100 %, and the function of the grafted parathyroid seems to begin after a minimum of 2 weeks to a maximum of 6 months [50].

Autotransplantation can be performed *selectively* when one or more than one parathyroid gland cannot be preserved or appear with a compromised vascularization or *routinely* for the prevention of hypoparathyroidism that could rise after the simple manipulation or dissection around the parathyroid glands. Selective autotransplantation is the most commonly adopted. It has been shown that increasing this employment in patients undergoing thyroidectomy determines a zero incidence of persistent hypoparathyroidism [51]. Some surgeons have advocated routine autotransplantation of at least one parathyroid gland for preventing a persistent hypoparathyroidism [11, 52–54]. They explain that the function of the autotransplanted parathyroid tissue is more predictable than that of parathyroid glands left in situ with a questionable blood supply. The routine autotransplantation of one or more parathyroid glands is usually accompanied by a high incidence of transient hypoparathyroidism but by very low or zero incidence of persistent hypoparathyroidism [20, 50, 52]. Also, intraoperative PTH assay has been proposed for guiding the surgeon to a selective parathyroid tissue autotransplantation adopting as criterion a level of PTH less than 10 pg/dl 10–20 min after completion of total thyroidectomy. In one recent experience none of the 21 patients selected with this criterion and submitted to parathyroid autotransplantation suffered a persistent hypoparathyroidism [22].

Impact of New Surgical Technologies

The introduction of the minimally invasive video-assisted thyroidectomy (MIVAT) or of new energy-based devices (LigaSure vessel sealing system, Harmonic scalpel) has not modified the incidence of persistent hypoparathyroidism. However, a lower incidence of transient hypoparathyroidism has been observed in some experiences with MIVAT [55] and with LigaSure [56].

Surgeon Expertise

It is controversial if the expertise of a surgeon has a determinant role in preventing postoperative hypoparathyroidism. The majority of the authors did not find any significant difference in transient and persistent hypoparathyroidism between patients undergoing operation by an experienced surgeon and those undergoing surgery by residents who are supervised by their superiors [10, 16].

22.4.2 Is the Early Prediction of Post-thyroidectomy Hypoparathyroidism Possible?

In the last decade, many efforts have been dedicated to the possibility of predicting postoperative hypoparathyroidism in order to prevent hypocalcemia with an early supplementation of vitamin D and calcium and/or allow a safe early dimission (same day or postoperative day 1). Various proposals have been reported: one of the most frequently adopted is the monitoring of serum calcium with one or more serial determination in order to design a slope of the calcemia. The serum calcium slope correlates well with the development of symptomatic hypocalcemia, but the results often are not useful until 12 or 24 h after operation. Therefore, there is a delay in initiating the therapy and the hospitalization cannot be avoided. More recently, the dosage of PTH has been indagated for a prompt identification of post-operative hypoparathyroidism. For contain the cost, a single value of PTH has been proposed. Either an intraoperative assay by a quick method at the end of surgery or an early postoperative determination (immediately after surgery or at 1, 2, 4, or 8 h after operation) has been indagated. The accuracy of this method in predicting post-thyroidectomy hypocalcemia was not very high, confirming that hypocalcemia can be multifactorial even if the most important cause remains the impaired parathyroid function [19]. Comparing different criteria of the intraoperative PTH (IOPTH) monitoring, it has been shown that the highest accuracy in predicting hypocalcemia after total thyroidectomy is a serum level of PTH less than 10 pg/dl at 4 h postoperatively [57, 58]. A positive predictive value of postoperative hypocalcemia around 100 % has been observed in the presence of a decrease of PTH in the first postoperative day of more than 75 % compared to the preoperative value [21, 59, 60]. Values of PTH less than 15 pg/mL on the first postoperative day and of serum calcium less than 1.9 mmol/L on the second postoperative day have a predictive value of 86 % for postoperative hypoparathyroidism [10]. In any case, a very high negative predictive value of postoperative hypocalcemia has been observed in patients with normal parathyroid hormone levels after thyroidectomy [10]. Prospective clinical trials have demonstrated the safety for same day discharge after thyroidectomy in the presence of a 4 h postoperative PTH level within the normal range [61] or the efficacy of prophylactic calcium and vitamin D administration based on the early postoperative PTH evaluation in minimizing the symptomatic hypocalcemia [62, 63].

22.5 Parathyroidectomy

Transient hypoparathyroidism is frequently observed in relationship to a functional suppression of the normal parathyroid glands after surgery for primary and secondary hyperparathyroidism; meanwhile, persistent hypoparathyroidism is essentially a surgical complication associated to a damage of parathyroid tissue, more frequently observed when it is necessary to explore bilaterally the neck, to remove more than 2 parathyroid glands, or in case of reoperation (Table 22.2).

22.5.1 Surgery for Sporadic Adenoma

22.5.1.1 Transient Hypoparathyroidism

Patients affected with PHPT who undergo parathyroidectomy have a rapid decrease in serum PTH and calcium levels after successful removal of one or more hyperactive parathyroid gland(s). In this occurrence the presence of hyperfunctioning parathyroid gland(s) suppresses the biochemical function of healthy parathyroid(s). The hypocalcemia is usually mild, has a peak on the 2–4 postoperative days, and becomes symptomatic in 15–30 % of the patients [77]. This complication seems independent of the size of pathological gland(s) and of surgical approach (focused parathyroidectomy, unilateral or bilateral neck exploration) [64, 65, 70, 71, 78–81].

The hypocalcemia lasts usually for few days and it is easily controlled by medical substitutive therapy. However, sometimes a prolonged (longer than 4 days postoperatively), symptomatic hypocalcemia characterized by a profound decrease of serum calcium (<2.1 mmol/l) that is difficult to treat can occur. In such patients in whom hyperparathyroidism is particularly severe, the preoperative indices of bone turnover are high (osteocalcin, bone alkaline phosphatase), and marked osteoporosis, osteitis fibrosa cystica, and *brown tumors* are present. *Hungry bone syndrome* is to be suspected. This syndrome persists until the normal parathyroid glands regain their full sensitivity and activity [82].

Some pre- or intraoperative biochemical parameters can predict transient postoperative hypocalcemia or hypoparathyroidism:

Normocalcemia: normal or only slightly elevated preoperative calcemia is an independent risk factor for postoperative transient hypocalcemia within the first 4 postoperative days [68].

Deficit of vitamin D: patients with PHPT and concurrent vitamin D deficiency seem to show a significantly higher preoperative PTH level and a greater incidence of late-onset symptomatic hypocalcemia after parathyroidectomy [83].

Intraoperative PTH: a drop of more than 80 % of IOPTH 10 min after removal of hyperfunctioning parathyroid tissue seems to be a significant factor for predicting postoperative hypoparathyroidism [77].

22.5.1.2 Persistent Hypoparathyroidism

PTH values lower than the normal range and not increasing after more than 4 days after parathyroidectomy may be due not to suppression of residual non-pathological parathyroid glands but to organic damage of the residual parathyroid glands [64, 71, 78, 84].

Data about surgery for PHPT showed 0–14 % persistent hypoparathyroidism after a first operation, with the lowest rates obtained in high-volume centers [85]. The experience of the surgeon and especially the choice of the surgical approach are extremely important in this setting, as postsurgical hypoparathyroidism is usually due to damage to the parathyroid gland(s) or to their vascular supply or to inadvertent (or

Table 22.2 Studies on the incidence of hypocalcemia (hypoparathyroidism) after parathyroidectomy for sporadic primary hyperparathyroidism

Author (year)	Study	N° PTS	Type HPT	Surgery	Diagnosis of hypocalcemia	Hypoparathyroidism % Transient	Hypoparathyroidism % Persistent
Kaplan (1982) [64]	Retrospective	107	PHPT	BNE	<7.9 mg/dL	37.4	0
Zamboni (1986) [65]	Retrospective	51	PHPT	BNE	NA	37	0
Thompson (1999) [66]	Retrospective	124	Recurrent PHPT	Reoperation	<2.5 mmol/L	NA	13
Westerdahl (2000) [67]	Prospective	86	PHPT	UNE	Symptoms	25.6	1.2
Bergenfelz (2002) [68]	Prospective-randomized	91	PHPT	BNE (44) UNE (47)	<2.5 mmol/L	49 28	2.2 0
Burkey (2002) [69]	Prospective	150	PHPT	BNE/UNE (50) IOPTH (50) γ PROBE (50)	<8.9 mg/dL	6 2 0	2 0 0
Lew (2006) [70]	Retrospective	43	PHPT (hypercalcemia crisis)	BNE (27) UNE (16)	NA	12	0
Norman (2007) [71]	Prospective	3,000 Consecutive	PHPT	NA	Symptoms	7	NA
Allendorf (2007) [72]	Retrospective-single surgeon	1,112 consecutive	PHPT	BNE (general 264; local 848)	Clinical with 1gx4/ dietary calcium	1.8	0
Richards (2008) [73]	Prospective	228	Recurrent PHPT	Reoperation With IOPTH(181) Without IOPTH (46)	<8.5 mg/dL	NA NA	2 9
Young (2010) [74]	Retrospective-many surgeons	687	PHPT	BNE and MIP	<8.5 mg/dL	2	1.5
Hessmann (2010) [75]	Prospective-randomized	143	PHPT	BNE (75) MIVAP (68)	NA	8.4	2.1
Udelsman (2011) [76]	Retrospective-single surgeon	1,650 Consecutive	PHPT	BNE 613 (general) MIP 1,037 (local)	NA	0.49 0.10	NA

PHPT primary hyperparathyroidism, *BNE* bilateral neck exploration, *UNE* unilateral neck exploration, *MIVAP* minimally invasive video-assisted parathyroidectomy, *MIP* minimally invasive parathyroidectomy, *NA* not available unavoidable) removal of the parathyroid gland(s) [85, 86]. Persistent hypoparathyroidism was observed when bilateral neck exploration, eventually performing biopsies of all parathyroid glands, was the treatment of choice for PHPT. Today, the localization modalities based on high-resolution ultrasonography [87–89] and sestamibi scintigraphy [90–92] allowed to indicate a safe and successful unilateral approach. The unilateral approach is based on the concept that if an enlarged parathyroid gland and a normal gland are found while exploring the first side of the neck, then this is an adenoma, and the second side should not be explored. Only if both glands on the initial side are recognized to be abnormal is the second side explored. The theoretical advantages of this unilateral approach are a decrease both in operative morbidity rates (hypoparathyroidism and nerve injuries) and in operative time.

Furthermore, the intraoperative measurement of PTH has been shown to be a valid method for confirming the complete removal of hyperfunctioning parathyroid tissue [93–99] and allows a focused surgery without exploring the other ipsilateral parathyroid gland. Good results have been claimed, with a decreased risk of hypocalcemia [67, 100–102] and vocal cord injury [7]. Bergenfelz et al. analyzed 91 patients affected by PHPT and randomized them for unilateral or bilateral neck exploration finding, after unilateral procedure, a lower incidence of persistent hypoparathyroidism compared with patients undergoing bilateral exploration (0 vs 2.5 %, respectively) [68].

22.5.2 Surgery for Persistent or Recurrent PHPT

The percentage of persistent hypoparathyroidism is high (>10 %) after a cervical reoperation to correct a recurrent or persistent PHPT, even if the operation is performed by an experienced surgical team [66]. In fact, a reoperation is characterized by high morbidity and complication rates, since an intensive scarring due to the previous operation sometimes makes it difficult to identify and to preserve the recurrent laryngeal nerve and the normal parathyroid gland(s) [86]. In the last years, the introduction of IOPTH monitoring seems to have reduced the incidence of postoperative hypoparathyroidism after reoperations for persistent or recurrent primary hyperparathyroidism changing from 9 % before to 2 % after the employment of IOPTH dosage [73]. However, it is to underline that the real contribution of IOPTH monitoring for the cure of persistent or recurrent PHPT is still controversial as Irvin et al. reported an increase in operative success from 76 to 94.3 %, while Sebag et al. found no statistically significant differences in success with (82 %) or without (87 %) the use of IOPTH monitoring [103, 104].

22.5.3 Surgery for PHPT: Familial Syndromes

The risk of transient or persistent hypoparathyroidism is clearly increased after surgical treatment of multiglandular familial PHPT; in the majority of cases, multiglandular PHPT needs to be treated with bilateral cervical exploration, and in familial syndromes all parathyroid glands are often genetically affected, even if apparently macroscopically healthy, since parathyroid gland involvement is asynchronous and asymmetrical with a variable volume enhancement.

Many data are now available about surgical results in multiple endocrine neoplasia type 1 syndrome (MEN1). According to the different surgical approach adopted, postoperative persistent hypoparathyroidism is variable: after less than subtotal parathyroidectomy is reported in 7.38 % of cases, while after subtotal (SPTX) and total parathyroidectomy (TPTX) with thymectomy and autograft, the rates are 11.1 and 25.2 %, respectively [105]. Analyzing our experience about MEN1 PHPT, we globally found after TPTX with autograft a persistent hypoparathyroidism in 25 % of cases, with a higher incidence in patients undergoing a second surgical cervical revision than in those who underwent a primary surgery (50% vs 22 %, respectively) [106]. In our recent experience, better results seem to be obtained after TPTX with autograft using 12–14 fragments of parathyroid tissue of about 1 mm^3 in volume each, obtained from the most normal-appearing gland, removed at surgery as the last one, and immediately placed in a sterile lactated Ringer or Wisconsin solution at 4 °C to be grafted at the end of the operation, after a time of ischemia <60 min, into separate pockets between the muscular fibers of the brachioradial muscle of the nondominant forearm.

Apparently higher rates of postoperative hypoparathyroidism are reported after surgical treatment of PHPT in MEN2A syndrome, with persistent hypoparathyroidism affecting 17.4, 30, and 13.6 % of patients who had undergone less than subtotal, subtotal, and total parathyroidectomy with autograft, respectively [107].

Only few data are still available about postoperative hypoparathyroidism in the other genetic form of PHPT as MEN4, FHH-NSHPT, ADMH, HPT-JT, and FIHPT [107].

22.5.4 Surgery for Secondary Hyperparathyroidism

TPTX with autograft of parathyroid tissue and SPTX are currently considered as the gold standard surgical procedures also for the treatment of secondary hyperparathyroidism (SHPT) [108, 109]. Transient hypocalcemia is extremely frequent after all surgical procedures for SHPT as it develops as a consequence of the lack of osteoclastic activity due to PTH decrease and *hungry bone syndrome* [110]. In the majority of patients, it can be easily treated by oral or intravenous supplementation of calcium and vitamin D. Otherwise, persistent hypoparathyroidism after surgery for SHPT seems to be extremely rare. Schneider et al. found this condition in only 1.19 % of patients who had undergone total parathyroidectomy with thymectomy and autotransplantation, in 0 % after total parathyroidectomy without thymectomy and without autotransplantation, and in 0 % after subtotal parathyroidectomy. However, paradoxically these authors described 14.29 % persistent hypoparathyroidism after less than subtotal parathyroidectomy underlining as the postoperative persistent hypoparathyroidism rate does not depend from the type of surgery adopted but from the function of the preserved parathyroid tissue [111]. For this reason, the IOPTH monitoring could be theoretically useful to guide the surgeon even if Roshan et al. reported that in the surgical management of SHPT, the IOPTH fails to predict postoperative hypoparathyroidism while it is useful in predicting the cure of hyperparathyroidism [112]. Considering that in uremic patients long-lasting decreased PTH levels lead to a suppression of bone turnover that may provoke the development of adynamic bone disease [113], adequate PTH levels should be maintained. Values ranging from 150 to 300 pg/ml after parathyroidectomy have been recently recommended for patients with stage V chronic kidney disease [114]. For this reason, persistent hypoparathyroidism should be treated by autotransplantation of autologous cryopreserved parathyroid tissue [115]. Because of the lack of reliable predicting factors for persistent postoperative hypoparathyroidism, cryopreservation of parathyroid tissue should be theoretically considered in any SHPT patient undergoing parathyroid surgery [111].

Levels of calcium and parathyroid hormone (PTH) in the immediate postoperative period (1–3 days after surgery) are informative about the function of the parathyroid remnant within the neck after SPTX [116, 117], while parathyroid fragments at the site of the autograft after TPTX may require up to 6 months to resume adequate function [116–118].

22.5.5 Laryngectomy and Hypopharyngo-esophagectomy

Transient and persistent hypoparathyroidism may also occur after surgical treatment for laryngeal and hypopharyngo-esophageal cancer as a consequence of neoplastic invasion of the parathyroid gland(s) and/or of its resection. Furthermore, during this kind of surgery, resection of all or part of the thyroid gland or ligature of thyroidal vessels is usually performed. The occurrence of transient hypoparathyroidism is variable as the incidence range reported in the literature is particularly wide (from 7.3 to 92 %) and it seems to be particularly increased when it is necessary to perform an associated total thyroidectomy [119–122]. Hypocalcemia is more common after total laryngectomy, in the postradiotherapy setting, and in patients undergoing bilateral neck dissection; meanwhile, the preservation of one thyroid lobe, when oncologically feasible [120, 122], the extent of pharyngectomy, and the preoperative tracheostomy are not significantly related to the postoperative development of hypoparathyroidism [120]. No considerable data is available about persistent hypoparathyroidism occurring in these operations because of the short-term prognosis of the majority of these patients.

References

1. Falk SA (1990) Complications of thyroid surgery. Hypocalcemia and hypoparathyroidism; hypocalcitonemia, Chapter 38. In: Falk SA (ed) Thyroid disease: endocrinology, surgery, nuclear medicine, and radiotherapy. Raven Press, Ltd, New York
2. Lo CY (2000) Parathyroid autotransplantation during thyroidectomy. ANZ J Surg 72:902–907
3. Rude RK, Singer FR (1981) Magnesium deficiency and excess. Annu Rev Med 32:245–259
4. Coburn JW, Massry SG, Kleeman CR (1970) The effect of calcium infusion on renal handling of magnesium with normal and reduced glomerulal filtration rate. Nephron 7:131–143
5. Wilkin TJ, Isles TE, Paterson CR et al (1977) Postthyroidectomy hypocalcemia: a feature of operation or the thyroid disorder? Lancet 1:621–623
6. Kirkby-Bott J, Markogiannakis H, Skandarajah A et al (2011) Preoperative vitamin D deficiency predicts postoperative hypocalcemia after total thyroidectomy. World J Surg 35:324–330
7. Erbil Y, Barbaros U, Temel B et al (2009) The impact of age, vitamin D(3) level, and incidental parathyroidectomy on postoperative hypocalcemia after total or near total thyroidectomy. Am J Surg 197:439–446
8. Hallgrimsson P, Nordenstrom E, Almquist M et al (2012) Risk factors for medically treated hypocalcemia after surgery for Graves's disease: a Swedish multicenter study of 1,157 patients. World J Surg 36:1933–1942
9. Lang BH, Wong KP, Cowling BJ et al (2013) Do low preoperative vitamin D levels reduce the accuracy of quick parathyroid hormone in predicting postthyroidectomy hypocalcemia? Ann Surg Oncol 20:739–745
10. Asari R, Passler C, Kaczireh K et al (2008) Hypoparathyroidism after total thyroidectomy. A prospective study. Arch Surg 143:132–137
11. Lo CY, Lam KY (2001) Routine parathyroid autotransplantation during thyroidectomy. Surgery 129:318–323
12. Claussen MS, Pehling GB, Kisken WA (1993) Delayed recovery from post-thyroidectomy hypoparathyroidism: a case report. Wis Med J 92:331–334
13. Sitges-Serra A, Ruiz S, Girvent M et al (2010) Outcome of protracted hypoparathyroidism after total thyroidectomy. Br J Surg 97:1687–1695
14. Cusano NE, Anderson L, Rubin MR et al (2013) Recovery of parathyroid hormone secretion and function in postoperative hypoparathyroidism: a case series. J Clin Endocrinol Metab 98:4285–4290
15. Pappalardo G, Guadalaxara A, Frattaroli FM et al (1998) Total compared with subtotal thyroidectomy in benign nodular disease: personal series and review of published reports. Eur J Surg 164:501–506
16. Thomusch O, Machens A, Sekulla C et al (2003) The impact of surgical technique on postoperative hypoparathyroidism in bilateral thyroid surgery: a multivariate analysis of 5846 consecutive patients. Surgery 133:180–185
17. RiosZambudo A, Rodriguez J, Riquelme J et al (2004) Prospective study of postoperative complications after total thyroidectomy for multinodular goiters by surgeons with experience in endocrine surgery. Ann Surg 240:18–25
18. Colak T, Akca T, Kanik A et al (2004) Total versus subtotal thyroidectomy for the management of benign multinodular goiter in an endemic region. ANZ J Surg 74:974–978
19. Lombardi CP, Raffaelli M, Princi P et al (2006) Parathyroid hormone levels 4 hours after surgery do not accurately predict post-thyroidectomy hypocalcemia. Surgery 140:1016–1025
20. Testini M, Rosato L, Avenia N et al (2007) The impact of single parathyroid gland autotransplantation during thyroid surgery on postoperative hypoparathyroidism: a multicenter study. Transplant Proc 39:225–230
21. Toniato A, Merante Boschin I, Piotto A et al (2008) Thyroidectomy and parathyroid hormone: tracing hypocalcemia-prone patients. Am J Surg 196:285–288
22. Barczynski M, Cichon S, Konturek A et al (2008) Applicability of intraoperative parathyroid hormone assay during total thyroidectomy as a guide for the surgeon to selective parathyroid tissue autotransplantation. World J Surg 32:822–828
23. Di Fabio F, Casella C, Bugari G et al (2006) Identification of patients at low risk for thyroidectomy-related hypocalcemia by intraoperative quick PTH. World J Surg 30:1428–1433
24. Berghenfelz A, Jansson S, Kristofferson A et al (2008) Complications to thyroid surgery: results as

reported in a database from a multicenter audit comprising 3,660 patients. Langenbeckes Arch Surg 393:667–673
25. Cayo AK, Yen T, Misustin SM et al (2012) Predicting the need for calcium and calcitriol supplementation after total thyroidectomy: Results of a prospective, randomized study. Surgery 152:1059–1567
26. De Andrade Sousa A, Porcaro Salles JM, Arantes Soares JM et al (2012) Predictors factors for post-thyroidectomy hypocalcemia. Rev Col Bras Cir 39:476–482
27. Salinger EM, Moore JT (2013) Perioperative indicators oh hypocalcemia in total thyroidectomy: the role of vitamin D and parathyroid hormone. Am J Surg 206:876–882
28. Rayes N, Steinmuller T, Schroder S et al (2013) Bilateral subtotal thyroidectomy versus hemithyroidectomy plus subtotal resection (Dunhill procedure) for benign goiter: long-term results of a prospective, randomized study. World J Surg 37:84–90
29. Julián MT, Balibrea JM, Granada ML et al (2013) Intact parathyroid hormone measurement at 24 hours after thyroid surgery as predictor of parathyroid function at long term. Am J Surg 206:783–789
30. Gonzales Botas JH, Lourido Pietrahita D (2013) Hypocalcemia after total thyroidectomy: incidence, control and treatment. Acta Otorrinolaringol Esp 64:102–107
31. Prim MP, de Diego JI, Hardisson D et al (2001) Factors related to nerve injury and hypocalcemia in thyroid gland surgery. Otolaryngol Head Neck Surg 124:111–114
32. Yamashita H, Noguchi S, Murakami T et al (2000) Calcium and its regulating hormones in patients with Graves disease: sex differences and relation to postoperative tetany. Eur J Surg 166:924–928
33. Mc Henry CR, Speroff T, Wintworth D et al (1994) Risk factors for postthyroidectomy hypocalcemia. Surgery 116:641–648
34. Rosato L, Avenia N, Bernante P et al (2004) Complications of thyroid surgery: analysis of a multicentric study on 14,934 patients operated on in Italy over 5 years. World J Surg 28:271–276
35. Tezelman S, Borucu I, Senyurek Y et al (2009) The change in surgical practice from subtotal to near-total or total thyroidectomy in the treatment of patients with benign multinodular goiter. World J Surg 33:400–405
36. Gulcelik MA, Kuru B, Dincer H et al (2012) Complications of completion versus total thyroidectomy. Asian Pac J Cancer Prev 13:5225–5229
37. Thompson NW, Harness JK (1970) Complications of total thyroidectomy for carcinoma. Surg Gynecol Obstet 131:861–868
38. Kim MK, Mandel SH, Baloch Z et al (2004) Morbidity following central compartment reoperation for recurrent or persistent thyroid cancer. Arch Otolaryngol Head Neck Surg 130:1214–1216
39. Paek SH, Lee YM, Min SY et al (2013) Risk factors of hypoparathyroidism following total thyroidectomy for thyroid cancer. World J Surg 37:94–101
40. Hammerstad SS, Norheim I, Paulsen T et al (2013) Excessive decrease in serum magnesium after total thyroidectomy for Graves's disease is related to development of permanent hypocalcemia. World J Surg 37:369–375
41. Palit TK, Miller CC, Miltenburg DM (2000) The efficacy of thyroidectomy for Graves's disease: a meta-analysis. J Surg Res 24:1303–1311
42. Al-Adhami A, Snaith AC, Craig WL et al (2013) Changing trends ion surgery for Graves's disease: a cohort comparison of those having surgery intended to preserve thyroid function with those having ablative surgery. J Otolaryngol Head Neck Surg 42:37
43. Pfleideree AG, Ahmad N, Dreaper MR et al (2009) The timing of calcium measurement in helping to predict temporary or permanent hypocalcaemia in patients having completion and total thyroidectomy. Ann R Coll Surg Engl 91:140–146
44. Kim YS (2012) Impact of preserving the parathyroid glands on hypocalcemia after total thyroidectomy with neck dissection. J Korean Surg Soc 83:75–82
45. Attie JN, Khafif RA (1975) Preservation of parathyroid glands during total thyroidectomy. Improved technic utilizing microsurgery. Am J Surg 130:399–404
46. Olson JA Jr, De Benedetti MK, Baumann DS et al (1996) Parathyroid autotransplantation during thyroidectomy. Results of long-term follow-up. Ann Surg 223:472–478
47. Halsted W (1907) Hypoparathyroidism, status parathyreoprivus, and transplantation of the parathyroid glands. Am J Med Sci 134:1–5
48. Carter WB, Uy K, Ward MD et al (2000) Parathyroid-induced angiogenesis is VEGF-dependent. Surgery 128:458–464
49. D'Avanzo A, Parangi S, Morita E et al (2000) Hyperparathyroidism after thyroid surgery and autotransplantation of histologically normal parathyroid glands. J Am Coll Surg 190:546–552
50. Lo CY, Lam SC (2001) Parathyroid autotransplantation during thyroidectomy: documentation of graft function. Arch Surg 136:1381–1385
51. Paloyan E, Lawrence AM, Paloyan D (1977) Successful autotransplantation of the parathyroid glands during total thyroidectomy for carcinoma. Surg Gynecol Obstet 145:364–368
52. Zedenius J, Wadstrom C, Delbridge L (1999) Routine autotransplantation of at least one parathyroid gland during total thyroidectomy may reduce permanent hypoparathyroidism to zero. Aust N Z J Surg 69:794–799
53. Gauger PG, Reeve TS, Wilkinson M et al (2000) Routine parathyroid autotransplantation during total thyroidectomy: the influence of technique. Eur J Surg 166:605–609
54. Kikumori T, Imai T, Tanaka Y et al (1999) Parathyroid autotransplantation with total thyroidectomy for thyroid carcinoma: long-term follow-up of grafted parathyroid function. Surgery 125:504–508

55. Gao W, Liu L, Ye G et al (2013) Application of minimally invasive video-assisted technique in papillary thyroid microcarcinoma. Surg Laparosc Endosc Percutan Tech 23:468–473
56. Kuboki A, Nakayama T, Konno W et al (2013) New technique using energy-based device versus conventional technique in open thyroidectomy. Auris Nasus Larynx 40:558–562
57. Lombardi CP, Raffaelli M, Princi P et al (2004) Early prediction of postthyroidectomy hypocalcemia by one single iPTH measurement. Surgery 136:1236–1241
58. Barczynski M, Cichon S, Konturek A (2007) Which criterion of intraoperative iPTH assay is the most accurate in prediction of true serum calcium levels after thyroid surgery? Langenbecks Arch Surg 392:693–698
59. Mc Leod IK, Arciero C, Noordzij JP et al (2006) The use of rapid parathyroid hormone assay in predicting postoperative hypocalcemia after total or completion thyroidectomy. Thyroid 16:259–265
60. Scurry WC, Beus KS, Hollenbeak CS et al (2005) Perioperative parathyroid hormone assay for diagnosis and management of postthyroidectomy hypocalcemia. Laryngoscope 115:1362–1366
61. Australian Endocrine Surgeons Guidelines AES06/01 (2007) Postoperative parathyroid hormone measurement and early discharge after total thyroidectomy. Analysis of Australian data and management recommendations. ANZ J Surg 77:199–202
62. Tartaglia F, Giuiani A, Sgueglia M et al (2005) Randomized study on oral administration of calcitriol to prevent symptomatic hypocalcemia after total thyroidectomy. Am J Surg 190:424–429
63. Carter Y, Chen H, Sippel RS (2014) An intact parathyroid hormone-based protocol for the prevention and treatment of symptomatic hypocalcemia after thyroidectomy. J Surg Res 186:23–28
64. Kaplan EL, Bartlett S, Sugimoto J et al (1982) Relation of postoperative hypocalcemia to operative techniques: deleterious effect of excessive use of parathyroid biopsy. Surgery 92:827–834
65. Zamboni WA, Folse R (1986) Adenoma weight: a predictor of transient hypocalcemia after parathyroidectomy. Am J Surg 152:611–615
66. Thompson GB, Grant CS, Perrier ND et al (1999) Reoperative parathyroid surgery in the era of sestamibi scanning and intraoperative parathyroid hormone monitoring. Arch Surg 134:699–704
67. Westerdal J, Lindblom P, Valdemarsson S et al (2000) Risk factors for postoperative hypocalcemia after surgery for primary hyperparathyroidism. Arch Surg 135:142–147
68. Bergenfelz A, Lindblom P, Tibblin S et al (2002) Unilateral versus bilateral neck exploration for primary hyperparathyroidism: a prospective randomized controlled trial. Ann Surg 236:543–551
69. Burkey SH, Van Heerden JA, Farley DR et al (2002) Will directed parathyroidectomy utilizing the gamma probe or intraoperative parathyroid hormone assay replace bilateral cervical exploration as the preferred operation for primary hyperparathyroidism? World J Surg 26:914–920
70. Lew JI, Solorzano CC, Irvin GL 3rd (2006) Long-term results of parathyroidectomy for hypercalcemic crisis. Arch Surg 141:696–700
71. Norman JG, Politz DE (2007) Safety of immediate discharge after parathyroidectomy: a prospective study of 3,000 consecutive patients. Endocr Pract 13:105–113
72. Allendorf J, DiGorgi M, Spanknebel K et al (2007) 1112 consecutive bilateral neck explorations for primary hyperparathyroidism. World J Surg 31:2075–2080
73. Richards ML, Thompson GB, Farley DR et al (2008) Reoperative parathyroidectomy in 228 patients during the era of minimal-access surgery and intraoperative parathyroid hormone monitoring. Am J Surg 196:937–943
74. Young VN, Osborne KM, Fleming MM et al (2010) Parathyroidectomy in the elderly population: does age really matter? Laryngoscope 120:247–252
75. Hessman O, Westerdahl J, Al-Suliman N et al (2010) Randomized clinical trial comparing open with video-assisted minimally invasive parathyroid surgery for primary hyperparathyroidism. Br J Surg 97:177–184
76. Udelsman R, Lin Z, Donovan P (2011) The superiority of minimally invasive parathyroidectomy based on 1650 consecutive patients with primary hyperparathyroidism. Ann Surg 253:585–591
77. Steen S, Rabeler B, Fisher T et al (2009) Predictive factors for early postoperative hypocalcemia after surgery for primary hyperparathyroidism. Proc (Bayl Univ Med Cent) 22:124–127
78. Brasier AR, Nussbaum SR (1988) Hungry bone syndrome: clinical and biochemical predictors of its occurrence after parathyroid surgery. Am J Med 84:654–660
79. Heath DA, Van't Hoff W, Barnes AD et al (1979) Value of 1-a-hydroxy vitamin D3 in treatment of primary hyperparathyroidism before parathyroidectomy. BMJ 1:450–452
80. Chia SH, Weisman RA, Tieu D et al (2006) Prospective study of perioperative factors predicting hypocalcemia after thyroid and parathyroid surgery. Arch Otolaryngol Head Neck Surg 132:41–45
81. Brossard JH, Garon J, Lepage R et al (1993) Inhibition of 1,25(OH)2D production by hypercalcemia in osteitis fibrosa cystica: influence on parathyroid hormone secretion and hungry bone disease. Bone Miner 23:15–26
82. Witteveen JE, van Thiel S, Romijn JA et al (2013) Hungry bone syndrome: still a challenge in the postoperative management of primary hyperparathyroidism: a systematic review of the literature. Eur J Endocrinol 168:45–53
83. Lang BH, Lo CY (2010) Vitamin D3 deficiency is associated with late-onset hypocalcemia after

minimally invasive parathyroidectomy in a vitamin D borderline area. World J Surg 34:1350–1355
84. Wong WK, Wong NA, Farndon JR (1996) Early postoperative plasma calcium concentration as a predictor of the need for calcium supplement after parathyroidectomy. Br J Surg 83:532–534
85. Kaplan EL, Yashiro T, Salti G (1992) Primary hyperparathyroidism in the 1990s. Choice of surgical procedures for this disease. Ann Surg 215:300–317
86. Karakas E, Müller HH, Schlosshauer T et al (2013) Reoperations for primary hyperparathyroidism – improvement of outcome over two decades. Langenbecks Arch Surg 398:99–106
87. Chapuis Y, Fulla Y, Bonnichon P et al (1996) Values of ultrasonography, sestamibi scintigraphy and intraoperative measurement of 1–84 PTH for unilateral neck exploration of primary hyperparathyroidism. World J Surg 20:835–840
88. Koslin DB, Adams J, Andersen P et al (1997) Preoperative evaluation of patients with primary hyperparathyroidism: role of high-resolution ultrasound. Laryngoscope 107:1249–1253
89. Inabnet WB, Fulla Y, Richard B et al (1999) Unilateral neck exploration under local anesthesia: the procedure of choice for asymptomatic primary hyperparathyroidism. Surgery 126:1004–1009
90. Tsukamoto E, Russell CF, Fergurson WR et al (1995) The role of preoperative thallium-technetium subtraction scintigraphy in the surgical management of patients with solitary parathyroid adenoma. Clin Radiol 50:677–680
91. Borley NR, Collins RE, O'Doherty M et al (1996) Technetium-99 m sestamibi is accurate enough for scan-directed unilateral neck exploration. Br J Surg 83:989–991
92. Carty SE, Worsey J, Virji MA et al (1997) Concise parathyroidectomy: the impact of preoperative SPECT 99mTc sestamibi scanning and intraoperative quick parathormone assay. Surgery 122:1107–1116
93. Bergenfelz A, Algotsson L, Ahrén B et al (1992) Surgery for primary hyperparathyroidism performed under local anaesthesia. Br J Surg 79:931–934
94. Chen H, Sokoll LJ, Udelsman R (1999) Outpatient minimally invasive parathyroidectomy: a combination of sestamibi-SPECT localization, cervical block anesthesia, and intraoperative parathyroid hormone assay. Surgery 126:1016–1021
95. Nussbaum S, Thompson A, Hutcheson K et al (1988) Intraoperative measurement of parathyroid hormone in the surgical management of hyperparathyroidism. Surgery 104:1121–1127
96. Bergenfelz A, Nordén NE, Ahrén B (1991) Intraoperative fall in plasma levels of intact parathyroid hormone after removal of one enlarged parathyroid gland in hyperparathyroid patients. Eur J Surg 157:109–112
97. Bergenfelz A, Isaksson A, Ahrén B (1994) Intraoperative monitoring of intact PTH during surgery for primary hyperparathyroidism. Langenbecks Arch Surg 379:178–181
98. Bergenfelz A, Isaksson A, Lindblom P et al (1998) Measurement of parathyroid hormone in patients with primary hyperparathyroidism undergoing first an reoperative surgery. Br J Surg 85:1129–1132
99. Irvin GL, Prudhomme DL, Deriso GT et al (1994) A new approach to parathyroidectomy. Ann Surg 219:574–579
100. Worsey MJ, Carty SE, Watson CG (1993) Success of unilateral neck exploration for sporadic primary hyperparathyroidism. Surgery 114:1024–1030
101. Tibblin S, Bondeson AG, Bondeson L et al (1984) Surgical strategy in hyperparathyroidism due to solitary adenoma. Ann Surg 200:776–784
102. Tibblin S, Bizard JP, Bondeson AG et al (1991) Primary hyperparathyroidism due to solitary adenoma. A comparative multicentre study of early and long-term results of different surgical regimens. Eur J Surg 157:511–515
103. Irvin GL III, Molinari AS, Figueroa C et al (1999) Improved success rate in reoperative parathyroidectomy with intraoperative PTH assay. Ann Surg 229:874–878
104. Sebag F, Shen W, Brunaud L et al (2003) Intraoperative parathyroid hormone assay and parathyroid reoperations. Surgery 134:1049–1055
105. Tonelli F, Giudici F, Cavalli T et al (2012) Surgical approach in patients with hyperparathyroidism in multiple endocrine neoplasia type 1: total versus partial parathyroidectomy. Clinics (Sao Paulo) 1:155–160
106. Tonelli F, Marcucci T, Fratini G et al (2007) Is total parathyroidectomy the treatment of choice for hyperparathyroidism in multiple endocrine neoplasia type 1? Ann Surg 246:1075–1082
107. Tonelli F, Marcucci T, Giudici F et al (2009) Surgical approach in hereditary hyperparathyroidism. Endocr J 56:827–841
108. Schlosser K, Veit JA, Witte S et al (2007) Comparison of total parathyroidectomy without autotransplantation and without thymectomy versus total parathyroidectomy with autotransplantation and with thymectomy for secondary hyperparathyroidism: TOPAR PILOT-Trial. Trials 8:22
109. Riss P, Asari R, Scheuba C et al (2013) Current trends in surgery for renal hyperparathyroidism (RHPT) – an international survey. Langenbecks Arch Surg 398:121–130
110. Saliba W, El-Haddad B (2009) Secondary hyperparathyroidism: pathophysiology and treatment. J Am Board Fam Med 22:574–581
111. Schneider R, Slater EP, Karakas E et al (2012) Initial parathyroid surgery in 606 patients with renal hyperparathyroidism. World J Surg 36:318–326
112. Roshan A, Kamath B, Roberts S et al (2006) Intraoperative parathyroid hormone monitoring in secondary hyperparathyroidism: is it useful? Clin Otolaryngol 31:198–203

113. Andress DL (2008) Adynamic bone in patients with chronic kidney disease. Kidney Int 73:1345–1354
114. National Kidney Foundation (2003) K/DOQI clinical practice guidelines for bone metabolism and disease in chronic kidney disease. Am J Kidney Dis 42:S1–S201
115. Rothmund M, Wagner PK (1984) Assessment of parathyroid graft function after autotransplantation of fresh and cryopreserved tissue. World J Surg 8:527–533
116. Wells SA Jr, Farndon JR, Dale JK et al (1980) Long-term evaluation of patients with primary parathyroid hyperplasia managed by total parathyroidectomy and heterotopic autotransplantation. Ann Surg 192:451–458
117. Wells SA Jr, Gunnells JC, Gutman RA et al (1977) The successful transplantation of frozen parathyroid tissue in man. Surgery 81:86–90
118. Shoback D (2008) Clinical practice. Hypoparathyroidism. N Engl J Med 359:391–403
119. Lo Galbo AM, Kuik DJ, Lips P et al (2013) A prospective longitudinal study on endocrine dysfunction following treatment of laryngeal or hypopharyngeal carcinoma. Oral Oncol 49:950–955
120. Basheeth N, O'Cathain E, O'Leary G et al (2014) Hypocalcemia after total laryngectomy: incidence and risk factors. Laryngoscope 124:1128–1133. doi:10.1002/lary.24429
121. Thorp MA, Levitt NS, Mortimore S et al (1999) Parathyroid and thyroid function five years after treatment of laryngeal and hypopharyngeal carcinoma. Clin Otolaryngol Allied Sci 24:104–108
122. Martins AS, Tincani AJ (2006) Thyroidectomy and hypoparathyroidism in patients with pharyngo-esophageal tumors. Head Neck 28:135–141

23 Hypoparathyroidism During Pregnancy, Lactation, and Fetal/Neonatal Development

Christopher S. Kovacs

23.1 Introduction

Pregnancy and lactation require that women deliver substantial calcium and other minerals to meet the needs of the rapidly growing fetus and neonate, respectively. The specific adaptations that are invoked to meet this demand for mineral lead to altered maternal serum chemistries and calciotropic hormone concentrations. In turn, the presentation, diagnosis, and management of hypoparathyroidism are also altered during pregnancy and lactation.

Fetal mineral metabolism is characterized by low levels of parathyroid hormone (PTH) and calcitriol. After birth the blood calcium falls, and this triggers an upregulation in parathyroid function and calcitriol synthesis. Fetal and neonatal parathyroids can be disturbed by genetic parathyroid disorders and by exposure in utero to maternal hypocalcemia or hypercalcemia.

This chapter begins with a review of mineral physiology in pregnant and lactating women, and in normal fetuses and neonates. The focus then turns to the presentation and management of hypoparathyroidism during pregnancy and lactation, and the effects that altered maternal mineral homeostasis and genetic parathyroid disorders can have on fetal and neonatal parathyroid function. Due to space limitations, other reviews by the author provide more detailed references for normal mineral physiology during reproduction and development [1–5].

23.2 Mineral Physiology During Pregnancy

Women deliver about 30 g of calcium to the average fetal skeleton by term, 80 % of which is transported across the placenta during the third trimester [6–8].

23.2.1 Serum Minerals and Calciotropic Hormones

Pregnancy is characterized by distinct changes in serum mineral and calciotropic hormone concentrations. The total serum calcium concentration falls in tandem with a decline in the serum albumin. This is physiologically unimportant and should not be interpreted to indicate true hypocalcemia. The ionized calcium (the physiologically important fraction) and the albumin-corrected serum calcium remain normal during pregnancy, and these should be measured in pregnant women whenever a disturbance in serum calcium is

C.S. Kovacs, MD, FRCPC, FACP, FACE (✉)
Faculty of Medicine – Endocrinology, Obstetrics and Gynecology, and BioMedical Sciences, Health Sciences Centre, Memorial University of Newfoundland, 300 Prince Philip Drive, St. John's, NL A1B 3 V6, Canada
e-mail: ckovacs@mun.ca

M.L. Brandi, E.M. Brown (eds.), *Hypoparathyroidism*,
DOI 10.1007/978-88-470-5376-2_23, © Springer-Verlag Italia 2015

suspected. Serum magnesium and phosphorus remain normal.

In pregnant North American and European women who consume diets adequate in calcium, intact PTH is usually in the low-normal range or suppressed below it during early pregnancy, after which it steadily increases back to the mid-normal range by term [9–14]. In contrast, several studies of women from Asia and Africa did not find suppressed PTH levels; instead, serum PTH concentrations may even increase above normal [15].

Serum calcitriol concentrations double or triple in the first trimester and remain there until after delivery, whereas free calcitriol levels may only be increased in the third trimester [1, 16]. PTH is evidently not responsible for this increased production of calcitriol because PTH is declining while calcitriol is increasing, and because mice lacking the *Pth* gene achieved a more than fourfold increase in calcitriol during pregnancy [17]. The mother's main source of calcitriol is her kidneys and not the placenta (as has often been assumed). This conclusion is supported by an anephric woman on hemodialysis whose endogenous calcitriol level was low before and during a pregnancy [18], and by the finding that *Cyp27b1* mRNA expression is 30-fold higher in maternal kidneys of pregnant mice as compared to their placentas [17]. Serum calcitonin is increased during pregnancy, and some of that may come from non-thyroidal sources, including the breasts and placenta.

PTH-related protein (PTHrP) shows a steady increase in the maternal circulation between the first and third trimesters [14], and it likely derives from the breasts, decidua, and placenta (see also Chap. 3). Two case reports confirmed that placental PTHrP reaches the maternal circulation in sufficient amounts to affect maternal calcium homeostasis. In the first case, a hypoparathyroid woman developed sudden and symptomatic hypocalcemia immediately after delivery, which resolved when lactation was established [19]. In the second case, a healthy woman developed a hypercalcemic crisis in the third trimester associated with a high (27 pmol/L) circulating level of PTHrP and undetectable (<5 pg/mL) PTH [20]. Within 6 h after an urgent C-section, she became hypocalcemic with undetectable (<1.1 pmol/L) PTHrP and high (110 pg/ml) PTH [20]. Other case reports confirmed that the breasts also contribute PTHrP to the maternal circulation during pregnancy, with high levels of breast-derived PTHrP causing maternal hypercalcemia (so-called pseudohyperparathyroidism of pregnancy) [21, 22]. In one such case, a bilateral mastectomy was needed to correct the hypercalcemia [23, 24].

Although PTHrP may achieve high levels in the maternal circulation by the third trimester, this is too late to explain the rise in calcitriol and suppression of PTH during the first trimester. Furthermore, PTHrP is much less potent than PTH in stimulating Cyp27b1, the enzyme that synthesizes calcitriol [25, 26]. On the other hand, PTHrP is a prohormone that is processed into amino-terminal, mid-region, and carboxyl-terminal forms. Each of these may have effects on maternal mineral and bone physiology. Carboxyl-terminal PTHrP, also called osteostatin, has been shown to suppress bone resorption in vitro [27–29]; conceivably, osteostatin may help prevent excessive resorption of the maternal skeleton during pregnancy.

23.2.2 Intestinal Absorption of Calcium

Intestinal calcium absorption in the mother doubles as early as the 12th week [1] and creates a positive calcium balance by midpregnancy [30]. The two- to threefold increased calcitriol concentrations likely contribute to the upregulation of intestinal calcium absorption. However, pregnant rodents made severely deficient in vitamin D or which lack the vitamin D receptor upregulate intestinal calcium absorption [31–33]; therefore, factors other than calcitriol must also stimulate calcium transport during pregnancy. Isotope studies revealed that over 90 % of calcium in the fetus is absorbed from the maternal diet during pregnancy [34], which agrees with other estimations that increased intestinal calcium absorption meets the fetal demand for calcium.

23.2.3 Renal Handling of Calcium

A consequence of increased intestinal calcium absorption is that urinary calcium excretion increases during pregnancy and hypercalciuric values can be reached. Low PTH and high calcitonin concentrations likely contribute to this increase. The fasting urine calcium remains normal or may be low, reflecting that this is absorptive hypercalciuria [1].

23.2.4 Skeletal Calcium Metabolism

There is evidence from clinical studies that skeletal resorption increases modestly during normal pregnancy. In 15 women who had elective first trimester abortions, iliac crest histomorphometry demonstrated an increase in indices of bone resorption as compared to biopsies from nonpregnant women [35]. Biochemical markers of bone resorption increase early in pregnancy, while markers of bone formation are low in the first trimester, and normal or slightly increased in the third trimester [1]. Use of these markers of bone resorption and formation has confounding problems that are discussed in detail elsewhere [36]. Total alkaline phosphatase in the mother increases markedly due to a placental fraction and does not indicate increased bone resorption.

Whether the modest increase in bone resorption causes net loss of the bone during pregnancy is less certain. Longitudinal studies used single- and/or dual-photon absorptiometry and found no significant change in cortical or trabecular bone density during pregnancy [1]. Dual x-ray absorptiometry (DXA) has been used 1–18 months before planned pregnancy and 1–6 weeks postpartum, but not during pregnancy [37–43]. These small studies have found no change to as much as a 5 % decrease in lumbar spine bone density between the two measurements. Serial ultrasound measurements of a peripheral site, the os calcis, have found small decreases in BMD; in one study, this was shown to be restored after pregnancy [44].

Although there have been no longitudinal DXA studies of hip or spine *during* pregnancy, the available data indicate that some resorption of the maternal skeleton may occur. Such resorption may be needed to supply mineral or to adapt the skeleton to the changing load that it bears during pregnancy. Does this resorption have consequences? Vertebral crush fractures have rarely occurred during or shortly after pregnancy, but whether excess loss of bone mass occurred prior to the fracture is unknown. So-called transient osteoporosis of the hip can also occur, but this appears to result from local factors that increase the water content of one or both femora, not from skeletal resorption [5]. For most women, these apparent small losses of bone mass during pregnancy are inconsequential in the long term. Several dozen studies have found no significant association of parity with bone density or fracture risk [1, 45]. Some studies have even found a protective effect of parity on adult bone mass, including a study of twins [46].

23.3 Mineral Physiology During Fetal Development

The fetus and placenta work together to actively pump mineral from the maternal circulation against concentration gradients, maintain serum mineral concentrations above simultaneous maternal values, and rapidly mineralize the skeleton during the final quarter of gestation.

23.3.1 Minerals and Calciotropic Hormones

Human babies are hypercalcemic relative to their mothers from as early as 15 to 20 weeks of gestation, the earliest time point studied. Serum phosphorus and magnesium are also higher than maternal values. This "fetal hypercalcemia" has been found among all mammalian fetuses and is robustly maintained even when the mother has hypocalcemia from various causes [1, 47, 48]. A high serum calcium appears to be necessary to achieve normal mineralization of the fetal skeleton, because a lower (normal adult) level of calcium is associated with a reduced skeletal mineral content at term in mice [49]. Studies in fetal rodents found that loss of the normally high serum calcium does not affect survival to term in utero [1, 50–52]; however, it may protect against neonatal mortality from hypocalcemia. The serum calcium undergoes an obligate 20–30 % fall after birth in humans [53–55] and by 40 % in rodents [56, 57], before increasing to adult values during the next 24–48 h. A lower fetal blood calcium may, therefore, predispose to an even lower trough level of blood calcium after birth, and a higher risk of tetany and death.

Intact PTH is normally low at term in human and other mammalian fetuses, but PTH remains important for fetal development because the loss of parathyroids or PTH in rodents causes hypocalcemia, hyperphosphatemia, and under-mineralized skeletons [47, 50, 51, 58]. 25-Hydroxyvitamin D (25(OH)D) readily crosses the placenta, but calcitriol does not [59–61], which means that calcitriol within the fetal circulation derives from fetal sources. Cord blood 25(OH)D is normally 75–100 % of the maternal 25(OH)D value, while calcitriol is 25–50 % of the mother's calcitriol level [62–66]. Such low calcitriol levels may result from suppression of Cyp27b1 by the fetal milieu of high serum calcium and phosphorus, and low PTH, combined with relatively high activity of Cyp24a1 that catabolizes 25(OH)D and calcitriol into inactive 24-hydroxylated forms. Fibroblast growth factor-23 (FGF23) reduces calcitriol by inhibiting Cyp27b1 and stimulating Cyp24a1, but the loss of FGF23 does not affect the circulating calcitriol level in fetal mice [67].

Clinical studies have found that severely vitamin D-deficient babies have normal blood calcium, phosphorus, PTH, and mineral content of skeletal ash at birth; however, in rare cases, vitamin D-deficient rickets has been diagnosed soon after birth [68]. Children with inactivating mutations of Cyp27b1 or the vitamin D receptor do not present with hypocalcemia or rickets until a year or two of age [68]. Animal studies have also found that despite severe vitamin D deficiency, the loss of Cyp27b1, or deletion of the vitamin D receptor, affected fetuses have normal serum calcium, phosphorus, and PTH and normally developed and mineralized skeletons at term [68]. These findings suggest that calcitriol may be nonessential for fetal mineral homeostasis, and this is likely because the fetal intestines are not a significant source of mineral. Fetal serum calcitonin levels are also increased above maternal values, possibly as a response to the increased serum calcium in fetal blood.

In human babies, cord blood PTHrP [1–86] is up to 15-fold higher than simultaneous intact PTH when expressed in equimolar units [49]. PTHrP is a significant regulator of fetal mineral homeostasis because fetal mice lacking PTHrP (*Pthrp* null fetuses) have abnormal endochondral bone development, hypocalcemia, hyperphosphatemia, and reduced placental calcium transfer [69, 70]. Serum PTH increases several-fold in *Pthrp* null fetuses [50], but their low calcium and high phosphorus indicate that PTH cannot fully compensate for the loss of PTHrP. Conversely, PTHrP does not compensate for the absence of PTH, since aparathyroid and *Pth* null fetuses have normal serum PTHrP despite significant hypocalcemia and hyperphosphatemia [47, 50, 51]. The absence of fetal parathyroids, PTH, or PTHrP each causes a similar degree of hyperphosphatemia [47, 50, 51, 71, 72]. Overall, despite PTH circulating at low levels, PTH and PTHrP are both important regulators of fetal mineral homeostasis.

23.3.2 Placental Mineral Transport

Studies in lambs and mice have shown that a mid-regional form of PTHrP stimulates placental calcium transfer [49, 72, 73]; it also stimulates placental magnesium transport in fetal sheep. PTH has recently been found to be expressed at low levels in murine placentas, where it may act locally to stimulate placental calcium transport [47, 71]. PTHrP and PTH had no effect on placental phosphorus transport in fetal lambs, and no other regulators of phosphate transport are known [74, 75].

23.3.3 Intestinal Mineral Absorption and Renal Mineral Handling

The placenta is the main source of mineral for the fetus. Fetuses also excrete mineral into urine, which in turn contributes much of the volume of amniotic fluid. When amniotic fluid is swallowed, voided mineral can then be absorbed by the fetal intestines. The extent to which this pathway contributes to fetal mineral homeostasis is unknown, but it may be a trivial route.

23.4 Mineral Physiology During Lactation

Breastfeeding women lose a mean 210 mg in milk daily during exclusive lactation; women who nurse twins lose even more.

23.4.1 Minerals and Calciotropic Hormones

The ionized calcium and albumin-corrected serum calcium are normal during lactation, whereas the serum phosphorus may exceed the normal range [1]. Intact or bio-intact PTH is normally low or undetectable during the first several months in North American and European women who exclusively breastfeed their babies. As solid foods are introduced, lactation becomes less intense, and PTH increases into the mid-normal range. In contrast, this decline in PTH during lactation may not occur in women from Asia and Africa, possibly due to their diets being lower in calcium and higher in phytate. Calcitriol increases two- to threefold during pregnancy but falls promptly to normal values during lactation, which may indicate that the factors that stimulate Cyp27b1 during pregnancy are lost at delivery.

The lactating breasts produce substantial amounts of PTHrP, such that human and cow's milk contain 1,000–10,000 times the concentration of PTHrP found in patients with hypercalcemia of malignancy and normal human controls. Breast-derived PTHrP also enters the maternal circulation with the highest levels reached after suckling [76, 77]. Data from lactating animals confirmed that mammary tissue is the main source of PTHrP. The venous drainage of mammary glands contains a higher PTHrP concentration than the arterial inflow [78], and selective deletion of the murine PTHrP gene from mammary tissue reduced the circulating PTHrP level [79].

Upon entering the maternal circulation from the breasts, PTHrP plays a central role to alter maternal mineral metabolism. It stimulates osteoclast-mediated bone resorption, renal tubular reabsorption of calcium, and (at least in rodents) osteocytic osteolysis. Higher plasma PTHrP levels in breastfeeding women correlate with greater loss of bone mineral density [80], higher serum calcium, and lower PTH [76, 81]. In mice, ablation of the PTHrP gene from the mammary tissue caused less bone to be resorbed during lactation [79]. In hypoparathyroid women (discussed in detail below), PTHrP normalizes mineral homeostasis during lactation and occasionally can cause hypercalcemia. PTHrP-mediated hypercalcemia has also occurred in normal women during lactation; it resolves with cessation of breastfeeding.

Calcitonin levels are elevated during the first 6 weeks of lactation in women, but whether this has physiological importance is unknown. Calcitonin protects the maternal skeleton of mice against excess resorption, since deletion of the calcitonin gene resulted in the loss of twice the normal amount of bone mineral content during lactation [82].

23.4.2 Intestinal Absorption of Calcium

Intestinal calcium absorption occurs at normal rates in breastfeeding women. Increased intake of calcium does not alter skeletal resorption or breast milk calcium [83–86]. Similarly, high intake of vitamin D (up to 6,400 IU daily in trials) has no effect on breast milk calcium, even with a mean maternal 25(OH)D level of 160 nmol/l (64 ng/ml) [87–89].

23.4.3 Renal Handling of Calcium

Twenty-four-hour renal calcium excretion drops to low-normal or overtly hypocalciuric values, due to the loss of calcium in breast milk and the effect of PTHrP to stimulate renal tubular reabsorption of calcium. The phosphaturic actions of PTHrP and the increased filtered load of phosphorus (which arises from resorbed bone) contribute to increased urine phosphorus excretion.

23.4.4 Skeletal Calcium Metabolism

Resorption of the skeleton during lactation provides calcium for milk production. In breastfeeding women, 3–10 % of trabecular bone mineral content will be resorbed over the first 2–6 months, with smaller losses occurring at cortical sites [1, 45, 90]. Randomized clinical trials and cohort studies have shown that these lactational bone losses are not minimized by high intake of calcium [83–86] or aggravated by low intake of calcium [91–94]. Instead, resorption of the bone appears to be programmed by hormonal and other changes that are invoked by breastfeeding, such that more intense lactation or greater breast milk output causes greater loss of the bone [95]. Bone turnover markers increase, and bone resorption markers increase more than markers of bone formation. Lactating rodents show increased osteoclast-mediated bone resorption, accompanied by enhanced resorption of mineral by osteocytes from their perilacunar matrices (osteocytic osteolysis).

Figure 23.1 depicts the known mechanisms that increase bone resorption during lactation. Prolactin and suckling both suppress ovarian function by acting on the GnRH pulse center in the hypothalamus, leading to low estradiol levels. Prolactin and

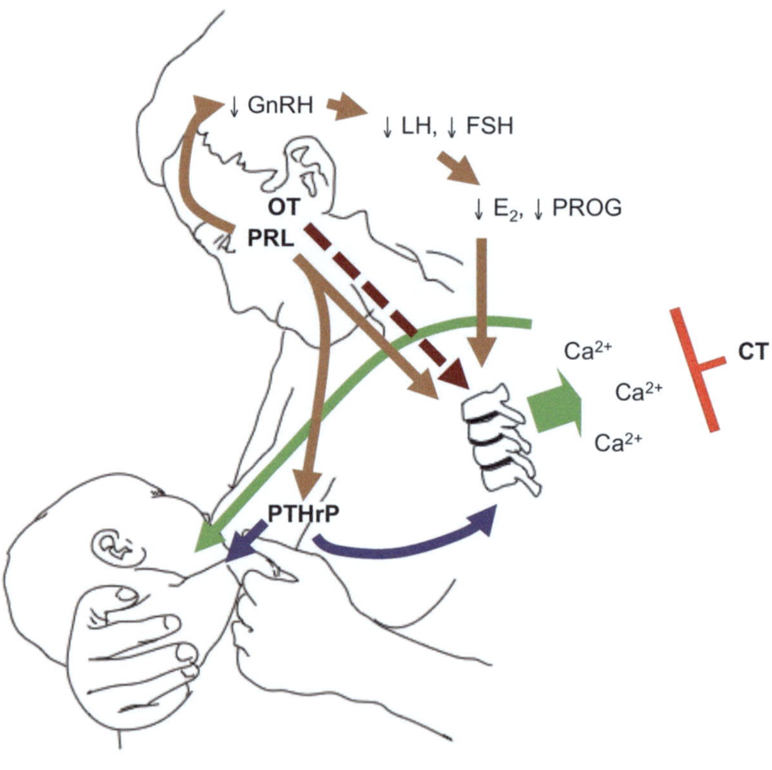

Fig. 23.1 Brain-breast-bone circuit. Suckling and prolactin [*PRL*] both inhibit the hypothalamic gonadotropin-releasing hormone (*GnRH*) pulse center, which in turn suppresses the gonadotropins (luteinizing hormone [*LH*] and follicle-stimulating hormone [*FSH*]), leading to low levels of the ovarian sex steroids (estradiol [E_2] and progesterone [*PROG*]). Prolactin may have direct effects on its receptor in bone cells. *PTHrP* production and release from the breast is controlled by several factors, including suckling, prolactin, low estradiol, and the calcium receptor. PTHrP enters the bloodstream and combines with systemically low estradiol levels to markedly upregulate bone resorption. Increased bone resorption releases calcium and phosphate into the bloodstream, which then reaches the breast ducts and is actively pumped into the breast milk. *PTHrP* also passes into milk at high concentrations, but whether swallowed *PTHrP* plays a role in regulating calcium physiology of the neonate is unknown. In addition to stimulating milk ejection, oxytocin (*OT*) may directly affect osteoblast and osteoclast function (*dashed line*). Calcitonin (*CT*) may inhibit skeletal responsiveness to *PTHrP* and low estradiol (Adapted from Kovacs [96]; ©2005 Springer Science and Business Media B.V.)

suckling also stimulate PTHrP production by the breasts. PTHrP and low estradiol have synergistic effects to upregulate bone resorption and osteocytic osteolysis, achieving more rapid bone loss than caused by estradiol deficiency alone. Other factors such as oxytocin and serotonin have also been proposed to contribute to the increased bone resorption during lactation.

After weaning, bone formation upregulates, and skeletal mineral content appears fully restored over the subsequent 6–12 months in most women [1, 45, 85]. Restoration of normal ovarian function facilitates bone recovery but cannot fully explain it. Animal studies have found that the classical calciotropic hormones PTH, PTHrP, calcitriol, calcitonin, and estradiol are not required for skeletal recovery to be fully achieved [1, 17, 33, 82, 97]. The identification of the factor(s) that stimulate post-weaning bone formation is the subject of active research [98]. Although skeletal recovery is complete as assessed by DXA, the trabecular microarchitecture may not be completely restored at all sites after weaning in women and rodents [99, 100]. The cross-sectional diameters of the long bones have increased by the end of lactation or post-weaning recovery, while cortical bone area is restored after weaning to prepregnant values [101–103]. The increase in bone volumes may compensate for the loss of trabecular microarchitecture and, thereby, maintain bone strength.

Vertebral compression fractures occasionally occur during lactation, but subsequent skeletal recovery makes lactational bone loss clinically unimportant in the long term. Several dozen epidemiological studies have found that a history of lactation has no effect, or even a protective effect, on peak bone mass, bone density, and risk of osteoporosis or hip fractures [1, 45].

23.5 Mineral Physiology in the Neonate

Cutting the umbilical cord causes the placental calcium infusion and placental hormones (especially PTHrP) to be lost, and the onset of breathing causes the blood pH to rise. These factors contribute to a sharp fall in the ionized calcium, which triggers a switch in how mineral metabolism is regulated.

23.5.1 Minerals and Calciotropic Hormones

In human babies, the serum calcium and ionized calcium each fall 20–30 % over the first 12 h, followed by an increase to adult values over the succeeding 24 h [53]. Phosphorus increases over the same interval and then declines as the serum calcium rises. These changes in calcium and phosphorus are indicators of a progressive stimulation of parathyroid function after birth. PTH rises from the suppressed fetal values during the first 24 h, and this is followed by a rise in calcitriol [53].

As noted earlier, at birth, the 25(OH)D value is typically 75–100 % of the maternal level [62–66]. Cord blood calcium is usually no different between babies born of vitamin D-replete and severely vitamin D-deficient mothers, but can be significantly lower after the first or second day. The serum calcium is also normal at birth in animal fetuses that have severe vitamin D deficiency or lack vitamin D receptors or Cyp27b1 [2]. In the days to months after birth, skeletal manifestations of rickets will develop if hypocalcemia and deficient mineral delivery persist [2, 68]. Breast milk normally contains little vitamin D or 25(OH)D, and so exclusively breastfed babies are at high risk for developing vitamin D-deficient rickets.

23.5.2 Intestinal Calcium Absorption and Renal Mineral Handling

The human neonate now requires the intestines and kidneys to provide the mineral delivery functions that the placenta previously performed. Intestinal calcium absorption is initially passive, non-saturable [104, 105], and facilitated by lactose in milk [106, 107]. As the neonate matures, the intestines become less permeable to passive calcium absorption, and the dominant mechanism of intestinal calcium delivery becomes active, saturable, and calcitriol dependent [104, 108, 109]. A similar developmentally programmed change from passive, lactose-facilitated to active, calcitriol-dependent mechanisms has also been observed in neonatal rodents [49, 110].

23.6 Hypoparathyroidism During Pregnancy

Most cases of maternal hypoparathyroidism are known prior to pregnancy, but some were not diagnosed until the newborn displayed severe secondary hyperparathyroidism, hypercalcemia, bone demineralization, and fractures [111, 112]. The normal adaptations in mineral homeostasis that occur during pregnancy, and the fetal demand for mineral, can contribute to two quite different outcomes that have been reported in pregnant, hypoparathyroid women.

First, hypoparathyroidism can become significantly improved during pregnancy, with fewer hypocalcemic symptoms, and reduced need for supplemental calcitriol and calcium. This likely occurs because the normal upregulation of calcitriol and intestinal calcium absorption, which begin in the first trimester, do not appear to require PTH. Instead, high levels of calcitriol, PTHrP, placental lactogen, prolactin, and other factors may upregulate intestinal calcium absorption despite the absence of PTH. Indeed, in several case reports of hypoparathyroidism during pregnancy, hypocalcemic symptoms lessened, serum calcium increased, and the requirement for supplemental calcium and calcitriol decreased [1, 19, 113]. Animal studies have confirmed that despite total parathyroidectomy or the absence of the *Pth* gene, calcitriol and intestinal calcium absorption increase normally during pregnancy [1, 17].

Second, hypoparathyroidism may worsen during pregnancy with the fetal demand for calcium overwhelming the hypoparathyroid woman's ability to maintain her own blood calcium. Several case reports have confirmed that higher doses of calcium and calcitriol were implemented during pregnancy [1, 113, 114]. One woman had a low calcitriol level at midpregnancy, indicating that the normal two- to threefold increase in calcitriol did not occur [114]. In other cases, it was not clinical symptoms but the normal pregnancy-related fall in serum calcium (uncorrected for albumin) that prompted the clinician to increase the dose of calcium and calcitriol [1]. The ionized calcium or albumin-corrected calcium must be used during pregnancy so that the artifactual fall in serum total but not ionized serum calcium is not misinterpreted as worsening hypocalcemia.

It is unknown why some hypoparathyroid women improve while others worsen, but variable responsiveness to the normal adaptations of pregnancy may be the cause. For example, the high estradiol concentrations of pregnancy may cause more marked suppression of bone turnover in some women, and more potent stimulation of Cyp27b1 in other women. The production of PTHrP by the breasts and placenta, and achieved level in the maternal circulation, may also vary. Animal models have confirmed that the fetal demand for calcium can cause maternal hypocalcemia during pregnancy when challenged by a low-calcium diet or a large litter size.

In nonpregnant hypoparathyroid adults, the treatment target is a serum calcium at or just below the lower end of the normal range, thereby balancing prevention of hypocalcemia against worsening hypercalciuria and nephrocalcinosis. There are no consensus guidelines for management of hypoparathyroidism during pregnancy, but this author contends that the target during pregnancy should be to maintain the ionized or albumin-corrected calcium *well within the normal range* in order to minimize the risk of fetal and neonatal complications. Maternal hypocalcemia increases the risk of premature birth and fetal and neonatal secondary hyperparathyroidism (see below) [115].

Due to the variability in presentation, management should be expectant, with the ionized or albumin-corrected calcium, and hypocalcemic symptoms, used as indicators as to whether the condition is improving (due to the pregnancy-related adaptations) or worsening (due to the fetal demand for calcium, especially in the third trimester). Calcitriol normally increases at least twofold starting in the first trimester, and so a higher dose of calcitriol may be the most appropriate method to raise the ionized or albumin-corrected calcium into the mid-normal range. As pregnancy progresses, the calcitriol dose should be adjusted based on the ionized or albumin-corrected calcium. Thiazides decrease urine calcium excretion and the dose of calcium or calcitriol required to maintain normocalcemia, but are usually avoided during pregnancy because they cross the placenta (FDA category C). However, hydrochlorothiazide was well tolerated in a pregnant hypoparathyroid woman whose severe hypocalcemia and hypercalciuria were unresponsive to calcium and calcitriol, but responded well to a thiazide [116].

Maternal hypercalcemia must also be avoided because it causes suppression of the fetal and neonatal parathyroids, and because higher doses of vitamin D analogs may increase the risk of teratogenicity [117, 118]. Calcitriol and 1α-calcidiol have shorter half-lives and lower risk of toxicity as compared to the older preparations, and are preferred over newer vitamin D analogs for use during pregnancy. High-dose vitamin D (cholecalciferol) is still used in many cases because it is cheap. Very high levels of 25(OH)D (>250 nmol/l or 100 ng/ml) are needed, and the risk of fetal adverse effects of such doses is uncertain.

Genetic resistance to PTH action, or pseudohypoparathyroidism, resembles hypoparathyroidism except that for the presence of high PTH levels (see also Chaps. 32, 33, 34, and 35). Pseudohypoparathyroidism may improve during pregnancy, such that the women have become normocalcemic and asymptomatic, with lower PTH levels and no need for supplemental calcium or vitamin D analogs [119]. Such improvement may reflect that the pregnancy-related increase in intestinal calcium absorption occurs independent of PTH.

But two other case reports found that pseudohypoparathyroidism symptomatically worsened during pregnancy, such that increased doses of calcium, calcitriol, or 1α-calcidiol were needed to maintain a normal serum calcium. In both cases, the worsening occurred in the third trimester, which is when the peak fetal demand for calcium occurs [120, 121].

In two women whose hypocalcemia improved during pregnancy, their calcitriol levels more than doubled (similar to normal pregnancy) during the second and third trimester [119]. In contrast, in a woman whose hypocalcemia worsened in the third trimester, calcitriol had increased during the first two trimesters but declined in the third [121]. An increase in calcitriol may result from actions of estradiol, placental lactogen, prolactin, or other factors to stimulate Cyp27b1, as has been shown in animal models. Also, higher estradiol levels and other hormonal changes of pregnancy could conceivably improve post-receptor signaling of the PTH receptor. Placental production of calcitriol has also been invoked to explain normalization of mineral homeostasis in four pseudohypoparathyroid women [122], but this explanation seems doubtful because of the evidence (cited earlier) that the placenta normally does not contribute a significant amount of calcitriol to the maternal circulation and has 30-fold lower expression of Cyp27b1.

The treatment goal should be to maintain normocalcemia in the mother, thereby minimizing the risk of fetal and neonatal secondary hyperparathyroidism. As with hypoparathyroidism, the approach needs to be expectant, anticipating that there may be either increased or decreased need for supplemental calcium and calcitriol.

23.7 Hypoparathyroidism During Lactation

Increased production of PTHrP by lactating breasts has clinically obvious effects in hypoparathyroid women. Years before PTHrP was identified, clinicians recognized that doses of calcium and calcitriol in hypoparathyroid women need to be decreased or stopped during lactation; otherwise, hypercalcemia occurs [123–127]. In fact, this effect of lactation to "normalize" hypoparathyroidism led to the correct deduction that breastfeeding induces a novel calcium-regulating hormone. High concentrations of PTHrP upregulate bone turnover, renal tubular calcium reabsorption, and endogenous calcitriol formation [126–128], thereby normalizing mineral homeostasis despite the absence of PTH. In clinical studies of nonpregnant adults, PTHrP is less potent than PTH in stimulating Cyp27b1, and that probably explains why calcitriol does not increase above normal [25, 26].

The management plan for hypoparathyroidism should include reducing or stopping calcium and calcitriol as lactation becomes established. Hypercalcemia has occurred as early as the first- or second-day postpartum when the doses were not decreased. When the baby later begins to take solid food, the implied reduction in milk output will be accompanied by a decline in PTHrP. Calcium and calcitriol will need to be reintroduced when the serum calcium drifts below normal and hypocalcemic symptoms resume. However, production of PTHrP can also be sustained well after lactation ceases. The author is aware of a hypoparathyroid woman who normalized mineral homeostasis during lactation and maintained this for over a year after the baby was weaned. Symptomatic hypocalcemia eventually recurred, and calcium and calcitriol had to be resumed. These observations underscore that there will be variability in when the calcium and calcitriol need to be restarted. It may be needed while lactation is still ongoing, and it is most commonly needed at weaning, but it may not be required for months after breastfeeding has ceased.

There are no case reports describing the clinical experience of pseudohypoparathyroid women during lactation (see also Chaps. 32, 33, 34, and 35). As occurs with hypoparathyroidism, lactation should lead to decreased calcium and calcitriol requirements when PTHrP and low estradiol combine to cause increased bone resorption. Skeletal responsiveness to PTH is normal in pseudohypoparathyroidism, and so it is conceivable that these patients may resorb more bone than normal, because they will have the effects of breast-derived PTHrP added to that of concurrently high levels of PTH.

23.8 Effects of Maternal Hypoparathyroidism on Fetal and Neonatal Parathyroid Function

Maternal hypocalcemia due to hypoparathyroidism limits the supply of mineral and provokes compensatory responses in the fetal-placental unit. The placenta upregulates the expression of factors known to be involved in active calcium transport (calbindins, calcium channels, Ca^{2+}-ATPase). Of greater concern is that the fetal parathyroids will enlarge (secondary hyperparathyroidism) and cause resorption of the fetal skeleton. Prolonged maternal hypocalcemia can result in a significantly demineralized fetal skeleton and fractures that occur in utero or during birth [129–133]. Spontaneous abortion, stillbirth, and neonatal death have also been reported [134–136]. The fetal parathyroids are enlarged and hyperplastic, while the cord blood calcium may be normal, low, or even increased. After birth, the increased parathyroid function can appear autonomous for days to weeks, resulting in hypercalcemia and progressive skeletal demineralization before eventually subsiding to normal [111]. These potential outcomes emphasize why significant maternal hypocalcemia must be avoided during pregnancy.

These outcomes are not normally associated with maternal hypocalcemia due to vitamin D deficiency; instead, PTH, calcium, phosphorus, and skeletal mineral content are normal in fetuses born of severely vitamin D-deficient mothers (reviewed in detail in [2]). The explanation is that secondary hyperparathyroidism blunts the fall in serum calcium caused by vitamin D deficiency, whereas hypocalcemia is usually more severe in hypoparathyroidism.

There are limited data examining whether the specific treatments used during pregnancy affect the fetal or neonatal outcomes. One report examined 12 hypoparathyroid women treated with calcitriol and calcium. Ten women delivered healthy babies, while serious adverse events occurred in two others, including premature closure of the frontal fontanelle and stillbirth [113]. It is unknown whether those were chance events or the result of over- or undertreatment of maternal hypoparathyroidism. In nine other women, use of high-dose cholecalciferol, calcitriol, or 1α-calcidiol was not associated with obvious teratogenicity or toxicity [113].

If a pregnant woman with pseudohypoparathyroidism remains hypocalcemic, parathyroid hyperplasia will occur in the fetus, and skeletal demineralization and fractures may result [137, 138]. The autonomous parathyroid function can cause neonatal hypercalcemia before the parathyroids eventually involute. As with hypoparathyroidism, normocalcemia must be maintained in the mothers during pregnancy in order to avoid these complications.

23.9 Fetal and Neonatal Hypoparathyroidism Secondary to Maternal Disturbances

23.9.1 Primary Hyperparathyroidism in the Mother

Maternal hypercalcemia due to primary hyperparathyroidism has caused fetal morbidity and mortality in up to 80 % of published cases [139]. These complications include severe outcomes such as stillbirth, miscarriage, and tetany. When case series from older and more recent decades are compared, the impression is that these outcomes have improved, with stillbirth declining from 13 to 2 %, neonatal death from 8 to 2 %, and neonatal tetany from 38 to 15 % [140]. Fetal death still occurs in modern case series, such that in 30 of 62 medically managed cases the babies were lost in the second or third trimester, which represented a 3.5-fold increased risk of fetal mortality that correlated with the increase in maternal serum calcium [141].

Fetal and neonatal hypoparathyroidism remains an important complication of maternal primary hyperparathyroidism during pregnancy. The pathophysiology may simply be that maternal hypercalcemia increases the flow of calcium across the placenta, which in turn suppresses the fetal parathyroids. The cord blood calcium may be higher than normal fetal values, after which it slowly declines [142]. The normal increase in PTH after birth may be delayed or not occur at all in the baby born of a hypercalcemic mother, thereby leading to hypocalcemia and tetany in the neonate. As much as 50 % of neonates have had some complication of maternal hypercalcemia (most commonly tetany), while 25–30 % of neonates died (presumably from hypocalcemia-induced tetany or arrhythmias) [143–146].

Neonatal hypoparathyroidism caused by maternal hypercalcemia usually resolves by 3–5 months after birth [140], but permanent hypoparathyroidism has also been reported [140, 145, 147]. The presentation may even be delayed, with symptomatic hypocalcemia not occurring until several weeks or months after birth [148–150]. Bottle-feeding increases the risk of hypocalcemia because the higher phosphate content of infant formulas and cow's milk binds calcium more tightly than breast milk does [140].

Fetal and neonatal hypoparathyroidism are not inevitable, since many neonates of hypercalcemic mothers do not show signs of hypocalcemia. The degree of elevation in maternal serum calcium is not an accurate predictor either because neonatal hypocalcemia and tetany have occurred after mild maternal primary hyperparathyroidism [151]. This variability and unpredictability is exemplified by primary hyperparathyroidism complicating a twin pregnancy: one neonate had hypocalcemic seizures while the other remained normocalcemic [152].

When primary hyperparathyroidism occurs during pregnancy, the neonate should be closely monitored for hypocalcemia. The cord blood calcium may be normal or increased, but it is after the normal postnatal fall in serum calcium that problems can develop. In an infant that is normocalcemic up to discharge from hospital, the parents should be advised to look for signs of hypocalcemia that may be delayed for days or weeks. Calcium and calcitriol are the usual treatments, but the latter will not be effective in premature infants due to low intestinal expression of the vitamin D receptor at that early stage of development. Formulas that are high in calcium and low in phosphate will minimize the risk of hypocalcemia. In the infant that does develop hypoparathyroidism, follow-up will reveal whether it is transient or permanent.

23.9.2 Familial Hypocalciuric Hypercalcemia (FHH) in the Mother

The comparatively mild maternal hypercalcemia of FHH is sufficient to cause suppression of fetal parathyroids followed by neonatal hypocalcemia and tetany [153–155]. Even the heterozygous neonate – who will later develop hypercalcemia and hypocalciuria – can present with hypocalcemia and tetany [111]. Hypocalcemia will occur unexpectedly if the mother is not known to have FHH. Animal models have confirmed that fetal

PTH is suppressed by the hypercalcemia in mothers with FHH [156], likely because hypercalcemia causes increased flux of calcium across the placenta. Surveillance and treatment considerations are the same as in babies born of mothers with primary hyperparathyroidism.

23.9.3 Maternal Magnesium Infusions (Tocolytic Therapy)

Intravenous magnesium sulfate ("tocolytic therapy") is used to treat preterm labor, preeclampsia, and eclampsia. Magnesium readily and actively crosses the placenta and can lead to fetal hypermagnesemia, suppressed PTH and parathyroid responsiveness, and variable effects on the total and ionized calcium of neonates [157, 158] (see also Chap. 7). Published reports have described beneficial, neutral, and adverse effects of tocolytic therapy on the fetus and neonate. Respiratory depression and hypotonia were more likely to occur if the mothers had received several days of tocolytic therapy [159, 160]. Prolonged magnesium exposure has caused defective ossification of the bone and enamel in the teeth [161] and abnormal mineralization within the metaphyses of long bones [162–164]. Severe hypermagnesemia (>7 mg/dl) is also more likely to cause hypotonia, respiratory depression, and bone abnormalities [157, 160, 164, 165].

Monitoring of fetal movements is advisable when tocolytic therapy is given for 2 days or longer. Neonates with hypotonia and respiratory depression may need to be ventilated for a day or two. Hypocalcemia may develop from parathyroid suppression. Intravenous calcium has also been used to reverse central nervous system depression and peripheral neuromuscular blockade caused by the high magnesium level [160].

23.9.4 Hypercalcemia of Malignancy (HoM) in the Mother

HoM is uncommon, but maternal hypercalcemia can be more severe and likely to cause fetal and (possibly permanent) neonatal hypoparathyroidism. Among limited data from babies born of mothers with HoM, hypercalcemia was present in cord blood and the first few postnatal samples [166–168]. It should be expected that after the serum calcium falls, the neonate will be at high risk for hypocalcemia, respiratory distress, tetany, and seizures [169, 170]. The baby died in one of four cases where the outcome was reported [169].

High doses of bisphosphonates have been used to treat affected women during pregnancy. Bisphosphonates cross the placenta and, at least theoretically, could impair endochondral bone development and reduce bone turnover, thereby lowering the blood calcium. Fetal and neonatal hypocalcemia have been reported in cases where women received bisphosphonates to treat HHM or other causes of hypercalcemia during pregnancy, but since maternal hypercalcemia itself causes fetal hypocalcemia, one cannot determine from those few cases whether bisphosphonates added to the risk of hypoparathyroidism in the baby. Pamidronate was used to treat HoM in two reported pregnancies, and no adverse effects were reported in the neonates [166, 168].

23.9.5 Pseudohyperparathyroidism in the Mother

Excess production of PTHrP by the breasts (pseudohyperparathyroidism) during pregnancy can cause maternal hypercalcemia. In one case, the baby was normal after the maternal condition was corrected by mastectomy during pregnancy [23, 24]. In another case, the baby was hypercalcemic for several days after birth and only mildly hypocalcemic thereafter [21]. As with other causes of maternal hypercalcemia during pregnancy, these babies are at increased risk of neonatal hypoparathyroidism and hypocalcemia.

23.9.6 Maternal Diabetes Causing Fetal and Neonatal Hypoparathyroidism

Poorly controlled maternal diabetes is a known risk factor for neonatal hypocalcemia, seizures, and tetany. Hyperphosphatemia is often present and provides an additional indication that the neonatal parathyroids are hypofunctioning. Why maternal diabetes can cause this outcome remains unclear. One case series found that, compared to controls, neonates born of diabetic mothers had higher ionized and total serum calcium in the cord blood and were more likely to have prolonged parathyroid suppression [171]. A high cord blood calcium can explain suppression of the fetal and neonatal parathyroids, but how maternal diabetes might cause fetal hypercalcemia is unknown. Another theory is that maternal hypomagnesemia results from glucosuria-induced wasting of magnesium in pregnant diabetic women, and in turn maternal hypomagnesemia causes fetal hypomagnesemia and parathyroid suppression. However, there are no data on cord blood magnesium in babies born of diabetic mothers, and the data cited earlier suggest that the cord blood calcium in these babies is increased, not decreased. Furthermore, postnatal magnesium supplementation had no effect on the incidence of neonatal hypocalcemia in infants of diabetic mothers [172]. Hypocalcemia also occurs in infants of diabetic mothers because they are at increased risk of preterm birth, lung immaturity, and asphyxia.

23.10 Fetal and Neonatal Hypoparathyroidism Due to Primary Parathyroid Disorders

23.10.1 Hypoparathyroidism

Hypoparathyroidism can result from genetic deletion of the parathyroids (DiGeorge and other 22q11.2 deletion syndromes, ablation of Gcm2, etc.), activating mutations of the calcium-sensing receptor, and other genetic mutations [173]. The absence of parathyroids does not necessarily cause symptomatic hypocalcemia; instead, development of hypocalcemia in utero may enable the affected neonate, infant, and child to be accustomed to a low level of calcium that a normal child would find intolerable. In one large series of 22q11.2 deletions, only 60 % of affected individuals ever had hypocalcemia [174]. In those who did develop hypocalcemia, most presented as neonates (some with seizures), but in others the hypocalcemia occurred in childhood or as late as 18 years of age [174]. In another series of 12 cases with confirmed hypocalcemia, only 4 were symptomatic, 10 were diagnosed before 1 month of age, and the remaining 2 presented at 3 months and 12 years of age, respectively [175]. In another series of 10 cases of hypoparathyroidism, the age at diagnosis ranged from 9 days to 13 years [176].

Numerous animal species (sheep, rats, mice) and models (loss of PTH, parathyroids, or the type 1 PTH receptor) have confirmed that the loss of PTH causes hypocalcemia in utero [47, 51]. Fetal mice lacking parathyroids (*Hoxa3* null) or the type 1 PTH receptor (*Pth1r* nulls) have the lowest blood calcium (well below the maternal level), whereas mice lacking PTH (*Pth* nulls) have a blood calcium equal to the maternal level [47, 51, 70]. The absence of PTH also causes hyperphosphatemia and hypomagnesemia in thyroparathyroidectomized fetal lambs and rats, and in all PTH-deficient fetal mouse models [47, 51]. The skeletons in these PTH-deficient fetuses were also under-mineralized. The long bones of aparathyroid or PTH-deficient mice showed either normal or slightly shortened lengths, depending upon the genetic background [47, 51, 58].

23.10.2 Pseudohypoparathyroidism

Clinical experience of pseudohypoparathyroidism is that hypocalcemia develops later in childhood, preceded by an interval of hyperphosphatemia and elevated PTH [177, 178]. Although no data have been specifically reported from affected newborns, it is anticipated that cord blood calcium will be normal and neonatal hypocalcemia unlikely to occur.

23.10.3 Deficient Production of PTHrP

Mice lacking the PTHrP gene have a hypoparathyroid phenotype in utero with hypocalcemia, hyperphosphatemia, reduced placental calcium transport, and dwarfism due to accelerated endochondral bone formation [69, 70]. They normally die at birth, but a few have survived for several days. Death may be due to a variety of causes, including hypocalcemia, impaired ventilation due to an abnormally calcified rib cage, and pulmonary abnormalities that include deficient type II alveolar cells and surfactant deficiency [69, 70, 179]. The human equivalent has not been reported but would presumably cause hypocalcemia, skeletal abnormalities, and lethality at or soon after birth. An autosomal dominant microdeletion in the PTHrP gene has been linked to the nonlethal human condition of brachydactyly type E. It is characterized by short stature and shortened metacarpals and metatarsals; calcium and phosphorus are normal [180].

23.10.4 Absent Type 1 PTH Receptor

Pth1r null fetal mice have a phenotype that is similar but more severe than the absence of PTHrP, including dwarfism, hypocalcemia, hyperphosphatemia, shortened limbs due to accelerated endochondral bone formation, and lethality at birth (embryonic lethality in some genetic backgrounds) [70, 181] (see also Chap. 9). The human equivalent is Blomstrand chondrodysplasia, which is characterized by accelerated endochondral ossification and dysplasia, and lethality in utero [182, 183]. Hypocalcemia and hyperphosphatemia are likely present but have not been measured.

Conclusions

Pregnancy invokes a doubling of intestinal calcium absorption to meet the fetal demand for calcium. In contrast, lactation programs increased bone resorption in order to provide calcium to the breast milk. These adaptations during normal pregnancy and lactation can lead to novel presentations and management issues for hypoparathyroid women.

Fetal calcium metabolism is regulated differently from that is the adult, but the loss of the placenta and decline in serum calcium after birth invoke a switch to adult regulatory mechanisms. Maternal hypoparathyroidism can cause fetal and neonatal hyperparathyroidism. Fetal and neonatal hypoparathyroidism can be caused by maternal disorders during pregnancy (hypercalcemia, diabetes, and use of tocolysis) and by disorders that directly impair the baby's parathyroid anatomy or function. If the mother has abnormal calcium homeostasis during pregnancy, then the fetus and neonate must be closely monitored for abnormalities. Similarly, if a neonate has an unexpected disorder of calcium homeostasis, the mother should be assessed for an undiagnosed disturbance of parathyroid function.

References

1. Kovacs CS, Kronenberg HM (1997) Maternal-fetal calcium and bone metabolism during pregnancy, puerperium and lactation. Endocr Rev 18:832–872
2. Kovacs CS (2012) The role of vitamin D in pregnancy and lactation: insights from animal models and clinical studies. Ann Rev Nutr 32:9.1–9.27
3. Kovacs CS (2012) Control of skeletal homeostasis during pregnancy and lactation – lessons from physiological models. In: Thakker RV, Whyte MP, Eisman JA, Igarashi T (eds) Genetics of bone biology and skeletal disease. Academic Press/Elsevier, San Diego, pp 221–240
4. Kovacs CS (2012) Fetal control of calcium and phosphate homeostasis – lessons from mouse models. In: Thakker RV, Whyte MP, Eisman JA, Igarashi T (eds) Genetics of bone biology and skeletal disease. Academic Press/Elsevier, San Diego, pp 205–220
5. Kovacs CS (2011) Calcium and bone metabolism disorders during pregnancy and lactation. Endocrinol Metab Clin North Am 40:795–826
6. Givens MH, Macy IC (1933) The chemical composition of the human fetus. J Biol Chem 102:7–17
7. Trotter M, Hixon BB (1974) Sequential changes in weight, density, and percentage ash weight of human skeletons from an early fetal period through old age. Anat Rec 179:1–18
8. Widdowson EM, Dickerson JW (1964) Chemical composition of the body. In: Comar CL, Bronner F (eds) Mineral metabolism: an advanced treatise, volume II, the elements, part A. Academic Press, New York, pp 1–247
9. Dahlman T, Sjoberg HE, Bucht E (1994) Calcium homeostasis in normal pregnancy and puerperium. A longitudinal study. Acta Obstet Gynecol Scand 73:393–398

10. Gallacher SJ, Fraser WD, Owens OJ, Dryburgh FJ, Logue FC, Jenkins A, Kennedy J, Boyle IT (1994) Changes in calciotrophic hormones and biochemical markers of bone turnover in normal human pregnancy. Eur J Endocrinol 131:369–374
11. Cross NA, Hillman LS, Allen SH, Krause GF, Vieira NE (1995) Calcium homeostasis and bone metabolism during pregnancy, lactation, and postweaning: a longitudinal study. Am J Clin Nutr 61:514–523
12. Rasmussen N, Frolich A, Hornnes PJ, Hegedus L (1990) Serum ionized calcium and intact parathyroid hormone levels during pregnancy and postpartum. Br J Obstet Gynaecol 97:857–859
13. Seki K, Makimura N, Mitsui C, Hirata J, Nagata I (1991) Calcium-regulating hormones and osteocalcin levels during pregnancy: a longitudinal study. Am J Obstet Gynecol 164:1248–1252
14. Ardawi MS, Nasrat HA, BA'Aqueel HS (1997) Calcium-regulating hormones and parathyroid hormone-related peptide in normal human pregnancy and postpartum: a longitudinal study. Eur J Endocrinol 137:402–409
15. Singh HJ, Mohammad NH, Nila A (1999) Serum calcium and parathormone during normal pregnancy in Malay women. J Matern Fetal Med 8:95–100
16. Bikle DD, Gee E, Halloran B, Haddad JG (1984) Free 1,25-dihydroxyvitamin D levels in serum from normal subjects, pregnant subjects, and subjects with liver disease. J Clin Invest 74:1966–1971
17. Kirby BJ, Ma Y, Martin HM, Favaro KL, Karaplis AC, Kovacs CS (2013) Upregulation of calcitriol during pregnancy and skeletal recovery after lactation do not require parathyroid hormone. J Bone Miner Res 28(9):1987–2000
18. Turner M, Barre PE, Benjamin A, Goltzman D, Gascon-Barre M (1988) Does the maternal kidney contribute to the increased circulating 1,25-dihydroxyvitamin D concentrations during pregnancy? Miner Electrolyte Metab 14:246–252
19. Sweeney LL, Malabanan AO, Rosen H (2010) Decreased calcitriol requirement during pregnancy and lactation with a window of increased requirement immediately post partum. Endocr Pract 16:459–462
20. Eller-Vainicher C, Ossola MW, Beck-Peccoz P, Chiodini I (2012) PTHrP-associated hypercalcemia of pregnancy resolved after delivery: a case report. Eur J Endocrinol 166:753–756
21. Sato K (2008) Hypercalcemia during pregnancy, puerperium, and lactation: review and a case report of hypercalcemic crisis after delivery due to excessive production of PTH-related protein (PTHrP) without malignancy (humoral hypercalcemia of pregnancy). Endocr J 55:959–966
22. Marx SJ, Zusman RM, Umiker WO (1977) Benign breast dysplasia causing hypercalcemia. J Clin Endocrinol Metab 45:1049–1052
23. Khosla S, van Heerden JA, Gharib H, Jackson IT, Danks J, Hayman JA, Martin TJ (1990) Parathyroid hormone-related protein and hypercalcemia secondary to massive mammary hyperplasia [letter]. N Engl J Med 322:1157
24. Jackson IT, Saleh J, van Heerden JA (1989) Gigantic mammary hyperplasia in pregnancy associated with pseudohyperparathyroidism. Plast Reconstr Surg 84:806–810
25. Horwitz MJ, Tedesco MB, Sereika SM, Hollis BW, Garcia-Ocana A, Stewart AF (2003) Direct comparison of sustained infusion of human parathyroid hormone-related protein-(1–36). J Clin Endocrinol Metab 88:1603–1609
26. Horwitz MJ, Tedesco MB, Sereika SM, Syed MA, Garcia-Ocana A, Bisello A, Hollis BW, Rosen CJ, Wysolmerski JJ, Dann P, Gundberg C, Stewart AF (2005) Continuous PTH and PTHrP infusion causes suppression of bone formation and discordant effects on 1,25(OH)2 vitamin D. J Bone Miner Res 20:1792–1803
27. Cornish J, Callon KE, Nicholson GC, Reid IR (1997) Parathyroid hormone-related protein-(107–139) inhibits bone resorption in vivo. Endocrinology 138:1299–1304
28. Fenton AJ, Kemp BE, Hammonds RG Jr, Mitchelhill K, Moseley JM, Martin TJ, Nicholson GC (1991) A potent inhibitor of osteoclastic bone resorption within a highly conserved pentapeptide region of parathyroid hormone-related protein; PTHrP[107–111]. Endocrinology 129:3424–3426
29. Fenton AJ, Kemp BE, Kent GN, Moseley JM, Zheng MH, Rowe DJ, Britto JM, Martin TJ, Nicholson GC (1991) A carboxyl-terminal peptide from the parathyroid hormone-related protein inhibits bone resorption by osteoclasts. Endocrinology 129:1762–1768
30. Heaney RP, Skillman TG (1971) Calcium metabolism in normal human pregnancy. J Clin Endocrinol Metab 33:661–670
31. Halloran BP, DeLuca HF (1980) Calcium transport in small intestine during pregnancy and lactation. Am J Physiol 239:E64–E68
32. Brommage R, Baxter DC, Gierke LW (1990) Vitamin D-independent intestinal calcium and phosphorus absorption during reproduction. Am J Physiol 259:G631–G638
33. Fudge NJ, Kovacs CS (2010) Pregnancy up-regulates intestinal calcium absorption and skeletal mineralization independently of the vitamin D receptor. Endocrinology 151:886–895
34. Wasserman RH, Comar CL, Nold MM, Lengemann FW (1957) Placental transfer of calcium and strontium in the rat and rabbit. Am J Physiol 189:91–97
35. Purdie DW, Aaron JE, Selby PL (1988) Bone histology and mineral homeostasis in human pregnancy. Br J Obstet Gynaecol 95:849–854
36. Kovacs CS (2001) Calcium and bone metabolism in pregnancy and lactation. J Clin Endocrinol Metab 86:2344–2348
37. Naylor KE, Iqbal P, Fledelius C, Fraser RB, Eastell R (2000) The effect of pregnancy on bone density and bone turnover. J Bone Miner Res 15:129–137

38. Black AJ, Topping J, Durham B, Farquharson RG, Fraser WD (2000) A detailed assessment of alterations in bone turnover, calcium homeostasis, and bone density in normal pregnancy. J Bone Miner Res 15:557–563
39. Ritchie LD, Fung EB, Halloran BP, Turnlund JR, Van Loan MD, Cann CE, King JC (1998) A longitudinal study of calcium homeostasis during human pregnancy and lactation and after resumption of menses. Am J Clin Nutr 67:693–701
40. Ulrich U, Miller PB, Eyre DR, Chesnut CH, Schlebusch H, Soules MR (2003) Bone remodeling and bone mineral density during pregnancy. Arch Gynecol Obstet 268:309–316
41. Kaur M, Pearson D, Godber I, Lawson N, Baker P, Hosking D (2003) Longitudinal changes in bone mineral density during normal pregnancy. Bone 32:449–454
42. Gambacciani M, Spinetti A, Gallo R, Cappagli B, Teti GC, Facchini V (1995) Ultrasonographic bone characteristics during normal pregnancy: longitudinal and cross-sectional evaluation. Am J Obstet Gynecol 173:890–893
43. Pearson D, Kaur M, San P, Lawson N, Baker P, Hosking D (2004) Recovery of pregnancy mediated bone loss during lactation. Bone 34:570–578
44. To WW, Wong MW (2011) Changes in bone mineral density of the os calcis as measured by quantitative ultrasound during pregnancy and 24 months after delivery. Aust N Z J Obstet Gynaecol 51:166–171
45. Sowers M (1996) Pregnancy and lactation as risk factors for subsequent bone loss and osteoporosis. J Bone Miner Res 11:1052–1060
46. Paton LM, Alexander JL, Nowson CA, Margerison C, Frame MG, Kaymakci B, Wark JD (2003) Pregnancy and lactation have no long-term deleterious effect on measures of bone mineral in healthy women: a twin study. Am J Clin Nutr 77:707–714
47. Simmonds CS, Karsenty G, Karaplis AC, Kovacs CS (2010) Parathyroid hormone regulates fetal-placental mineral homeostasis. J Bone Miner Res 25:594–605
48. Kovacs CS (2003) Fetal mineral homeostasis. In: Glorieux FH, Pettifor JM, Jüppner H (eds) Pediatric bone: biology and diseases. Academic Press, San Diego, pp 271–302
49. Kovacs CS (2011) Fetal mineral homeostasis. In: Glorieux FH, Pettifor JM, Jüppner H (eds) Pediatric bone: biology and diseases, 2nd edn. Elsevier/Academic Press, San Diego, pp 247–275
50. Kovacs CS, Chafe LL, Fudge NJ, Friel JK, Manley NR (2001) PTH regulates fetal blood calcium and skeletal mineralization independently of PTHrP. Endocrinology 142:4983–4993
51. Kovacs CS, Manley NR, Moseley JM, Martin TJ, Kronenberg HM (2001) Fetal parathyroids are not required to maintain placental calcium transport. J Clin Invest 107:1007–1015
52. Suzuki Y, Kovacs CS, Takanaga H, Peng JB, Landowski CP, Hediger MA (2008) Calcium TRPV6 is involved in murine maternal-fetal calcium transport. J Bone Miner Res 23:1249–1256
53. Loughead JL, Mimouni F, Tsang RC (1988) Serum ionized calcium concentrations in normal neonates. Am J Dis Child 142:516–518
54. David L, Anast CS (1974) Calcium metabolism in newborn infants. The interrelationship of parathyroid function and calcium, magnesium, and phosphorus metabolism in normal, sick, and hypocalcemic newborns. J Clin Invest 54:287–296
55. Schauberger CW, Pitkin RM (1979) Maternal-perinatal calcium relationships. Obstet Gynecol 53:74–76
56. Garel JM, Barlet JP (1976) Calcium metabolism in newborn animals: the interrelationship of calcium, magnesium, and inorganic phosphorus in newborn rats, foals, lambs, and calves. Pediatr Res 10:749–754
57. Krukowski M, Smith JJ (1976) pH and the level of calcium in the blood of fetal and neonatal albino rats. Biol Neonate 29:148–161
58. Miao D, He B, Karaplis AC, Goltzman D (2002) Parathyroid hormone is essential for normal fetal bone formation. J Clin Invest 109:1173–1182
59. Haddad JG Jr, Boisseau V, Avioli LV (1971) Placental transfer of vitamin D_3 and 25-hydroxycholecalciferol in the rat. J Lab Clin Med 77:908–915
60. Noff D, Edelstein S (1978) Vitamin D and its hydroxylated metabolites in the rat. Placental and lacteal transport, subsequent metabolic pathways and tissue distribution. Horm Res 9:292–300
61. Hillman LS, Haddad JG (1974) Human perinatal vitamin D metabolism. I. 25-Hydroxyvitamin D in maternal and cord blood. J Pediatr 84:742–749
62. Wieland P, Fischer JA, Trechsel U, Roth HR, Vetter K, Schneider H, Huch A (1980) Perinatal parathyroid hormone, vitamin D metabolites, and calcitonin in man. Am J Physiol 239:E385–E390
63. Fleischman AR, Rosen JF, Cole J, Smith CM, DeLuca HF (1980) Maternal and fetal serum 1,25-dihydroxyvitamin D levels at term. J Pediatr 97:640–642
64. Seki K, Furuya K, Makimura N, Mitsui C, Hirata J, Nagata I (1994) Cord blood levels of calcium-regulating hormones and osteocalcin in premature infants. J Perinat Med 22:189–194
65. Hollis BW, Pittard WB (1984) Evaluation of the total fetomaternal vitamin D relationships at term: evidence for racial differences. J Clin Endocrinol Metab 59:652–657
66. Viljakainen HT, Saarnio E, Hytinantti T, Miettinen M, Surcel H, Makitie O, Andersson S, Laitinen K, Lamberg-Allardt C (2010) Maternal vitamin D status determines bone variables in the newborn. J Clin Endocrinol Metab 95:1749–1757
67. Ma Y, Samaraweera M, Cooke-Hubley S, Kirby BJ, Karaplis AC, Lanske B, Kovacs CS (2014) Neither absence nor excess of FGF23 disturbs murine fetal-placental phosphorus homeostasis or prenatal

skeletal development and mineralization. Endocrinology 155:1596–1605
68. Kovacs CS (2011) Fetus, neonate and infant. In: Feldman D, Pike WJ, Adams JS (eds) Vitamin D, 3rd edn. Academic Press, New York, pp 625–646
69. Karaplis AC, Luz A, Glowacki J, Bronson RT, Tybulewicz VL, Kronenberg HM, Mulligan RC (1994) Lethal skeletal dysplasia from targeted disruption of the parathyroid hormone-related peptide gene. Genes Dev 8:277–289
70. Kovacs CS, Lanske B, Hunzelman JL, Guo J, Karaplis AC, Kronenberg HM (1996) Parathyroid hormone-related peptide (PTHrP) regulates fetal-placental calcium transport through a receptor distinct from the PTH/PTHrP receptor. Proc Natl Acad Sci U S A 93:15233–15238
71. Simmonds CS, Kovacs CS (2010) Role of parathyroid hormone (PTH) and PTH-related protein (PTHrP) in regulating mineral homeostasis during fetal development. Crit Rev Eukaryot Gene Expr 20:235–273
72. Kovacs CS (2011) Bone development in the fetus and neonate: role of the calciotropic hormones. Curr Osteoporos Rep 9:274–283
73. Care AD (1997) Fetal calcium homeostasis. Equine Vet J Suppl 24:59–61
74. Barlet JP, Davicco MJ, Rouffet J, Coxam V, Lefaivre J (1994) Short communication: parathyroid hormone-related peptide does not stimulate phosphate placental transport. Placenta 15:441–444
75. Stulc J, Stulcova B (1996) Placental transfer of phosphate in anaesthetized rats. Placenta 17:487–493
76. Dobnig H, Kainer F, Stepan V, Winter R, Lipp R, Schaffer M, Kahr A, Nocnik S, Patterer G, Leb G (1995) Elevated parathyroid hormone-related peptide levels after human gestation: relationship to changes in bone and mineral metabolism. J Clin Endocrinol Metab 80:3699–3707
77. Yamamoto M, Duong LT, Fisher JE, Thiede MA, Caulfield MP, Rosenblatt M (1991) Suckling-mediated increases in urinary phosphate and 3′,5′-cyclic adenosine monophosphate excretion in lactating rats: possible systemic effects of parathyroid hormone-related protein. Endocrinology 129:2614–2622
78. Ratcliffe WA, Thompson GE, Care AD, Peaker M (1992) Production of parathyroid hormone-related protein by the mammary gland of the goat. J Endocrinol 133:87–93
79. VanHouten JN, Dann P, Stewart AF, Watson CJ, Pollak M, Karaplis AC, Wysolmerski JJ (2003) Mammary-specific deletion of parathyroid hormone-related protein preserves bone mass during lactation. J Clin Invest 112:1429–1436
80. Sowers MF, Hollis BW, Shapiro B, Randolph J, Janney CA, Zhang D, Schork A, Crutchfield M, Stanczyk F, Russell-Aulet M (1996) Elevated parathyroid hormone-related peptide associated with lactation and bone density loss. JAMA 276:549–554
81. Kovacs CS, Chik CL (1995) Hyperprolactinemia caused by lactation and pituitary adenomas is associated with altered serum calcium, phosphate, parathyroid hormone (PTH), and PTH-related peptide levels. J Clin Endocrinol Metab 80:3036–3042
82. Woodrow JP, Sharpe CJ, Fudge NJ, Hoff AO, Gagel RF, Kovacs CS (2006) Calcitonin plays a critical role in regulating skeletal mineral metabolism during lactation. Endocrinology 147:4010–4021
83. Cross NA, Hillman LS, Allen SH, Krause GF (1995) Changes in bone mineral density and markers of bone remodeling during lactation and postweaning in women consuming high amounts of calcium. J Bone Miner Res 10:1312–1320
84. Kalkwarf HJ, Specker BL, Bianchi DC, Ranz J, Ho M (1997) The effect of calcium supplementation on bone density during lactation and after weaning. N Engl J Med 337:523–528
85. Polatti F, Capuzzo E, Viazzo F, Colleoni R, Klersy C (1999) Bone mineral changes during and after lactation. Obstet Gynecol 94:52–56
86. Kolthoff N, Eiken P, Kristensen B, Nielsen SP (1998) Bone mineral changes during pregnancy and lactation: a longitudinal cohort study. Clin Sci (Colch) 94:405–412
87. Hollis BW, Wagner CL (2004) Vitamin D requirements during lactation: high-dose maternal supplementation as therapy to prevent hypovitaminosis D for both the mother and the nursing infant. Am J Clin Nutr 80:1752S–1758S
88. Basile LA, Taylor SN, Wagner CL, Horst RL, Hollis BW (2006) The effect of high-dose vitamin D supplementation on serum vitamin D levels and milk calcium concentration in lactating women and their infants. Breastfeed Med 1:27–35
89. Wagner CL, Hulsey TC, Fanning D, Ebeling M, Hollis BW (2006) High-dose vitamin D3 supplementation in a cohort of breastfeeding mothers and their infants: a 6-month follow-up pilot study. Breastfeed Med 1:59–70
90. Laskey MA, Prentice A (1997) Effect of pregnancy on recovery of lactational bone loss [letter]. Lancet 349:1518–1519
91. Prentice A (2000) Calcium in pregnancy and lactation. Annu Rev Nutr 20:249–272
92. Prentice A, Jarjou LM, Cole TJ, Stirling DM, Dibba B, Fairweather-Tait S (1995) Calcium requirements of lactating Gambian mothers: effects of a calcium supplement on breast-milk calcium concentration, maternal bone mineral content, and urinary calcium excretion. Am J Clin Nutr 62:58–67
93. Prentice A, Jarjou LM, Stirling DM, Buffenstein R, Fairweather-Tait S (1998) Biochemical markers of calcium and bone metabolism during 18 months of lactation in Gambian women accustomed to a low calcium intake and in those consuming a calcium supplement. J Clin Endocrinol Metab 83:1059–1066
94. Prentice A, Yan L, Jarjou LM, Dibba B, Laskey MA, Stirling DM, Fairweather-Tait S (1997) Vitamin D status does not influence the breast-milk calcium concentration of lactating mothers accustomed to a low calcium intake. Acta Paediatr 86:1006–1008

95. Laskey MA, Prentice A, Hanratty LA, Jarjou LM, Dibba B, Beavan SR, Cole TJ (1998) Bone changes after 3 mo of lactation: influence of calcium intake, breast-milk output, and vitamin D-receptor genotype. Am J Clin Nutr 67:685–692
96. Kovacs CS (2005) Calcium and bone metabolism during pregnancy and lactation. J Mammary Gland Biol Neoplasia 10:105–118
97. Kirby BJ, Ardeshirpour L, Woodrow JP, Wysolmerski JJ, Sims NA, Karaplis AC, Kovacs CS (2011) Skeletal recovery after weaning does not require PTHrP. J Bone Miner Res 26:1242–1251
98. Collins JN, Kirby BJ, Woodrow JP, Gagel RF, Rosen CJ, Sims NA, Kovacs CS (2013) Lactating Ctcgrp nulls lose twice the normal bone mineral content due to fewer osteoblasts and more osteoclasts, whereas bone mass is fully restored after weaning in association with up-regulation of Wnt signaling and other novel genes. Endocrinology 154:1400–1413
99. Liu XS, Ardeshirpour L, VanHouten JN, Shane E, Wysolmerski JJ (2012) Site-specific changes in bone microarchitecture, mineralization, and stiffness during lactation and after weaning in mice. J Bone Miner Res 27:865–875
100. Bjørnerem A, Ghasem-Zadeh A, Vu T, Seeman E (2010) Bone microstructure during and after lactation [abstract]. J Bone Miner Res 25(Suppl 1):S121
101. Wiklund PK, Xu L, Wang Q, Mikkola T, Lyytikainen A, Volgyi E, Munukka E, Cheng SM, Alen M, Keinanen-Kiukaanniemi S, Cheng S (2012) Lactation is associated with greater maternal bone size and bone strength later in life. Osteoporos Int 23:1939–1945
102. Bowman BM, Miller SC (2001) Skeletal adaptations during mammalian reproduction. J Musculoskelet Neuronal Interact 1:347–355
103. Vajda EG, Bowman BM, Miller SC (2001) Cancellous and cortical bone mechanical properties and tissue dynamics during pregnancy, lactation, and postlactation in the rat. Biol Reprod 65:689–695
104. Giles MM, Fenton MH, Shaw B, Elton RA, Clarke M, Lang M, Hume R (1987) Sequential calcium and phosphorus balance studies in preterm infants. J Pediatr 110:591–598
105. Barltrop D, Oppe TE (1973) Calcium and fat absorption by low birthweight infants from a calcium-supplemented milk formula. Arch Dis Child 48:580–582
106. Kobayashi A, Kawai S, Obe Y, Nagashima Y (1975) Effects of dietary lactose and lactase preparation on the intestinal absorption of calcium and magnesium in normal infants. Am J Clin Nutr 28:681–683
107. Kocian J, Skala I, Bakos K (1973) Calcium absorption from milk and lactose-free milk in healthy subjects and patients with lactose intolerance. Digestion 9:317–324
108. Shaw JC (1976) Evidence for defective skeletal mineralization in low-birthweight infants: the absorption of calcium and fat. Pediatrics 57:16–25
109. Senterre J, Salle B (1982) Calcium and phosphorus economy of the preterm infant and its interaction with vitamin D and its metabolites. Acta Paediatr Scand Suppl 296:85–92
110. Buchowski MS, Miller DD (1991) Lactose, calcium source and age affect calcium bioavailability in rats. J Nutr 121:1746–1754
111. Thomas AK, McVie R, Levine SN (1999) Disorders of maternal calcium metabolism implicated by abnormal calcium metabolism in the neonate. Am J Perinatol 16:515–520
112. Demirel N, Aydin M, Zenciroglu A, Okumus N, Cetinkaya S, Yildiz YT, Ipek MS (2009) Hyperparathyroidism secondary to maternal hypoparathyroidism and vitamin D deficiency: an uncommon cause of neonatal respiratory distress. Ann Trop Paediatr 29:149–154
113. Callies F, Arlt W, Scholz HJ, Reincke M, Allolio B (1998) Management of hypoparathyroidism during pregnancy–report of twelve cases. Eur J Endocrinol 139:284–289
114. Krysiak R, Kobielusz-Gembala I, Okopien B (2011) Hypoparathyroidism in pregnancy. Gynecol Endocrinol 27:529–532
115. Landing BH, Kamoshita S (1970) Congenital hyperparathyroidism secondary to maternal hypoparathyroidism. J Pediatr 77:842–847
116. Kurzel RB, Hagen GA (1990) Use of thiazide diuretics to reduce the hypercalciuria of hypoparathyroidism during pregnancy. Am J Perinatol 7:333–336
117. Taussig HB (1966) Possible injury to the cardiovascular system from vitamin D. Ann Intern Med 65:1195–1200
118. Friedman WF, Mills LF (1969) The relationship between vitamin D and the craniofacial and dental anomalies of the supravalvular aortic stenosis syndrome. Pediatrics 43:12–18
119. Breslau NA, Zerwekh JE (1986) Relationship of estrogen and pregnancy to calcium homeostasis in pseudohypoparathyroidism. J Clin Endocrinol Metab 62:45–51
120. Saito H, Saito M, Saito K, Terauchi A, Kobayashi T, Tominaga T, Hosoi E, Senoo M, Saito T (1989) Subclinical pseudohypoparathyroidism type II: evidence for failure of physiologic adjustment in calcium metabolism during pregnancy. Am J Med Sci 297:247–250
121. Seki K, Osada H, Yasuda T, Sekiya S (1999) Pseudohypoparathyroidism type 1b in pregnancy. Gynecol Obstet Invest 47:278–280
122. Zerwekh JE, Breslau NA (1986) Human placental production of 1a,25-dihydroxyvitamin D_3: biochemical characterization and production in normal subjects and patients with pseudohypoparathyroidism. J Clin Endocrinol Metab 62:192–196

123. Salle BL, Berthezene F, Glorieux FH, Delvin EE, Berland M, David L, Varenne JP, Putet G (1981) Hypoparathyroidism during pregnancy: treatment with calcitriol. J Clin Endocrinol Metab 52:810–813
124. Sadeghi-Nejad A, Wolfsdorf JI, Senior B (1980) Hypoparathyroidism and pregnancy. Treatment with calcitriol. JAMA 243:254–255
125. Cathebras P, Cartry O, Sassolas G, Rousset H (1996) Hypercalcemia induced by lactation in 2 patients with treated hypoparathyroidism. Rev Med Interne 17:675–676
126. Shomali ME, Ross DS (1999) Hypercalcemia in a woman with hypoparathyroidism associated with increased parathyroid hormone-related protein during lactation. Endocr Pract 5:198–200
127. Caplan RH, Beguin EA (1990) Hypercalcemia in a calcitriol-treated hypoparathyroid woman during lactation. Obstet Gynecol 76:485–489
128. Mather KJ, Chik CL, Corenblum B (1999) Maintenance of serum calcium by parathyroid hormone-related peptide during lactation in a hypoparathyroid patient. J Clin Endocrinol Metab 84:424–427
129. Bronsky D, Kiamko RT, Moncada R, Rosenthal IM (1968) Intra-uterine hyperparathyroidism secondary to maternal hypoparathyroidism. Pediatrics 42:606–613
130. Stuart C, Aceto T Jr, Kuhn JP, Terplan K (1979) Intrauterine hyperparathyroidism. Postmortem findings in two cases. Am J Dis Child 133:67–70
131. Loughead JL, Mughal Z, Mimouni F, Tsang RC, Oestreich AE (1990) Spectrum and natural history of congenital hyperparathyroidism secondary to maternal hypocalcemia. Am J Perinatol 7:350–355
132. Sann L, David L, Thomas A, Frederich A, Chapuy MC, Francois R (1976) Congenital hyperparathyroidism and vitamin D deficiency secondary to maternal hypoparathyroidism. Acta Paediatr Scand 65:381–385
133. Aceto T Jr, Batt RE, Bruck E, Schultz RB, Perz YR (1966) Intrauterine hyperparathyroidism: a complication of untreated maternal hypoparathyroidism. J Clin Endocrinol Metab 26:487–492
134. Eastell R, Edmonds CJ, de Chayal RC, McFadyen IR (1985) Prolonged hypoparathyroidism presenting eventually as second trimester abortion. Br Med J (Clin Res Ed) 291:955–956
135. Anderson GW, Musselman L (1942) The treatment of tetany in pregnancy. Am J Obstet Gynecol 43:547–567
136. Kehrer E (1913) Die geburtschilflich-gynäkologische bedeutung der tetanie. Arch Gynaek 99:372–447
137. Glass EJ, Barr DG (1981) Transient neonatal hyperparathyroidism secondary to maternal pseudohypoparathyroidism. Arch Dis Child 56:565–568
138. Vidailhet M, Monin P, Andre M, Suty Y, Marchal C, Vert P (1980) Neonatal hyperparathyroidism secondary to maternal hypoparathyroidism. Arch Fr Pediatr 37:305–312
139. Schnatz PF, Curry SL (2002) Primary hyperparathyroidism in pregnancy: evidence-based management. Obstet Gynecol Surv 57:365–376
140. Shangold MM, Dor N, Welt SI, Fleischman AR, Crenshaw MC Jr (1982) Hyperparathyroidism and pregnancy: a review. Obstet Gynecol Surv 37:217–228
141. Norman J, Politz D, Politz L (2009) Hyperparathyroidism during pregnancy and the effect of rising calcium on pregnancy loss: a call for earlier intervention. Clin Endocrinol (Oxf) 71:104–109
142. Shani H, Sivan E, Cassif E, Simchen MJ (2008) Maternal hypercalcemia as a possible cause of unexplained fetal polyhydramnion: a case series. Am J Obstet Gynecol 199(410):e411–e415
143. Kelly TR (1991) Primary hyperparathyroidism during pregnancy. Surgery 110:1028–1033; discussion 1033–1034
144. Wagner G, Transhol L, Melchior JC (1964) Hyperparathyroidism and pregnancy. Acta Endocrinol (Copenh) 47:549–564
145. Ludwig GD (1962) Hyperparathyroidism in relation to pregnancy. N Engl J Med 267:637–642
146. Delmonico FL, Neer RM, Cosimi AB, Barnes AB, Russell PS (1976) Hyperparathyroidism during pregnancy. Am J Surg 131:328–337
147. Bruce J, Strong JA (1955) Maternal hyperparathyroidism and parathyroid deficiency in the child, with account of effect of parathyroidectomy on renal function, and of attempt to transplant part of tumor. Q J Med 24:307–319
148. Ip P (2003) Neonatal convulsion revealing maternal hyperparathyroidism: an unusual case of late neonatal hypoparathyroidism. Arch Gynecol Obstet 268:227–229
149. Tseng UF, Shu SG, Chen CH, Chi CS (2001) Transient neonatal hypoparathyroidism: report of four cases. Acta Paediatr Taiwan 42:359–362
150. Naru T, Khan RS, Khan MA (2011) Primary hyperparathyrodism in pregnancy and review of literature. J Pak Med Assoc 61:401–403
151. Pieringer H, Hatzl-Griesenhofer M, Shebl O, Wiesinger-Eidenberger G, Maschek W, Biesenbach G (2007) Hypocalcemic tetany in the newborn as a manifestation of unrecognized maternal primary hyperparathyroidism. Wien Klin Wochenschr 119:129–131
152. McDonnell CM, Zacharin MR (2006) Maternal primary hyperparathyroidism: discordant outcomes in a twin pregnancy. J Paediatr Child Health 42:70–71
153. Powell BR, Buist NR (1990) Late presenting, prolonged hypocalcemia in an infant of a woman with

hypocalciuric hypercalcemia. Clin Pediatr (Phila) 29:241–243
154. Thomas BR, Bennett JD (1995) Symptomatic hypocalcemia and hypoparathyroidism in two infants of mothers with hyperparathyroidism and familial benign hypercalcemia. J Perinatol 15:23–26
155. Thomas BR, Bennett JD (1997) Late-onset neonatal hypocalcemia as an unusual presentation in an offspring of a mother with familial hypocalciuric hypercalcemia. Clin Pediatr (Phila) 36:547–550
156. Kovacs CS, Ho-Pao CL, Hunzelman JL, Lanske B, Fox J, Seidman JG, Seidman CE, Kronenberg HM (1998) Regulation of murine fetal-placental calcium metabolism by the calcium-sensing receptor. J Clin Invest 101:2812–2820
157. Donovan EF, Tsang RC, Steichen JJ, Strub RJ, Chen IW, Chen M (1980) Neonatal hypermagnesemia: effect on parathyroid hormone and calcium homeostasis. J Pediatr 96:305–310
158. Cruikshank DP, Pitkin RM, Reynolds WA, Williams GA, Hargis GK (1979) Effects of magnesium sulfate treatment on perinatal calcium metabolism. I. Maternal and fetal responses. Am J Obstet Gynecol 134:243–249
159. Lipsitz PJ, English IC (1967) Hypermagnesemia in the newborn infant. Pediatrics 40:856–862
160. Lipsitz PJ (1971) The clinical and biochemical effects of excess magnesium in the newborn. Pediatrics 47:501–509
161. Lamm CI, Norton KI, Murphy RJ, Wilkins IA, Rabinowitz JG (1988) Congenital rickets associated with magnesium sulfate infusion for tocolysis. J Pediatr 113:1078–1082
162. Santi MD, Henry GW, Douglas GL (1994) Magnesium sulfate treatment of preterm labor as a cause of abnormal neonatal bone mineralization. J Pediatr Orthop 14:249–253
163. Cumming WA, Thomas VJ (1989) Hypermagnesemia: a cause of abnormal metaphyses in the neonate. AJR Am J Roentgenol 152:1071–1072
164. Malaeb SN, Rassi AI, Haddad MC, Seoud MA, Yunis KA (2004) Bone mineralization in newborns whose mothers received magnesium sulphate for tocolysis of premature labour. Pediatr Radiol 34:384–386
165. Savory J, Monif GR (1971) Serum calcium levels in cord sera of the progeny of mothers treated with magnesium sulfate for toxemia of pregnancy. Am J Obstet Gynecol 110:556–559
166. Culbert EC, Schfirin BS (2006) Malignant hypercalcemia in pregnancy: effect of pamidronate on uterine contractions. Obstet Gynecol 108:789–791
167. Usta IM, Chammas M, Khalil AM (1998) Renal cell carcinoma with hypercalcemia complicating a pregnancy: case report and review of the literature. Eur J Gynaecol Oncol 19:584–587
168. Illidge TM, Hussey M, Godden CW (1996) Malignant hypercalcaemia in pregnancy and antenatal administration of intravenous pamidronate. Clin Oncol (R Coll Radiol) 8:257–258
169. Montoro MN, Paler RJ, Goodwin TM, Mestman JH (2000) Parathyroid carcinoma during pregnancy. Obstet Gynecol 96:841
170. Abraham P, Ralston SH, Hewison M, Fraser WD, Bevan JS (2002) Presentation of a PTHrP-secreting pancreatic neuroendocrine tumour, with hypercalcaemic crisis, pre-eclampsia, and renal failure. Postgrad Med J 78:752–753
171. Tsang RC, Chen I, Friedman MA, Gigger M, Steichen J, Koffler H, Fenton L, Brown D, Pramanik A, Keenan W, Strub R, Joyce T (1975) Parathyroid function in infants of diabetic mothers. J Pediatr 86:399–404
172. Mehta KC, Kalkwarf HJ, Mimouni F, Khoury J, Tsang RC (1998) Randomized trial of magnesium administration to prevent hypocalcemia in infants of diabetic mothers. J Perinatol 18:352–356
173. Brandi ML (2011) Genetics of hypoparathyroidism and pseudohypoparathyroidism. J Endocrinol Invest 34:27–34
174. Ryan AK, Goodship JA, Wilson DI, Philip N, Levy A, Seidel H, Schuffenhauer S, Oechsler H, Belohradsky B, Prieur M, Aurias A, Raymond FL, Clayton-Smith J, Hatchwell E, McKeown C, Beemer FA, Dallapiccola B, Novelli G, Hurst JA, Ignatius J, Green AJ, Winter RM, Brueton L, Brondum-Nielsen K, Scambler PJ et al (1997) Spectrum of clinical features associated with interstitial chromosome 22q11 deletions: a European collaborative study. J Med Genet 34:798–804
175. Brauner R, Harivel L, de Gonneville A, Kindermans C, Le Bidois J, Prieur M, Lyonnet S, Souberbielle JC (2003) Parathyroid function and growth in 22q11.2 deletion syndrome. J Pediatr 142:504–508
176. Adachi M, Tachibana K, Masuno M, Makita Y, Maesaka H, Okada T, Hizukuri K, Imaizumi K, Kuroki Y, Kurahashi H, Suwa S (1998) Clinical characteristics of children with hypoparathyroidism due to 22q11.2 microdeletion. Eur J Pediatr 157:34–38
177. Mantovani G (2011) Clinical review: pseudohypoparathyroidism: diagnosis and treatment. J Clin Endocrinol Metab 96:3020–3030
178. Ngai YF, Chijiwa C, Mercimek-Mahmutoglu S, Stewart L, Yong SL, Robinson WP, Gibson WT (2010) Pseudohypoparathyroidism type 1a and the GNAS p.R231H mutation: somatic mosaicism in a mother with two affected sons. Am J Med Genet A 152A:2784–2790
179. Rubin LP, Kovacs CS, De Paepe ME, Tsai SW, Torday JS, Kronenberg HM (2004) Arrested pulmonary alveolar cytodifferentiation and defective surfactant synthesis in mice missing the gene for

parathyroid hormone-related protein. Dev Dyn 230:278–289
180. Klopocki E, Hennig BP, Dathe K, Koll R, de Ravel T, Baten E, Blom E, Gillerot Y, Weigel JF, Kruger G, Hiort O, Seemann P, Mundlos S (2010) Deletion and point mutations of PTHLH cause brachydactyly type E. Am J Hum Genet 86:434–439
181. Lanske B, Karaplis AC, Lee K, Luz A, Vortkamp A, Pirro A, Karperien M, Defize L, Ho C, Abou-Samra AB, Jüppner H, Segre GV, Kronenberg HM (1996) PTH/PTHrP receptor in early development and Indian hedgehog-regulated bone growth. Science 273:663–666
182. Karaplis AC, He B, Nguyen MT, Young ID, Semeraro D, Ozawa H, Amizuka N (1998) Inactivating mutation in the human parathyroid hormone receptor type 1 gene in Blomstrand chondrodysplasia. Endocrinology 139:5255–5258
183. Oostra RJ, van der Harten JJ, Rijnders WP, Scott RJ, Young MP, Trump D (2000) Blomstrand osteochondrodysplasia: three novel cases and histological evidence for heterogeneity. Virchows Arch 436:28–35

Rare Causes of Acquired Hypoparathyroidism

24

Jean-Louis Wémeau

24.1 Introduction

The main causes of hypoparathyroidism are congenital anomalies of parathyroid organogenesis, autoimmune diseases, and surgery [1, 2]. However, infiltrative disorders, radioisotopic destruction, and burn injuries also cause lesions to the parathyroid glands and acute or permanent parathyroid gland hypofunction (see also Chaps. 14 and 15). Transient or sometimes prolonged functional parathyroid dysfunction can also develop in the neonatal period in relation to maternal hypercalcemia that is usually linked to undiagnosed primary hyperparathyroidism (see also Chap. 23).

Knowledge of these rarer causes is warranted, since hypoparathyroidism may be the presenting sign. In this review, we will identify, as far as possible, the prevalence of these conditions, their mechanisms, the specific features of their management and their prognosis.

J.-L. Wémeau
Department of Endocrinology and Metabolic Disease, Marc Linquette Endocrinological Clinic, Claude Huriez Hospital, CHRU, Lille Cedex 59 037, France
e-mail: Jean-Louis.WEMEAU@chru-lille.fr; jl-wemeau@hotmail.fr

24.2 Radiation-Induced Hypoparathyroidism

External irradiation, such as that delivered for benign cervical diseases, is a traditional risk factor for thyroid cancers as well as parathyroid adenomas, which are responsible for primary hyperparathyroidism [3, 4]. Hyperparathyroidism can also be observed after iodine 131 treatment for hyperthyroidism [5]. Paradoxically, the role of ionizing radiation must also be considered in the presence of parathyroid hypofunction, even if the irradiation occurred long ago and has been forgotten.

The first evaluation of parathyroid functional reserve was carried out in 1965 by Adams and Chalmers in 60 subjects who had received a therapeutic dose of iodine 131 for hyperthyroidism [6]. The assessment, done over a variable time period, consisted of a calcium deprivation test provoked by the infusion of a calcium-chelating agent (disodium EDTA = disodium hydrogen ethylenediaminetetracetic acid), followed by serum calcium measurements after 2, 6, 12, and 24 h. Ten percent of patients had persistent hypocalcemia, while normal subjects had a restoration of serum calcium levels to at least 90 % of pre-EDTA levels by 12 h. Another investigation was done in 19 asymptomatic subjects 6 months after receiving a therapeutic dose of iodine 131 (100–150 mCi) for thyroid cancer remnant ablation. The study results showed low serum calcium levels (<84 mg/l) with inappropriately

M.L. Brandi, E.M. Brown (eds.), *Hypoparathyroidism*,
DOI 10.1007/978-88-470-5376-2_24, © Springer-Verlag Italia 2015

Table 24.1 Review of post-RAI therapy hypoparathyroidism

References	Age (years)	Gender	Total RAI dose (mCi)	Time to onset of symptoms	Duration of treatment for	Calcium at time of symptoms mg/dl
TIGHE	14	M	4	72 days	6 months	n/a
CHATTERJEE	36	F	5	2 months	More than 1 year	6.4
FULOP	45	F	14	9 months	10 years	6.5
ORME	57	M	40	10 months		5
WINSLOW	38	F	150	1 month	More than 1 year	5.7
TOWNSEND	52	M	10	4 months	More than 1 year	n/a
JIALAL	46	F	5	6 months	2 years	5
BURCH	22	F	4	10 months	12 years	7.1
EIPE	58	M	15.7	5 months	More than 1 year	4.5
FREEMAN	58	F	3.5	5 days		7.7
KORAKOVSKIY	12	F	11.1	1 month	More than 6 months	6.6

From Komarovskiy [7]

normal parathyroid hormone PTH levels in 7 of 19 patients. None of them had symptoms of hypocalcemia.

In contrast, reported cases of severe hypocalcemia related to radiation-induced hypoparathyroidism are rare. The excellent report by Komarovskiy and Raghavan in 2012 found 11 cases in the literature, including one personal case, which occurred after radioactive iodine treatment and in the absence of parathyroid disease [7]. The female to male ratio was 7:4. Hypocalcemia related to hypoparathyroidism occurred within 5 days to 10 months of the treatment. The situation required prolonged calcium supplementation with vitamin D therapy for at least 1 year (up to 12 years) in the majority of cases (Table 24.1). Animal studies showed the development of parathyroid gland lesions caused by iodine 131 β-radiation, which could penetrate surrounding structures to a depth of up to 2.5 cm [8, 9].

Potential additional contributors to hypocalcemia in this setting include acceleration of bone formation during recovery from the prior hyperthyroidism (i.e., "hungry bones"), a shortage or deficiency of vitamin D, and other factors such as corticosteroid therapy for asthma [7]. The possibility of autoimmune hypoparathyroidism should also be kept in mind (see Chap. 17), although its association with Graves' disease is very rare.

24.3 Infiltrative Hypoparathyroidism

Parathyroid gland hypofunction can result in infiltration related to inflammatory (granulomatosis, thyroiditis) or neoplastic processes or as a reaction to metal overload (hemochromatosis, Wilson's disease) (Table 24.2).

24.3.1 Infiltrative Granulomatosis Hypoparathyroidism

Sarcoidosis, tuberculosis, and some other granulomatous diseases are well-known causes of hypercalcemia and hypercalciuria [10]. They are explained by the extrarenal production of 1, 25-hydroxyvitamin D in the granuloma. However, there are few reports of hypoparathyroidism in patients with sarcoidosis, which are usually not well documented. Possible infiltration of the parathyroid glands by sarcoid granulomas was suspected by some authors [11–13].

Some histologically described inflammatory and granulomatous parathyroid disorders (96 cases in 27 articles) have been reported using the terms "parathyroiditis," "inflammation of the parathyroid glands," "granulomatous inflammation," "sarcoidosis," or "tuberculosis." There has been a surprising absence of associations

Table 24.2 Infiltrative hypoparathyroidisms

Granulomatosis: sarcoidosis, tuberculosis, syphilis
Amyloidosis
Riedel's thyroiditis
Metastasis
Metal deposition
Hemochromatosis
Wilson's diseases
Aluminum deposit in patients with renal failure

between histologically proven parathyroiditis and hypoparathyroidism [14].

Hypoparathyroidism has also been reported in the endocrine expression of POEMS syndrome (polyneuropathy, organomegaly, endocrinopathy, monoclonal protein, skin changes) [15, 16]. There is still a lack of knowledge concerning the nosology, pathogenesis, and medical approach to this condition [17]

24.3.2 Amyloidosis and the Parathyroid Glands

Autopsy investigations have found a very high frequency of amyloid infiltration of the parathyroid glands, which increases with age and with atherosclerosis and occurs without evidence of altered endocrine function [18]. In a Scottish study, it was found to be more prevalent in the parathyroid glands removed at necropsy (46 %) than in those that were surgically removed [19]. Parathyroid gland infiltration is also seen in systemic amyloidosis, as in other endocrine glands [20]. It has also been observed in amyloidosis related to the relapsing disease and familial Mediterranean fever [21, 22], and it is a possible accompanying sign of amyloid goiter [23]. In these conditions, a reduction in parathyroid hormone production is possible but rare.

24.3.3 Hypoparathyroidism Associated with Riedel's Thyroiditis

Riedel's thyroiditis is a rare, if not exceptional, condition, which is defined as an invading fibrosis of the thyroid gland. It infiltrates and extends beyond the capsule, then become an invasive fibrosis of the neck [24, 25]. As a result, a rock-hard goiter eventually leads to severe compressive signs; this process may include hypoparathyroidism, as well as possibly hypothyroidism [26, 27].

Interestingly, hypoparathyroidism is sometimes an early presenting sign of a fibrous process forming in the neck, which has been little known until now [28]. These conditions should be distinguished from hypoparathyroid lesions formed during attempts at debulking surgery for thyroiditis [29]. Riedel's thyroiditis is actually discovered during thyroidectomies in 0.06 % of cases [30].

The mechanisms behind the formation of these fibrous processes are poorly understood. In rare cases, it is a reaction to an autoimmune disorder corresponding to a fibrous variant of Hashimoto thyroiditis, with extracapsular extension of the processes marking the transition to Riedel's thyroiditis [31]. The fibrosis is usually isolated, however, showing no evidence of preexistent autoimmunity, giant-cell granulomatous reaction, or neoplastic invasion [32]. This fibrotic process is, however, likely to spread to the mediastinum, the hepatobiliary tract, the retroperitoneum, and the eye socket. The fibroblastic proliferation appears to be induced by cytokines from the B and T lymphocytes [24, 33].

Means of limiting the spread of fibrosis include corticosteroid therapy, thyroid hormone therapy, tamoxifen (antiestrogen agent that stimulates TGF-β production and reduces the maturation of immature fibroblasts) and, as far as possible, partial surgical resection. Therapeutic success is modest, however [24, 33, 34].

According to the published cases, 14 cases of hypoparathyroidism accompanying Riedel's thyroiditis have been reported, 10 of them in women. The mean age at occurrence was 45 years (range 36–66 years). Hypoparathyroidism was a presenting diagnosis in one case; it occurred at the time of the discovery of the thyroiditis in four cases and occurred separately with an interval between the diagnoses of the two conditions of between 4 and 32 months in the other cases.

Hypoparathyroidism was present in 3 of the 21 cases of Riedel's thyroiditis studied at the Mayo Clinic between 1976 and 2008 [35].

24.3.4 Hypoparathyroidism Related to Metastasis

There are very few reports of secondary tumors developing in the parathyroid glands, sometimes with hypoparathyroidism [36].

24.3.4.1 Hypoparathyroidism Related to Metal Storage

Hypoparathyroidism is rare but constitutes a classical feature of hemochromatosis, mainly observed in secondary iron overload following blood transfusions. However, parathyroid insufficiency can also be observed in relation to copper (Wilson's disease) and aluminum (in patients with renal failure).

There are well-documented cases of hypoparathyroidism occurring in patients who received multiple blood transfusions for thalassemia major, Diamond-Blackfan anemia, or other varieties of chronic transient aplastic anemia. These cases particularly involved subjects treated from a young age. The onset of hypocalcemia was most often after the age of 10 years. It usually was associated with few symptoms and was discovered on laboratory testing. As a general rule, it occurred simultaneously with massive tissue overload of iron, a determinant of liver involvement, pituitary insufficiency (hypogonadism with eunuchoidism, growth hormone insufficiency resulting in short stature), and hypothyroidism. With iron chelation therapy, the hypoparathyroidism rarely appeared to regress [37–43].

In an Iranian prospective study of 220 patients with thalassemia, hypoparathyroidism was present in 7.6 % (while short stature was seen in 39.3 %, hypogonadism in 22.9 % of boys and 12.8 % of girls, and hypothyroidism in 7.7 %). The patients' mean age at the time of diagnosis of hypoparathyroidism was 16.9 ± 3.7 years, and the highest prevalence was seen at the age of 20, with a distribution of 81.8 % in males and 18.2 % in females. The mean serum ferritin level (1,444 ± 798 μg/l) was not significantly different from other patients [44]. In other reports, the prevalence of hypoparathyroidism was lower: 3.6–7 % [45–47], and there was a lower male to female ratio [46, 48]. There are no reports showing a potential protective effect of hypoparathyroidism with regard to osteoporosis and osteopenia, which were observed in the lower back and femoral neck in 10–50 % of individuals [44, 49]. In a patient with multiple endocrinopathies and secondary hemochromatosis due to multiple blood transfusions, admitted due to adrenal crisis, serum calcium decreased to 6.4 mg/dl; it was presumed that administration of glucocorticoid and alendronate therapy, prescribed for fractures and osteoporosis, had unmasked a latent hypoparathyroidism [50]. In contrast, hypoparathyroidism has rarely been reported with primary hemochromatosis caused by a mutation of the HFE gene (<1 %) [51]. This difference in the expression of the disease is surprising. It shows that hypoparathyroidism in the setting of hemochromatosis may not only result from iron overload but also from the ability of the disease to generate sclerosis and tissue lesions.

Observations of long-term hemodialysis patients suggest that multiple transfusions and iron overload causing hemochromatosis play a causative role in hypoparathyroidism [52]. The authenticity of staining of bone for iron was confirmed in a series of 48 patients on dialysis. It occurred with osteomalacia and with bone aluminum staining. In patients with iron overload in the bones and other tissue such as bone marrow, the observation of lower levels of iPTH in overloaded patients raises the possibility that iron overload induces a state of relative hypoparathyroidism [53].

Hypoparathyroidism is a possible manifestation of Wilson's disease, which involves congenital anomalies of copper metabolism and affects the liver and nervous system. Its prevalence is rare (<1/100,000) [51]. Hypoparathyroidism has been attributed to copper deposits in the parathyroid glands [54, 55]. It has also been observed in the setting of intestinal lymphangiectasia with protein-losing enteropathy and toxic copper accumulation [56].

Hypoparathyroidism has also been described in adrenoleukodystrophy of Schilder's disease, which involves an accumulation of long-chain triglycerides, with a particular affinity for the central nervous system, the adrenal glands, and testicles [57]. It has also been found in Fabry's disease [58], and finally during mitochondrial cytopathies that alter the respiratory chain [59, 60]. Transient hypoparathyroidism has also been reported in cases of acute decompensation of fatty acid oxidation disorders (FAO), which also have involvement of muscle, heart, and liver [61].

24.4 Hypoparathyroidism Related to Burn Injuries

It is now well known that extensive burns are likely to alter calcium and bone metabolism. This has been particularly well established in children. The survey of Klein et al. showed that in middle-school-aged children (9.6±4.7), burns affecting ±57 % of the body surface area clearly lower the total and ionized calcium. At the same time, the serum iPTH concentrations are not increased, which is inappropriate given the hypocalcemia, and appear to be weakly reactive during the calcium deprivation test provoked by EDTA [62]. Due to simultaneous hypomagnesemia and the well-known role of magnesium deficiency on the production and activity of PTH [63], the role of the magnesium deficiency had been suggested, but oral and parenteral magnesium supplementation proved ineffective in correcting the hypocalcemia and the hypoparathyroidism [64]. In contrast, experimental studies on sheep suggest rather an upregulation of the parathyroid calcium-sensing receptor and a related decrease in the set point for calcium suppression of the parathyroid hormone, thus contributing to post-burn hypoparathyroidism and hypocalcemia [65]. At the same time, there is a decrease in the parameters of bone formation and bone mass as assessed by bone mineral density testing, which may be contributed to by the increase in endogenous glucocorticoid production, functional growth hormone deficiency, immobilization, and the proinflammatory cytokines interleukin 1β and 6. As a consequence, the risk for post-burn fractures increases [66]. The administration of the bisphosphonate pamidronate did not exacerbate post-burn hypocalcemia and effectively preserved bone mass, as attested by a double-blind randomized controlled study [67].

24.5 Hypoparathyroidism in Children in Relation to Maternal Hypercalcemia

Primary hyperparathyroidism is the main cause of hypercalcemia, and its prevalence is estimated to be 1/1,000. Fortunately the condition is rare in women of childbearing age, with an incidence during pregnancy estimated at 8 cases per 100,000 and per year [68]. Hyperparathyroidism has profound consequences for pregnant women (vomiting, kidney stones, nephrocalcinosis, pancreatitis, and pre-eclampsia), as well as fetuses (miscarriage, mental retardation) [69–72] (see also Chap. 23). Consequently, surgical intervention is usually recommended during the second trimester of pregnancy and particularly if the serum calcium exceeds 2.85 mmol/L. The intervention and successful cure of the hyperparathyroidism very significantly reduce the maternal and fetal risk [72, 73]. Cinacalcet, a calcimimetic agent was used during pregnancy [74], but its potential effects on calcium transport in the placenta and the fetus require that its risks and benefits be carefully considered.

Maternal hyperparathyroidism may go undetected during pregnancy and present as neonatal tetany. The chronic exposure of the fetus to hypercalcemia from the mother in fact causes inhibition of parathyroid function in the fetus, which continues into the neonatal period. This effect is ordinarily transient, but prolonged parathyroid hypofunction has been reported [72].

Due to the placental transfer of calcium to the fetus, neonatal hypocalcemia is delayed and is usually observed from the second or third week with a dramatic presentation of convulsions in some cases [75–77] (Table 24.3).

> All neonatal hypocalcemia requires the measurement of serum calcium in the mother.

Table 24.3 Causes of neonatal hypocalcemia

Soon after birth	
	Low birth weight
	Intrauterine growth retardation
	Diabetic mother
	Prolonged deliveries
Late after postnatal 72 h	
	Increased phosphate load
	Hypomagnesemia
	Vitamin D deficiency
	PTH resistance
	Primary hypoparathyroidism
	Maternal hyperparathyroidism

From Cakir et al. [77]

Neonatal hypocalcemia may be related to other causes of maternal hypercalcemia. The diagnosis of familial hypocalciuric hypercalcemia should be considered here in conjunction with an inactivating mutation of the calcium-sensing receptor expressed in the parathyroid glands and kidneys [78]. These situations are usually asymptomatic and have no consequences in adults, particularly for the pregnant woman. However, the situation of the fetus needs to be very carefully considered. If the child were to inherit a homozygous mutation from the mother and father, he/she would be at risk of developing severe neonatal hyperparathyroidism, with dramatic consequences and therapeutic management that is difficult [78]. Some cases have also been reported of severe neonatal hypercalcemia simply from inheritance of a single heterozygous mutation of the gene [79]. If the fetus is unaffected, as in half of the cases with a Mendelian pattern of inheritance for an autosomal dominant condition, the exposure to maternal hypercalcemia is inconsequential, since it is usually moderate (less than 115 mg/L). Cases of functional inhibition and convulsions during the neonatal period have been reported, however [76].

References

1. Shoback D (2008) Clinical practice. Hypoparathyroidism. N Engl J Med 359:391–403
2. Al-Azem H, Khan AA (2012) Hypoparathyroidism. Best Pract Res Clin Endocrinol Metab 26:517–522
3. Tisell LE, Hansson G, Lindberg S, Ragnhult I (1977) Hyperparathyroidism in persons treated with X-rays for tuberculous cervical adenitis. Cancer 40:846–854
4. Christmas TJ, Chapple CR, Noble JG et al (1988) Hyperparathyroidism after neck irradiation. Br J Surg 75:873–874
5. Colaço SM, Ming S, Reiff E (2007) Hyperparathyroidism after radioactive iodine therapy. Am J Surg 194:323–327
6. Adams PH, Chalmers TM (1965) Parathyroid function after I131 therapy for hyperthyroidism. Clin Sci 29:391–395
7. Komarovskiy K, Raghavan S (2012) Hypocalcemia following treatment with radioiodine in a child with Graves' disease. Thyroid 22:218–222. doi:10.1089/thy.2011.0094
8. Gorbman A (1950) Functional and structural changes consequent to high dosages of radioactive iodine. J Clin Endocrinol Metab 10:1177–1191
9. Soley MH, Foreman N (1949) Radioiodine therapy in Graves' disease; a review. J Clin Invest 28:1367–1374
10. Sharma OP (1996) Vitamin D, calcium, and sarcoidosis. Chest 109:535–539
11. Brinkane A, Peschard S, Leroy-Terquem E et al (2001) Rare association of hypoparathyroidism and mediastinal-pulmonary sarcoidosis. Ann Med Interne (Paris) 152:63–64
12. Saeed A, Khan M, Irwin S, Fraser A (2011) Sarcoidosis presenting with severe hypocalcaemia. Ir J Med Sci 180:575–577. doi:10.1007/s11845-009-0277-9
13. Badell A, Servitje O, Graells J et al (1998) Hypoparathyroidism and sarcoidosis. Br J Dermatol 138:915–917
14. Talat N, Diaz-Cano S, Schulte KM (2011) Inflammatory diseases of the parathyroid gland. Histopathology. doi:10.1111/j.1365-2559.2011.04001
15. Cabezas-Agricola JM, Lado-Abeal JJ, Otero-Anton E et al (1996) Hypoparathyroidism in POEMS syndrome. Lancet 347:701–712
16. Belkhribchia MR, El Moutawakil B, Amarti Riffi A (2013) POEMS syndrome revealed unusually early with unexplained hepatic and peritoneal granulomas: to an associated malignant lymphoma? Rev Neurol (Paris). doi:10.1016/j.neurol.2012.05.012
17. Soubrier MJ, Dubost JJ, Sauvezie BJ (1994) POEMS syndrome: a study of 25 cases and a review of the literature. French Study Group on POEMS Syndrome. Am J Med 97:543–553
18. Thiele J, Ries P, Georgii A (1975) Special and functional pathomorphology of parathyroid glands as revealed in non-selected autopsies (589 cases). Virchows Arch A Pathol Anat Histol 12:367
19. Anderson TJ, Ewen SW (1974) Amyloid in normal and pathological parathyroid glands. J Clin Pathol 27:656–663
20. Ozdemir D, Dagdelen S, Erbas T (2010) Endocrine involvement in systemic amyloidosis. Endocr Pract 16:1056–1063. doi:10.4158/EP10095.RA

21. Arutiunian VM, Eganian GA, Grigorian GA (1986) Structural and functional changes in the parathyroid glands in patients with periodic disease. Ter Arkh 58:123–125
22. Keven K, Oztas E, Aksoy H et al (2001) Polyglandular endocrine failure in a patient with amyloidosis secondary to familial Mediterranean fever. Am J Kidney Dis 38:E39
23. Ellis HA, Mawhinney WH (1984) Parathyroid amyloidosis. Arch Pathol Lab Med 108:689–690
24. Perimenis P, Marcelli S, Leteurtre E et al (2008) Riedel's thyroiditis: current aspects. Presse Med 37:1015–1021
25. Hennessey JV (2011) Clinical review: Riedel's thyroiditis: a clinical review. J Clin Endocrinol Metab 96:3031–3041. doi:10.1210/jc.2011-0617
26. Woolner LB, Mcconahey WM, Beahrs OH (1957) Granulomatous thyroiditis (De Quervain's thyroiditis). J Clin Endocrinol Metab 17:1202–1221
27. Marín F, Araujo R, Páramo C et al (1989) Riedel's thyroiditis associated with hypothyroidism and hypoparathyroidism. Postgrad Med J 65:381–383
28. Nazal EM, Belmatoug N, de Roquancourt A et al (2003) Hypoparathyroidism preceding Riedel's thyroiditis. Eur J Intern Med 14:202–204
29. Lorenz K, Gimm O, Holzhausen HJ et al (2007) Riedel's thyroiditis: impact and strategy of a challenging surgery. Langenbecks Arch Surg 392:405–412
30. Hay ID (1985) Thyroiditis: a clinical update. Mayo Clin Proc 60:836–843
31. Best TB, Munro RE, Burwell S et al (1991) Riedel's thyroiditis associated with Hashimoto's thyroiditis, hypoparathyroidism, and retroperitoneal fibrosis. J Endocrinol Invest 14:767–772
32. Schwaegerle SM, Bauer TW, Esselstyn CB Jr (1988) Riedel's thyroiditis. Am J Clin Pathol 90:715–722
33. Yasmeen T, Khan S, Patel SG et al (2002) Clinical case seminar: Riedel's thyroiditis: report of a case complicated by spontaneous hypoparathyroidism, recurrent laryngeal nerve injury, and Horner's syndrome. J Clin Endocrinol Metab 87:3543–3547
34. Lo JC, Loh KC, Rubin AL et al (1998) Riedel's thyroiditis presenting with hypothyroidism and hypoparathyroidism: dramatic response to glucocorticoid and thyroxine therapy. Clin Endocrinol (Oxf) 48:815–818
35. Fatourechi MM, Hay ID, McIver B (2011) Invasive fibrous thyroiditis (Riedel thyroiditis): the Mayo Clinic experience, 1976–2008. Thyroid 21:765–772. doi:10.1089/thy.2010.0453
36. Horwitz CA, Myers WP, Foote FW Jr (1972) Secondary malignant tumors of the parathyroid glands. Report of two cases with associated hypoparathyroidism. Am J Med 52:797–808
37. Sherman LA, Pfefferbaum A, Brown EB Jr (1970) Hypoparathyroidism in a patient with longstanding iron storage disease. Ann Intern Med 73:259–261
38. Mautalen CA, Kvicala R, Perriard D et al (1978) Case report: hypoparathyroidism and iron storage disease. Treatment with 25-hydroxy-vitamin D3. Am J Med Sci 276:363–368
39. Himoto Y, Kanzaki S, Nomura H (1995) Hypothyroidism and hypoparathyroidism in an 11 year old boy with hemochromatosis secondary to aplastic anemia. Acta Paediatr Jpn 37:534–536
40. Lanes R, Muller A, Palacios A (2000) Multiple endocrine abnormalities in a child with Blackfan-Diamond anemia and hemochromatosis. Significant improvement of growth velocity and predicted adult height following growth hormone treatment despite liver damage. J Pediatr Endocrinol Metab 13:325–328
41. Chern JP, Lin KH (2002) Hypoparathyroidism in transfusion-dependent patients with beta-thalassemia. J Pediatr Hematol Oncol 24:291–293
42. Angelopoulos NG, Goula A, Rombopoulos G et al (2006) Hypoparathyroidism in transfusion-dependent patients with beta-thalassemia. J Bone Miner Metab 24:138–145
43. Gamberini MR, De Sanctis V, Gilli G (2008) Hypogonadism, diabetes mellitus, hypothyroidism, hypoparathyroidism: incidence and prevalence related to iron overload and chelation therapy in patients with thalassaemia major followed from 1980 to 2007 in the Ferrara Centre. Pediatr Endocrinol Rev 6(Suppl 1): 158–169
44. Shamshirsaz AA, Bekheirnia MR, Kamgar M (2003) Metabolic and endocrinologic complications in beta-thalassemia major: a multicenter study in Tehran. BMC Endocr Disord 3:4
45. Vullo C, De Sanctis V, Katz M (1990) Endocrine abnormalities in thalassemia. Ann N Y Acad Sci 612:293–310
46. Italian Working Group on Endocrine Complications in Non-endocrine Diseases (1995) Multicentre study on prevalence of endocrine complications in thalassaemia major. Italian Working Group on Endocrine Complications in Non-endocrine Diseases. Clin Endocrinol (Oxf) 42:581–586
47. De Sanctis V, Vullo C, Katz M, Wonke B, Hoffbrand VA, Di Palma A, Bagne B (1989) Endocrine complication in thalassemia major in advances and controversies. In: Buckner CD, Gale RP, Lucarreli G (eds) Thalassemia therapy. Alan Liss, New York, pp 77–83
48. De Sanctis V, Wonke B (1998) Growth and endocrine complications. In: Mediprint (ed) Growth and endocrine complications in thalassemia. Roma, pp 17–30
49. Jensen CE, Tuck SM, Agnew JE et al (1998) High prevalence of low bone mass in thalassaemia major. Br J Haematol 103:911–915
50. Tanimoto K, Okubo Y, Harada C et al (2008) Latent hypoparathyroidism in an osteoporotic patient with multiple endocrinopathies and secondary hemochromatosis due to multiple blood transfusions, unmasked by alendronate and glucocorticoid at adrenal crisis. Intern Med 47:515–520
51. Vantyghem MC, Dobbelaere D, Mention K et al (2012) Endocrine manifestations related to inherited metabolic diseases in adults. Orphanet J Rare Dis 7:11. doi:10.1186/1750-1172-7-11
52. Shirota T, Shinoda T, Aizawa T (1992) Primary hypothyroidism and multiple endocrine failure in

association with hemochromatosis in a long-term hemodialysis patient. Clin Nephrol 38:105–109
53. McCarthy JT, Hodgson SF, Fairbanks VF et al (1991) Clinical and histologic features of iron-related bone disease in dialysis patients. Am J Kidney Dis 17:551–561
54. Carpenter TO, Carnes DL Jr, Anast CS (1983) Hypoparathyroidism in Wilson's disease. N Engl J Med 309:873–877
55. Okada M, Higashi K, Enomoto S et al (1998) A case of Wilson's disease associated with hypoparathyroidism and amenorrhea. Nihon Shokakibyo Gakkai Zasshi 95:445–449
56. O'Donnell D, Myers AM (1990) Intestinal lymphangiectasia with protein losing enteropathy, toxic copper accumulation and hypoparathyroidism. Aust N Z J Med 20:167–169
57. Faggiano A, Pisani A, Milone F et al (2006) Endocrine dysfunction in patients with Fabry disease. J Clin Endocrinol Metab 91:4319–4325
58. Misery L, Gregoire M, Prieur F et al (2002) Fabry's disease and hypoparathyroidism. Ann Med Interne (Paris) 153:283–285
59. Papadimitriou A, Hadjigeorgiou GM, Divari R (1996) The influence of Coenzyme Q10 on total serum calcium concentration in two patients with Kearns-Sayre Syndrome and hypoparathyroidism. Neuromuscul Disord 6:49–53
60. Tengan CH, Kiyomoto BH, Rocha MS (1998) Mitochondrial encephalomyopathy and hypoparathyroidism associated with a duplication and a deletion of mitochondrial deoxyribonucleic acid. J Clin Endocrinol Metab 83:125–129
61. Baruteau J, Levade T, Redonnet-Vernhet I et al (2009) Hypoketotic hypoglycemia with myolysis and hypoparathyroidism: an unusual association in medium chain acyl-CoA desydrogenase deficiency (MCADD). J Pediatr Endocrinol Metab 22:1175–1177
62. Klein GL, Nicolai M, Langman CB et al (1997) Dysregulation of calcium homeostasis after severe burn injury in children: possible role of magnesium depletion. J Pediatr 131:246–251
63. Rude RK, Oldham SB, Singer FR (1976) Functional hypoparathyroidism and parathyroid hormone end-organ resistance in human magnesium deficiency. Clin Endocrinol (Oxf) 5:209–224
64. Klein GL, Langman CB, Herndon DN (2000) Persistent hypoparathyroidism following magnesium repletion in burn-injured children. Pediatr Nephrol 14:301–304
65. Murphey ED, Chattopadhyay N, Bai M et al (2000) Up-regulation of the parathyroid calcium-sensing receptor after burn injury in sheep: a potential contributory factor to postburn hypocalcemia. Crit Care Med 28:3885–3890
66. Klein GL, Herndon DN (1999) The role of bone densitometry in the diagnosis and management of the severely burned patient with bone loss. J Clin Densitom 2:11–15
67. Klein GL, Wimalawansa SJ, Kulkarni G et al (2005) The efficacy of acute administration of pamidronate on the conservation of bone mass following severe burn injury in children: a double-blind, randomized, controlled study. Osteoporos 16:631–635
68. Pieringer H, Hatzl-Griesenhofer M, Shebl O (2007) Hypocalcemic tetany in the newborn as a manifestation of unrecognized maternal primary hyperparathyroidism. Wien Klin Wochenschr 119:129–131
69. Kristoffersson A, Dahlgren S, Lithner F, Järhult J (1985) Primary hyperparathyroidism in pregnancy. Surgery 97:326–330
70. Shangold MM, Dor N, Welt SI et al (1982) Hyperparathyroidism and pregnancy: a review. Obstet Gynecol Surv 37:217–228
71. Kelly TR (1991) Primary hyperparathyroidism during pregnancy. Surgery 110:1028–1033
72. Cooper MS (2011) Disorders of calcium metabolism and parathyroid disease. Best Pract Res Clin Endocrinol Metab 25:975–983
73. Norman J, Politz D, Politz L (2009) Hyperparathyroidism during pregnancy and the effect of rising calcium on pregnancy loss: a call for earlier intervention. Clin Endocrinol (Oxf) 71:104–109. doi:10.1111/j.1365-2265.2008.03495.x
74. Horjus C, Groot I, Telting D (2009) Cinacalcet for hyperparathyroidism in pregnancy and puerperium. J Pediatr Endocrinol Metab 22:741–749
75. Jaafar R, Yun Boo N, Rasat R (2004) Neonatal seizures due to maternal primary hyperparathyroidism. J Paediatr Child Health 40:329
76. Thomas BR, Bennett JD (1995) Symptomatic hypocalcemia and hypoparathyroidism in two infants of mothers with hyperparathyroidism and familial benign hypercalcemia. J Perinatol 15:23–26
77. Cakir U, Alan S, Erdeve O et al (2013) Late neonatal hypocalcemic tetany as a manifestation of unrecognized maternal primary hyperparathyroidism. Turk J Pediatr 55:438–440
78. Pollak MR, Brown EM, Chou YH et al (1993) Mutations in the human Ca(2+)-sensing receptor gene cause familial hypocalciuric hypercalcemia and neonatal severe hyperparathyroidism. Cell 75:1297–1303
79. Pearce SH, Trump D, Wooding C et al (1995) Calcium-sensing receptor mutations in familial benign hypercalcemia and neonatal hyperparathyroidism. J Clin Invest 96:2683–2692

25 Refractory Hypoparathyroidism

Laura Masi

25.1 Introduction

Hypoparathyroidism, a hormonal insufficiency state, is characterized by hypocalcemia and hyperphosphatemia which are the result of a deficiency in parathyroid hormone (PTH) secretion or action [1]. The prevalence of this disorder in the population remains unknown. Postsurgical hypoparathyroidism has been observed in 5–25 % of patients in different series in the literature, and it is the most frequent form of hypoparathyroidism. Primary hypoparathyroidism may also be caused by developmental defects in the parathyroid glands, resulting from agenesis (e.g., the Di George syndrome) or destruction of the parathyroid glands (e.g., in autoimmune diseases) or due to reduction of PTH secretion (e.g., neonatal hypocalcemia or hypomagnesemia). Hypoparathyroidism may occur as an inherited disorder and finally as an impaired regulation of PTH secretion, as in CaR mutations [2, 3].

The majority of cases of hypoparathyroidism are well controlled under conventional treatment with calcium and vitamin D analogs.

However, this treatment may be difficult to manage, especially in the following situations: (1) in the context of autoimmune polyendocrinopathy-candidiasis-ectodermal dystrophy, (2) in all conditions of hypoparathyroidism associated with malabsorption (celiac disease), and (3) activating mutations in the calcium-sensing receptor [4]. These situations are indicated as *refractory hypoparathyroidism*.

Some conditions of hypocalcemia which are not always follow-on hypoparathyroidism may be resistant to administration of oral calcium and vitamin D and will be here indicated as *refractory hypocalcemia* due to: (1) hungry bone syndrome, (2) abdominal surgery, and (3) hypomagnesemia.

The biochemical diagnosis of hypoparathyroidism is made with the coexistence of inappropriately low serum PTH with hypocalcemia and hyperphosphatemia. Primary renal failure must be excluded by determination of serum creatinine or urea, because it may produce similar plasma mineral changes. The differentiation of hypoparathyroidism and pseudohypoparathyroidism, a hormonal resistance state, can be made by simultaneous measurement of serum Ca and PTH concentrations. Serum 25-hydroxyvitamin D (25 OH D) should also be measured to exclude vitamin D deficiency. In addition, mutational analysis of genes involved in the pathogenesis of hypocalcemia is important to identify mutations that have been described in the literature and to identify new mutations responsible for the various forms of hypoparathyroidism.

Conditions with high bone turnover can induce a hungry bone syndrome resulting in increased

L. Masi, MD, PhD
Metabolic Bone Diseases Unit, University Hospital AOU-Careggi, University of Florence,
CTO 25 Building – II Floor, Largo Palagi, 1, Florence 50134, Italy
e-mail: l.masi@dmi.unifi.it

M.L. Brandi, E.M. Brown (eds.), *Hypoparathyroidism*,
DOI 10.1007/978-88-470-5376-2_25, © Springer-Verlag Italia 2015

bone demand for calcium with hypocalcemia that can be resistant to administration of oral calcium and vitamin D [5]. Symptoms of hypocalcemia difficult to treat with oral therapy may be due to malabsorption, possibly as a result of intestinal hurry in subjects after abdominal surgery treatment [6]. In addition, ionized hypocalcemia accompanied by significant elevation of intact PTH has been described during surgical procedures of varying severity [7]. Indeed, Lapage R. et al. showed that an important part of this fall in ionized calcium was associated with falls in albumin resulting from acute hemodilution by physiological saline [7].

Mg depletion is often secondary to another disease process or to a therapeutic agent; the features of the primary disease process may complicate or mask the Mg depletion. A common laboratory feature of Mg depletion is hypokalemia due to a loss of potassium from the cell with intracellular potassium depletion and an inability of the kidney to conserve potassium. Hypocalcemia is also a common manifestation of moderate to severe Mg depletion.

Treatment of hypoparathyroidism is dependent on many factors including the presenting symptoms and the severity and rapidity that these symptoms developed. The aim of therapy is to maintain the serum calcium at or around the lower limit of the normal concentration range (8–9 mg/dl), so that, on the one hand, hypocalcemic manifestations are limited to the mildest symptoms and, on the other hand, harmful hypercalcemia and hypercalciuria are avoided. If therapy is successful, symptoms associated with hypoparathyroidism should not disturb the patient's daily life, and long-term complications should be avoided.

Transient hypercalcemia should be avoided because recurrent episodes may cause irreparable kidney damage.

The conventional treatment for hypoparathyroidism is *vitamin D and its analogs*. Table 25.1 indicated the calciferol steroid therapy used in the therapy of hypoparathyroidism.

Some forms of hypocalcemia with or without hypoparathyroidism require special attention and monitoring, and patients become refractory to oral steroid therapy [8].

Table 25.1 Calciferol steroid therapy used in the therapy of hypoparathyroidism

Sterol	Average dosage (μg/kg-day)	Average T1/2 (days)	Comments
1,25 $(OH)_2D_3$	0.03	1	Risk of cumulative action
1α (OH) D	0.06	2	
DHT	20	7	
25 (OH) D	4	15	
D2 and D3	50	30	

From Masi and Brandi [8]

Magnesium deficiency can be managed with $MgCl_2$ supplementation in daily doses of 2 mmol/kg divided into four doses.

25.2 Refractory Hypoparathyroidism

25.2.1 Refractory Hypoparathyroidism Associated with Autoimmune Polyendocrine Syndrome (APS)

This refractoriness is most common in patients with hypoparathyroidism associated with autoimmune polyendocrine syndrome (APS) [9, 10]. There are four different types of APS, types I–IV (Table 25.2) [11]. Their presentation and manifestations are quite varied, and therefore careful attention to clinical and laboratory evaluation is important. The term "polyendocrine" itself may be a misnomer because some patients have multiple endocrine disorders while some have many nonendocrine issues [11, 12]. Prompt recognition of APS is crucial because it may require that a patient or family member undergo further evaluation for certain genetic syndromes or autoimmune disorders. APS type I often appears early in life, typically in infants with chronic candidal infections, hypoparathyroidism, and autoimmune AD 8 [11]. The APS type I disorder has also been referred to as either the autoimmune polyendocrinopathy-candidiasis-ectodermal dystrophy (APECED) [13]. It is characterized by

Table 25.2 Types of APS

Type	Characteristic
I	Chronic candidiasis, chronic hypoparathyroidism, autoimmune Addison's disease (AD) (at least 2 present)
II	Autoimmune AD + autoimmune thyroid diseases and/or type 1 diabetes
III	Autoimmune thyroid disease + other autoimmune diseases (excluding autoimmune AD)
IV	≥2 autoimmune diseases

the triad hypoparathyroidism, mucocutaneous candidiasis, and adrenal insufficiency and two or three of the following: insulin-dependent diabetes primary hypogonadism, autoimmune thyroid disease, pernicious anemia, chronic active hepatitis, steatorrhea (malabsorption), alopecia (totalis or areata), and vitiligo [13]. It is caused by a mutation in the autoimmune regulator gene (AIRE), located in locus 21q22.3, which produces a protein that functions as a transcription regulator [14–15].

In these patients, one contributing factor to their chronic hypocalcemia and refractoriness to vitamin D therapy is fat malabsorption. Gastrointestinal (GI) dysfunction is commonly induced by infection, allergy, gluten-sensitive enteropathy, inflammatory bowel disease, or eosinophilic gastroenteropathy. In APECED, GI symptoms such as malabsorption, constipation, watery diarrhea, or steatorrhea are part of the syndrome in around 24 % of the patients [16, 17]. In contrast to the major organ-specific autoimmune symptoms of APECED, the GI symptoms and their underlying pathogenesis are poorly understood. Yet isolated case reports and small series depict severe intestinal involvement in children, leading to malabsorption, multiple deficiencies, growth impairment, and possible death [11]. Possible explanations include intestinal candidiasis, mucosal atrophy, intestinal lymphangiectasia, pancreatic insufficiency, bile salt deficiency, and hypoparathyroidism leading to hypocalcemia [17–20]. Some studies proposed the importance of enteroendocrine (EE) cells in APECED-associated diarrhea [16, 17]. Posovszky C. et al. demonstrated that APECED with GI dysfunction is associated with severe or complete loss of enteroendocrine (EE) cells. The author showed that GI symptoms together with a loss of EE cells preceded the onset of the typical diagnostic features of APECED. This has enormous clinical implications because GI dysfunction is the first manifestation of the syndrome in approximately10 % of APECED patients [17]. In these patients, dietary and supplemental calcium is poorly absorbed. These patients are prone to vitamin D deficiency which further exacerbates their tendency to hypocalcemia. Magnesium deficiency is also common which can be managed with MgCl2 supplementation in daily doses of 2 mmol/kg divided into 4 doses [21].

25.2.2 Refractory Hypoparathyroidism Associated with Celiac Disease

Hypoparathyroidism can coexist with celiac disease and can lead to dramatic fluctuations in plasma calcium levels. Cases of idiopathic hypoparathyroidism coexisting with celiac disease are described in a handful of cases in the literature [22–27], although this situation occurs very rarely and mainly in patients with long-standing disease and therefore lengthy exposure to antibodies [28]. Celiac disease has a high prevalence of 1:300 in white Caucasians of northern European ancestry [28]. Both hypoparathyroidism and celiac disease lead to hypocalcemia. When calcium levels in a previously stable treated patient with hypoparathyroidism decrease or begin to fluctuate significantly, the differential diagnosis includes prolonged use of laxatives or anticonvulsant therapy, chronic renal failure, decreased dietary intake of calcium and vitamin D or malabsorption such as occurs with celiac disease. The possibility of celiac disease should be considered in patients with hypoparathyroidism that seems unduly difficult to treat, and this should be evaluated even in the absence of gastrointestinal symptoms [27]. The hypoparathyrodism causes hypocalcemia through a fall of PTH levels and in celiac disease, the mechanism leading to hypocalcemia is chronic inflammation of the intestinal mucosa leading to malabsorption of vitamin D and calcium [28, 29]. Hypocalcemia is often exacerbated by a decreased intake of dairy products because of lactose intolerance, as lactase cannot be produced by degenerated intestinal epithelial cells and is not available to cleave the lactose in dairy products [30]. Enteric bacteria switch to lactose metabolism, and the resultant fermentation produces large amounts of gas, which can cause painful abdominal bloating. In both conditions oral or intravenous calcium (if calcium levels fall rapidly and severe symptoms occur) as well as vitamin D supplements are given. The institution of and adherence to a gluten-free diet in a newly diagnosed patient with celiac disease on vitamin D may lead to rapid improvement of intestinal absorption with prompt increases in plasma calcium levels in 70 % of patients [28]. Kohler S. et al. described two patients diagnosed with hypoparathyroidism, both of whom went on to develop celiac disease at a later stage. Profound alterations in calcium balance occurred before and after the diagnosis of celiac disease and illustrate the changes in calcium levels that may result from the combination of hypoparathyroidism and celiac disease and to alert them to the potential complications of this combination of pathologies. On the other hand, the authors suggested that given that the response to a gluten-free diet is rapid, this must be considered when a patient taking calcium and vitamin D supplements starts the diet. Calcium levels need to be monitored carefully, as dangerously high plasma calcium levels and acute renal failure can develop [28].

25.2.3 Refractory Hypoparathyroidism Associated with Activating Mutation of Calcium-Sensing Receptor

This particular form of hypoparathyroidism is insert in the "refractory hypoparathyroidism" because require a particular attention in the management of the therapy. Indeed, this form of hypocalcemia is due to the activating mutation of calcium-sensing receptor (CaSR) where there is a tendency to excessive hypercalciuria even at low or below normal serum calcium levels. Congenital isolated hypoparathyroidism caused by activating mutations in the *CaSR* gene is identified as autosomal dominant hypocalcemia (ADH) [31]. Lienhardt et al. identified activating CaSR mutations in 8 (42 %) of 19 unrelated probands with isolated hypoparathyroidism. The severity of hypocalcemic symptoms at diagnosis was independent of age, mutation type, or mode of inheritance but was related to the degree of hypocalcemia [31]. The authors underlined that treatment of hypocalcemia in these patients need to be optimized, because the use of 1-hydroxylated vitamin D3 derivatives can cause hypercalciuria and nephrocalcinosis [31]. The prevalence of activating mutations of the CaSR as a cause of isolated hypoparathyroidism is unknown, making it difficult to identify those patients with hypoparathyroidism in whom mutational analysis is warranted. However, the diagnosis of ADH should be suspected in any case with isolated, autosomal hypocalcemia or in any sporadic case of idiopathic hypoparathyroidism, particularly those who have had complications, such as marked hypercalciuria and nephrocalcinosis during treatment with oral calcium and vitamin D metabolites [32]. It is important, therefore, to determine the optimal mode of treatment for this condition, which minimizes the risk of hypercalciuria and resultant complications, particularly nephrocalcinosis and impaired renal function [31–33].

The hypocalcemia in the families with hypocalcemia and hypercalciuria was initially attributed to hypoparathyroidism [15, 33–39] because it was associated with serum parathyroid hormone concentrations in the low-normal range [33]. However, it is important to differentiate patients with familial hypocalcemic hypercalciuria from those with hypoparathyroidism, because treatment with vitamin D to correct the hypocalcemia in the former may lead to hypercalciuria, nephrocalcinosis, and renal impairment. In addition, polyuria and polydipsia develop at normal serum calcium concentrations in some subjects with hypocalcemia hypercalciuria, perhaps due to increased activity of the mutant receptors in the collecting duct and thus, the combined effects of hypercalciuria and dehydration may make subjects with hypocalcemic hypercalciuria particularly susceptible to nephrocalcinosis and renal impairment [33] Asymptomatic patients with familial hypocalcemic hypercalciuria should not routinely receive vitamin D; such treatment should be reserved for symptomatic patients and given to them with the aim not of restoring normocalcemia, but of maintaining a serum calcium concentration just sufficient to alleviate the symptoms. Familial hypocalcemic hypercalciuria may be difficult to distinguish from hypoparathyroidism on the basis of measurements of serum parathyroid hormone and urinary calcium. However, the identification of mutations in the calcium-sensing receptor gene will help in making this distinction and in facilitating early recognition of patients with hypocalcemic hypercalciuria, but the mutational diversity of the gene makes screening for the disorder arduous and time consuming [33].

25.3 Refractory Hypocalcemia with or Without Hypoparathyroidism

25.3.1 Refractory Hypocalcemia due to Hungry Bone Syndrome

Postsurgical hypoparathyroidism is one of the common postsurgical complications following thyroidectomy and/or parathyroidectomy. However, serum calcium concentration can be controlled by oral administration of vitamin D_3 and calcium or venous infusion of calcium gluconate hydrate. Hungry bone syndrome is recognized caused of hypocalcemia following thyroidectomy for thyrotoxicosis since 1958 [40]. Indeed, severe Graves' disease (GD) and/or severe primary hyperparathyroidism (PHPT) may induce a high bone turnover that can be the cause of a hungry bone syndrome. In these cases, after surgical treatment for hyperthyroidism and/or hyperparathyroidism, the bone metabolism is dramatically changed because of decreased bone resorption and increased bone formation. If hungry bone syndrome is complicated by postsurgical hypoparathyroidism, hypocalcemia after surgery is exacerbated. In brief, bone formation becomes greater than bone resorption after surgery, resulting in increased demand for calcium that can be resistant to administration of oral calcium and vitamin D [5].

In the literature, there have been some reports of hungry bone syndrome or bone loss associated with PHPT [41, 42]. In addition, Yamashita et al. reported that GD patients often show secondary hyperparathyroidism because of a relative deficiency in calcium and vitamin D due to increased demand for bone restoration after preoperative medical therapy [43].

Recently, Tachibana et al. [5] described a case with severe hypocalcemia after total parathyroidectomy and thyroidectomy in a multiple endocrine neoplasia type 1 (MEN1) patient with PHPT, GD, and acromegaly (AC) indicating that all the three conditions are associated with high bone turnover resulting in severe bone loss and their surgery treatment can induce a hungry bone syndrome with refractory hypokalemia [5]. Finally, severe hypocalcemia due to a hungry bone syndrome is also present in parathyroid carcinoma, a rare and severe entity, with marked clinical and laboratory manifestations at diagnosis. Hungry bone syndrome observed reflects the rapid mineralization after correction of hyperparathyroidism and is related to bone disease severity prior to surgery [44].

The hypoparathyroidism and hungry bone syndrome may induce a critical hypocalcemia that can be resistant to the oral calcium and vitamin D therapy.

25.3.2 Refractory Hypocalcemia due to Abdominal Surgery

A poor response to treatment of hypoparathyroidism following thyroidectomy has been reported as due to malabsorption as a result of abdominal surgery [45–47].

Recently Etheridge et al. [6] described a case of hypocalcemia unresponsive to oral therapy in a patient with hypoparathyroidism following a thyroidectomy. The patient had a history of panproctocolectomy that concurred to realize refractories to oral therapy.

25.3.3 Refractory Hypocalcemia due to Hypomagnesemia

Magnesium (Mg) is required both for the synthesis and release and the peripheral action of PTH. Because Mg depletion is often secondary to another diseases process or to a therapeutic agent, the features of the primary disease process may complicate or mask the Mg depletion. Table 25.3 indicates the major causes of Mg depletion [48].

The hypocalcemia may be a major contributing factor to the increased neuromuscular excitability in these conditions. The pathogenesis of hypocalcemia is multifactorial. In normal subjects, acute change in the serum Mg concentration can influence PTH secretion in a manner similar to calcium through binding to the calcium-sensing receptor. During chronic Mg depletion, however, PTH secretion is impaired. Impaired PTH secretion seems to be a major factor in hypomagnesemia-induced hypocalcemia. Patients with hypocalcemia caused by Mg depletion have both skeletal and renal resistance to exogenously administrated PTH. Clinically, patients with hypocalcemia caused by Mg depletion are resistant not only to PTH but also to calcium and vitamin D therapy. The vitamin D resistance may be caused by impaired metabolism of vitamin D, because serum concentration of 1,25-dihydroxyvitamin D is low [49].

Table 25.3 Major causes of hypomagnesemia

Gastrointestinal losses
Disorders of the small bowel
Small bowel bypass surgery
Primary intestinal hypomagnesemia (X-linked recessive inheritance or autosomally recessive with linkage to chromosome 9q)
Acute pancreatitis
Renal losses
Loop and thiazide-type diuretics (inhibition of Mg absorption)
Volume expansion
Alcohol
Hypercalcemia
Nephrotoxic drugs (aminoglycoside antibiotics, amphotericin B, cisplatin, pentamidine, and cyclosporine)
Loop of Henle or distal tubule dysfunction (tubular necrosis following renal transplantation, during a postobstructive diuresis, or in patients with Bartter's syndrome)
Primary renal magnesium wasting
(a) Associated with hypercalciuria
(b) Associated with hypocalciuria and hypokalemia (Gitelman's syndrome)
(c) Isolated magnesium wasting with both an autosomal dominant and recessive mode of inheritance
Miscellaneous
Diabetes
Hungry bone syndrome

References

1. Bilezikian JP, Khan A, Potts JT Jr (2009) Guidelines for the management of asymptomatic primary hyperparathyroidism: summary statement f rom the third international workshop. J Clin Endocrinol Metab 94:335–339
2. Pollack ME, Brown EM, Estep HL, McLaine PN, Kifor O, Park J, Hebert SC, Seidman CE, Seidman JG (1994) Autosomal dominant hypocalcemia caused by Ca+−sensing receptor gene mutation. Nat Genet 8:303–307
3. Takker R (2013) Chapter 26. Hypoparathyroidism. In: Thakker RV, Ehyte MP, Eisman JA, Igarashi T (eds) Genetic of bone biology and skeletal diseases. Academic, London, pp 409–422
4. Linglart A, Rothenbuhler A, Gueorgieva I, Lucchini P, Silve C, Bougneres P (2011) Long-term results of continuous subcutaneous recombinant PTH (1–34) infusion in children with refractory hypoparathyroidism. J Clin Endocrinol Metab 96:3308–3312
5. Tachibana S, Sato S, Yokoi T, Nagaishi R, Akehi Y, Yanase T, Yamashita H (2012) Severe hypocalcemia complicated by postsurgical hypoparathyroidism and hungry bone syndrome in a patient with primary hyperparathyroidism, Graves' disease, and acromegaly. Intern Med 51:1869–1873
6. Etherige ZC, Schofield C, Prinsloo PJJ, Sturrock NDC (2014) Hypocalcaemia following thyroidectomy unresponsive to oral therapy. Hormones 13:286–289
7. Lepage R, Légaré P, Racicot C, Brossard J-H, Lapointe R, Dagenais M, D'Amour P (1999) Hypocalcemia induced during major and minor

abdominal surgery in humans. J Clin Endocrinol Metab 84:2654–2658
8. Masi L, Brandi ML (2012) Chapter 15. Hypoparathyroidism and hypocalcemic states. In: Khan AA, Clark OH (eds) Handbook of parathyroid diseases. Springer, New York, pp 245–256
9. Dent CE, Harper CM, Morgans ME, Philpot GR, Trotter WR (1955) Insensitivity to vitamin D developing during treatment of postoperative tetany. Its specificity as regards the form of vitamin D taken. Lancet 2:687–690
10. Harrison HE, Lifshitz F, Blizzard RM (1967) Comparison between crystalline dihydrotachyferol and calciferol in patients requiring pharmacologic vitamin D therapy. N Engl J Med 276:894–900
11. Schneller C, Finkel L, Wise M, Hageman JR, Littlejohn E (2013) Autoimmune polyendocrine syndrome: a case-based review. Pediatr Ann 42: 203–208
12. Eisenbarth G, Gottlieb P (2004) Autoimmune polyendocrine syndromes. N Engl J Med 350:2068–2079
13. Ahonen P, Myllarniemi S, Sipila I, Perheentupa J (1990) Clinical variation of autoimmune polyendocrinopaty-candidiasis-ectordermal dystrophy (APECED) in a series of 68 patients. N Engl J Med 322:1829–1836
14. Maeda SS, Fortes EM, Oliveira UM, Borba VCZ, Lazaretti-Castro M (2006) Hypoparathyroidism and pseudohypoparathyroidism. Arq Bras Endocrinol Metabol 50:664–673
15. Aaltonen J, Bjorses P, Sandkuijl L, Perheentupa J, Peltonen L (1994) An autosomal locus causing autoimmune disease: autoimmune polyglandular disease type 1 assigned to chromosome 21. Nat Genet 8:83–87
16. Ekwall O, Hedstrand H, Grimelius L, Haavik J, Perheentupa J, Gustafsson J, Husebye E, Ka¨mpe O, Rorsman F (1998) Identification of tryptophan hydroxylase as an intestinal autoantigen. Lancet 352:279–283
17. Posovszky C, Lahr G, von Schnurbein J, Buderus S, Findeisen A, der Schro C, Schu¨ tz C, Schulz A, Debatin KM, Wabitsch M, Barth TF (2012) Loss of enteroendocrine cells in autoimmune-polyendocrine-candidiasis-ectodermal-dystrophy (APECED) syndrome with gastrointestinal dysfunction. J Clin Endocrinol Metab 97:E292–E300
18. Scire G, Magliocca FM, Cianfarani S, Scalamandre A, Petrozza V, Bonamico M (1991) Autoimmune polyendocrine candidiasis syndrome with associated chronic diarrhea caused by intestinal infection and pancreas insufficiency. J Pediatr Gastroenterol Nutr 13:224–227
19. Bereket A, Lowenheim M, Blethen SL, Kane P, Wilson TA (1995) Intestinal lymphangiectasia in a patient with autoimmune polyglandular disease type I and steatorrhea. J Clin Endocrinol Metab 80:933–935
20. Heubi JE, Partin JC, Schubert WK (1983) Hypocalcemia and steatorrhea—clues to etiology. Dig Dis Sci 28:124–128
21. Masi L, Winer KK, Potts JP, Brandi ML (2004) Management of hypoparathyroidism. Clin Cases Min Bone Metab 2:127–128
22. Khandwala HM, Chibbar R, Bedia A (2006) Celiac disease occurring in a patient with hypoparathyroidism and autoimmune thyroid disease. South Med J 99:290–292
23. Matsueda K, Rosenberg IH (1982) Malabsorption with idiopathic hypoparathyroidism responding to treatment for coincident celiac sprue. Dig Dis Sci 27:269–273
24. Sari R, Yildirim B, Sevinc A, Buyukberber S (2000) Idiopathic hypoparathyroidism and celiac disease in two patients with previous history of cataract. Indian J Gastroenterol 19:31–32
25. Isaia GC, Casalis S, Grosso I, Molinatti PA, Tamone C, Sategna-Guidetti CJ (2004) Hypoparathyroidism and co-existing celiac disease. J Endocrinol Invest 27:778–781
26. Frysák Z, Hrcková Y, Rolinc Z, Hermanová Z, Lukl J (2000) Idiopathic hypoparathyroidism with celiac disease – diagnostic and therapeutic problem. Vnitr Lek 46:408–412
27. Marcondes JAM, Seferian PJ, da Silveria Mitteldorf CAP (2009) Resistance to vitamin D treatment as an indication of celiac disease in a patient with primary hypoparathyroidism. Clinics 64:259–261
28. Kohler S, Wass JAH (2009) Hypoparathyroidism and coeliac disease: a potentially dangerous combination. J R Soc Med 102:311–314
29. Van Heel DA, West J (2000) Recent advances in coeliac disease. Gut 55:1037–1046
30. Selby PL, Davies M, Adams JE, Mawer EB (1999) Bone loss in celiac disease is related to secondary hypoparathyroidism. J Bone Miner Res 14:652–765
31. Lienhardt A, Bai M, Lagarde JP, Rigaud M, Zhang Z, Jiang Y, Kottler ML, Brown EM, Garabedian M (2001) Activating mutations of the calcium-sensing receptor: management of hypocalcemia. J Clin Endocrinol Metab 86:5313–5323
32. Egbuna OI, Brown EM (2008) Hypercalcaemic and hypocalcaemic conditions due to calcium- sensing receptor mutations. Best Pract Res Clin Rheumatol 22:129–148
33. Pearce SH, Williamson C, Kifor O, Bai M, Coulthard MG, Davies M, Lewis-Barned N, McCredie D, Powell H, Kendall-Taylor P, Brown EM, Thakker RV (1996) A familial syndrome of hypocalcemia with hypercalciuria due to mutations in the calcium-sensing receptor. N Engl J Med 335:1115–1122
34. Thakker RV (1994) Molecular genetics of hypoparathyroidism. In: Bilezikian JP, Marcus R, Levine MA (eds) The parathyroids: basic and clinical concepts. Raven, New York, pp 765–779
35. Carey AH, Kelly D, Halford S et al (1992) Molecular genetic study of the frequency of monosomy 22q11 in DiGeorge syndrome. Am J Hum Genet 51:964–970
36. Thakker RV, Davies KE, Whyte MP, Wooding C, O'Riordan JL (1990) Mapping the gene causing X-linked recessive idiopathic hypoparathyroidism to

Xq26-Xq27 by linkage studies. J Clin Invest 86:40–45
37. Arnold A, Horst SA, Gardella TJ, Baba H, Levine MA, Kronenberg HM (1990) Mutations of the signal peptide-encoding region of the preproparathyroid hormone gene in familial isolated hypoparathyroidism. J Clin Invest 86:1084–1087
38. Parkinson DB, Thakker RV (1992) A donor splice site mutation in the parathyroid hormone gene is associated with autosomal recessive hypoparathyroidism. Nat Genet 1:149–152
39. Ahn TG, Antonarakis SE, Kronenberg HM, Igarashi T, Levine MA (1986) Familial isolated hypoparathyroidism: a molecular genetic analysis of 8 families with 23 affected persons. Medicine (Baltimore) 65:73–81
40. Dent CE, Harper CM (1958) Hypoparathyroid tetany (following thyroidectomy) apparently resistant to vitamin D. Proc R Soc Med 51:489–490
41. Brasier AR, Nussbaum SR (1988) Hungry bone syndrome: clinical and biochemical predictors of its occurrence after parathyroid surgery. Am J Med 84:654–660
42. Smith D, Murray BF, McDermott E, O'Shea D, McKenna MJ, McKenna TJ (2005) Hungry bones without hypocalcaemia following parathyroidectomy. J Bone Miner Metab 23:514–515
43. Ymashita H, Murakami T, Noguchi S et al (1999) Postoperative tetany in Graves disease important role of vitamin D metabolites. Ann Surg 229:237–245
44. Ohe NM, Santos RO, Hojaij F, Neves MC, Kunii IS, Orlandi D, Valle L, Martins C, Janovsky C, Ferreira R, Delcelo R, Domingos AM, Abrahão M, Cervantes O, Lazaretti-Castro M, Vieira JGH (2013) Parathyroid carcinoma and hungry bone syndrome. Arq Bras Endocrinol Metabol 57:79–86
45. Pietras S, Holick M (2009) Refractory hypocalcemia following near-total thyroidectomy in a patient with a prior Roux-en-Y gastric bypass. Obes Surg 19:524–526
46. Hylander E, Madsen S (1979) 1 alpha-hydroxyvitamin D3 treatment of therapy-resistant symptomatic hypocalcemia in a hypoparathyroid patient with intestinal malabsorption. Acta Med Scand 205:603–605
47. Seki T, Yamamoto M, Ohwada R et al (2010) Successful treatment of postsurgical hypoparathyroidism by intramuscular injection of vitamin D3 in a patient associated with malabsorption syndrome due to multiple abdominal surgeries. J Bone Miner Metab 28:227–232
48. Agus Z (1999) Disease of the month. Hypomagnesemia. J Am Soc Nephrol 10:1616–1622
49. Rude RK Chapter 70. Hypomagnesemia In: Primer and metabolic bone diseases and disorders of mineral metabolism, 7th edn. ASBMR, pp 325–328

Bone Histomorphometry in Hypoparathyroidism

26

David W. Dempster

26.1 Introduction

PTH is one of the key regulators of skeletal homeostasis, and the effects of chronic PTH deficiency on the human skeleton are dramatic. These effects have been well characterized by histomorphometric analysis of iliac crest bone biopsy specimens. Under normal circumstances, a delicate balance between bone resorption and bone formation, in the process termed bone remodeling, maintains bone mass and structure. Reduced circulating concentrations of PTH initially cause a decrease in bone resorption followed by a coupled decrease in bone formation. Over time, bone mass increases indicating that the bone balance in each remodeling cycle is positive, i.e., more bone is replaced than is removed.

26.2 The Bone Biopsy in Hypoparathyroidism Treated with Vitamin D

The first histomorphometric study of hypoparathyroidism was conducted by Langdahl and colleagues [1]. They analyzed biopsies from 8 women and 4 men with vitamin D-treated hypoparathyroidism and compared them with 13 age- and sex-matched controls. The duration of the disease ranged from 2 to 53 years. Cancellous bone volume was higher in the hypoparathyroid patients, although given the small sample size, this difference was not statistically significant and other structural indices (marrow star volume, trabecular star volume, and trabecular thickness) were also not different from the controls. In the hypoparathyroid subjects, mineralizing surface, bone formation rate, and remodeling activation frequency were all significantly reduced by 58, 80, and 54 %, respectively. Resorption depth was reduced and the total resorption period was extended from 26 to 80 days. Figure 26.1 shows the reconstructed remodeling cycles from the hypoparathyroid and control subjects in this study. Note that there was a slightly positive balance of approximately 5 μm between the resorption depth and wall thickness of cancellous bone packets in the hypoparathyroid subjects compared to the controls. In other words, slightly more bone was being replaced than was removed in each remodeling transaction.

More recently, our group performed a larger histomorphometric study on 33 subjects (24 women and 9 men) with vitamin D-treated hypoparathyroidism and compared the results with 33 age- and sex-matched controls [2]. The etiologies of the hypoparathyroidism were post-thyroid surgery (n = 18), autoimmune ($n = 13$), and DiGeorge syndrome ($n = 2$), and the mean duration of the disease was 17 ± 13 (SD) years. Vitamin D intake

D.W. Dempster (✉)
Department of Pathology and Cell Biology, College of Physicians and Surgeons of Columbia University, 630 W 168th St, New York, NY 10032, USA

Department of Regional Bone Center, Helen Hayes Hospital, West Haverstraw, NY, USA
e-mail: ddempster9@aol.com

M.L. Brandi, E.M. Brown (eds.), *Hypoparathyroidism*,
DOI 10.1007/978-88-470-5376-2_26, © Springer-Verlag Italia 2015

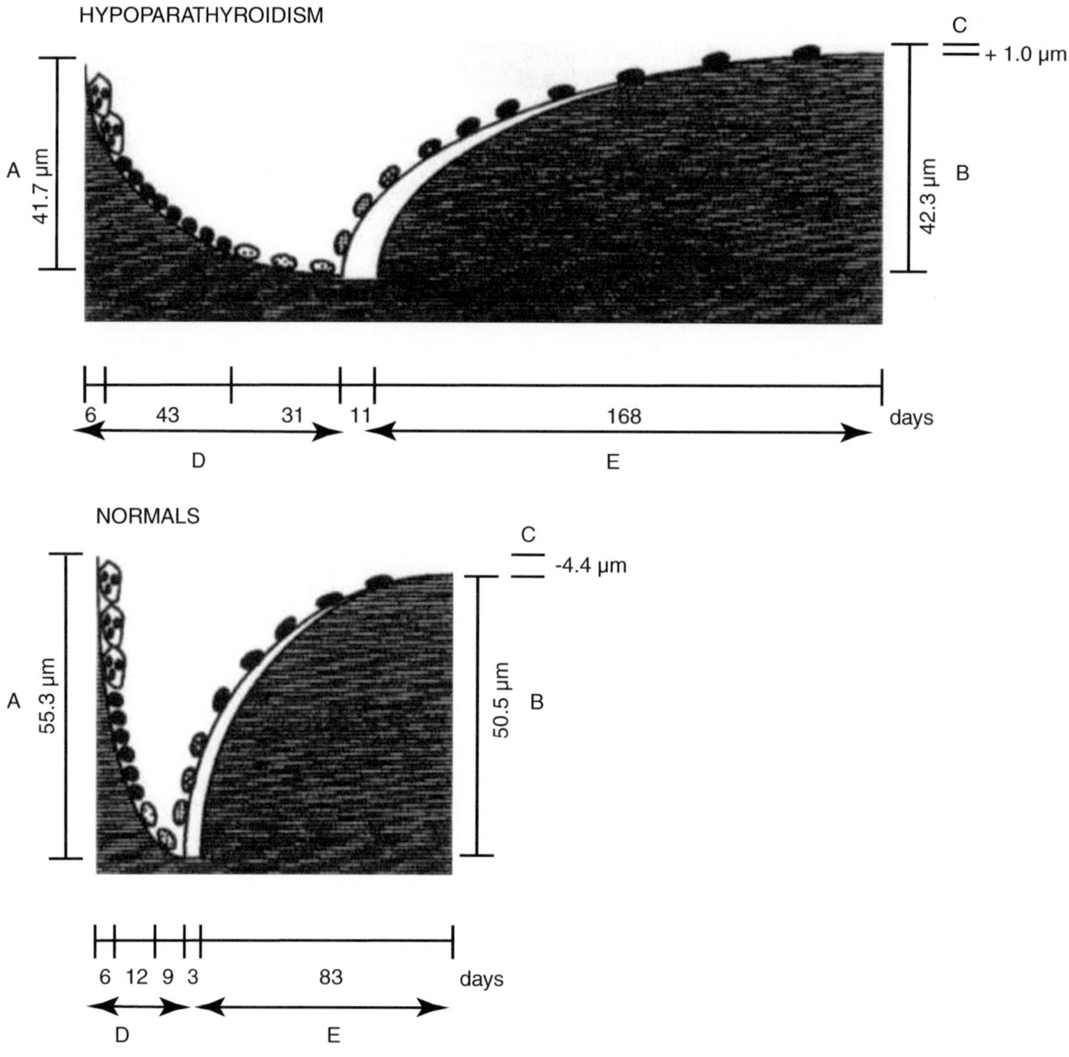

Fig. 26.1 Bone remodeling cycles in hypoparathyroid (*upper*) and normal (*lower*) subjects. All phases of the remodeling cycle are elongated in hypoparathyroidism (Reproduced with permission [1])

ranged between 400 and 100,000 IU/day, calcitriol intake was between 0 and 3 µg/day, and calcium supplementation ranged between 0 and 9 g/day. Ten of the 33 hypoparathyroid subjects were receiving thiazide diuretics. In contrast to the study described above [1], we found that cancellous bone volume was elevated in the hypoparathyroid subjects (Figs. 26.2 and 26.3). The structural basis for this was higher trabecular width, with trabecular number and trabecular spacing being similar to those in control subjects. The hypoparathyroid subjects also displayed higher cortical width than the controls, and cortical porosity was slightly, but not significantly, lower. We assessed remodeling separately on cancellous, endocortical, and intracortical skeletal envelopes. On all three envelopes, bone formation rate, osteoid surface, and width were reduced in the hypoparathyroid subjects relative to controls with the greatest decrease in bone formation rate (>5-fold) occurring on the cancellous envelope (Fig. 26.4). Decreases in both the mineralizing surface and the mineral apposition rate contributed to the marked decrease in bone formation rate.

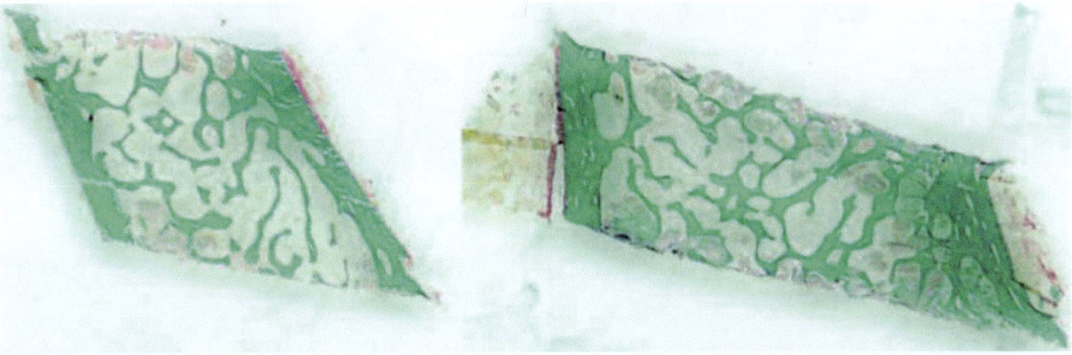

Fig. 26.2 Iliac crest bone biopsies from a control subject (*left*) and a hypoparathyroid subject (*right*). Goldner trichrome stain. Note the higher cortical thickness and cancellous bone volume in the hypoparathyroid subject (Reproduced with permission [2])

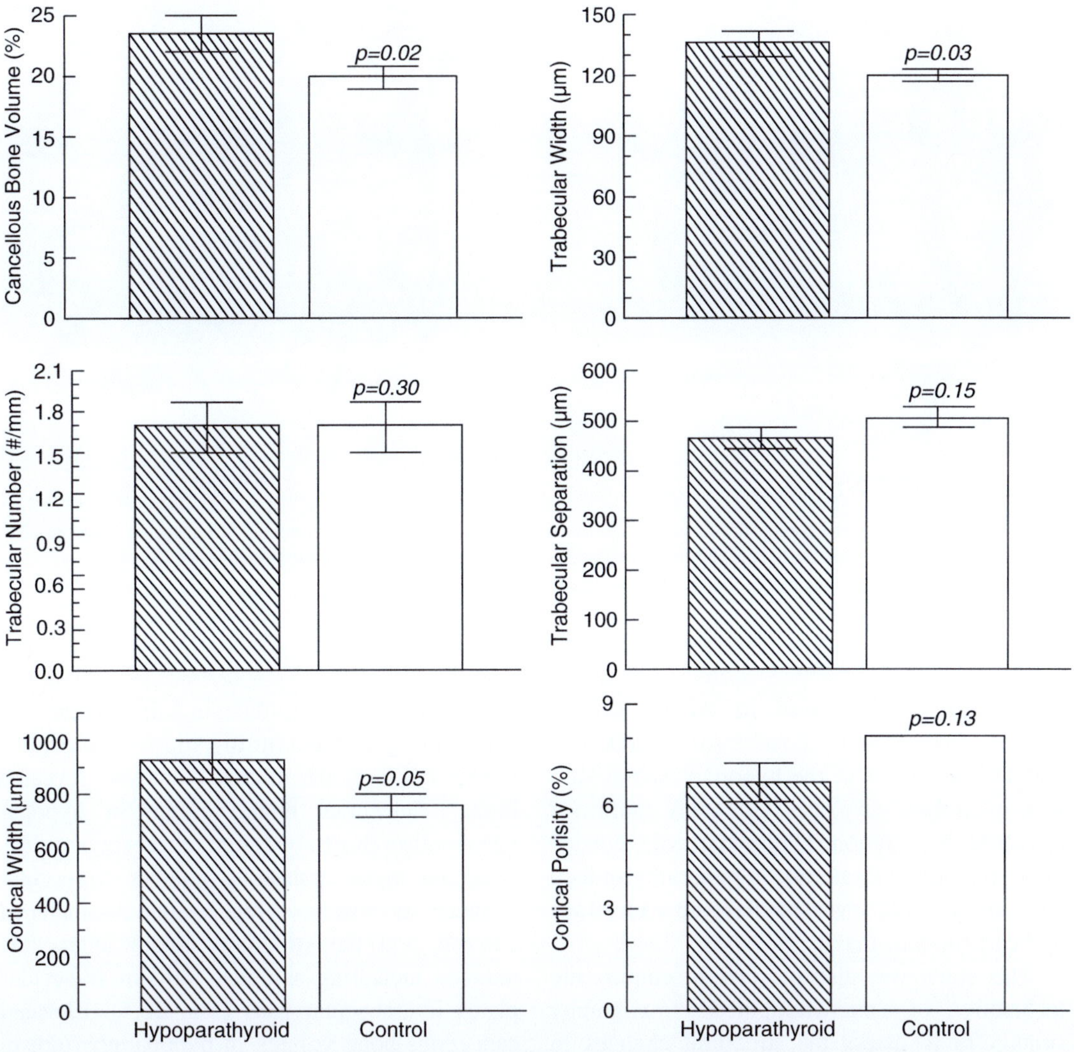

Fig. 26.3 Histomorphometric parameters reflecting cancellous and cortical bone structure in subjects with hypoparathyroidism (*hatched bars*) and controls (*open bars*). Values are mean ± SD (Drawn from data from Rubin et al. [2])

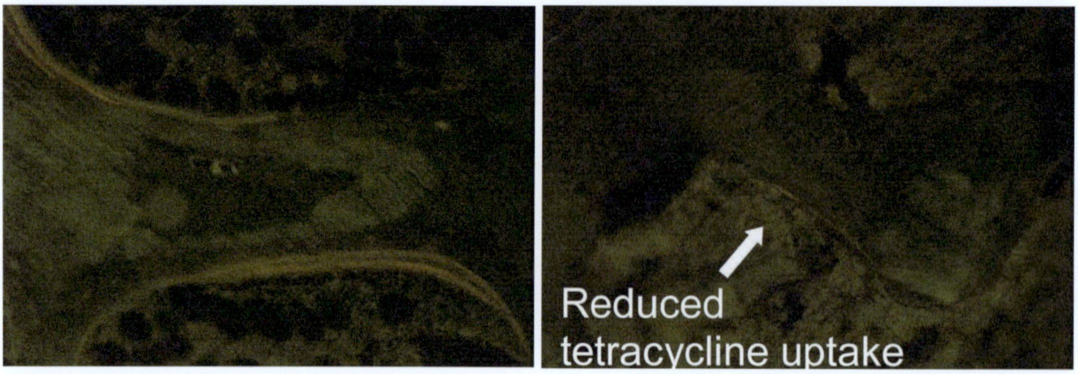

Fig. 26.4 Tetracycline labels in a hypoparathyroid (*left*) and control subject (*right*). Note reduction in tetracycline uptake in the hypoparathyroid subject reflecting reduced bone turnover (Reproduced with permission [2])

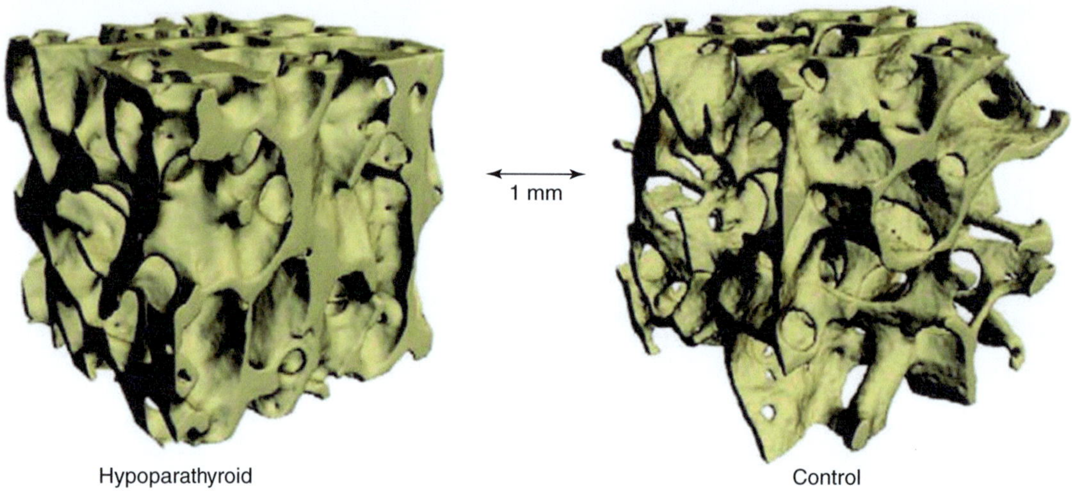

Fig. 26.5 Microcomputed tomographic images of cancellous bone from a hypoparathyroid subject (*left*) and a control subject (*right*). Note the higher cancellous bone volume and dense trabecular structure in hypoparathyroidism (Reproduced with permission [3])

There was no difference in eroded surface between groups, but the bone resorption rate was significantly lower in the hypoparathyroid subjects in all three envelopes. Our study confirmed the earlier observation [1] of a markedly reduced bone remodeling rate in hypoparathyroidism and, in addition, revealed elevated cancellous and cortical bone mass.

This study was also the first to employ the technique of microcomputed tomography (microCT) to assess the structural changes in bone in hypoparathyroidism in 3 dimensions [3]. Confirming and extending the histomorphometric findings, the microCT analysis revealed higher cancellous bone volume in hypoparathyroidism due to both higher trabecular thickness and higher trabecular number. Trabecular connectivity was also higher than in matched controls and the structural model index was reduced indicating a higher ratio of trabecular plates to trabecular rods (Fig. 26.5). Elevated cancellous bone volume in hypoparathyroidism

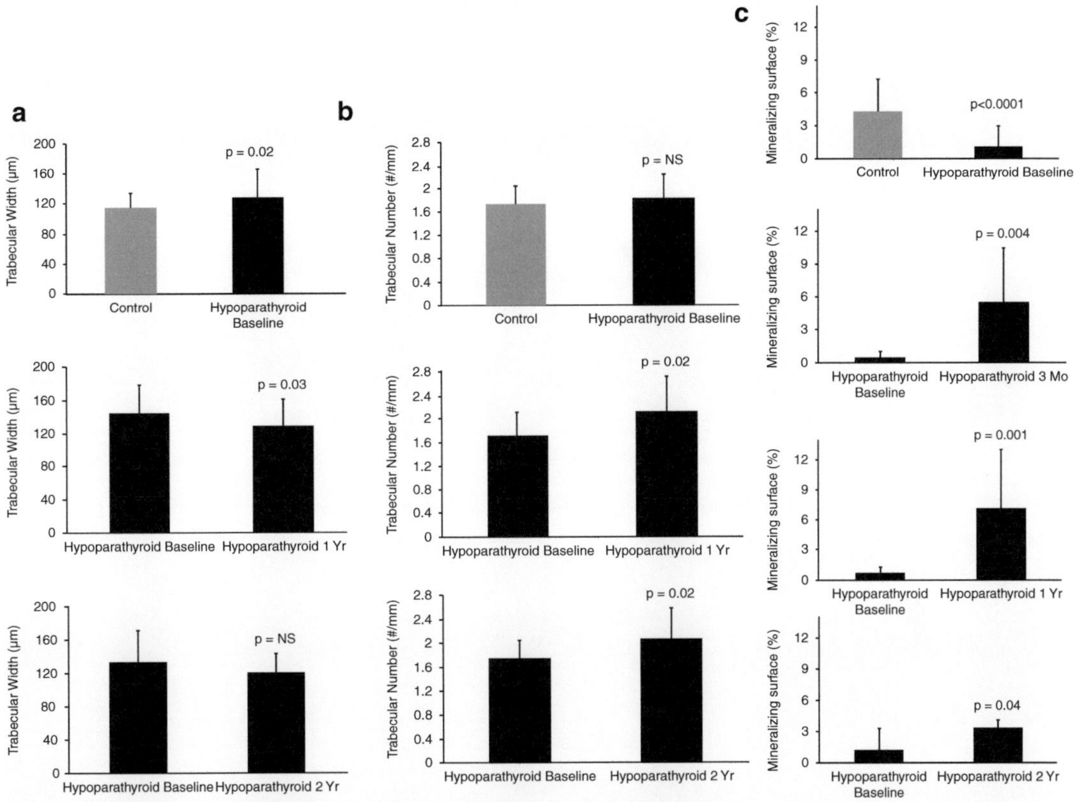

Fig. 26.6 Temporal changes in trabecular width (**a**), trabecular number (**b**), and mineralizing surface (**c**) in cancellous bone following treatment with PTH(1–84) 100 μg every other day for the indicated durations (Reproduced with permission [6])

has also been confirmed by high-resolution pQCT and microfinite element analysis indicates that bone strength is also improved [4].

26.3 The Bone Biopsy in Hypoparathyroidism Treated with PTH

There have now been several studies in which bone histomorphometry has been used to characterize the skeletal effects of treating hypoparathyroidism with PTH (see also Chaps. 30 and 31). In our own study, we used PTH(1–84) at a dose of 100 μg every 2 days [5]. We performed a longitudinal study in which we examined the effects of 3 months of treatment using a quadruple fluorochrome labeling technique and the effects of 1 or 2 years of treatment with a paired biopsy design. The principal findings are summarized in Figs. 26.6 and 26.7. There was no overall change in cancellous bone volume with PTH(1–84) treatment, but trabecular width decreased relative to baseline at 12 months and was still lower than baseline at 24 months, although the latter difference was non-significant. On the other hand, trabecular number was significantly increased at both 12 and 24 months of treatment compared to baseline. These structural changes in cancellous bone were accompanied by an increase in cancellous mineralizing surface, which was significantly elevated as early as 3 months and remained high through 2 years of treatment, indicating a rapid and persistent increase in bone turnover. Tunneling resorption was also seen within trabeculae, explaining the increase in trabecular number. Similar increments

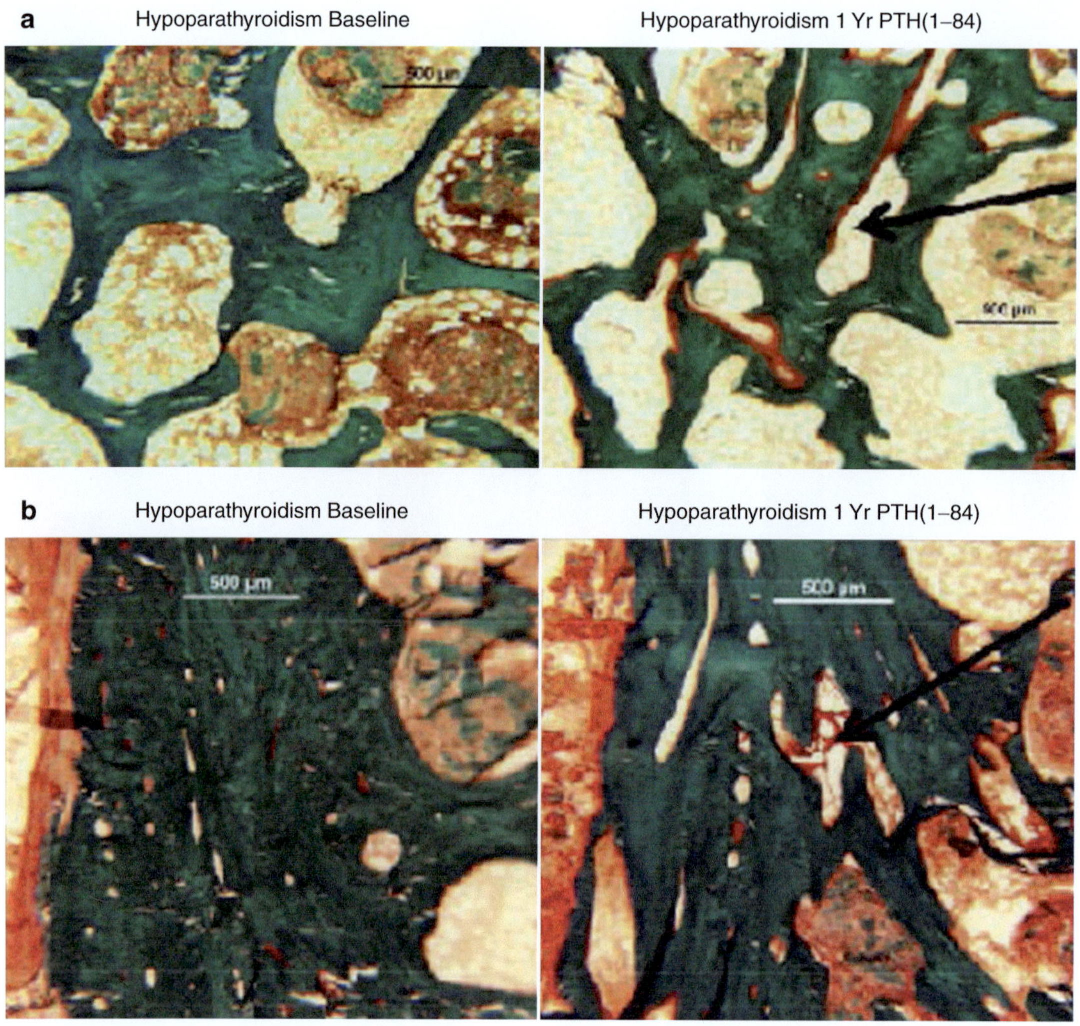

Fig. 26.7 Changes in cancellous (**a**) and cortical (**b**) bone structure following 1 year of treatment with PTH(1–84) in a subject with hypoparathyroidism. *Arrow* indicates tunneling in cancellous and cortical bone (Reproduced with permission [6])

in mineralizing surface, bone formation rate, and bone resorption rate were also seen in the endocortical and intracortical envelopes. The histomorphometric evidence of increased turnover was mirrored by contemporaneous increases in biochemical markers of bone turnover. Interestingly, PTH(1–84) treatment was also associated with an increase in circulating osteogenic precursor cells, which were highly correlated with bone formation indices in the biopsy and showed the same temporal pattern as biochemical markers of bone formation [6]. Figure 26.7 illustrates the changes in cancellous and cortical bone structure and turnover in a patient who was treated with PTH(1–84) for 1 year. This study was the largest histomorphometric study of the skeletal effects of PTH replacement in hypoparathyroidism to date, involving 64 subjects, ranging in age from 18 to 71 years.

Gafni et al. [7] studied the effects of PTH(1–34) replacement in 2 adults and 3 adolescents with hypoparathyroidism. Subjects were injected 2–3 times daily with doses of PTH(1–34) that were titrated to maintain total serum calcium

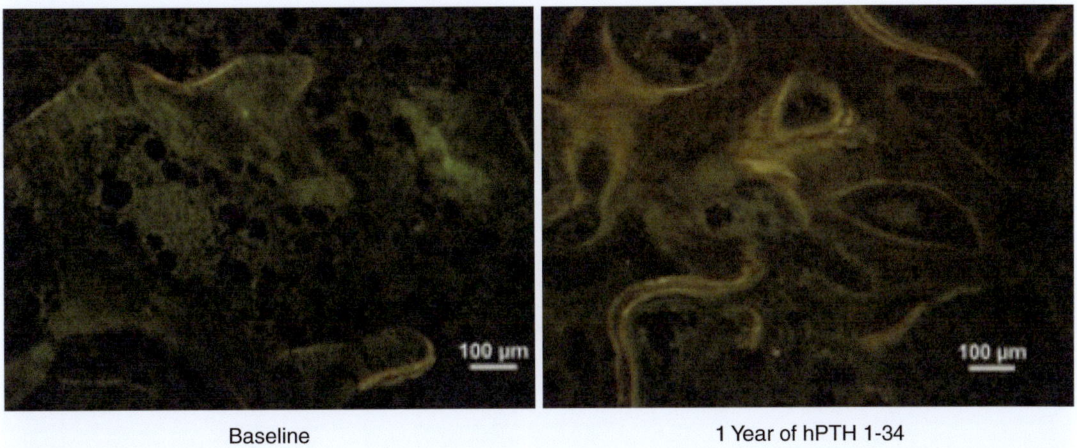

Fig. 26.8 Tetracycline labels in cancellous bone from a 46-year-old woman with hypoparathyroidism before (*left*) and after (*right*) treatment with daily injections of PTH(1–34) (Reproduced with permission [7])

concentration in the range 1.9–2.25 mmol/L. Paired biopsies were performed before and after 1 year of treatment. Unlike the studies described above, cancellous bone volume was dramatically increased by an average of 58 % with PTH(1–34) treatment. This was due to an increase in trabecular number and a decrease in trabecular separation. As previously seen, trabecular remodeling was stimulated (Fig. 26.8) and intra-trabecular tunneling was observed, leading to variable changes in trabecular width. Cortical width did not change but cortical porosity was increased. Similar changes in remodeling were seen on cancellous, endocortical, and intracortical envelopes.

Sikjaer and colleagues [8] have used microCT to assess the effects of PTH(1–84) 100 μg/day on three-dimensional cortical and cancellous bone structure in hypoparathyroid patients. This was a randomized control study in which 23 PTH(1–84)-treated subjects were compared with 21 placebo-treated subjects. The treatment period was 6 months. Compared to the placebo group, PTH(1–84) treatment lowered trabecular bone tissue density by 4 % and trabecular thickness by 27 %, whereas connectivity density was increased by 34 %. Trabecular tunneling was seen in 48 % of the PTH(1–84)-treated subjects (Fig. 26.9), and those with tunneling had higher levels of biochemical markers of bone resorption and formation than those without. In cortical bone, there was a 139 % increase in the Haversian canal number per unit area and a strong trend towards an increase in cortical porosity in the PTH(1–84)-treated group.

The longest reported duration of PTH treatment of a patient with hypoparathyroidism is 13.75 years [9]. The patient was a 6-year-old girl with a sporadic calcium-sensing receptor mutation who was treated continuously with PTH(1–34) until age 20. A bone biopsy obtained after 13.5 years of treatment revealed dramatically elevated cancellous bone volume with increased trabecular number and intra-trabecular tunneling. Tetracycline labeling was not performed.

26.4 The Mechanism Underlying Elevated Bone Mass in Hypoparathyroidism

While bone histomorphometry has revealed much insight into the effects of hypoparathyroidism on bone metabolism and structure, it has so far provided little information on the cellular mechanisms underlying the marked increase in bone mass in this disease. This is partly because the biopsies have generally been performed at a point when the disease is well established rather

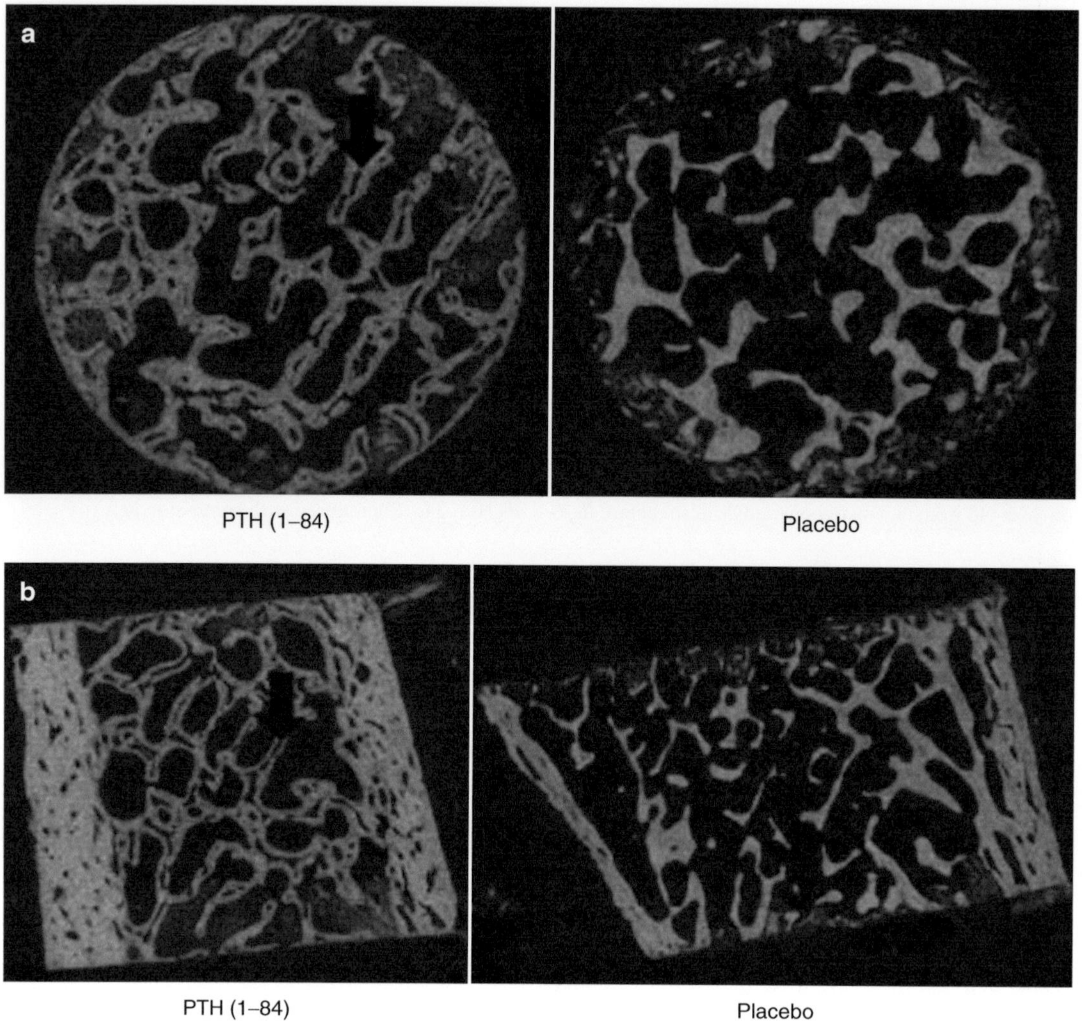

Fig. 26.9 Microcomputed tomography images of a control subject (*right*) and a subject with hypoparathyroidism (*left*) after 6 months of treatment with PTH(1–84) 100 μg/day. The upper images show cross-sectional views and the lower images show longitudinal views. Note the tunneling resorption of trabeculae following treatment with PTH(1–84) (*arrows*) (Reproduced with permission [8])

than when it is developing. The biopsy has revealed low turnover, and in adults, this should certainly preserve bone mass, but not necessarily increase it to the extent seen in hypoparathyroidism. The one study [1] that reported bone balance at the bone remodeling unit level found a modest positive balance of 5 μm. This is unlikely to account for the magnitude of the increases in bone mass, especially since turnover is low and, therefore, the activation frequency of bone remodeling units in which bone formation exceeds resorption is limited. Christen and colleagues [10] have taken a theoretical approach to this problem. These authors used a load adaptive bone modeling and remodeling simulation model that is able to predict changes in microarchitecture due to changes in mechanical loading or cellular activity. They applied the simulation to iliac crest biopsies from 7 healthy subjects and validated the outcome by comparing the models they generated with biopsies from 13 subjects with hypoparathyroidism. Their model

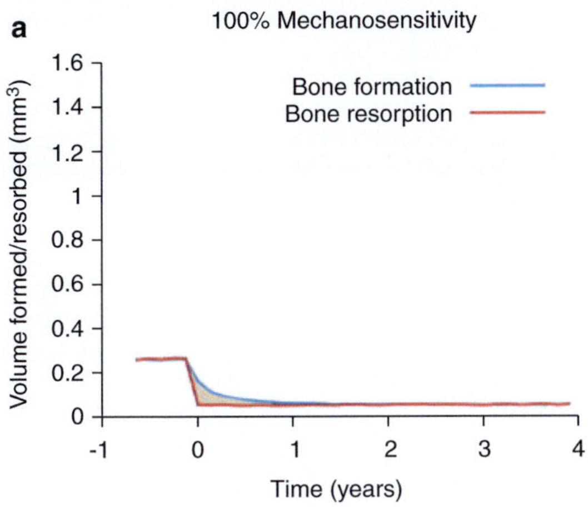

Fig. 26.10 Bone formation and resorption in a simulation of the onset of hypoparathyroidism. On the left, osteocyte mechanosensitivity is set to 100 %, whereas on the right, it is set to 140 % (Reproduced with permission [10])

predicted that, in addition to lowering turnover, the hypoparathyroid state must also cause increased mechanosensitivity of the osteocytes, which leads to a dramatic increase in bone formation during the first year after disease onset (Fig. 26.10).

Conclusion

In conclusion, histomorphometric analysis of iliac crest bone biopsies has contributed significantly to our understanding of the pathogenesis of hypoparathyroidism and the effects of hormone replacement therapy on bone remodeling and structure. The biopsies collected so far will continue to yield new information as researchers apply other techniques to determine the effects of the disease and its treatment on the material properties of bone matrix, such as mineralization density and collagen cross-links. More studies are needed on the long-term effects of injectable PTH therapy and on the effects of continuous infusion of PTH peptides.

Acknowledgment I wish to acknowledge the seminal contributions of my colleagues and collaborators in the studies reviewed in this chapter, in particular Drs. John Bilezikian, Michaela Rubin, and Hua Zhou.

References

1. Langdahl BL, Mortensen L, Vesterby A, Eriksen EF, Charles P (1996) Bone histomorphometry in hypoparathyroid patients treated with vitamin D. Bone 18:103–108
2. Rubin MR, Dempster DW, Zhou H, Shane E, Nickolas T, Sliney J Jr, Silverberg SJ, Bilezikian JP (2008) Dynamic and structural properties of the skeleton in hypoparathyroidism. J Bone Miner Res 23:2018–2024
3. Rubin MR, Dempster DW, Kohler T, Stauber M, Zhou H, Shane E, Nickolas T, Stein E, Sliney J Jr, Silverberg SJ, Bilezikian JP, Müller R (2010) Three dimensional cancellous bone structure in hypoparathyroidism. Bone 46:190–195
4. Cohen A, Dempster DW, Müller R, Guo XE, Nickolas TL, Liu XS, Zhang XH, Wirth AJ, van Lenthe GH, Kohler T, McMahon DJ, Zhou H, Rubin MR, Bilezikian JP, Lappe JM, Recker RR, Shane E (2010) Assessment of trabecular and cortical architecture and mechanical competence of bone by high-resolution peripheral computed tomography: comparison with transiliac bone biopsy. Osteoporos Int 21:263–273
5. Rubin MR, Dempster DW, Sliney J Jr, Zhou H, Nickolas TL, Stein EM, Dworakowski E, Dellabadia M, Ives R, McMahon DJ, Zhang C, Silverberg SJ, Shane E, Cremers S, Bilezikian JP (2011) PTH(1–84) administration reverses abnormal bone-remodeling dynamics and structure in hypoparathyroidism. J Bone Miner Res 26(11):2727–2736
6. Rubin MR, Manavalan JS, Dempster DW, Shah J, Cremers S, Kousteni S, Zhou H, McMahon DJ, Kode A, Sliney J, Shane E, Silverberg SJ, Bilezikian JP (2011) Parathyroid hormone stimulates circulating

osteogenic cells in hypoparathyroidism. J Clin Endocrinol Metab 96(1):176–186
7. Gafni RI, Brahim JS, Andreopoulou P, Bhattacharyya N, Kelly MH, Brillante BA, Reynolds JC, Zhou H, Dempster DW, Collins MT (2012) Daily parathyroid hormone 1–34 replacement therapy for hypoparathyroidism induces marked changes in bone turnover and structure. J Bone Miner Res 27:1811–1820
8. Sikjaer T, Rejnmark L, Thomsen JS, Tietze A, Brüel A, Andersen G, Mosekilde L (2012) Changes in 3-dimensional bone structure indices in hypoparathyroid patients treated with PTH(1–84): a randomized controlled study. J Bone Miner Res 27:781–788
9. Theman TA, Collins MT, Dempster DW, Zhou H, Reynolds JC, Brahim JS, Roschger P, Klaushofer K, Winer KK (2009) PTH(1–34) replacement therapy in a child with hypoparathyroidism caused by a sporadic calcium receptor mutation. J Bone Miner Res 24:964–973
10. Christen P, Ito K, Müller R, Rubin MR, Dempster DW, Bilezikian JP, van Rietbergen B (2012) Patient-specific bone modelling and remodelling simulation of hypoparathyroidism based on human iliac crest biopsies. J Biomech 45:2411–2416

27 Management of Acute Hypocalcemia

Mark Stuart Cooper and Katherine Benson

27.1 Introduction

The spectrum of hypocalcemia ranges from an asymptomatic biochemical abnormality to a life-threatening medical emergency with manifestations such as paresthesias, carpopedal spasm, tetany, and seizures. Its management will depend on factors such as rapidity of onset, the extent of the fall in serum calcium, the presence of signs and symptoms, and its likely cause. The aims of acute management are to safely raise the serum calcium to a level at which symptoms dissipate and to prevent serious cardiac disturbance. Symptomatic patients and those with an acute decrease in serum calcium to less than 1.9 mmol/L (7.6 mg/dL) usually require the intravenous administration of calcium salts. Asymptomatic patients with milder degrees of hypocalcemia can often be managed with oral calcium preparations, with or without the addition of vitamin D (or one of its analogues). The optimal management of acute hypocalcemia has not been examined extensively in clinical trials, and thus, there is not a well-developed evidence base. There are, however, long-standing treatment regimens that are regarded as effective and safe [1–5].

M.S. Cooper (✉) • K. Benson
Department of Endocrinology, Concord Repatriation General Hospital, Hospital Road, Sydney, NSW 2073, Australia
e-mail: mark.cooper@sydney.edu.au; katherinebenson@iinet.net.au

27.2 The Biochemical Diagnosis of Acute Hypocalcemia

The serum calcium concentration must be interpreted in relation to the serum albumin concentration as approximately 50 % of the calcium within the serum is bound to albumin or to other small anions (e.g., citrate), with the remaining fraction being unbound ionized calcium. It is the ionized calcium that is biologically active. An abnormal serum albumin concentration will alter the ratio of ionized calcium to bound calcium and must be corrected for. Various formulae are available for estimation of the corrected serum calcium. Most commonly, the serum calcium concentration is corrected to a reference albumin concentration of 40 g/L (4.0 g/dL) [1, 3]. For every 1 g/L (0.1 g/dL) of albumin above or below this value, the calcium is adjusted by decreasing or increasing it by 0.02 mmol/L (0.08 mg/dL), respectively. For example, a calcium level of 2.05 mmol/L (8.2 mg/dL) with an albumin concentration of 35 g/L (3.5 g/dL) would be corrected to 2.15 mmol/L (8.6 mg/dL).

There are, however, limitations to the use of these correction formulae. The extent of binding of calcium to albumin is influenced by acid–base balance, with alkalosis being associated with an increase in protein binding and a reduction in the fraction of ionized calcium. In patients with critical illness or a very low serum albumin concentration, the estimates of total calcium based on albumin correction can be particularly unreliable [6, 7].

M.L. Brandi, E.M. Brown (eds.), *Hypoparathyroidism*,
DOI 10.1007/978-88-470-5376-2_27, © Springer-Verlag Italia 2015

In these situations, the ionized calcium should be measured directly. Other factors that can interfere with serum calcium measurement and give a falsely low estimate include contamination of blood collection tubes with EDTA and recent use of gadolinium-containing contrast materials [8].

In addition to the measurement of serum calcium, patients presenting with acute hypocalcemia should have urgent measurement of renal function, serum phosphate, and magnesium. Serum should also be collected for measurement of PTH and 25-hydroxyvitamin D.

27.3 The Clinical Presentation of Acute Hypocalcemia

Hypocalcemia causes neuromuscular excitability. Classical symptoms of hypocalcemia are skeletal muscle twitching, carpopedal spasm, and both perioral and peripheral tingling and numbness. In severe cases, life-threatening generalized tetany, laryngeal spasm, and seizures can occur.

In hypocalcemic subjects who are asymptomatic, neuromuscular excitability can frequently be unmasked by use of the Chvostek or Trousseau test [9]. The Chvostek test involves gently tapping the parotid gland over the facial nerve (approximately 2 cm anterior to the earlobe) (Fig. 27.1) to induce facial muscle spasm. However, this test lacks both sensitivity and specificity as 10 % of normocalcemic individuals display a positive Chvostek response and a substantial proportion of patients with clinically significant hypocalcemia have a negative response. Trousseau test involves inflating a blood pressure cuff placed on the arm above arterial pressure for up to 3 min (this induces ischemia of the muscle which is thought to increase its sensitivity to low calcium levels). A positive test suggesting neuromuscular excitability is the development of carpal spasm in the limb (Fig. 27.2). Trousseau test is positive in approximately 95 % of patients with significant hypocalcemia and in only 1 % of normocalcemic individuals and is thus a more discriminating test [9].

The typical ECG finding in significant hypocalcemia is prolongation of the QT interval, but cardiac dysrhythmias can also occur. An important but infrequent feature of hypocalcemia is cardiac failure. This condition is poorly responsive to inotropic therapy unless the hypocalcemia is rectified. Long-standing hypocalcemia can occasionally present with neuropsychiatric symptoms, cataract formation, or raised intracranial pressure/optic nerve damage in the absence of the more typical neuromuscular symptoms.

Depending on the cause of hypocalcemia, symptoms can develop slowly over many months or rapidly over hours. The development

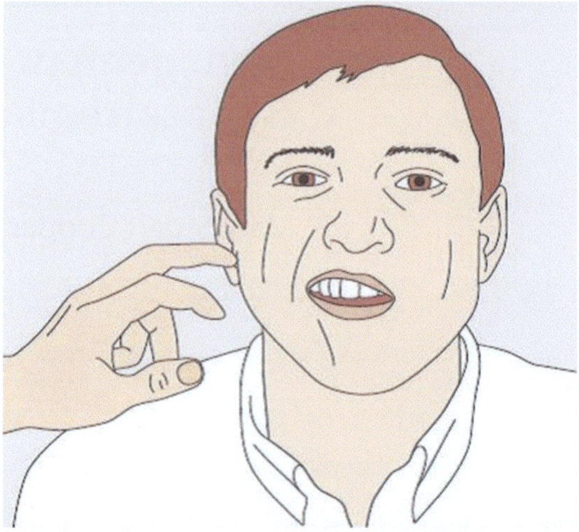

Fig. 27.1 Chvostek sign. Tapping the lateral aspect of the face over the facial nerve induces a facial twitch in the presence of hypocalcemia. This sign lacks specificity and sensitivity for the accurate diagnosis of significant hypocalcemia when used in isolation

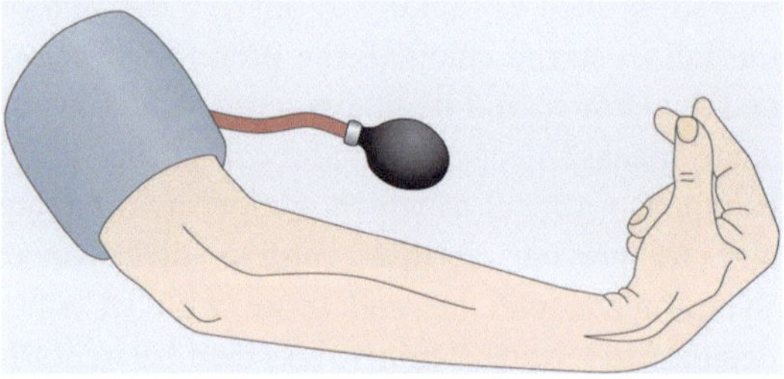

Fig. 27.2 Trousseau sign. Inflation of a blood pressure cuff at the level of the forearm to above arterial pressure for up to 3 min induces carpal spasm in the presence of hypocalcemia. This sign has greater sensitive and specificity for the diagnosis of significant hypocalcemia compared to Chvostek sign

of neuromuscular excitability depends on both the absolute level of serum calcium and how rapidly the serum calcium falls. When hypocalcemia develops gradually, e.g., in autoimmune hypoparathyroidism, the serum calcium can fall to as low as 1.1 mmol/L (4.4 mg/dL) before the patient becomes symptomatic. By contrast, in patients with postsurgical hypoparathyroidism, symptoms commonly arise when the serum calcium falls below 1.9 mmol/L (7.6 mg/dL). In patients who develop hypoparathyroidism gradually, there is frequently an intercurrent event that reduces calcium levels acutely and precipitates the development of symptoms. Examples include viral infections, hyperventilation, or use of glucocorticoids (which reduce calcium absorption and stimulate renal calcium loss).

27.4 The Immediate Management of Acute Hypocalcemia

The following management plan relates to the management of hypocalcemia in adults, but the principles of management are similar for children (discussed further below). Neuromuscular irritability due to hypocalcemia necessitates prompt admission to hospital and treatment with intravenous calcium. This should also be considered in asymptomatic patients with a corrected serum calcium of less than 1.9 mmol/L (7.6 mg/dL) as there is a risk of serious complications developing. The approach to management of acute hypocalcemia is outlined in Fig. 27.3. This is based on clinical experience and expert recommendations [1, 4, 5, 10] (but not randomized trials).

Both calcium gluconate and calcium chloride are available for intravenous use, but calcium gluconate should be used preferentially as the chloride form can cause serious local irritation if extravasation occurs. With either preparation, care should be taken to avoid extravasation.

One to two 10-mL ampules of 10 % calcium gluconate (each of which contains 1 g of calcium gluconate which equates to approximately 94 mg of elemental calcium) should be diluted in 50–100 mL of 5 % dextrose in water and infused slowly over 10 min. ECG monitoring is recommended as dysrhythmias can occur if correction is too rapid. This regimen can be repeated until acute symptoms have resolved. Possible side effects of intravenous calcium infusion include the development of a chalky taste, hot flushes, and peripheral vasodilatation. Boluses of intravenous

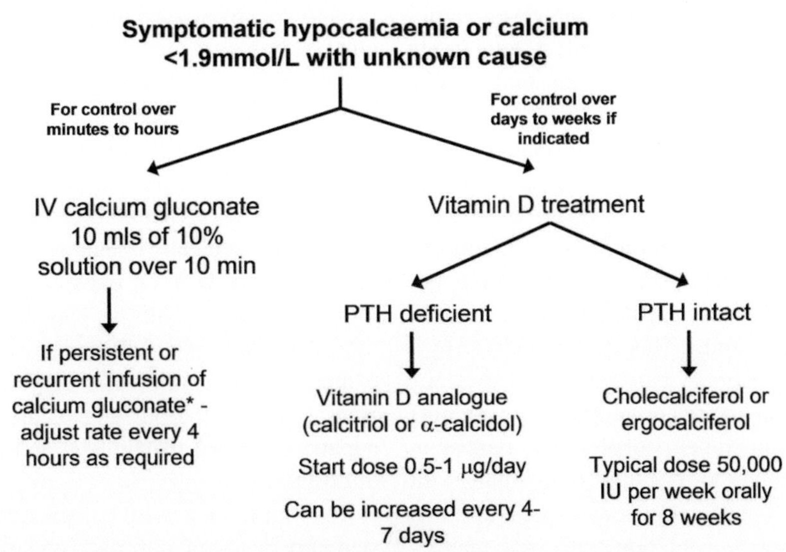

Fig. 27.3 Algorithm for the management of acute hypocalcemia (adults)

calcium will often only provide temporary relief of symptoms, and continuous infusion of a dilute calcium solution may be required to prevent recurrent hypocalcemia. A typical approach would involve using 10 ampules of 10 mL of 10 % calcium gluconate in 1 L of 5 % dextrose in water or 0.9 % saline. This is infused at an initial rate of 50 mL/h with subsequent frequent monitoring of the serum calcium, aiming to maintain it at the lower end of the reference range [5]. An infusion of 10 mL/kg of this solution over 4–6 h will be expected to increase serum calcium by approximately 0.3–0.5 mmol/L (1.2–2.0 mg/dL). Particular care is needed when administering intravenous calcium to subjects taking digoxin as it is associated with increased cardiac sensitivity to changes in extracellular calcium. Slower infusion rates and ECG monitoring are required in this situation.

27.5 Concurrent Hypomagnesemia

Hypomagnesemia is an important cause of hypocalcemia, both by inducing resistance to PTH and diminishing its secretion (see also Chap. 7). When hypocalcemia is induced by hypomagnesemia, the level of magnesium in typically less than 0.5 mmol/L. However, serum levels of magnesium poorly reflect total body (and primarily intracellular) magnesium levels, and thus, any level of magnesium below the normal range needs to be considered as a possible contributory factor to hypocalcemia. Hypomagnesemia should be rectified with an infusion of magnesium sulfate 1–2 g in 100 mL 5 % dextrose in water over 10 min (1 g of magnesium sulfate contains approximately 93 mg of elemental magnesium). This can be followed by a continuous infusion of between 4 and 6 g/day. Lower infusion rates are required in patients with significant renal impairment. Persistent hypomagnesemia can occur in individuals with ongoing gastrointestinal or renal losses, and this necessitates supplementation with oral magnesium. The most frequently recommended form of oral magnesium in this situation is magnesium oxide 400 mg. Each tablet contains 242 mg of elemental magnesium and should be given two to three times per day.

Serum phosphate should be measured prior to administration of intravenous calcium, as precipitation of calcium phosphate can occur in hyperphosphatemic subjects. In clinical practice, this situation has been encountered in patients with tumor lysis syndrome and renal failure but is unlikely to occur in uncomplicated hypoparathyroidism. Caution is necessary if the serum phosphate exceeds 2.0 mmol/L (6 mg/dL).

27.6 Oral Calcium and Vitamin D

Almost all subjects with significant hypocalcemia will require supplementation with oral calcium and/or vitamin D or one of its analogues. Oral calcium supplementation is preferable for those with mild acute hypocalcemia (1.9–2.0 mmol/L) or for chronic hypocalcemia. Various preparations of oral calcium salts are available. The most commonly used are calcium carbonate and calcium citrate. The carbonate form has the highest amount of elemental calcium at about 40% of its total weight, with the citrate form being around 21%. The typical starting dose is 1,000–1,200 mg of calcium two to three times per day. There are limited data to suggest that the absorption of calcium carbonate is reduced in patients with achlorhydria or those on proton pump inhibitors. In this situation, it is reasonable to use calcium citrate.

Several vitamin D preparations are available for the treatment of hypocalcemia due to hypoparathyroidism or vitamin D deficiency. Vitamin D deficiency is typically treated with oral ergocalciferol (vitamin D2) or cholecalciferol (vitamin D3). In some countries, parenteral forms are available. The major advantage of vitamin D in comparison to calcitriol (1,25-dihydroxyvitamin D) is its low cost. Disadvantages are the need for hepatic and renal metabolism and slow duration of onset.

In patients with hypoparathyroidism, standard doses of non-hydroxylated vitamin D (cholecalciferol or ergocalciferol) are unlikely to be effective as PTH is required for the renal conversion of 25-hydroxyvitamin D to calcitriol (1,25-dihydroxyvitamin D). The most appropriate treatment is with calcitriol, initially 0.25–0.5 mcg twice daily. The half-life of this preparation is approximately 5–8 h, and the dose can be adjusted every 48 h.

Pharmacological doses of vitamin D, e.g., 10,000 to 50,000 units per day (0.25–1.25 mg/day), can be used, but the half-life of these preparations is prolonged, and they have a slower onset of action and a greater risk of serious toxicity (hypervitaminosis D) than calcitriol.

As with intravenous calcium administration, the goal of oral therapy is to increase the level of serum calcium to a level at which symptoms of hypocalcemia resolve and the risk of their recurrence is minimized. The major risks of therapy are hypercalcemia and hypercalciuria, which can lead to nephrolithiasis and nephrocalcinosis. Hypercalciuria is most likely to occur in subjects with hypoparathyroidism, as PTH stimulates renal calcium reabsorption. The goal of therapy in hypoparathyroid subjects is maintenance of serum calcium in the low to normal range 2.0–2.1 mmol/L. Serum and urine calcium should be monitored regularly.

27.7 Special Situations

The treatment of neonates, infants, and children with acute hypocalcemia follows similar principles to those described above, but the doses of medications are based on patient weight rather than absolute amounts. Acute hypocalcemia in the context of recent neck surgery is treated in a similar fashion to that outlined above. Where hypocalcemia might be anticipated, patients are often treated preemptively with oral calcium supplements. The decision to initiate calcitriol or alfacalcidol (1-alpha-hydroxycholecalciferol) is often based on a measurement of serum PTH six to twelve hours after the operation. An undetectable level indicates a need for active vitamin D preparations. Depending on the nature of the surgery, the hypoparathyroidism could be temporary or permanent, but this often only becomes clear in the months following surgery when withdrawal of calcium and vitamin D supplements is attempted. An alternative approach to the management of acute hypoparathyroidism is the injection of PTH or one of its analogues (e.g., teriparatide) as a replacement therapy. The effectiveness of these agents has been evaluated in the context of chronic hypoparathyroidism and is discussed in depth in chapters 30 and 31. However, the use of PTH in acute hypoparathyroidism has not been studied. Another therapeutic approach that is commonly used in patients with chronic hypoparathyroidism is the administration of thiazide diuretics. These drugs reduce urinary calcium excretion and can maintain a higher level of serum calcium for the same amount of vitamin D analogue therapy. In the context of acute hypoparathyroidism, the use of thiazides to increase serum calcium has not been reported. As such, both PTH and thiazide therapy are inappropriate in the acute setting.

References

1. Cooper MS, Gittoes NJ (2008) Diagnosis and management of hypocalcemia. BMJ 336(7656): 1298–1302
2. Compston JE (1995) Investigation of hypocalcemia. Clin Endocrinol (Oxf) 42(2):195–8
3. Bushinsky DA, Monk RD (1998) Electrolyte quintet: calcium. Lancet 352(9124):306–311
4. Fong J, Khan A (2012) Hypocalcemia: updates in diagnosis and management for primary care. Can Fam Physician 58(2):158–62
5. Thakker RV (2003) Parathyroid disorders and diseases altering calcium metabolism. In: Warrall D (ed) Oxford textbook of medicine. Oxford University Press, Oxford
6. Dickerson RN et al (2004) Accuracy of methods to estimate ionized and "corrected" serum calcium concentrations in critically ill multiple trauma patients receiving specialized nutrition support. JPEN J Parenter Enteral Nutr 28(3):133–41
7. Slomp J et al (2003) Albumin-adjusted calcium is not suitable for diagnosis of hyper- and hypocalcemia in the critically ill. Crit Care Med 31(5):1389–1393
8. Doorenbos CJ, Ozyilmaz A, van Wijnen M (2003) Severe pseudohypocalcemia after gadolinium-enhanced magnetic resonance angiography. N Engl J Med 349(8):817–8
9. Urbano FL (2000) Signs of Hypocalcemia: Chvostek's and Trousseau's. Hosp Physician 36:43–45
10. Shoback D (2008) Clinical practice hypoparathyroidism. N Engl J Med 359(4):391–403

Conventional Treatment of Chronic Hypoparathyroidism

28

Lars Rejnmark

28.1 Introduction

Hypoparathyroidism is characterized by hypocalcemia due to inappropriately low levels of PTH. Conventional treatment of hypoparathyroidism is a combination of calcium supplementation and vitamin D analogues. Although serum calcium levels are (near-)normalized in response to this therapy, it does not restore the normal physiological regulation of calcium homeostasis. Despite normocalcemia, the lack of PTH causes an abnormally low bone turnover, and bone mineral density is often markedly increased [1]. Moreover, renal calcium losses are increased with a concomitantly low renal phosphate clearance. Due to reduced renal clearance, serum phosphate levels are often high normal or above the upper level of the reference interval causing a relatively high calcium–phosphorous product, which may increase the risk of extraskeletal calcifications. Compared with the general background population, patients with postoperative hypoparathyroidism have an almost fourfold increased risk of renal complications [2]. In addition to these well-known effects of the lack of PTH, an increasing number of studies suggest that PTH may exert effects beyond its classical effects on kidney and bone. Receptors for PTH are expressed by cells in several tissues, including the central nervous system [3]. It is possible that the lack of PTH in the central nervous system may account for the neuropsychological complains often reported by patients [4]. Despite normal serum calcium levels, patients with hypoparathyroidism on conventional therapy often complain of reduced quality of life (QoL), mood disorders, cognitive dysfunction, and numerous nonspecific symptoms [4–6]. Accordingly, although calcium levels are restored in response to conventional therapy, it should be emphasized that a number of physiological and neuropsychological deficits are present in hypoparathyroidism.

28.2 Therapeutic Targets

Table 28.1 shows the therapeutic targets during conventional treatment of chronic hypoparathyroidism. Treatment should be titrated in such a manner that patients are without major symptoms of hypocalcemia. Preferably, serum calcium levels are maintained slightly below or in the lower range of the reference interval. This is of importance in order to lower urinary calcium excretion and to keep the calcium–phosphorous product as low as possible. Magnesium levels should be monitored, as magnesium homeostasis may be disturbed with a tendency towards hypomagnesemia [7]. Moreover, if magnesium levels are very low, PTH cannot be secreted by the parathyroid

L. Rejnmark, PhD, DMSc
Department of Endocrinology and Internal Medicine, Aarhus University Hospital, Tage-Hansens Gade 2, Aarhus C DK-8000, Denmark
e-mail: rejnmark@post6.tele.dk

M.L. Brandi, E.M. Brown (eds.), *Hypoparathyroidism*,
DOI 10.1007/978-88-470-5376-2_28, © Springer-Verlag Italia 2015

glands, and severe hypomagnesemia may thereby blunt a recovery of the parathyroid function [8]. Finally, hypomagnesemia may cause symptoms similar to hypocalcemia, including an increased neuromuscular excitability.

28.3 Treatment with Calcium and Vitamin D

Calcium supplements in combination with vitamin D analogues are the mainstay in the treatment of hypoparathyroidism. Different traditions seem to exist on how to titrate the dose of calcium and vitamin D, as some institutions seem to prefer a relative high dose of calcium (3–5 g/d) in combination with a relative low dose of vitamin D, whereas other institutions are using a relatively high dose of vitamin D analogues in combination with a lower dose of calcium (800–1,200 mg/d). The two strategies have not been formally compared, but is does somehow seem reasonable to keep calcium intake relatively low in order to lower urinary calcium and thereby (probably) lower risk of renal stone formation [9].

28.3.1 Calcium Supplements

Doses of calcium supplements should be spread throughout the day, as the fraction of calcium absorbed decreases as the ingested amount of calcium increases. Normally, a dose of 400–500 mg of elementary calcium should be given twice a day. If patients experience symptoms of hypocalcemia despite normal serum calcium levels, it may be difficult to increase the dose of vitamin D, as this often will result in hypercalcemia. If so, patients may have relief of their hypocalcemic symptoms by using calcium supplements more than twice a day. If symptoms of hypocalcemia occur intermittently, but less frequently than on a daily basis, patients may be instructed to use additional calcium supplements PRN (pro re nata). Taking a dose of calcium at bedtime may be favorable, as nighttime is a state of fasting. Normally, PTH levels and bone turnover increase during nighttime causing an efflux of calcium from the bone to the extracellular fluid [10]. Although this has not been thoughtfully investigated, it seems likely that these homeostatic mechanisms do not function properly in hypoparathyroidism. Evening administration of calcium supplements may be a solution to patients with symptoms of hypocalcemia during nighttime or in the early morning hours.

Calcium is bound to carbonate in most calcium supplements, but it may as well be bound to citrate or an organic molecule, such as malate in others. Calcium carbonate is often preferred, as it is less expensive than other calcium supplements. It should, however, be taken with food, as its absorption in the intestine depends on low pH levels [11]. Despite impaired acidification, patients with achlorhydria seem to have proper absorption calcium from calcium carbonate as long as it is taken together with a meal [11]. Calcium citrate and calcium malate can be taken without a meal as calcium from these compounds is better absorbed on an empty stomach. Apparently, the absorption of calcium from supplements is similar to the absorption from dairy products [12]. Although there are no data comparing intake from supplements with intake from dairy products in hypoparathyroidism, a high intake of calcium from dairy products may not be favorable, as dairy products are rich in phosphate. As renal phosphate clearance is impaired in hypoparathyroidism, it seems reasonable to limit phosphate intake.

Discrepant results have been reported on the effects of use of calcium supplements on cardiovascular events [13, 14]. However, in a recent epidemiological study on comorbidity in hypoparathyroidism, risk of cardiovascular diseases was not increased in patients with hypoparathyroidism [2].

Use of calcium supplements may cause gastrointestinal discomfort, especially constipation. Such side effects may be avoided by taking the supplements with food and concomitant use of magnesium supplements [15].

A common cause of hypoparathyroidism is accidental removal of the parathyroid glands during thyroid surgery or other operations in the neck such as radical neck dissections, and many

Table 28.1 Therapeutic targets and monitoring of patients with chronic hypoparathyroidism on conventional therapy

	Therapeutical target	Monitoring interval
Symptomatic hypocalcemia	None or infrequently	Every 3rd month
Serum calcium	Slightly below or in the lower range of the reference interval	Every 3rd month
Calcium–phosphorus product	<4.4 $mmol^2/L^2$ (<55 mg^2/dL^2)	Every 3rd–6th month
Serum magnesium	Within the reference interval	Every 3rd–6th month
24-h urinary calcium	Within the sex-specific reference interval	Once a year
Serum PTH		Following surgery causing hypoparathyroidism, PTH levels should be measured once a month for the first 6 months. In chronic hypoparathyroidism, PTH should be measured once a year in order to detect whether there is any recovery of parathyroid function
Eye examination	No cataract	Awareness of symptoms at clinical follow-up (once a year). If symptoms occur, an eye examination should be performed

patients with postsurgical hypoparathyroidism are on substitution therapy with thyroid hormones. As calcium supplements may impair the absorption of levothyroxine, patients should be instructed not to take levothyroxine and calcium simultaneously or near-simultaneously [16]. Similarly, calcium may interfere with the absorption of other drugs such as ciprofloxacin and tetracycline [17]. Care should be taken to advise patients with hypoparathyroidism appropriately on such "food"-drug interactions.

28.3.2 Vitamin D Treatment

Before the development of active (1α-hydroxylated) vitamin D analogues, calciferol was used in the treatment of hypoparathyroidism, either as cholecalciferol (vitamin D3) or ergocalciferol (vitamin D2) [18]. Although the affinity of $1{,}25(OH)_2D$ to the vitamin D receptor (VDR) is much higher than the affinity of calciferol, calciferol do, nevertheless, bind to the VDR and exert biological effects if the concentrations are high enough. Moreover, despite reduced activity of the 1α-hydroxylase if PTH is missing, a small amount of 25-hydroxyvitamin D (25OHD) is nevertheless converted into its active metabolite. In hypoparathyroidism, serum levels of $1{,}25(OH)_2D$ are subnormal, but a positive correlation is present between levels of 25OHD and $1{,}25(OH)_2D$ [19, 20]. If treated with calciferol, very high doses are needed to achieve normocalcemia resulting in markedly elevated serum 25OHD levels. Typically, patients receive a daily dose of 25,000–200,000 IU of vitamin D2 or D3 resulting in serum 25OHD levels between 500 and 1,000 nmol/l [1, 19, 21].

Since the development of 1α-hydroxylated vitamin D analogues, most patients with hypoparathyroidism have been shifted to treatment with either alfacalcidol or calcitriol, as the serum half-life of activated vitamin D metabolites is much shorter (approximately 3–6 h) than the biological half-life of vitamin D2 or D3 (approximately 3 weeks) [18, 19, 22–25]. The advantage of the shorter serum half-life is that a new equilibrium is obtained at a much faster rate [7]. Dose titration may be performed every 2–3 days if activated vitamin D metabolites are used, whereas 2–3 months have to pass before a new equilibrium is obtained in response to dose titration of calciferol. Similarly, if intoxication occurs, this has a much longer duration if caused by too high levels of calciferol rather than too high levels of calcitriol.

The synthetic vitamin D analogue *dihydrotachysterol* may also be used. Similar to alfacalcidol, dihydrotachysterol is activated in the liver, but does not require renal hydroxylation [26, 27].

Although used frequently in the past, most institutions now prefer the use of more specific vitamin D analogues (alfacalcidol or calcitriol).

The dose of alfacalcidol needed to maintain normocalcemia is normally 1–3 μg/day, whereas calcitriol typically is administrated in a daily dose of 0.5–1.5 μg. However, large interindividual variations exist in dose needed, and the dose required to maintain serum calcium in the desired range may vary with time within patients. No formal comparison has been performed on whether differences exist on long-term treatment outcomes between alfacalcidol, calcitriol, and dihydrotachysterol. However, it has been reported that hypercalcemia following treatment with dihydrotachysterol may take slightly longer to resolve (3–14 days) than following treatment with either alfacalcidol (5–10 days) or calcitriol (2–10 days) [27, 28].

Although active vitamin D analogues are the mainstay of treatment of hypoparathyroidism, a sufficient vitamin D status in terms of 25OHD levels >80 nmol/l should be ensured, as calciferol may be of importance to extraskeletal tissues [29].

28.4 How to Handle Hypercalciuria?

In hypoparathyroidism, a linear relationship has been reported between serum and urinary calcium [18, 30]. In order to lower urinary calcium, serum calcium levels should be in the lower part (or slightly below the lower limit) of the reference interval. Moreover, salt intake should be restricted [31].

Treatment of patients with hypoparathyroidism with thiazide diuretics has been shown to lower urinary calcium by increasing the renal tubular reabsorption of calcium [32–34]. A low-salt diet should be advocated, as a high-salt intake may abolish the hypocalciuric action of thiazides [34].

There are no data available on long-term effects of treatment with thiazides in hypoparathyroidism. However, as thiazides have been shown to lower risk of renal stones in idiopathic hypercalciuria, it seems likely that a similar beneficial effect may apply to patients with hypoparathyroidism [35].

To obtain a sustained hypocalciuric effect throughout the day, most thiazide diuretics (e.g., hydrochlorothiazide and bendroflumethiazide) need to be administrated twice a day due to their relatively short serum half-life. The hypocalciuric effect is dose dependent, why a relative high dose may be used if tolerated by patients [36]. Hydrochlorothiazide may be given as 25–50 mg twice a day and bendroflumethiazide as 2.5–5.0 mg twice a day. Thiazide-like diuretics such as chlorthalidone have a longer serum half-life and may be administrated only once a day [34].

Effects on 24-h urinary calcium should be assessed 2–3 weeks after initiation of treatment or following dose increment. In addition, blood pressure should be measured to exclude hypotensive side effects. Potassium supplementation is needed in most patients. It is of importance to avoid hypokalemia, as this is associated with metabolic alkalosis. In serum, approximately 50 % of calcium is bound to proteins, and the binding of calcium is pH dependent. The fraction of calcium bound to proteins increases with an increase in serum pH (0.04–0.05 mmol/L change in ionized calcium per 0.1 unit change in pH) [37]. Accordingly, the free (physiological active) fraction of calcium decreases if a hypokalemic metabolic alkalosis develops [38]. This is in contrast to the common finding of a slight increase in total serum calcium levels in response to therapy with thiazide diuretics. Importantly, the increase in total levels does not necessarily correspond with an increase in the free fraction if metabolic alkalosis develops. Moreover, if ionized serum calcium levels are measured, the result is often reported as a value adjusted to a pH value of 7.4 and does therefore not provide information on serum level of free calcium at the actual pH.

As thiazide diuretics may increase renal magnesium excretion, magnesium supplements or co-administration of a potassium-sparing diuretic may be considered (see below). Treatment with loop diuretics should be avoided as this will increase renal calcium excretion [36, 39].

28.5 How to Handle Hyperphosphatemia?

In addition to an increased risk of renal calcifications, long-standing hypoparathyroidism is associated with extraskeletal calcifications at other organs such as the eye (cataract) and the brain (basal ganglia calcifications) [40–42]. Serum phosphorus levels are often high normal or elevated in hypoparathyroidism which is attributable to a reduced renal clearance (the phosphaturic effect of PTH is lacking) and an enhanced intestinal phosphate absorption caused by vitamin D treatment.

As high serum phosphorus levels and a high serum calcium–phosphorus product may increase the risk of soft tissue precipitation of calcium phosphate salts, it seems reasonable to aim at lowering phosphorus level. In a case series with 145 patients with idiopathic hypoparathyroidism, basal ganglia calcification was present in 74 % of the patients, and the progression of basal ganglia calcifications was associated with the calcium–phosphorus ratio. For each 1 % increase in the ratio, risk of progression decreased by 5 % [42].

There are no available data from clinical trials on whether soft tissue calcifications or renal stones are preventable with aggressive therapy (e.g., use of phosphate binders) in hypoparathyroidism [43]. If the calcium–phosphorus product is elevated, it seems, however, reasonable to prescribe a low-phosphorus diet.

28.6 How to Handle Hypomagnesemia?

Magnesium is essential for PTH secretion and activation of the PTH receptor. If magnesium stores are depleted, the parathyroid glands are unable to secrete PTH, and renal and skeletal responses to PTH are blunted causing hypocalcemia (functional hypoparathyroidism) [8, 44]. The hypocalcaemia is unlikely to be corrected without first correcting the low magnesium levels [45]. Accordingly, if magnesium levels are low, this may blunt a potential recovery of the parathyroid glands in postsurgical hypoparathyroidism. Moreover, a number of case reports have suggested that patients with postsurgical and idiopathic hypoparathyroidism may develop resistance to conventional treatment with vitamin D if magnesium levels are low and hypomagnesemia may cause symptoms similar to hypocalcemia [46–48].

In chronic hypoparathyroidism, not caused by magnesium depletion, mild hypomagnesemia often occur during conventional therapy [7, 49]. It seems that hypoparathyroidism may influence magnesium status by affecting the renal magnesium handling as well as its intestinal absorption [7]. PTH has been shown to increase magnesium reabsorption in the renal distal tubule [50, 51].

In patients with hypomagnesemia, other causes than hypoparathyroidism should of course be considered, including treatment with proton pump inhibitors [52]. Magnesium supplements may be used to correct hypomagnesemia. Supplementation with magnesium does not affect serum calcium levels in patients with hypoparathyroidism [53]. Common side effects include stomach upset, nausea, and diarrhea. If magnesium supplements are not tolerated, a potassium-sparing diuretic may be an option, as these diuretics also lower renal magnesium excretion [54]. In patients with intact parathyroid function treated with hydrochlorothiazide, amiloride (but not spironolactone) has been shown to cause a dose-dependent increase in serum magnesium levels [55]. There are no specific data available on effects of amiloride in hypoparathyroidism.

28.7 Is Estrogen Status of Importance to Treatment Responses?

In a few small case series and case reports, estrogen status has been reported to influence vitamin D requirement in women with hypoparathyroidism, although discrepant results have been reported on the effect of estrogen on calcium levels [56]. Hypercalcemia necessitating a reduced dose of active vitamin D analogue drugs has been reported in women with hypoparathyroidism entering the menopause [56], starting treatment with danazol (a hypoestrogenic drug) [57] or stopping use of oral contraceptive pills [58] or hormone replacement therapy [56]. In accordance with these observations, hypocalcemia has been reported in women starting estrogen therapy [59]. However, changes in serum calcium levels in relation to estrogen therapy is not a universal finding, as no changes in serum calcium in women stopping oral contraceptive pills also have been reported [60]. In addition, some women of childbearing age may experience symptoms of hypocalcemia at time of menses or at the time of withdrawal bleeding (hormonal contraceptive users), which may or may not be related to measurable changes in serum calcium levels [60]. The mechanism of action by which estrogen status affects calcium homeostasis is unclear. Apparently, the menses-associated symptoms of hypocalcemia are related to an acute decline in estrogen levels. However, a decrease in estrogen levels also occurs if hormone replacement therapy is stopped, which has been associated with the development of hypercalcemia. Although estrogen increases the synthesis of vitamin D binding protein and thereby the total concentration of 25OHD and $1,25(OH)_2D$, the calculated free index is not altered in women using hormonal contraceptives [61]. It has been suggested that the effects of estrogens on calcium homeostasis in hypoparathyroidism may be due to effects on bone resorption, but further studies are needed to clarify such mechanisms [56].

Hypercalcemia may develop immediately after delivery, necessitating a reduced dose of active vitamin D as long as the woman is lactating [62–64]. It is unclear whether this is due to the hypoestrogenic state associated with breast feeding or due to an increased level of parathyroid hormone-related peptide (PTHrP) synthesized by the mammary glands [65]. In one study, it was suggested that a window may exist during which there are increased needs for active vitamin D immediately following delivery [66].

28.8 How to Manage Acute Hypocalcemia?

Whether hypocalcemia causes symptoms depends on both the absolute serum calcium level and how rapidly it falls. A rapid fall in serum calcium is often associated with marked symptoms, whereas patients who are born with hypoparathyroidism or who develop hypoparathyroidism slowly can be almost free of symptoms (see also Chap. 27). Moreover, hypoparathyroidism represent a spectrum extending from mild degrees of PTH insufficiency to the complete absence of circulating PTH. Some patients may only experience symptoms when they are in a state of emotional or physiological stress. Indication for acute treatment should therefore not be based on serum calcium levels alone. The decision on whether to initiate treatment and/or to change the dose of already initiated treatment should be based on a combination of biochemical measurements and symptoms experienced by the patient.

Most patients with hypocalcemia can be managed by initiating treatment with oral calcium supplementation and a 1α-hydroxylated vitamin analogue. Typical starting doses are 1–2 μg of alfacalcidol or 0.5–1.0 μg of calcitriol each day. According to measurement of serum calcium levels, dose can be increased every 4–7 days, as appropriate. If the hypoparathyroidism is newly developed, for example, following surgical damage to the parathyroid glands, the initial dose of 1α-hydroxylated vitamin analogue needed is often higher than the long-term maintenance dose. It seems that an initial refractoriness of hypocalcemia to vitamin D and oral calcium supplementation may exist during initiation of therapy. In such patients, it may be appropriate to use a relatively high starting dose of active vitamin D analogues, i.e., a daily dose of 2–4 μg of alfacalcidol or 1–2 μg of calcitriol, while patients are monitored closely.

If severe symptoms (including tetany) are present, these may be controlled by a bolus intravenous infusion (IV) of calcium. Calcium gluconate is often preferred, as calcium chloride is more likely to cause local irritation. Ten to twenty milliliter of 10 % calcium gluconate diluted in

50–100 ml of isotonic sodium chloride, which should be infused slowly over 5–10 min. A bolus IV infusion can be repeated until symptoms have weaned. If repeated infusions are needed, therapy may be changed to a continuous calcium infusion. If the hypocalcemia is likely to persist, therapy with oral calcium supplements and an active vitamin D analogue should also be started. During IV infusion of calcium, serum calcium levels should be monitored every 4–6 h. If hypomagnesemia is present, this needs to be corrected before the hypocalcemia will resolve. Other electrolyte abnormalities (potassium, sodium, etc.) should be corrected as well. Electrocardiographic monitoring is recommended because dysrhythmias can occur if correction is too rapid. This is of particular importance to patients on treatment with a cardiac glycoside as rapid changes in serum calcium levels may cause arrhythmias.

28.9 How to Treat Episodes of Hypercalcemia?

Patients with chronic hypoparathyroidism may develop episodes of hypercalcemia that may occur unpredictably without prior changes in dose of calcium supplements or vitamin D analogue. The hypercalcemia is most often due to vitamin D intoxication, although levels of $1,25(OH)_2D$ may not always be elevated.

In most instances, serum calcium levels are only slightly elevated with no or only mild symptoms of hypercalcemia. Mild hypercalcemia can often be handled by encouraging a high oral fluid intake and by decreasing dose of calcium supplement and active vitamin D analogues by 25–50 %. If severe hypercalcemia is present with deterioration in renal function, calcium supplements and vitamin D analogues should be discontinued until serum calcium concentration returns to the desired level, and patients should be treated by IV isotonic saline infusions (3–4 L per day). Renal calcium excretion can be increased by treatment with loop diuretics. However, their use is not supported by strong clinical evidence, and the combination of aggressive fluid therapy and loop diuretics may cause hypokalemia and hypomagnesemia and increase the risk of precipitation of calcium phosphate crystals in the renal tissue. The use of forced diuresis should therefore be reserved for the management of volume overload [67].

Infusion with IV calcitonin may be of value if a rapid reduction of serum calcium is warranted. Unlike other hypocalcemic agents, calcitonin is effective within 2 h after infusion. As the half-life of calcitonin is short, infusions may need to be repeated every 12–24 h. However, tolerance often develops within 72–96 h with disappearance of the calcium-lowering effects [68]. If additional treatment is needed, glucocorticoids have long been used in the treatment of hypercalcemia associated with vitamin D intoxication. The rational for this is that glucocorticoids partly function as a vitamin D antagonist by decreasing intestinal calcium absorption and increasing renal calcium excretion. Normally, it takes 2–3 days before the hypocalcemic effect of glucocorticoids emerges. The duration of glucocorticoid administration should be short as possible to minimize long-term complications of these agents.

In order to avoid episodes of severe hypercalcemia, it is of importance that patients with hypoparathyroidism are aware on symptoms of hypercalcemia and are told to contact their health-care providers in case of such symptoms, including nausea, vomiting, constipation, diarrhea, dry mouth with polydipsia, polyuria, lethargy, generalized malaise, or drowsiness. Case series have been published suggesting that the requirement for vitamin D is diminished after an episode of hypercalcemia [69, 70]. This may especially apply to those treated with calciferol due to its long serum half-life.

28.10 Therapeutic Challenges

Although serum calcium levels are stable for most of the time in patients with chronic hypoparathyroidism managed by conventional therapy, it is obvious that this treatment regime does not restore the normal physiology of calcium homeostasis and that patients often have a number of complaints in terms of musculoskeletal symptoms, a reduced quality of life, and an increased risk of renal complications. The daily dose of vitamin D or serum measurement of 25OHD or 1,25(OH)$_2$D levels does not correlate well with serum calcium levels, as large (unexplained) differences exist between individuals in their needs for vitamin D [71]. Moreover, fluctuations in calcium levels may occur without changes in treatment in any given individual. Patients may suddenly become *overly sensitive* or *insensitive* to a given dose of vitamin D after months or years of good control on a given therapeutic regimen [19, 29, 70]. Although a few case series have suggested that such changes in some women may be due to changes in estrogen status [56], the change in vitamin D sensitivity is largely unexplained in most patients. No associations have been found between changes in vitamin D requirements and the season of the year or the state of thyroid function [72]. Emphasis should also be paid to the fact that although serum calcium levels are measured as being within the target range, calcium homeostasis exhibits diurnal variations. Patients may experience symptoms due to diurnal fluctuations in calcium levels that are not disclosed by a single measurement. In subjects with intact parathyroid function, there is a minute-to-minute regulation of PTH release; moreover, PTH is secreted in a diurnal pattern, which helps to maintain serum calcium levels within very narrow limits [73, 74].

As hypoparathyroidism represents a spectrum extending from mild degrees of PTH insufficiency to the complete absence of PTH, there is a need for studies focusing on how to classify the severity of the disease. It may be that patients with some residual PTH secretion do better and should be treated differentially from patients with a complete lack of PTH [75]. There is also a need for data that may shed light on whether serial measurements of calcium levels during the day provide a better measure for assessing calcium homeostasis in hypoparathyroid patients than a single measurement. Moreover, the inherited and idiopathic forms of hypoparathyroidism may need to be treated differently from postsurgical hypoparathyroidism. Finally, data are needed on how to protect patients from renal complications and how to improve their quality of life.

References

1. Sikjaer T, Rejnmark L, Rolighed L, Heickendorff L, Mosekilde L, the hypoparathyroid study group (2011) The effect of adding PTH (1–84) to conventional treatment of hypoparathyroidism – A randomized, placebo controlled study. J Bone Miner Res 26:2358–2370
2. Underbjerg L, Sikjaer T, Mosekilde L, Rejnmark L (2013) Cardiovascular and renal complications to postsurgical hypoparathyroidism: a Danish nationwide controlled historic follow-up study. J Bone Miner Res 28:2277–2285
3. Bagó AG, Dimitrov E, Saunders R, Seress L, Palkovits M, Usdin TB, Dobolyi A (2009) Parathyroid hormone 2 receptor and its endogenous ligand tuberoinfundibular peptide of 39 residues are concentrated in endocrine, viscerosensory and auditory brain regions in macaque and human. Neuroscience 162:128–147
4. Hadker N, Egan J, Sanders J, Lagast H, Clarke B (2014) Understanding the burden of illness associated with hypoparathyroidism reported among patients in the paradox study. Endocr Pract 20(7):671–679
5. Arlt W, Fremerey C, Callies F, Reincke M, Schneider P, Timmermann W, Allolio B (2002) Well-being, mood and calcium homeostasis in patients with hypoparathyroidism receiving standard treatment with calcium and vitamin D. Eur J Endocrinol 146:215–222
6. Cusano NE, Rubin MR, McMahon DJ, Irani D, Tulley A, Sliney J, Bilezikian JP (2013) The effect of PTH(1–84) on quality of life in hypoparathyroidism. J Clin Endocrinol Metab 98:2356–2361
7. Mortensen L, Hyldstrup L, Charles P (1997) Effect of vitamin D treatment in hypoparathyroid patients: a study on calcium, phosphate and magnesium homeostasis. Eur J Endocrinol 136:52–60
8. Rude RK, Oldham SB, Singer FR (1976) Functional hypoparathyroidism and parathyroid hormone end-organ resistance in human magnesium deficiency. Clin Endocrinol (Oxf) 5:209–224
9. Liebman SE, Taylor JG, Bushinsky DA (2006) Idiopathic hypercalciuria. Curr Rheumatol Rep 8:70–75
10. Blumsohn A, Herrington K, Hannon RA, Shao P, Eyre DR, Eastell R (1994) The effect of calcium supplementation on the circadian rhythm of bone resorption. J Clin Endocrinol Metab 79:730–735
11. Heaney RP, Smith KT, Recker RR, Hinders SM (1989) Meal effects on calcium absorption. Am J Clin Nutr 49:372–376
12. Mortensen L, Charles P (1996) Bioavailability of calcium supplements and the effect of Vitamin D: comparisons between milk, calcium carbonate, and calcium carbonate plus vitamin D. Am J Clin Nutr 63:354–357
13. Reid IR, Bolland MJ, Grey A (2010) Does calcium supplementation increase cardiovascular risk? Clin Endocrinol (Oxf) 73:689–695
14. Heaney RP, Kopecky S, Maki KC, Hathcock J, MacKay D, Wallace TC (2012) A review of calcium

supplements and cardiovascular disease risk. Adv Nutr 3:763–771
15. Lembo A, Camilleri M (2003) Chronic constipation. N Engl J Med 349:1360–1368
16. Singh N, Singh PN, Hershman JM (2000) Effect of calcium carbonate on the absorption of levothyroxine. JAMA 283:2822–2825
17. Frost RW, Lasseter KC, Noe AJ, Shamblen EC, Lettieri JT (1992) Effects of aluminum hydroxide and calcium carbonate antacids on the bioavailability of ciprofloxacin. Antimicrob Agents Chemother 36:830–832
18. Davies M, Taylor CM, Hill LF, Stanbury SW (1977) 1,25-dihydroxycholecalciferol in hypoparathyroidism. Lancet 309:55–59
19. Lund B, Sorensen OH, Lund B, Bishop JE, Norman AW (1980) Vitamin D metabolism in hypoparathyroidism. J Clin Endocrinol Metab 51:606–610
20. Brumbaugh PF, Haussler DH, Bressler R, Haussler MR (1974) Radioreceptor assay for 1,25-dihydroxyvitamin D3. Science 183:1089–1091
21. Avioli LV (1974) The therapeutic approach to hypoparathyroidism. Am J Med 57:34–42
22. Russell RGG, Walton RJ, Smith R, Preston C, Basson R, Henderson RG, Norman AW (1974) 1,25-dihydroxycholecalciferol and 1alpha-hydroxycholecalciferol in hypoparathyroidism. Lancet 304:14–17
23. Hill LF, Davies M, Taylor CM, Standbury SW (1976) Treatment of hypoparathyroidism with 1,25-dihydroxycholecalciferol. Clin Endocrinol (Oxf) 5(Suppl):167S–173S
24. Jorgensen H, Vogt JH (1977) 1alpha-hydroxycholecalciferol in the treatment of hypoparathyroidism. Acta Med Scand 201:3–7
25. Neer RM, Holick MF, DeLuca HF, Potts J (1975) Effects of 1a-hydroxy-vitamin D3 and 1,25-dihydroxy-vitamin D3 on calcium and phosphorus metabolism in hypoparathyroidism. Metabolism 24:1403–1413
26. Kanis JA, Russell RGG, Smith R (1977) Physiological and therapeutic differences between vitamin D, its metabolites and analogues. Clin Endocrinol (Oxf) 7:191s–201s
27. Haussler MR, Cordy PE (1982) Metabolites and analogues of vitamin d: which for what? JAMA 247:841–844
28. Quack I, Zwernemann C, Weiner SM, Sellin L, Henning BF, Waldherr R, Büchner NJ, Stegbauer J, Vonend O, Rump LC (2005) Dihydrotachysterol therapy for hypoparathyroidism: consequences of inadequate monitoring. Five cases and a review. Exp Clin Endocrinol Diabetes 113:376–380
29. Holick MF (2007) Vitamin D deficiency. N Engl J Med 357:266–281
30. Mitchell DM, Regan S, Cooley MR, Lauter KB, Vrla MC, Becker CB, Burnett-Bowie SA, Mannstadt M (2012) Long-term follow-up of patients with hypoparathyroidism. J Clin Endocrinol Metab 97:4507–4514
31. Massey LK, Whiting SJ (1996) Dietary salt, urinary calcium, and bone loss. J Bone Miner Res 11:731–736
32. Rizzoli R, Hugi K, Fleisch H, Bonjour JP (1981) Effect of hydrochlorothiazide on 1,25-dihydroxyvitamin D3-induced changes in calcium metabolism in experimental hypoparathyroidism in rats. Clin Sci (Lond) 60:101–107
33. Santos F, Smith MJV, Chan JCM (1986) Hypercalciuria associated with long-term administration of calcitriol (1,25-dihydroxyvitamin D3) action of hydrochlorothiazide. Am J Dis Child 140:139–142
34. Porter RH, Cox BG, Heaney D, Hostetter TH, Stinebaugh BJ, Suki WN (1978) Treatment of hypoparathyroid patients with chlorthalidone. N Engl J Med 298:577–581
35. Xu H, Zisman AL, Coe FL, Worcester EM (2013) Kidney stones: an update on current pharmacological management and future directions. Expert Opin Pharmacother 14:435–447
36. Rejnmark L, Vestergaard P, Pedersen AR, Heickendorff L, Andreasen F, Mosekilde L (2003) Dose-effect relations of loop- and thiazide-diuretics on calcium homeostasis: a randomized, double-blinded Latin-square multiple cross-over study in postmenopausal osteopenic women. Eur J Clin Invest 33:41–50
37. Wang S, McDonnell EH, Sedor FA, Toffaletti JG (2002) pH effects on measurements of ionized calcium and ionized magnesium in blood. Arch Pathol Lab Med 126:947–950
38. Rejnmark L, Vestergaard P, Heickendorff L, Andreasen F, Mosekilde L (2001) Effects of thiazide- and loop-diuretics, alone or in combination, on calcitropic hormones and biochemical bone markers: a randomized controlled study. J Intern Med 250:144–153
39. Rejnmark L, Vestergaard P, Heickendorff L, Andreasen F, Mosekilde L (2006) Loop diuretics increase bone turnover and decrease BMD in osteopenic postmenopausal women: results from a randomized controlled study with bumetanide. J Bone Miner Res 21:163–170
40. Tambyah PA, Ong BK, Lee KO (1993) Reversible parkinsonism and asymptomatic hypocalcemia with basal ganglia calcification from hypoparathyroidism 26 years after thyroid surgery. Am J Med 94:444–445
41. Pohjola S (1962) Ocular manifestations of idiopathic hypoparathyroidism. Acta Ophthalmol 40:255–265
42. Goswami R, Sharma R, Sreenivas V, Gupta N, Ganapathy A, Das S (2012) Prevalence and progression of basal ganglia calcification and its pathogenic mechanism in patients with idiopathic hypoparathyroidism. Clin Endocrinol (Oxf) 77:200–206
43. Malberti F (2013) Hyperphosphataemia: treatment options. Drugs 73:673–688
44. Levi J, Massry SG, Coburn JW, Llach F, Kleeman CR (1974) Hypocalcemia in magnesium-depleted dogs: evidence for reduced responsiveness to parathyroid hormone and relative failure of parathyroid gland function. Metabolism 23:323–335
45. Cholst IN, Steinberg SF, Tropper PJ, Fox HE, Segre GV, Bilezikian JP (1984) The influence of hypermagnesemia on serum calcium and parathyroid hormone levels in human subjects. N Engl J Med 310:1221–1225

46. Rösler A, Rabinowitz D (1973) Magnesium-induced reversal of vitamin-D resistance in hypoparathyroidism. Lancet 301:803–805
47. Jones KH, Fourman P (1966) Effects of infusions of magnesium and of calcium in parathyroid insufficiency. Clin Sci 30:139–150
48. Dent CE, Harper CM, Morgans M, Philpot GR, Trotter WR (1955) Insensitivity of vitamin D developing during the treatment of postoperative tetany; its specificity as regards the form of vitamin D taken. Lancet 266:687–690
49. Elin RJ, Hosseini JM, Fitzpatrick L, Bliziotes MM, Marx SJ (1990) Blood magnesium status of patients with parathyroid disease. Magnes Trace Elem 9:119–123
50. Quamme GA, Carney SL, Wong NL, Dirks JH (1980) Effect of parathyroid hormone on renal calcium and magnesium reabsorption in magnesium deficient rats. Pflugers Arch 386:59–65
51. Quamme GA (1997) Renal magnesium handling: new insights in understanding old problems. Kidney Int 52:1180–1195
52. Hoorn EJ, van der Hoek J, de Man RA, Kuipers EJ, Bolwerk C, Zietse R (2010) A case series of proton pump inhibitor – induced hypomagnesemia. Am J Kidney Dis 56:112–116
53. Lubi M, Tammiksaar K, Matjus S, Vasar E, Volke V (2012) Magnesium supplementation does not affect blood calcium level in treated hypoparathyroid patients. J Clin Endocrinol Metab 97:E2090–E2092
54. Ryan MP (1986) Magnesium and potassium-sparing diuretics. Magnesium 5:282–292
55. Murdoch DL, Forrest G, Davies DL, McInnes GT (1993) A comparison of the potassium and magnesium-sparing properties of amiloride and spironolactone in diuretic-treated normal subjects. Br J Clin Pharmacol 35:373–378
56. McIlroy J, Dryburgh F, Hinnie J, Dargie R, Al-Rawi A (1999) Oestrogen and calcium homeostasis in women with hypoparathyroidism. BMJ 319:1252–1253
57. Hepburn NC, Abdul-Aziz LA, Whiteoak R (1989) Danazol-induced hypercalcaemia in alphacalcidol-treated hypoparathyroidism. Postgrad Med J 65:849–850
58. Verbeelen D, Fuss M (1979) Hypercalcaemia induced by oestrogen withdrawal in vitamin D-treated hypoparathyroidism. BMJ 1:522–523
59. de Nagant DC (1979) Oestrogen-induced hypocalcaemia in hypoparathyroidism. BMJ 1:1563–1564
60. Mallette LE (1992) Case report: hypoparathyroidism with menses-associated hypocalcemia. Am J Med Sci 304:32–37
61. Moller UK, Streym S, Jensen LT, Mosekilde L, Schoenmakers I, Nigdikar S, Rejnmark L (2013) Increased plasma concentrations of vitamin D metabolites and vitamin D binding protein in women using hormonal contraceptives: a cross-sectional study. Nutrients 5:3470–3480
62. Cundy T, Haining SA, Guilland-Cumming DF, Butler J, Kanis JA (1987) Remission of hypoparathyroidism during lactation: evidence for a physiological role for prolactin in the regulation of vitamin D metabolism. Clin Endocrinol (Oxf) 26:667–674
63. Rude RK, Haussler MR, Singer FR (1984) Postpartum resolution of hypocalcemia in a lactating hypoparathyroid patient. Endocrinol Jpn 31:227–233
64. Wright AD, Joplin GF, Dixon HG (1969) Post-partum hypercalcaemia in treated hypoparathyroidism. BMJ 1:23–25
65. Lepre F, Grill V, Ho PWM, Martin TJ (1993) Hypercalcemia in pregnancy and lactation associated with parathyroid hormone-related protein. N Engl J Med 328:666–667
66. Sato K (2008) Hypercalcemia during pregnancy, puerperium, and lactation: review and a case report of hypercalcemic crisis after delivery Due to excessive production of PTH-related protein (PTHrP) without malignancy (humoral hypercalcemia of pregnancy). Endocr J 55:959–966
67. LeGrand SB, Leskuski D, Zama I (2008) Narrative review: furosemide for hypercalcemia: an unproven yet common practice. Ann Intern Med 149:259–263
68. Wisneski LA (1990) Salmon calcitonin in the acute management of hypercalcemia. Calcif Tissue Int 46(Suppl):S26–S30
69. Spaulding WB, Yendt ER (1964) Prolonged vitamin D intoxication in a patient with hypoparathyroidism. Can Med Assoc J 90:1049–1054
70. Hossain M (1970) Vitamin-D intoxication during treatment of hypoparathyroidism. Lancet 295:1149–1151
71. Stamp TCB (1981) Calcitriol dosage in osteomalacia, hypoparathyroidism and attempted treatment of myositis ossificans progressiva. Curr Med Res Opin 7:316–336
72. Ireland AW, Clubb JS, Neale FC, Posen S, Reeve TS (1968) The calciferol requirements of patients with surgical hypoparathyroidism. Ann Intern Med 69:81–89
73. Kitamura N, Shigeno C, Shiomi K, Lee K, Ohta S, Sone T, Katsushima S, Tadamura E, Kousaka T, Yamamoto I (1990) Episodic fluctuation in serum intact parathyroid hormone concentration in men. J Clin Endocrinol Metab 70:252–263
74. Shrestha RP, Hollot CV, Chipkin SR, Schmitt CP, Chait Y (2010) A mathematical model of parathyroid hormone response to acute changes in plasma ionized calcium concentration in humans. Math Biosci 226:46–57
75. Promberger R, Ott J, Kober F, Karik M, Freissmuth M, Hermann M (2011) Normal parathyroid hormone levels do not exclude permanent hypoparathyroidism after thyroidectomy. Thyroid 21:145–150

29 Follow-up in Chronic Hypoparathyroidism

Michael Mannstadt and Deborah M. Mitchell

29.1 Introduction

Patients with hypoparathyroidism are at risk of many complications, both from the disease itself as well as from adverse effects of currently available treatment regimens. Ongoing care with a provider familiar with the treatment of this disorder is critical in order to meet the complex needs of patients and to optimize their outcomes (see also Chap. 15). The overarching goals of follow-up care for hypoparathyroidism are to ensure that effective care is in place and that the manifestations of the disease are as well controlled as possible [1].

A key aspect of ongoing follow-up is engaging the patient as a partner in his or her medical care. Patients with hypoparathyroidism need to have a basic understanding of the underlying pathophysiology, the rationale for treatment, and signs and symptoms of complications of the disorder. This is particularly true due to the fact that it is an extremely rare disorder, and other medical providers seen by the patient may be less familiar with the potential manifestations of the disease. Patient education and encouragement are therefore important parts of follow-up visits. For example, the patient who understands the importance of preventing kidney damage due to excessive urinary calcium excretion might be more accepting of the inconvenient 24-h urine collections.

There are no data from clinical trials available that would determine the optimal follow-up intervals or the optimal frequency of laboratory and imaging tests. As a result, there are currently no guidelines for the management of the disease available from professional societies. The following chapter therefore is a reflection mainly based on personal preference and experience.

29.2 Control of Symptoms

In acute, severe hypocalcemia, the goal of treatment is the rapid correction of the biochemical abnormalities and accompanying symptoms, often through intravenous infusion of calcium gluconate (see Chap. 27 for more details). For chronic management, the goal of treatment is to control symptoms while minimizing complications of overtreatment. Symptoms of hypocalcemia to be sought include the more common paresthesias and tetany, but also rare symptoms which may not be readily recognized as hypocalcemic, such as stridor, bronchospasm, and symptoms of heart failure [2, 3].

The generally accepted target serum calcium concentration is in the low-normal range, a state in which symptoms of hypocalcemia are generally rare. Higher serum levels of calcium, even

M. Mannstadt, MD (✉) • D.M. Mitchell, MD
Endocrine Unit, Massachusetts General Hospital, Harvard Medical School, Thier 10, 50 Blossom St, Boston, MA 02114, USA
e-mail: mannstadt@mgh.harvard.edu; dmmitchell@mgh.harvard.edu

M.L. Brandi, E.M. Brown (eds.), *Hypoparathyroidism*,
DOI 10.1007/978-88-470-5376-2_29, © Springer-Verlag Italia 2015

within the normal range, are to be avoided. This is because urinary calcium excretion, an important determinant of long-term renal complications, is directly correlated with the filtered load of calcium, which in turn is correlated with the serum calcium level. Because of the lack of parathyroid hormone (PTH), urinary excretion of calcium at any given level of serum calcium is greater than normal. It follows therefore logically that the lower the serum calcium level, the lower the urinary calcium excretion. Although no prospective data exist proving that this approach minimizes long-term complications, retrospective data does support the notion that keeping serum calcium at a low-normal or slightly hypocalcemic level has beneficial effects in the long term. For example, Mitchell et al. reported that relative hypercalcemia over time was inversely associated with estimated glomerular filtration rate, a measure of renal function [4]. In certain cases, such as hypoparathyroidism due to activating mutations of the calcium-sensing receptor, maintenance of a serum calcium level just high enough to prevent symptomatic hypocalcemia may be the goal.

Occasional mild symptoms of hypocalcemia can often be tolerated, since they indicate that the patient's serum calcium level is within, or close to, the target range. By contrast, serum calcium concentrations in the high-normal or slightly elevated range are often asymptomatic or associated with nonspecific symptoms such as fatigue or abdominal pain, which may be missed. However, keeping serum calcium in the low to low-normal range comes at a price: the lower the serum calcium level, the more likely a patient is to develop symptomatic hypocalcemia. This is particularly relevant in the setting of a stressor such as an intercurrent illness or increase in physical activity.

When inquiring about the frequency and intensity of hypocalcemic symptoms, it is important to also identify triggers of hypocalcemia. In some patients, strenuous physical activity can lead to a drop in serum calcium, as can gastroenteritis. In addition, some women report an association of hypocalcemia with the onset of menstruation [5, 6]. Once triggers are identified, strategies to mitigate the symptoms can be instituted, for example, taking extra 500–1,000 mg oral calcium half an hour to one hour before engaging in strenuous physical activity. Monitoring serum calcium levels is recommended when feasible.

29.3 Minimizing Complications

29.3.1 Biochemical Measurements

For the typical patient with chronic hypoparathyroidism, serum calcium levels should be monitored every 3–6 months. Additional serum calcium checks are indicated after changes in oral calcium or vitamin D therapy, or if unexplained symptoms of hypo- or hypercalcemia occur. We typically measure serum albumin, creatinine, and phosphate simultaneously to ensure that albumin-corrected serum calcium can be calculated, renal function is stable, and the calcium-phosphate product is not elevated. The goals of therapy include control of symptoms (see above), albumin-corrected serum calcium level at or slightly below the lower end of the normal range (about 8.0–8.5 mg/dL), and a serum calcium-phosphate product well below 55 mg^2/dL^2.

In clinical practice, we typically measure total serum calcium concentrations, despite the fact that it is the ionized portion that is both regulated and biologically active. In typical physiologic states, approximately 50 % of the total calcium is ionized, 10 % is bound to anions such as bicarbonate, and 40 % is protein bound, mostly to albumin. Formulas exist which correct the measured calcium concentration if the serum albumin is significantly different from 4.0 mg/dL. The most commonly used formula is:

$$\begin{aligned}&\text{Corrected calcium concentration } = \\ &\text{Total calcium concentration } + \\ &0.8 \times \left(4.0 - \text{Albumin}\right)\end{aligned}$$

Additional complexities arise in the setting of altered acid-base status, as alkalosis increases the fraction of total serum calcium bound to albumin, decreasing the ionized fraction. In addition, the above formula does not perform well in patients with chronic kidney disease (CKD), due at least in part to the high prevalence of metabolic acidosis among patients with CKD. In certain situations, it is therefore best to check an ionized calcium level, but the handling of the specimen is demanding [7, 8].

As the serum calcium concentration is often lowest in the morning, we typically adjust calcium and vitamin D therapy to achieve a morning fasting serum calcium concentration that is in the low-normal range. Once the fasting serum calcium is in the target range, we occasionally check serum calcium values during the day. If the serum calcium level or the calcium-phosphate product is unacceptably high during the day, changes in dosing or frequency of medication might be helpful. Examples of changes include spreading out calcium tablets more evenly throughout the day, or splitting a once-daily calcitriol dose into two doses.

24-h urinary calcium excretion is measured at least once yearly to detect hypercalciuria. A simultaneous urinary creatinine measurement ensures adequate sampling. The goal is a 24-h urinary calcium excretion of well below 300 mg for a typical adult or 4 mg/kg for a pediatric patient. Decreasing the urinary calcium excretion can often be achieved by decreasing serum calcium. If this cannot be achieved, thiazide diuretics may be given as they can significantly decrease urinary calcium excretion [9, 10] with or without potassium bicarbonate [11]. Hydrochlorothiazide or the longer-acting chlorthalidone is generally started at 25 mg/d and uptitrated to 50–100 mg/day. Hydrochlorothiazide at 50 mg or higher is typically split into two doses. As described in other chapters, either synthetic PTH(1–34) or recombinant PTH(1–84) represents additional treatment modalities that may have advantages over the traditional treatment of hypoparathyroidism with calcium and activated vitamin D supplementation.

29.3.2 Imaging

Nephrocalcinosis, nephrolithiasis, and impaired renal function are important potential complications of hypoparathyroidism. Urinary calcium excretion is directly proportional to serum calcium levels [12], and since PTH increases renal calcium reabsorption, PTH deficiency leads to higher urinary calcium losses for any given serum calcium level [13, 14]. Standard therapy with calcium supplementation and active vitamin D can therefore cause renal calcifications [15], and great care has to be given to prevent these outcomes.

Estimates of the prevalence of nephrolithiasis range from 2 to 15 % [13, 16, 17]. For example, a Danish population-based case-control study involving 688 patients with hypoparathyroidism and 2,064 controls reported an almost fivefold increased risk of renal stones and renal insufficiency in hypoparathyroidism [17]. Reported rates of medullary nephrocalcinosis range from 12 to 57 %, with the higher rates observed in patients with calcium-sensing receptor mutations [18–20].

In addition to symptomatic renal colic, calcification may also lead to impaired renal function. Of 27 participants in a trial of PTH replacement, two thirds had creatinine clearance rates below the normal range, and participants with nephrocalcinosis had, on average, lower GFRs than those without nephrocalcinosis [20]. In a cohort of 120 patients seen in a single hospital system, 41 % had estimated Glomerular filtration rate (GFR) less than 60 mL/min/1.73 m^2, consistent with chronic kidney disease stage 3 or higher [4]. The prevalence of decreased GFR among hypoparathyroid patients was 2–35-fold higher than in an age-matched population-based cohort [21]. Predictors of lower GFR in this study included age, duration of disease, and relative hypercalcemia. In addition, two patients in this study (1.7 %) progressed to severe renal failure requiring transplant.

There is no consensus regarding modality and frequency of renal imaging. Options for detecting nephrocalcinosis and nephrolithiasis include plain radiographs of the abdomen, computed tomography (CT) without contrast, and renal ultrasound. Ultrasound has been shown in one recent study to be more sensitive than CT for the detection of calcification [22]. We typically check for occult kidney stones or nephrocalcinosis when initiating treatment to establish a baseline. We then obtain an ultrasound of the abdomen every several years to investigate whether nephrocalcinosis or kidney stones are developing.

Brain calcifications, especially basal ganglia calcifications (BGC), are a well-known complication of hypoparathyroidism [4, 19, 23]. The reported prevalence of BGC varies widely. In one cohort of 33 patients, 12 % had BGC [16]. In another cohort of 31 patients who had head imaging, 52 % had BGC [4]. In a large study of 145 patients with hypoparathyroidism, all of whom were imaged by CT, 74 % had BGC [23]. The prevalence in the general population is not well established, but estimates are significantly lower [24, 25]. The clinical significance of BGC in hypoparathyroid patients is unclear. It is therefore unknown at this time whether head imaging has any role in the care of patients with hypoparathyroidism.

29.3.3 Slit-Lamp and Ophthalmoscopic Examinations

Cataracts are opacities in the lens of the eye and are a potential complication of hypoparathyroidism. In several case series, cataracts have consistently been found in approximately 50 % of hypoparathyroid patients [26–28]. Arlt et al. conducted slit-lamp eye exams in 20 women with postsurgical hypoparathyroidism and reported that 11 (55 %) had cataracts, the majority bilaterally [13]. Older age and longer duration of illness were more common among patients with cataracts [13, 27]. Annual slit-lamp and ophthalmoscopic examinations are therefore recommended to monitor for the development of cataracts in all patients.

29.3.4 Dental Abnormalities

A wide variety of dental manifestations of hypoparathyroidism have been reported including missing or hypoplastic teeth, abnormal tooth spacing, and "stumpy" roots [29, 30]. Enamel hypoplasia, often considered to be the most characteristic dental finding in hypoparathyroidism, has been reported at a prevalence ranging from 20 to 80 % vs. 3–15 % of normal controls [31, 32]. We do not modify the recommendations for general dental care solely because the patient has hypoparathyroidism. More research is needed to define dental abnormalities and their implication for management.

29.3.5 Other

The clinician should check an electrocardiogram or an echocardiogram if symptoms of heart failure or arrhythmia occur [2, 33]. While bone mineral density as measured by dual-energy x-ray absorptiometry (DXA) is often increased in patients with hypoparathyroidism, bone microarchitecture is abnormal [16, 34]. However, currently no data are available with which to assess whether risk of fracture is altered in hypoparathyroidism. Therefore, the presence of hypoparathyroidism does not change our approach to screening for osteoporosis for the general population. In patients who develop myalgia or muscle weakness, we measure creatinine phosphokinase (CPK) in the blood, as hypocalcemia is associated with a nonspecific myopathy and elevation in circulating levels of this muscle enzyme [35, 36].

29.4 Practical Considerations for Follow-up

In contrast to patients with conditions such as hypertension and diabetes mellitus, who can self-monitor blood pressure or blood sugar, patients with hypoparathyroidism cannot easily measure blood calcium levels themselves. It is therefore important for the patient to have quick access to laboratory measurements of serum calcium. It can be useful, especially for newly diagnosed patients, to have several lab slips at their disposal that are already filled out (or standing laboratory orders) in order to check calcium and albumin as needed. Combined with clear instructions for when to use them, these can empower the patient and significantly increase the efficiency of responding to symptoms of hypo- or hypercalcemia.

In certain situations, a patient diary can also be of tremendous help. For example, in a patient with frequent symptoms and significant fluctuations in serum calcium levels, it can be helpful to have detailed notes available recording the timing of medication, doses, meals, physical activity, etc., as a means of determining the reason(s) for the fluctuations and optimizing treatment.

Several excellent educational resources for patients are available, most of them through the Internet. The Hypoparathyroidism Association (www.hypopara.org) is an independent nonprofit organization that provides educational material, including videos, for patients (see also Chap. 38). Other resources include disease websites maintained by academic institutions such as the Mayo Clinic (www.mayoclinic.org).

References

1. Shoback D (2008) Clinical practice. Hypoparathyroidism. N Engl J Med 359(4):391–403
2. Newman DB, Fidahussein SS, Kashiwagi DT, Kennel KA, Kashani KB, Wang Z et al (2014) Reversible cardiac dysfunction associated with hypocalcemia: a systematic review and meta-analysis of individual patient data. Heart Fail Rev 19(2):199–205
3. Chakrabarty AD (2013) Adult primary hypoparathyroidism: A rare presentation. Indian J Endocrinol Metab 17(Suppl 1):S201–2
4. Mitchell DM, Regan S, Cooley MR, Lauter KB, Vrla MC, Becker CB et al (2012) Long-term follow-up of patients with hypoparathyroidism. J Clin Endocrinol Metab 97(12):4507–14
5. Mallette LE (1992) Case report: hypoparathyroidism with menses-associated hypocalcemia. Am J Med Sci 304(1):32–7
6. Thys-Jacobs S, McMahon D, Bilezikian JP (2007) Cyclical changes in calcium metabolism across the menstrual cycle in women with premenstrual dysphoric disorder. J Clin Endocrinol Metab 92(8):2952–9
7. Gauci C, Moranne O, Fouqueray B, de la Faille R, Maruani G, Haymann JP et al (2008) Pitfalls of measuring total blood calcium in patients with CKD. J Am Soc Nephrol 19(8):1592–8
8. Nordenstrom E, Katzman P, Bergenfelz A (2011) Biochemical diagnosis of primary hyperparathyroidism: analysis of the sensitivity of total and ionized calcium in combination with PTH. Clin Biochem 44(10–11):849–52
9. Brickman AS, Coburn JW, Koppel MH, Peacock M, Massry SG (1971) The effect of hydrochlorothiazide administration on serum and urinary calcium in normal, hypoparathyroid and hyperparathyroid subjects. Studies on mechanisms. Isr J Med Sci 7(3):518–9
10. Lemann J Jr, Gray RW, Maierhofer WJ, Cheung HS (1985) Hydrochlorothiazide inhibits bone resorption in men despite experimentally elevated serum 1,25-dihydroxyvitamin D concentrations. Kidney Int 28(6):951–8
11. Frassetto LA, Nash E, Morris RC Jr, Sebastian A (2000) Comparative effects of potassium chloride and bicarbonate on thiazide-induced reduction in urinary calcium excretion. Kidney Int 58(2):748–52
12. Macfadyen IJ, Nordin BE, Smith DA, Wayne DJ, Rae SL (1965) Effect of variation in dietary calcium on plasma concentration and urinary excretion of calcium. Br Med J 1(5428):161–4
13. Arlt W, Fremerey C, Callies F, Reincke M, Schneider P, Timmermann W et al (2002) Well-being, mood and calcium homeostasis in patients with hypoparathyroidism receiving standard treatment with calcium and vitamin D. Eur J Endocrinol 146(2):215–22
14. Peacock M, Robertson WG, Nordin BE (1969) Relation between serum and urinary calcium with particular reference to parathyroid activity. Lancet 1(7591):384–6
15. Hossain M (1970) Vitamin-D intoxication during treatment of hypoparathyroidism. Lancet 1(7657):1149–51
16. Rubin MR, Dempster DW, Zhou H, Shane E, Nickolas T, Sliney J Jr et al (2008) Dynamic and structural properties of the skeleton in hypoparathyroidism. J Bone Miner Res 23(12):2018–24
17. Underbjerg L, Sikjaer T, Mosekilde L, Rejnmark L (2013) Cardiovascular and renal complications to postsurgical hypoparathyroidism: a Danish nationwide controlled historic follow-up study. J Bone Miner Res 28(11):2277–85
18. Lienhardt A, Bai M, Lagarde JP, Rigaud M, Zhang Z, Jiang Y et al (2001) Activating mutations of the calcium-sensing receptor: management of hypocalcemia. J Clin Endocrinol Metab 86(11):5313–23
19. Raue F, Pichl J, Dorr HG, Schnabel D, Heidemann P, Hammersen G et al (2011) Activating mutations in the calcium-sensing receptor: genetic and clinical spectrum in 25 patients with autosomal dominant hypocalcaemia – a German survey. Clin Endocrinol (Oxf) 75(6):760–5
20. Winer KK, Ko CW, Reynolds JC, Dowdy K, Keil M, Peterson D et al (2003) Long-term treatment of hypoparathyroidism: a randomized controlled study comparing parathyroid hormone-(1–34) versus calcitriol and calcium. J Clin Endocrinol Metab 88(9):4214–20
21. Levey AS, Stevens LA, Schmid CH, Zhang YL, Castro AF 3rd, Feldman HI et al (2009) A new equa-

tion to estimate glomerular filtration rate. Ann Intern Med 150(9):604–12
22. Boyce AM, Shawker TH, Hill SC, Choyke PL, Hill MC, James R et al (2013) Ultrasound is superior to computed tomography for assessment of medullary nephrocalcinosis in hypoparathyroidism. J Clin Endocrinol Metab 98(3):989–94
23. Goswami R, Sharma R, Sreenivas V, Gupta N, Ganapathy A, Das S (2012) Prevalence and progression of basal ganglia calcification and its pathogenic mechanism in patients with idiopathic hypoparathyroidism. Clin Endocrinol (Oxf) 77(2):200–6
24. Fenelon G, Gray F, Paillard F, Thibierge M, Mahieux F, Guillani A (1993) A prospective study of patients with CT detected pallidal calcifications. J Neurol Neurosurg Psychiatry 56(6):622–5
25. Gomille T, Meyer RA, Falkai P, Gaebel W, Konigshausen T, Christ F (2001) Prevalence and clinical significance of computerized tomography verified idiopathic calcinosis of the basal ganglia. Radiologe 41(2):205–10
26. Steinberg H, Waldron BR (1952) Idiopathic hypoparathyroidism; an analysis of fifty-two cases, including the report of a new case. Medicine (Baltimore) 31(2):133–54
27. Ireland AW, Hornbrook JW, Neale FC, Posen S (1968) The crystalline lens in chronic surgical hypoparathyroidism. Arch Intern Med 122(5):408–11
28. Pohjola S (1962) Ocular manifestations of idiopathic hypoparathyroidism. Acta Ophthalmol 40:255–65
29. Kalb RE, Grossman ME (1986) Ectodermal defects and chronic mucocutaneous candidiasis in idiopathic hypoparathyroidism. J Am Acad Dermatol 15(2 Pt 2):353–6
30. Pisanty S, Garfunkel A (1977) Familial hypoparathyroidism with candidiasis and mental retardation. Oral Surg Oral Med Oral Pathol 44(3):374–83
31. Jensen SB, Illum F, Dupont E (1981) Nature and frequency of dental changes in idiopathic hypoparathyroidism and pseudohypoparathyroidism. Scand J Dent Res 89(1):26–37
32. Nikiforuk G, Fraser D, Poyton HG, McKendry JB (1981) Calcific bridging of dental pulp caused by iatrogenic hypercalcemia. Report of a case. Oral Surg Oral Med Oral Pathol 51(3):317–9
33. Vered I, Vered Z, Perez JE, Jaffe AS, Whyte MP (1989) Normal left ventricular performance documented by Doppler echocardiography in patients with long-standing hypocalcemia. Am J Med 86(4):413–6
34. Rubin MR, Dempster DW, Kohler T, Stauber M, Zhou H, Shane E et al (2010) Three dimensional cancellous bone structure in hypoparathyroidism. Bone 46(1):190–5
35. Policepatil SM, Caplan RH, Dolan M (2012) Hypocalcemic myopathy secondary to hypoparathyroidism. WMJ 111(4):173–5
36. Roca B, Minguez C, Saez-Royuela A, Simon E (1995) Dementia, myopathy, and idiopathic hypoparathyroidism. Postgrad Med J 71(841):702

Treatment of Hypoparathyroidism with Parathyroid Hormone 1–34

30

Karen K. Winer and Gordon B. Cutler Jr.

30.1 Introduction

The clinical presentation of hypoparathyroidism includes inappropriately low or undetectable serum parathyroid hormone (PTH), hypocalcemia, hyperphosphatemia, and frequently, hypomagnesemia. Chronic symptoms include neuromuscular irritability causing tetany, muscle cramping, spasms, and seizures. In adults, the disorder is usually a complication of neck surgery due to an excision of a goiter or thyroid cancer. In children, the condition is most often due to inherited disorders such as autoimmune polyglandular syndrome type I (APS1) or an activating mutation in the calcium-sensing receptor (CaSR). Children with hypoparathyroidism pose a particular therapeutic dilemma, because recurrent episodes of hypocalcemia, if associated with seizures, may adversely affect brain development.

Disclosure Statement Authors KKW, GBC have nothing to disclose. Neither of the authors received an honorarium, grant, or other form of payment to produce the manuscript.

K.K. Winer, MD (✉)
National Institutes of Health, Eunice Kennedy Shriver National Institute of Child Health and Human Development/PGNB, Bldg 6100, Rm4B11, Bethesda, MD 20892-7510, USA
e-mail: winerk@mail.nih.gov

G.B. Cutler Jr., MD
Gordon Cutler Consultancy, LLC, Deltaville, VA 23043-2172, USA
e-mail: gbcutler@gmail.com

30.2 Why Is Conventional Therapy Not Adequate?

Unlike most other hormonal insufficiencies, hypoparathyroidism is not treated by replacing the missing hormone, PTH [1–3]. Instead, conventional treatment of this disorder consists of daily doses of vitamin D analogues and calcium and magnesium supplementation (see also Chap. 28). But, this therapy does not fully restore normal mineral ion homeostasis, because it relies entirely on calcium transport across the intestinal epithelium to normalize blood calcium (see also Chap. 25). The objective of standard therapy is to flood the GI tract with calcium, which is given with oral supplements throughout the day, and to enhance the absorption of calcium across the intestinal epithelium by providing the active form of vitamin D (calcitriol). Although calcitriol is usually effective in raising serum calcium, this therapy is a poor surrogate for replacement of the missing hormone as it entirely bypasses the kidney, where much of the sensing and regulation of blood calcium occurs, and the bone, the main calcium reservoir for the body. In the distal tubule, PTH and the calcium-sensing receptor actively regulate calcium reabsorption stimulating and inhibiting, respectively, the paracellular reabsorption of calcium in the cortical thick ascending limb, while PTH also stimulates transcellular reabsorption in the distal convoluted tubule. Passive calcium reabsorption also occurs in the proximal tubule, where reabsorption occurs along

M.L. Brandi, E.M. Brown (eds.), *Hypoparathyroidism*,
DOI 10.1007/978-88-470-5376-2_30, © Springer-Verlag Italia 2015

with the flow of sodium and water. Without the renal calcium-retaining effect of PTH, excessive urinary calcium losses eventually lead to calcium deposits in the kidneys. This condition may progress over time to irreversible kidney damage, renal insufficiency, and failure [4–9]. This progression is particularly evident in patients with hypoparathyroidism due to activating mutations of the calcium-sensing receptor (CaSR), where the molecular defect inappropriately activates the receptor even at low serum calcium concentrations, signaling an apparent hypercalcemic state (leading to suppression of PTH secretion and increased renal calcium excretion) [10, 11] (see also Chap. 25).

To minimize hypercalciuria-induced renal damage, the calcitriol dosage can be adjusted to maintain serum calcium at the lowest tolerated level, but for most hypoparathyroid patients, this leads to varying symptoms of hypocalcemia. Frequent monitoring of urine and serum calcium levels is required to properly attain the therapeutic goals of avoiding symptomatic hypocalcemia while minimizing the risk of renal damage. Furthermore, many patients do not tolerate the many pills, often more than a dozen calcium, calcitriol, and magnesium pills daily, that are required by the conventional therapy regimen. Moreover, patients with autoimmune polyglandular syndrome type 1 (APS-1), also called autoimmune polyendocrinopathy candidiasis ectodermal dystrophy (APECED), often have intermittent malabsorption, which interferes with the absorption of the multiple conventional therapy pills and can lead to hypocalcemic crisis [12].

Additionally, various adverse effects of PTH deficiency are difficult to measure such as fatigue, muscle weakness, exercise intolerance, and cognitive deficits including memory loss [13, 14]. Patients with hypoparathyroidism often complain of an increased level of anxiety, malaise, and depression. These symptoms may lead to a significant reduction in quality of life after the onset of hypoparathyroidism. Conventional therapy often fails to restore a sense of well-being.

30.3 Past History of Experimental PTH Replacement Therapy

In 1925, J.B. Collip was the first to show that bovine PTH extract can be used as a replacement therapy when injected into parathyroidectomized dogs [15, 16]. A few years later, Fuller Albright treated a 14-year-old boy with postsurgical hypoparathyroidism with bovine PTH for 27 days [17]. In 1967, an attempt to use bovine PTH extract [18] in the treatment of a woman with postsurgical hypoparathyroidism resulted in neutralizing antibodies within days after injections were initiated [19]. The investigators concluded that, although this result was disappointing, vitamin D analogues remained the appropriate treatment for the disease. Over the subsequent decade, the sequencing of parathyroid hormone 1-34 [20] and PTH 1-37 [21] led to the use of synthetic parathyroid hormone 1-34 fragment for studies in animals and humans to characterize its effects on the kidney and bone. PTH(1–34) was also used experimentally as a therapy for osteoporosis [22–24]. These studies demonstrated dose-dependent effects of PTH(1–34) on bone in animals and humans. In 1990, Strogmann et al. [25] were the first to report successful PTH replacement therapy for hypoparathyroidism using synthetic human PTH 1-38 in 2 adolescents who were refractory to conventional therapy with calcitriol and calcium. One child had APS-1 and the other had idiopathic hypoparathyroidism. Although given just every other day, this replacement therapy was successful in one of the adolescents, but treatment was stopped due to lack of drug availability. This study and the studies demonstrating the safety and efficacy of PTH(1–34) in adults with osteoporosis [22–24] provided the foundation for our studies. In 1991, we initiated studies of the replacement therapy of hypoparathyroidism with synthetic human PTH(1–34) [6].

Six decades have passed from the first successful use of bovine PTH in the treatment of parathyroidectomized animals and more than four decades since the identification of the human PTH amino acid sequence before the first studies of synthetic human PTH(1–34) replacement were initiated in patients with hypoparathyroidism. What concerns led to the hesitation to explore PTH as a replacement therapy and allowed this vital area to remain dormant? First, the pharmacokinetic profile of PTH(1–34) given subcutaneously did not appear to be well suited for disease management. Peak plasma concentrations of PTH occur approximately 30 min after injection, and the serum elimination half-life of PTH is approximately 2 h after injection [26]. Second, PTH(1–34) must be injected, whereas for most individuals with hypoparathyroidism, oral calcitriol and calcium can maintain serum calcium levels within an acceptable range. Third, only a small subset of patients were considered so refractory to conventional therapy that they were at high risk of recurrent hypocalcemic crises. Fourth, an additional impediment occurred in 1999, with the emergence of Eli Lilly carcinogenicity data in rats demonstrating dose-dependent occurrence of osteosarcoma with chronic rhPTH(1–34) administration. In 2002, when recombinant human PTH(1–34) (Forteo®; teriparatide; rhPTH(1–34)) received FDA approval for treatment of osteoporosis, a black box warning was included because of the rodent osteosarcoma data. According to the warning, Forteo was contraindicated in growing children, and the duration of treatment in adults with osteoporosis was limited to 2 years, which was the duration of the osteoporosis trials at the time that rat osteosarcomas were detected and the trials were halted as a precaution. During the more than 10 years since FDA approval, however, there has been no osteosarcoma signal from human or primate exposure to rhPTH(1–34) [27–35].

Over the past two decades, studies of replacement therapy for hypoparathyroidism sought to determine the optimal PTH(1–34) regimen, in terms of dose and frequency of administration that best restores normal calcium homeostasis. During these studies it has become apparent that administration of the missing hormone can achieve more physiologic calcium homeostasis than conventional therapy. Furthermore, recent studies suggest that PTH replacement may remediate the quality of life impairments that have been associated with conventional treatment [13, 14].

30.4 Evidence of Safety and Efficacy of Synthetic Human PTH(1–34) Replacement in the Treatment of Hypoparathyroidism

Hypoparathyroidism is one of the few hormonal deficiencies for which replacement therapy with the missing hormone is not available. Based upon the demonstrable need for an alternative to conventional therapy, we have studied synthetic human parathyroid hormone 1-34 over the past two decades as a potential replacement therapy for hypoparathyroidism [6, 7, 36–40]. Our studies of PTH(1–34) replacement of hypoparathyroidism produced results that one might have predicted from the known physiologic effects of PTH and calcitriol. For many protocol subjects, the replacement of the missing hormone was as effective as conventional therapy in maintaining serum calcium within the desired range. For a majority of patients, however, replacing the missing hormone was better than conventional therapy, as PTH replacement reduced urinary calcium excretion and improve quality of life. For patients who were refractory to conventional treatment due to APS-1 and malabsorption, PTH replacement was a potentially life-saving measure.

We investigated different doses, dose regimens, and delivery modalities in adults and children ages 4–70 years. The simultaneous normalization and minimal fluctuation of serum and urine calcium, phosphorus, magnesium, and markers of bone turnover were the main goals of this hormone replacement therapy. No study participant received diuretics or phosphate binders. Patients who received PTH did not concurrently receive calcitriol or calcium supplements. PTH is a peptide and, therefore, cannot be given as an oral medication because it is broken down by proteolytic enzymes in the digestive tract. To bypass the gastrointestinal tract, we administered PTH by subcutaneous injection. Synthesized human PTH(1–34) was formulated as previously described at the NIH Clinical Center Pharmacy in vials containing 50 mcg/mL and 200 mcg/mL synthetic human PTH(1–34) [6].

30.4.1 Once-Daily Injection Therapy

The initial study with synthetic human parathyroid hormone 1-34 included 10 adults with hypoparathyroidism [6]. At study entry, 80 % of subjects had evidence of renal insufficiency and half had radiographic evidence of renal calcifications. A 27-year-old woman with APS-1 and recurrent episodes of hypocalcemia in the past requiring intravenous calcium infusions had stage 3 chronic kidney disease with a creatinine clearance of 47 mL/min (normal: 90–125 mL/min). We compared daily treatment with subcutaneous PTH to conventional therapy in a 20-week randomized crossover trial. During the conventional treatment arm, subjects received oral twice-daily calcitriol and four times daily calcium supplementation. With the initiation of PTH therapy, calcitriol and calcium supplements were abruptly discontinued. None of the patients received thiazides or cholecalciferol during either treatment arm. The patients were instructed to consume a daily diet containing approximately 1–2 g of calcium. Consistent with our hypothesis, PTH replacement simultaneously normalized both serum and urinary calcium levels and demonstrated an improvement over conventional therapy in control of mineral metabolism.

Within this cohort of 10 subjects, one patient ([6]; Table 1, patient H) required higher mean PTH (120 mcg/day; 1.4 mcg/kg/day) and calcitriol doses (3 mcg/day) compared to the other patients. Despite high doses of medication, serum calcium remained low and urine calcium remained above the normal range. This subject's markers of bone turnover in response to PTH therapy were also elevated (alkaline phosphatase: 196 U/L; normal 37–116). We discovered a mutation in the calcium-sensing receptor [10] and concluded that she could not be well managed on a single daily dose of PTH(1–34). The relatively large dose of PTH(1–34) that she required in this initial study produced too much stimulation of bone turnover, and urine calcium could not be normalized. This case was the first of our patients to illustrate a phenomenon that we later observed in other subjects, namely, that the rise in bone markers (which reflects the level of PTH stimulation of bone) is directly proportional to the PTH(1–34) dose. This observation led us to focus on the use of smaller, more frequent doses of PTH(1–34).

30.4.2 Twice-Daily Injection Therapy

We studied 17 adults (age 19–64 years) and 14 children (age 4–17 years) comparing once-daily with twice-daily PTH(1–34) injections in a randomized crossover design over a 28-week period. Adults and children yielded similar results [7, 36]. The study included five subjects with APS-1 (four children), five subjects with activating mutations of the CaSR (one child), ten subjects with postsurgical hypoparathyroidism (1 child), and the remaining subjects had idiopathic hypoparathyroidism. Of the 17 adult patients, 14 (80 %) had renal insufficiency at the start of the study. Four patients with CaSR from the same family, had stage 3–4 chronic kidney disease (creatinine clearance ranging from 21.6 to 44.7 mL/min).

We compared the safety and efficacy of these two treatment schedules between adults with congenital hypoparathyroidism due to a mutation of the CaSR and adults with acquired hypoparathyroidism due to APS-1 or thyroid surgery. We found that the responses to PTH therapy differed significantly between the CaSR group and the group with postsurgical or autoimmune hypoparathyroidism. For the adult subjects with postsurgical or autoimmune hypoparathyroidism, single or twice-daily subcutaneous injections of PTH(1–34) restored both serum and urine calcium to the normal or near-normal range. This was achieved with a significantly lower total daily dose during the twice-daily PTH(1–34) regimen than during once-daily administration.

This study was the first to highlight contrasting physiologic responses and resultant management issues in the patients with CaSR mutations compared to patients with other forms of hypoparathyroidism. As we had observed in the prior study, a single daily PTH(1–34) injection was inadequate to maintain serum calcium in the near-normal range throughout the day in patients with CaSR mutations. To maintain normal serum calcium, this regimen required excessively large doses and caused fluctuations in both serum and urine calcium. Furthermore, once-daily PTH(1–34) produced lower extremity bone pain and nausea post-injection in one subject with a CaSR mutation.

Twice-daily injections of PTH(1–34) lowered the daily dose and improved metabolic control in the CaSR patients, but the average urinary calcium remained in the high or above-normal range. We concluded that smaller, more frequent PTH(1–34) doses would be needed for this group of patients. We implemented administration of more frequent doses in the 3-year long-term study (twice-daily injections) and in a more recent long-term observational study (unpublished data) in which most patients with CaSR mutations were treated with thrice-daily PTH injections. Optimal therapy was achieved when PTH replacement through an insulin pump was initiated. Investigators in Spain and France were the first to report the advantages of pump delivery of rhPTH(1–34) (Forteo) over conventional therapy. Their work shows improved control of refractory hypoparathyroidism for periods of 1–3 years in one adult [41] and in 3 children [42] while they were receiving PTH replacement therapy delivered by an insulin pump (Medtronic).

30.4.3 PTH Replacement by Continuous Subcutaneous Infusion

To improve further the metabolic response to PTH, we initiated studies using an insulin pump (OmniPod by Insulet) to deliver PTH(1–34). For the first time, in all patients, regardless of disease etiology, PTH delivery by insulin pump simultaneously normalized serum and urine calcium and markers of bone turnover. We used 200 mcg/mL PTH(1–34) vials. Calcitriol and calcium supplements were discontinued at baseline, and all patients received cholecalciferol 1,000 IU daily. This method enabled continuous delivery of fixed subcutaneous microbolus doses at varying time intervals, determined by the basal rate [39, 40].

During pump delivery, subjects filled the pump-pod device with PTH(1–34), attached it to the abdomen or back, and changed it every 72 h. Seven basal rates were programmed into a wireless device that controlled the pump delivery rate. Initial basal rates were estimated for each subject based upon body weight (0.2 mcg/kg/day) and prior calcitriol or PTH(1–34) dose requirements. Basal rates ranged from 3 to 14 pulses per hour, with each pulse delivering 0.1 mcg of PTH(1–34). Two of the 7 basal options included a 4-h or 8-h nighttime dosage step-up of 1 pulse per hour from midnight to 0400 or from MN to 0800 to mimic the known circadian variation in circulating PTH [43, 44].

We compared PTH(1–34) delivered by an insulin pump with twice-daily injections in a randomized crossover study, first in 8 adults with postsurgical hypoparathyroidism [39] and subsequently in 12 children with APS-1 or CaSR mutations [40]. Subjects were randomized to either pump therapy or to twice-daily injections at the beginning of the study and crossed over to the alternate PTH delivery system (injections vs. pump) at the conclusion of the initial 3-month treatment period. There were three inpatient admissions, baseline, 3 months, and 6 months.

Compared to twice-daily injection, pump delivery in adult patients with postsurgical hypoparathyroidism resulted in a 65 % reduction in the mean $\pm$ SD daily PTH dose (13 $\pm$ 4 μg/day [0.17 $\pm$ 0.03 μg/kg/day] vs. 37 $\pm$ 14 μg/day [0.47 $\pm$ 0.13 μg/kg/day], $P < 0.001$). Pump delivery in children with APS-1 or CaSR mutations resulted in a 62 % reduction in their daily PTH dose during pump therapy compared to treatment with injections (0.32 $\pm$ 0.04 vs. 0.85 $\pm$ 0.11 mcg/

kg/day, $P<0.001$). Additionally, by raising serum magnesium levels, pump delivery of PTH permitted reduction in mean magnesium supplement dose. Most importantly, PTH delivered by pump produced normal urinary calcium excretion and normal bone turnover markers.

30.4.3.1 Vitamin D Supplementation

Initially, PTH replacement was not supplemented with cholecalciferol in our studies unless patients developed vitamin D deficiency. At dose-study baseline [7], adult subjects had normal mean serum 25-hydroxyvitamin D (25(OH)D) levels (40±29 ng/mL) that dropped by 27 % (to 29±22 ng/mL) and 45 % (to 22±14 ng/mL) while on once- and twice-daily PTH, respectively. The children's dose study also showed a reduction of 25(OH)D levels during PTH therapy, from 32±3.3 ng/mL at baseline to 30 % less (22±2.5 ng/mL) and 15 % less (27±2.2 ng/mL) in response to once- and twice-daily PTH injections, respectively [36]. In a subsequent three-year study in children [38], serum 1,25-dihydroxyvitamin D_3 (1,25$(OH)_2D_3$) levels were significantly higher during PTH therapy despite lower levels of 25(OH)D (23±1 [when receiving PTH(1–34)] vs. 34±2 ng/mL [when receiving calcitriol], $P<0.01$).

The reductions in 25(OH)D during PTH treatment suggested that it might be desirable to supplement vitamin D_3 intake. Initially, treatment with ergocalciferol or cholecalciferol was provided in response to decreasing 25(OH)D levels [37, 38]. Later, during the pump studies, prophylactic cholecalciferol (1,000 IU daily) was given to all patients to avoid a drop in 25(OH)D levels [39, 40]. Because a major effect of PTH is to regulate conversion of 25(OH)D precursor to the active 1,25-dihydroxyvitamin D_3 product by the enzyme 1-alpha hydroxylase, it seemed prudent to ensure that normal levels of 25(OH)D were maintained during PTH treatment. The intent was to ensure that PTH replacement would have the desired effect of providing exogenous PTH and endogenous calcitriol (1,25$(OH)_2D_3$), which would in turn act synergistically on the kidney, gut, and bone to maintain calcium homeostasis.

30.4.3.2 Magnesium Supplementation

Hypomagnesemia and the requirement for magnesium supplementation are most common in patients with autoimmune hypoparathyroidism or with CaSR mutations. In our study, we divided the daily magnesium supplement into 3 or 4 doses to bring the target level of magnesium to just below the normal range. An attempt to normalize serum magnesium levels may result in large magnesium losses in the urine which, along with excess calcium excretion, can damage the kidney. Additionally, large magnesium doses may cause diarrhea which can result in malabsorption.

For patients who received magnesium supplementation, the dose of elemental magnesium was similar during both treatment arms in the study comparing once-daily with twice-daily PTH replacement therapy [7, 36]. The delivery of PTH by pump,, however, significantly reduced the need for magnesium supplementation [39, 40]. Pump therapy allowed 4 of the 5 adult patients who were magnesium deficient to discontinue their supplements and still maintain normal serum magnesium levels. In children, pump delivery increased serum magnesium and permitted a reduction in mean±SE magnesium supplement (532±105 [pump] vs. 944±158 mg/day [injections], $P<0.001$).

30.4.3.3 Impact of PTH on Bone

PTH can be either anabolic or catabolic to the skeleton. The effects on bone are dose and frequency dependent. Our studies of long-term twice-daily replacement therapy, did not result in an increase or decrease in bone mineral density over time in adult subjects [37]. The 3-year study in children demonstrated a normal rise in bone density—nearly identical to that of age-matched growing children [38].

The PTH stimulation of bone, with increases in bone turnover markers, was apparent from the initial study comparing once-daily PTH with twice-daily calcitriol [6]. A single daily dose of PTH with no additional calcium or vitamin D supplements caused a rise in markers of bone turnover. In response to these observations, the second study [7] was undertaken, in part, to ascertain whether administering PTH twice-daily in smaller doses would result in lower markers of bone turnover. In this second study, we found that twice-daily injections produced far less stimulation of bone turnover and resulted in lower bone markers than once-daily injections. At baseline, subjects had normal mean total alkaline phosphatase levels (66 ± 21 U/L), which increased by 270 % (243 ± 272 ng/mL) and 120 % (146 ± 51 U/L) while on once- and twice-daily PTH, respectively [7]. The same study performed in children produced similar but less remarkable increases in bone markers [36]. At baseline, children with hypoparathyroidism had normal mean alkaline phosphatase levels (207 ± 20 U/L), which increased by 60 % (330 ± 29.4 U/L) and 44 % (298 ± 26.4 U/L) while on once- or twice-daily PTH, respectively.

We then investigated whether more frequent microboluses of PTH, through an insulin pump, would be more physiologic for the bone and avoid increases in markers of bone turnover beyond the normal range. PTH(1–34) replacement delivery with an insulin pump in both adults and children produced consistently lower markers of bone turnover compared to twice-daily PTH injections and maintained such markers within the normal range [39, 40].

30.4.3.4 PTH Dose

Most adults with postsurgical hypoparathyroidism respond well to a total daily dose of PTH(1–34) of 0.5 mcg/kg/day divided into two daily doses. Children with congenital hypoparathyroidism require a larger dose of 0.6 mg/kg/day divided into two or three daily doses. The mean total daily PTH dose required to maintain serum calcium in the normal or near-normal range varied from one study to the next due to the different etiologies and severity of hypoparathyroidism within the subject groups studied. The evidence suggests that more frequent, smaller doses of PTH result in lower total daily PTH doses that are needed to normalize serum calcium and less stimulation to the bone. The mean daily PTH dose was significantly lower for all adult subjects receiving twice-daily PTH (46 ± 32 μg/day [0.62 ± 0.45 μg/kg/day]) than once-daily PTH (97 ± 60 μg/day [1.48 ± 1.29 μg/kg/day], $P < 0.001$) [7]. Children required half the PTH(1–34) dose during the twice-daily regimen compared to the once-daily regimen (twice daily, 25 ± 15 mcg/day vs. once daily, 58 ± 28 mcg/day, $P < 0.001$).

When PTH was delivered by insulin pump to adults with postsurgical hypoparathyroidism, the mean ± SD daily PTH(1–34) dose (0.17 ± 0.03 μg/kg/day) was 65 % less than twice-daily delivery (0.47 ± 0.13 μg/kg/day). Compared to the results observed in the earlier study of pump delivery of PTH(1–34) in adults with postsurgical hypoparathyroidism [39], pediatric patients with APS-1 or CaSR mutations [40] required nearly twice the per kilogram PTH(1–34) pump dosage (0.17 ± 0.03 [adults] vs. 0.32 ± 0.11 mcg/kg/day).

30.4.3.5 Long-Term Replacement Therapy

We studied PTH(1–34) replacement in both adults and children in a 3-year randomized, parallel trial comparing twice-daily PTH(1–34) to twice-daily calcitriol with supplemental calcium [37, 38]. In addition to the outcome measures in the adults [37], the study in growing children included observations of linear growth, weight gain, and bone mineral accrual [38].

Our studies have shown that PTH(1–34) therapy is safe and effective in maintaining for up to three years stable calcium homeostasis and renal function in adults and children with hypoparathyroidism. PTH(1–34) was able to maintain mean serum calcium in the low or just below the normal range with normal concurrent urine calcium excretion. Markers of bone turnover remained elevated, but PTH did not produce, in the adults, significant longitudinal changes in A-P spine or whole body bone mineral density (BMD) or bone mineral content (BMC) as measured by dual-energy x-ray absorptiometry (DXA). The children participating in our studies had normal linear growth and bone accrual; there were no differences in BMD *Z* scores or mean height percentiles over time between the PTH and calcitriol treatment groups.

30.4.3.6 Treatment of Children

The study of PTH(1–34) replacement therapy in children began with comparing once-daily and twice-daily PTH [36]. Results of this study of 14 children ages 4–17 years demonstrated that twice-daily dosing effectively reduced urine calcium excretion and maintained normal serum calcium with half the total daily PTH dose needed in the once-daily arm (twice daily, 25 ± 15 mcg/day vs. once daily, 58 ± 28 mcg/day).

A subsequent three-year study of PTH vs. calcitriol and calcium in 12 children, ages 5–14 years, demonstrated that twice-daily PTH maintained normal serum calcium and reduced urine calcium levels compared to conventional therapy. Both PTH(1–34) and conventional treatment maintained normal skeletal development, linear growth, weight gain, and renal function. PTH therapy in children led to higher markers of bone remodeling than was seen in those treated with calcitriol. Bone mineral density *Z* scores, reflecting rates of bone mineral accrual, did not differ across time or between treatment groups (PTH vs. calcitriol). As one would expect in growing children, BMC and BMD showed a consistent upward trend over the three-year study period. Unpublished data from a ten-year observational study of 13 children and adolescents treated with PTH also demonstrated normal linear growth, weight gain, and bone mass accrual.

30.4.3.7 Treatment of Infants

Severe neonatal hypocalcemia due to hypoparathyroidism may be refractory to calcitriol and calcium treatment and can lead to life-threatening seizures. Emergency therapy with intravenous calcium, often administered through a central line, has risks of thrombosis and infection. Cho et al. [45] describe a preterm infant with a rare autosomal recessive form of hypoparathyroidism, Sanjad-Sakati syndrome, which included perinatal growth retardation and dysmorphic facial features. This infant developed hypocalcemia and hypomagnesemia on day 3 of life with concurrent malabsorption. After 2 weeks of hypocalcemia refractory to calcitriol and calcium supplementation, she received a 1 mcg/kg rhPTH(1–34) (Forteo) subcutaneous injection followed by the same dose divided in twice-daily injections. Serum calcium and phosphorus normalized after 6 days of therapy and there was an improvement in weight gain. PTH replacement was continued for an additional week. The child was discharged on conventional therapy and required 9 hospitalizations for hypocalcemia during her first year of life. During this time, linear growth and weight gain were poor. Newfield [46] described a 17-day-old infant who received successful rhPTH(1–34) (Forteo) therapy during a hypocalcemic crisis with seizures. After 2 days of unsuccessful treatment with intravenous calcium and oral calcitriol, a single 1 mcg/kg subcutaneous PTH injection raised the serum calcium to the normal range within 4 hours. Mittleman et al. [47] described a 14-month-old boy with poorly controlled hypoparathyroidism due to CaSR mutation diagnosed at 3 weeks old after hypocalcemic seizures. He remained hospitalized for a month due to refractory hypocalcemia. Twice-daily rhPTH(1–34) (Forteo) injections were initiated when the child was 14 months old at a daily dose of 0.5 mcg/kg/day and continued for 17 months with symptomatic relief and improved urine and serum calcium levels.

Large boluses of intravenous calcium may lead to fluctuations in plasma ionized calcium which cause further neuromuscular irritability and increased risk of seizures. Furthermore, large intravenous doses of calcium may lead to renal damage if administered repeatedly. Subcutaneous PTH is a rapid, safe, and more physiologic therapy for severe hypocalcemia due to hypoparathyroidism. Further study is needed in both the acute hypocalcemic setting as well as in long-term maintenance therapy during infancy. Additionally, early intervention with PTH replacement therapy would theoretically avoid kidney damage that can be evident from a young age in patients with congenital hypoparathyroidism treated with conventional therapy.

30.5 Management of the Patient with Hypoparathyroidism: Synopsis

In the last 20 years, we have witnessed major changes in the approach to the treatment of hypoparathyroidism. The commercial availability of recombinant human PTH(1–34) as an approved drug for osteoporosis has provided the option for off-label use in hypoparathyroidism, which was not available prior to 2002. The improvement in drug manufacturing techniques allowed for both rhPTH(1–84) and synthetic human PTH(1–34) peptides to be available for investigational use in the treatment of hypoparathyroidism. These preparations replaced the animal extracted preparations from decades ago that were associated with antibody formation. New PTH analogues, with distinct pharmacokinetic profiles, are in development and may provide advantages over treatment with PTH(1–34). The use of the insulin pump to deliver PTH(1–34) represents an important breakthrough as it provides a potential for true physiologic replacement that mimics endogenous secretion of parathyroid hormone.

A successful therapeutic approach to the patient with hypoparathyroidism requires attention to disease etiology to determine individual PTH replacement requirements. These requirements also depend upon age, level of activity, nutrition, and GI tract integrity.

Dietary management, a key component of successful therapy, is often ignored. The use of a low-phosphorus diet does not reduce the serum phosphorus level. This diet also forces the patient to avoid dairy products, a key source of calcium. The use of calcium supplements to raise serum calcium levels can increase the risk of renal calcifications [48]. Providing additional calcium in the diet is the optimal way for the patient to consume the calcium needed to manage the disease. The amount of dietary calcium intake should be no more than the recommended levels for age. 25-Hydroxyvitamin D levels should be maintained between 30 and 50 ng/mL, which is readily accomplished with 1,000 IU daily supplemental cholecalciferol. Patients with malabsorption, as in APS-1, need two to four times this daily dose of vitamin D_3.

Until recently, there has been a lack of guidelines for treatment of hypoparathyroidism with conventional therapy or with PTH. Much of the practice in this area is based on anecdotal information rather than evidence-based recommendations. Several common misconceptions: include the following:

1. Supplementing conventional therapy with thiazides offers the same protection to renal function as PTH to decrease urine calcium excretion.

Hypoparathyroidism patients may develop renal disease and nephrocalcinosis due to chronic elevated urinary calcium levels. Thiazide diuretics are often added to the conventional treatment regimen as an off-label therapy to block sodium chloride cotransport and raise the distal tubular calcium reabsorption, which should theoretically decrease calcium excretion. However, when Parfitt compared the reaction to thiazide diuretics of control subjects with hypercalciuria to that of patients with hypoparathyroidism, he observed distinctly different responses between the two groups [49]. The effect of a thiazide diuretic on urinary calcium in patients with hypoparathyroidism differed from its effect on urine calcium in subjects with hypercalciuria and intact parathyroid function. Urine calcium increased initially in the hypoparathyroid group but did not exhibit the expected decrease observed in the control group. Both groups experienced a rise in urine magnesium, phosphorus, potassium, and sodium excretion with concurrent fall in plasma magnesium and potassium. Parfitt concluded that PTH is necessary to achieve the usual hypocalciuric effects of thiazides. These results were supported by a second study demonstrating that thiazide diuretics reduce urine calcium excretion in patients with intact parathyroid glands but not in patients with hypoparathyroidism, despite equivalent sodium losses between the two groups [50]. Although the reasons why thiazides were ineffective in patients with hypoparathyroidism remained unknown, the investigators from both studies hypothesized that thiazides may potentiate the renal calcium-retaining action of parathyroid hormone and, in its absence, are not effective.

There is no evidence that thiazides are effective in avoiding long-term adverse renal outcomes for patients with hypoparathyroidism, especially in the treatment of more severe forms of hypoparathyroidism associated with a CaSR or APS-1. One could argue that the off-label use of these drugs should be avoided in hypoparathyroid patients who have concurrent hypomagnesemia or who are prone to other electrolyte imbalances as in Addison's disease, a common feature of APS-1. To be effective in treating hypercalciuria, thiazides should be administered with a low-salt diet [51, 52], which would be contraindicated in a patient with Addison's disease and difficult to implement in most other forms of hypoparathyroidism. Eknoyan et al. [52] demonstrated, in hypoparathyroid dogs given chlorothiazide, a progressive increase in the fractional clearance of calcium, magnesium, and phosphate as the clearance of sodium increased. We observed, in patients with hypoparathyroidism, a rise in magnesium excretion and a decrease in serum magnesium with thiazide therapy. This was particularly exaggerated in patients with CaSR. Furthermore, the use of thiazides may lead to potassium depletion which requires adding potassium supplements to the regimen.

2. Patients prefer conventional therapy because they find oral medication far more practical than injections.

   When given the choice, patients will choose the therapy that provides the optimal metabolic and health advantages. Ultimately, patients choose whatever regimen improves their quality of life. Parenteral treatment was chosen over oral conventional therapy by all but one patient in the first study comparing PTH with conventional therapy and the majority of patients we treated in subsequent studies. In all of our studies, we discontinued calcitriol and all calcium supplements. For many, this was, on average, 15 pills daily. Substituting two subcutaneous injections was a relief to most of the patients and the preferred therapy, especially if the metabolic advantages were evident in their individual response to the drug.

3. FDA's black box warning implies that osteosarcoma risk heightens with each year of therapy with PTH.

   In 1998, as part of obtaining approval for rhPTH(1–34) (Forteo) as a treatment for osteoporosis in adults, Eli Lilly released to the FDA their 2-year rat carcinogenicity data demonstrating a dose-dependent risk of osteosarcoma. These experiments later prompted an FDA black box warning against rhPTH(1–34) use in children with open epiphyses or in anyone for more than 2 years [27]. The experimental rats did not have parathyroid hormone insufficiency and were given daily supraphysiologic subcutaneous doses (5, 30, or 75 mcg/kg/dose) of rhPTH(1–34) from the time of weaning throughout most of their natural lifespan. The appearance of osteosarcomas appeared to be dose and duration dependent and was most evident in the highest dose group [28]. Compared to adults with hypoparathyroidism, the rhPTH(1–34) doses administered during the rat carcinogenicity studies were 20, 125, and 330 times greater than those administered during twice-daily injections (0.23 μg/kg/dose) and were 33, 166, and 500 times greater than the daily PTH replacement dose during pump administration (0.17 μg/kg/day) [39].

   Additional data have emerged further demonstrating that the effect of PTH on bone is dose dependent and that, thus far, no osteosarcoma signal has been observed during clinical use of PTH(1–34) in humans [29–33]. Furthermore, in nonhuman primate studies, daily high-dose PTH(1–34) (5 mcg/kg), administered to ovariectomized monkeys with normal parathyroid function for 18 months, produced a significant rise in bone density but no bone proliferative lesions or microscopic osteosarcomas – either during treatment or during 3 years of observation after stopping treatment [34]. These results suggest that the potential osteosarcoma risk of PTH(1–34) is far greater in rats than in nonhuman primates. A possible explanation for the greater oncogenic sensitivity of rat skeleton is that differences in skeletal physiology between rodents and primates lead to an exaggerated anabolic response to PTH treatment in the rat. An additional observation is that osteosarcoma is not a feature of long-standing hyperparathyroidism in humans despite chronically elevated endogenous serum PTH blood levels [35].

30.6 Future Directions

Continuous monitoring devices with alarms to signal hypoglycemia have revolutionized diabetes management and served to reduce the anxiety associated with not knowing the glycemic status. Similar technologic advances could be implemented for monitoring blood calcium. One of the major difficulties for patients with hypoparathyroidism is their inability to monitor their own calcium levels. When symptoms develop, doses may be changed without knowing the calcemic status. Alternatively, patients are forced to ignore their symptoms or seek emergency care. Monitoring devices would enable more fine-tuned dose adjustments based on real-time serum calcium levels.

It would also be useful to further explore PTH replacement therapy in infants and young children, particularly the long-term safety, efficacy, and tolerability of PTH(1–34) delivered by pump. Pump therapy from early life in children with congenital hypoparathyroidism offers the best current prospect of avoiding the renal damage that often develops in these patients with conventional treatment.

References

1. Horowitz MJ, Stewart AF (2008) Hypoparathyroidism: is it time for replacement therapy? J Clin Endocrinol Metab 93(9):3307–3309
2. Romijn JA, Smit JWA, Lamberts SW (2003) Intrinsic imperfections of endocrine replacement therapy. Eur J Endocrinol 149:91–97
3. Rejnmark L, Sikjaer T, Underbjerg L, Mosekilde L (2013) PTH replacement therapy of hypoparathyroidism. Osteoporos Int 24:1529–1536
4. Weber G, Cazzuffi MA, Frisone F, De Angelis M, Pasolini D, Tomaselli V et al (1998) Nephrocalcinosis in children and adolescents: sonographic evaluation during long-term treatment with 1, 25-dihydrocholecalciferol. Child Nephrol Urol 9:273–276
5. Santos F, Smith MJ, Chan JC (1986) Hypercalciuria associated with long term administration of calcitriol (1.25- dihydroxyvitamin D). Am J Dis Child 140:139–142
6. Winer KK, Yanovski JA, Cutler GB Jr (1996) Synthetic human parathyroid hormone 1-34 vs calcitriol and calcium in the treatment of hypoparathyroidism: results of a randomized crossover trial. JAMA 276:631–636
7. Winer KK, Yanovsiki JA, Sarani B, Cutler GB Jr (1998) A randomized, crossover trial of once-daily versus twice-daily human parathyroid hormone 1-34 in the treatment of hypoparathyroidism. J Clin Endocrinol Metab 83:3480–3486
8. Underbjerg L, Sikjaer T, Mosekilde L, Rejnmark L (2013) Cardiovascular and renal complications to postsurgical hypoparathyroidism: a Danish nationwide controlled historic follow-up study. J Bone Miner Res 28(11):2277–2285
9. Mitchell DM, Regan S, Cooley MR, Lauter KB, Vrla MC, Becker CB, Burnett-Bowie SA, Mannstadt M (2012) Long-term follow-up of patients with hypoparathyroidism. J Clin Endocrinol Metab 97(12):4507–4514
10. Mancilla EE, De Luca F, Ray K, Winer KK, Fan GF (1997) A Ca(2+) sensing receptor mutation causes hypoparathyroidism by increasing receptor sensitivity to CA2+ and maximal signal transduction. Pediatr Res 42(4):443–447
11. Pearce SH, Williamson C, Kifor O, Bai M, Coulthard MG, Davies M, Lewis-Barned N, McCredie D, Powell H, Kendall-Taylor P, Brown EM, Thakker RV (1996) Familial Syndrome of hypocalcemia with hypercalciuria due to mutations in the calcium-sensing receptor. N Engl J Med 335(15):1115–1122
12. Husebye ES, Perheentupa J, Rautemaa R, Kämpe O (2009) Clinical manifestations and management of patients with autoimmune polyendocrine syndrome type I. J Intern Med 265(5):514–529
13. Arlt W, Fremerey C, Callies F, Reincke M, Schneider P, Timmermann W, Allolio B (2002) Well-being, mood and calcium homeostasis in patients with hypoparathyroidism receiving standard treatment with calcium and vitamin D. Eur J Endocrinol 146(2):215–222
14. Cusano NE, Rubin MR, McMahon DJ, Irani D, Tulley A, Sliney J Jr, Bilezikian JP (2013) The effect of PTH(1-84) on quality of life in hypoparathyroidism. J Clin Endocrinol Metab 98(6):2356–2361
15. Collip JB (1925) The extraction of parathyroid hormone which will prevent or control parathyroid tetany and which regulates the level of blood calcium. J Biol Chem 63:395–438
16. Collip JB (1925) Clinical use of parathyroid hormone. Can Med Assoc J 15(11):1158
17. Albirght F, Ellsworth R (1929) Studies on the physiology of the parathyroid glands; calcium and phosphorus studies on a case of idiopathic hypoparathyroidism. J Clin Invest 7(2):183–201
18. Aurbach GD (1959) Isolation of parathyroid hormone after extraction with phenol. J Biol Chem 234:3179–3181
19. Melick RA, Gill JR Jr, Berson SA, Yalow RS, Bartter FC, Potts JT Jr, Aurbach GD (1967) Antibodies and clinical resistance to parathyroid hormone. N Engl J Med 276(3):144–147
20. Brewer HB, Fairwell R, Ronan R, Sizemore GW, Arnaud CD (1972) Human parathyroid hormone amino acid sequence of the amino- terminal residues 1-34. Proc Natl Acad Sci U S A 69:3585–3588

21. Niall HD, Sauer RT, Jacobs JW, Keutmann HT, Segre GV, O'Riordan JL, Aurbach GD, Potts JT Jr (1974) The amino-acid sequence of the amino-terminal 37 residues of human parathyroid hormone. Proc Natl Acad Sci U S A 71(2):384–388
22. Slovik D, Neer R, Potts J (1981) Short term effects of synthetic human parathyroid hormone on bone mineral metabolism in osteoporotic patients. J Clin Invest 68(5):1261–1271
23. Slovik DM, Rosenthal DI, Doppelt SH, Potts JT, Daly MA, Campbell JA, Neer RM (1986) Restoration of spinal bone of osteoporotic men by treatment with human parathyroid hormone (1-34) and 1, 25-dihydroxyvitamin D. J Bone Miner Res 1(4):377–381
24. Reeve J, Meunier P, Parson J, Bernat M, Bijvoet OL, Courpron P, Edouard C, Klenerman L, Neer RM, Renier JC, Slovik D, Vismans FJ, Potts JT Jr (1980) Anabolic effect of human parathyroid hormone fragment on trabecular bone in involutional osteoporosis: a multicenter trail. Br Med J 280:1340–1344
25. Strogmann W, Bohrn E, Woloszczuk W (1990) First experiences in the substitution treatment of hypoparathyroidism with synthetic human parathyroid hormone. Monatasschr Kinderheilkd 138:141–146
26. Chu NN, Li XN, Chen WL, Xu HR (2007) Pharmacokinetics and safety of recombinant human parathyroid hormone (1-34) (teriparatide) after single ascending doses in Chinese healthy volunteers. Pharmazie 62(11):869–871
27. Vahle JL, Sato M, Long GG, Young JK, Francis PC, Engelhardt JA, Westmore MS, Linda Y, Nold JB (2002) Skeletal changes in rats given daily subcutaneous injections of recombinant human parathyroid hormone (1-34) for 2 years and relevance to human safety. Toxicol Pathol 30(3):312–321
28. Vahle JL, Long GG, Sandusky G, Westmore M, Ma YL, Sato M (2004) Bone Neoplasms in F344 rats given teriparatide [rhPTH(1-34)] are dependent on duration of treatment and dose. Toxicol Pathol 32(4):426–438
29. Vahle JL, Sato M, Long GG (2007) Variations in animal populations over time and differences in diagnostic thresholds used can impact tumor incidence data. Toxicol Pathol 35(7):1045–1046
30. Turner RT, Evans GL, Lotinun S, Lapke PD, Iwaniec UT, Morey-Holton E (2007) Dose-response effects of intermittent PTH on cancellous bone in hindlimb unloaded rats. J Bone Miner Res 22(1):64–71
31. Tashijan AH Jr, Gagel RF (2006) Teriparatide: 2.5 years of experience on the use and safety of the drug for the treatment of osteoporosis. J Bone Miner Res 21:354–365
32. Tashjian AH Jr, Golzman D (2008) Perspective on the interpretation of Rat carcinogenicity studies for human PTH 1-34 and human PTH 1–84. J Bone Miner Res 23(6):803–811
33. Haseman JK, Hailey JR, Morris RW (1998) Spontaneous neoplasm incidences in Fischer 344 rats and B6C3F1 mice in two-year carcinogenicity studies: a National Toxicology Program Update. Toxicol Pathol 26(3):428–441
34. Vahle JL, Zuehlke U, Schmidt A, Westmore M, Chen P, Sato M (2008) Lack of bone neoplasms and persistence of bone efficacy in cynomolgus macaques after long-term treatment with teriparatide [rhPTH(1-34)]. J Bone Miner Res 23(12):2033–2039
35. Silverberg SJ, Shane E, Jacobs TP et al (1999) A 10 year prospective study of primary hyperparathyroidism with or without parathyroid surgery. N Engl J Med 341:1249–1255
36. Winer KK, Sinaii N, Peterson D, Sainz B Jr, Cutler GB Jr (2008) Effects of once-daily versus twice-daily parathyroid hormone 1-34 in children with hypoparathyroidism. J Clin Endocrinol Metab 93(9):3389–3395
37. Winer KK, Ko CW, Reynolds J, Dowdy K, Keil M, Peterson D et al (2003) Long-term treatment of hypoparathyroidism: a randomized controlled study comparing parathyroid hormone 1–34 and calcitriol and calcium. J Clin Endocrinol Metab 88:4214–4220
38. Winer KK, Sinaii N, Reynolds J, Peterson D, Dowdy K, Cutler GB Jr (2010) Long-term treatment of 12 children with chronic hypoparathyroidism: a randomized trial comparing synthetic human parathyroid hormone 1-34 versus calcitriol and calcium. J Clin Endocrinol Metab 95(6):2680–2688
39. Winer KK, Zhang B, Shrader JA, Peterson D, Smith M, Albert PS et al (2012) Synthetic human parathyroid hormone 1-34 replacement therapy: a randomized crossover trial comparing pump versus injections in the treatment of chronic hypoparathyroidism. J Clin Endocrinol Metab 97(2):391–399
40. Winer KK, Fulton K, Albert PS, Cutler GB Jr (2014) Twice-daily subcutaneous injections vs. pump delivery of PTH 1-34 in the Treatment of Children with Severe Congenital Hypoparathyroidism. J Pediatr 165(3):556–563.
41. Puig-Domingo M, Diaz G, Nicolau J, Fernandez C, Rueda S, Halperin I (2008) Successful treatment of vitamin D unresponsive hypoparathyroidism with multipulse subcutaneous infusion of teriparatide. Eur J Endocrinol 159:653–657
42. Linglart A, Rothernbuhler A, Gueorgieva I, Lucchini P, Silve C, Bougneres P (2011) Long-term results of continuous subcutaneous recombinant PTH 1–34 infusion in children with refractory hypoparathyroidism. J Clin Endocrinol Metab 96(11):3308–3312
43. Samuels MH, Veldhuis JD, Cawley C, Urban RJ, Luther M, Bauer R et al (1993) Pulsatile secretion of parathyroid hormone in normal young subjects: assessment by deconvolution analysis. J Clin Endocrinol Metab 77(2):399–403
44. Harms HS, Kaptiana U, Kulpmann WR, Brabant G, Hesch RD (1989) Pulse amplitude and frequency modulation of parathyroid hormone in plasma. J Clin Endocrinol Metab 69:843–851
45. Cho YH, Tchan M, Roy B, Halliday R, Wilson M, Dutt S, Siew S, Munns C, Howard N (2012) Recombinant parathyroid hormone therapy for severe neonatal hypoparathyroidism. J Pediatr 160(2):345–348

46. Newfield RS (2007) Recombinant PTH for initial management of neonatal hypocalcemia. N Engl J Med 356(16):1687–1688
47. Mittleman SD, Hendy GN, Fefferman RA, Canaff L, Mosesova I, Cole DE, Burkett L, Geffner ME (2006) A Hypocalcemic child with a novel activating mutation of the calcium-sensing receptor gene: successful treatment with recombinant human parathyroid hormone. J Clin Endocrinol Metab 91(7):2474–2479
48. Jackson RD, LaCroix AZ, Gass M, Wallace RB, Robbins J, Lewis CE, et al. Women's Health Initiative Investigators (2006) Calcium plus vitamin D supplementation and the risk of fractures. N Engl J Med 354(7):669–683
49. Parfitt AM (1972) The interactions of thiazide Diuretics with parathyroid hormone and vitamin D studies in patients with hypoparathyroidism. J Clin Invest 51(7):1879–1888
50. Middler S, Pak CY, Murad F, Bartter FC (1973) Thiazide diuretics and calcium metabolism. Metabolism 22(2):139
51. Porter RH, Cox BG, Heaney D, Hostetter TH, Stinebaugh BJ, Suki WN (1978) Treatment of hypoparathyroid patients with chlorthalidone. N Engl J Med 298(11):577–581
52. Eknoyan G, Suki WN, Martinez-Maldonado M (1970) Effect of diuretics on urinary excretion of phosphate, calcium and magnesium, in thyroparathyroidectomized dogs. J Lab Clin Med 76:257–266

31 Replacement Therapy with PTH(1–84)

Mishaela R. Rubin, Natalie E. Cusano, and John P. Bilezikian

31.1 Introduction

Hypoparathyroidism, a rare disorder of mineral metabolism, is characterized biochemically by low serum calcium and low or undetectable parathyroid hormone (PTH) levels. In adults, it is most often caused by inadvertent removal of all parathyroid glands during neck surgery, while in children it is associated with rare genetic disorders, including familial-isolated hypoparathyroidism, DiGeorge syndrome, autoimmune polyglandular syndrome type 1, and autosomal dominant hypocalcemia [1]. The most recent estimates place the prevalence of hypoparathyroidism to be approximately 58,700 in the United States [2]. In the absence of PTH, well-characterized, biochemical abnormalities develop. A normal serum calcium concentration cannot be maintained and hypocalcemia often ensues with associated symptoms of neuromuscular irritability, such as muscle spasms, numbness, and paresthesias of the extremities. At its worst, laryngeal spasm and seizures occur. Moreover, in the absence of PTH, the filtered calcium load at the renal tubule cannot be conserved, especially during treatment of the condition with calcium and vitamin D supplementation, leading to hypercalciuria. Nephrocalcinosis, nephrolithiasis, and renal dysfunction can follow.

Conventional therapy of hypoparathyroidism attempts to address these abnormalities by treatment with large, pharmacologic dosages of oral calcium, parent vitamin D (cholecalciferol), and/or active vitamin D (1,25-dihydroxyvitamin D) (see also Chap. 28). This therapeutic approach is associated with variable success in regulating serum calcium levels. It also does not address many other key management issues. For example, impaired quality of life, specifically with regard to cognition and mood, a nearly universal finding in hypoparathyroidism, is not improved by calcium and vitamin D. Furthermore, without PTH, bone turnover is abnormally low, leading to markedly altered microarchitectural and biomechanical properties of the skeleton (see also Chap. 26). Calcium and vitamin D cannot improve these structural and dynamic skeletal abnormalities. Moreover, in addition to renal calcifications, other extraskeletal calcifications can develop in hypoparathyroidism, such as in the basal ganglia, other areas in the brain, and in the vasculature itself. Thus, there has been a clear need to improve the management of hypoparathyroidism with an approach that goes beyond the use of pharmacologic amounts of calcium and vitamin D.

M.R. Rubin • N.E. Cusano • J.P. Bilezikian, MD (✉)
Metabolic Bone Disease Unit, Columbia University Medical Center, College of Physicians and Surgeons, Columbia University, PH8W-864, 630W. 168th St., New York, NY 10032, USA
e-mail: jpb2@columbia.edu

M.L. Brandi, E.M. Brown (eds.), *Hypoparathyroidism*,
DOI 10.1007/978-88-470-5376-2_31, © Springer-Verlag Italia 2015

31.2 Treatment of Hypoparathyroidism with Injections of PTH(1–84)

Hypoparathyroidism is the only classic hormone deficiency state for which approved replacement therapy is still not available. To address this gap, studies have investigated the utility of PTH replacement therapy in hypoparathyroidism over the past two decades. PTH(1–34) has been studied in a series of randomized and controlled studies, in which it was titrated to fully abolish the need for supplemental calcium and vitamin D therapy [3–7]. It was able to maintain normocalcemia for up to 3 years in hypoparathyroidism adults and children, although because of the short half-life of PTH(1–34), twice daily and sometimes more frequent dosing was required [3–7].

Other studies, conducted over the past decade, have investigated the use of PTH(1–84), the full length, native PTH molecule, as a potentially more natural therapy of hypoparathyroidism [8–10]. The pharmacokinetic profile of PTH(1–84) differs from that of PTH(1–34). With PTH(1–34), PTH peaks in 30 min and calcium levels reach their peak within 4–6 h [11]. In contrast, PTH(1–84) reaches its peak 1–2 h after subcutaneous administration with calcium peaking between 6 and 8 h [12]. Although no head-to-head comparisons with PTH(1–34) exist, the longer half-life of PTH(1–84), particularly when injected in the thigh [13], theoretically offers an advantage in comparison to PTH(1–34) by having a more prolonged biologic effect and requiring less frequent injections.

To date, three studies of PTH(1–84) therapy in hypoparathyroidism have been conducted. One study, reported by Sikjaer et al. was a randomized clinical trial (RCT) with a fixed dose of 100 mcg daily as add-on therapy to conventional treatment for 24 weeks [8]. The participants included 62 subjects with hypoparathyroidism, mostly women with postsurgical disease. Replacement therapy included alfacalcidol (1-hydroxycholecalciferol) for treatment of the hypoparathyroidism. A second study, the largest to date, was a multicenter randomized control trial (RCT) sponsored by NPS Pharmaceuticals, of 134 hypoparathyroid patients who were randomized in a 2:1 fashion to PTH(1–84) or placebo for 24 weeks [9]. Treatment with PTH(1–84) was started at 50 mcg/d, with the option to increase to either 75 or 100 mcg/d. A third study, the longest trial to date, is an ongoing open-label cohort study in the metabolic bone diseases group at Columbia University Medical Center in which PTH(1–84) was administered initially at 100 mcg every other day. The 4-year results of 27 subjects (postsurgical $n=16$, idiopathic $n=10$, DiGeorge $n=1$) have been reported [10]. Overall, the data from these studies indicate that PTH(1–84) addresses many of the biochemical, renal, skeletal, and neuropsychological abnormalities of hypoparathyroidism to a greater extent than conventional treatment.

31.3 Effects of PTH(1–84) on Serum Calcium and Requirement for Calcium Supplementation

In each of the PTH(1–84) studies, supplemental calcium and vitamin D dosages could be reduced while normal serum calcium levels were maintained. In the RCT of Sikjaer et al., calcium supplementation fell by 75 % and active vitamin D by 73 %; seven patients were able to stop calcium and vitamin D altogether [8]. In the face of these reductions in supplemental calcium and vitamin D, ionized calcium levels with PTH(1–84) treatment actually increased, although at 24 weeks the ionized calcium level was not significantly higher than baseline [8]. Pharmacodynamic studies showed that the rise in PTH levels after injection into the thigh was rapid, peaking at 15 min postinjection, with ionized calcium peaking at 7 h and PTH levels returning to baseline levels over 16 h [14].

Similar effects on the control of serum calcium were observed in the NPS randomized control study (RCT). Calcium and/or calcitriol dosages could be reduced by at least 50 % while maintaining the serum calcium level in the normal range (the triple primary endpoint of the study) in 53 % of the PTH group, as compared with 2 % in the placebo group [9]. Moreover, 41 % of patients in

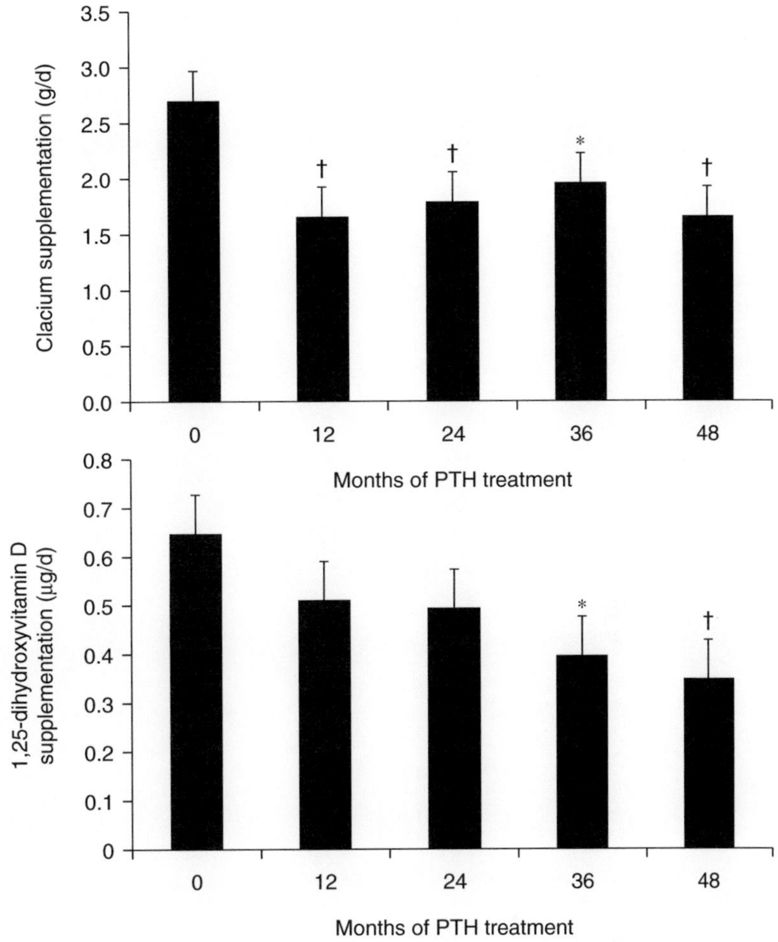

Fig. 31.1 Changes in calcium and 1,25-dihydroxyvitamin D supplementation. Calcium requirements decreased by 6 months after baseline, whereas 1,25-dihydroxyvitamin D requirements decreased by 36 months. Data are expressed as mean ± SE. *, $P<0.05$ compared with baseline; †, $P<0.01$ compared with baseline (Reproduced with permission from Cusano et al. [10])

the PTH group completely stopped vitamin D and decreased their calcium dosages to <500 mg/d. Despite these large reductions in active vitamin D and calcium doses, as in the Sikjaer study [8], serum calcium levels remained at or above baseline in the PTH-treated patients [9].

Data from the Columbia cohort suggest that these beneficial effects of PTH(1–84) to maintain normocalcemia in hypoparathyroidism, with reduced need for treatment with calcium and vitamin D, can be maintained long term [10]. With 4 years of PTH(1–84) treatment, supplemental calcium was reduced by 37 % and vitamin D by 45 % (Fig. 31.1), with seven subjects completely stopping active vitamin D [10]. Serum calcium levels remained in the low-normal range throughout the 4 years [10], suggesting that PTH(1–84) may be effective for long-term treatment of chronic hypoparathyroidism.

31.4 Effects of PTH(1–84) on Quality of Life

Hypoparathyroid patients treated with calcium and vitamin D have been found to have deficits in mental and physical functioning as measured by the RAND 36-Item Health Survey (SF36) tool [15], suggesting that the absence of PTH, even in the presence of eucalcemia, is accompanied by compromised quality of life. Data from the Columbia cohort suggest that 1 year of PTH(1–84) treatment is associated with an improvement in both mental and physical health domains

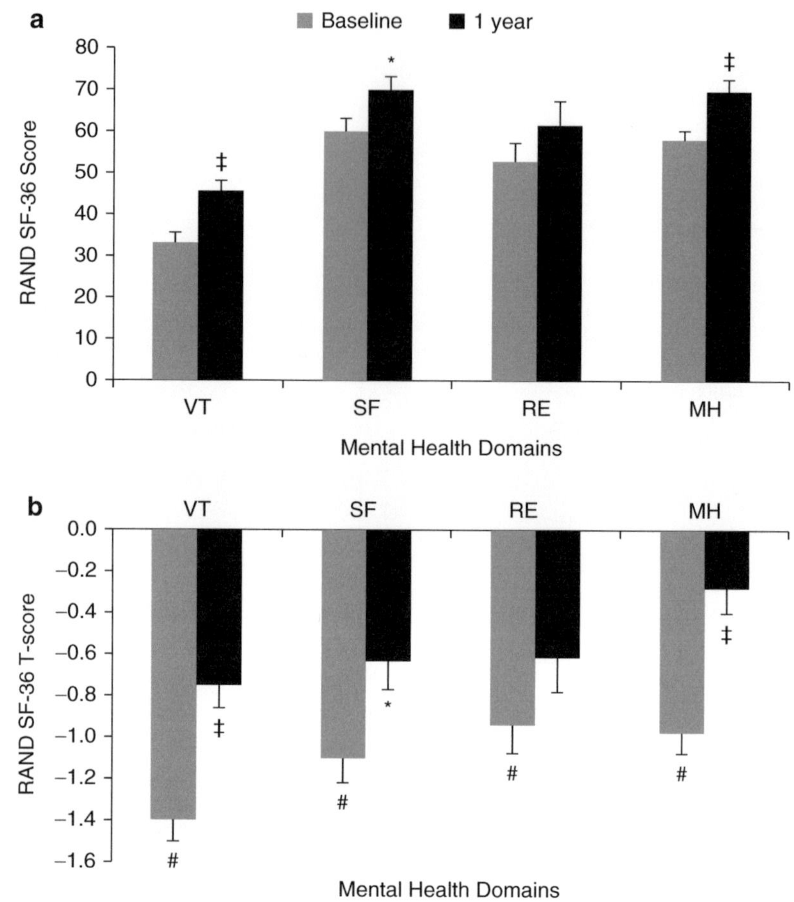

Fig. 31.2 Changes in the mental health domains with PTH(1–84) therapy. (**a**) Change in RAND 36-Item Health Survey domain scores from baseline to 1 year. (**b**) Change in RAND 36-Item Health Survey domain T-scores from baseline to 1 year. Values are mean ± SE. #, $P<.05$ compared with normal population; *, $P<.05$ compared with baseline; ‡, $P<.001$ compared with baseline

(Fig. 31.2) [15]. The PTH dose for subjects in this study was titrated to maintain serum calcium in the intended low-normal range. Although these data are limited by the uncontrolled, open-label design of the study, they nevertheless indicate a marked improvement in quality of life with PTH(1–84). Sikjaer et al. published the results of their randomized clinical trial investigating the effect of a fixed dose of PTH(1–84) 100 μg daily vs. placebo on quality of life over 24 weeks [16]. There was improvement in quality of life measures in both the placebo and PTH arms but no between-group differences. In their study, mean serum calcium levels in the PTH-treated group were significantly increased, and there was a relatively high incidence of hypercalcemia, which the investigators noted as asymptomatic. The authors posited that the large fluctuations in serum calcium might have negated the possible benefits of PTH therapy. These results may indicate that the reference range for serum calcium in normal subjects may not be suitable for all hypoparathyroid subjects with regard to their mental and physical health considering that they are acclimated to relatively lower serum calcium values. Further randomized and controlled data will be necessary to investigate this outcome.

31.5 Effect of PTH(1–84) on Urinary Calcium Excretion

Without PTH, hypoparathyroid patients are unable to conserve normally the filtered calcium load at the renal tubule, predisposing them to hypercalciuria and renal damage. Nephrocalcinosis, nephrolithiasis, and renal failure can ensue [17]. Recent data suggest that nearly half of hypoparathyroid subjects have eGFR levels that are consistent with stage 3 CKD or worse [17], along with a nearly fourfold risk of renal complications [18]. Preliminary data suggest that PTH(1–84) might address this issue. In the Sikjaer study, urinary calcium excretion initially increased, possibly due to the increased filtered calcium load, but after week 12 was not different from baseline levels [8]. In the NPS RCT, there was no change in urinary calcium excretion [9], while in the Columbia cohort there was a decrease at 3 but not at 4 years [10]. Most likely, a dosing regimen which maximizes the exposure of PTH to the renal tubule and thus enhances renal calcium absorption, without increasing the filtered calcium load, would optimally address the hypercalciuria.

31.6 Effects of PTH(1–84) on Extraskeletal Calcifications

In each of the PTH(1–84) studies, serum phosphate levels fell, but the calcium-phosphate product did not decrease [8–10]. It remains to be seen whether the fall in the serum phosphate, with a possible reordering of the calcium-phosphate product, will have a beneficial effect on extraskeletal calcifications.

31.7 Effects of PTH(1–84) on Bone Turnover

Biochemical markers of bone turnover initially increase dramatically with PTH(1–84) treatment [8, 10]. Sikjaer et al. found that biochemical markers of bone turnover rose dramatically (P1NP by 1,315 % and s-CTx increased by 1,209 %), although the levels of osteocalcin and s-CTx appeared to plateau between 20 and 24 weeks [8]. Similarly, in the 27 subjects treated in the Columbia study with PTH (1–84) for 4 years, bone turnover markers increased significantly, reaching a threefold peak from baseline values at 6–12 months and subsequently declining to steady-state levels at 30 months, with P1NP and tartrate-resistant acid phosphatase (TRAP) remaining statistically higher than at baseline [10] (Fig. 31.3). Taken together, these data suggest that PTH has an initial exuberant effect to increase biochemical markers of bone turnover, with subsequent tempering over time to a new, steady-state, more euparathyroid level.

31.8 Effects of PTH(1–84) Treatment on Bone Mineral Density (BMD)

When PTH(1–84) was given by Sikjaer et al. to adults at 100 μg/d for 6 months, BMD decreased at the whole body, spine, hip, and femoral neck, but not at the forearm [8]; the BMD decreases correlated with the increases in biochemical markers of bone turnover [8]. Quantitative computed tomography (QCT) analysis of this cohort showed that vBMD in cancellous bone increased, despite the decrease in a BMD at the lumbar spine, while cortical vBMD decreased [8]. These data suggest that the relative distribution of trabecular and cortical bone might differ with PTH treatment at specific skeletal sites. In the 4-year Columbia treatment study of PTH(1–84), lumbar spine BMD increased by 5.5 %, while the femoral neck and total hip BMD remained stable; the distal radial BMD decreased, but at 4 years was not different from baseline [10] (see also Chap. 26).

31.9 Effects of PTH(1–84) Treatment on Histomorphometric Indices

Iliac crest bone biopsies were performed by Sikjaer et al. in 51 patients in the 6-month RCT of treatment with PTH(1–84) (PTH group $n=26$; placebo group $n=25$) [19]. MicroCT analysis demonstrated lower trabecular thickness with PTH treatment, with an increase in the bone surface and the presence of a more complex trabecular network, suggesting the development of thinner and better connected trabeculae [19]. Intratrabecular tunneling, or the longitudinal splitting of single trabeculae into two thinner new

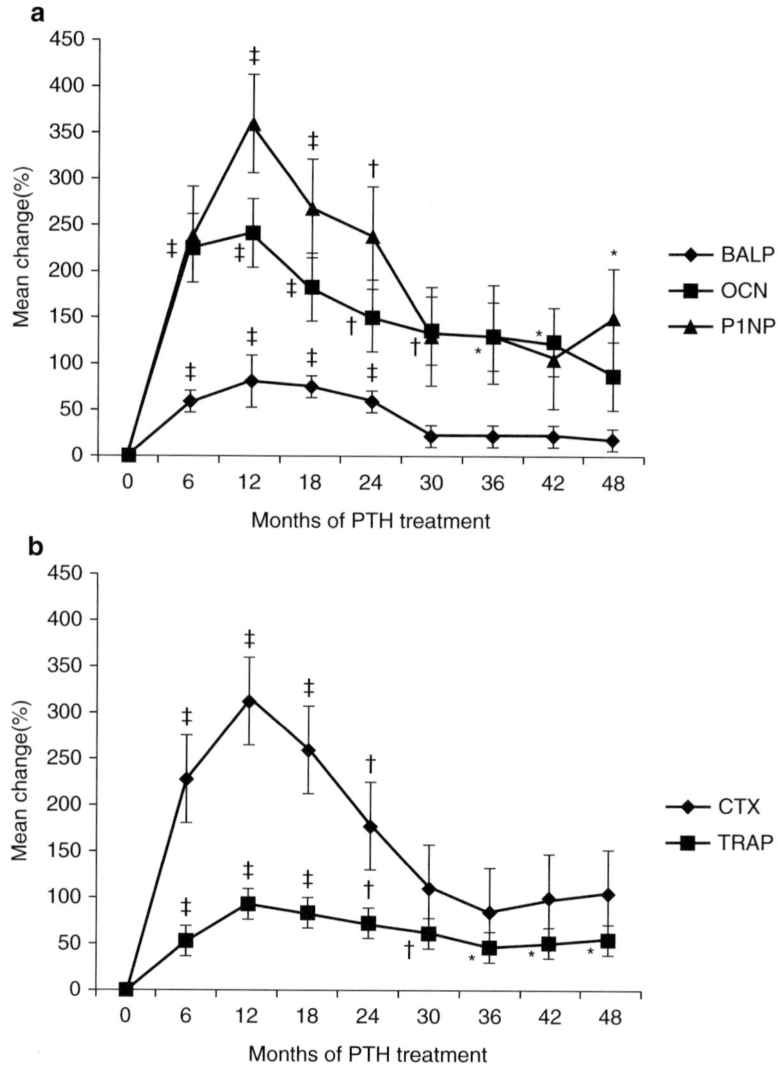

Fig. 31.3 Changes in markers of bone formation (P1NP, BALP, OCN; **a**) and resorption (CTX, TRAP; **b**) over 4 years of PTH(1–84). With PTH(1–84) treatment, all bone turnover markers increased significantly, peaking at levels of up to threefold above baseline values at 6–12 months and subsequently declining to steady-state levels at 30 months. *BALP* bone-specific alkaline phosphatase, *OCN* N-mid osteocalcin, *CTX* collagen type 1 cross-linked C-telopeptide. Data are expressed as mean ± SE. *, $P<0.05$ compared with baseline; †, $P<0.01$ compared with baseline; ‡, $P<0.0001$ compared with baseline (Reproduced with permission from Cusano et al. [10])

trabeculae (Fig. 31.4), was observed in the PTH(1–84) group. The presence of intratrabecular tunneling was associated with greater calcium mobilization, as evidenced by higher bone turnover and a tendency toward a greater decrease in calcium and active vitamin D supplementation [8, 19]. With regard to cortical bone, at 6 months, more Haversian canals per unit area were observed, with a trend toward increased cortical porosity, although cortical bone tissue density was not different [19](see also Chap. 26).

In a 2-year study of open-label PTH(1–84) treatment in the Columbia cohort, paired iliac crest bone biopsies were obtained before and after PTH(1–84) treatment at 1 year ($n=14$) and at 2 years ($n=16$); a separate group had an early "quadruple-label" biopsy [20] at 3 months ($n=16$) [21]. An immediate anabolic effect was apparent, with an early increase in the mineralizing surface (MS), osteoid surface, and bone formation rate at 3 months (MS at baseline: 0.39 ± 0.6 % vs. MS at 3 months: 5.47 ± 6.0 %; $p=0.004$), which peaked at 12 months (MS at baseline: 0.7 ± 0.6 % vs. MS at 1 year: 7.1 ± 6.0 %, $p=0.001$) and was similar to euparathyroid levels at 2 years (MS at baseline: 1.18 ± 2.2 % vs. MS at 2 years: 3.34 ± 0.8 %; $p=0.04$; MS in healthy controls: 4.33 ± 3.2 %). The remodeling changes were most pronounced in

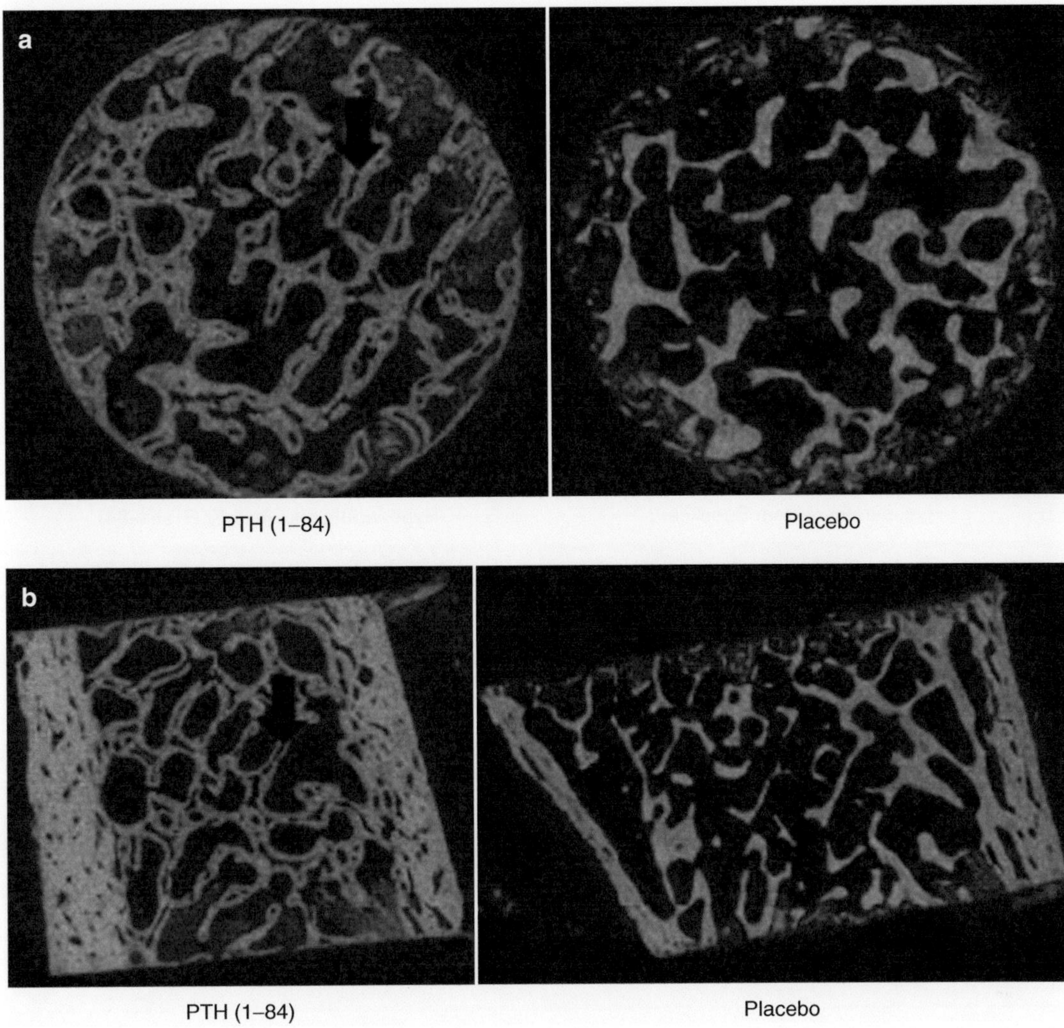

Fig. 31.4 Iliac crest biopsies, one with intratrabecular tunneling from a patient treated with PTH(1–84) 100 mcg/day for 24 weeks and one without tunneling from a placebo-treated patient. (**a**) Cross-sectional view; (**b**) longitudinal sectional view (Reproduced with permission from Ref. [19])

the cancellous envelope at 1 year; within 2 years, with the exception of osteoid surface, the differences were no longer significant at the endocortical and intracortical envelopes [21]. Structural changes after 2 years of PTH(1–84) treatment included reduced trabecular width (144±34 to 128±34 μm, $p=0.03$) and increases in trabecular number (1.74±0.34 to 2.07±0.50/mm, $p=0.02$). As in the study of Sikjaer et al., intratrabecular tunneling was apparent (Fig. 31.5). Cortical porosity increased at 2 years (7.4±3.2 % to 9.2±2.4 %, $p=0.03$), although cortical width did not change. Recent data employing longitudinal 3-D analysis of the biopsies by microcomputed tomography (microCT) confirm that the microstructural changes, including decreased trabecular thickness and increased connectivity density, occur relatively early with PTH treatment and are detectable to a greater extent at 1 than at 2 years [22]. Overall, the histomorphometric data suggest that administration of PTH improves abnormal dynamic and structural skeletal properties in hypoparathyroidism, restoring bone metabolism toward normal euparathyroid levels.

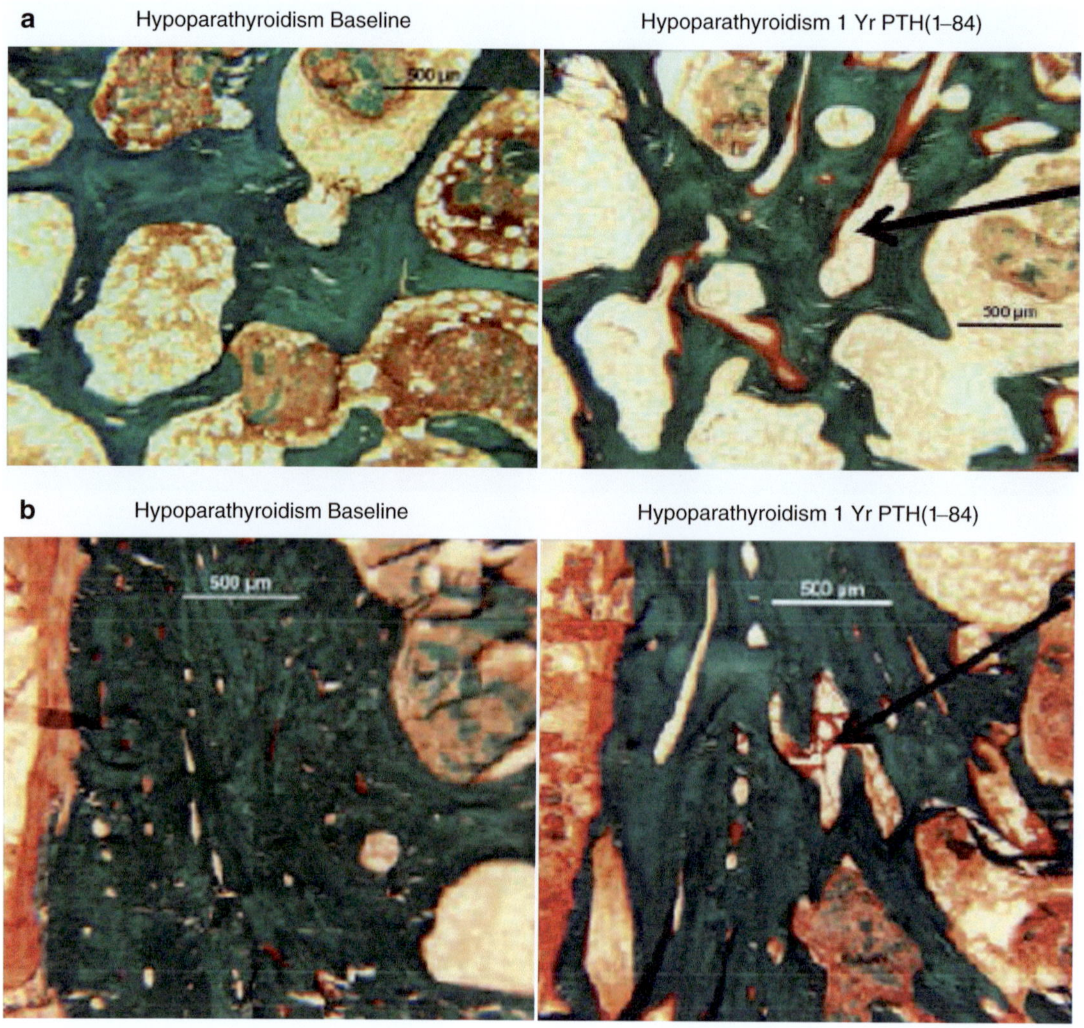

Fig. 31.5 Iliac crest biopsy illustrating changes in trabecular (**a**) and cortical (**b**) structure before and after 1 year of PTH(1–84) treatment in a hypoparathyroid subject. Note the increases in trabecular tunneling and cortical porosity (*arrows*) in the posttreatment biopsy (Reproduced with permission from Rubin et al. [21])

31.10 Effects of PTH(1–84) Treatment on Fractures

Data are not available on the effects of PTH(1–84) treatment in hypoparathyroidism on fracture risk. Given the well-characterized improvements in skeletal properties, a decrease in fracture risk would be anticipated, but this expectation awaits confirmation from larger and longer studies.

31.11 Safety Concerns: Hypercalcemia with PTH (1–84) Treatment

Hypercalcemia occurred, albeit with low frequency, in the three PTH(1–84) studies. In the Sikjaer study, 11 patients had a total of 17 episodes of symptomatic hypercalcemia [8]. In the NPS study, there was 1 hospitalization because of hypercalcemia [9], while in the Columbia cohort there were 11 episodes of hypercalcemia in 8 subjects over 4 years [10].

31.12 Safety Concerns: Osteosarcoma Risk with PTH(1–84) Treatment

PTH(1–84) was found to increase osteosarcoma risk in rats [23]. However, the risk is dose and duration related, and the noncarcinogenic doses for PTH(1–84) (10 μg/kg/d) in the studies in rats are markedly above that used to treat hypoparathyroidism in humans [24]. Most reassuring of all, no increased risk of osteosarcoma has emerged, since recombinant human PTH(1–34) (Forteo) was approved in 2002 [25] (see also Chap. 26).

Conclusion

In a disorder characterized by absent or inadequate parathyroid function, accumulating data suggest that PTH(1–84) is able to address many of the biochemical, renal, skeletal, and neuropsychological features of hypoparathyroidism to a greater extent than conventional treatment. One looks forward to a time when the missing hormone, namely, PTH(1–84), will become the standard option for therapy of this disease.

References

1. Bilezikian JP, Khan A, Potts JT Jr et al (2011) Hypoparathyroidism in the adult: epidemiology, diagnosis, pathophysiology, target-organ involvement, treatment, and challenges for future research. J Bone Miner Res 26(10):2317–2337
2. Powers J, Joy K, Ruscio A, Lagast H (2013) Prevalence and incidence of hypoparathyroidism in the United States using a large claims database. J Bone Miner Res 28(12):2570–2576
3. Winer KK, Yanovski JA, Cutler GB Jr (1996) Synthetic human parathyroid hormone 1–34 vs calcitriol and calcium in the treatment of hypoparathyroidism. JAMA 276(8):631–636
4. Winer KK, Yanovski JA, Sarani B, Cutler GB Jr (1998) A randomized, cross-over trial of once-daily versus twice-daily parathyroid hormone 1–34 in treatment of hypoparathyroidism. J Clin Endocrinol Metab 83(10):3480–3486
5. Winer KK, Ko CW, Reynolds JC et al (2003) Long-term treatment of hypoparathyroidism: a randomized controlled study comparing parathyroid hormone-(1–34) versus calcitriol and calcium. J Clin Endocrinol Metab 88(9):4214–4220
6. Winer KK, Sinaii N, Peterson D, Sainz B Jr, Cutler GB Jr (2008) Effects of once versus twice-daily parathyroid hormone 1–34 therapy in children with hypoparathyroidism. J Clin Endocrinol Metab 93(9):3389–3395
7. Winer KK, Sinaii N, Reynolds J, Peterson D, Dowdy K, Cutler GB Jr (2010) Long-term treatment of 12 children with chronic hypoparathyroidism: a randomized trial comparing synthetic human parathyroid hormone 1–34 versus calcitriol and calcium. J Clin Endocrinol Metab 95(6):2680–2688
8. Sikjaer T, Rejnmark L, Rolighed L, Heickendorff L, Mosekilde L (2011) The effect of adding PTH(1–84) to conventional treatment of hypoparathyroidism: a randomized, placebo-controlled study. J Bone Miner Res 26(10):2358–2370
9. Mannstadt M, Clarke BL, Vokes T et al (2013) Efficacy and safety of recombinant human parathyroid hormone (1–84) in hypoparathyroidism (REPLACE): a double-blind, placebo-controlled, randomised, phase 3 study. Lancet Diabetes Endocrinol 1(4):275–283
10. Cusano NE, Rubin MR, McMahon DJ et al (2013) Therapy of hypoparathyroidism with PTH(1–84): a prospective four-year investigation of efficacy and safety. J Clin Endocrinol Metab 98(1):137–144
11. Product Monograph: Forteo (2004) Eli Lilly, Canada
12. Sikjaer T, Rejnmark L, Mosekilde L (2011) PTH treatment in hypoparathyroidism. Curr Drug Saf 6(2):89–99
13. Fox J, Wells D, Garceau R (2011) Relationships between pharmacokinetic profile of human PTH(1–84) and serum calcium response in postmenopausal women following 4 different methods of administration. J Bone Miner Res 26(Suppl 1)
14. Sikjaer T, Amstrup AK, Rolighed L, Kjaer SG, Mosekilde L, Rejnmark L (2013) PTH(1–84) replacement therapy in hypoparathyroidism: a randomized controlled trial on pharmacokinetic and dynamic effects after 6 months of treatment. J Bone Miner Res 28(10):2232–2243
15. Cusano NE, Rubin MR, McMahon DJ et al (2013) The effect of PTH(1–84) on quality of life in hypoparathyroidism. J Clin Endocrinol Metab 98(6):2356–2361
16. Sikjaer T, Rolighed L, Hess A, Fuglsang-Frederiksen A, Mosekilde L, Rejnmark L (2014) Effects of PTH(1–84) therapy on muscle function and quality of life in hypoparathyroidism: results from a randomized controlled trial. Osteoporos Int 25(6):1717–1726
17. Mitchell DM, Regan S, Cooley MR et al (2012) Long-term follow-up of patients with hypoparathyroidism. J Clin Endocrinol Metab 97(12):4507–4514
18. Underbjerg L, Sikjaer T, Mosekilde L, Rejnmark L (2013) Cardiovascular and renal complications to postsurgical hypoparathyroidism: a Danish nationwide controlled historic follow-up study. J Bone Miner Res 28(11):2277–2285

19. Sikjaer T, Rejnmark L, Thomsen JS et al (2012) Changes in 3-dimensional bone structure indices in hypoparathyroid patients treated with PTH(1–84): a randomized controlled study. J Bone Miner Res 27(4):781–788
20. Lindsay R, Cosman F, Zhou H et al (2006) A novel tetracycline labeling schedule for longitudinal evaluation of the short-term effects of anabolic therapy with a single iliac crest bone biopsy: early actions of teriparatide. J Bone Miner Res 21(3):366–373
21. Rubin MR, Dempster DW, Sliney J Jr et al (2011) PTH(1–84) administration reverses abnormal bone-remodeling dynamics and structure in hypoparathyroidism. J Bone Miner Res 26(11):2727–2736
22. Rubin M, Zwahlen A, Dempster D et al (2013) PTH(1–84) administration alters three dimensional cancellous bone structure in hypoparathyroidism. J Bone Miner Res (Supp 1). http://www.asbmr.org/asbmr-2013-abstract-detail?aid=ed685e42-6df2-47f7-8955-5fd09bc92a3b. Accessed Oct 2013
23. Watanabe A, Yoneyama S, Nakajima M et al (2012) Osteosarcoma in Sprague-Dawley rats after long-term treatment with teriparatide (human parathyroid hormone (1–34)). J Toxicol Sci 37(3):617–629
24. Jolette J, Wilker CE, Smith SY et al (2006) Defining a noncarcinogenic dose of recombinant human parathyroid hormone 1–84 in a 2-year study in Fischer 344 rats. Toxicol Pathol 34(7):929–940
25. Andrews EB, Gilsenan AW, Midkiff K et al (2012) The US postmarketing surveillance study of adult osteosarcoma and teriparatide: study design and findings from the first 7 years. J Bone Miner Res 27(12):2429–2437

Part III

Functional Hypoparathyroidism

Classification of Pseudohypoparathyroidism and Differential Diagnosis

32

Giovanna Mantovani and Francesca M. Elli

32.1 Introduction

Pseudohypoparathyroidism was described in 1942 by Fuller Albright and colleagues, as they reported patients with normal renal function in whom significantly reduced levels of plasma calcium with hyperphosphatemia were associated with parathyroid overactivity. This contrasted with patients affected with primary hypoparathyroidism, in whom low-normal or frankly low PTH levels are insufficient to maintain normocalcemia. Due to clinical and biochemical similarities with hypoparathyroidism, the disease was called Pseudohypoparathyroidism (PHP) and the presence of target tissue resistance (in proximal renal tubules) to PTH action was hypothesized as the pathogenetic mechanism.

Moreover, these individuals presented with a constellation of specific somatic and developmental abnormalities referred to as Albright's hereditary osteodystrophy (AHO) that included short stature, centripetal obesity, rounded face, short neck, short and low-set nasal bridge, and brachydactyly due to early closure of the epiphyses with resultant shortening of one or more metacarpals or metatarsals [1]. Further reports also described ectopic subcutaneous calcifications (now better defined as true ossifications) and cognitive abnormalities of varying degrees, from learning disabilities to severe mental retardation, as additional features found in the majority of AHO cases [2, 3]. Subsequent studies demonstrated that there was no urinary cyclic AMP (cAMP) generation and reduced calcemic and phosphaturic responses in PHP patients after the administration of parathyroid tissue extracts, thus confirming that PTH resistance was indeed the underlying defect [4, 5]. A decade after their first report of PHP, Albright and colleagues reported patients showing all physical features of AHO without any evidence of PTH resistance and termed this new syndrome pseudopseudohypoparathyroidism (PPHP) [6]. As more cases came to be described, it appeared that this disease might present either as a sporadic or as a familial defect and that in familial cases it was inherited in an autosomal dominant manner [7–9].

The identification of the PTH receptor and its signal transduction cascade enhanced our understanding of PHP pathophysiology and the underlying molecular defect, i.e., inactivating mutations in the gene encoding for the alpha subunit of the stimulatory G protein (Gsα), now known as *GNAS* [10–15]. GNAS is a complex imprinted locus mapping to chromosome 20q13.2–13.3 coding for several different transcripts besides Gsα (see Chap. 10) [16].

G. Mantovani, MD, PhD (✉) • F.M. Elli
Endocrinology and Diabetology Unit, Department of Clinical Sciences and Community Health, University of Milan, Fondazione IRCCS Ca' Granda Ospedale Maggiore Policlinico, Via F. Sforza, 35, Milan 20122, Italy
e-mail: giovanna.mantovani@unimi.it; francesca.elli@unimi.it

M.L. Brandi, E.M. Brown (eds.), *Hypoparathyroidism*,
DOI 10.1007/978-88-470-5376-2_32, © Springer-Verlag Italia 2015

Several variants of this disorder have been defined since then, and different forms have been classified based on the absence or presence of AHO, as well as of resistance to other hormones in addition to PTH. Indeed, the term PHP now encompasses a heterogeneous group of rare, related metabolic disorders with proven genetic components. The most frequently encountered variants of PHP, caused by molecular alterations within or upstream of the GNAS locus, include PHP type Ia (PHP-Ia), pseudopseudohypoparathyroidism (PPHP), and PHP type Ib (PHP-Ib).

Patients with PHP-Ia have features of AHO and present with hypocalcemia and hyperphosphatemia despite elevated serum PTH levels. Hormone resistance is usually not limited to PTH, as affected individuals may show evidence also for resistance to thyroid-stimulating hormone (TSH), gonadotropins, growth hormone-releasing hormone (GHRH), and calcitonin, even if the frequencies of these abnormalities remain poorly defined so far [17–24]. Most PHP-Ia carry heterozygous maternally derived mutations in *GNAS* exons 1–13 and show a partial deficiency (about 50 %) of Gs alpha activity in red blood cells [10].

In contrast, patients with PPHP have the typical features of AHO but do not show evidence for resistance to PTH or other hormones. PPHP is also caused by heterozygous inactivating mutations in Gs alpha coding exons, but the phenotype (AHO only) is associated with paternal transmission of the genetic defect. As two sides of the same coin, PHP-Ia and PPHP can be found in the same kindred but not in the same sibship, because clinical features result from the gender of the parent transmitting the mutation, thus reflecting the imprinted nature of *GNAS* [13].

On the other hand, the majority of PHP-Ib patients present with signs and symptoms of PTH resistance but lack features of AHO, and hormone resistance seems to be confined to the renal actions of PTH. Only recently, mild TSH resistance has been observed in some PHP-Ib patients, which raises the possibility that additional endocrine systems may also be affected in this PHP variant. Relatively recent studies showed a normal/slightly reduced Gs alpha activity in erythrocytes and fibroblasts and identified as underlying molecular cause methylation defects in the imprinted GNAS cluster [25–29].

During the last decade, incoming data on both clinical and molecular aspects of these complex disorders have challenged the distinction between different *GNAS*-related diseases. In particular, in a subset of patients with PHP and variable degrees of AHO, GNAS epigenetic defects similar to those classically found in PHP-Ib patients have been detected by independent groups, suggesting a molecular overlap between PHP-Ia and PHP-Ib. These findings confirm the complexity in establishing an accurate diagnosis of PHP, as sometimes clinical and molecular data do not fully discriminate between the main variants of the disease [30–34] (Table 32.1).

32.2 Pathophysiology of PTH Resistance/Molecular Basis

Deficiency of a given hormone leads to decreased activity at its target organs, often leading to clinical manifestations. Similar abnormalities may arise when a target organ becomes nonresponsive to the hormone itself, despite sufficient or even elevated hormone levels. Endocrine disorders deriving from target organ resistance are mostly caused by inactivating mutations affecting hormone receptors. PHP is therefore an unusual form of hormone resistance as the molecular defect affects a downstream effector of PTH, the Gs alpha protein (alpha subunit of the stimulatory guanine nucleotide-binding protein) [35–39].

Heterotrimeric guanine nucleotide-binding proteins, called G proteins, are a superfamily of heterotrimers composed of three distinct subunits (alpha, beta, and gamma), which are involved in the intracellular signal transduction from seven transmembrane receptors, called G-protein-coupled receptors (GPCRs), and their functional specificity depends on the alpha subunit. The interaction between the agonist and its specific GPCR activates alpha subunit-triggered effectors, enzymes, and ion channels that induce both short-term effects on hormone secretion, neurotransmission, and muscle contraction and long-term effects on gene transcription [40].

Table 32.1 PHP diagnostic criteria

Laboratory findings
Major:
Hypocalcemia, hyperphosphatemia, and raised serum PTH levels in the absence of vitamin D deficiency (PTH resistance)
Additional:
Raised serum TSH levels, in the absence of antithyroid antibodies and in the presence of normal thyroid scan (TSH resistance)
Elevated LH and FSH levels, together with low estradiol/testosterone levels (resistance to gonadotropins)
Blunted GH response to provocative tests (GHRH resistance)
Clinical findings
Major (associated with acute or chronic hypocalcemia):
Nervous hyperexcitability with paresthesias, cramps, tetany, hyperreflexia, convulsions, and tetanic crisis
Cataracts
Basal ganglia calcifications
Additional:
Secondary amenorrhea and/or infertility
Reduced growth velocity (in children)
AHO manifestations (at least brachydactyly and/or heterotopic ossifications are required for the definition of AHO):
Brachydactyly (shortening of fourth and/or fifth metacarpals defined as the metacarpal sign and/or shortening below -2SDS at the metacarpophalangeal profile pattern in at least one metacarpal bone or distal phalanx)
Ectopic ossifications (either clinically evident or at X-ray)
Short stature (height below the third percentile for chronological age)
Obesity (BMI >30 kg/m^2 in adults and >97th centile in children)
Round face
Mental retardation, defined in case of history of delayed motor and/or speech milestones or need of extra help in preschool or mainstream school

Characterization of the molecular bases of PHP began with the observation that cAMP is the mediator of various PTH actions in different cell types and organs. PTH is the primary regulator of serum calcium, mainly acting on kidney and bone via its Gs alpha-coupled receptor PTHR1 [41, 42]. The alpha subunit of Gs is a primary modulator of the adenylyl cyclase, promoting the intracellular formation of cyclic adenosine monophosphate (cAMP), which rapidly activates protein kinase A (PKA), and a cascade of intracellular responses (Fig. 32.1). At the kidney level, PTH enhances active reabsorption of calcium and magnesium from the distal tubules and the thick ascending limb. It also decreases the reabsorption of phosphate from the proximal tubule, further increasing free calcium in the circulation [43]. In the intestine it promotes the absorption of calcium (as Ca^{2+} ions) via TRPV6, an apical Ca^{2+} channel, calbindin, a putative ferry to the basolateral membrane, and membrane pumps and transporters on the latter membrane, all of which are stimulated by activated vitamin D. The concentration of this metabolite in blood is, by necessity, tightly regulated, and the most important stimulus for renal 1,25-(OH)2D synthesis is PTH through the upregulation of 25-hydroxyvitamin D 1-alpha-hydroxylase mRNA and protein. Moreover, PTH indirectly enhances the release of calcium from bone by osteoblast-mediated stimulation of bone resorption by osteoclasts [44].

As a result of PTH resistance, patients with PHP develop hypocalcemia and hyperphosphatemia have diminished serum concentrations of 1,25-(OH)2D, and exogenous administration of PTH fails to induce an appropriate increase in urinary phosphate and cAMP excretion when compared with normal controls or patients with other forms of hypoparathyroidism [5, 45, 46]. Because of the different sites of action of the anti-calciuric (thick ascending limb and distal convoluted tubule) and the phosphaturic (proximal tubule) effects of PTH, the anti-calciuric action of this hormone seems to remain intact in PHP patients. As a result of these unimpaired functions, PHP patients may present with prolonged periods of normocalcemia maintained by an elevated PTH concentration, interrupted by episodes of hypocalcemia with associated clinical manifestations (i.e., seizures or muscle spasm), and normo-/hypocalciuria with, however, conserved renal handling of calcium in the absence of kidney stones. This cell-specific defect in PHP is in accordance with the demonstration of the cell-specific imprinting of GNAS. The extent to which PTH signaling in bone, which is

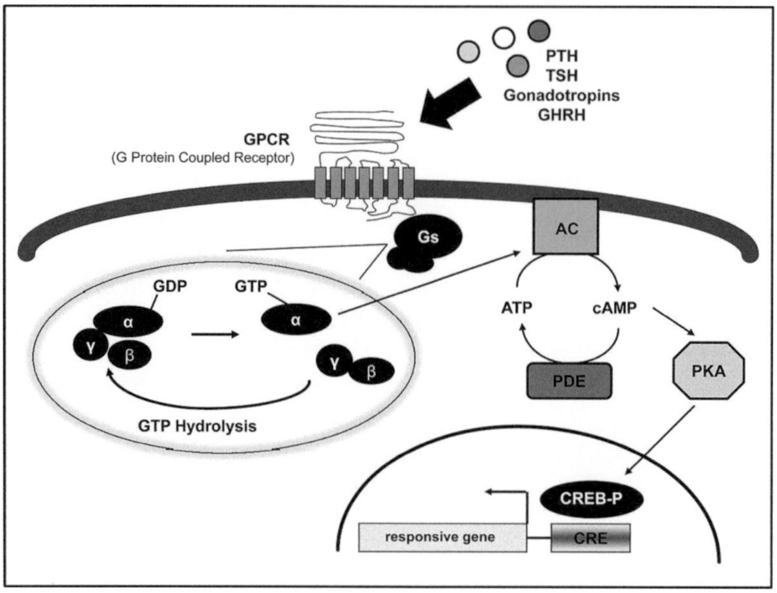

Fig. 32.1 The hormone-activated cAMP intracellular transduction pathway

independent of vitamin D action, is defective in PHP patients is less clear, as bone remodeling in response to PTH appears to be intact in most patients, although clinical manifestations in the skeleton, ranging from decreased bone density (BMD) to overt osteitis fibrosa cystica to osteosclerosis, have been reported [47–54].

32.3 Classification and Differential Diagnosis

The distinction between PHP and primary hypoparathyroidism is usually easy, given the significantly elevated PTH levels that characterize PHP. Moreover, unlike patients with primary hypoparathyroidism, PHP patients are not prone to hypercalciuria, thus maintaining normal renal calcium handling and indicating that the anticalciuric action of PTH in the thick ascending limb and DCT is unaffected. On the contrary, the differential diagnosis among the various forms of the disorders is sometimes challenging, as outlined below.

32.4 PHP Type I Versus Type II

Since the first description of PHP, different clinical variants of this disorder have been reported (see also Chaps. 33, 34 and 35), and today PHP is divided into two main subtypes according to the response to the administration of exogenous PTH. PTH infusion remains the most consistent test for the distinction between the two forms, as PHP type I patients show blunted nephrogenous cAMP generation and phosphate excretion following this administration [1, 5, 18, 45, 46]. In contrast, PHP type II patients retain a normal generation of nephrogenous cAMP in response to PTH, with impaired urinary excretion of phosphate. Only few cases of PHP type II have been described, and the molecular determinant underlying this PHP variant is still elusive, although the observed deficient phosphaturic response suggests a defect distal to cAMP generation in the PTH-mediated transduction pathway in target cells [55]. It is now believed that in most of these cases, PHP-II may be an acquired defect secondary to vitamin D deficiency, as calcium and vitamin D replacement often normalizes the phosphaturic response to PTH in these subjects.

32.5 PHP Type I Subtypes

PHP type I, besides PTH resistance associated with blunted cAMP and phosphaturic responses to exogenous PTH, is classically further differentiated according to the presence (PHP-Ia, MIM 103580; PHP-Ic, MIM 612462) or absence (PHP-Ib, MIM 603233) of AHO features (Table 32.2).

32.5.1 PHP-Ia, PPHP, and PHP-Ic

In PHP-Ia, PTH resistance usually develops over the first years of life, with hyperphosphatemia and elevated PTH generally preceding hypocalcemia, even if some patients remain normocalcemic throughout life [17]. Typically, in PHP-Ia patients hormone resistance is not restricted to PTH, as they also display resistance to other hormones that act via GPCRs, such as TSH, gonadotropins, and GHRH. Resistance to these additional hormones may develop with interindividual variability in severity and time course [11, 17, 56, 57]. Indeed, most PHP-Ia patients become clinically resistant to TSH over childhood or adolescence, but hypothyroidism may be detected at neonatal screening as well [58–61]. Generally, TSH resistance is mild, with normal or slightly low thyroid hormone levels, no goiter, and absence of antithyroid antibodies.

Clinical evidence of hypogonadism, particularly in females, is usually manifested as delayed or incomplete sexual maturation, amenorrhea or oligomenorrhea, and/or infertility. Laboratory findings showed that PHP-Ia women are slightly hypoestrogenic, but the relation with increased basal or GnRH-stimulated levels of circulating gonadotropins is still unclear [62]. More recently, Mantovani and Germain-Lee reported deficiency in GH secretion due to resistance to GHRH in a large subset of these patients [63–65]. Although the mechanism is unknown, prolactin deficiency has also been documented in some PHP-Ia patients. In PHP-Ia patients plasma cAMP responses to glucagon and isoproterenol can also be reduced, but the rise in serum glucose is normal, suggesting that even a blunted cAMP response is able to induce the physiological response. Moreover, patients are not resistant to vasopressin, ACTH, or CRH, whereas resistance to calcitonin has been occasionally described [24, 63, 66–68]. Although conclusive data are lacking, a recent report described normal region-specific bone mineral density (BMD) together with increased total body BMD in a quite large series of patients, with consequent normal or even reduced risk of fracture, confirming the apparently intact skeletal responsiveness to PTH in PHP-Ia and PHP-Ib [69].

As previously stated, PPHP patients may coexist with PHP-Ia within the same family but never in the same sibship and, similarly to their relatives with PHP-Ia, have an approximately 50 % deficiency in Gs activity in cell membranes, but show a normal response of urinary cAMP to exogenous PTH [70, 71]. Clinical features of PPHP can also be found in sporadic cases, making the diagnosis more difficult. In fact, some of the typical AHO features, especially when

Table 32.2 PHP classification

PHP subtype	AHO	Hormone resistance	PTH infusion	Gs activity	GNAS defect
Type Ia	Yes	Yes	No response	Reduced	Genetic (maternal transmission)
PPHP	Yes	No	Normal response	Reduced	Genetic (paternal transmission)
Type Ib	No	Yes	No response	Normal/slightly reduced	Epigenetic
Type Ic	Yes	Yes	No response	Normal	Few cases reported

considered individually, may be present within the normal population as well as in other disorders, some of which ascribed to specific chromosomal defects such as deletions involving chromosome 2q37 associated with the brachydactyly-mental retardation syndrome (BDMR) [72, 73]. Moreover, in clinical practice AHO is often difficult to diagnose as some clinical features are not obvious at birth or shortly after and may be very heterogeneous later. PHP-Ia and PPHP are caused by heterozygous inactivating mutations located within *GNAS* coding exons inherited from the mother or the father, respectively, and this pattern of inheritance reflects the tissue-specific imprinting of GNAS.

In 1990, Patten and colleagues identified the first GNAS mutation responsible for PHP-Ia, and since then numerous different mutations (about 130 unique mutations reported in the literature) distributed throughout the entire gene have been identified [15, 74]. Missense mutations and small insertions/deletions predominate, but nonsense mutations, splice mutations, and macrodeletions have also been documented. Most of these genetic defects are private mutations, confined to one or few patients, and the observation of few recurring mutations in unrelated patients suggests that the presence of identical de novo mutations probably derives from the presence of a common molecular mechanism rather than a founder effect. No genotype-phenotype correlations have been observed, as neither the mutation type nor its location correlate with the onset of the disease (as marker of disease precocity and severity), with the severity of endocrine resistances, or with the number of AHO signs [74]. The detection of a *GNAS* mutation is associated with 50 % risk of recurrence, thus providing the possibility of predictive genetic testing in relatives, as well as prenatal diagnosis. In case of negative *GNAS* mutational screening, patients should be also tested for the presence of GNAS imprinting defects and *GNAS* macrodeletions. Screening of mutations in the *PRKAR1A* and *PDE4D* genes may be also considered in those cases showing a phenotype suggestive of acrodysostosis.

From the clinical point of view, PHP-Ic is indistinguishable from PHP-Ia, being characterized by the presence of multi-hormone resistance and AHO. Nevertheless, it is possible to differentiate between these two PHP variants by means of Gs activity measurements in the membranes of various cell types (erythrocytes, fibroblasts, platelets): the partial deficiency (about 50 %) demonstrated in patients with PHP-Ia is usually absent in patients with PHP-Ic although exceptions to these rules have been reported [75, 76].

32.5.2 PHP-Ib

PHP-Ib is typically characterized by renal resistance to PTH in the absence of other endocrine or physical abnormalities and a normal/slightly reduced Gs activity in red blood cells [77–80]. Recent studies have also reported resistance to the action of TSH, with conserved GH secretion in a large subset of patients [20, 29]. In all PHP-Ib cases, a defect in the signaling pathway proximal to cAMP generation is documented by a blunted urinary cAMP response to exogenous PTH. For this reason, originally, the PTH receptor type 1 (PTHR1) gene was suggested as a candidate to explain the molecular basis of the disease. We now know that PHP-Ib is associated with methylation alterations affecting GNAS differentially methylated regions (DMRs), streches of DNA showing an allele-specific methylation pattern according to the parental inheritance. As in PHP-Ia, hormonal resistance develops only after maternal inheritance of the disease, whereas paternal inheritance of the same defect is not associated with endocrine abnormalities [27, 28, 81].

These imprinting defects are often sporadic (spor-PHP-Ib), but the disease may occasionally present as familial, with an autosomal dominant, maternally inherited pattern of transmission, known as AD-PHP-Ib. The molecular findings in the two forms are different and will be specifically described in another chapter [82–87] Nevertheless, it is important to note that, despite different underlying pathogenetic mechanisms, no clinical differences have been observed between the sporadic and the familial forms of the disease [88]. Finally, a partial clinical overlap between PHP type Ia and type Ib has been recently demonstrated, as patients with GNAS imprinting defects may occasionally present with variable degrees of AHO. These findings indicate the need to investigate GNAS methylation status in all PHP patients with associated AHO signs who are found to be negative for classical *GNAS* mutations [76].

32.6 AHO Versus AHO-Like Syndromes

The term AHO encompasses heterogeneous and nonspecific clinical findings, including rounded face, short stature, central obesity, and variable degrees of mental retardation; accordingly, brachydactyly and heterotopic ossifications are considered as the most specific features of AHO phenotype. However, the diagnosis may remain unclear in the absence of a specific molecular diagnosis, because of the occasional detection within the normal population of some AHO features (i.e., rounded face, obesity, or shortening of hand bones as well). Moreover, most of AHO features may be detected in other genetically determined diseases outlined thereafter. Mental retardation is included among the clinical characteristics of AHO, despite the fact that its frequency and severity are not well established, with an apparent discrepancy between the adult (27 %) and the pediatric populations (64 %) [76]. It is well known that mental retardation is included in a wide variety of genetic syndromes.

Recently, mutations in *PRKAR1A* and *PDE4D* genes, both encoding proteins crucial for cAMP-mediated signaling, have been detected in a small subset of patients with PHP-Ia or PPHP features, showing a phenotypic overlap with acrodysostosis (ACRDYS). Acrodysostosis is a rare congenital malformation syndrome characterized by skeletal dysplasia presenting with short stature, severe brachydactyly, facial dysostosis, nasal hypoplasia, and often advanced bone age and obesity. In some patients, laboratory findings show resistance to multiple hormones (including PTH, TSH, calcitonin, GHRH, and gonadotropins) [89, 90].

Finally, brachydactyly-mental retardation syndrome (BDMR) is a contiguous gene syndrome, as most patients described to date display a large deletion of chromosome 2q37.2. Patients show a phenotype resembling the physical anomalies found in AHO, such as short stature, mental retardation, and brachymetaphalangia, but classically no soft tissue ossifications and no abnormalities in calcium metabolism are observed [72]. In conclusion, patients negative for GNAS defects with suggestive clinical presentations should be also screened for chromosomal regions and genes associated with diseases that undergo differential diagnosis with PHP.

References

1. Albright F, Burnett CH, Smith CH et al (1942) Pseudohypoparathyroidism: an example of "Seabright-Bantam syndrome". Endocrinology 30:922–932
2. Eyre WG, Reed WB (1971) Albright's hereditary osteodystrophy with cutaneous bone formation. Arch Dermatol 104:634–642
3. Farfel Z, Friedman E (1986) Mental deficiency in pseudohypoparathyroidism type I is associated with Ns-protein deficiency. Ann Intern Med 105(2): 197–199
4. Tashjian AH Jr, Frantz AG, Lee JB (1966) Pseudohypoparathyroidism: assays of parathyroid hormone and thyrocalcitonin. Proc Natl Acad Sci U S A 56(4):1138–1142
5. Chase LR, Melson GL, Aurbach GD (1969) Pseudohypoparathyroidism: defective excretion of 3',5'-AMP in response to parathyroid hormone. J Clin Invest 48(10):1832–1844
6. Albright F, Forbes AP, Henneman PH (1952) Pseudo-pseudohypoparathyroidism. Trans Assoc Am Physicians 65:337–350
7. Farfel Z, Brothers VM, Brickman AS et al (1981) Pseudohypoparathyroidism: inheritance of deficient receptor-cyclase coupling activity. Proc Natl Acad Sci U S A 78(5):3098–3102
8. Fitch N (1982) Albright's hereditary osteodystrophy: a review. Am J Med Genet 11(1):11–29
9. Weinberg AG, Stone RT (1971) Autosomal dominant inheritance in Albright's hereditary osteodystrophy. J Pediatr 79(6):996–999
10. Levine MA, Downs RW, Singer M et al (1980) Deficient activity of guanine nucleotide regulatory protein in erythrocytes from patients with pseudohypoparathyroidism. Biochem Biophys Res Commun 94:1319–1324
11. Levine MA, Downs RW Jr, Moses AM et al (1983) Resistance to multiple hormones in patients with pseudohypoparathyroidism. Association with deficient activity of guanine nucleotide regulatory protein. Am J Med 74(4):545–556

12. Weinstein LS, Gejman PV, Friedman E et al (1990) Mutations of the Gs alpha-subunit gene in Albright hereditary osteodystrophy detected by denaturing gradient gel electrophoresis. Proc Natl Acad Sci U S A 87(21):8287–8290
13. Davies SJ, Hughes HE (1993) Imprinting in Albright's hereditary osteodystrophy. J Med Genet 30(2): 101–103
14. Levine MA, Modi WS, O'Brien SJ (1991) Mapping of the gene encoding the alpha subunit of the stimulatory G protein of adenylyl cyclase (GNAS1) to 20q13.2–q13.3 in human by in situ hybridization. Genomics 11(2):478–479
15. Patten JL, Johns DR, Valle D et al (1990) Mutation in the gene encoding the stimulatory G protein of adenylate cyclase in Albright's hereditary osteodystrophy. N Engl J Med 322(20):1412–1419
16. Weinstein LS, Liu J, Sakamoto A et al (2004) Minireview: GNAS: normal and abnormal functions. Endocrinology 145(12):5459–5464
17. Weinstein LS, Yu S, Warner DR et al (2001) Endocrine manifestations of stimulatory G protein α-subunit mutations and the role of genomic imprinting. Endocr Rev 22:675–705
18. Levine MA (2002) Pseudohypoparathyroidism. In: Bilezikian JP, Raisz LG, Rodan GA (eds) Principles of bone biology. Academic, New York, pp 1137–1163
19. Shima M, Nose O, Shimizu K et al (1988) Multiple associated endocrine abnormalities in a patient with pseudohypoparathyroidism type 1a. Eur J Pediatr 147:536–538
20. Liu J, Erlichman B, Weinstein LS (2003) The stimulatory G protein α-subunit $G_s\alpha$ is imprinted in human thyroid glands: implications for thyroid function in pseudohypoparathyroidism types 1A and 1B. J Clin Endocrinol Metab 88:4336–4341
21. Kaji M, Umeda K, Ashida M et al (2001) A case of pseudohypoparathyroidism type la complicated with growth hormone deficiency: recovery of growth hormone secretion after vitamin D therapy. Eur J Pediatr 160:679–681
22. Scott DC, Hung W (1995) Pseudohypoparathyroidism type Ia and growth hormone deficiency in two siblings. J Pediatr Endocrinol Metab 8(3):205–207
23. Zwermann O, Piepkorn B, Engelbach M et al (2002) Abnormal pentagastrin response in a patient with pseudohypoparathyroidism. Exp Clin Endocrinol Diabetes 110:86–91
24. Vlaeminck-Guillem V, D'Herbomez M, Pigny P et al (2001) Pseudohypoparathyroidism Ia and hypercalcitoninemia. J Clin Endocrinol Metab 86:3091–3096
25. Bastepe M, Lane AH, Jüppner H (2001) Paternal uniparental isodisomy of chromosome 20q (patUPD20q)–and the resulting changes in GNAS1 methylation–as a plausible cause of pseudohypoparathyroidism. Am J Hum Genet 68:1283–1289
26. Bastepe M, Pincus JE, Sugimoto T et al (2001) Positional dissociation between the genetic mutation responsible for pseudohypoparathyroidism type Ib and the associated methylation defect at exon A/B: evidence for a long-range regulatory element within the imprinted GNAS1 locus. Hum Mol Genet 10: 1231–1241
27. Liu J, Litman D, Rosenberg MJ et al (2000) A GNAS imprinting defect in pseudohypoparathyroidism type Ib. J Clin Invest 106:1167–1174
28. Kelsey G (2010) Imprinting on chromosome 20: tissue-specific imprinting and imprinting mutations in the GNAS locus. Am J Med Genet C Semin Med Genet 154C:377–386
29. Mantovani G, Bondioni S, Linglart A et al (2007) Genetic analysis and evaluation of resistance to thyrotropin and growth hormone-releasing hormone in pseudohypoparathyroidism type Ib. J Clin Endocrinol Metab 92:3738–3742
30. de Nanclares GP, Fernández-Rebollo E, Santin I et al (2007) Epigenetic defects of GNAS in patients with pseudohypoparathyroidism and mild features of Albright hereditary osteodystrophy. J Clin Endocrinol Metab 92:2370–2373
31. Mariot V, Maupetit-Méhouas S, Sinding C et al (2008) A maternal epimutation of GNAS leads to Albright osteodystrophy and parathyroid hormone resistance. J Clin Endocrinol Metab 93:661–665
32. Unluturk U, Harmanci A, Babaoglu M et al (2008) Molecular diagnosis and clinical characterization of pseudohypoparathyroidism type-Ib in a patient with mild Albright hereditary osteodystrophy-like features, epileptic seizures, and defective renal handling of uric acid. Am J Med Sci 336:84–90
33. Mantovani G, deSanctis L, Barbieri AM et al (2010) Pseudohypoparathyroidism and *GNAS* epigenetic defects: clinical evaluation of Albright hereditary osteodystrophy and molecular analysis in 40 patients. J Clin Endocrinol Metab 95:651–658
34. Elli FM, DeSanctis L, Bollati V et al (2014) Quantitative analysis of methylation defects and correlation with clinical characteristics in patients with Pseudohypoparathyroidism type I and GNAS epigenetic alterations. J Clin Endocrinol Metab 99(33): E508–E517
35. Spiegel AM, Shenker A, Weinstein LS (1992) Receptor-effector coupling by G proteins: implications for normal and abnormal signal transduction. Endocr Rev 13(3):536–565
36. Spiegel AM (1996) Mutations in G proteins and G protein-coupled receptors in endocrine disease. J Clin Endocrinol Metab 81(7):2434–2442
37. Farzel Z, Bourne HR, Iiri T (1999) The expanding spectrum of G protein diseases. N Engl J Med 340(13):1012–1020
38. Lania A, Mantovani G, Spada A (2001) G protein mutations in endocrine diseases. Eur J Endocrinol 145(5):543–559
39. Mantovani G, Spada A (2006) Mutations in the Gs alpha gene causing hormone resistance. Best Pract Res Clin Endocrinol Metab 20(4):501–513
40. Wettschureck N, Offermanns S (2005) Mammalian G proteins and their cell type specific functions. Physiol Rev 85(4):1159–1204

41. Potts JT (2005) Parathyroid hormone: past and present. J Endocrinol 187(3):311–325
42. Gensure RC, Gardella TJ, Juppner H (2005) Parathyroid hormone and parathyroid hormone-related peptide and their receptors. Biochem Biophys Res Commun 328:666–678
43. Gardner D, Shoback D (2011) Greenspan's basic & clinical endocrinology, 9th edn. McGraw Hill, New York, p 232
44. Poole K, Reeve J (2005) Parathyroid hormone – a bone anabolic and catabolic agent. Curr Opin Pharmacol 5(6):612–617
45. Breslau NA, Weinstock RS (1988) Regulation of 1,25 (OH)2D synthesis in hypoparathyroidism and pseudohypoparathyroidism. Am J Physiol 255:E730–E736
46. Drezner MK, Neelon FA, Haussler M et al (1976) 1,25-Dihydroxycholecalciferol deficiency: the probable cause of hypocalcemia and metabolic bone disease in pseudohypoparathyroidism. J Clin Endocrinol Metab 42(4):621–628
47. Stone MD, Hosking DJ, Garcia-Himmelstine C et al (1993) The renal response to exogenous parathyroid hormone in treated pseudohypoparathyroidism. Bone 14(5):727–735
48. Kidd GS, Schaaf M, Adler RA et al (1980) Skeletal responsiveness in pseudohypoparathyroidism: a spectrum of clinical disease. Am J Med 68:772–781
49. Murray TM, Rao LG, Wong MM et al (1993) Pseudohypoparathyroidism with osteitis fibrosa cystica: direct demonstration of skeletal responsiveness to parathyroid hormone in cells cultured from bone. J Bone Miner Res 8:83–91
50. Eubanks PJ, Stabile BE (1998) Osteitis fibrosa cystica with renal parathyroid hormone resistance: a review of pseudohypoparathyroidism with insight into calcium homeostasis. Arch Surg 133:673–676
51. CohenRD, VinceFP(1969)Pseudohypoparathyroidism with raised plasma alkaline phosphatase. Arch Dis Child 44:96–101
52. Kolb FO, Steinbach HL (1962) Pseudohypoparathyroidism with secondary hyperparathyroidism and osteitis fibrosa. J Clin Endocrinol Metab 22:59–70
53. Tollin SR, Perlmutter S, Aloia JF (2000) Serial changes in bone mineral density and bone turnover after correction of secondary hyperparathyroidism in a patient with pseudohypoparathyroidism type Ib. J Bone Miner Res 15:1412–1416
54. Ish-Shalom S, Rao LG, Levine MA et al (1996) Normal parathyroid hormone responsiveness of bone-derived cells from a patient with pseudohypoparathyroidism. J Bone Miner Res 11:8–14
55. Drezner M, Neelon FA, Lebovitz HE (1973) Pseudohypoparathyroidism type II: a possible defect in the reception of the cyclic AMP signal. N Engl J Med 289:1056–1060
56. Wemeau JL, Balavoine AS, Ladsous M et al (2006) Multihormonal resistance to parathyroid hormone, thyroid stimulating hormone, and other hormonal and neurosensory stimuli in patients with pseudohypoparathyroidism. J Pediatr Endocrinol Metab 19(Suppl 2):653–661
57. Germain-Lee EL (2006) Short stature, obesity, and growth hormone deficiency in pseudohypoparathyroidism type Ia. Pediatr Endocrinol Rev 3:318–327
58. Levine MA, Jap TS, Hung W (1985) Infantile hypothyroidism in two sibs: an unusual presentation of pseudohypoparathyroidism type Ia. J Pediatr 107:919–922
59. Pohlenz J, Ahrens W, Hiort O (2003) A new heterozygous mutation (L338N) in the human Gsα (GNAS1) gene as a cause for congenital hypothyroidism in Albright's hereditary osteodystrophy. Eur J Endocrinol 148:463–468
60. Riepe FG, Ahrens W, Krone N et al (2005) Early manifestation of calcinosis cutis in pseudohypoparathyroidism type Ia associated with a novel mutation in the GNAS gene. Eur J Endocrinol 152:515–519
61. Pinsker JE, Rogers W, McLean S et al (2006) Pseudohypoparathyroidism type 1a with congenital hypothyroidism. J Pediatr Endocrinol Metab 19:1049–1052
62. Mantovani G, Spada A (2006) Resistance to growth hormone releasing hormone and gonadotropins in Albright's hereditary osteodystrophy. J Pediatr Endocrinol Metab 19:663–670
63. Mantovani G, Maghnie M, Weber G et al (2003) Growth hormone-releasing hormone resistance in pseudohypoparathyroidism type Ia: new evidence for imprinting of the Gsα gene. J Clin Endocrinol Metab 88:4070–4074
64. Germain-Lee EL, Groman J, Crane JL et al (2003) Growth hormone deficiency in pseudohypoparathyroidism type 1a: another manifestation of multihormone resistance. J Clin Endocrinol Metab 88:4059–4069
65. de Sanctis L, Bellone J, Salerno M et al (2007) GH secretion in a cohort of children with pseudohypoparathyroidism type Ia. J Endocrinol Invest 30:97–103
66. Moses AM, Weinstock RS, Levine MA et al (1986) Evidence for normal antidiuretic responses to endogenous and exogenous arginine vasopressin in patients with guanine nucleotide-binding stimulatory protein-deficient pseudohypoparathyroidism. J Clin Endocrinol Metab 62:221–224
67. Faull CM, Welbury RR, Paul B et al (1991) Pseudohypoparathyroidism: its phenotypic variability and associated disorders in a large family. Q J Med 78:251–264
68. Tsai KS, Chang CC, Wu DJ et al (1989) Deficient erythrocyte membrane Gsα activity and resistance to trophic hormones of multiple endocrine organs in two cases of pseudohypoparathyroidism. Taiwan Yi Xue Hui Za Zhi 88:450–455
69. Long DN, Levine MA, Germain-Lee EL (2010) Bone mineral density in pseudohypoparathyroidism type 1a. J Clin Endocrinol Metab 95:4465–4475

70. Mann JB, Alterman S, Hills AG (1962) Albright's hereditary osteodystrophy comprising pseudohypoparathyroidism and pseudopseudohypoparathyroidism with a report of two cases representing the complete syndrome occurring in successive generations. Ann Intern Med 56:315–342
71. Levine MA, Jap TS, Mauseth RS et al (1986) Activity of the stimulatory guanine nucleotide-binding protein is reduced in erythrocytes from patients with pseudohypoparathyroidism and pseudopseudohypoparathyroidism: biochemical, endocrine, and genetic analysis of Albright's hereditary osteodystrophy in six kindreds. J Clin Endocrinol Metab 62:497–502
72. Wilson LC, Leverton K, Oude Luttikhuis ME et al (1995) Brachydactyly and mental retardation: an Albright's hereditary osteodystrophy-like syndrome localized to 2q37. Am J Hum Genet 56:400–407
73. Phelan MC, Rogers RC, Clarkson KB et al (1995) Albright's hereditary osteodystrophy and del(2)(q37.3) in four unrelated individuals. Am J Med Genet 58:1–7
74. Elli FM, deSanctis L, Ceoloni B (2013) Pseudohypoparathyroidism type Ia and pseudo-pseudohypoparathyroidism: the growing spectrum of GNAS inactivating mutations. Hum Mutat 34(3):411–416
75. Aldred MA (2006) Genetics of pseudohypoparathyroidism types Ia and Ic. J Pediatr Endocrinol Metab 19(2):635–640
76. Mantovani G (2011) Pseudohypoparathyroidism: diagnosis and treatment. J Clin Endocrinol Metab 96(10):3020–3030
77. Barret D, Breslau NA, Wax MB et al (1989) New form of pseudohypoparathyroidism with abnormal catalytic adenylate cyclase. Am J Physiol 257:E277–E283
78. Winter JS, Hughes IA (1986) Familial pseudohypoparathyroidism without somatic anomalies. Can Med Assoc J 123:26–31
79. Nusynowitz ML, Frame B, Kolb FO (1976) The spectrum of the hypoparathyroid states: a classification based on physiologic principles. Medicine 55:105–119
80. Zazo C, Thiele S, Martín C et al (2011) Gsα activity is reduced in erythrocyte membranes of patients with pseudohypoparathyroidism due to epigenetic alterations at the GNAS locus. J Bone Miner Res 26(8):1864–1870
81. Schipani E, Weinstein LS, Bergwitz C et al (1995) Pseudohypoparathyroidism type Ib is not caused by mutations in the coding exons of the human parathyroid hormone (PTH)/PTH-related peptide receptor gene. J Clin Endocrinol Metab 80(5):1611–1621
82. Juppner H, Bastepe M (2006) Different mutations within or upstream of the GNAS locus cause distinct forms of pseudohypoparathyroidism. J Pediatr Endocrinol Metab 19(Suppl 2):641–646
83. Bastepe M, Frohlich LF, Hendy GN et al (2003) Autosomal dominant pseudohypoparathyroidism type Ib is associated with a heterozygous microdeletion that likely disrupts a putative imprinting control element of GNAS. J Clin Invest 112(8):1255–1263
84. Bastepe M, Frohlich LF, Linglart A et al (2005) Deletion of the NESP55 differentially methylated region causes loss of maternal GNAS imprints and pseudohypoparathyroidism type Ib. Nat Genet 37(1):25–27
85. Linglart A, Gensure RC, Olney RC et al (2005) A novel STX16 deletion in autosomal dominant pseudohypoparathyroidism type Ib redefines the boundaries of a cis-acting imprinting control element of GNAS. Am J Hum Genet 76:804–814
86. Chillambhi S, Turan S, Hwang D et al (2010) Deletion of the noncoding GNAS antisense transcript causes pseudohypoparathyroidism type Ib and biparental defects of GNAS methylation in cis. J Clin Endocrinol Metab 95(8):3993–4002
87. Richard N, Abeguilè G, Coudray N et al (2012) A new deletion ablating NESP55 causes loss of maternal imprint of A/B GNAS and autosomal dominant pseudohypoparathyroidism type Ib. J Clin Endocrinol Metab 97(5):E863–E867
88. Linglart A, Bastepe M, Jüppner H (2007) Similar clinical and laboratory findings in patients with symptomatic autosomal dominant and sporadic pseudohypoparathyroidism type Ib despite different epigenetic changes at the GNAS locus. Clin Endocrinol (Oxf) 67:822–831
89. Linglart A, Menguy C, Couvineau A et al (2011) Recurrent PRKAR1A mutation in acrodysostosis with hormone resistance. N Engl J Med 364(23):2218–2226
90. Lee H, Graham JM Jr, Rimoin DL et al (2012) Exome sequencing identifies PDE4D mutations in acrodysostosis. Am J Hum Genet 90:746–751

33 Pseudohypoparathyroidism Type 1a, Pseudopseudohypoparathyroidism, and Albright Hereditary Osteodystrophy

Lee S. Weinstein

33.1 Introduction

Albright hereditary osteodystrophy (AHO) is a congenital disorder caused by heterozygous inactivating mutations of the $G_s\alpha$ gene (*GNAS*) on chromosome 20 that is characterized by the presence of short stature, brachydactyly, subcutaneous ossifications, centripetal facial abnormalities, including depressed nasal bridge or hypertelorism, and mental deficits or developmental delay [1] (Fig. 33.1) (see also Chaps. 32, 34, and 35). Patients who inherit the *GNAS* mutation from their mother (or have a de novo maternal *GNAS* mutation) also develop resistance to parathyroid hormone (PTH) and other hormones as well as early-onset obesity (also known as pseudohypoparathyroidism type 1a; PHP1a). In contrast, patients who inherit their mutation paternally only have the physical and neurobehavioral features of AHO (also known as pseudopseudohypoparathyroidism; PPHP). Rarely patients with *GNAS* mutations will develop a more severe form of ectopic ossification referred to as progressive osseous heteroplasia (POH). This chapter will summarize the clinical features, genetics, pathogenesis, diagnosis, and treatment for these disorders associated with *GNAS* mutations.

L.S. Weinstein
Metabolic Diseases Branch, National Institute of Diabetes and Digestive and Kidney Diseases, National Institutes of Health, Bldg 10 Rm 8C101, Bethesda, MD 20892-1752, USA
e-mail: leew@mail.nih.gov

33.2 Clinical Features

The extent and severity of AHO features found in individual patients (both with PHP1a and PPHP) is very variable. Almost all AHO patients present with short stature and most have brachydactyly (shortening and widening of the long bones in the hands and feet), most often the distal thumb and third, fourth, and fifth metacarpals and metatarsals. The extent of brachydactyly is highly variable and is often asymmetric. Brachydactyly is primarily due to premature closure of the growth plate and is associated with coning of the epiphysis [2]. Other musculoskeletal abnormalities associated with AHO include spinal cord compression [3] and carpal tunnel syndrome [4]. AHO patients often, although not always, present with neurocognitive abnormalities, including developmental delay, mental retardation, and emotional disorders. A recent report suggests that these neurocognitive features may be more prominent in PHP1a than in PPHP [5].

The most specific feature of AHO is the presence of ectopic ossifications (osteoma cutis), which are generally limited to the dermis and subcutaneous tissues and may present as palpable hard nodules or calcifications on radiographs. While short stature and brachydactyly are more frequently observed in AHO patients, ectopic ossification is a more specific manifestation and therefore its presence is more diagnostically useful. In rare cases the lesions coalesce to form plate- or cast-like structures and invade into deep

M.L. Brandi, E.M. Brown (eds.), *Hypoparathyroidism*,
DOI 10.1007/978-88-470-5376-2_33, © Springer-Verlag Italia 2015

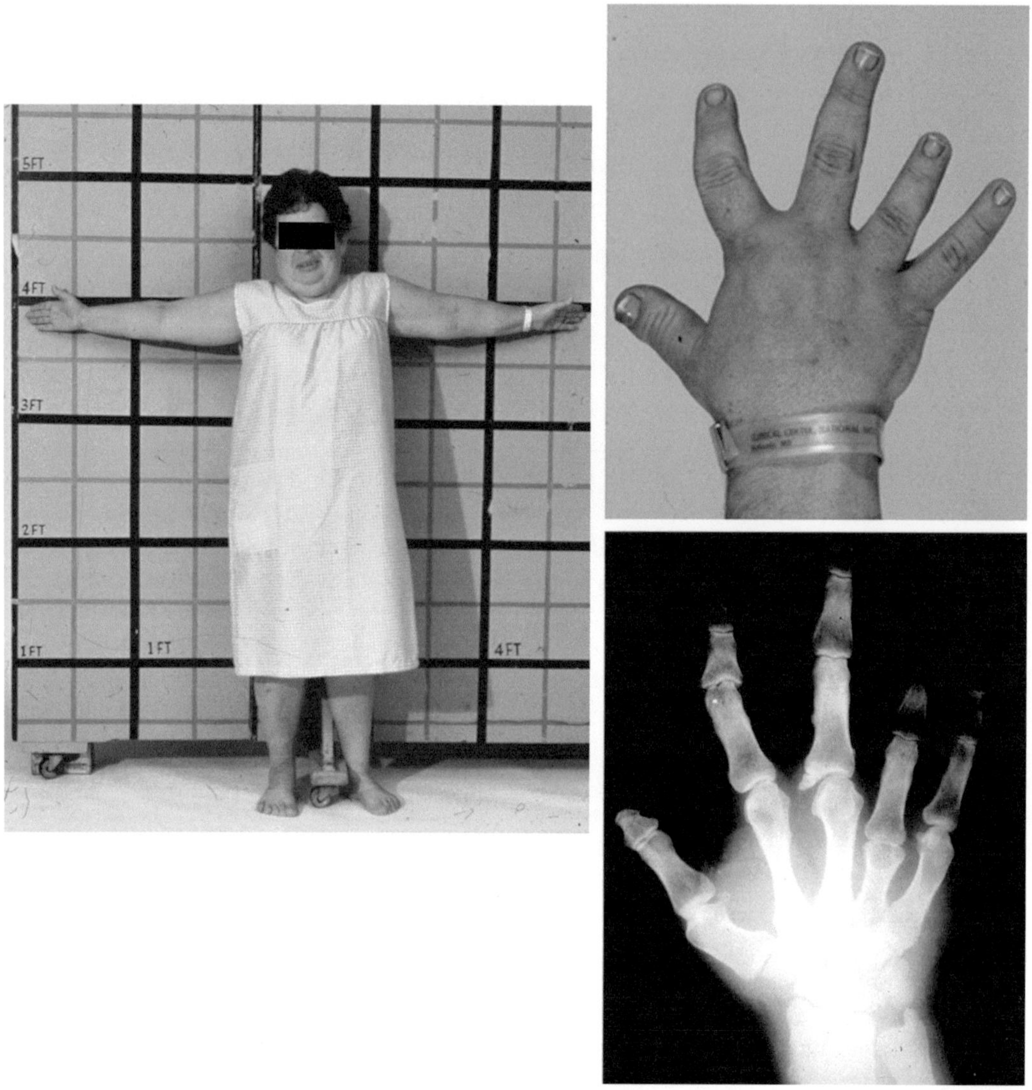

Fig. 33.1 Albright hereditary osteodystrophy. *Left*: AHO patient with short stature, obesity, and rounded face. *Right*: Photograph (*above*) and radiograph (*below*) showing brachydactyly of the fourth metacarpal (Reproduced from: Thakkar et al. [39])

soft tissues leading to joint stiffness and bone deformity, and this presentation is referred to as POH [6].

In addition to the AHO features described above, PHP1a patients also have other additional features not present in PPHP, including multihormonal resistance and early-onset obesity. The most prominent hormonal resistance seen in PHP1a is renal resistance to the actions of PTH in the proximal tubule, leading to impaired generation of 1,25-dihydroxyvitamin D and increased urinary phosphate reabsorption. The biochemical hallmarks of PHP1a (and other forms of PHP) are hypocalcemia, hyperphosphatemia, and elevated serum PTH levels in the absence of renal failure or 25-hydroxyvitamin D deficiency. Serum levels of 1,25-dihydroxyvitamin D are typically low or low normal despite PTH levels being elevated. Elevated PTH levels and hyperphosphatemia usually develop in early childhood prior to the development of hypocalcemia. Typically patients present with hypocalcemic symptoms (paresthesias, tetany, seizures) during late childhood, although some patients never develop hypocalcemic symptoms or may even remain eucalcemic. Patients often develop features in common with primary hypoparathyroidism, such as basal ganglia calcifications and cataracts, and rarely will develop rachitic bone changes. One study showed that bone mineral density is maintained in PHP1a [7].

PHP1a patients also develop resistance to other hormones that also activate $G_s\alpha$ in their target tissues, including thyrotropin (TSH), gonadotropins, and growth hormone-releasing hormone (GHRH). TSH resistance is often detected during perinatal screening with elevated TSH levels and typically leads to mild to moderate nongoitrous hypothyroidism. Gonadotropin resistance presents primarily in females as delayed or incomplete sexual development, oligomenorrhea, and/or infertility, and typically estrogen levels are low, although gonadotropins are not uniformly elevated [8]. Another feature in many, but not all, PHP1a patients is growth hormone deficiency due to pituitary GHRH resistance. However this may not be the major factor leading to short stature, as short stature primarily results from premature closure of growth plates in the axial skeleton [9]. It should be noted that in PHP1a, there is no clinical resistance to other hormones that activate $G_s\alpha$, such as vasopressin, glucagon, and ACTH. Another feature associated with PHP1a but not PPHP is severe, early-onset obesity [10]. Adult PHP1a patients also develop insulin resistance independent of their level of adiposity and are probably more prone to the development of type 2 diabetes [11]. Prolactin deficiency and impaired olfaction have also been reported in PHP1a patients.

33.3 Genetics

AHO is an autosomal dominant disorder resulting from heterozygous mutations in the *GNAS* gene at 20q13 resulting in loss of expression or function of the ubiquitously expressed G protein α-subunit $G_s\alpha$ [1]. $G_s\alpha$ couples many seven-transmembrane receptors for hormones, neurotransmitter, and other signals to the enzyme adenylyl cyclase and is required for receptor-stimulated intracellular cAMP generation. $G_s\alpha$ mutations on the paternal allele (de novo or inherited from the father) lead to PPHP (AHO alone), while the same mutations on the maternal allele lead to PHP1a (AHO plus multihormone resistance and obesity). This parent-of-origin effect of $G_s\alpha$ mutations is due to genomic imprinting leading to tissue-specific effects on $G_s\alpha$ expression from each parental allele. While $G_s\alpha$ is biallelically expressed in most tissues (i.e., from both copies of the gene), it is expressed primarily from the maternal allele in some tissues, including renal proximal tubules, thyroid, gonad, pituitary somatotrophs, and certain brain regions [1, 12]. Tissue-specific $G_s\alpha$ imprinting has been confirmed in mice [13]. $G_s\alpha$ imprinting is associated with and caused by differences in DNA methylation within the $G_s\alpha$ gene *GNAS* between the two parental alleles. Pseudohypoparathyroidism type 1b (PHP type 1b), a form of renal PTH resistance without the features of AHO, is associated with a loss of DNA methylation at a specific region on the maternal allele which leads to both parental alleles having a paternal methylation pattern. Recently it has been reported that patients with the *GNAS* methylation defect typically associated with PHP type 1b may occasionally have features of AHO, particularly brachydactyly [14, 15], and this may correlate with the extent of tissue-specific $G_s\alpha$ imprinting [16].

Most $G_s\alpha$ mutations associated with AHO are complete null mutations (splice junction, nonsense, or frameshift mutations) and there is no clear genotype-phenotype correlation. One specific 4 base pair deletion in exon 7 has been identified in many families [17]. A mutation within the alternatively spliced exon 3 resulting in loss of expression of the long but not short form of $G_s\alpha$ was associated with a mild form of PHP1a without hypocalcemia or obesity [18]. Several specific missense mutations leading to single amino acid substitutions have been identified and shown to have specific effects on $G_s\alpha$ function [1]. Mutation of either

Arg231 or Glu259 results in a receptor-activation defect, probably by disrupting interactions required to stabilize the active conformation. The Ala366Ser mutation results in PHP1a plus gonadotropin-independent precocious puberty (testotoxicosis) in males. At core body temperature, the mutant protein is thermolabile resulting in PHP1a, while at the lower testicular temperature, the mutant protein is stable but is constitutively activated due to increased basal GDP release leading to more $G_s\alpha$ in the active GTP-bound conformation.

POH (severe ectopic ossification) is associated with the same mutations as AHO and may present with or without other features of AHO or PHP1a. There is a predilection for patients who present with POH alone to inherit their mutations from the father, although even within these families, female POH patients may have affected offspring with classic AHO [19].

33.4 Pathogenesis

The differences in clinical manifestations between PHP1a and PPHP (presence or absence of multihormonal resistance and obesity) result from tissue-specific $G_s\alpha$ imprinting, which results from differences in DNA methylation between the two parental alleles of *GNAS* [1, 20]. $G_s\alpha$ is biallelically expressed in most tissues but primarily expressed from the maternal allele in specific tissues, including those that are targets of hormone action (e.g., renal proximal tubules for PTH, thyroid, pituitary, and gonad). In these tissues an inactivating mutation on the active maternal allele disrupts $G_s\alpha$ expression and hormone signaling, while the same mutation on the inactive paternal allele has little effect on $G_s\alpha$ expression or hormone signaling (Fig. 33.2). This is consistent with the markedly reduced PTH-stimulated urinary cAMP observed in PHP1a but not PPHP patients.

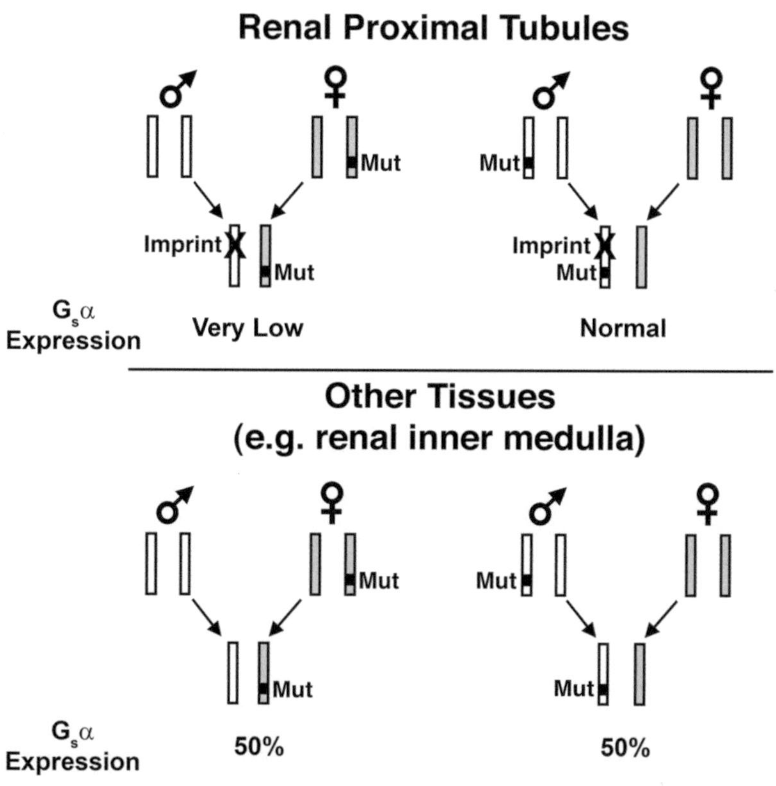

Fig. 33.2 Role of tissue-specific $G_s\alpha$ imprinting in the pathogenesis of PHP1a. In renal proximal tubules (*above*) $G_s\alpha$ is silenced from the paternal allele due to genomic imprinting (denoted with *X*). Mutation (*Mut*) on the active maternal allele in the setting of PHP1a leads to loss of $G_s\alpha$ expression and PTH signaling, while mutation on the inactive paternal allele in the setting of PPHP has little effect on $G_s\alpha$ expression or PTH action. In other tissues, including renal inner medulla (*below*), $G_s\alpha$ is not imprinted and therefore both maternal and paternal $G_s\alpha$ mutations lead to similar ~50 % loss of $G_s\alpha$ expression (Reproduced from: Thakkar et al. [39])

In PHP1a, loss of $G_s\alpha$ in renal proximal tubules leads to impaired PTH signaling in renal proximal tubules, leading to lower conversion of 25-hydroxyvitamin D to 1,25-dihydroxy vitamin D and lower reabsorption of phosphate. Low 1,25-dihydroxy vitamin D levels result in low intestinal calcium absorption and skeletal calcium mobilization. Decreased phosphate excretion leads to hyperphosphatemia, which further inhibits 1,25-dihydroxy vitamin D production. All of these effects combine to produce hypocalcemia, hyperphosphatemia, and secondary hyperparathyroidism. In PHP1a patients PTH resistance is not present at birth but develops over the first 2–3 years of life. A recent study in mice suggests that this is due to the early postnatal onset of $G_s\alpha$ imprinting in renal proximal tubules [21], which may relate to the postnatal maturation of proximal tubular cells [22].

Hypothyroidism, hypogonadism, and growth hormone deficiency in PHP1a result from TSH, gonadotropin, and GHRH resistance due to partial $G_s\alpha$ deficiency in thyroid, gonads, and pituitary somatotrophs, respectively [1]. Because gonadotropin levels are not clearly elevated in PHP1a, it has proposed that PHP1a patients have a partial gonadotropin resistance that allows for follicular development and estrogen production but does not allow for the high level of gonadotropin signaling required for ovulation [8].

The absence of clinical resistance to other hormones that also activate $G_s\alpha$ in their target tissues, such as ACTH and vasopressin, in PHP1a patients may be related to the absence of $G_s\alpha$ imprinting in their respective target tissues [20]. In these tissues heterozygous mutations lead to only a 50 % reduction in $G_s\alpha$, which may still allow enough cAMP signaling to elicit a normal physiological response. Lack of $G_s\alpha$ imprinting in the thick ascending limb of the nephron may also explain why the anticalciuric action of PTH is maintained in PHP1a patients [23].

Severe early-onset obesity associated with PHP1a (but not PPHP) likely results from $G_s\alpha$ imprinting in one or more metabolically active tissues leading to severe $G_s\alpha$ deficiency. Maternal (but not paternal) germline $G_s\alpha$ mutation in mice is also associated with this parent-of-origin effect on obesity, as well as insulin-resistant diabetes and hyperlipidemia, and this phenotype is prevented when $G_s\alpha$ imprinting is lost [24]. The liver, muscle, and adipose tissue are unlikely to mediate these effects as there is no $G_s\alpha$ imprinting in these tissues and mice with $G_s\alpha$ knockout in these specific tissues do not mimic the germline phenotype [24]. However maternal $G_s\alpha$ mutation limited to the central nervous system (CNS) recapitulated the metabolic phenotype observed in the germline knockout mice, indicating that the metabolic effects are related to $G_s\alpha$ imprinting in one or more regions of the CNS [12]. Although $G_s\alpha$ is imprinted in the paraventricular nucleus of the hypothalamus (PVH), $G_s\alpha$ knockout limited to this region resulted in only a minimal metabolic phenotype indicating that the metabolic effects of $G_s\alpha$ mutations are the consequence of impaired $G_s\alpha$ signaling in a brain region outside of the PVH [25].

The obesity in maternal $G_s\alpha$ knockout mice (both germline and CNS-specific) is associated with reduced sympathetic nervous system (SNS) activity and energy expenditure but not with an increase in food intake [12, 24]. This is consistent with the finding of severely reduced plasma norepinephrine levels in PHP1a patients [26] and the report of a PHP1a infant who developed early-onset obesity in the absence of hyperphagia [27]. However two recent studies failed to show reduced metabolic rate in PHP1a patients [11, 28]. Obesity in maternal CNS-specific $G_s\alpha$ knockout mice was associated with an impaired ability of a central melanocortin agonist to stimulate energy expenditure, while the ability of the agonist to acutely reduce food intake was maintained [12], indicating selective resistance to the effects of central melanocortins (which mediate their actions via $G_s\alpha$) on SNS activity and energy expenditure. Brain $G_s\alpha$ imprinting may also account for the greater severity of neurocognitive problems in PHP1a patients as compared to PPHP patients [5]. It is likely that the increased insulin resistance observed in PHP1a patients [11] and in $G_s\alpha$ knockout mouse models [12] is also related to impaired melanocortin actions on peripheral glucose metabolism.

AHO features common to both PHP1a and PPHP are likely the consequence of $G_s\alpha$

haploinsufficiency in tissues where $G_s\alpha$ is not imprinted. Brachydactyly appears to be the result of impaired local action of PTHrP (which activates $G_s\alpha$ via the type 1 PTH receptor that mediates most the actions of both PTH and PTHrP) on growth plate chondrocytes leading to accelerated differentiation, premature growth plate closure, and long bone shortening [1, 29, 30]. Ectopic ossification in AHO and POH, which has also been reported in a mouse AHO model [31], occurs by intramembranous ossification as reduced cAMP promotes osteoblast differentiation and expression of the osteoblast-specific factor Cbfa1/RUNX2 [1]. Hedgehog signaling has been shown to be upregulated in ectopic ossifications from POH patients and manipulations of Hedgehog signaling in mouse models indicate that $G_s\alpha$ deficiency also promotes ectopic ossification by altering both Wnt-β-catenin and Hedgehog signaling in mesenchymal cells [32]. As AHO and POH patients have similar complete null mutations of $G_s\alpha$, the difference in severity of the ossifications in these two disorders likely relates to genetic or environmental factors leading to differences in $G_s\alpha$ expression from the unaffected allele, in another component of the $G_s\alpha$ signaling pathway, or in another pathway involved in osteoblast differentiation.

33.5 Diagnosis

PHP1a (as well as PHP1b) typically presents with hypo- or eucalcemia, hyperphosphatemia, and elevated PTH in the absence of renal insufficiency or vitamin D deficiency. Renal PTH resistance can be confirmed by showing an impaired urinary cAMP response to exogenous PTH analogue (Ellsworth-Howard test). However the analogue is not commercially available in a form to easily perform this test and this test is generally not clinically necessary. Brachydactyly, obesity, and neurocognitive deficits are not specific for AHO and are seen in other genetic disorders such as Prader-Willi syndrome, brachydactyly syndromes, Turner's syndrome, Rubinstein-Taybi syndrome, 2q37 deletion, and acrodysostosis. Therefore the presence of only these features is not enough to make the specific diagnosis of AHO or PPHP. Osteoma cutis and PTH resistance are much more specific for AHO and PHP1a and therefore are more useful for establishing this diagnosis. The coexistence of PTH resistance with other hormone resistance (TSH, gonadotropins) and early-onset obesity makes the diagnosis of PHP1a highly likely. Acrodysostosis is another syndrome associated with severe short stature, brachydactyly, and neurocognitive impairment that has been shown to be associated with mutations affecting cAMP action (protein kinase A regulatory-1α subunit; PRKAR1A) or cAMP degradation (phosphodiesterase 4D) [33, 34]. Patients with PRKAR1A mutations also develop multihormonal resistance. Mutations in patients who present with severe ectopic ossifications typical of POH should be examined for features of AHO and biochemically screened for multihormonal resistance.

The presence of the features of AHO in the absence of a clear family history or hormone resistance requires biochemical or genetic confirmation of a $G_s\alpha$ defect to establish the diagnosis. *GNAS* mutation screening can be obtained from commercial laboratories but is only ~70 % sensitive. The diagnosis of AHO can also be confirmed by demonstrating an ~50 % loss of erythrocyte $G_s\alpha$ bioactivity or expression levels, but these tests are only performed in research laboratories. For the biochemical assays, patient erythrocyte membranes are mixed with membranes from a $G_s\alpha$-deficient cell line and reconstitution of $G_s\alpha$ signaling is quantified. Historically, these assays were performed by stimulating $G_s\alpha$ with a non-hydrolyzable GTP analogue. However this method will not identify a defect in patients whose mutation leads to a disruption in receptor-Gsα coupling, and this could lead to a misdiagnosis of PHP1c (clinical PHP1a in the absence of a $G_s\alpha$ defect) [35]. Therefore these assays should be performed using a receptor ligand such as isoproterenol. *GNAS* methylation analysis can be performed in patients with AHO features and PTH resistance with no apparent $G_s\alpha$ mutation. However one must consider that an apparent loss of methylation typical of PHP1b may in fact be due to genetic loss of the differentially methylated region on the maternal allele.

33.6 Treatment/Counseling

There is no specific therapy for the physical and neurocognitive manifestations of AHO. Ectopic ossifications do not require surgical excision unless they are causing discomfort or disfigurement. PTH resistance should be treated aggressively with oral calcium and vitamin D (either high-dose ergo- or cholecalciferol or calcitriol at more physiologic doses) to normalize both calcium and PTH, if possible. In contrast to patients with primary hypoparathyroidism, PHP patients generally do not develop hypercalciuria on treatment as the calcium-reabsorbing effect of PTH in the renal distal tubule is not affected [23]. However urine calcium should be periodically monitored. Normalizing PTH is important for preventing the skeletal consequences of high circulating PTH levels and preventing the development of tertiary hyperparathyroidism, although these complications are more typically present in PHP1b rather than PHP1a [36]. TSH and gonadotropin resistance in PHP1a is treated with levothyroxine and oral contraceptives (in females) or testosterone (in males), respectively. Growth hormone replacement may improve short stature in PHP1a patients [37].

AHO patients (both PHP1a and PPHP) should be counseled that each offspring has a 50 % chance of inheriting AHO, and offspring of affected females (both with PHP1a and PPHP) will also develop multihormone resistance, obesity, and potentially significant neurocognitive problems. Offspring of affected males will almost certainly not develop multihormone resistance or severe obesity, although neurocognitive problems may be present. The AHO phenotype is variable and it is impossible to predict its severity in offspring. Even patients with mild features need to be told that their affected offspring may develop severe physical and neurocognitive manifestations, including POH. Male POH patients should be counseled that each of their offspring has a 50 % chance of having PPHP or possibly POH. Female POH patients should be counseled that each of their offspring has a 50 % chance of developing PHP1a and a chance of developing POH. *GNAS* mutation screening may be useful in identifying further affected family members and possibly for prenatal testing [38].

References

1. Weinstein LS (2008) Guanine nucleotide-binding protein $G_s\alpha$ (*GNAS1*): fibrous dysplasia, McCune-Albright syndrome, Albright hereditary osteodystrophy, and pseudohypoparathyroidism. In: Epstein CJ, Erickson RP, Wynshaw-Boris A (eds) Inborn errors of development: the molecular basis of clinical disorders of morphogenesis, 2nd edn. Oxford University Press, New York, pp 1277–1288
2. de Sanctis L, Vai S, Andreo MR et al (2004) Brachydactyly in 14 genetically characterized pseudohypoparathyroidism type Ia patients. J Clin Endocrinol Metab 89:1650–1655
3. Roberts TT, Khasnavis S, Papaliodis DN et al (2013) Spinal cord compression in pseudohypoparathyroidism. Spine J 13:e15–e19
4. Joseph AW, Shoemaker AH, Germain-Lee EL (2011) Increased prevalence of carpal tunnel syndrome in Albright hereditary osteodystrophy. J Clin Endocrinol Metab 96:2065–2073
5. Mouallem M, Shaharabany M, Weintrob N et al (2008) Cognitive impairment is prevalent in pseudohypoparathyroidism type Ia, but not in pseudopseudohypoparathyroidism: possible cerebral imprinting of Gsα. Clin Endocrinol (Oxf) 68:233–239
6. Kaplan FS, Shore EI (2000) Progressive osseous heteroplasia. J Bone Miner Res 15:2084–2094
7. Long DN, Levine MA, Germain-Lee EL (2010) Bone mineral density in pseudohypoparathyroidism type 1a. J Clin Endocrinol Metab 95:4465–4475
8. Namnoum AB, Merriam GR, Moses AM et al (1998) Reproductive dysfunction in women with Albright's hereditary osteodystrophy. J Clin Endocrinol Metab 83:824–829
9. Germain-Lee EL, Groman J, Crane JL et al (2003) Growth hormone deficiency in pseudohypoparathyroidism type 1a: another manifestation of multihormone resistance. J Clin Endocrinol Metab 88: 4059–4069
10. Long DN, McGuire S, Levine MA et al (2007) Body mass index differences in pseudohypoparathyroidism type 1a versus pseudopseudohypoparathyroidism may implicate paternal imprinting of $G\alpha_s$ in the development of human obesity. J Clin Endocrinol Metab 92:1073–1079
11. Muniyappa R, Warren MA, Zhao X et al (2013) Reduced insulin sensitivity in adults with pseudohypoparathyroidism type 1a. J Clin Endocrinol Metab 98:E1796–E1801
12. Chen M, Wang J, Dickerson KE et al (2009) Central nervous system imprinting of the G protein $G_s\alpha$ and its role in metabolic regulation. Cell Metab 9:548–555

13. Weinstein LS, Xie T, Zhang QH et al (2007) Studies of the regulation and function of the $G_s\alpha$ gene *Gnas* using gene targeting technology. Pharmacol Ther 115:271–291
14. Mariot V, Maupetit-Mehouas S, Sinding C et al (2008) A maternal epimutation of *GNAS* leads to Albright osteodystrophy and parathyroid hormone resistance. J Clin Endocrinol Metab 93:661–665
15. Mantovani G, de Sanctis L, Barbieri AM et al (2010) Pseudohypoparathyroidism and *GNAS* epigenetic defects: clinical evaluation of Albright hereditary osteodystrophy and molecular analysis in 40 patients. J Clin Endocrinol Metab 95:651–658
16. Zazo C, Thiele S, Martin C et al (2011) $G_s\alpha$ activity is reduced in erythrocyte membranes of patients with pseudohypoparathyroidism due to epigenetic alterations at the *GNAS* locus. J Bone Miner Res 26:1864–1870
17. Yu S, Yu D, Hainline BE et al (1995) A deletion hotspot in exon 7 of the $G_s\alpha$ gene (*GNAS1*) in patients with Albright hereditary osteodystrophy. Hum Mol Genet 4:2001–2002
18. Thiele S, Werner R, Ahrens W et al (2007) A disruptive mutation in exon 3 of the *GNAS* gene with Albright hereditary osteodystrophy, normocalcemic pseudohypoparathyroidism, and selective long transcript variant Gsα-L deficiency. J Clin Endocrinol Metab 92:1764–1768
19. Shore EM, Ahn J, Jan de Beur S et al (2002) Paternally inherited inactivating mutations of the *GNAS1* gene in progressive osseous heteroplasia. N Engl J Med 346: 99–106
20. Yu S, Yu D, Lee E et al (1998) Variable and tissue-specific hormone resistance in heterotrimeric G_s protein α-subunit $G_s\alpha$ knockout mice is due to tissue-specific imprinting of the $G_s\alpha$ gene. Proc Natl Acad Sci U S A 95:8715–8720
21. Turan S, Fernandez-Rebollo E, Aydin C et al (2014) Postnatal establishment of allelic $G\alpha_s$ silencing as a plausible explanation for delayed onset of parathyroid hormone resistance owing to heterozygous $G\alpha_s$ disruption. J Bone Miner Res 29:749–760
22. Weinstein LS (2001) The stimulatory G protein α-subunit gene: mutations and imprinting lead to complex phenotypes. J Clin Endocrinol Metab 86:4622–4626
23. Stone MD, Hosking DJ, Garcia-Himmelstine C et al (1993) The renal response to exogenous parathyroid hormone in treated pseudohypoparathyroidism. Bone 14:727–735
24. Xie T, Chen M, Gavrilova O et al (2008) Severe obesity and insulin resistance due to deletion of the maternal $G_s\alpha$ allele is reversed by paternal deletion of the $G_s\alpha$ imprint control region. Endocrinology 149: 2443–2450
25. Chen M, Berger A, Kablan A et al (2012) $G_s\alpha$ deficiency in the paraventricular nucleus of the hypothalamus partially contributes to obesity associated with $G_s\alpha$ mutations. Endocrinology 153:4256–4265
26. Carel JC, Le Stunff C, Condamine L et al (1999) Resistance to the lipolytic action of epinephrine: a new feature of protein G_s deficiency. J Clin Endocrinol Metab 84:4127–4131
27. Dekelbab BH, Aughton DJ, Levine MA (2009) Pseudohypoparathyroidism type 1A and morbid obesity in infancy. Endocr Pract 15:249–253
28. Shoemaker AH, Lomenick JP, Saville BR et al (2013) Energy expenditure in obese children with pseudohypoparathyroidism type 1a. Int J Obes (Lond) 37: 1147–1153
29. Sakamoto A, Chen M, Kobayashi T et al (2005) Chondrocyte-specific knockout of the G protein $G_s\alpha$ leads to epiphyseal and growth plate abnormalities and ectopic chondrocyte formation. J Bone Miner Res 20:663–671
30. Bastepe M, Weinstein LS, Ogata N et al (2004) Stimulatory G protein directly regulates hypertrophic differentiation of growth plate cartilage *in vivo*. Proc Natl Acad Sci U S A 101:14794–14799
31. Huso DL, Edie S, Levine MA et al (2011) Heterotopic ossifications in a mouse model of Albright hereditary osteodystrophy. PLoS One 6:e21755
32. Regard JB, Malhotra D, Gvozdenovic-Jeremic J et al (2013) Activation of Hedgehog signaling by loss of *GNAS* causes heterotopic ossification. Nat Med 19:1505–1512
33. Linglart A, Fryssira H, Hiort O et al (2012) PRKAR1A and PDE4D mutations cause acrodysostosis but two distinct syndromes with or without GPCR-signaling hormone resistance. J Clin Endocrinol Metab 97: E2328–E2338
34. Linglart A, Menguy C, Couvineau A et al (2011) Recurrent PRKAR1A mutation in acrodysostosis with hormone resistance. N Engl J Med 364:2218–2226
35. Thiele S, de Sanctis L, Werner R et al (2011) Functional characterization of *GNAS* mutations found in patients with pseudohypoparathyroidism type Ic defines a new subgroup of pseudohypoparathyroidism affecting selectively $G_s\alpha$-receptor interaction. Hum Mutat 32:653–660
36. Neary NM, El-Maouche D, Hopkins R et al (2012) Development and treatment of tertiary hyperparathyroidism in patients with pseudohypoparathyroidism type 1B. J Clin Endocrinol Metab 97:3025–3030
37. Mantovani G, Ferrante E, Giavoli C et al (2010) Recombinant human GH replacement therapy in children with pseudohypoparathyroidism type Ia: first study on the effect on growth. J Clin Endocrinol Metab 95:5011–5017
38. Lietman SA, Goldfarb J, Desai N et al (2008) Preimplantation genetic diagnosis for severe Albright hereditary osteodystrophy. J Clin Endocrinol Metab 93:901–904
39. Weinstein LS, Collins MT, Spiegel AM (2013) $G_s\alpha$, pseudohypoparathyroidism, fibrous dysplasia, and McCune-Albright syndrome. In: Thakkar RV, Whyte MP, Eisman JA, Igarashi T (eds) (2013) Genetics of bone biology and skeletal disease. Academic, London, pp 425–440

Pseudohypoparathyroidism Type Ib (PHP-Ib): PTH-Resistant Hypocalcemia and Hyperphosphatemia Due to Abnormal *GNAS* Methylation

34

Harald Jüppner

34.1 Introduction

This chapter will focus primarily on the forms of pseudohypoparathyroidism type Ib (PHP-Ib) that are caused by different genetic and epigenetic abnormalities at the *GNAS* locus. To better comprehend the underlying molecular mechanisms leading to these PHP-Ib variants, two closely related disorders that are caused by different *GNAS* mutations, namely, PHP type Ia (PHP-Ia) and pseudo-pseudohypoparathyroidism (PPHP), will be reviewed first (see also Chaps. 32 and 33).

34.2 Parathyroid Hormone and the Regulation of Calcium Homeostasis

An important role of the parathyroids in the regulation of calcium and phosphate homeostasis was established approximately 100 years ago (see also Chap. 1). Since the early 1900s, it was furthermore known that parathyroid extracts can correct the hypocalcemia in parathyroidectomized animals and in patients with postsurgical or idiopathic hypoparathyroidism and increase urinary phosphate excretion, thereby improving serum phosphate levels. The biologically active principle in the parathyroids, namely, parathyroid hormone (PTH), could thus be used for investigations exploring some of the mechanisms contributing to the regulation of mineral ion homeostasis.

34.3 PTH Resistance Causes Hypocalcemia and Hyperphosphatemia

Once biologically active PTH had become available, Fuller Albright and his colleagues were able to show that certain patients with hypocalcemia and hyperphosphatemia failed to increase urinary phosphate excretion upon treatment with parathyroid extracts [1]. It was therefore hypothesized that these patients had end-organ resistance to PTH rather than a deficiency of this hormone, which led to the term "pseudohypoparathyroidism" (PHP). These PTH-resistant patients showed, besides the abnormalities in mineral ion homeostasis, a combination of different physical features, including short stature, brachydactyly, and ectopic ossifications, as well as obesity and various degrees of intellectual and cognitive impairment. Largely because of the skeletal findings, these physical stigmata are now referred to as Albright hereditary osteodystrophy (AHO), although a different term reflecting the nonskeletal aspects of the syndrome might be more appropriate.

H. Jüppner
Endocrine Unit and Pediatric Nephrology Unit, Massachusetts General Hospital and Harvard Medical School, Boston, MA 02114, USA
e-mail: hjueppner@partners.org

M.L. Brandi, E.M. Brown (eds.), *Hypoparathyroidism*,
DOI 10.1007/978-88-470-5376-2_34, © Springer-Verlag Italia 2015

34.4 Multiple Variants of Pseudohypoparathyroidism

To add to the complexity of the PHP syndrome, Albright and colleagues described patients who presented with AHO features but without evidence for an abnormal regulation of calcium and phosphate homeostasis; this disorder is referred to as pseudo-pseudohypoparathyroidism (PPHP) [2].

Consistent with the conclusion that PHP-Ia patients are resistant toward the actions of PTH rather than suffering from hormonal deficiency, Tashjian et al. subsequently showed that affected patients have elevated concentrations of immunoreactive PTH [3]. Subsequently, it was furthermore shown that PHP-Ia patients have, besides elevated PTH levels, resistance toward other hormones that are now known to mediate their actions through G protein-coupled receptors (GPCRs) (see Chap. 33). Thus, although PTH resistance is usually the most prominent feature of the disease, involvement of multiple hormonal systems raises the question of whether pseudohypoparathyroidism is the most appropriate term.

After the discovery of cAMP as a second messenger, Aurbach and colleagues demonstrated that PTH increases the formation of cAMP in kidney- and bone-derived tissue and that the increase in the urinary excretion of phosphate is preceded by a striking increase in urinary cAMP excretion [4, 5]. These authors furthermore showed that patients with PHP and obvious AHO features, i.e., individuals affected by PHP-Ia, failed to respond to a PTH challenge with an increase in urinary cAMP excretion [6]. This indicated that the lack of PTH-induced phosphaturia in PHP-Ia, initially described by Albright et al. [1], is associated with a severely impaired production of this second messenger. However, Marcus et al. had been unable to detect a gross deficiency in PTH-dependent cAMP formation in renal cortical tissue obtained from a PHP-Ia patient [7], and the authors therefore speculated correctly that the underlying defect may occur only in certain cells of the renal cortex. It was subsequently discovered that tissues readily accessible from PHP-Ia patients (erythrocytes, skin fibroblasts, and platelets) showed an about 50 % reduction of G protein activity (see Chap. 33). This indicated that the hormonal resistance observed in these patients is caused by an abnormal coupling between PTH/PTHrP receptor and adenyl cyclase, rather than deficiency of the hormone receptor or the enzyme catalyzing the formation of cAMP. Reduction in G protein activity, combined with the presence or absence of urinary cAMP and phosphate excretion in response to PTH, led to the current classification of the different disorders of PTH resistance, namely, PHP type I in which the PTH-induced urinary excretions of cAMP and phosphate are both impaired and PHP type II in which phosphate excretion, but not cAMP excretion, is blunted [8]. PHP type I was further subdivided, based on reduced or normal G protein activity, into PHP type Ia or PHP type Ib, respectively; patients with PHP type Ic show normal G protein activity in complementation assays using non-hydrolyzable GTP analogs but present with clinical and laboratory features that are indistinguishable from those of PHP type Ia.

34.5 PHP-Ia Is Caused by Mutations in Those *GNAS* Exons That Encode the α-Subunit of the Stimulatory G (Gsα)

Consistent with the reduction in G protein activity, subsequent studies identified mutations in *GNAS*, the gene encoding the stimulatory G protein (Gsα), i.e., the signaling protein that couples the adenylate cyclase to a large variety of different GPCRs, thereby stimulating the formation of cAMP and the subsequent activation of PKA [9, 10]. Numerous Gsα mutations (point mutations, deletions/insertions, intronic or constitutional deletions) have been described to date that can affect any of the 13 exons, intervening sequences, or splice sites, most of which cause the previously observed reduction in G protein activity (see Chap. 33). It remained unclear, however, why these Gsα mutations were all heterozygous and why the loss of Gsα protein from one parental allele should lead to PTH resistance at all. This conundrum was partially resolved when Davies and Hughes revealed that PHP-Ia and the associated PTH resistance become apparent only when the genetic defect is inherited maternally [11].

34.6 Gsα Is Expressed in Some Tissues Only from the Maternal Allele

It is now well established that Gsα is ubiquitously expressed and couples adenylate cyclase to numerous GPCRs, including the PTH/PTHrP receptor. Gsα is derived from the *GNAS* locus, a complex imprinted genomic region located on chromosome 20q13, which encodes besides Gsα several other alternatively spliced transcripts (see Chap. 10). In some tissues such as the proximal renal tubules (PTH target), the thyroid (TSH target), or the pituitary gland (CRF, GHRH target), Gsα is predominantly or exclusively expressed from the maternal allele, while expression from the paternal allele is silenced through as-of-yet unknown mechanisms (Fig. 34.1). This mechanism appears to be particularly efficient in the proximal renal tubules, thus explaining the PTH resistance observed in patients with different PHP variants that are caused by maternally inherited *GNAS* mutations. PTH resistance develops gradually after birth in humans, and this delay is consistent with recent findings in mice showing that the silencing of paternal Gsα expression in the proximal tubule develops gradually after early postnatal stages [12, 13]. Gsα expression is biallelic in most other tissues, including distal renal tubules [14–16] and bone [17, 18], i.e., two tissues in which patients with PHP-Ia and PHP-Ib show no evidence for PTH resistance. Silencing of Gsα expression from the paternal allele in the proximal renal tubules is thus of critical importance for the development of PTH-resistant hypocalcemia and hyperphosphatemia that is encountered in the *GNAS*-related forms of PHP.

34.7 Pseudohypoparathyroidism Type Ib (PHP-Ib)

In contrast to PHP-Ia, which was resolved at the molecular level following the identification of heterozygous *GNAS* mutations that lead to an inactive Gsα protein [9, 10], the mechanisms leading to PHP-Ib remained to a large extent unknown. Fuller Albright and others had already described some patients with PTH-resistant hypocalcemia and hyperphosphatemia, who did not show typical AHO features [19–21]; in one family, this form of PHP followed an autosomal dominant mode of inheritance [22]. Subsequent studies showed that similar patients without apparent AHO features have normal G protein activity in peripheral blood cells or skin fibroblasts [23, 24]; this disease variant is therefore referred to as PHP-Ib. Fibroblasts from some of these patients showed evidence for decreased responsiveness to PTH, which suggested that this PHP variant could be caused by mutations in the receptor for PTH, i.e., the PTH/PTHrP receptor [25]. However, mutations in this GPCR were subsequently excluded at the genomic and the mRNA level [26–30]. It was furthermore shown that PTH/PTHrP receptor mutations lead to diseases that are much more severe than PHP-Ib, namely, Jansen's and Blomstrand's diseases that are caused by heterozygous activating and homozygous/compound heterozygous inactivating mutations, respectively, in this GPCR (for review, see [31]).

34.8 Genetic Causes of Autosomal Dominant PHP-Ib (AD-PHP-Ib)

After mutations had been excluded in the gene encoding the PTH/PTHrP receptor, genetic linkage studies were pursued to identify the disease-associated locus. DNA was available from a large family, in which the disease followed an autosomal dominant trait, albeit with incomplete penetrance since two daughters of the index case were healthy, yet had affected children and grandchildren [32].

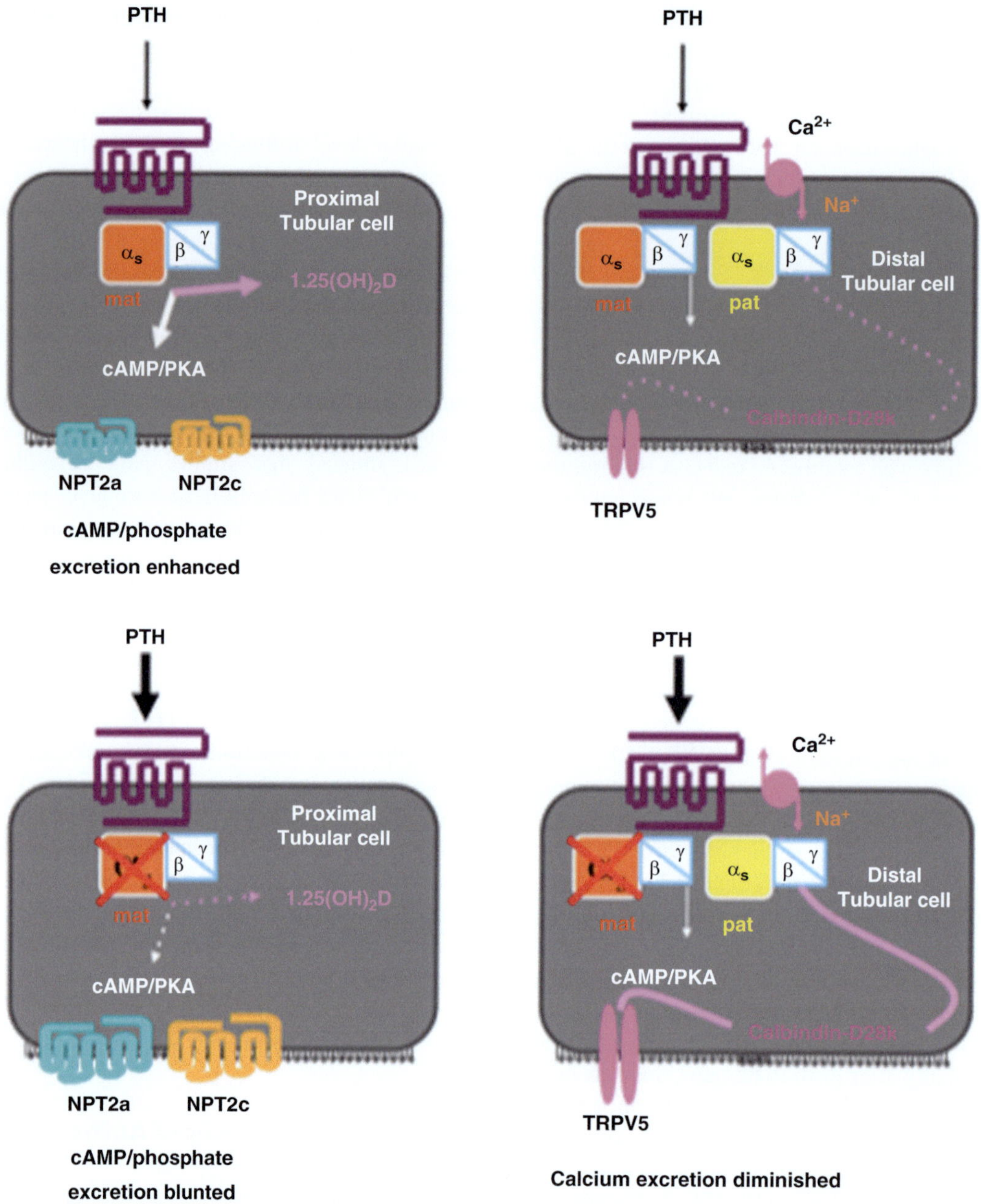

Fig. 34.1 PTH actions in proximal and distal renal tubular cells. *Upper panels:* Proximal (*left*) and distal (*right*) tubular cells express the PTH/PTHrP receptor. Stimulation by PTH enhances the formation of cAMP (Ca^{2+}/IP3 signaling pathway not shown) resulting in PKA activation. Second messenger production is mediated by the stimulatory G protein (Gsα), which is expressed in the distal tubules from both parental alleles, while this signaling protein is derived in the proximal tubules predominantly from the maternal allele. *Lower panels:* A loss of Gsα expression from the maternal allele is observed in patients affected either by PHP-Ia (maternally inherited mutations in the *GNAS* exons encoding Gsα) or PHP-Ib (maternally inherited deletions within or upstream of *GNAS* that are associated with loss of the maternal methylation imprints). Consequently, the PTH-stimulated increase in urinary cAMP and phosphate excretion by the proximal renal tubules is blunted (*left*). In the distal tubules (*right*), however, Gsα expression from the paternal allele remains intact, thus enhancing PTH-stimulated calcium reabsorption which results in diminished urinary calcium excretion

A single linkage peak was observed for chromosome 20q13.3, e.g., the genomic region comprising the *GNAS* locus. Several additional unrelated kindreds mapped to the same locus suggesting that AD-PHP-Ib could be caused by a genetic defect involving *GNAS*, although the exons encoding Gsα had been excluded [32, 33]. In addition, it became apparent that PTH-resistant hypocalcemia and hyperphosphatemia develop only when the disease-associated allele is inherited from a female who is either affected by PHP-Ib or is an obligate carrier; affected males never had affected children. This imprinted mode of inheritance was identical to the findings in patients with Gsα mutations, who develop hormonal resistance only when the *GNAS* mutation is inherited from a female affected either by PHP-Ia or PPHP [11]. Subsequently, Weinstein et al. showed that all familial and most sporadic PHP-Ib cases show methylation changes involving one or several of the four differentially methylated regions (DMRs) within the *GNAS* locus [34]. However, additional mapping of the linked region on chromosome 20q13.3 showed that the mutation which causes AD-PHP-Ib is located centromeric of the *GNAS* locus itself [33]. The disease-causing mutation was eventually identified as a maternally inherited 3-kb deletion within *STX16*, the gene encoding syntaxin 16, and that the deletion is associated with a loss of *GNAS* methylation affecting only exon A/B [35]. This heterozygous deletion removes *STX16* exons 4–6 and is the most frequent cause of AD-PHP-Ib, which has been identified thus far in almost 50 unrelated families [36]. Two subsequently identified *STX16* deletions that are also associated with only limited *GNAS* methylation change remove either exons 2–4 [12] or exons 2–8 [37]. Another deletion of about 18.9 kb was discovered in a kindred in which the affected members show an isolated loss of exon A/B methylation. This novel deletion removes most of the genomic region between *GNAS* antisense exons 4 and 5, including exon NESP55 [38]. In contrast, deletions extending from exon NESP55 to AS exon 3 or comprising only AS exons 3–4 are associated with loss of methylation of all maternal methylation imprints [39, 40].

34.9 Genetic Causes of Sporadic PHP-Ib

The sporadic variant of PHP-Ib remains unresolved at the molecular level for most patients. These sporadic cases have broad *GNAS* methylation changes that usually involve all four DMRs; incomplete loss of methylation has been observed in several of these patients [34, 41–43]. Analysis of microsatellites and single nucleotide polymorphisms of the *GNAS* region for siblings and parents of affected individuals showed, for numerous families, that the healthy siblings shared the same maternally inherited allele as the affected patients; furthermore, the healthy mothers of these patients revealed no evidence for apparent methylation changes at *GNAS* exon NESP55, which would be expected for a *GNAS* deletion comprising this exon. These data excluded an inherited deletion or point mutation involving this chromosomal region, but a de novo mutation remained plausible [42]. However, more recent data made such an event unlikely, at least in several females with sporadic PHP-Ib, who passed either the maternally or the paternally inherited *GNAS* allele to their children, yet these children were healthy and showed no *GNAS* methylation changes [44, 45]. The findings made linkage to the *GNAS* region unlikely and raised the possibility that the sporadic variant of PHP-Ib is caused by a recessive mutation elsewhere in the genome. However, certain findings argue against this possibility. Thus far, only a single family with two affected individuals has been described that does not appear to be linked to the *GNAS* locus [46]. Furthermore, all other unsolved sporadic PHP-Ib cases have, in our experience, only healthy siblings (unpublished observations), thus raising doubts that dominant or recessive mutations in an as-of-yet unknown gene are a frequent cause of this disease variant. Moreover, mutations involving a gene that is involved in establishing or maintaining *GNAS* methylation would be expected to lead to changes at other imprinted genomic loci; however, only few sporadic PHP-Ib cases revealed epigenetic changes outside the *GNAS* locus [47, 48].

A clearly established cause of sporadic PHP-Ib is paternal uniparental isodisomy (patUPD) involving either the long arm of chromosome 20, large segments thereof, or the entire chromosome 20 [46, 49–52] (Fig. 34.2). Because these regions comprise only paternally inherited DNA that is not methylated at exons A/B, XL, and AS, Gsα expression is predicted to be markedly reduced in the proximal renal tubules and other tissues in which Gsα expression from the paternal allele is silenced. The known cases with patUPD20q comprise large chromosomal segments, but such extensive duplications were excluded in numerous sporadic PHP-Ib patients, thus raising the possibility that other causes are responsible for this disorder.

34.10 Hypotheses Regarding the Cause of Unresolved Sporadic PHP-Ib Cases

The lack of multiple families with more than one affected child yet the apparent exclusion of the *GNAS* locus raises the question of whether mechanisms other than patUPD20q or a recessive mutation might be responsible for the sporadic PHP-Ib variant. Such an alternative cause for sporadic PHP-Ib could include interallelic gene conversion [53, 54] or stochastic *GNAS* methylation changes [55]. Most sporadic PHP-Ib patients show a loss of the three maternal methylation

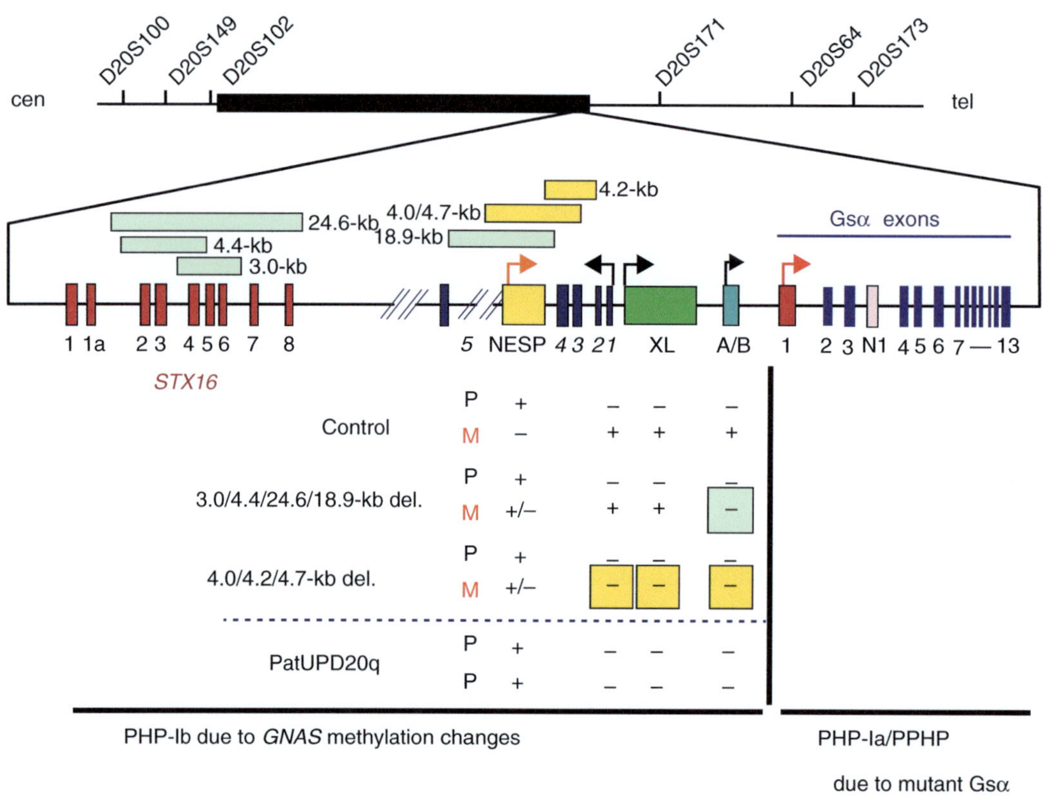

Fig. 34.2 *GNAS* locus: parent-specific methylation and locations of heterozygous deletions that cause autosomal dominant PHP-Ib. Maternally inherited mutations in the Gsα-encoding exons 1–13 cause PHP-Ia, while paternally inherited mutations lead to the related disorders PPHP and POH. Autosomal dominant PHP-Ib is caused either by maternal deletions within *STX16* or within *GNAS*; these disease variants are associated either with loss of *GNAS* methylation restricted to exon A/B alone (*light green horizontal bars*) or with methylation changes at multiple *GNAS* exons (*dark yellow horizontal bars*). Paternal uniparental isodisomy for chromosome 20q (patUPD20q) is a rare cause of sporPHP-Ib, but most sporadic cases have not yet been defined at the molecular level. *Boxes* exons, *connecting lines* introns, *P* paternal, *M* maternal, *methylated* (+), *non-methylated* (–), transcriptional direction (*arrows*) (Modified from Turan et al. [36])

imprints (A/B, XL, and AS) and a gain of methylation at the NESP55 DMR, yet no evidence for a loss of the maternal allele. Duplication of the paternal DNA through gene conversion would thus be expected to comprise all DMRs at the *GNAS* locus, unless all methylation imprints at this locus are established and/or maintained through a small, as-of-yet undefined region within this locus.

34.11 Laboratory and Radiographic Abnormalities in the Different PHP-Ib Variants

Autosomal dominant PHP-Ib and patUPD20q, and other sporadic variants of PHP-Ib, are all associated with *GNAS* methylation changes. Despite having distinct epigenetic abnormalities at the *GNAS* locus (i.e., isolated loss of methylation at exon A/B vs. broad methylation defects involving exon A/B and at least one other *GNAS* DMR), most PHP-Ib patients typically present with similar laboratory abnormalities early in the second decade of life [42]; some patients become symptomatic much earlier, and others do not present with symptoms until much later in life or they develop only mild or no laboratory abnormalities. The delay in the development of symptoms is most likely related to the lack of imprinted Gsα expression in the bone [56], which allows maintaining normocalcemia for extended periods of time through increased PTH-dependent bone resorption. In fact, PHP-Ib patients with long-standing elevations in PTH levels can show an impressive increase in bone turnover resulting in hyperparathyroid bone disease, which led to the introduction of the term "pseudohypo-hyperparathyroidism" [57–60]. Besides PTH resistance, PHP-Ib patients can show evidence for TSH and calcitonin resistance, while GHRH resistance leading to growth hormone deficiency is not a frequent occurrence, which is different from the findings in PHP-Ia patients [49, 61].

Initially PHP-Ib patients were thought to lack AHO features. However, several recent reports identified patients who carry genetic and epigenetic defects associated with PHP-Ib yet present with mild AHO features, particularly shortness of metacarpal bones [43, 62–64]. In one large family in which PHP-Ib is caused by the frequently observed 3-kb deletion within *STX16*, skeletal abnormalities vary from brachydactyly to Madelung-like deformity despite the same underlying genetic defect [65]. These skeletal abnormalities occurred only when the genetic mutation was maternally inherited.

34.12 Pseudohypoparathyroidism Type II

Patients with PTH-induced nephrogenous cAMP formation but a blunted phosphaturic effect without skeletal abnormalities are referred to as having pseudohypoparathyroidism type II (PHP-II). This rare form of PHP is typically sporadic, but a case with a familial form of PHP-II type has been reported [66], and several reports describe evidence for a self-limited form of this disease in newborns, which could indicate that it is transient in nature [67–70]; the underlying molecular defect was postulated to involve a defect downstream of cAMP generation [8]. In fact, mutations in the regulatory subunit of protein kinase A have been identified in some patients with a form of acrodysostosis that is associated with normal PTH-stimulated urinary cAMP excretion but with a blunted phosphaturic response. However, unlike individuals with transient PHP-II, these patients show resistance toward multiple hormones, and they present with characteristic skeletal abnormalities [16, 71]. Alternatively, Gq- or G_{11}-dependent signaling by the PTH/PTHrP receptor may be defective in PHP-II patients, since PKC signaling appears to be important for sustaining the phosphaturic actions of PTH, as recently shown for mice expressing a mutant PTH/PTHrP receptor that fails to activate IP3/PKC signaling [72, 73]. Furthermore, vitamin D deficiency has been associated with PTH-resistant hyperphosphatemia [74–76]. PHP-II may thus be caused by different genetic and non-genetic defects.

References

1. Albright F, Burnett CH, Smith PH, Parson W (1942) Pseudohypoparathyroidism – an example of "Seabright-Bantam syndrome". Endocrinology 30:922–932
2. Albright F, Forbes AP, Henneman PH (1952) Pseudo-pseudohypoparathyroidism. Trans Assoc Am Physicians 65:337–350
3. Tashjian AH Jr, Frantz AG, Lee JB (1966) Pseudohypoparathyroidism: assays of parathyroid hormone and thyrocalcitonin. Proc Natl Acad Sci U S A 56:1138–1142
4. Chase LR, Aurbach GD (1967) Parathyroid function and the renal excretion of 3′5′-adenylic acid. Proc Natl Acad Sci U S A 58:518–525
5. Chase LR, Fedak SA, Aurbach GD (1969) Activation of skeletal adenyl cyclase by parathyroid hormone in vitro. Endocrinology 84:761–768
6. Chase LR, Melson GL, Aurbach GD (1969) Pseudohypoparathyroidism: defective excretion of 3′,5′-AMP in response to parathyroid hormone. J Clin Invest 48:1832–1844
7. Marcus R, Wilber JF, Aurbach GD (1971) Parathyroid hormone-sensitive adenyl cyclase from the renal cortex of a patient with pseudohypoparathyroidism. J Clin Endocrinol Metab 33:537–541
8. Drezner M, Neelon FA, Lebovitz HE (1973) Pseudohypoparathyroidism type II: a possible defect in the reception of the cyclic AMP signal. N Engl J Med 289:1056–1060
9. Patten JL, Johns DR, Valle D, Eil C, Gruppuso PA, Steele G, Smallwood PM, Levine MA (1990) Mutation in the gene encoding the stimulatory G protein of adenylate cyclase in Albright's hereditary osteodystrophy. N Engl J Med 322:1412–1419
10. Weinstein LS, Gejman PV, Friedman E, Kadowaki T, Collins RM, Gershon ES, Spiegel AM (1990) Mutations of the Gs a-subunit gene in Albright hereditary osteodystrophy detected by denaturing gradient gel electrophoresis. Proc Natl Acad Sci U S A 87:8287–8290
11. Davies AJ, Hughes HE (1993) Imprinting in Albright's hereditary osteodystrophy. J Med Genet 30:101–103
12. Linglart A, Gensure RC, Olney RC, Jüppner H, Bastepe M (2005) A novel *STX16* deletion in autosomal dominant pseudohypoparathyroidism type Ib redefines the boundaries of a cis-acting imprinting control element of *GNAS*. Am J Hum Genet 76:804–814
13. Turan S, Fernandez-Rebollo E, Aydin C, Zoto T, Reyes M, Bounoutas G, Chen M, Weinstein LS, Erben RG, Marshansky V et al (2014) Postnatal establishment of allelic galphas silencing as a plausible explanation for delayed onset of parathyroid hormone resistance owing to heterozygous galphas disruption. J Bone Miner Res 29:749–760
14. Yamamoto M, Takuwa Y, Masuko S, Ogata E (1988) Effects of endogenous and exogenous parathyroid hormone on tubular reabsorption of calcium in pseudohypoparathyroidism. J Clin Endocrinol Metab 66: 618–625
15. Stone M, Hosking D, Garcia-Himmelstine C, White D, Rosenblum D, Worth H (1993) The renal response to exogenous parathyroid hormone in treated pseudohypoparathyroidism. Bone 14:727–735
16. Linglart A, Fryssira H, Hiort O, Holterhus PM, Perez de Nanclares G, Argente J, Heinrichs C, Kuechler A, Mantovani G, Leheup B et al (2012) PRKAR1A and PDE4D mutations cause acrodysostosis but two distinct syndromes with or without GPCR-signaling hormone resistance. J Clin Endocrinol Metab 97:E2328–E2338
17. Murray T, Gomez Rao E, Wong MM, Waddell JP, McBroom R, Tam CS, Rosen F, Levine MA (1993) Pseudohypoparathyroidism with osteitis fibrosa cystica: direct demonstration of skeletal responsiveness to parathyroid hormone in cells cultured from bone. J Bone Miner Res 8:83–91
18. Ish-Shalom S, Rao LG, Levine MA, Fraser D, Kooh SW, Josse RG, McBroom R, Wong MM, Murray TM (1996) Normal parathyroid hormone responsiveness of bone-derived cells from a patient with pseudohypoparathyroidism. J Bone Miner Res 11:8–14
19. Elrick H, Albright F, Bartter FC, Forbes AP, Reeves JD (1950) Further studies on pseudo-hypoparathyroidism: report of four new cases. Acta Endocrinol (Copenh) 5:199–225
20. Peterman MG, Garvey JL (1949) Pseudohypoparathyroidism; case report. Pediatrics 4:790
21. Reynolds TB, Jacobson G, Edmondson HA, Martin HE, Nelson CH (1952) Pseudohypoparathyroidism: report of a case showing bony demineralization. J Clin Endocrinol Metab 12:560
22. Winter JSD, Hughes IA (1980) Familial pseudohypoparathyroidism without somatic anomalies. Can Med Assoc J 123:26–31
23. Farfel Z, Brickman AS, Kaslow HR, Brothers VM, Bourne HR (1980) Defect of receptor-cyclase coupling protein in pseudohypoparathyroidism. N Engl J Med 303:237–242
24. Levine MA, Downs RW Jr, Singer M, Marx SJ, Aurbach GD, Spiegel AM (1980) Deficient activity of guanine nucleotide regulatory protein in erythrocytes from patients with pseudohypoparathyroidism. Biochem Biophys Res Commun 94:1319–1324
25. Silve C, Santora A, Breslau N, Moses A, Spiegel A (1986) Selective resistance to parathyroid hormone in cultured skin fibroblasts from patients with pseudohypoparathyroidism type Ib. J Clin Endocrinol Metab 62:640–644
26. Schipani E, Weinstein LS, Bergwitz C, Iida-Klein A, Kong XF, Stuhrmann M, Kruse K, Whyte MP, Murray T, Schmidtke J et al (1995) Pseudohypoparathyroidism type Ib is not caused by mutations in the coding exons of the human parathyroid hormone (PTH)/PTH-related peptide receptor gene. J Clin Endocrinol Metab 80:1611–1621
27. Suarez F, Lebrun JJ, Lecossier D, Escoubet B, Coureau C, Silve C (1995) Expression and modulation of the parathyroid hormone (PTH)/PTH-related peptide receptor messenger ribonucleic acid in skin fibroblasts from patients with type Ib pseudohypoparathyroidism. J Clin Endocrinol Metab 80:965–970

28. Fukumoto S, Suzawa M, Takeuchi Y, Nakayama K, Kodama Y, Ogata E, Matsumoto T (1996) Absence of mutations in parathyroid hormone (PTH)/PTH-related protein receptor complementary deoxyribonucleic acid in patients with pseudohypoparathyroidism type Ib. J Clin Endocrinol Metab 81:2554–2558
29. Bettoun JD, Minagawa M, Kwan MY, Lee HS, Yasuda T, Hendy GN, Goltzman D, White JH (1997) Cloning and characterization of the promoter regions of the human parathyroid hormone (PTH)/PTH-related peptide receptor gene: analysis of deoxyribonucleic acid from normal subjects and patients with pseudohypoparathyroidism type Ib. J Clin Endocrinol Metab 82:1031–1040
30. Jan de Beur S, Ding C, LaBuda M, Usdin T, Levine M (2000) Pseudohypoparathyroidism 1b: exclusion of parathyroid hormone and its receptors as candidate disease genes. J Clin Endocrinol Metab 85:2239–2246
31. Silve C, Jüppner H (2015) Genetic disorders caused by mutations in the PTH/PTHrP receptor and downstream effector molecules. In: Bilezikian J, Marcus R, Levine MA (eds) The Parathyroids (3rd edition). Academic Press, San Diego, CA 587–605.
32. Jüppner H, Schipani E, Bastepe M, Cole DEC, Lawson ML, Mannstadt M, Hendy GN, Plotkin H, Koshiyama H, Koh T et al (1998) The gene responsible for pseudohypoparathyroidism type Ib is paternally imprinted and maps in four unrelated kindreds to chromosome 20q13.3. Proc Natl Acad Sci U S A 95:11798–11803
33. Bastepe M, Pincus JE, Sugimoto T, Tojo K, Kanatani M, Azuma Y, Kruse K, Rosenbloom AL, Koshiyama H, Jüppner H (2001) Positional dissociation between the genetic mutation responsible for pseudohypoparathyroidism type Ib and the associated methylation defect at exon A/B: evidence for a long-range regulatory element within the imprinted *GNAS1* locus. Hum Mol Genet 10:1231–1241
34. Liu J, Litman D, Rosenberg M, Yu S, Biesecker L, Weinstein L (2000) A GNAS1 imprinting defect in pseudohypoparathyroidism type IB. J Clin Invest 106:1167–1174
35. Bastepe M, Fröhlich LF, Hendy GN, Indridason OS, Josse RG, Koshiyama H, Korkko J, Nakamoto JM, Rosenbloom AL, Slyper AH et al (2003) Autosomal dominant pseudohypoparathyroidism type Ib is associated with a heterozygous microdeletion that likely disrupts a putative imprinting control element of *GNAS*. J Clin Invest 112:1255–1263
36. Turan S, Ignatius J, Moilanen J, Kuismin O, Stewart H, Mann N, Linglart A, Bastepe M, Jüppner H (2012) De novo STX16 deletions: an infrequent cause of pseudohypoparathyroidism type Ib that should be excluded in sporadic cases. J Clin Endocrinol Metab 97(12):E2314–E2319
37. Elli FM, de Sanctis L, Peverelli E, Bordogna P, Pivetta B, Miolo G, Beck-Peccoz P, Spada A, Mantovani G (2014) Autosomal Dominant Pseudohypoparathyroidism type Ib: a novel inherited deletion ablating STX16 causes Loss of Imprinting at the A/B DMR. J Clin Endocrinol Metab 99(4):E724–E728, jc20133704
38. Richard N, Abeguile G, Coudray N, Mittre H, Gruchy N, Andrieux J, Cathebras P, Kottler ML (2012) A new deletion ablating NESP55 causes loss of maternal imprint of A/B GNAS and autosomal dominant pseudohypoparathyroidism type Ib. J Clin Endocrinol Metab 97:E863–E867
39. Bastepe M, Fröhlich LF, Linglart A, Abu-Zahra HS, Tojo K, Ward LM, Jüppner H (2005) Deletion of the NESP55 differentially methylated region causes loss of maternal *GNAS* imprints and pseudohypoparathyroidism type Ib. Nat Genet 37:25–27
40. Chillambhi S, Turan S, Hwang DY, Chen HC, Jüppner H, Bastepe M (2010) Deletion of the noncoding *GNAS* antisense transcript causes pseudohypoparathyroidism type Ib and biparental defects of *GNAS* methylation in cis. J Clin Endocrinol Metab 95:3993–4002
41. Liu J, Nealon J, Weinstein L (2005) Distinct patterns of abnormal GNAS imprinting in familial and sporadic pseudohypoparathyroidism type IB. Hum Mol Genet 14:95–102
42. Linglart A, Bastepe M, Jüppner H (2007) Similar clinical and laboratory findings in patients with symptomatic autosomal dominant and sporadic pseudohypoparathyroidism type Ib despite different epigenetic changes at the *GNAS* locus. Clin Endocrinol (Oxf) 67:822–831
43. Mantovani G, de Sanctis L, Barbieri AM, Elli FM, Bollati V, Vaira V, Labarile P, Bondioni S, Peverelli E, Lania AG et al (2010) Pseudohypoparathyroidism and GNAS epigenetic defects: clinical evaluation of albright hereditary osteodystrophy and molecular analysis in 40 patients. J Clin Endocrinol Metab 95:651–658
44. Cavaco BM, Tomaz RA, Fonseca F, Mascarenhas MR, Leite V, Sobrinho LG (2010) Clinical and genetic characterization of Portuguese patients with pseudohypoparathyroidism type Ib. Endocrine 37:408–414
45. Fernández-Rebollo E, Pérez de Nanclares G, Lecumberri B, Turan S, Anda E, Pérez de Nanclares G, Feig D, Nik-Zainal S, Bastepe M, Jüppner H (2011) Exclusion of the GNAS locus in PHP-Ib patients with broad GNAS methylation changes: evidence for an autosomal recessive form of PHP-Ib? J Bone Miner Res 26:1854–1863
46. Fernández-Rebollo E, Lecumberri B, Garin I, Arroyo J, Bernal-Chico A, Goni F, Orduna R, Castano L, Pérez de Nanclares G (2011) New mechanisms involved in paternal 20q disomy associated with pseudohypoparathyroidism. Eur J Endocrinol 163:953–962
47. Pérez-Nanclares G, Romanelli V, Mayo S, Garin I, Zazo C, Fernandez-Rebollo E, Martinez F, Lapunzina P, Pérez de Nanclares G (2012) Detection of hypomethylation syndrome among patients with epigenetic alterations at the GNAS locus. J Clin Endocrinol Metab 97:E1060–E1067
48. Maupetit-Mehouas S, Azzi S, Steunou V, Sakakini N, Silve C, Reynes C, Perez de Nanclares G, Keren B, Chantot S, Barlier A et al (2013) Simultaneous hyper- and hypomethylation at imprinted loci in a subset of patients with GNAS epimutations underlies a complex and different mechanism of multilocus methylation defect in pseudohypoparathyroidism type 1b. Hum Mutat 34:1172–1180

49. Bastepe M, Lane AH, Jüppner H (2001) Paternal uniparental isodisomy of chromosome 20q (patUPD20q) – and the resulting changes in *GNAS1* methylation – as a plausible cause of pseudohypoparathyroidism. Am J Hum Genet 68:1283–1289
50. Lecumberri B, Fernandez-Rebollo E, Sentchordi L, Saavedra P, Bernal-Chico A, Pallardo LF, Bustos JM, Castano L, de Santiago M, Hiort O et al (2009) Coexistence of two different pseudohypoparathyroidism subtypes (Ia and Ib) in the same kindred with independent Gs{alpha} coding mutations and GNAS imprinting defects. J Med Genet 47:276–280
51. Bastepe M, Altug-Teber O, Agarwal C, Oberfield SE, Bonin M, Jüppner H (2011) Paternal uniparental isodisomy of the entire chromosome 20 as a molecular cause of pseudohypoparathyroidism type Ib (PHP-Ib). Bone 48:659–662
52. Dixit A, Chandler KE, Lever M, Poole RL, Bullman H, Mughal MZ, Steggall M, Suri M (2013) Pseudohypoparathyroidism type 1b due to paternal uniparental disomy of chromosome 20q. J Clin Endocrinol Metab 98:E103–E108
53. Chuzhanova N, Chen JM, Bacolla A, Patrinos GP, Ferec C, Wells RD, Cooper DN (2009) Gene conversion causing human inherited disease: evidence for involvement of non-B-DNA-forming sequences and recombination-promoting motifs in DNA breakage and repair. Hum Mutat 30:1189–1198
54. Chen JM, Cooper DN, Chuzhanova N, Ferec C, Patrinos GP (2007) Gene conversion: mechanisms, evolution and human disease. Nat Rev Genet 8:762–775
55. Jan de Beur S, Ding C, Germain-Lee E, Cho J, Maret A, Levine M (2003) Discordance between genetic and epigenetic defects in pseudohypoparathyroidism type 1b revealed by inconsistent loss of maternal imprinting at GNAS1. Am J Hum Genet 73:314–322
56. Mantovani G, Bondioni S, Locatelli M, Pedroni C, Lania AG, Ferrante E, Filopanti M, Beck-Peccoz P, Spada A (2004) Biallelic expression of the Gsalpha gene in human bone and adipose tissue. J Clin Endocrinol Metab 89:6316–6319
57. Costello JM, Dent CE (1963) Hypo-hyperparathyroidism. Arch Dis Child 38:397–407
58. Allen EH, Millard FJC, Nassim JR (1968) Hypo-hyperparathyroidism. Arch Dis Child 43:295–301
59. Frame B, Hanson CA, Frost HM, Block M, Arnstein AR (1972) Renal resistance to parathyroid hormone with osteitis fibrosa: "pseudohypohyperparathyroidism". Am J Med 52:311–321
60. Farfel Z (1999) Pseudohypohyperparathyroidism-pseudohypoparathyroidism type Ib. J Bone Miner Res 14:1016
61. Mantovani G, Bondioni S, Linglart A, Maghnie M, Cisternino M, Corbetta S, Lania AG, Beck-Peccoz P, Spada A (2007) Genetic analysis and evaluation of resistance to thyrotropin and growth hormone-releasing hormone in pseudohypoparathyroidism type Ib. J Clin Endocrinol Metab 92:3738–3742
62. Pérez de Nanclares G, Fernández-Rebollo E, Santin I, Garcia-Cuartero B, Gaztambide S, Menendez E, Morales MJ, Pombo M, Bilbao JR, Barros F et al (2007) Epigenetic defects of GNAS in patients with pseudohypoparathyroidism and mild features of Albright's hereditary osteodystrophy. J Clin Endocrinol Metab 92:2370–2373
63. Mariot V, Maupetit-Mehouas S, Sinding C, Kottler ML, Linglart A (2008) A maternal epimutation of GNAS leads to Albright osteodystrophy and parathyroid hormone resistance. J Clin Endocrinol Metab 93:661–665
64. Unluturk U, Harmanci A, Babaoglu M, Yasar U, Varli K, Bastepe M, Bayraktar M (2008) Molecular diagnosis and clinical characterization of pseudohypoparathyroidism type-Ib in a patient with mild Albright's hereditary osteodystrophy-like features, epileptic seizures, and defective renal handling of uric acid. Am J Med Sci 336:84–90
65. Sanchez J, Perera E, Jan de Beur S, Ding C, Dang A, Berkovitz GD, Levine MA (2011) Madelung-like deformity in pseudohypoparathyroidism type 1b. J Clin Endocrinol Metab 96:E1507–E1511
66. van Dop C (1989) Pseudohypoparathyroidism: clinical and molecular aspects. Semin Nephrol 9:168–178
67. Kruse K, Kustermann W (1987) Evidence for transient peripheral resistance to parathyroid hormone in premature infants. Acta Paediatr Scand 76:115–118
68. Narang M, Salota R, Sachdev SS (2006) Neonatal pseudohypoparathyroidism. Indian J Pediatr 73:97–98
69. Lee CT, Tsai WY, Tung YC, Tsau YK (2008) Transient pseudohypoparathyroidism as a cause of late-onset hypocalcemia in neonates and infants. J Formos Med Assoc 107:806–810
70. Manzar S (2001) Transient pseudohypoparathyroidism and neonatal seizure. J Trop Pediatr 47:113–114
71. Linglart A, Menguy C, Couvineau A, Auzan C, Gunes Y, Cancel M, Motte E, Pinto G, Chanson P, Bougneres P et al (2011) Recurrent PRKAR1A mutation in acrodysostosis with hormone resistance. N Engl J Med 364:2218–2226
72. Guo J, Liu M, Yang D, Bouxsein ML, Thomas CC, Schipani E, Bringhurst FR, Kronenberg HM (2010) Phospholipase C signaling via the parathyroid hormone (PTH)/PTH-related peptide receptor is essential for normal bone responses to PTH. Endocrinology 151:3502–3513
73. Guo J, Song L, Liu M, Segawa H, Miyamoto K, Bringhurst FR, Kronenberg HM, Jüppner H (2013) Activation of a non-cAMP/PKA signaling pathway downstream of the PTH/PTHrP receptor is essential for a sustained hypophosphatemic response to PTH infusion in male mice. Endocrinology 154:1680–1689
74. Rao DS, Parfitt AM, Kleerekoper M, Pumo BS, Frame B (1985) Dissociation between the effects of endogenous parathyroid hormone on adenosine 3′,5′-monophosphate generation and phosphate reabsorption in hypocalcemia due to vitamin D depletion: an acquired disorder resembling pseudohypoparathyroidism type II. J Clin Endocrinol Metab 61:285–290
75. Srivastava T, Alon US (2002) Stage I vitamin D-deficiency rickets mimicking pseudohypoparathyroidism type II. Clin Pediatr (Phila) 41:263–268
76. Shriraam M, Bhansali A, Velayutham P (2003) Vitamin D deficiency masquerading as pseudohypoparathyroidism type 2. J Assoc Physicians India 51:619–620

Genetic Testing in Pseudohypoparathyroidism

35

Agnès Linglart and Susanne Thiele

35.1 Introduction

When Fuller Albright described in 1942 the concept of *pseudohypoparathyroidism* (PHP) in three patients who showed no improvement in their serum calcium or their serum phosphate concentrations in response to parathyroid extract, he could not imagine the future extent of the field he was creating [1]. More than ten different phenotypes, a similar number of causative mechanisms and/or genes and major biological functions like *cAMP* signaling, epigenetics, development, and cell differentiation are concealed behind the term "pseudohypoparathyroidism." This explains the difficulty for experts to produce a consensual classification of PHP and the rapid evolution of the recommendations in both care and genetic/epigenetic testing of affected patients.

A. Linglart (✉)
Department of Pediatric Endocrinology, Center of Reference for Rare Disorders of the Calcium and Phosphorus Metabolism, APHP, Paris-Sud Hospital and Paris-Sud University, 82 avenue du Général Leclerc, Le Kremlin Bicêtre 94270, France
e-mail: agnes.linglart@inserm.fr

S. Thiele
Division of Pediatric Endocrinology and Diabetes, Department of Pediatrics and Adolescent Medicine, University of Luebeck, Ratzeburger Allee 160, Luebeck, Schleswig-Holstein, 23538, Germany
e-mail: thiele@paedia.ukl.mu-luebeck.de

The genetic testing of a patient affected with PHP or a PHP-like phenotype requires knowing the following elements:

- The peculiarity of *parental imprinting*
- The overlap among clinical presentations of PHP and their underlying molecular mechanisms
- The ongoing discovery of new genes

It is therefore of utmost importance to delineate the symptoms and biochemical findings of the affected patient in order to guide the genetic investigation and identify the disease-causing mechanism (see also Chaps. 32, 33, and 34).

35.2 Genetics, Epigenetics, and PTH Signaling

Under physiological conditions, upon ligand binding (e.g., parathyroid hormone (PTH)) to the parathyroid hormone (PTH), Gsα, the alpha-stimulatory subunit of the G protein, encoded by the *GNAS* gene, dissociates from the ßγ-subunits of the G protein, activates adenylate cyclase, and triggers cAMP synthesis, which acts as a second messenger to elicit the effects on target cells. Binding of cAMP to the regulatory subunits (R1A encoded by PRKAR1A) of protein kinase A (PKA) unlocks catalytic subunits and unleashes a cascade of events, including phosphorylation of PDE4D and CREB. Phosphodiesterases (PDEs), including PDE4D, degrade cAMP to maintain intracellular concentrations. Other cAMP targets have been identified, such as cyclic nucleotide-gated cation

M.L. Brandi, E.M. Brown (eds.), *Hypoparathyroidism*,
DOI 10.1007/978-88-470-5376-2_35, © Springer-Verlag Italia 2015

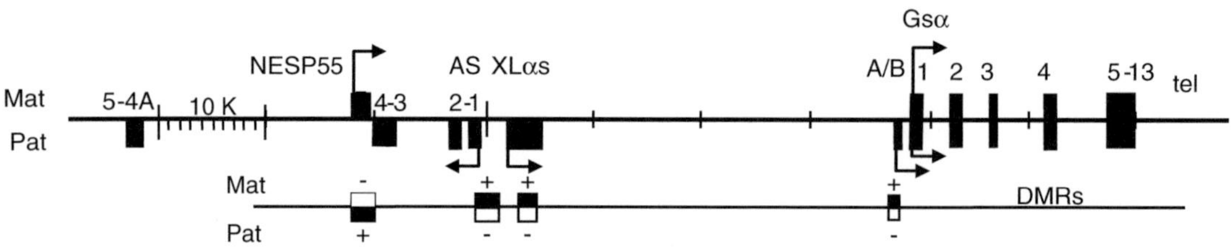

Fig. 35.1 Schematic representation of the *GNAS* locus. The *GNAS* locus is scaled, based on HG19. The four DMRs of *GNAS* are represented below the genomic line by *black boxes* (+ or methylated) or *white boxes* (– or unmethylated) for the paternal (*Pat*) or maternal (*Mat*) allele. Exons are indicated as *black rectangles*, and the allelic origin of transcription is indicated by *broken arrows* on the paternal (*Pat*) or maternal (*Mat*) allele

channels (CNG) and the exchange proteins 1 and 2 activated by cAMP (Epac1 and Epac2) [2, 3].

Pseudohypoparathyroidism is the consequence of the lack of the downstream, cAMP-mediated signaling activated via the PTH receptor (PTHR1) following its activation by its ligand, PTH. So far, most of the identified causes of PHP affect the signaling of the PTHR1 through this Gsα/cAMP/PKA pathway. Molecular alterations of this pathway also affect, although with different severities or at different times of development, the signaling of numerous other hormones and their receptors, including TSH, PTHrP (PTH-related peptide, a ligand of the PTHR1 specifically acting on chondrocyte differentiation) catecholamine, and calcitonin that signal through GPCRs and the same Gsα/cAMP/PKA pathway [4].

Genomic imprinting is an epigenetic mechanism whereby expression of a subset of genes is restricted to a single parental allele. The *GNAS* locus is a complex imprinted locus that encodes Gsα and four additional alternative transcripts: XLαs, NESP55 (neurosecretory protein 55), the A/B transcript, and the antisense transcript (AS). AS, XLαs, and A/B are paternally derived transcripts, while NESP55 is a maternally derived transcript [5, 6]. Consistent with their monoallelic expression, the promoters of these imprinted transcripts are located within differentially methylated regions (DMRs) (Fig. 35.1). In contrast, the promoter giving rise to Gsα is not methylated, and, accordingly, transcripts are derived in most tissues from both parental *GNAS* alleles. In some tissues, including proximal renal tubules, brown adipose tissue, pituitary, gonads, and thyroid, however, Gsα transcripts are predominantly derived from the maternal allele [7]. Gsα and its splice variant XLαs are identical in their C-termini encoded by exons 2–13 of *GNAS* and differ in their N-termini encoded by different first exons [8].

35.3 Classifications of PHP and Guidance for Genetic Testing

An update on the classification of PHP has been given in Chap. 32, and diseases have been described in detail in Chaps. 33 and 34. The phenotype of the patients will remain the primary guide for the genetic testing and will determine the order of investigations. Juxtaposing phenotypes and molecular findings in PHP provides the grounds for the choice of genetic/epigenetic testing in patients (Table 35.1).

35.4 Patients Affected with PHP1A

Patients presenting a collection of features, including Albright hereditary osteodystrophy (AHO) and obesity, and end-organ resistance to hormones that signal through G-protein-coupled receptors (GPCRs) are categorized under the term pseudohypoparathyroidism 1A (PHP1A). *Heterozygous loss-of-function mutations of* Gsα are the main cause of PHP1A [9, 12, 26–28]. Because of the parental imprinting of *GNAS*, these mutations

Table 35.1 The different classifications of PHPs and their overlapping phenotypes and molecular mechanisms

	PHP type 1A	PHP type 1C	PHP type 1B	Pseudo PHP	Progressive osseous heteroplasia	Acrodysostosis with hormonal resistance	Acrodysostosis without hormonal resistance	Brachydactyly type E
OMIM	PHP1A # 103580	PHP1C # 612462	PHP1B # 603233	PPHP # 612463	POH # 166350	ACRDYS1 # 101800	ACRDYS2 # 614613	BDE # 113300
Phenotype	Albright hereditary osteodystrophy and hormonal resistance	Albright hereditary osteodystrophy and hormonal resistance	Hormonal resistance, few or absence of clinical phenotype	Albright hereditary osteodystrophy	Ectopic ossifications	Acrodysostosis[a]	Acrodysostosis[a]	Brachymetacarpy, short stature
Main molecular mechanism(s)	Maternal *GNAS* mutations, coding regions	Maternal *GNAS* mutations	LOI at the *GNAS* DMRs	Paternal *GNAS* mutations, coding regions	Paternal *GNAS* mutations, coding regions	Heterozygous *PRKAR1A* mutations	Heterozygous *PDE4D* mutations	Heterozygous *PTHLH* mutations
Other identified mechanisms	LOI at the *GNAS* DMRs Heterozygous *PRKAR1A* mutations		*STX16* deletions patUPD of chromosome 20q Maternal deletion of NESP and/or AS Maternal heterozygous mutation in the coding sequence of *GNAS* (DelIle382)	Cytogenetic deletion at 2q37- Heterozygous *PTHLH* mutations				*HOXD13* mutations
References	[9–11]	[12, 13]	[14–18]	[9, 24]	[19]	[20]	[21, 22]	[23]

(continued)

Table 35.1 (continued)

	PHP type 1A	PHP type 1C	PHP type 1B	Pseudo PHP	Progressive osseous heteroplasia	Acrodysostosis with hormonal resistance	Acrodysostosis without hormonal resistance	Brachydactyly type E

Mutation/epimutation	Heterozygous *PTHLH* mutations	Heterozygous *PRKAR1A* mutations	Heterozygous *PDE4D* mutations	Paternal *GNAS* mutations, coding regions	Maternal *GNAS* mutations, coding regions	LOI at the *GNAS* DMRs
Related diseases	BDE2	ACRDYS1	ACRDYS2	PPHP POH IUGR and failure to thrive	PHP1A PHP1B PHP1C ACRDYS1	PHP1B PHP1A
References				[9, 19, 25]	[9, 12, 13, 18]	

LOI is for loss of imprinting. The table is divided in two parts. In the top table, the diseases are classified according to their clinical presentation and the actual classification of PHP; in the bottom table, the diseases are classified according to their molecular mechanisms

[a]Acrodysostosis could be considered as an extremely severe form of Albright hereditary osteodystrophy

always lie on the maternal allele of *GNAS*. As would be expected, these mutations are either maternally inherited or occur *de novo* on the maternal *GNAS* allele [12]. Deletions, insertions, amino-acid substitutions, or stop codons have been found in coding exons and exons-introns boundaries, with three hot spots located in exons 6, 7, and 13 of *GNAS*; mutations may also lie in the alternatively spliced exon 3 [29]. *In vitro* analyses, in transfected cells that endogenously lack both Gsα and XLαs (*Gnas*$^{-/-}$ cells), have demonstrated that mutations impair Gsα transcript expression or cAMP generation upon agonist stimulation [8, 30]. These mutations are identified in about 60–70 % of the PHP1A patients [12, 27]. The sequencing of *GNAS*, whatever the technique used, is therefore the primary investigation to consider in these patients. However, some issues have been identified. Firstly, as for many genes, exon 1 of *GNAS* is highly GC enriched and consequently is difficult to amplify and sequence. Secondly, *large deletions* removing the entire *GNAS* gene or several exons may cause PHP1A [31, 32]. If the sequencing technique, i.e., Sanger method, does not measure the gene dosage (copy number), such large deletions cannot be detected and require specific quantitative strategies such as multiplex ligation-dependent probe amplification (MLPA), a custom-made CGH array encompassing the *GNAS* locus (an approach utilized by us), quantitative genomic PCR [33], or next-generation sequencing (the list is not exhaustive).

As mentioned above, about 30 % of patients with a phenotype compatible with PHP1A do not display mutations in the coding regions of *GNAS*. There is now enough evidence that several different molecular defects present as phenocopies of PHP1A. They constitute the second line of investigation in these patients:

- We and others have shown that abnormal methylation of the *GNAS* DMRs is the cause, in some patients, of PTH resistance and Albright hereditary osteodystrophy (24 out of 40 patients in the series of Mantovani et al. [10, 33, 34]). It is therefore mandatory to investigate the imprinting pattern of the *GNAS* locus in these patients as described below in patients identified as PHP1B.
- Mutations in *PRKAR1A*, encoding the regulatory subunit of PKA, leading to a decreased affinity of the subunit for cAMP hence decreased PKA activity, have also been identified in patients first characterized as PHP1A by their physicians [11, 35] (our experience, Fig. 35.2).

35.5 Patients Affected with PHP1C

Pseudohypoparathyroidism 1C (PHP1C) differs from PHP1A by the preserved Gsα activity measured in patients' blood cells. Our group has demonstrated through functional *in vitro* analysis that some of these patients carry a maternal loss-of-function mutation of the coding regions of *GNAS*. Interestingly, these mutations (1) are located within the carboxy-terminus of Gsα, (2) impair the ability of the protein to interact with GPCRs, but (3) do not prevent the generation of cAMP through stimulation of adenylate cyclase which explains the normal *in vitro* Gsα bioactivity [12, 13]. All considerations regarding genetic testing for PHP1A, therefore, also apply to PHP1C.

35.6 Patients Affected with PHP1B

The molecular diagnosis of PHP1B is a three-step strategy:

First step: document that the patient's phenotype is compatible with PHP1B

In most cases, the phenotype of patients affected with PHP1B is restricted to the renal resistance to the action of PTH, i.e., the association of normal or low calcium levels, elevated serum phosphate, elevated PTH, and normal kidney function. Careful attention should be paid to the biochemical characterization of PTH resistance as other conditions may lead to increased PTH levels and normal calcium levels (normocalcemic hyperparathyroidism) or increased PTH levels and hypocalcemia (vitamin D deficiency). As mentioned above, in Chap. 34 and illustrated

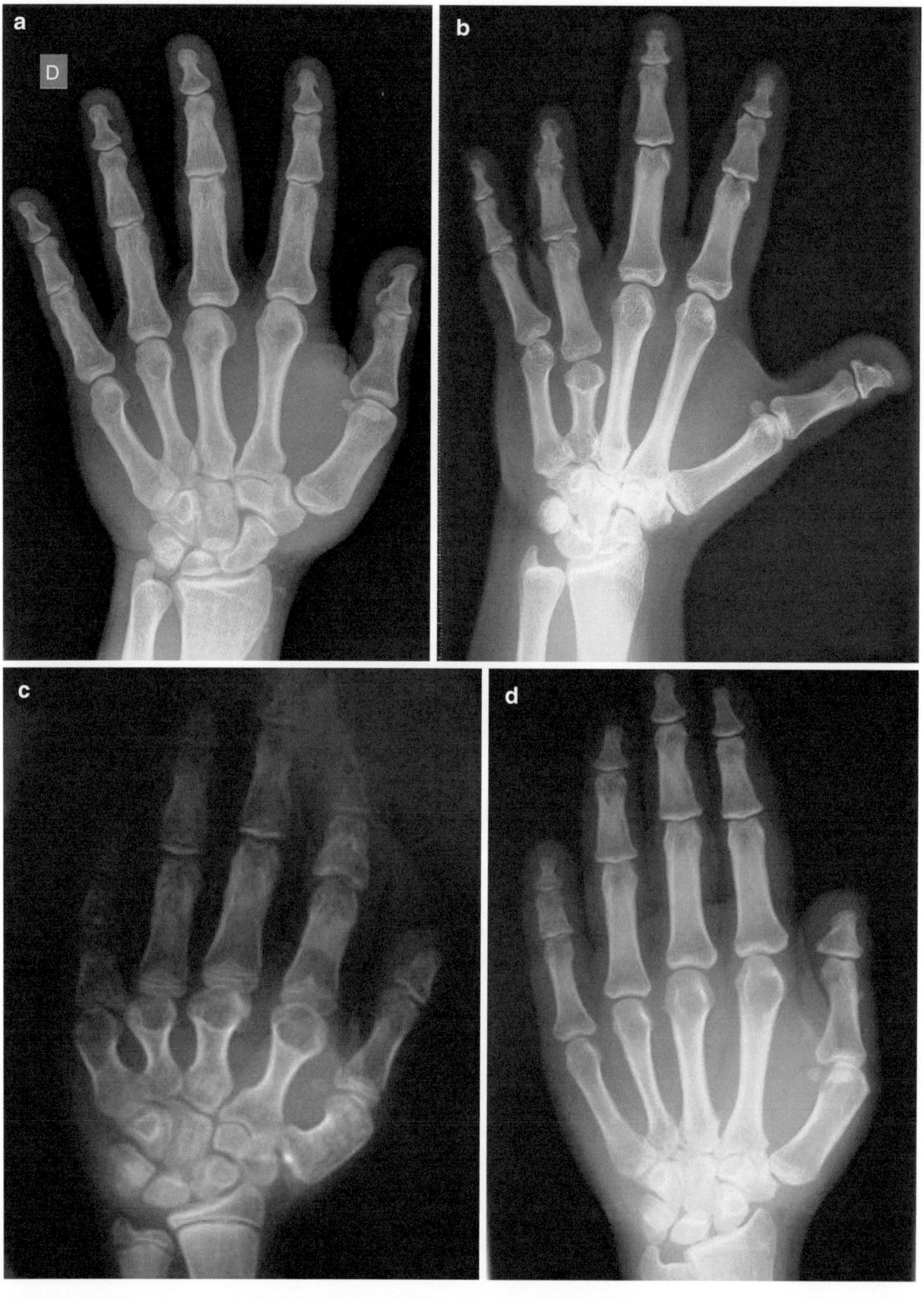

Fig. 35.2 Phenotypic overlap in patients affected with PHP. (**a–c**) are X-rays of the hand from a 30-year-old male, a 13-year-old girl, and an 11-year-old girl, respectively, all affected with PHP1A and a maternal loss-of-function mutation of *GNAS*. (**d**, **e**) are hand X-rays from a 24-year-old male and a 13-year-old girl, respectively, both affected with PHP1B and broad loss of imprinting at the *GNAS* DMRs. (**f**, **g**) are hand X-rays from a 9-year-old girl and an 11-year-old girl, respectively, both affected with a bone phenotype and PTH and TSH resistance and who carry a heterozygous mutation in PRKAR1A (OMIM ACRDYS1)

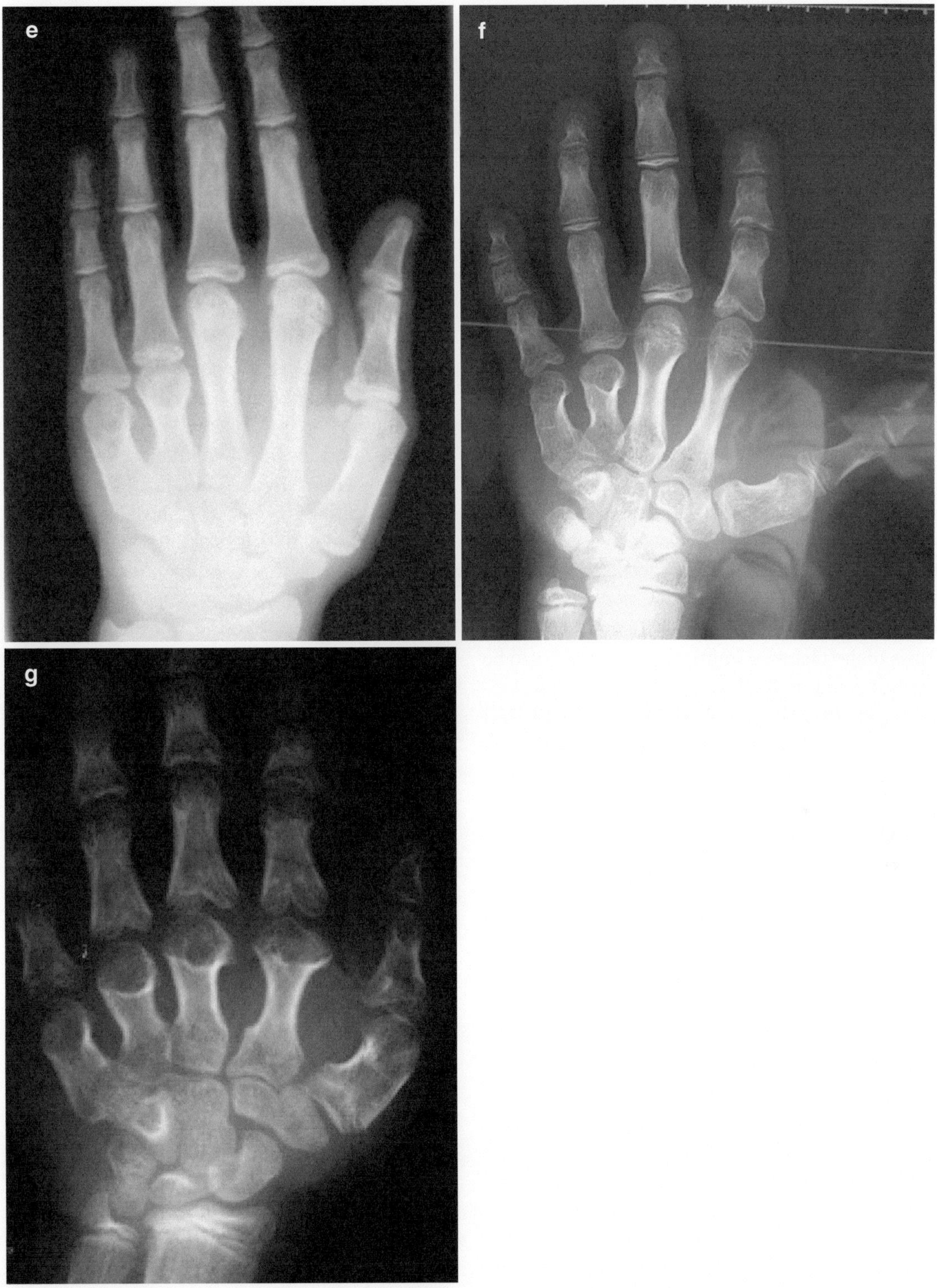

Fig. 35.2 (continued)

in Fig. 35.2, the diagnosis should also be considered in patients with resistance to hormones that signal through GPCRs and exhibit features of Albright hereditary osteodystrophy in the absence of loss-of-function mutations in the *GNAS* gene.

Second step: prove the loss of imprinting at the A/B promoter of *GNAS*

All patients affected with PHP1B share the loss of methylation at the A/B promoter of *GNAS* (Fig. 35.1), which likely results in suppressed Gsα transcription in imprinted tissues. The association between the decreased methylation of cytosines at the A/B promoter of *GNAS* and the increased expression of the A/B transcript (and hence the suppression of Gsα transcript expression) is based on a few experimental observations [36, 37], mostly from animal studies. We have shown that patients with LOI at the A/B promoter of *GNAS* display a significant decrease in erythrocyte Gsα activity compared to normal controls [13] and that the percent of methylation at A/B correlates with serum PTH in these patients [38]. Loss of the maternal imprint at A/B appears to be the molecular defect mandatory for the development of PHP1B. In about 80 % of PHP1B patients, loss of imprinting (LOI) encompasses at least another DMR of *GNAS*, i.e., XL, AS, and/or NESP [14, 39]. DMRs' methylation along the *GNAS* locus might be altered unevenly among patients [38].

Because imprinted DMRs have opposite methylation profiles, e.g., methylated on the paternal allele and unmethylated on the maternal allele, quantitative measure of DNA methylation at the A/B, XL, AS, and NESP DMRs of *GNAS* is more or less 50 % in samples from healthy individuals. The range of methylation indices is very narrow [38] and constant throughout life at these DMRs. In patients, the profound loss of methylation at A/B, XL, and AS and the gain of methylation at NESP indicate the loss of maternal imprinting. However, in some patients, the methylation defect is incomplete, suggesting that (1) different cell populations coexist, cells with abnormal methylation at *GNAS* and cells with a normal methylation pattern, and (2) the epimutation occurred postzygotically in these patients. There is a need to define the boundaries of normal/abnormal methylation patterns at the different DMRs of *GNAS*; the sharing of human samples and the results obtained should facilitate this.

Several methods have been developed to assess the *methylation* pattern at imprinted regions of *GNAS*. All techniques necessitate discriminating methylated and unmethylated DNA. The first description of LOI at *GNAS* was done by digesting the genomic DNA of patients extracted from leukocytes by methyl-sensitive enzymes, followed by probe hybridization onto the A/B DMR of *GNAS* (also called 1A). Southern blotting was then rapidly replaced by less DNA- and time-consuming methods. Today, most techniques differentiate the methylated and unmethylated alleles through chemical modification of unmethylated cytosines by sodium bisulfite followed by DNA amplification. Qualitative, semiquantitative, or quantitative assessment of the methylated cytosines (not modified by the sodium bisulfite) or unmethylated cytosines (converted into thymidines) may be done through numerous tactics including enzymatic restriction (COBRA or combined bisulfite restriction enzymatic analysis), Sanger sequencing, pyrosequencing, MethylQuant, or EpiTYPER. Alternatively, other methods, such as methylation-specific MLPA or MS-MLPA, profile the methylation semiquantitatively using methyl-sensitive restriction of the DNA combined with copy number detection (see below). The diagnosis of PHP1B should include characterization of the methylation defect at the A/B DMR of *GNAS*, which we know is associated with impaired Gsα expression.

Third step: identify the cause of the loss of maternal imprint at the A/B promoter of *GNAS*

Genomic imprinting is mainly due to the differential allelic DNA methylation of regulatory regions or promoters that occurs under the control of *imprinting control elements* (ICEs). The integrity of the ICEs and the factors necessary for the establishment of imprinting and its maintenance, together with a proper environment, are necessary to ensure a 50 % methylation index in all DMRs of *GNAS*. We and others propose that the LOI at the A/B promoter of *GNAS* is the consequence of a cytogenetic, genetic, or environmental assault that needs to be identified.

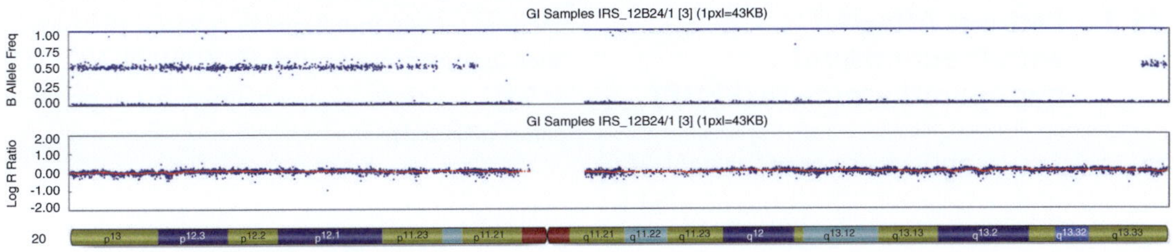

Fig. 35.3 Isodisomy encompassing most of chromosome 20q in a PHP1B patient with broad LOI at *GNAS*. Results from a SNP array

The known causes of loss of maternal imprint at *GNAS* are as follows:

- Deletion of an ICE of *GNAS*

In most familial cases of PHP1B, the LOI is restricted to A/B and associated with a maternal heterozygous deletion of *STX16* [14, 40, 41], located 220 kb telomeric of A/B, or NESP [15]. In a few familial cases with LOI extending along the whole *GNAS* locus, deletions of the entire maternal NESP or of the AS DMRs have been identified [16, 42]. The deleted regions are considered as ICEs of *GNAS*. In theory, maternal deletion of the entire *GNAS* locus should also lead to PHP.

Identification of ICE deletions has been carried out through various methods including Southern blot [14], quantitative genomic PCR [33], comparative genomic hybridization, multiplex long-range PCR [14], and MS-MLPA [41]. The latter is now widely used as it assesses both the methylation profile and copy number of *GNAS* alleles in a single experimental run and it is commercially available. However, the large size of the imprinted *GNAS* locus and the likelihood of necessary long-acting regulator elements of *GNAS* prevent the complete search for ICE deletions with these methods.

- Cytogenetic errors involving chromosome 20q

Uniparental disomy (UPD) arises usually from the failure of the two members of a chromosome pair to separate properly during meiosis in the parent's germline (nondisjunction). Rescue events such as duplication of a single chromosome in a monosomy or nondisjunction of chromosomes causing a trisomy will then result in UPD. UPD can also occur post fertilization. Consequently, the imbalance of paternal and maternal *GNAS* alleles leading to the loss of maternal imprinting is associated with decreased Gsα expression. Paternal UPD of chromosome 20q comprises isodisomy (identical copies of the paternal chromosome 20q – or segments containing *GNAS* – and loss of the maternal chromosome), heterodisomy (lack of transmission – or loss – of the maternal chromosome 20q and transmission of both paternal chromosome 20q), or a mixture of isodisomy and heterodisomy. Overall, patUPD represents 10–25 % of investigated PHP1B patients [17, 43, 44]. Because the discovery of patUPD will make it possible to inform the patient of the near-absent risk of transmission, we propose to investigate patients with broad LOI at the *GNAS* locus, especially if they also display LOI at the nearby imprinted locus *L3MBTL1* [44].

Isodisomy is characterized by perfect homozygosity along segments or the entire chromosome. All methods used to demonstrate isodisomy feature loss of heterozygosity (LOH), such as microsatellite analysis or single nucleotide polymorphism (SNP) arrays (Fig. 35.3). The search for heterodisomy requires analyzing parents' and proband's samples and performing haplotype analysis.

Identification of LOI will therefore trigger genetic and cytogenetic investigations depending on the *GNAS* methylation profile. Loss of methylation restricted to the A/B promoter of *GNAS* leads to a thorough analysis of the *STX16* and NESP regions. Broad LOI at *GNAS* implies the need to look for patUPD, deletion of the locus, and deletion of an ICE of *GNAS*. However, in about 70 % of PHP1B patients with broad LOI at *GNAS*, the underlying defect accounting for the loss of maternal imprint at A/B remains to be discovered.

35.7 Patients Affected with Pseudopseudo-hypoparathyroidism (PPHP)

The diagnosis of PPHP is remarkably challenging because of the nonspecific symptoms of Albright hereditary osteodystrophy. Brachydactyly may reveal PPHP as well as numerous developmental bone disorders [11, 45]. Several situations direct the genetic testing in these patients and need to be recognized.

35.7.1 PPHP with Ectopic Subcutaneous Ossifications or Osteoma Cutis

The presence of ectopic bone in the dermis or subcutaneous fat is essential to the diagnosis of PPHP. Very few diseases, especially in children, lead to the formation of ectopic bone. The main differential diagnosis is fibrodysplasia ossificans progressiva (FOP), a rare disorder characterized by bone growing from deep tissues, not from the surface. Altogether, when ectopic bone is present – in the absence of hormonal resistance – one should search for paternal mutations of the coding region of *GNAS*. Of note, in young children affected with PHP1A, including ectopic subcutaneous ossifications and a maternal mutation of *GNAS*, the hormonal resistance may be absent.

35.7.2 PPHP in a Familial Context of PHP1A

The genetic testing will aim at identifying a loss-of-function mutation in the coding region of *GNAS*.

35.7.3 "Isolated" PPHP

This represents the most problematic situation for genetic testing, which will depend on the involvement and expertise of the physician and availability and cost of gene analyses in order to trigger and orient the investigations. Besides paternal coding mutations in *GNAS*, PPHP could uncover mutations in PDE4D, PTHLH, HOXD3, or 2q37 deletions (non-exhaustive list of possible genes affected).

35.8 Patients Affected with POH

Mutations in the coding sequence of Gsα found in patients with POH are also found in patients with PHP1A or PPHP, although they are exclusively located on the paternal allele [19], and mostly severely affect protein function [46].

35.9 Patients Affected with Acrodysostosis

Acrodysostosis is a developmental chondrodysplasia presenting some phenotypic similarities to PHP1A. Patients typically present with facial and peripheral dysostosis characterized by severe brachydactyly, indicating a defect in impairment of PTHrP actions on endochondral bone development [20–22, 47]. In some patients, resistance to PTH and other hormones signaling through GPCRs is present, although it is less pronounced than in PHP1A. This resistance is associated with elevated cAMP levels measured in blood and urine, as expected with cAMP resistance. These patients are named ACRDYS1 in contrast with ACRDYS2 (acrodysostosis without hormonal phenotype). Heterozygous mutations of *PRKAR1A* are responsible for ACRDYS1 [20]. These mutations are mainly missense mutations, except for very few mutations, like the hot-spot mutation, R368X, that induces frameshifts and produces R1A proteins lacking the very last amino acids at the carboxy-terminus. So far, all identified mutations (more than 30 including those reported and our unpublished results) lie within the cAMP-binding domain A or B of the regulatory subunit 1A of the PKA. Functional studies are consistent with the *in vivo* hormonal resistance and cAMP-signaling impairment [20, 35]. Noteworthy, none of these 30 mutations were inherited; hence, they were all considered sporadic.

Heterozygous mutations in *PDE4D*, the gene encoding one of the numerous phosphodiesterases, were found in patients with acrodysostosis, yet there was no hormonal resistance (ACRDYS2). Vertical transmission of the mutation is possible. To date, there is no report of *in vitro* functional analysis of mutated PDE4D. It is therefore almost impossible to anticipate the consequences of the mutations on the protein function as (1) mutations have been identified in all domains of the protein, (2) they are almost exclusively missense mutations, (3) blood and urine cAMP are found normal in patients, and (4) resistance to PTHrP should result from impaired cAMP signaling and, therefore, increased activity of phosphodiesterases in chondrocytes.

Even if the phenotype/genotype correlation of ACRDYS1/*PRKAR1A* and ACRDYS2/*PED4D* is very convincing, physicians should be reminded that:

- ACRDYS1 and PHP1A have overlapping phenotypes.
- Mental retardation is more frequent in patients with *PDE4D* mutations.
- TSH resistance has been reported in few patients with ACRDYS2 and *PED4D* mutations, although not in any detail.
- *PRKAR1A* and *PDE4D* mutations do not account for all patients with acrodysostosis, and more genes are very likely involved in the pathogenesis of this disorder.

35.10 The Place of Complementary Biochemical Investigation to Guide Genetic Testing

The above paragraphs show the challenge in identifying the molecular cause of PHP in any given patient. It is of utmost importance to delineate the clinical and biochemical phenotype of the patients in order to adjust genetic exploration toward gene sequencing, methylation analysis, or cytogenetic arrays.

The renal response to the infusion of exogenous PTH (formerly Ellsworth-Howard test, which has been replaced by the *infusion of recombinant PTH(1–34)* [48, 49]) may be used to document the impairment of the cAMP-signaling pathway in vivo. PHP1A and PHP1B are typically associated with absent urinary cAMP production and an increase in urinary phosphate excretion; in contrast, patients with ACRDYS1 display elevated basal urine cAMP levels and a normal rise of cAMP upon agonist stimulation. Normal basal cAMP levels have been measured in PHP1A, PHP1B, and ACRDYS2, and, therefore, this finding should not exclude any of these diagnoses.

The erythrocyte bioassay for measuring Gsα *activity* allows an appraisal of the defect in the cAMP-signaling pathway. With the easy access of modern-day genetics, this assay is no longer used to discriminate PHP1A and PHP1B. However, it is valuable in phenocopies of PHP, as it may orient research toward a specific factor in the pathway [13].

35.11 The Place of Complementary Genetic/Epigenetic Analyses to Refine Diagnosis and Prognosis

The peculiarity of genomic imprinting adds a layer of complexity to the genetic testing and prediction of disease inheritance. Several laboratories have developed strategies to further characterize the molecular defects of PHPs.

35.11.1 Allelic Localization of Loss-of-Function Mutations Within the Coding Regions of *GNAS*

The parental origin of the *GNAS* mutations outlines the phenotype of the associated disease. Paternal mutations (PPHP) are associated with absent hormonal resistance, low risk of obesity, but increased occurrence of ectopic bone formation. By contrast, maternal mutations (PHP1A) are associated with pronounced hormonal resistance, greater risk of cognitive dysfunction, and obesity. Both diseases share similar phenotypes early in life as the Albright hereditary osteodystrophy and PTH resistance develop gradually over time. Through the allelic localization of the

GNAS mutations, genetics will predict the phenotype of the patient and have a significant impact on the subsequent clinical follow-up.

When the *GNAS* mutation occurs sporadically, the parental allele carrying the *GNAS* mutation can be recognized through various methods, such as analysis of linkage disequilibrium between the mutation and parental haplotypes [12], or the amplification and sequencing of the monoallelic transcripts [46, 50].

35.11.2 Identification of Multilocus Molecular Defects (MLMD)

In imprinting disorders different from PHP1B, i.e., Russell-Silver syndrome, Beckwith-Wiedemann syndrome, or transient neonatal diabetes mellitus, it has been shown that, in a subset of patients, the LOI may not be restricted to one imprinted locus but may affect other imprinted loci [51, 52]. We have demonstrated similarly that about 10 % of PHP1B patients display LOI at *PEG1/MEST*, *L3MBTL1*, and *DLK1/GTL2* DMRs located on chromosomes 7q32, 20, and 14q32, respectively. These MLMDs could account for the variable phenotype of some patients; they may have been underestimated by the targeted investigation of a few imprinted loci in a cohort of PHP1B patients with broad LOI at *GNAS* [44]. We propose that they may result from molecular alteration of factors necessary for the establishment/maintenance of genomic imprint marks or the integrity of the imprinted gene-coregulated network [53]. So far, no mutations have been identified in PHP1B in factors involved in maternal hypomethylation syndromes like *ZFP57* and *NLRP2* [54].

35.11.3 Involvement/Exclusion of *GNAS*

At the interface of clinical genetics and research, thorough haplotype analysis of chromosome 20q may help to determine in some patients if *GNAS* is involved in the development of the disease. It requires analyzing genotypes of patients, parents, and, as possible, affected and unaffected siblings.

35.12 Genetic Counseling and Prenatal Diagnosis

The identification of PHP or PHP phenocopies (acrodysostosis, PPHP, POH) in a patient implies the need to look for an underlying genetic/epigenetic mechanism. Depending on the molecular mechanisms, symptoms can be latent or develop over time, which renders essential identification of the molecular mechanism whenever possible. Once the genetic cause has been identified, familial screening is performed in coordination with physicians and geneticists. In puzzling situations without any hint of the molecular defect, determination of PTH, TSH, or calcitonin resistance may help to screen other family members.

Disease inheritance is a major concern for patients and an important trigger for the molecular investigations. We have summarized the main situations encountered in Table **35.2**. Schematically, PHP1A and PHP1B occur through maternal transmission, whereas PPHP and POH are paternally transmitted. UPD should be erased through germinal transmission. Acrodysostosis syndromes are compatible with autosomal dominant inheritance.

Ante- and prenatal diagnoses have been done in patients affected with *GNAS* coding mutations and PHP1A/PPHP willing to give birth to unaffected children [55] (and our experience). Cognitive development and ectopic bone are the main concerns of future parents. Their concern, however, will be impacted by the severity of the gonadotropin resistance and the individual countries' ethical laws. At birth, mutations or epigenetic anomalies may be searched for using cord blood [40].

Conclusion

Pseudohypoparathyroidism clearly is not one disorder but, for many, with different causes and mechanisms, different prognoses, and different risks of recurrence. It requires an integrated view of the patient's phenotype, epigenotype, and genotype to deliver the most appropriate care and counsel.

Despite our efforts, a significant number of patients with PHP1B and broad LOI at *GNAS*

Table 35.2 Evaluation of the transmission and recurrence of the diseases in the most common situations

Molecular defect/disease in the affected parent	Parental transmission	Risk of genetic transmission	Expected molecular defect for the affected baby	Expected phenotype for the baby
Maternal mutation of *GNAS*/ PHP1A	Through the mother	1/2	Maternal mutation of *GNAS*	PHP1A
Maternal mutation of *GNAS*/ PHP1A	Through the father	1/2	Paternal mutation of *GNAS*	PPHP or POH
Paternal mutation of *GNAS*/ PPHP or POH	Through the mother	1/2	Maternal mutation of *GNAS*	PHP1A
Paternal mutation of *GNAS*/ PPHP or POH	Through the father	1/2	Paternal mutation of *GNAS*	PPHP or POH
LOI restricted to the A/B DMR and maternal *STX16* deletion/PHP1B	Through the mother	1/2	LOM restricted to A/B DMR and maternal *STX16* deletion	PHP1B
LOM restricted to the A/B DMR and maternal *STX16* deletion/PHP1B	Through the father	1/2	Normal methylation at *GNAS* and paternal *STX16* deletion	None
Normal methylation at *GNAS* and paternal *STX16* deletion/ none	Through the mother	1/2	LOI restricted to A/B DMR and maternal *STX16* deletion	PHP1B
Normal methylation at *GNAS* and paternal *STX16* deletion/ none	Through the father	1/2	Normal methylation at *GNAS* and paternal *STX16* deletion	None
PatUPD encompassing *GNAS*/ PHP1B	Through the mother or the father	None	Normal methylation at *GNAS*	None
LOI at the whole *GNAS* locus/ sporPHP1B	Through the mother or the father	Unknown	Unpredictable	Unpredictable
PRKAR1A mutation/ ACRDYS1	Through the mother or the father	In theory ½ (never reported)	In theory *PRKAR1A* mutation	In theory ACRDYS1
PDE4D mutation/ACRDYS2	Through the mother or the father	1/2	*PDE4D* mutation	ACRDYS2

do not as yet have a complete molecular characterization of their disease. Research is very active in trying to identify the mechanisms leading to the loss of maternal imprints. Candidate gene strategies based on thorough phenotypic investigation of the patients as well as whole genome/methylome approaches are currently undergoing. It is therefore important to fully understand and complete the molecular characterization of these patients as it will (1) allow proper genetic counseling and clinical guidance and (2) promote research in this area.

References

1. Albright F, Burnett CH, Smith PH, Parson W (1942) Pseudohypoparathyroidism – an example of "Seabright-Bantam syndrome". Endocrinology 30:922–932
2. Tasken K, Aandahl EM (2004) Localized effects of cAMP mediated by distinct routes of protein kinase A. Physiol Rev 84(1):137–167
3. Breckler M, Berthouze M, Laurent AC, Crozatier B, Morel E, Lezoualc'h F (2011) Rap-linked cAMP signaling Epac proteins: compartmentation, functioning and disease implications. Cell Signal 23(8):1257–1266
4. Turan S, Bastepe M (2013) The GNAS complex locus and human diseases associated with loss-of-function mutations or epimutations within this imprinted gene. Horm Res Paediatr 80(4):229–241

5. Hayward B, Bonthron D (2000) An imprinted antisense transcript at the human GNAS1 locus. Hum Mol Genet 9:835–841
6. Hayward BE, Moran V, Strain L, Bonthron DT (1998) Bidirectional imprinting of a single gene: GNAS1 encodes maternally, paternally, and biallelically derived proteins. Proc Natl Acad Sci U S A 95: 15475–15480
7. Mantovani G, Ballare E, Giammona E, Beck-Peccoz P, Spada A (2002) The Gsa gene: predominant maternal origin of transcription in human thyroid gland and gonads. J Clin Endocrinol Metab 87(10):4736–4740
8. Linglart A, Mahon MJ, Kerachian MA, Berlach DM, Hendy GN, Juppner H et al (2006) Coding GNAS mutations leading to hormone resistance impair in vitro agonist- and cholera toxin-induced adenosine cyclic 3′,5′-monophosphate formation mediated by human XLalphas. Endocrinology 147(5):2253–2262
9. Weinstein LS, Gejman PV, Friedman E, Kadowaki T, Collins RM, Gershon ES et al (1990) Mutations of the Gs a-subunit gene in Albright hereditary osteodystrophy detected by denaturing gradient gel electrophoresis. Proc Natl Acad Sci U S A 87:8287–8290
10. de Nanclares GP, Fernandez-Rebollo E, Santin I, Garcia-Cuartero B, Gaztambide S, Menendez E et al (2007) Epigenetic defects of GNAS in patients with pseudohypoparathyroidism and mild features of Albright's hereditary osteodystrophy. J Clin Endocrinol Metab 92(6):2370–2373
11. Linglart A, Fryssira H, Hiort O, Holterhus PM, Perez de Nanclares G, Argente J et al (2012) PRKAR1A and PDE4D mutations cause acrodysostosis but two distinct syndromes with or without GPCR-signaling hormone resistance. J Clin Endocrinol Metab 97(12)): E2328–38
12. Linglart A, Carel JC, Garabedian M, Le T, Mallet E, Kottler ML (2002) GNAS1 lesions in pseudohypoparathyroidism Ia and Ic: genotype phenotype relationship and evidence of the maternal transmission of the hormonal resistance. J Clin Endocrinol Metab 87(1):189–197
13. Thiele S, de Sanctis L, Werner R, Grotzinger J, Aydin C, Juppner H et al (2011) Functional characterization of GNAS mutations found in patients with pseudohypoparathyroidism type Ic defines a new subgroup of pseudohypoparathyroidism affecting selectively Gsalpha-receptor interaction. Hum Mutat 32(6): 653–660
14. Bastepe M, Frohlich LF, Hendy GN, Indridason OS, Josse RG, Koshiyama H et al (2003) Autosomal dominant pseudohypoparathyroidism type Ib is associated with a heterozygous microdeletion that likely disrupts a putative imprinting control element of GNAS. J Clin Invest 112(8):1255–1263
15. Richard N, Abeguile G, Coudray N, Mittre H, Gruchy N, Andrieux J et al (2012) A new deletion ablating NESP55 causes loss of maternal imprint of A/B GNAS and autosomal dominant pseudohypoparathyroidism type Ib. J Clin Endocrinol Metab 97(5): E863–E867
16. Bastepe M, Frohlich LF, Linglart A, Abu-Zahra HS, Tojo K, Ward LM et al (2005) Deletion of the NESP55 differentially methylated region causes loss of maternal GNAS imprints and pseudohypoparathyroidism type Ib. Nat Genet 37(1):25–27
17. Fernandez-Rebollo E, Lecumberri B, Garin I, Arroyo J, Bernal-Chico A, Goni F et al (2010) New mechanisms involved in paternal 20q disomy associated with pseudohypoparathyroidism. Eur J Endocrinol 163(6):953–962
18. Wu WI, Schwindinger WF, Aparicio LF, Levine MA (2001) Selective resistance to parathyroid hormone caused by a novel uncoupling mutation in the carboxyl terminus of Gas: A cause of pseudohypoparathyroidism type Ib. J Biol Chem 276(1):165–171
19. Shore EM, Ahn J, Jan de Beur S, Li M, Xu M, Gardner RJ et al (2002) Paternally inherited inactivating mutations of the GNAS1 gene in progressive osseous heteroplasia. N Engl J Med 346(2):99–106
20. Linglart A, Menguy C, Couvineau A, Auzan C, Gunes Y, Cancel M et al (2011) Recurrent PRKAR1A mutation in acrodysostosis with hormone resistance. N Engl J Med 364(23):2218–26
21. Michot C, Le Goff C, Goldenberg A, Abhyankar A, Klein C, Kinning E et al (2012) Exome sequencing identifies PDE4D mutations as another cause of acrodysostosis. Am J Hum Genet 90(4):740–745
22. Lee H, Graham JM Jr, Rimoin DL, Lachman RS, Krejci P, Tompson SW et al (2012) Exome sequencing identifies PDE4D mutations in acrodysostosis. Am J Hum Genet 90(4):746–751
23. Klopocki E, Hennig BP, Dathe K, Koll R, de Ravel T, Baten E, Blom E et al (2010) Deletion and point mutations of PTHLH cause brachydactyly type E. Am J Hum Genet 86(3):434–439
24. Wilson LC, Leverton K, Oude Luttikhuis ME, Oley CA, Flint J et al (1995) Brachydactyly and mental retardation: an Albright hereditary osteodystrophy-like syndrome localized to 2q37. Am J Hum Genet 56(2):400–407
25. Genevieve D, Sanlaville D, Faivre L, Kottler ML, Jambou M, Gosset P et al (2005) Paternal deletion of the GNAS imprinted locus (including Gnasxl) in two girls presenting with severe pre- and post-natal growth retardation and intractable feeding difficulties. Eur J Hum Genet 13(9):1033–1039
26. Ahmed SF, Dixon PH, Bonthron DT, Stirling HF, Barr DG, Kelnar CJ et al (1998) GNAS1 mutational analysis in pseudohypoparathyroidism. Clin Endocrinol (Oxf) 49(4):525–531
27. Elli FM, deSanctis L, Ceoloni B, Barbieri AM, Bordogna P, Beck-Peccoz P et al (2013) Pseudohypoparathyroidism type Ia and pseudo-pseudohypoparathyroidism: the growing spectrum of GNAS inactivating mutations. Hum Mutat 34(3):411–6
28. Ahrens W, Hiort O, Staedt P, Kirschner T, Marschke C, Kruse K (2001) Analysis of the GNAS1 gene in Albright's hereditary osteodystrophy. J Clin Endocrinol Metab 86(10):4630–4634

29. Thiele S, Werner R, Ahrens W, Hoppe U, Marschke C, Staedt P et al (2007) A disruptive mutation in exon 3 of the GNAS gene with albright hereditary osteodystrophy, normocalcemic pseudohypoparathyroidism, and selective long transcript variant Gsalpha-L deficiency. J Clin Endocrinol Metab 92(5): 1764–1768
30. Bastepe M, Gunes Y, Perez-Villamil B, Hunzelman J, Weinstein LS, Juppner H (2002) Receptor-mediated adenylyl cyclase activation through XLalpha(s), the extra-large variant of the stimulatory G protein alpha-subunit. Mol Endocrinol 16(8):1912–1919
31. Fernandez-Rebollo E, Garcia-Cuartero B, Garin I, Largo C, Martinez F, Garcia-Lacalle C et al (2010) Intragenic GNAS deletion involving exon A/B in pseudohypoparathyroidism type 1A resulting in an apparent loss of exon A/B methylation: potential for misdiagnosis of pseudohypoparathyroidism type 1B. J Clin Endocrinol Metab 95(2):765–771
32. Mitsui T, Nagasaki K, Takagi M, Narumi S, Ishii T, Hasegawa T (2012) A family of pseudohypoparathyroidism type Ia with an 850-kb submicroscopic deletion encompassing the whole GNAS locus. Am J Med Genet A 158A(1):261–264
33. Mariot V, Maupetit-Mehouas S, Sinding C, Kottler ML, Linglart A (2008) A maternal epimutation of GNAS leads to Albright osteodystrophy and parathyroid hormone resistance. J Clin Endocrinol Metab 93(3):661–665
34. Mantovani G, de Sanctis L, Barbieri AM, Elli FM, Bollati V, Vaira V et al (2010) Pseudohypoparathyroidism and GNAS epigenetic defects: clinical evaluation of albright hereditary osteodystrophy and molecular analysis in 40 patients. J Clin Endocrinol Metab 95(2):651–658
35. Nagasaki K, Iida T, Sato H, Ogawa Y, Kikuchi T, Saitoh A et al (2012) PRKAR1A mutation affecting cAMP-mediated G protein-coupled receptor signaling in a patient with acrodysostosis and hormone resistance. J Clin Endocrinol Metab 97(9):E1808–E1813
36. Freson K, Hoylaerts MF, Jaeken J, Eyssen M, Arnout J, Vermylen J et al (2001) Genetic variation of the extra-large stimulatory G protein alpha-subunit leads to Gs hyperfunction in platelets and is a risk factor for bleeding. Thromb Haemost 86(3):733–738
37. Frohlich LF, Mrakovcic M, Steinborn R, Chung UI, Bastepe M, Juppner H (2010) Targeted deletion of the Nesp55 DMR defines another Gnas imprinting control region and provides a mouse model of autosomal dominant PHP-Ib. Proc Natl Acad Sci U S A 107(20):9275–9280
38. Maupetit-Mehouas S, Mariot V, Reynes C, Bertrand G, Feillet F, Carel JC et al (2011) Quantification of the methylation at the GNAS locus identifies subtypes of sporadic pseudohypoparathyroidism type Ib. J Med Genet 48(1):55–63
39. Liu J, Litman D, Rosenberg M, Yu S, Biesecker L, Weinstein L (2000) A GNAS1 imprinting defect in pseudohypoparathyroidism type Ib. J Clin Invest 106:1167–1174
40. Linglart A, Gensure RC, Olney RC, Juppner H, Bastepe M (2005) A novel STX16 deletion in autosomal dominant pseudohypoparathyroidism type Ib redefines the boundaries of a cis-acting imprinting control element of GNAS. Am J Hum Genet 76(5):804–814
41. Elli FM, de Sanctis L, Peverelli E, Bordogna P, Pivetta B, Miolo G et al (2014) Autosomal dominant pseudohypoparathyroidism type Ib: a novel inherited deletion ablating STX16 causes loss of imprinting at the A/B DMR. J Clin Endocrinol Metab 99(4):E724–E728, jc20133704
42. Chillambhi S, Turan S, Hwang DY, Chen HC, Juppner H, Bastepe M (2010) Deletion of the noncoding GNAS antisense transcript causes pseudohypoparathyroidism type Ib and biparental defects of GNAS methylation in cis. J Clin Endocrinol Metab 95(8):3993–4002
43. Bastepe M, Lane AH, Juppner H (2001) Paternal uniparental isodisomy of chromosome 20q–and the resulting changes in GNAS1 methylation–as a plausible cause of pseudohypoparathyroidism. Am J Hum Genet 68(5):1283–1289
44. Maupetit-Mehouas S, Azzi S, Steunou V, Sakakini N, Silve C, Reynes C et al (2013) Simultaneous hyper- and hypomethylation at imprinted loci in a subset of patients with GNAS epimutations underlies a complex and different mechanism of multilocus methylation defect in pseudohypoparathyroidism type 1b. Hum Mutat 34(8):1172–1180
45. Mundlos S (2009) The brachydactylies: a molecular disease family. Clin Genet 76(2):123–136
46. Lebrun M, Richard N, Abeguile G, David A, Coeslier Dieux A, Journel H et al (2010) Progressive osseous heteroplasia: a model for the imprinting effects of GNAS inactivating mutations in humans. J Clin Endocrinol Metab 95(6):3028–3038
47. Lynch DC, Dyment DA, Huang L, Nikkel SM, Lacombe D, Campeau PM et al (2013) Identification of novel mutations confirms PDE4D as a major gene causing acrodysostosis. Hum Mutat 34(1):97–102
48. Linglart A, Menguy C, Couvineau A, Auzan C, Gunes Y, Cancel M et al (2011) Recurrent PRKAR1A mutation in acrodysostosis with hormone resistance. N Engl J Med 364(23):2218–2226
49. Ellsworth R, Howard JE (1934) Studies on the physiology of the parathyroid glands. VII. Some responses of normal human kidneys and blood to intravenous parathyroid extract. Bull Johns Hopkins Hosp 55:296
50. Mariot V, Wu JY, Aydin C, Mantovani G, Mahon MJ, Linglart A et al (2011) Potent constitutive cyclic AMP-generating activity of XLalphas implicates this imprinted GNAS product in the pathogenesis of McCune-Albright syndrome and fibrous dysplasia of bone. Bone 48(2):312–320
51. Azzi S, Rossignol S, Steunou V, Sas T, Thibaud N, Danton F et al (2009) Multilocus methylation analysis in a large cohort of 11p15-related foetal growth disorders (Russell Silver and Beckwith Wiedemann syndromes) reveals simultaneous loss of methylation at paternal and maternal imprinted loci. Hum Mol Genet 18(24):4724–4733

52. Bliek J, Verde G, Callaway J, Maas SM, De Crescenzo A, Sparago A et al (2009) Hypomethylation at multiple maternally methylated imprinted regions including PLAGL1 and GNAS loci in Beckwith-Wiedemann syndrome. Eur J Hum Genet 17(5):611–619
53. Varrault A, Gueydan C, Delalbre A, Bellmann A, Houssami S, Aknin C et al (2006) Zac1 regulates an imprinted gene network critically involved in the control of embryonic growth. Dev Cell 11(5):711–722
54. Perez-Nanclares G, Romanelli V, Mayo S, Garin I, Zazo C, Fernandez-Rebollo E et al (2012) Detection of hypomethylation syndrome among patients with epigenetic alterations at the GNAS locus. J Clin Endocrinol Metab 97(6):E1060–E1067
55. Lietman SA, Goldfarb J, Desai N, Levine MA (2008) Preimplantation genetic diagnosis for severe albright hereditary osteodystrophy. J Clin Endocrinol Metab 93(3):901–904

Blomstrand's Chondrodysplasia

36

Francesca Giusti, Luisella Cianferotti, Laura Masi, and Maria Luisa Brandi

36.1 Introduction

Blomstrand's chondrodysplasia (BOCD) (OMIM phenotype number 215045) is an autosomal recessive disorder, caused by homozygous or compound heterozygous inactivating mutations in the parathyroid hormone receptor-1 gene (*PTH1R*) (OMIM gene number 168468) [1, 2]. Defects in this receptor are also known to be the cause of other disorders. Activating heterozygous mutations of PTH1R have been detected in Jansen's metaphyseal chondrodysplasia (JMC) (OMIM 156400) [3, 4], while recessive inactivating mutations have been found in isolated cases of multiple enchondromatosis (ENCHOM) (OMIM 166000) [5], Eiken skeletal dysplasia (EISD) (OMIM 600002) [6], and primary failure of tooth eruption (PFE) (OMIM 125350) [7]. All five disorders are characterized by various defects in skeletal development.

F. Giusti, MD, PhD • M.L. Brandi, MD, PhD (✉)
Bone Metabolic Diseases Unit, Department of Surgery and Translational Medicine, University of Florence, Largo Palagi 1, Florence, FI, 50139, Italy
e-mail: marialuisa.brandi@unifi.it

L. Cianferotti, MD, PhD • L. Masi, MD, PhD
Department of Surgery and Translational Medicine, University of Florence, Florence, Italy
e-mail: luisella.cianferotti@unifi.it; masi@dmi.unifi.it

36.2 Clinical Features

BOCD is a rare disease (prevalence: <1/1,000,000), with few cases reported in the literature to date. It is characterized by multiple malformations, including very short limbs and dwarfism; a narrow thorax; facial anomalies such as macroglossia, micrognathia, and depressed nasal bridge (Fig. 36.1a–c); as well as polyhydramnios, hydrops fetalis, hypoplastic lungs, protruding eyes showing cataracts, and internal malformations such as preductal aortic coarctation [1]. Fetuses show increased bone mineral density and advanced bone maturation (Fig. 36.2a–d). Signs of the disease are present at birth, and it leads to neonatal death. Although an assessment of mineral metabolism has not been performed in the reported cases due to the early lethality of the disease, it is likely that a triad of hypocalcemia, hyperphosphatemia, and elevated PTH levels, configuring a syndrome of PTH resistance, is present. Normal birth weight can be overestimated because most infants are hydroptic and the placenta can be immature and edematous. Recently, defects in mammary gland and tooth development have also been demonstrated [8].

The first case of a female Finnish neonate who died shortly after birth with clinical and radiologic features related to dysplasia was described in 1985 by Blomstrand, after whom the disorder was named [9]. In 1990, Spranger and Maroteaux [10] described another similar case with increased bone

M.L. Brandi, E.M. Brown (eds.), *Hypoparathyroidism*,
DOI 10.1007/978-88-470-5376-2_36, © Springer-Verlag Italia 2015

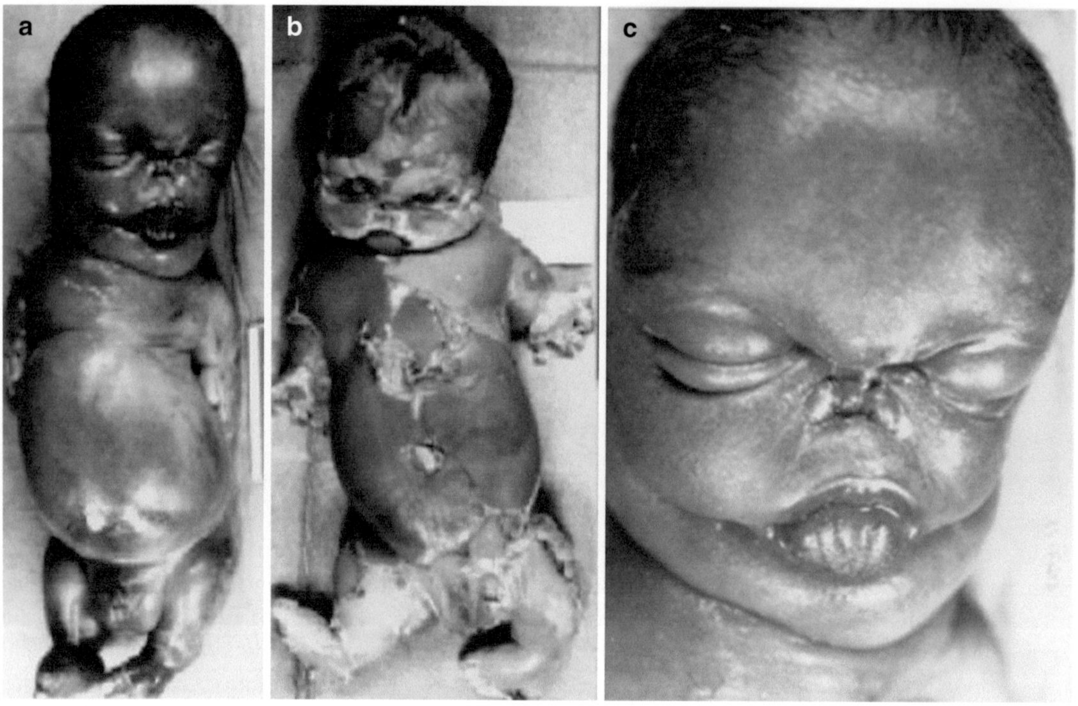

Fig. 36.1 (**a**) General appearance of the male fetus (at 26 weeks of gestation). (**b**) Postmortem appearance of the female fetus (at 33 weeks of gestation) with the same syndrome. (**c**) Detail of the face of the male fetus; note swollen tongue and depressed nasal bridge (Reproduced with permission from Ref. [1])

density. In 1993, Young et al. [11] reported a patient with advanced dentition, general ossification, and sclerotic bones. Autopsy revealed that the patient had a very small foramen magnum and larynx and pulmonary hypoplasia. In all these cases, the patient was born to consanguineous parents.

In 1996, Leroy et al. [12] described two fetuses who were the offspring of consanguineous parents and exhibited edema of the face, hypoplastic nose, hypoplastic and narrow thoracic cage, prominent abdomen, and extremely short limbs and were most likely affected by BOCD.The fact that the disease is most common in infants born to consanguineous parents, both in females and males, and with no ethnic differences [1], suggests that BOCD manifests as an autosomal recessive trait. In 1997, Loshkajian et al. [1] described a case of a 26-year-old woman who was referred for a routine ultrasound examination at 26 weeks of gestation. The exam showed a hydropic fetus with very short limbs and multiple anomalies and a normal 46, XY pattern. The patient opted for an elective termination of the pregnancy. A few years later, her fourth pregnancy terminated spontaneously at 33 weeks of gestation with the delivery of a dead girl with the same syndrome. These two fetuses were born to a non-consanguineous couple. At histology, the epiphyseal cartilage appeared markedly reduced with fusiform and occasionally vacuolated chondrocytes and erratic distribution of chondrocytes. The epiphyseal-metaphyseal junction was wide and irregular; the zone of proliferating cartilage was narrow and irregular with irregular columnization in the hypertrophic zone. Osteoclasts were rare with deficient bone remodeling. The growth cartilage showed dilated segments of the rough endoplasmic reticulum containing amorphous material, a relatively high number of cells with shrunken or pyknotic nuclei, and a matrix with fine unequal fibers, some with visible cross-striations, small granules, and fine filaments irregularly arranged between the fibers.

Den Hollander et al. [13] reported a family in which 2 female fetuses presented with a severe

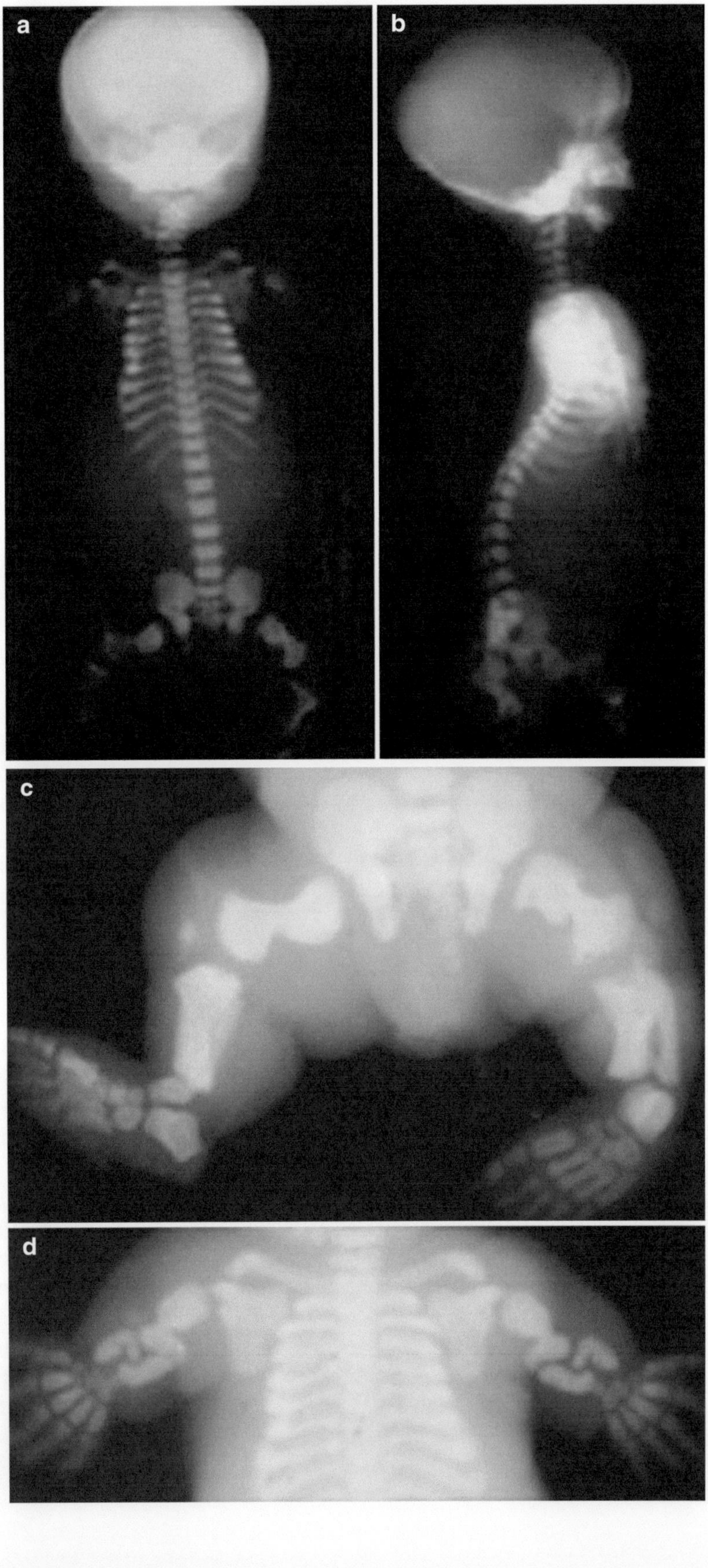

Fig. 36.2 (**a**) Radiograph of the male fetus, frontal view; overall bone density is increased, the tubular bones are short with wide ends and shafts of humeri and femora bowed. (**b**) Radiograph of the same fetus, lateral view. (**c**) Radiograph of the female fetus; most tarsal bones are ossified. (**d**) Radiograph of the upper limbs, showing advanced bone maturation; most all the carpal bones are present at 33 weeks of gestation (Reproduced with permission from Ref. [1])

Table 36.1 Signs associated with Blomstrand's chondrodysplasia

Skeletal anomalies	
	Short stature and/or nanism
	Very short limbs
	Micromelia
	Advanced bone maturation
	Osteosclerosis/osteopetrosis/bone condensation/ synostosis
	Abnormal vertebral size and shape
	Metacarpal anomalies (Archibald's sign)
	Metaphyseal anomaly
	Narrow and short rib cage
	Rib structure anomalies
	Clavicle absent or abnormal
Internal anomalies	
	Hypoplastic or agenesis lung
	Preductal aortic coarctation
	Defects in mammary gland
	Small larynx
Facial anomalies	
	Low and posteriorly rotated ears
	Macroglossia (prominent and hypertrophic tongue)
	Micrognathia
	Retrognathia
	Hypoplasia and/or arhinia
	Depressed nasal bridge
	Premature eruption of teeth or natal teeth
	Exophthalmos
	Cataracts
	Lens opacification
Intrauterine fetal anomalies	
	Polyhydramnios
	Hydrops fetalis

skeletal dysplasia during prenatal screening (at 18.5 and 12 weeks of gestation, respectively). The pregnancy was terminated and the diagnosis of BOCD was confirmed at autopsy. In 2000, Oostra et al. [14] described 3 novel cases (2 isolated and a sib-pair). The above reported clinical signs are the consequence of extremely accelerated skeletal maturation and mineralization at sites of endochondral bone formation. Metaphyseal growth plates are undetectable, and an increased bone mineral density is present in radiological examinations. Signs associated with BOCD are summarized in Table 36.1.

36.3 Molecular Pathogenesis

BOCD is caused by inactivating mutations of the *PTH1R* gene, which encodes for the receptor of PTH and PTH-related peptide (PTH/PTHrP) [15, 16]. The *PTH1R* gene is located on the short (p) arm of chromosome 3 between positions 22 and 21.1, and more precisely, it is located from base pair 46,877,745 to base pair 46,903,798 on chromosome 3 (Fig. 36.1) [15, 16]. The PTH1R is a member of the G protein-coupled receptor family 2; its activity is mediated by G proteins that activate adenylate (AC)/protein kinase A (PKA) and the phospholipase C beta (PLCβ)/protein kinase C (PKC) signaling pathway [15, 16] (see Chap. 9 for further details). The PTH receptor is expressed in most tissues but is found at particularly high levels in bone, kidneys, and growth plate (see Chap. 11 for further details).

Recent studies have demonstrated that signaling through the PTH/PTHrP receptor, in addition to its role in regulating mineral metabolism, plays an essential role in fetal development due to the important regulatory effects of PTHrP on the development of cartilage and bone [17–19].

Indeed, while during postnatal life, it regulates calcium and phosphate homeostasis mediating the endocrine actions of PTH in the bone and kidney; during fetal life, it is a critical component in endochondral bone formation as part of the PTHrP-dependent autocrine/paracrine regulation of chondrocyte growth and differentiation [2, 20]. In bone, it is expressed on the surface of osteoblasts. It is activated on osteoblasts when it binds PTH, which causes upregulation of RANKL expression (receptor activator of nuclear factor kB ligand). RANKL, in turn, binds to RANK (receptor activator of nuclear factor kB) on osteoclasts. This turns on osteoclasts to ultimately increase both their formation and resorption rate [2, 20].

In growth plate, when the receptor is activated through cAMP-dependent mechanisms, it stimulates proliferation of the fetal growth plate chondrocytes and inhibits their differentiation into hypertrophic chondrocytes [2, 20]. Mice with disruptions of either the PTH/PTHrP receptor or PTHrP genes exhibit multiple, severe skeletal defects characterized by an advanced endochondral bone formation that prove lethal in utero or shortly after birth [17–19], resembling individuals with BOCD. Indeed, the majority of the genetically ascertained cases of BOCD have been proven to be caused by homozygous inactivating mutations of the PTH1R.

In 1998, Jobert reported a case of BOCD, born to non-consanguineous parents, whose genetic assessment showed a heterozygous point mutation in the *PTH1R* gene (G→A substitution at nucleotide 1176) inherited from the mother [2]. This point mutation caused the deletion of the first 11 amino acids of exon M5 (encoding the fifth transmembrane domain of the receptor), resulting from the use of a novel splice site created by the base substitution [2]. In vitro studies showed that this altered receptor, although well expressed, was not capable of binding PTH nor PTHrP and failed to induce detectable stimulation of either cAMP or inositol phosphate production in response to these ligands [2]. The paternal allele was not expressed in the patient's chondrocytes, and only the abnormal and nonfunctional PTH/PTHrP receptor encoded by the maternal allele was expressed, indicating a dominant negative mode of inheritance [2].

After this initial report, all other ensuing cases of BOCD have been shown to follow a recessive mode of inheritance and to be determined by homozygous inactivating mutations of *PTH1R*.

An infant born to consanguineous parents, who show alteration of a single homozygous nucleotide changing a strictly conserved proline residue at position 132 in the receptor's amino terminal extracellular domain to leucine. An in vitro functional study showed that COS-1 cells expressing the mutant receptor did not accumulate cyclic adenosine 3′,5′-monophosphate in response to PTH or PTH-related peptide (PTHrP) and did not bind the radiolabeled ligand [20].

In another case, a homozygous deletion of G at position 1,122 (exon EL2) was identified [21]. This missense mutation resulted in a shift in the open reading frame, leading to a truncated protein, lacking transmembrane domains 5, 6, and 7, the connecting intra- and extracellular loops, and the cytoplasmic tail [21]. Functional analysis of the mutant receptor in COS-7 cells and of dermal fibroblasts obtained from the case proved that the mutation was indeed inactivating [21].

Recently, a P132L mutation, which inactivates the PTH/PTHrP receptor incompletely, was identified in two additional patients affected by BOCD characterized by a less severe phenotype [22]. This suggests that BOCD can be classified in two different forms, BPCD type I and BOCD type II, according to the degree of severity of the disease and the associated PTH1R abnormality. The more serious BOCD type I is determined by completely inactivating mutation in the *PTH1R* gene, while the milder BOCD type II phenotype is caused by incomplete inactivation of PTH1R [14].

As previously stated, mutations of the *PTH1R* gene can be found in additional disorders characterized by skeletal abnormalities not resembling BOCD. Jansen's metaphyseal chondrodysplasia is caused by constitutively active heterozygous mutations in the PTH1R gene on chromosome 3p21. JMC is a rare autosomal dominant disorder characterized by a short-limbed dwarfism associated with hypercalcemia and normal or low serum concentrations of the two parathyroid hormones [3, 4].

ENCHOM is a condition characterized by multiple formations of enchondromas, benign neoplasms derived from mesodermal cells that form cartilage, without abnormalities in mineral metabolism [5]. A recessive mutation in PTH1R, leading to a decrease in signal transduction, has been found in one case of ENCHOM. Indeed, a knock-in mouse model expressing this mutation displays enchondroma-like lesions.

EISD is a rare skeletal dysplasia characterized by severely retarded ossification, principally of the epiphyses, pelvis, and hands and feet, as well as by abnormal modeling of the bones in hands and feet, abnormal persistence of cartilage in the pelvis, and mild growth retardation [6]. In one case, a homozygous mutation in the C-terminal cytoplasmic tail of the *PTH1R* gene has been shown.

Primary failure of tooth eruption can be caused by heterozygous mutation in the *PTH1R* gene. PFE is a rare condition that has high penetrance and variable expressivity and in which failure of tooth eruption occurs without evidence of any obvious mechanical interference. Instead, malfunction of the eruptive mechanism itself appears to cause nonankylosed permanent teeth to fail to erupt, although the eruption pathway has been cleared by bone resorption [7].

Conclusions

Despite the fact that BOCD is a very rare and lethal disease, its study, determined by the absence or reduction of a functional PTH1R in humans, has helped further define the multiple roles of PTH1R in skeletal homeostasis. Although a proper assessment of mineral metabolism has not been undertaken in infants affected by BOCD because of early lethality, it is probable that it represents a syndrome of PTH resistance leading to functional hypoparathyroidism, and for this reason, it has been included in this volume.

References

1. Loshkajian A, Roume J, Stanescu V, Delezoide A, Stampf F, Maroteaux P (1997) Familial Blomstrand chondrodysplasia with advanced skeletal maturation: further delineation. Am J Med Genet 71:283–288
2. Jobert A-S, Zhang P, Couvineau A, Bonaventure J, Roume J, Le Merrer M, Silve C (1998) Absence of functional receptors for parathyroid hormone and parathyroid hormone-related peptide in Blomstrand chondrodysplasia. J Clin Invest 102:34–40
3. Jupper H, Schipani E (1997) The parathyroid hormone/parathyroid hormone-related peptide receptor in Jansen's metaphyseal chondrodysplasia. Curr Opin Endocrinol Diabetes 4:433–442
4. Charrow J, Poznanski AK (1984) The Jansen type of metaphyseal chondrodysplasia: confirmation of dominant inheritance and review of radiographic manifestations in the newborn and adult. Am J Med Genet 18:321–327
5. Couvineau A, Wouters V, Bertrand G, Rouyer C, Gerard B, Boon LM, Grandchamp B, Vikkula M, Silve C (2008) PTH1R mutations associated with Ollier disease result in receptor loss of function. Hum Mol Genet 17:2766–2775
6. Duchatelet S, Ostergaard E, Cortes D, Lemainque A, Julier C (2005) Recessive mutations in PTH1R cause contrasting skeletal dysplasias in Eiken and Blomstrand syndromes. Hum Mol Genet 14:1–5
7. Decker E, Stellzig-Eisenhauer A, Fiebig BS, Rau C, Kress W, Saar K, Ruschendorf F, Hubner N, Grimm T, Weber BHF (2008) PTH1R loss-of-function mutations in familial, nonsyndromic primary failure of tooth eruption. Am J Hum Genet 83:781–786
8. Wysolmerski JJ, Cormier S, Philbrick WM, Dann P, Zhang J-P, Roume J, Delezoide A-L, Silve C (2001) Absence of functional type 1 parathyroid hormone (PTH)/PTH-related protein receptors in humans is associated with abnormal breast development and tooth impaction. J Clin Endocrinol Metab 86:1788–1794
9. Blomstrand S, Claesson I, Save-Soderbergh J (1985) A case of lethal congenital dwarfism with accelerated skeletal maturation. Pediatr Radiol 15:141–143
10. Spranger J, Maroteaux P (1990) The lethal osteochondrodysplasias. In: Harris H, Hirschhorn K (eds) Advances in human genetics. Plenum Press (pub.), New York, pp 1–103
11. Young ID, Zuccollo JM, Broderick NJ (1993) A lethal skeletal dysplasia with generalised sclerosis and advanced skeletal maturation: Blomstrand chondrodysplasia? J Med Genet 30:155–157
12. Leroy JG, Keersmaeckers G, Coppens M, Dumon JE, Roels H (1996) Blomstrand lethal osteochondrodysplasia. Am J Med Genet 63:84–89
13. Den Hollander NS, van der Harten HJ, Vermeij-Keers C, Niermeijer MF, Wladimiroff JW (1997) First-trimester diagnosis of Blomstrand lethal osteochondrodysplasia. Am J Med Genet 73:345–350
14. Oostra RJ, van der Harten HJ, Rijnders WPHA, Scott RJ, Young MPA, Trump D (2000) Blomstrand osteochondrodysplasia: three novel cases and histological evidence for heterogeneity. Virchows Arch 436:28–35
15. Mannstadt M, Jüppner H, Gardella TJ (1999) Receptors for PTH and PTHrP: their biological importance and functional properties. Am J Physiol 277(5 Pt 2):F665–F675
16. Offermanns S, Iida-Klein A, Segre GV, Simon MI (1996) G alpha q family members couple parathyroid hormone (PTH)/PTH-related peptide and calcitonin receptors to phospholipase C in COS-7 cells. Mol Endocrinol 10(5):566–574
17. Karaplis AC, Luz A, Glowacki J, Bronson RT, Tybulewicz VLJ, Kronenberg HM, Mulligan RC (1994) Lethal skeletal dysplasia from targeted disruption of the parathyroid hormone-related peptide gene. Genes Dev 8:277–289
18. Weir EC, Philbrick WM, Amling M, Neff LA, Baron R, Broadus AE (1996) Targeted overexpression of parathyroid hormone-related peptide in chondrocytes causes chondrodysplasia and delayed endochondral bone formation. Proc Natl Acad Sci U S A 93:10240–10245
19. Lanske B, Karaplis AC, Lee K, Luz A, Vortkamp A, Pirro A, Karperien M, Defize LH, Ho C, Mulligan

RC, Abou-Samra AB, Juppner H, Segre GV, Kronenberg HM (1996) PTH/PTHrP receptor in early development and Indian hedgehog-regulated bone growth. Science 273:663–666

20. Karaplis AC, He B, Nguyen MT, Young ID, Semeraro D, Ozawa H, Amizuka N (1998) Inactivating mutation in the human parathyroid hormone receptor type 1 gene in Blomstrand chondrodysplasia. Endocrinology 139:5255–5258
21. Karperien M, van der Harten HJ, van Schooten R, Farih-Sips H, Den Hollander NS, Kneppers SL, Nijweide P, Papapoulos SE, Lowik CW (1999) A frame-shift mutation in the type I parathyroid hormone (PTH)/PTH-related peptide receptor causing Blomstrand lethal osteochondrodysplasia. J Clin Endocrinol Metab 84:3713–3720
22. Hoogendam J, Farih-Sips H, Wynaendts LC, Lowik CWGM, Wit JM, Karperien M (2007) Novel mutations in the parathyroid hormone (PTH)/PTH-related peptide receptor type 1 causing Blomstrand osteochondrodysplasia types I and II. J Clin Endocrinol Metab 92:1088–1095

Hypoparathyroidism During Magnesium Deficiency or Excess

37

René Rizzoli

37.1 Introduction

In humans, body magnesium content amounts to 25 g, with 66 % located in the bone, 33 % within cells, and only 1 % in the extracellular fluid (ECF), including blood [1–5] (see also Chap. 7). In the bone, magnesium as the divalent cation is adsorbed on the hydroxyapatite crystal and is in equilibrium with magnesium in the ECF. It is the most abundant intracellular cation together with potassium, reaching a concentration of approximately 0.5 mmol/l, which is thus close to that of magnesium in the ECF. Free cytosolic magnesium accounts for 5–10 % of total cellular magnesium. It binds to various organelles, 60 % of which is within mitochondria, where it is involved in phosphate transport, ATP synthesis, and utilization. ATP is synthesized by a magnesium-dependent oxidative phosphorylation process. Magnesium is a cofactor and regulator of a large series of enzymatic reactions, particularly those utilizing magnesium-ATP (glycolysis, oxidative phosphorylation), but also of DNA transcription and protein synthesis [1, 6]. In serum, 30 % of magnesium is protein bound. Circulating magnesium, which is between 0.7 and 1.0 mmol/l, is determined by bidirectional fluxes taking place at the levels of the intestine, kidney, and bone. Ionic magnesium interacts with the calcium-sensing receptor (CaSR) on parathyroid cells, and also on renal tubular cells, with a potency lower than calcium [7].

Magnesium is present in all nutrients of cellular origin. The recommended dietary allowance is 420 and 320 mg/day for men and women, respectively [8]. Inadequate dietary intake is rare. Net absorption is proportional to intake, usually representing 35–40 % of dietary magnesium [5, 9]. Phosphate and cellulose phosphate form complexes with this divalent cation, thereby impairing its absorption. A low pH is important to displace magnesium bound to dietary fiber and to make it available to the absorptive processes. Calcitriol does not stimulate intestinal magnesium absorption [10]. Bidirectional fluxes are voltage dependent. A paracellular pathway plays an important role in intestinal magnesium absorption. The cation transporter TRMP6 is present in the apical membrane of gut epithelial cells of the small intestine. It appears that magnesium absorption takes place through two processes, a saturable transcellular pathway mediated by TRPM6 when magnesium intake is low and passive paracellular diffusion when it is high [9]. TRPM6 is expressed in the small intestine, while paracellular magnesium absorption takes place along the whole small and large intestine [11, 12]. Inactivating mutations in the gene coding for the TRPM6 channel are associated with impaired intestinal magnesium absorption, hypomagnesemia, and hypocalcemia [13, 14] in a rare autosomal recessive disease.

R. Rizzoli, MD
Division of Bone Diseases, Faculty of Medicine,
Geneva University Hospitals,
1211, Geneva 14, Switzerland
e-mail: rene.rizzoli@unige.ch

M.L. Brandi, E.M. Brown (eds.), *Hypoparathyroidism*,
DOI 10.1007/978-88-470-5376-2_37, © Springer-Verlag Italia 2015

Table 37.1 Factors influencing renal tubular reabsorption of magnesium

Renal tubular reabsorption	
Increased	Decreased
ECF volume contraction	ECF volume expansion
Hypocalcemia	Hypercalcemia
Hypomagnesemia	Hypermagnesemia
Phosphate administration	Phosphorus deprivation
Metabolic alkalosis	Metabolic acidosis
PTH/PTHrP	Loop diuretics
	Cyclosporin A, tacrolimus
	Cinacalcet

Of serum magnesium, 70 % is ultrafiltrable, and 95 % of this is reabsorbed (15 % in the proximal tubule, 70 % in the cortical thick ascending limb of Henle's loop, 10 % in the distal convoluted tubule) [15–17]. TRMP6 is present in the apical membrane of the distal convoluted tubule [18]. Magnesium interacts with CaSR in the basolateral membrane and lowers tubular reabsorption of both calcium and magnesium [15]. Various factors control renal tubular magnesium reabsorption (Table 37.1). PTH and/or PTHrP stimulates renal tubular reabsorption of magnesium in the loop of Henle and in the distal convoluted tubule [15, 19], through mechanisms independent of TRMP6 expression [20]. The latter is not affected by $1,25(OH)_2$-vitamin D. ECF expansion, hypercalcemia, and hypermagnesemia (through the interaction with CaSR), loop diuretics, systemic metabolic acidosis, and alcohol decrease this transport, whereas an increase is detected in the thick ascending limb of the loop of Henle during hypomagnesemia, or during magnesium depletion in the distal convoluted tubule [20]. A series of inherited disorders of renal tubular reabsorption of magnesium cause renal wasting [15]. Some of them can be associated with hypercalciuria as well.

37.2 Role of Magnesium in PTH Secretion

Calcium is the main agonist of CaSR, but other divalent or even trivalent cations are also able to activate this receptor [7]. Among them, the divalent cation magnesium is also able to acutely modulate PTH secretion. Indeed, hypermagnesemia inhibits PTH secretion [21]. However, the efficacy of magnesium in controlling acute PTH secretion is lower than that of calcium [7, 22]. It is well recognized that chronic magnesium deficiency is associated with major perturbations of calcium and phosphate metabolism [5, 6] (Fig. 37.1). Indeed, magnesium depletion is accompanied by hypocalcemia, without a concomitant increase of PTH, producing a state of functional hypoparathyroidism [5, 23]. Low or inappropriately normal PTH levels are found in chronic magnesium deficiency despite hypocalcemia [5, 24]. Conversely, acute administration of magnesium to magnesium-depleted subjects leads to a rapid increase in PTH [24, 25]. This is quite different from the situation seen in magnesium-replete normal subjects in whom magnesium decreases PTH secretion [26]. To reconcile these clinical observations, one could assume that the PTH response is adequate when the magnesium concentration decreases, as expected when a divalent cation interacts with CaSR. However, with depletion of intracellular magnesium stores, as a result, for example, of long-term exposure to magnesium deficiency, PTH secretion becomes impaired, hence hypocalcemia and the inadequate PTH response. The rapid rise of PTH upon magnesium administration in this latter condition would indicate that impaired production of PTH in magnesium-deficient patients is related to an altered secretion rather to an inhibition of protein synthesis [5, 7].

Low circulating levels of calcitriol have been reported in hypocalcemic, magnesium-deficient patients [27]. This observation could be related to the low PTH, though hydroxylation of the 1-alpha position of 25-hydroxyvitamin D to form 1,25-dihydroxyvitamin D [$1,25(OH)_2$ vitamin D] in response to hypocalcemia could also occur through a PTH-independent mechanism and/or by a direct inhibition of calcitriol synthesis by magnesium deficiency [26].

37.3 Role of Magnesium in PTH Action

Though the response of PTH to magnesium administration is rapid in hypocalcemic patients with magnesium deficiency, the restoration of

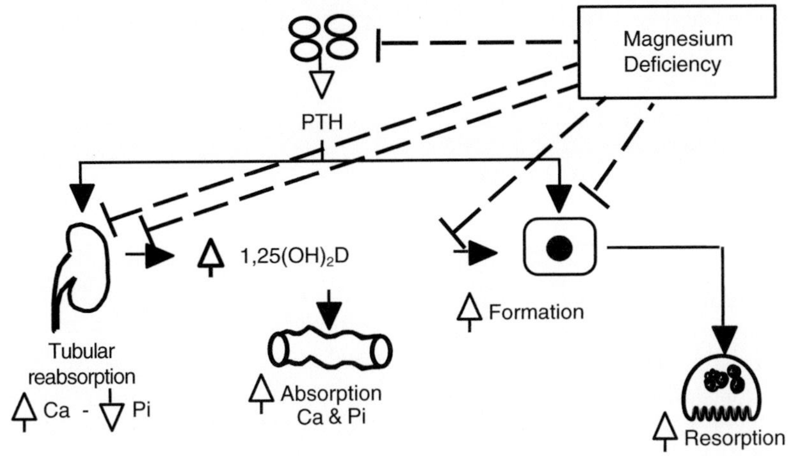

Fig. 37.1 Effects of magnesium deficiency on PTH-regulated calcium and phosphate homeostasis. *Dashed lines* illustrate the various effects of magnesium deficiency on PTH secretion and action

normocalcemia is delayed in time, suggesting some resistance to the effects of PTH [23] (Fig. 37.1). This resistance could be detected at the two major target organs for PTH, that are, the kidneys, as shown by a blunted urinary cyclic AMP excretion following PTH infusion in magnesium-deficient patients [23, 28], and bone [29]. One main intracellular mediator of PTH action is cyclic AMP. Magnesium is required for the stimulation of adenylyl cyclase and thus cAMP production. These observations are compatible with the hypothesis that there is impaired cyclic nucleotide metabolism in cells depleted of magnesium, both in terms of magnesium-induced enzyme stimulation and of magnesium constituting a component of the substrate magnesium-ATP [6]. However, in a child with primary hypomagnesemia, the urinary cyclic AMP and phosphate responses to PTH were found to be similar in normomagnesemic or hypomagnesemic states, suggesting a normal end-organ responsiveness in these conditions [30].

A reduced response to vitamin D in hypoparathyroidism with magnesium depletion together with a greater responsiveness upon magnesium administration has also been reported [31]. This suggests the presence of resistance to vitamin D during magnesium deficiency as well. A possible mechanism is an impaired conversion of 25-OH-vitamin D to $1,25(OH)_2$-vitamin D during magnesium depletion, but preclinical data and case reports have also suggested some primary resistance to active vitamin D metabolites [32].

37.4 Causes of Magnesium Deficiency

Magnesium deficiency is relatively frequent in hospitalized patients, particularly in intensive care units [33, 34]. It occurs when dietary intake of magnesium cannot compensate for gastrointestinal or renal losses of magnesium (Table 37.2). Prolonged nasogastric suction and chronic diarrhea are risk factors for magnesium depletion. Indeed upper GI tract fluid contains 2 mmol/l of magnesium, whereas in diarrheal fluids, magnesium concentration may be as high as 30 mmol/l. Regarding renal wasting, kidney tubule disorders and a large series of drugs or conditions, such as osmotic diuresis in the setting of diabetes mellitus or hypercalcemia, can contribute to the development of magnesium deficiency (Table 37.2). Chronic alcoholism affects both the intestinal component of magnesium homeostasis through the undernutrition often observed in subjects with chronic high consumption of alcoholic beverages and thus low magnesium intakes and the kidney by an alcohol-dependent impairment of tubular magnesium

Table 37.2 Causes of magnesium deficiency

Gastrointestinal tract loss	Malabsorption (including steatorrhea)
	Chronic diarrhea (including laxatives use)
	Bypass surgery, bowel resection
	Pancreatitis
	Abdominal irradiation
	Gastric suction
	Proton pump inhibitors[a]
	Inborn errors of metabolism
Renal loss	Loop diuretics
	Osmotic diuresis (diabetes mellitus)
	Alcohol consumption
	Nephrotoxics (aminoglycosides, pentamidine, amphotericin B)
	Cyclosporin A, tacrolimus
	Chemotherapy (cisplatinum derivatives)
	Hypercalcemia
	Metabolic acidosis
	Renal tubule disorders (pyelonephritis, renal tubular acidosis)
	Bartter's syndrome, Gitelman's syndrome
	cinacalcet[a]

[a]Limited evidence, mostly case reports

reabsorption. Among the most frequent causes of magnesium deficiency, several drugs are at the forefront. These include loop diuretics, aminoglycosides, cisplatinum derivatives, amphotericin B, cyclosporin A, and pentamidine. EGF appears to directly regulate the TRMP6 channel in the distal convoluted tubule. This could explain the high prevalence of hypomagnesemia in patients treated with the anti-EGF receptor antibodies cetuximab or panitunumab [35]. Although the underlying mechanisms are unclear, proton pump inhibitors, but not H2-blockers, may be associated with magnesium deficiency [36, 37]. By interacting with the renal tubule CaSR, cinacalcet, in a way similar to hypercalcemia or hypermagnesemia, can reduce renal tubular magnesium reabsorption [38].

37.5 Features and Diagnosis of Magnesium Deficiency

Because of impaired PTH secretion and action, magnesium deficiency can cause hypocalcemia and its associated neuromuscular hyperexcitability, including paresthesia, spasms, seizures and depression, and cardiac arrhythmia [5, 6]. The latter is further aggravated by hypokalemia secondary to a magnesium deficiency-dependent renal potassium wasting [39]. Under this condition, potassium therapy may be totally ineffective without magnesium repletion. The same may be true for hypocalcemia, and its correction may sometimes only be achieved with magnesium treatment, which restores adequate PTH secretion and action.

Circulating electrolyte concentrations are often poor reflections of body stores. This is even more relevant as far as magnesium is concerned. Indeed, magnesium is mainly an intracellular ion, and less than 1 % is present in ECF. However, serum magnesium concentration is commonly measured, and serum levels below 0.7 mmol/l are considered to be suggestive of magnesium deficiency (Table 37.3). Cellular magnesium can be evaluated by measuring lymphocyte magnesium content. This approach has been used in various studies, and the assay is available in commercial laboratories. Large variability (the ratio of lymphocyte magnesium

Table 37.3 Diagnosis of magnesium deficiency

1. Plasma magnesium level	Poor reflection of magnesium stores
2. Lymphocytes magnesium content	Large variability (ratio of 2 measurements, i.e., magnesium and protein)
3. Magnesium retention test[a]	Retention >0.5 -> magnesium deficiency
	>0.25 -> Possible magnesium deficiency

[a]Two 24 h urine collections for magnesium excretion, infusion of 8 mmol magnesium on the second day
Retention = 1 – ([U-Mg/24 h (day 2) – U-Mg/24 h (day 1)]/Infused Mg)

to protein is measured) and poor discriminatory values preclude the use of this determination as a standard diagnostic procedure for documenting magnesium deficiency. In the presence of magnesium deficiency, repletion of bodily stores is associated with a higher retention of an administered dose than in subjects with normal magnesium homeostasis. This provides the rationale for the magnesium retention test [40] (Table 37.3). A magnesium retention of more than 50 % indicates magnesium deficiency. The interpretation of this test may be limited in the case of renal magnesium leak or of drugs associated with renal magnesium wasting.

37.6 Management of Magnesium Deficiency

In the presence of hypocalcemia of unknown origin, magnesium deficiency should be suspected and magnesium repletion rapidly undertaken, like a therapeutic test (Table 37.4) (see also Chaps. 7 and 28). This is achieved by the intravenous administration of a magnesium salt [6]. For rapid magnesium repletion, the oral route is not recommended because of the limited amount of magnesium tolerated, before diarrhea occurs, because of the cathartic properties of magnesium. An effective regimen includes 8 mmol magnesium (200 mg of elemental Mg) over 1–2 h, followed by 20–24 mmol (500–600 mg of elemental Mg) over 24 h intravenously (Table 37.4). Though this treatment may transiently normalize serum magnesium concentration, possibly with correction of hypocalcemia and hypokalemia, this does not mean that repletion of bodily magnesium stores is complete. Intravenous therapy should be continued for 3–5 days, since deficiency may be as high as 100 mmol. Dietary sources of magnesium include almonds, soybeans, seeds, wheat germs, wheat brans, millets, dark green vegetables, fruits, and seafood. If magnesium losses from the intestine or kidney are persistent, dietary repletion may not be sufficient. A large variety of magnesium preparations are available (sulfate, lactate, hydroxide, chloride, and glycerophosphate). Daily doses should be between 12 and 24 mmol (300–600 of elemental Mg). Three to four divided daily doses may help to prevent diarrhea.

Table 37.4 Management of magnesium deficiency

1. In case of hypocalcemia of unknown origin: 8 mmol magnesium (aspartate or sulfate) in 100 ml 5 % glucose solution infused over 2 h
2. Magnesium repletion: 24 mmol by intravenous infusion over 4–6 h, for 3–5 days (magnesium deficit may be as high as 100 mmol)
3. Prevention of magnesium deficiency: oral magnesium supplementation (sulfate, lactate, chloride, glycerophosphate), up to 10 mmol/day in 3–4 divided doses to avoid diarrhea
Caution should be taken for patients with renal failure

Table 37.5 Major risk factors for magnesium deficiency

Stay in intensive care unit
Chemotherapy
Loop diuretics (with or without laxatives)
Chronic alcohol consumption
Malabsorption syndromes (including intestine resection or bypass)

Conclusion

Functional hypoparathyroidism under magnesium deficiency is a well-recognized clinical entity. Attention should be paid to risk factors for magnesium depletion (Table 37.5). In the case of hypocalcemia of unknown origin, a therapeutic test of magnesium administration should be undertaken.

References

1. Walser M (1961) Ion association. VI. Interactions between calcium, magnesium, inorganic phosphate, citrate and protein in normal human plasma. J Clin Invest 40:723–730
2. Wacker WE, Parisi AF (1968) Magnesium metabolism. N Engl J Med 278(14):772–776 concl, Epub 1968/04/04
3. Wacker WE, Parisi AF (1968) Magnesium metabolism. N Engl J Med 278(13):712–717, Epub 1968/03/28
4. Wacker WE, Parisi AF (1968) Magnesium metabolism. N Engl J Med 278(12):658–663, Epub 1968/03/21
5. Rude RK, Singer FR, Gruber HE (2009) Skeletal and hormonal effects of magnesium deficiency. J Am Coll Nutr 28(2):131–141, Epub 2009/10/16

6. Rude RK (1998) Magnesium deficiency: a cause of heterogeneous disease in humans. J Bone Miner Res 13(4):749–758, Epub 1998/04/29
7. Brown EM, MacLeod RJ (2001) Extracellular calcium sensing and extracellular calcium signaling. Physiol Rev 81(1):239–297, Epub 2001/01/12
8. Board IoMFaN (1997) Dietary references intakes: calcium, phosphorus, magnesium, vitamin D and fluoride. National Academic Press, Washington, DC
9. Fine KD, Santa Ana CA, Porter JL, Fordtran JS (1991) Intestinal absorption of magnesium from food and supplements. J Clin Invest 88(2):396–402, Epub 1991/08/01
10. Schmulen AC, Lerman M, Pak CY, Zerwekh J, Morawski S, Fordtran JS et al (1980) Effect of 1,25-(OH)2D3 on jejunal absorption of magnesium in patients with chronic renal disease. Am J Physiol 238(4):G349–G352, Epub 1980/04/01
11. Karbach U (1989) Cellular-mediated and diffusive magnesium transport across the descending colon of the rat. Gastroenterology 96(5 Pt 1):1282–1289, Epub 1989/05/01
12. Brannan PG, Vergne-Marini P, Pak CY, Hull AR, Fordtran JS (1976) Magnesium absorption in the human small intestine. Results in normal subjects, patients with chronic renal disease, and patients with absorptive hypercalciuria. J Clin Invest 57(6):1412–1418
13. Schlingmann KP, Weber S, Peters M, Niemann Nejsum L, Vitzthum H, Klingel K et al (2002) Hypomagnesemia with secondary hypocalcemia is caused by mutations in TRPM6, a new member of the TRPM gene family. Nat Genet 31(2):166–170, Epub 2002/05/29
14. Lainez S, Schlingmann KP, van der Wijst J, Dworniczak B, van Zeeland F, Konrad M et al (2014) New TRPM6 missense mutations linked to hypomagnesemia with secondary hypocalcemia. Eur J Hum Genet 22(4):497–504, Epub 2013/08/15
15. Houillier P (2014) Mechanisms and regulation of renal magnesium transport. Annu Rev Physiol 76:411–430, Epub 2014/02/12
16. Agus ZS (1999) Hypomagnesemia. J Am Soc Nephrol 10(7):1616–1622, Epub 1999/07/15
17. de Rouffignac C, Quamme G (1994) Renal magnesium handling and its hormonal control. Physiol Rev 74(2):305–322, Epub 1994/04/01
18. Voets T, Nilius B, Hoefs S, van der Kemp AW, Droogmans G, Bindels RJ et al (2004) TRPM6 forms the Mg2+ influx channel involved in intestinal and renal Mg2+ absorption. J Biol Chem 279(1):19–25, Epub 2003/10/25
19. Bailly C, Roinel N, Amiel C (1985) Stimulation by glucagon and PTH of Ca and Mg reabsorption in the superficial distal tubule of the rat kidney. Pflugers Arch 403(1):28–34, Epub 1985/01/01
20. Groenestege WM, Hoenderop JG, van den Heuvel L, Knoers N, Bindels RJ (2006) The epithelial Mg2+ channel transient receptor potential melastatin 6 is regulated by dietary Mg2+ content and estrogens. J Am Soc Nephrol 17(4):1035–1043, Epub 2006/03/10
21. Cholst IN, Steinberg SF, Tropper PJ, Fox HE, Segre GV, Bilezikian JP (1984) The influence of hypermagnesemia on serum calcium and parathyroid hormone levels in human subjects. N Engl J Med 310(19):1221–1225, Epub 1984/05/10
22. Mayer GP, Hurst JG (1978) Comparison of the effects of calcium and magnesium on parathyroid hormone secretion rate in calves. Endocrinology 102(6):1803–1814, Epub 1978/06/01
23. Rude RK, Oldham SB, Singer FR (1976) Functional hypoparathyroidism and parathyroid hormone end-organ resistance in human magnesium deficiency. Clin Endocrinol (Oxf) 5(3):209–224, Epub 1976/05/01
24. Fuss M, Cogan E, Gillet C, Karmali R, Geurts J, Bergans A et al (1985) Magnesium administration reverses the hypocalcaemia secondary to hypomagnesaemia despite low circulating levels of 25-hydroxyvitamin D and 1,25-dihydroxy vitamin D. Clin Endocrinol (Oxf) 22(6):807–815, Epub 1985/06/01
25. Rude RK, Oldham SB, Sharp CF Jr, Singer FR (1978) Parathyroid hormone secretion in magnesium deficiency. J Clin Endocrinol Metab 47(4):800–806, Epub 1978/10/01
26. Fatemi S, Ryzen E, Flores J, Endres DB, Rude RK (1991) Effect of experimental human magnesium depletion on parathyroid hormone secretion and 1,25-dihydroxyvitamin D metabolism. J Clin Endocrinol Metab 73(5):1067–1072, Epub 1991/11/01
27. Rude RK, Adams JS, Ryzen E, Endres DB, Niimi H, Horst RL et al (1985) Low serum concentrations of 1,25-dihydroxyvitamin D in human magnesium deficiency. J Clin Endocrinol Metab 61(5):933–940, Epub 1985/11/01
28. Estep H, Shaw WA, Watlington C, Hobe R, Holland W, Tucker SG (1969) Hypocalcemia due to hypomagnesemia and reversible parathyroid hormone unresponsiveness. J Clin Endocrinol Metab 29(6):842–848, Epub 1969/06/01
29. Freitag JJ, Martin KJ, Conrades MB, Bellorin-Font E, Teitelbaum S, Klahr S et al (1979) Evidence for skeletal resistance to parathyroid hormone in magnesium deficiency. Studies in isolated perfused bone. J Clin Invest 64(5):1238–1244, Epub 1979/11/01
30. Suh SM, Tashjian AH Jr, Matsuo N, Parkinson DK, Fraser D (1973) Pathogenesis of hypocalcemia in primary hypomagnesemia: normal end-organ responsiveness to parathyroid hormone, impaired parathyroid gland function. J Clin Invest 52(1):153–160, Epub 1973/01/01
31. Rosler A, Rabinowitz D (1973) Magnesium-induced reversal of vitamin-D resistance in hypoparathyroidism. Lancet 1(7807):803–804, Epub 1973/04/14

32. Medalle R, Waterhouse C, Hahn TJ (1976) Vitamin D resistance in magnesium deficiency. Am J Clin Nutr 29(8):854–858, Epub 1976/08/01
33. Wong ET, Rude RK, Singer FR, Shaw ST Jr (1983) A high prevalence of hypomagnesemia and hypermagnesemia in hospitalized patients. Am J Clin Pathol 79(3):348–352, Epub 1983/03/01
34. Ryzen E, Wagers PW, Singer FR, Rude RK (1985) Magnesium deficiency in a medical ICU population. Crit Care Med 13(1):19–21, Epub 1985/01/01
35. Petrelli F, Borgonovo K, Cabiddu M, Ghilardi M, Barni S (2012) Risk of anti-EGFR monoclonal antibody-related hypomagnesemia: systematic review and pooled analysis of randomized studies. Expert Opin Drug Saf 11(Suppl 1):S9–S19, Epub 2011/08/17
36. Furlanetto TW, Faulhaber GA (2011) Hypomagnesemia and proton pump inhibitors: below the tip of the iceberg. Arch Intern Med 171(15):1391–1392, Epub 2011/05/11
37. Hoorn EJ, van der Hoek J, de Man RA, Kuipers EJ, Bolwerk C, Zietse R (2010) A case series of proton pump inhibitor-induced hypomagnesemia. Am J Kidney Dis 56(1):112–116, Epub 2010/03/02
38. Drueke TB (2004) Modulation and action of the calcium-sensing receptor. Nephrol Dial Transplant 19(Suppl 5):V20–V26, Epub 2004/07/31
39. Whang R, Flink EB, Dyckner T, Wester PO, Aikawa JK, Ryan MP (1985) Magnesium depletion as a cause of refractory potassium repletion. Arch Intern Med 145(9):1686–1689, Epub 1985/09/01
40. Ryzen E, Elbaum N, Singer FR, Rude RK (1985) Parenteral magnesium tolerance testing in the evaluation of magnesium deficiency. Magnesium 4(2–3): 137–147, Epub 1985/01/01

Part IV

Advocating for Hypoparathyroidism

Advocacy and Hypoparathyroidism in the Twenty-First Century

38

James E. Sanders and Jim Sliney Jr.

38.1 Introduction

In the realm of hypoparathyroidism, one major way advocacy is supported is through the HypoPARAthyroidism Association, Inc. The focus of the HypoPARAthyroidism Association is to improve lives touched by hypoparathyroidism through awareness and support [1]. Encouraging hypoparathyroidism patients to become their own advocate has been a part of the HypoPARAthyroidism Association itself and is specifically relevant in the story of its founder James E. Sanders. His story serves to illustrate not only the vital role of the patient as self-advocate but also of the importance of networking within one's greater community.

38.2 James Sanders' Story

38.2.1 Misdiagnosis

When I was about 8 years old, I experienced an attack of severe tetany. This was my introduction to hypoparathyroidism, though I did not know it at the time. My father and brother had to take me to the hospital at Chateauroux Air Force Base in France where my father was stationed. The emergency room physicians there were unable to find anything medically wrong with me and told my father that this episode was probably "all in my head" and not a medical problem. So I lived with this diagnosis of "psychosomatic illness" for the next several years – a difficult time where I had to learn on my own how to deal with the many and, at times, complex symptoms as they presented themselves.

The symptoms I remember most vividly included tetany, severe muscle and joint pain, and facial cramps, but the worst were the laryngospasms where I would suddenly find myself unable to breathe or talk, or even be able to let anyone know I was in trouble.

The first time I experienced a *laryngospasm*, it was terrifying. I didn't know at the time that they usually last less than 60 s (it felt like an eternity to me); all I knew is that when they occurred I could not talk and often I couldn't even breathe! Panicking...desperate for air...breathing just didn't work...but within a few seconds...I was able to catch some breath...and soon after... I could breathe again, the panic response subsiding. Experience taught me to find ways to relax which was the only remedy available to subdue a laryngospasm – though it did not always help.

In the setting of the 1960s when "PE" (Physical Education) was considered an

J.E. Sanders (✉)
Hypoparathyroidism Association,
Idaho Falls, ID, USA
e-mail: jsanders@hypopara.org

J. Sliney Jr.
Division of Endocrinology, Columbia University
Medical Center, New York, USA

M.L. Brandi, E.M. Brown (eds.), *Hypoparathyroidism*,
DOI 10.1007/978-88-470-5376-2_38, © Springer-Verlag Italia 2015

important aspect of public education, I found the exertion to be a real problem for me. Strenuous exercising, like running laps, was commonplace, and whenever I exerted myself, I experienced muscle and joint pain. My muscles became uncontrollable and my joints would feel like they were tight and swollen, so much so that any movement felt like I was tearing or damaging the tissues. The only relief I was able to find was to lie on a cold bathroom floor next to a toilet. Eventually I found ways to avoid exercising so I could in turn avoid the pain I experienced. Naturally the PE teachers did not appreciate my "laziness," nor did it occur to them to understand the underlying cause.

I would also experience severe peripheral tetany. In the absence of blood flow, similar to having your blood pressure checked during a routine physical exam or if my hand fell asleep, my hypocalcemia, and subsequent neuromuscular irritability, would induce spasm in the muscles of my hand and forearm. My wrists would flex, the joints in my fingers would extend, and my fingers would adduct. Of course, I didn't understand the mechanisms at the time (and had never heard of Trousseau); I just knew my hand would sometimes "claw up." These spasms, so severe that adults were unable to separate my fingers or straighten my hands, seemed spontaneous at the time, though I would later learn that you could recreate them in just a few seconds by inflating a blood pressure cuff on my forearm. Talk of symptoms wouldn't be complete without what we refer today as "brain fog," which has been a part of most of my life.

I had to deal with all of these symptoms by myself since no one else, not even my parents, understood the causes, or how severe the pain could be, or how to correct any of it – a heavy, and at times frightening, burden for a boy not yet 10 years old. The symptoms remained a part of my life, in varying degrees, until I was diagnosed with "hypoparathyroidism" and began treatment at the age of 22. All the while, physicians could not find anything medically to account for them, and my medical records still bore the cruel indication that the problem was "in my head," "psychosomatic."

38.2.2 Road to Answers

When I was about 19 years old, I had returned from a work study program in the Philippine Islands and went to the hospital at Holloman Air Force Base in New Mexico, where my father was stationed at the time. I had been struggling with fatigue and wanted to find out why. The lab work and x-rays done there were puzzling and inconclusive, which led the doctors to consult with other physicians at William Beaumont Regional Medical Center (WBRMC), which was the major military hospital in the Southwest (in El Paso, Texas). After some testing, Dr. Martin Nusynowitz, a WBRMC department head, concluded that the lab results for serum calcium must be wrong and advised to redo the test. The subsequent blood test showed results even lower than the first.

I was then referred to Dr. Nusynowitz, at the WBRMC, who concluded that I had hypocalcemia and that it was probably caused by hypoparathyroidism. What is that? However, while he suspected hypoparathyroidism, I did not fit the physical and clinical characteristics for the disease as they were understood in 1969. Dr. Nusynowitz, thankfully, proved a tenacious investigator [2].

Many years later I asked him why he went the extra mile with me; after all he could have just treated my hypocalcemia and left it at that. He told me

> (my) motivation was a combination of (1) the fact that something was wrong with you on the basis of your history, physical, and lab work, and I felt it was my responsibility as your physician to make the diagnosis and initiate treatment to correct the condition, and (2) my innate intellectual curiosity. [3]

He went on to explain that the extensive diagnostic tests and treatment were covered by the military medical system enabling him to go beyond a simple diagnosis and treatment to try and find what the correct diagnosis should be. Over the next couple of years, I would spend months in the WBRMC at El Paso and at Holloman Air Force Base, as he looked for answers to questions which were missed earlier in my life.

38.2.3 Finally, a Diagnosis

Using the calcium homeostasis feedback loop as a basis for his research, Dr. Nusynowitz was, after several months, able to describe what had been "wrong" with me for most of my life. He diagnosed me with pseudo-idiopathic hypoparathyroidism. My parents were devastated. They had accepted for all these years that these problems were in my head but now learned that, in fact, I had been sick all of those years. Of course, we were also relieved because now we finally had a diagnosis and a treatment which offered some hope to us.

Dr. Nusynowitz published his findings in an article in the *American Journal of Medicine* [2], and followed with a second paper describing a new category of hypoparathyroidism disorders [4].

My quest for answers was well underway and things were changing. Within several years of my diagnosis, Dr. Michael A. Levine, then of Johns Hopkins University School of Medicine, contacted me. He was just beginning his molecular studies into the causes of hypoparathyroidism and had read the papers by Dr. Nusynowitz. I began working with him to see if he could find the origin of my hypoparathyroidism in my family.

38.2.4 The Path to a More Effective Treatment

In the early 1990s, Dr. Karen K. Winer, a research fellow at the National Institutes of Health (NIH) in Bethesda, Maryland, had begun investigating the use of parathyroid hormone as an effective means of treating hypoparathyroidism [5].

To recruit patients for her clinical trials, she got in touch with Dr. Levine and other physicians to see if they had any patients who would be interested in participating in her clinical trials. Though the 1990s may not seem like so long ago, it was long enough that recruiting patients for a rare-disease trial took more time, money, and patience than it might now (and it's still not easy). But like all successful investigators, Dr. Winer was tenacious. Dr. Winer contacted me in 1992 and asked if I would be interested in participating in her clinical trials! *I was!* I was able to obtain permission from my employer to get the time off, with full pay, without which I would not have been able to participate in the clinical trials. It was during my participation in Dr. Winer's clinical trials that the idea of the HypoPARAthyroidism Association was born.

38.2.5 Birth of an Idea

I was held over for a few days on one of my periodic trips to NIH, when I had an opportunity to actually meet one of Dr. Winer's other hypoparathyroidism patients. It was the first time either of us had ever met another patient, and, needless to say, it was an emotional meeting for us both. We shared common misadventures and reveled in the realization that neither of us was alone anymore as a patient with hypoparathyroidism. It was something neither of us would ever forget.

I discussed this transformative encounter with a psychiatrist I was seeing for depression (yet another comorbidity of hypoparathyroidism), when he asked me what I wanted to do with this new discovery. My own answer surprised me. I explained that I wanted others to be able to experience the same great liberation of not feeling alone while dealing with this "thing" that had invaded our lives. His encouragement was priceless and gave me the strength I needed to overcome my doubts and my shyness and take action.

And so it was that in August of 1994 I wrote the first issue of the Hypoparathyroidism Newsletter [6]. I sent it to nearly 100 medical schools across the United States and my relatives (a captive audience) and a few friends. By December 31, 1998, we had evolved into the tax-exempt 501(3) (c) HypoPARAthyroidism Association, Inc., and were moving forward. Fast forward over 20 years and we have over 4500 members in 70 different countries. As an association for a rare medical disorder, we have been able to make a difference by helping physicians understand the disorder and the impact it has on the lives of those suffering with it. We have also been able to make a difference in the lives of individual patients and treating physicians, all in ways not possible prior to 1994.

Though I've come a long way, I am still learning about this disease; however, now I do it through the eyes of my extended hypoparathyroidism family.

38.3 Patient Advocacy and Hypoparathyroidism

Patient advocacy, as a policy, is a "concept that generally refers to efforts to support patients and their general interests within the context of the health care system. A more specific or applied definition of patient advocacy is difficult to articulate, in part because the term has been used in many difference ways" [7]. But here is how we see it.

As the medical community grows and oceans of medical knowledge become vaster, it is becoming increasingly important for patients to have an advocate within the medical system. Advocacy goes from important to essential when the patient is dealing with a rare medical disorder. So how are we going to define patient advocacy?

While patient advocacy can involve groups that develop policies that help patients, or civil committees that develop legislation, the real starting point for advocacy is always *activities that benefit patients.*

Patient advocacy is a relatively new and evolving field, but then, so is the medical industrial complex. The collective experience of the HypoPARAthyroidism Association shows us that advocacy must begin from the bottom up (starting with the patient), instead of the top down (starting with government legislation). Advocacy ends up playing a role between patients, doctors, researchers, and the many people impacted by a patient's disease, so it enhances every part of medical care. That advocate might be a sibling, a neighbor, a member of your parish, your nurse, a researcher assistant, your doctor, your parent, your spouse…essentially anyone, oneself included, who shares in the impact of the disease and has the will to take action or play a role.

Advocacy is necessary because *patients have needs that are not being properly met.* Given the underserved, underfunded, and misunderstood nature of rare diseases, its patients are often left to their own devices to seek the best possible help and information. Fear of consequences and the impact on loved ones often motivates the patient to become active in seeking solutions to their needs. Those needs may be fear of being sick, fear of what will happen to loved ones in a worst case scenario, the consequences of not having a diagnosis (such as years of misguided treatment and random testing), the burden of the cost of testing, the impact of radiation, the wasted time, or simply the ongoing suffering with the symptoms of the disease. So we can say that the advocate is born out of motivation to address the unmet needs of the patient.

As we return to considering the hypoparathyroidism patient as a model for any rare disease patient, we begin with the question:

What are the needs of the hypoparathyroidism patient?

38.3.1 Communication

The hypoparathyroidism patient needs to know that they are not alone. Alienation is a terrible thing, especially when stacked on top of a debilitating disease. When James Sanders met another hypoparathyroidism patient for the first time, it was a transformative experience. Suddenly the ongoing struggle with hypoparathyroidism in an individual could be communicated between two people using ideas and concepts that had never quite made sense to other outsiders. Imagine visiting a place that doesn't speak the language you speak – what a relief to find someone you can talk to who understands what you are saying and who you in turn can understand! Being able to share "war stories" between members of the Association has been fuel for the overall progress of the group. It takes a patient's internalized struggle and brings it into the light to be confronted. It allows patients to share hard-earned wisdom on how to live with the disease.

A patient must also be able to share their experiences with loved ones. It is a sad truth that some relationships cannot withstand a partner becoming ill. When a partner in a codependent or dependent

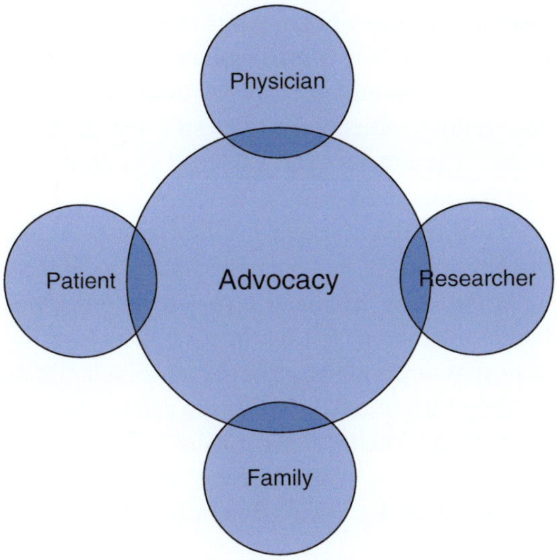

Fig. 38.1 Advocacy can be performed by multiple people in the patient's life and should not be limited just to the patient

relationship suddenly undergoes a drastic change, it feels like a violation of a pact. The young spouse suddenly becomes the unwilling caregiver, or the child gets less time and attention from the parent, or previously enjoyed activities are now limited or eliminated. The impact can do great damage to the unprepared. A sense of betrayal (you don't act the way you used to), unfairness (why did your disease happen to me?), or burden (years of emergency room visits) can fester.

If rescue from such despair were to be possible, above all else the partners would *need to communicate*. They must convey their experiences, revelations, fears, and hopes, not just from the mouth of the diagnosed but by all parties profoundly affected. By understanding, respecting, and accepting the impact of the disease, the illness need not be the dominant feature of the relationship.

The advocate must, therefore, nurture a comfortable arena where *communication* can flourish (Fig. 38.1).

38.3.2 Understanding the Disease

Patients also need to understand the nature of what they are experiencing. Communicating between partners is not likely to be effective without at least a rudimentary understanding of the nature of the disease. Fear of the uncontrollable numbness around their mouth might be diminished by understanding the physiologic principles behind it. Anxiety and depression (so prominent in hypoparathyroidism) only serve to magnify the gravity and turmoil of each involuntary tingle, twitch, and ache. Rather than being a victim of these symptoms, a person can take control of the symptom if they *understand* its nature. To obtain such understanding, a patient/advocate can turn to sources of information starting with their doctors, but turning also to reliable online sources, medical texts, and medically verified support groups.

The first step any hypoparathyroidism patients should take should be to educate themselves and their loved ones about the disorder. By doing so, not only will they be able to understand the disorder and the reasons for their symptoms but they will be able to communicate better and be in a better position to cooperate with their physicians. Self-education is important but, if done poorly, has serious downsides. For the patient who cannot find the time to read the medical texts, search the pertinent academic publications or scour the internet for expertise; one could turn to the resources available through the HypoPARAthyroidism Association, which strives to be a reliable source of accurate and easily digestible information.

38.3.3 Stay Informed/Stay Involved

Understanding the disease is not complete without further understanding of the treatment options that are available for the disease or those emerging on the horizon. Living with hypoparathyroidism becomes more manageable when one understands the causes and the impacts of the disease. Of equal importance is how one addresses those impacts and that includes staying informed on what advances are emerging.

Every hypoparathyroidism patient has been told at one time or another that they don't have problems that can't be fixed by taking calcium. If that were so, texts like this one would not need to exist. By simply asking a few basic questions, we can see the problem is deeper than taking calcium. What are the consequences of the disease, what are the treatments for the disease, and what are the consequences of the treatments? Treatment has consisted of calcium and vitamin D analogs for longer than a lifetime, and these tools, though potentially effective, do come with many caveats and secondary effects. The capabilities of calcium and vitamin D may be many, but it can be argued that they cannot, for instance, correct the quality of life issues that the hypoparathyroidism patient faces [8], and what of the threat to the kidneys and soft tissues from the overuse of calcium? In other words, "simply" taking calcium is not so simple. Fortunately, we live in an exciting time for hypoparathyroidism because we are only a few steps away from seeing recombinant human parathyroid hormone (1–84) get approval as a treatment for hypoparathyroidism which will alter the lives of patients profoundly.

Patients need to be cautious about the financial and physiologic consequences of available treatments, explore alternatives to those treatments, and contribute in any way possible to the evolution of drugs and treatments that they and their loved ones will benefit from. What was the leading therapy when the patient was first diagnosed may now be well outdated. Treatments cannot evolve without participation and interaction between patient, physician, and investigator. Therefore, it is prudent for patients to not only *stay informed* of new therapies and approaches to wellness but to *stay involved* in the investigation of these approaches. The best ways to stay involved are by providing quality feedback to the treating physician, by communicating with fellow patients, hopefully through an advocacy group, and by participating in research studies.

In summary, the patient advocate, or self-advocate, or advocacy organization must foster communication, understanding, and ongoing education. Maintaining this level of advocacy takes much effort which is why the more successful patient is the one who involves their families, friends or community, physician-infrastructure, and fellow patients and doesn't try to go it alone. Therein lays the value of a central support group.

"How different might my life have been had the physicians in the emergency room in France, or in the other hospitals I visited before I was finally diagnosed, felt it was *their* responsibility as *my* physician to make the diagnosis and initiate treatment to correct the condition and *maintained their* innate intellectual curiosity." Therein lays the foundation of "Advocacy and Hypoparathyroidism in the Twenty-First Century."

38.3.4 The Argument for a Patient Advocacy Group

Many patients diagnosed with hypoparathyroidism had a long and tortuous journey leading to their diagnosis, involving several physicians, years of misdiagnosis and mistreatment, and feeling isolated from their families and society because they "do not look sick," when, in fact, they have a serious medical disorder. Their journey was made difficult for them because they experienced symptoms that others could not see, and which were not fully understood until recently. It wasn't until the collective experience of hundreds of patients over several years found a forum that the impact of hypoparathyroidism on people began to be more fully understood by the medical community.

The HypoPARAthyroidism Association is, at its core, an advocacy generator. It attracts advocates, like its Board of Directors, Medical Adviser's, and volunteers. It is also there to provide the tools to others who wish to advocate for hypoparathyroidism. We communicate with our members and help them communicate with each other. We keep them informed of our activities and educate them about their disease. We try to keep them apprised of new studies and new ideas and encourage them to share their own. We also aim to teach them how to better care for themselves, to improve the care with their doctors, to gain independence from this disease, to improve their quality of life, and to ultimately have fewer unmet needs. That is, after all, what advocacy is all about.

A good support group is not complete until it makes information available to physicians as well. By providing easy access to precisely the information that physicians need to administer the best possible care, we help them as well.

There are several organizations that can provide legitimate medical information on hypoparathyroidism, or rare diseases in general. These are both built by patients and then vetted by medical experts or come directly from physician or commercial sources:

National Organization for Rare Disorders (NORD) http://www.rarediseases.org/

Office of Rare Diseases Research (ORDR) http://rarediseases.info.nih.gov/

HypoPARAthyroidism Association https://www.hypopara.org/

Mayo Clinic, United States http://www.mayoclinic.org/diseases-conditions/hypoparathyroidism/basics/definition/con-20030780

Hypoparathyroidism Answers – Getting Clarity on a Complex Disorder http://hypoparathyroidism.com/

38.4 The Physician's Role

Few people look to their physician as one of their advocates, but in reality the physician should be the most important advocate! The physician is the medical expert who brings all of the pieces of the puzzle together which results in holistic care for individual patients. In the theater of rare diseases, this may require extra work for the physician, but the rewards far outweigh the labor thanks to the growing availability of information on rare diseases.

In a successful doctor-patient relationship, the patient must come to terms with the doctor's expertise and ability to prognosticate, while the doctor must come to terms with what he "does not yet know," namely, the impacts and limitations brought on by the disease for that particular patient. Therefore, if the doctor and patient choose to enter into this special relationship, the physician's part of the agreement includes accepting the responsibility of becoming educated about the patient's specific medical disorder and how it may best be managed.

One way the doctor can accomplish this is to allow the patient to become a partner in their care. The patient who is practicing good self-advocacy can bring in pertinent medical information (i.e., medical publications or information from medically vetted sources) and discuss this information with the doctor. With such open exchange, the doctor becomes more familiar with both the overt and subtle impacts of hypoparathyroidism, and the patient is able to gain more targeted insights from the medical professional.

Given the value of a well-informed patient, the physician who has rare disease patients serves both himself and his patient best by encouraging them to participate in advocacy groups like the HypoPARAthyroidism Association. From such sources patients can get many of their needs met and get good quality education on their disease, allowing the doctor-patient relationship to focus on more immediate medical needs.

The physician ultimately has a responsibility to help his patient become well, and so if, for any reason, the physician is unable or unwilling to take on the level of commitment necessary for a rare disease patient, that doctor should help the patient find another physician who will.

In the HypoPARAthyroidism Association's experience, their medical advisers are extremely willing to help other physicians who are looking for help and/or information (including referrals to other doctors) in order to help their patients.

Conclusion

By putting the well-educated patient and doctor together in a partnership based on mutual respect and understanding, it becomes possible to achieve a paradigm of healthcare. Let that patient and that doctor be advocates for improvements in holistic care of hypoparathyroidism and they can overcome much of the secondary suffering caused by the disease. With the addition of support groups, clinical research, and extended advocates, there is virtually no limit to the achievements possible. Together they can make a difference for each other and for all people with hypoparathyroidism.

References

1. 2014 (Online). Available: https://www.hypopara.org/about-us/mission-statement.html. Accessed 18 Nov 2014
2. Nusynowitz ML (1973) Pseudoidiopathic hypoparathyroidism. Hypoparathyroidism with ineffective parathyroid hormone. Am J Med 725(7):677–686
3. Nusynowitz (2013) Dr. Martin L Nusynowitz, Colonel, US Army, retired. Email to James Sanders
4. Nusynowitz ML (1976) The spectrum of the hypoparathyroid states: a classification based on physiologic principles. Medicine 55(2):105–119
5. Winer KK (1996) Synthetic human parathyroid hormone 1-34 vs calcitriol and calcium in the treatment of hypoparathyroidism. JAMA 276(8):631–636
6. Sanders J (1994) Hypoparathyroidism Association Inc. [Online]. Available: https://www.hypopara.org/wwwroot/userfiles/newsletters/August1994.pdf. Accessed 26 Aug 2013
7. Gilkey MB (2009) Defining patient advocacy in the post-quality chasm era. N C Med J 70(2):120–124
8. Cusano NE (2013) The effect of PTH(1-84) on quality of life in hypoparathyroidism. J Clin Endocrinol Metabol 98(6):2356–2361

Index

M.L. Brandi, E.M. Brown (eds.), *Hypoparathyroidism*,
DOI 10.1007/978-88-470-5376-2, © Springer-Verlag Italia 2015

F

Q

R

MIX
Papier aus verantwortungsvollen Quellen
Paper from responsible sources
FSC® C105338

If you have any concerns about our products, you can contact us on
ProductSafety@springernature.com

In case Publisher is established outside the EU, the EU authorized representative is:
Springer Nature Customer Service Center GmbH
Europaplatz 3, 69115 Heidelberg, Germany

Printed by Libri Plureos GmbH
in Hamburg, Germany